D1406588

Customer Support Information

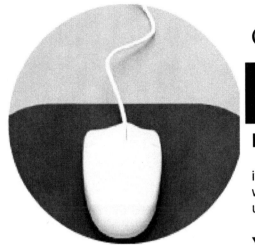

Plunkett's Biotech & Genetics Industry Almanac 2011

Please register your book immediately...

if you did not purchase it directly from Plunkett Research, Ltd. This will enable us to fulfill your requests for assistance. Also, it will enable us to notify you of future editions.

Your purchase includes access to Book Data and Exports online

As a book purchaser, you can register for free, 1-year, 1-seat online access to the latest data for your book's industry trends, statistics and company profiles. This includes tools to export company data. We are migrating from our former CD-ROMs, for supplemental data and export tools, to the web. Simply send us this registration form, and we will send you a user name and password. In this manner, you will have access to our continual updates during the year. Certain restrictions apply.

_____ YES, please register me as a purchaser of the book.
I did not buy it directly from Plunkett Research, Ltd.

_____ YES, please register me for free online access. I am the actual, original purchaser. (Proof of purchase may be required.)

Customer Name _____

Title_____

Organization _____

Address_____

City_____State_____Zip_____

Country (if other than USA) _____

Phone_____Fax _____

E-mail _____

Return to: ## Plunkett Research, Ltd.

Attn: Registration
P.O. Drawer 541737, Houston, TX 77254-1737 USA
713.932.0000 · Fax 713.932.7080 · www.plunkettresearch.com
customersupport@plunkettresearch.com

* Purchasers of used books are not eligible to register. Use of online access is subject to the terms of the end user license agreement.

Stem cell transplantation
will have a predictable
side effects
and thus ~~easier to avoid~~
more efficient to develop
before clinical Trials

It is a transplantation !

PLUNKETT'S BIOTECH & GENETICS INDUSTRY ALMANAC 2011

The only comprehensive guide to biotechnology and genetics companies and trends

Jack W. Plunkett

Published by:
Plunkett Research, Ltd., Houston, Texas
www.plunkettresearch.com

PLUNKETT'S
BIOTECH & GENETICS
INDUSTRY ALMANAC
2011

Editor and Publisher:
Jack W. Plunkett

Executive Editor and Database Manager:
Martha Burgher Plunkett

Senior Editors and Researchers:
Brandon Brison
Michael Esterheld
Addie K. FryeWeaver
Christie Manck

Editors, Researchers and Assistants:
Kalonji Bobb
Elizabeth Braddock
Austin Bunch
Michelle Dotter
Jeremy Faulk
Lucinda Gaines
Leandra Hernandez
Andrew Olsen
Jill Steinberg
Suzanne Zarosky

Enterprise Accounts Manager
Emily Hurley

Information Technology Manager:
Wenping Guo

E-Commerce Managers:
Alejandra Avila
Kelly Burke
Geoffrey Trudeau

Cover Design:
Kim Paxson, Just Graphics
Junction, TX

Special Thanks to:
Burrill & Company
Centers for Medicare & Medicaid Services (CMS)
Ernst & Young
IMS Health
ISAAA
National Science Foundation
Pharmaceutical Research & Manufacturers
Association (PhRMA)
Tufts Center for the Study of Drug Development
U.S. Department of Labor, Bureau of Labor Statistics
U.S. Food & Drug Administration (FDA)
U.S. National Institutes of Health
U.S. National Science Foundation
U.S. Patent & Trademark Office

Plunkett Research, Ltd.
P. O. Drawer 541737, Houston, Texas 77254 USA
Phone: 713.932.0000   Fax: 713.932.7080
www.plunkettresearch.com

Published by:

Plunkett Research, Ltd.

P.O. Drawer 541737

Houston, Texas 77254-1737

Phone: 713.932.0000

Fax: 713.932.7080

Internet: www.plunkettresearch.com

ISBN13 # 978-1-59392-179-8

(eBook Edition # 978-1-59392-522-2)

Disclaimer of liability
for use and results of use:

PLUNKETT'S BIOTECH & GENETICS INDUSTRY ALMANAC 2011

CONTENTS

Continued on next page

Continued from previous page

A Short Biotech & Genetics Industry Glossary

10-K: An annual report filed by publicly held companies. It provides a comprehensive overview of the company's business and its finances. By law, it must contain specific information and follow a given form, the "Annual Report on Form 10-K." The U.S. Securities and Exchange Commission requires that it be filed within 90 days after fiscal year end. However, these reports are often filed late due to extenuating circumstances. Variations of a 10-K are often filed to indicate amendments and changes. Most publicly held companies also publish an "annual report" that is not on Form 10-K. These annual reports are more informal and are frequently used by a company to enhance its image with customers, investors and industry peers.

510 K: An application filed with the FDA for a new medical device to show that the apparatus is "substantially equivalent" to one that is already marketed.

Abbreviated New Drug Application (ANDA): An application filed with the FDA showing that a substance is the same as an existing, previously approved drug (i.e., a generic version).

Absorption, Distribution, Metabolism and Excretion (ADME): In clinical trials, the bodily processes studied to determine the extent and duration of systemic exposure to a drug.

Accelerated Approval: A process at the FDA for reducing the clinical trial length for drugs designed for certain serious or life-threatening diseases.

ADME: See "Absorption, Distribution, Metabolism and Excretion (ADME)."

Adverse Event (AE): In clinical trials, a condition not observed at baseline or worsened if present at baseline. Sometimes called Treatment Emergent Signs and Symptoms (TESS).

AE: See "Adverse Event (AE)."

Agricultural Biotechnology (AgriBio): The application of biotechnology methods to enhance agricultural plants and animals.

Amino Acid: Any of a class of 20 molecules that combine to form proteins. Also, see "Codon."

ANDA: See "Abbreviated New Drug Application (ANDA)."

Angiogenesis: Blood vessel formation, typically in the growth of malignant tissue.

Angioplasty: The re-opening of a blood vessel by non-surgical techniques such as balloon dilation or laser, or through surgery.

Antibody: A protein produced by white blood cells in response to a foreign substance. Each antibody can bind only to one specific antigen. See "Antigen."

Antigen: A foreign substance that causes the immune system to create an antibody. See "Antibody."

Antisense Technology: The use of RNA-like oligonucleotides that bind to RNA and inhibit the expression of a gene.

APAC: Asia Pacific Advisory Committee. A multi-country committee representing the Asia and Pacific region.

Apoptosis: A normal cellular process leading to the termination of a cell's life.

Applied Research: The application of compounds, processes, materials or other items discovered during basic research to practical uses. The goal is to move discoveries along to the final development phase.

Array: An orderly arrangement, such as a rectangular matrix of data. In some laboratory systems, such as microarrays, multiple detectors (probes) are positioned in an array in order to best perform research. See "Microarray."

Artificial Life (AL): See "Synthetic Biology."

Assay: A laboratory test to identify and/or measure the amount of a particular substance in a sample. Types of assays include endpoint assays, in which a single measurement is made at a fixed time; kinetic assays, in which increasing amounts of a product are formed with time and are monitored at multiple points; microbiological assays, which measure the concentration of antimicrobials in biological material;

and immunological assays, in which analysis or measurement is based on antigen-antibody reactions.

Astrobiology: A field focused on the study of the origin, evolution and future of life on Earth and beyond. Participants include scientists in physics, chemistry, astronomy, biology, geology and computer science.

Baby Boomer: Generally refers to people born in the U.S. and Western Europe from 1946 to 1964. In the U.S., the initial number of Baby Boomers totaled about 78 million. The term evolved to include the children of soldiers and war industry workers who were involved in World War II.

Baseline: A set of data used in clinical studies, or other types of research, for control or comparison.

Basic Research: Attempts to discover compounds, materials, processes or other items that may be largely or entirely new and/or unique. Basic research may start with a theoretical concept that has yet to be proven. The goal is to create discoveries that can be moved along to applied research. Basic research is sometimes referred to as "blue sky" research.

Big Pharma: The top tier of pharmaceutical companies in terms of sales and profits (e.g., Pfizer, Merck, Johnson & Johnson).

Bioaccumulation: A process in which chemicals are retained in fatty body tissue and increased in concentration over time.

Bioavailability: In pharmaceuticals, the rate and extent to which a drug is absorbed or is otherwise available to the treatment site in the body.

Biochemical Engineering: A sector of chemical engineering that deals with biological structures and processes. Biochemical engineers may be found in the pharmaceutical, biotechnology and environmental fields, among others.

Biochemicals: Chemicals that either naturally occur or are identical to naturally occurring substances. Examples include hormones, pheromones and enzymes. Biochemicals function as pesticides through non-toxic, non-lethal modes of action, disrupting insect mating patterns, regulating growth or acting as repellants. They tend to be environmentally desirable, and may be produced by

industry from organic sources such as plant waste (biomass). Biochemicals also may be referred to as bio-based chemicals, green chemicals or plant-based chemicals.

Biodiesel: A fuel derived when glycerin is separated from vegetable oils or animal fats. The resulting byproducts are methyl esters (the chemical name for biodiesel) and glycerin which can be used in soaps and cleaning products. It has lower emissions than petroleum diesel and is currently used as an additive to that fuel since it helps with lubricity.

Bioengineering: Engineering principles applied when working in biology and pharmaceuticals.

Bioequivalence: In pharmaceuticals, the demonstration that a drug's rate and extent of absorption are not significantly different from those of an existing drug that is already approved by the FDA. This is the basis upon which generic and brand name drugs are compared.

Bioethanol: A fuel produced by the fermentation of plant matter such as corn. Fermentation is enhanced through the use of enzymes that are created through biotechnology. Also, see "Ethanol."

Biogeneric: See "Follow-on Biologics."

Biogenerics: Genetic versions of drugs that have been created via biotechnology. Also, see "Follow-on Biologics."

Bioinformatics: Research, development or application of computational tools and approaches for expanding the use of biological, medical, behavioral or health data, including those to acquire, store, organize, archive, analyze or visualize such data. Bioinformatics is often applied to the study of genetic data. It applies principles of information sciences and technologies to make vast, diverse and complex life sciences data more understandable and useful.

Biologics: Drugs that are synthesized from living organisms. That is, drugs created using biotechnology, sometimes referred to as biopharmaceuticals. Specifically, biologics may be any virus, therapeutic serum, toxin, antitoxin, vaccine, blood, blood component or derivative, allergenic or analogous product, or arsphenamine or one of its derivatives used for the prevention, treatment or cure of disease. Also, see "Biologics

License Application (BLA)," "Follow-on Biologics," and "Biopharmaceuticals."

Biologics License Application (BLA): An application to be submitted to the FDA when a firm wants to obtain permission to market a novel, new biological drug product. Specifically, these are drugs created through the use of biotechnology. It was formerly known as Product License Application (PLA). Also see "Biologics."

Biomagnification: The increase of tissue accumulation in species higher in the natural food chain as contaminated food species are eaten.

Biomass: Organic, non-fossil material of biological origin constituting a renewable energy source. The biomass can be burnt as fuel in a system that creates steam to turn a turbine, generating electricity. For example, biomass can include wood chips and agricultural crops.

BioMEMS: MEMS used in medicine. See "Microelectromechanical Systems (MEMS)."

Biomimetic: Mimicking, imitating, copying or learning from nature.

Biopharmaceuticals: That portion of the pharmaceutical industry focused on the use of biotechnology to create new drugs. A biopharmaceutical can be any biological compound that is intended to be used as a therapeutic drug, including recombinant proteins, monoclonal and polyclonal antibodies, antisense oligonucleotides, therapeutic genes, and recombinant and DNA vaccines. Also, see "Biologics."

Biopolymeroptoelectromechanical Systems (BioPOEMS): Combination of MEMS and optics used in biological applications.

Biorefinery: A refinery that produces fuels from biomass. These fuels may include bioethanol (produced from corn or other plant matter) or biodiesel (produced from plant or animal matter).

Biosensor: A sensor based on the use of biological materials or that targets biological analytes.

Biosimilar: See "Follow-on Biologics."

Biotechnology: A set of powerful tools that employ living organisms (or parts of organisms) to make or modify products, improve plants or animals (including humans) or develop microorganisms for specific uses. Biotechnology is most commonly thought of to include the development of human medical therapies and processes using recombinant DNA, cell fusion, other genetic techniques and bioremediation.

BLA: See "Biologics License Application (BLA)."

Blastocyst: A fertilized embryo, aged four to 11 days, which consists of multiplying cells both outside and inside a cavity. It is the blastocyst that embeds itself in the uterine wall and ultimately develops into a fetus. Blastocysts are utilized outside the womb during the process of stem cell cultivation.

BPO: See "Business Process Outsourcing (BPO)."

Branding: A marketing strategy that places a focus on the brand name of a product, service or firm in order to increase the brand's market share, increase sales, establish credibility, improve satisfaction, raise the profile of the firm and increase profits.

BRIC: An acronym representing Brazil, Russia, India and China. The economies of these four countries are seen as some of the fastest growing in the world. A 2003 report by investment bank Goldman Sachs is often credited for popularizing the term; the report suggested that by 2050, BRIC economies will likely outshine those countries which are currently the richest in the world.

B-to-B, or B2B: See "Business-to-Business."

B-to-C, or B2C: See "Business-to-Consumer."

Business Process Outsourcing (BPO): The process of hiring another company to handle business activities. BPO is one of the fastest-growing segments in the offshoring sector. Services include human resources management, billing and purchasing and call centers, as well as many types of customer service or marketing activities, depending on the industry involved. Also, see "Knowledge Process Outsourcing (KPO)."

Business-to-Business: An organization focused on selling products, services or data to commercial customers rather than individual consumers. Also known as B2B.

Business-to-Consumer: An organization focused on selling products, services or data to individual consumers rather than commercial customers. Also known as B2C.

CANDA: See "Computer-Assisted New Drug Application (CANDA)."

Captive Offshoring: Used to describe a company-owned offshore operation. For example, Microsoft owns and operates significant captive offshore research and development centers in China and elsewhere that are offshore from Microsoft's U.S. home base. Also see "Offshoring."

Carcinogen: A substance capable of causing cancer. A suspected carcinogen is a substance that may cause cancer in humans or animals but for which the evidence is not conclusive.

Case Report Form (CRF): In clinical trials, a standard document used by clinicians to record and report subject data pertinent to the study protocol.

Case Report Tabulation (CRT): In clinical trials, a tabular listing of all data collected on study case report forms.

CBER: See "Center for Biologics Evaluation and Research (CBER)."

CDC: See "Centers for Disease Control and Prevention (CDC)."

CEM: Contract electronic manufacturing. See "Contract Manufacturer."

Center for Biologics Evaluation and Research (CBER): The branch of the FDA responsible for the regulation of biological products, including blood, vaccines, therapeutics and related drugs and devices, to ensure purity, potency, safety, availability and effectiveness. www.fda.gov/cber

Center for Devices and Radiological Health (CDRH): The branch of the FDA responsible for the regulation of medical devices. www.fda.gov/cdrh

Center for Drug Evaluation and Research (CDER): The branch of the FDA responsible for the regulation of drug products. www.fda.gov/cder

Centers for Disease Control and Prevention (CDC): The federal agency charged with protecting the public health of the nation by providing leadership and direction in the prevention and control of diseases and other preventable conditions and responding to public health emergencies. Headquartered in Atlanta, it was established as an operating health agency within the U.S. Public Health Service on July 1, 1973. See www.cdc.gov.

Centers for Medicare and Medicaid Services (CMS): A federal agency responsible for administering Medicare and monitoring the states' operations of Medicaid. See www.cms.hhs.gov.

Chromosome: A structure in the nucleus of a cell that contains genes. Chromosomes are found in pairs.

CIS: See "Commonwealth of Independent States (CIS)."

Class I Device: An FDA classification of medical devices for which general controls are sufficient to ensure safety and efficacy.

Class II Device: An FDA classification of medical devices for which performance standards and special controls are sufficient to ensure safety and efficacy.

Class III Device: An FDA classification of medical devices for which pre-market approval is required to ensure safety and efficacy, unless the device is substantially equivalent to a currently marketed device. See "510 K."

Climate Change (Greenhouse Effect): A theory that assumes an increasing mean global surface temperature of the Earth caused by gases (sometimes referred to as greenhouse gases) in the atmosphere (including carbon dioxide, methane, nitrous oxide, ozone and chlorofluorocarbons). The greenhouse effect allows solar radiation to penetrate the Earth's atmosphere but absorbs the infrared radiation returning to space.

Clinical Research Associate (CRA): An individual responsible for monitoring clinical trial data to ensure compliance with study protocol and FDA GCP regulations.

Clinical Trial: See "Phase I Clinical Trials," along with definitions for Phase II, Phase III and Phase IV.

Clone: A group of identical genes, cells or organisms derived from one ancestor. A clone is an identical copy. "Dolly" the sheep is a famous case of a clone of an animal. Also see "Cloning (Reproductive)" and "Cloning (Therapeutic)."

Cloning (Reproductive): A method of reproducing an exact copy of an animal or, potentially, an exact copy of a human being. A scientist removes the nucleus from a donor's unfertilized egg, inserts a nucleus from the animal to be copied and then stimulates the nucleus to begin dividing to form an embryo. In the case of a mammal, such as a human, the embryo would then be implanted in the uterus of a host female. Also see "Cloning (Therapeutic)."

Cloning (Therapeutic): A method of reproducing exact copies of cells needed for research or for the development of replacement tissue or organs. A scientist removes the nucleus from a donor's unfertilized egg, inserts a nucleus from the animal whose cells are to be copied and then stimulates the nucleus to begin dividing to form an embryo. However, the embryo is never allowed to grow to any significant stage of development. Instead, it is allowed to grow for a few hours or days, and stem cells are then removed from it for use in regenerating tissue. Also see "Cloning (Reproductive)."

CMS: See "Centers for Medicare and Medicaid Services (CMS)."

Codon: Groupings of three letters each (for example, AGT) that represent the amino acid molecules that form the components of proteins. The letters used are limited to A, C, G and T.

Combinatorial Chemistry: A chemistry in which molecules are found that control a pre-determined protein. This advanced computer technique enables scientists to use automatic fluid handlers to mix chemicals under specific test conditions at extremely high speed. Combinatorial chemistry can generate thousands of chemical compound variations in a few hours. Previously, traditional chemistry methods could have required several weeks to do the same work.

Committee on Proprietary Medicinal Products (CPMP): A committee, composed of two people

from each EU Member State (see "European Union (EU)"), that is responsible for the scientific evaluation and assessment of marketing applications for medicinal products in the EU. The CPMP is the major body involved in the harmonization of pharmaceutical regulations within the EU and receives administrative support from the European Medicines Evaluation Agency. See "European Medicines Evaluation Agency (EMEA)."

Commonwealth of Independent States (CIS): An organization consisting of 11 former members of the Soviet Union: Russia, Ukraine, Armenia, Moldova, Georgia, Belarus, Kazakhstan, Uzbekistan, Azerbaijan, Kyrgyzstan and Tajikistan. It was created in 1991. Turkmenistan recently left the Commonwealth as a permanent member, but remained as an associate member. The Commonwealth seeks to coordinate a variety of economic and social policies, including taxation, pricing, customs and economic regulation, as well as to promote the free movement of capital, goods, services and labor.

Complementary-DNA (cDNA): A sequence acquired by copying a messenger RNA (mRNA) molecule back into DNA. In contrast to the original DNA, mRNA codes for an expressed protein without non-coding DNA sequences (introns). Therefore, a cDNA probe can also be used to find the specific gene in a complex DNA sample from another organism with different non-coding sequences. Also called Copy DNA.

Computational Biology: The development and application of data-analytical and theoretical methods, mathematical modeling and computational simulation techniques to the study of biological, behavioral, and social systems. Computational biology uses mathematical and computational approaches to address theoretical and experimental questions in biology.

Computed Tomography (CT): An imaging method that uses x-rays to create cross-sectional pictures of the body. The technique is frequently referred to as a "CAT Scan." A patient lies on a narrow platform while the machine's x-ray beam rotates around him or her. Small detectors inside the scanner measure the amount of x-rays that make it through the part of the body being studied. A computer takes this information and uses it to create several individual images, called slices. These images can be stored, viewed on a monitor, or printed on film. Three-dimensional models of organs can be created by stacking the individual slices together. The newest machines are capable of operating at 256 slice levels, creating very high resolution images in a short period of time.

Computer-Assisted New Drug Application (CANDA): An electronic submission of a new drug application (NDA) to the FDA.

Consumer Price Index (CPI): A measure of the average change in consumer prices over time in a fixed market basket of goods and services, such as food, clothing and housing. The CPI is calculated by the U.S. Federal Government and is considered to be one measure of inflation.

Contract Manufacturer: A company that manufactures products that will be sold under the brand names of its client companies. For example, a large number of consumer electronics, such as laptop computers, are manufactured by contract manufacturers for leading brand-name computer companies such as Dell. Many other types of products are made under contract manufacturing, from apparel to pharmaceuticals. Also see "Original Equipment Manufacturer (OEM)" and "Original Design Manufacturer (ODM)."

Contract Manufacturing: See "Contract Manufacturer."

Contract Research Organization (CRO): An independent organization that contracts with a client to conduct part of the work on a study or research project. For example, drug and medical device makers frequently outsource clinical trials and other research work to CROs.

Coordinator: In clinical trials, the person at an investigative site who handles the administrative responsibilities of the trial, acts as a liaison between the investigative site and the sponsor, and reviews data and records during a monitoring visit.

CPMP: See "Committee on Proprietary Medicinal Products (CPMP)."

CRA: See "Clinical Research Associate (CRA)."

CRF: See "Case Report Form (CRF)."

CT: See "Computed Tomography (CT)."

Data and Safety Monitoring Board (DSMB): See "Data Monitoring Board (DMB)."

Data Monitoring Board (DMB): A committee that monitors the progress of a clinical trial and carefully observes the safety data.

Deoxyribonucleic Acid (DNA): The carrier of the genetic information that cells need to replicate and to produce proteins.

Development: The phase of research and development (R&D) in which researchers attempt to create new products from the results of discoveries and applications created during basic and applied research.

Device: In medical products, an instrument, apparatus, implement, machine, contrivance, implant, in vitro reagent or other similar or related article, including any component, part or accessory, that 1) is recognized in the official National Formulary or United States Pharmacopoeia or any supplement to them, 2) is intended for use in the diagnosis of disease or other conditions, or in the cure, mitigation, treatment or prevention of disease, in man or animals or 3) is intended to affect the structure of the body of man or animals and does not achieve any of its principal intended purposes through chemical action within or on the body of man or animals and is not dependent upon being metabolized for the achievement of any of its principal intended purposes.

Diagnostic Radioisotope Facility: A medical facility in which radioactive isotopes (radiopharmaceuticals) are used as tracers or indicators to detect an abnormal condition or disease in the body.

Distributor: An individual or business involved in marketing, warehousing and/or shipping of products manufactured by others to a specific group of end users. Distributors do not sell to the general public. In order to develop a competitive advantage, distributors often focus on serving one industry or one set of niche clients. For example, within the medical industry, there are major distributors that focus on providing pharmaceuticals, surgical supplies or dental supplies to clinics and hospitals.

DMB: See "Data Monitoring Board (DMB)."

DNA: See "Deoxyribonucleic Acid (DNA)."

DNA Chip: A revolutionary tool used to identify mutations in genes like BRCA1 and BRCA2. The chip, which consists of a small glass plate encased in plastic, is manufactured using a process similar to the one used to make computer microchips. On the surface, each chip contains synthetic single-stranded DNA sequences identical to a normal gene.

Drug Utilization Review: A quantitative assessment of patient drug use and physicians' patterns of prescribing drugs in an effort to determine the usefulness of drug therapy.

DSMB: See "Data and Safety Monitoring Board (DSMB)."

Ecology: The study of relationships among all living organisms and the environment, especially the totality or pattern of interactions; a view that includes all plant and animal species and their unique contributions to a particular habitat.

Efficacy: A drug or medical product's ability to effectively produce beneficial results within a patient.

EFGCP: See "European Forum for Good Clinical Practices (EFGCP)."

ELA: See "Establishment License Application (ELA)."

Electronic Data Interchange (EDI): An accepted standard format for the exchange of data between various companies' networks. EDI allows for the transfer of e-mail as well as orders, invoices and other files from one company to another.

Electroporation: A health care technology that uses short pulses of electric current (DC) to create openings (pores) in the membranes of cancerous cells, thus leading to death of the cells. It has potential as a treatment for prostate cancer. In the laboratory, it is a means of introducing foreign proteins or DNA into living cells.

EMEA: The region comprised of Europe, the Middle East and Africa.

Endpoint: A clinical or laboratory measurement used to assess safety, efficacy or other trial objectives of a test article in a clinical trial.

Enterprise Resource Planning (ERP): An integrated information system that helps manage all aspects of a business, including accounting, ordering and human resources, typically across all locations of a major corporation or organization. ERP is considered to be a critical tool for management of large organizations. Suppliers of ERP tools include SAP and Oracle.

Enzyme: A protein that acts as a catalyst, affecting the chemical reactions in cells.

Epigenetics: A branch of biology focused on gene "silencers." Scientists involved in epigenetics are studying the function within a gene that regulates whether a gene is operating a full capacity or is toned down to a lower level. The level of operation of a given gene may lead to a higher risk of disease, such as certain types of cancer, within a patient. Epigenetics may be very effective at correcting abnormal gene expressions that cause cancer.

Epigenomics: The study of epigenetics. See "Epigenetics."

ERP: See "Enterprise Resource Planning (ERP)."

EST: See "Expressed Sequence Tags (EST)."

Establishment License Application (ELA): Required for the approval of a biologic (see "Biologics"). It permits a specific facility to manufacture a biological product for commercial purposes. Compare to "Product License Agreement (PLA)."

ESWL: See "Extracorporeal Shock Wave Lithotripter (ESWL)."

Ethanol: A clear, colorless, flammable, oxygenated hydrocarbon, also called ethyl alcohol. In the U.S., it is used as a gasoline octane enhancer and oxygenate in a 10% blend called E10. Ethanol can be used in higher concentrations (such as an 85% blend called E85) in vehicles designed for its use. It is typically produced chemically from ethylene or biologically from fermentation of various sugars from carbohydrates found in agricultural crops and cellulose residues from crops or wood. Grain ethanol production is typically based on corn or sugarcane. Cellulosic ethanol production is based on agricultural waste, such as wheat stalks, that has been treated with enzymes to break the waste down into component sugars.

Etiology: The study of the causes or origins of diseases.

EU: See "European Union (EU)."

EU Competence: The jurisdiction in which the EU can take legal action.

European Community (EC): See "European Union (EU)."

European Forum for Good Clinical Practices (EFGCP): The organization dedicated to finding common ground in Europe on the implementation of Good Clinical Practices. See "Good Clinical Practices (GCP)." www.efgcp.org

European Medicines Evaluation Agency (EMEA): The European agency responsible for supervising and coordinating applications for marketing medicinal products in the European Union (see "European Union (EU)" and "Committee on Proprietary Medicinal Products (CPMP)"). The EMEA is headquartered in the U.K. www.eudraportal.eudra.org

European Union (EU): A consolidation of European countries (member states) functioning as one body to facilitate trade. Previously known as the European Community (EC), the EU expanded to include much of Eastern Europe in 2004, raising the total number of member states to 25. In 2002, the EU launched a unified currency, the Euro. See europa.eu.int.

Expressed Sequence Tags (EST): Small pieces of DNA sequence (usually 200 to 500 nucleotides long) that are generated by sequencing either one or both ends of an expressed gene. The idea is to sequence bits of DNA that represent genes expressed in certain cells, tissues or organs from different organisms and use these tags to fish a gene out of a portion of chromosomal DNA by matching base pairs. See "Gene Expression."

Extracorporeal Shock Wave Lithotripter (ESWL): A medical device used for treating stones in the kidney or urethra. The device disintegrates kidney stones noninvasively through the transmission of acoustic shock waves directed at the stones.

Fast Track Development: An enhanced process at the FDA for rapid approval of drugs that treat certain life-threatening or extremely serious conditions. Some Fast Track drugs come to market in very short periods of time. The Fast Track designation is intended for drugs that address an unmet medical need, but is independent of Priority Review and Accelerated Approval. The benefits of Fast Track include scheduled meetings to seek FDA input into development as well as the option of submitting a New Drug Application in sections rather than all components at once.

FD&C Act: See "Federal Food Drug and Cosmetic Act (FD&C Act)."

FDA: See "Food and Drug Administration (FDA)."

Federal Food, Drug and Cosmetic Act (FD&C Act): A set of laws passed by the U.S. Congress, which controls, among other things, residues in food and feed.

Follow-on Biologics: A term used to describe generic versions of drugs that have been created using biotechnology. Because biotech drugs ("biologics") are made from living cells, a generic version of a drug probably won't be biochemically identical to the original branded version of the drug. Consequently, they are described as "follow-on" biologics to set them apart. Since these drugs won't be exactly the same as the originals, there are concerns that they may not be as safe or effective unless they go through clinical trials for proof of quality. In Europe, these drugs are referred to as "biosimilars." See "Biologics."

Food and Drug Administration (FDA): The U.S. government agency responsible for the enforcement of the Federal Food, Drug and Cosmetic Act, ensuring industry compliance with laws regulating products in commerce. The FDA's mission is to protect the public from harm and encourage technological advances that hold the promise of benefiting society. www.fda.gov

Formulary: A preferred list of drug products that typically limits the number of drugs available within a therapeutic class for purposes of drug purchasing, dispensing and/or reimbursement. A government body, third-party insurer or health plan, or an institution may compile a formulary. Some institutions or health plans develop closed (i.e.

restricted) formularies where only those drug products listed can be dispensed in that institution or reimbursed by the health plan. Other formularies may have no restrictions (open formulary) or may have certain restrictions such as higher patient cost-sharing requirements for off-formulary drugs.

Functional Foods: Food products that contain nutrients, such as vitamins, associated with certain health benefits. The nutrients may occur naturally, or the foods may have been enhanced with them.

Functional Genomics: The process of attempting to convert the molecular information represented by DNA into an understanding of gene functions and effects. To address gene function and expression specifically, the recovery and identification of mutant and over-expressed phenotypes can be employed. Functional genomics also entails research on the protein function (proteomics) or, even more broadly, the whole metabolism (metabolomics) of an organism.

Functional Imaging: The uses of PET scan, MRI and other advanced imaging technology to see how an area of the body is functioning and responding. For example, brain activity can be viewed, and the reaction of cancer tumors to therapies can be judged using functional imaging.

Functional Proteomics: The study of the function of all the proteins encoded by an organism's entire genome.

GCP: See "Good Clinical Practices (GCP)."

GDP: See "Gross Domestic Product (GDP)."

Gene: A working subunit of DNA; the carrier of inheritable traits.

Gene Chip: See "DNA Chip."

Gene Expression: The term used to describe the transcription of the information contained within the DNA (the repository of genetic information) into messenger RNA (mRNA) molecules that are then translated into the proteins that perform most of the critical functions of cells. Scientists study the kinds and amounts of mRNA produced by a cell to learn which genes are expressed, which in turn provides insights into how the cell responds to its changing needs.

Gene Knock-Out: The inhibition of gene expression through various scientific methods.

Gene Therapy: Treatment based on the alteration or replacement of existing genes. Genetic therapy involves splicing desired genes insolated from one patient into a second patient's cells in order to compensate for that patient's inherited genetic defect, or to enable that patient's body to better fight a specific disease.

Genetic Code: The sequence of nucleotides, determining the sequence of amino acids in protein synthesis.

Genetically Modified (GM) Foods: Food crops that are bioengineered to resist herbicides, diseases or insects; have higher nutritional value than non-engineered plants; produce a higher yield per acre; and/or last longer on the shelf. Additional traits may include resistance to temperature and moisture extremes. Agricultural animals also may be genetically modified organisms.

Genetically Modified Organism (GMO): An organism that has undergone genome modification by the insertion of a foreign gene. The genetic material of a GMO is not found through mating or natural recombination.

Genetically-Modified (GM): See "Genetically Modified (GM) Foods" and "Genetically Modified Organism (GMO)."

Genetics: The study of the process of heredity.

Genome: The genetic material (composed of DNA) in the chromosomes of a living organism.

Genomics: The study of genes, their role in diseases and our ability to manipulate them.

Genotype: The genetic constitution of an organism.

Global Warming: An increase in the near-surface temperature of the Earth. Global warming has occurred in the distant past as the result of natural influences, but the term is most often used to refer to a theory that warming occurs as a result of increased use of hydrocarbon fuels by man. See "Climate Change (Greenhouse Effect)."

Globalization: The increased mobility of goods, services, labor, technology and capital throughout the world. Although globalization is not a new development, its pace has increased with the advent of new technologies, especially in the areas of telecommunications, finance and shipping.

GLP: See "Good Laboratory Practices (GLP)."

GM: See "Genetically-Modified (GM) Foods."

GMO: See "Genetically Modified Organism (GMO)."

GMP: See "Good Manufacturing Practices (GMP)."

Good Clinical Practices (GCP): FDA regulations and guidelines that define the responsibilities of the key figures involved in a clinical trial, including the sponsor, the investigator, the monitor and the Institutional Review Board. See "Institutional Review Board (IRB)."

Good Laboratory Practices (GLP): A collection of regulations and guidelines to be used in laboratories where research is conducted on drugs, biologics or devices that are intended for submission to the FDA.

Good Manufacturing Practices (GMP): A collection of regulations and guidelines to be used in manufacturing drugs, biologics and medical devices.

Gross Domestic Product (GDP): The total value of a nation's output, income and expenditures produced with a nation's physical borders.

Gross National Product (GNP): A country's total output of goods and services from all forms of economic activity measured at market prices for one calendar year. It differs from Gross Domestic Product (GDP) in that GNP includes income from investments made in foreign nations.

HESC: See "Human Embryonic Stem Cell (HESC)."

HHS: See "U.S. Department of Health and Human Services (HHS)."

High-Throughput Screening (HTP): Makes use of techniques that allow for a fast and simple test on the presence or absence of a desirable structure, such as a specific DNA sequence. HTP screening often uses DNA chips or microarrays and automated data processing for large-scale screening, for instance, to identify new targets for drug development.

Human Embryonic Stem Cell (HESC): See "Stem Cells."

IEEE: The Institute of Electrical and Electronic Engineers. The IEEE sets global technical standards and acts as an authority in technical areas including computer engineering, biomedical technology, telecommunications, electric power, aerospace and consumer electronics, among others. www.ieee.org.

Imaging: In medicine, the viewing of the body's organs through external, high-tech means. This reduces the need for broad exploratory surgery. These advances, along with new types of surgical instruments, have made minimally invasive surgery possible. Imaging includes MRI (magnetic resonance imaging), CT (computed tomography or CAT scan), MEG (magnetoencephalography), improved x-ray technology, mammography, ultrasound and angiography.

Imaging Contrast Agent: A molecule or molecular complex that increases the intensity of the signal detected by an imaging technique, including MRI and ultrasound. An MRI contrast agent, for example, might contain gadolinium attached to a targeting antibody. The antibody would bind to a specific target, a metastatic melanoma cell for example, while the gadolinium would increase the magnetic signal detected by the MRI scanner.

Immunoassay: An immunological assay. Types include agglutination, complement-fixation, precipitation, immunodiffusion and electrophoretic assays. Each type of assay utilizes either a particular type of antibody or a specific support medium (such as a gel) to determine the amount of antigen present.

In Vitro: Laboratory experiments conducted in the test tube, or otherwise, without using live animals and/or humans.

In Vivo: Laboratory experiments conducted with live animals and/or humans.

IND: See "Investigational New Drug Application (IND)."

Indication: Refers to a specific disease, illness or condition for which a drug is approved as a treatment. Typically, a new drug is first approved for one indication. Then, an application to the FDA is later made for approval of additional indications.

Induced Pluripotent State Cell (IPSC): A human stem cell produced without human cloning or the use of human embryos or eggs. Adult cells are drawn from a skin biopsy and treated with four reprogramming factors, rendering cells that can produce all human cell types and grow indefinitely.

Industrial Biotechnology: The application of biotechnology to serve industrial needs. This is a rapidly growing field on a global basis. The current focus on industrial biotechnology is primarily on enzymes and other substances for renewable energy such as biofuels; chemicals such as pharmaceuticals, food additives, solvents and colorants; and bioplastics. Industrial biotech attempts to create synergies between biochemistry, genetics and microbiology in order to develop exciting new substances.

Informatics: See "Bioinformatics."

Informed Consent: Must be obtained in writing from people who agree to be clinical trial subjects prior to their enrollment in the study. The document must explain the risks associated with the study and treatment and describe alternative therapy available to the patient. A copy of the document must also be provided to the patient.

Information Technology (IT): The systems, including hardware and software, that move and store voice, video and data via computers and telecommunications.

Initial Public Offering (IPO): A company's first effort to sell its stock to investors (the public). Investors in an up-trending market eagerly seek stocks offered in many IPOs because the stocks of newly public companies that seem to have great promise may appreciate very rapidly in price, reaping great profits for those who were able to get the stock at the first offering. In the United States, IPOs are regulated by the SEC (U.S. Securities Exchange Commission) and by the state-level regulatory agencies of the states in which the IPO shares are offered.

Insertion Mutants: Mutants of genes that are obtained by inserting DNA, for instance through mobile DNA sequences. In plant research, the capacity of the bacterium Agrobacterium to introduce DNA into the plant genome is employed to induce mutants. In both cases, mutations lead to lacking or changing gene functions that are revealed by aberrant phenotypes. Insertion mutant isolation, and subsequent identification and analysis are employed in model plants such as Arabiopsis and in crop plants such as maize and rice.

Institutional Review Board (IRB): A group of individuals usually found in medical institutions that is responsible for reviewing protocols for ethical consideration (to ensure the rights of the patients). An IRB also evaluates the benefit-to-risk ratio of a new drug to see that the risk is acceptable for patient exposure. Responsibilities of an IRB are defined in FDA regulations.

Interactomics (Interactome): The study of the interactions between genes, RNA, proteins and metabolites within the cell.

Interferon: A type of biological response modifier (a substance that can improve the body's natural response to disease).

Investigational New Device Exemption (IDE): A document that must be filed with the FDA prior to initiating clinical trials of medical devices considered to pose a significant risk to human subjects.

Investigational New Drug Application (IND): A document that must be filed with the FDA prior to initiating clinical trials of drugs or biologics.

Investigator: In clinical trials, a clinician who agrees to supervise the use of an investigational drug, device or biologic in humans. Responsibilities of the investigator, as defined in FDA regulations, include administering the drug, observing and testing the patient, collecting data and monitoring the care and welfare of the patient.

Iontophoresis: The transfer of ions of medicine through the skin using a local electric current.

IRB: See "Institutional Review Board (IRB)."

ISO 9000, 9001, 9002, 9003: Standards set by the International Organization for Standardization. ISO 9000, 9001, 9002 and 9003 are the highest quality certifications awarded to organizations that meet exacting standards in their operating practices and procedures.

IT: See "Information Technology (IT)."

IT-Enabled Services (ITES): The portion of the Information Technology industry focused on providing business services, such as call centers, insurance claims processing and medical records transcription, by utilizing the power of IT, especially the Internet. Most ITES functions are considered to be back-office procedures. Also, see "Business Process Outsourcing (BPO)."

ITES: See "IT-Enabled Services (ITES)."

Kinase (Protein Kinase): Enzymes that influence certain basic functions within cells, such as cell division. Kinases catalyze the transfer of phosphates from ATP to proteins, thus causing changes in protein function. Defective kinases can lead to diseases such as cancer.

Knowledge Process Outsourcing (KPO): The use of outsourced and/or offshore workers to perform business tasks that require judgment and analysis. Examples include such professional tasks as patent research, legal research, architecture, design, engineering, market research, scientific research, accounting and tax return preparation. Also, see "Business Process Outsourcing (BPO)."

LAC: An acronym for Latin America and the Caribbean.

LDCs: See "Least Developed Countries (LDCs)."

Least Developed Countries (LDCs): Nations determined by the U.N. Economic and Social Council to be the poorest and weakest members of the international community. There are currently 50 LDCs, of which 34 are in Africa, 15 are in Asia Pacific and the remaining one (Haiti) is in Latin America. The top 10 on the LDC list, in descending order from top to 10th, are Afghanistan, Angola, Bangladesh, Benin, Bhutan, Burkina Faso, Burundi, Cambodia, Cape Verde and the Central African Republic. Sixteen of the LDCs are also Landlocked Least Developed Countries (LLDCs) which present them with additional difficulties often due to the high cost of transporting trade goods. Eleven of the LDCs are Small Island Developing States (SIDS), which are often at risk of extreme weather phenomenon

(hurricanes, typhoons, Tsunami); have fragile ecosystems; are often dependent on foreign energy sources; can have high disease rates for HIV/AIDS and malaria; and can have poor market access and trade terms.

Lifestyle Drug: Lifestyle drugs target a variety of medical conditions ranging from the painful to the inconvenient, including obesity, impotence, memory loss and depression, rather than illness or disease. Drug companies continue to develop lifestyle treatments for hair loss and skin wrinkles in an effort to capture their share of the huge anti-aging market aimed at the baby-boomer generation.

Ligand: Any atom or molecule attached to a central atom in a complex compound.

Liposome: A micro or nanoscale lipid or phospholipid layer enclosing a liquid core used for transport for particular molecules or biological structures or as a model for membranes.

M3 (Measurement): Cubic meters.

Marketing: Includes all planning and management activities and expenses associated with the promotion of a product or service. Marketing can encompass advertising, customer surveys, public relations and many other disciplines. Marketing is distinct from selling, which is the process of sell-through to the end user.

Mass Spectrometry: Usage of analytical devices that can determine the mass (or molecular weight) of proteins and nucleic acids, the sequence of protein molecules, the chemical organization of almost all substances and the identification of gram-negative and gram-positive microorganisms.

Medical Device: See "Device."

MEMS: See "Microelectromechanical Systems (MEMS)."

Metabolomics: The study of low-molecular-weight materials produced during genomic expression within a cell. Such studies can lead to a better understanding of how changes within genes and proteins affect the function of cells.

Metagenomics: An advanced form of genomics that increases the understanding of complex microbial systems.

Micrelectromechanical Systems (MEMS): Micron scale structures that transduce signals between electronic and mechanical forms. MEMS typically combine electronic sensors or switches with mechanical features such as gears, pumps or motors. A common use is the accelerometer found in electronic game playing machines and automobile airbags.

Microarray: A DNA analysis tool consisting of a microscopic ordered array that enables parallel analysis of complex biochemical samples; used to analyze how large numbers of genes interact with each other; used for genotyping, mapping, sequencing, sequence detection; usually constructed by applying biomolecules onto a slide or chip and then scanning with microscope or other imaging equipment.

Microfluidics: Refers to the manipulation of microscopic amounts of fluid, generally for analysis in microarrays using high throughput screening. The volume of the liquid involved is on the nanolitre scale.

Molecular Imaging: An emerging field in which advanced biology on the molecular level is combined with noninvasive imaging to determine the presence of certain proteins and other important genetic material.

Monoclonal Antibodies (mAb, Human Monoclonal Antibody): Antibodies that have been cloned from a single antibody and massed produced as a therapy or diagnostic test. An example is an antibody specific to a certain protein found in cancer cells.

Nanocantilever: The simplest micro-electro-mechanical system (MEMS) that can be easily machined and mass-produced via the same techniques used to make computer chips. The ability to detect extremely small displacements make nanocantilever beams an ideal device for detecting extremely small forces, stresses and masses. Nanocantilevers coated with antibodies, for example, will bend from the mass added when substrate binds to its antibody, providing a detector capable of sensing the presence of single molecules of clinical importance.

Nanoparticle: A nanoscale spherical or capsule-shaped structure. Most, though not all, nanoparticles are hollow, which provides a central reservoir that can be filled with anticancer drugs, detection agents, or chemicals, known as reporters, that can signal if a drug is having a therapeutic effect. The surface of a nanoparticle can also be adorned with various targeting agents, such as antibodies, drugs, imaging agents, and reporters. Most nanoparticles are constructed to be small enough to pass through blood capillaries and enter cells.

Nanopharmaceuticals: Nanoscale particles that modulate drug transport in drug uptake and delivery applications.

Nanoshell: A nanoparticle composed of a metallic shell surrounding a semiconductor. When nanoshells reach a target cancer cell, they can be irradiated with near-infrared light or excited with a magnetic field, either of which will cause the nanoshell to become hot, killing the cancer cell.

Nanotechnology: The science of designing, building or utilizing unique structures that are smaller than 100 nanometers (a nanometer is one billionth of a meter). This involves microscopic structures that are no larger than the width of some cell membranes.

Nanowires: A nanometer-scale wire made of metal atoms, silicon, or other materials that conduct electricity. Nanowires are built atom by atom on a solid surface, often as part of a microfluidic device. They can be coated with molecules such as antibodies that will bind to proteins and other substances of interest to researchers and clinicians. By the very nature of their nanoscale size, nanowires are incredibly sensitive to such binding events and respond by altering the electrical current flowing through them, and thus can form the basis of ultra sensitive molecular detectors.

National Drug Code (NDC): An identifying drug number maintained by the FDA.

National Institutes of Health (NIH): A branch of the U.S. Public Health Service that conducts biomedical research. www.nih.gov

NCE: See "New Chemical Entity (NCE)."

NDA: See "New Drug Application (NDA)."

New Chemical Entity (NCE): See "New Molecular Entity (NME)."

New Drug Application (NDA): An application requesting FDA approval, after completion of the all-important Phase III Clinical Trials, to market a new drug for human use in the U.S. The drug may contain chemical compounds that were previously approved by the FDA as distinct molecular entities suitable for use in drug trials. See "New Molecular Entity (NME)."

New Molecular Entity (NME): Defined by the FDA as a medication containing chemical compound that has never before been approved for marketing in any form in the U.S. An NME is sometimes referred to as a New Chemical Entity (NCE). Also, see "New Drug Application (NDA)."

NIH: See "National Institutes of Health (NIH)."

NME: See "New Molecular Entity (NME)."

Nonclinical Studies: In vitro (laboratory) or in vivo (animal) pharmacology, toxicology and pharmacokinetic studies that support the testing of a product in humans. Usually at least two species are evaluated prior to Phase I clinical trials. Nonclinical studies continue throughout all phases of research to evaluate long-term safety issues.

Nucleic Acid: A large molecule composed of nucleotides. Nucleic acids include RNA, DNA and antisense oligonucleotides.

Nutraceutical: Nutrient + pharmaceutical – a food or part of a food that has been isolated and sold in a medicinal form and claims to offer benefits such as the treatment or prevention of disease.

Nutraceuticals: Food products and dietary supplements that may have certain health benefits. Nutraceuticals may offer specific vitamins or minerals. Also see "Functional Foods."

Nutrigenomics: The study of how food interacts with genes.

ODM: See "Original Design Manufacturer (ODM)."

OECD: See "Organisation for Economic Co-operation and Development (OECD)."

OEM: See "Original Equipment Manufacturer (OEM)."

Offshoring: The rapidly growing tendency among U.S., Japanese and Western European firms to send knowledge-based and manufacturing work overseas. The intent is to take advantage of lower wages and operating costs in such nations as China, India, Hungary and Russia. The choice of a nation for offshore work may be influenced by such factors as language and education of the local workforce, transportation systems or natural resources. For example, China and India are graduating high numbers of skilled engineers and scientists from their universities. Also, some nations are noted for large numbers of workers skilled in the English language, such as the Philippines and India. Also see "Captive Offshoring" and "Outsourcing."

Oncogene: A unit of DNA that normally directs cell growth, but which can also promote or allow the uncontrolled growth of cancer.

Oncology: The diagnosis, study and treatment of cancer.

Onshoring: The opposite of "offshoring." Providing or maintaining manufacturing or services within or nearby a company's domestic location. Sometimes referred to as reshoring.

Organisation for Economic Co-operation and Development (OECD): A group of 31 countries that are strongly committed to the market economy and democracy. Some of the OECD members include Japan, the U.S., Spain, Germany, Australia, Korea, the U.K., Canada and Mexico. Although not members, Chile, Estonia, Israel, Russia and Slovenia are invited to member talks; and Brazil, China, India, Indonesia and South Africa have enhanced engagement policies with the OECD. The Organisation provides statistics, as well as social and economic data; and researches social changes, including patterns in evolving fiscal policy, agriculture, technology, trade, the environment and other areas. It publishes over 250 titles annually; publishes a corporate magazine, the OECD Observer; has radio and TV studios; and has centers in Tokyo, Washington, D.C., Berlin and Mexico City that distributed the Organisation's work and organizes events.

Original Design Manufacturer (ODM): A contract manufacturer that offers complete, end-to-end design, engineering and manufacturing services. ODMs design and build products, such as consumer electronics, that client companies can then brand and sell as their own. For example, a large percentage of laptop computers, cell phones and PDAs are made by ODMs. Also see "Original Equipment Manufacturer (OEM)" and "Contract Manufacturer."

Original Equipment Manufacturer (OEM): A company that manufactures a product or component for sale to a customer that will integrate the component into a final product or assembly. The OEM's customer will distribute the end product or resell it to an end user. For example, a personal computer made under a brand name by a given company may contain various components, such as hard drives, graphics cards or speakers, manufactured by several different OEM "vendors," but the firm doing the final assembly/manufacturing process is the final manufacturer. Also see "Original Design Manufacturer (ODM)" and "Contract Manufacturer."

Orphan Drug: A drug or biologic designated by the FDA as providing therapeutic benefit for a rare disease affecting less than 200,000 people in the U.S. Companies that market orphan drugs are granted a period of market exclusivity in return for the limited commercial potential of the drug.

OTC: See "Over-the-Counter Drugs (OTC)."

Outsourcing: The hiring of an outside company to perform a task otherwise performed internally by the company, generally with the goal of lowering costs and/or streamlining work flow. Outsourcing contracts are generally several years in length. Companies that hire outsourced services providers often prefer to focus on their core strengths while sending more routine tasks outside for others to perform. Typical outsourced services include the running of human resources departments, telephone call centers and computer departments. When outsourcing is performed overseas, it may be referred to as offshoring. Also see "Offshoring."

Over-the-Counter Drugs (OTC): FDA-regulated products that do not require a physician's prescription. Some examples are aspirin, sunscreen, nasal spray and sunglasses.

Patent: A property right granted by the U.S. government to an inventor to exclude others from making, using, offering for sale, or selling the invention throughout the U.S. or importing the invention into the U.S. for a limited time in exchange for public disclosure of the invention when the patent is granted.

Pathogen: Any microorganism (e.g., fungus, virus, bacteria or parasite) that causes a disease.

PCR: See "Polymerase Chain Reaction (PCR)."

Peer Review: The process used by the scientific community, whereby review of a paper, project or report is obtained through comments of independent colleagues in the same field.

Pharmacodynamics (PD): The study of reactions between drugs and living systems. It can be thought of as the study of what a drug does to the body.

Pharmacoeconomics: The study of the costs and benefits associated with various drug treatments.

Pharmacogenetics: The investigation of the different reactions of human beings to drugs and the underlying genetic predispositions. The differences in reaction are mainly caused by mutations in certain enzymes responsible for drug metabolization. As a result, the degradation of the active substance can lead to harmful by-products, or the drug might have no effect at all.

Pharmacogenomics: The use of the knowledge of DNA sequences for the development of new drugs.

Pharmacokinetics (PK): The study of the processes of bodily absorption, distribution, metabolism and excretion of compounds and medicines. It can be thought of as the study of what the body does to a drug. See "Absorption, Distribution, Metabolism and Excretion (ADME)."

Pharmacology: The science of drugs, their characteristics and their interactions with living organisms.

Pharmacy Benefit Manager (PBM): An organization that provides administrative services in processing and analyzing prescription claims for pharmacy benefit and coverage programs. Many PBMs also operate mail order pharmacies or have arrangements to include prescription availability through mail order pharmacies.

Phase I Clinical Trials: Studies in this phase include initial introduction of an investigational drug into humans. These studies are closely monitored and are usually conducted in healthy volunteers. Phase I trials are conducted after the completion of extensive nonclinical or pre-clinical trials not involving humans. Phase I studies include the determination of clinical pharmacology, bioavailability, drug interactions and side effects associated with increasing doses of the drug.

Phase II Clinical Trials: Include randomized, masked, controlled clinical studies conducted to evaluate the effectiveness of a drug for a particular indication(s). During Phase II trials, the minimum effective dose and dosing intervals should be determined.

Phase III Clinical Trials: Consist of controlled and uncontrolled trials that are performed after preliminary evidence of effectiveness of a drug has been established. They are conducted to document the safety and efficacy of the drug, as well as to determine adequate directions (labeling) for use by the physician. A specific patient population needs to be clearly identified from the results of these studies. Trials during Phase III are conducted using a large number of patients to determine the frequency of adverse events and to obtain data regarding intolerance.

Phase IV Clinical Trials: Conducted after approval of a drug has been obtained to gather data supporting new or revised labeling, marketing or advertising claims.

Phenomics: The study of how an organism's structure responds to such things as toxins or drugs.

Phenotype: Observable characteristics of an organism produced by the organism's genotype interacting with the environment.

Phylogenetic Systematics: The field of biology that deals with identifying and understanding the evolutionary relationships among the many different kinds of life on earth, both living (extant) and dead (extinct).

Pivotal Studies: In clinical trials, a Phase III trial that is designed specifically to support approval of a product. These studies are well-controlled (usually by placebo) and are generally designed with input from the FDA so that they will provide data that is adequate to support approval of the product. Two pivotal studies are required for drug product approval, but usually only one study is required for biologics.

PLA: See "Product License Agreement (PLA)."

Plant Patent: A plant patent may be granted by the U.S. Patent and Trademark Office to anyone who invents or discovers and asexually reproduces any distinct and new variety of plant.

Platform or Technology Platform Companies: Firms hoping to profit by providing information systems, software, databases and related support to biopharmaceutical companies.

PMA: See "Pre-Market Approval (PMA)."

PMCs: Postmarketing study commitments. PMCs are clinical studies that are not required for FDA initial approval of a drug, but the FDA nonetheless feels these studies will provide important data on a newly marketed drug. Consequently, the drug firm makes a commitment for continuing studies.

Polymerase Chain Reaction (PCR): In molecular biology, PCR is a technique used to reproduce or amplify small, selected sections of DNA or RNA for analysis. It enables researchers to create multiple copies of a given sequence.

Positional Cloning: The identification and cloning of a specific gene, with chromosomal location as the only source of information about the gene.

Post-Marketing Surveillance: The FDA's ongoing safety monitoring of marketed drugs.

PPP: See "Purchasing Power Parity (PPP) or Point-to-Point Protocol (PPP)."

Preclinical Studies: See "Nonclinical Studies."

Pre-Market Approval (PMA): Required for the approval of a new medical device or a device that is to be used for life-sustaining or life-supporting purposes, is implanted in the human body or presents potential risk of illness or injury.

Priority Reviews: The FDA places some drug applications that appear to promise "significant improvements" over existing drugs for priority approval, with a goal of returning approval within six months.

Product License Agreement (PLA): See "Biologics License Application (BLA)."

Protein: A large, complex molecule made up of amino acids. Proteins are essential functional or structural components of living cells, and they may include vital enzymes, antibodies or hormones. Genes provide the codes for synthesis of specific proteins.

Proteome: The genetic material (composed of amino acids) in the chromosomes of a living organism.

Proteomics: The study of gene expression at the protein level, by the identification and characterization of proteins present in a biological sample.

Public Health Service (PHS): May stand for the Public Health Service Act, a law passed by the U.S. Congress in 1944. PHS also may stand for the Public Health Service itself, a U.S. government agency established by an act of Congress in July 1798, originally authorizing hospitals for the care of American merchant seamen. Today, the Public Health Service sets national health policy; conducts medical and biomedical research; sponsors programs for disease control and mental health; and enforces laws to assure the safety and efficacy of drugs, foods, cosmetics and medical devices. The FDA (Food and Drug Administration) is part of the Public Health Service, as are the Centers for Disease Control and Prevention (CDC).

Purchasing Power Parity (PPP): Currency conversion rates that attempt to reflect the actual purchasing power of a currency in its home market, as opposed to examining price levels and comparing an exchange rate. PPPs are always given in the national currency units per U.S. dollar.

Qdots: See "Quantum Dots (Qdots)."

QOL: See "Quality of Life (QOL)."

Quality of Life (QOL): In medicine, an endpoint of therapeutic assessment used to adjust measures of effectiveness for clinical decision-making. Typically, QOL endpoints measure the improvement of a patient's day-to-day living as a result of specific therapy.

Quantum Dots (Qdots): Nanometer sized semiconductor particles, made of cadmium selenide (CdSe), cadmium sulfide (CdS) or cadmium telluride (CdTe) with an inert polymer coating. The semiconductor material used for the core is chosen based upon the emission wavelength range being targeted: CdS for UV-blue, CdSe for the bulk of the visible spectrum, CdTe for the far red and near-infrared, with the particle's size determining the exact color of a given quantum dot. The polymer coating safeguards cells from cadmium toxicity but also affords the opportunity to attach any variety targeting molecules, including monoclonal antibodies directed to tumor-specific biomarkers. Because of their small size, quantum dots can function as cell- and even molecule-specific markers that will not interfere with the normal workings of a cell. In addition, the availability of quantum dots of different colors provides a powerful tool for following the actions of multiple cells and molecules simultaneously.

R&D: Research and development. Also see "Applied Research" and "Basic Research."

Radioisotope: An object that has varying properties that allows it to penetrate other objects at different rates. For example, a sheet of paper can stop an alpha particle, a beta particle can penetrate tissues in the body and a gamma ray can penetrate concrete. The varying penetration capabilities allow radioisotopes to be used in different ways. (Also called radioactive isotope or radionuclide.)

Receptor: Proteins in or on a cell that selectively bind a specific substance called a ligand. See "Ligand."

Recombination: The natural process of breaking and rejoining DNA strands to produce new combinations of genes.

Reporter Gene: A gene that is inserted into DNA by researchers in order to indicate when a linked gene is successfully expressed or when signal transduction has taken place in a cell.

Return on Investment (ROI): A measure of a company's profitability, expressed in percentage as net profit (after taxes) divided by total dollar investment.

Ribonucleic Acid (RNA): A macromolecule found in the nucleus and cytoplasm of cells; vital in protein synthesis.

RNAi (RNA interference): A biological occurrence where double-stranded RNA is used to silence genes.

Safe Medical Devices Act (SMDA): An act that amends the Food, Drug and Cosmetic Act to impose additional regulations on medical devices. The act became law in 1990.

Semiconductor: A generic term for a device that controls electrical signals. It specifically refers to a material (such as silicon, germanium or gallium arsenide) that can be altered either to conduct electrical current or to block its passage. Carbon nanotubes may eventually be used as semiconductors. Semiconductors are partly responsible for the miniaturization of modern electronic devices, as they are vital components in computer memory and processor chips. The manufacture of semiconductors is carried out by small firms, and by industry giants such as Intel and Advanced Micro Devices.

Single Nucleotide Polymorphisms (SNPs): Stable mutations consisting of a change at a single base in a DNA molecule. SNPs can be detected by HTP analyses, such as gene chips, and they are then mapped by DNA sequencing. They are the most common type of genetic variation.

SMDA: See "Safe Medical Devices Act (SMDA)."

SNP: See "Single-Nucleotide Polymorphisms (SNPs)."

Sponsor: The individual or company that assumes responsibility for the investigation of a new drug, including compliance with the FD&C Act and regulations. The sponsor may be an individual, partnership, corporation or governmental agency and may be a manufacturer, scientific institution or

investigator regularly and lawfully engaged in the investigation of new drugs. The sponsor assumes most of the legal and financial responsibility of the clinical trial.

Standard Drugs: NCEs (New Chemical Entities) that the FDA feels offer few advantages over existing drugs that are therefore given lower status for review.

Stem Cells: Cells found in human bone marrow, the blood stream and the umbilical cord that can be replicated indefinitely and can turn into any type of mature blood cell, including platelets, white blood cells or red blood cells. Also referred to as pluripotent cells.

Study Coordinator: See "Coordinator."

Subsidiary, Wholly-Owned: A company that is wholly controlled by another company through stock ownership.

Supply Chain: The complete set of suppliers of goods and services required for a company to operate its business. For example, a manufacturer's supply chain may include providers of raw materials, components, custom-made parts and packaging materials.

Synthetic Biology: Synthetic biology can be defined as the design and construction of new entities, including enzymes and cells, or the reformatting of existing biological systems. This science capitalizes on previous advances in molecular biology and systems biology, by applying a focus on the design and construction of unique core components that can be integrated into larger systems in order to solve specific problems.

Systems Biology: The use of combinations of advanced computer hardware, software and database technologies to take a systemic approach to biological research. Advanced technologies will enable scientists to view genetic predisposition by integrating information about entire biological systems, from DNA to proteins to cells to tissues.

T Cell (T-Cell): A white blood cell that carries out immune system responses. The T cell originates in the bone marrow and matures in the thymus.

Targets: The proteins involved in a specific disease. Drug compounds concentrate on specific targets in order to have the greatest positive effect and cut down on the incidence of side effects.

Taste Masking: The creation of a barrier between a drug molecule and taste receptors so the drug is easier to take. It masks bitter or unpleasant tastes.

TESS: See "Adverse Event (AE)."

Toxicogenomics: The study of the relationship between responses to toxic substances and the resulting genetic changes.

Trial Coordinator: See "Coordinator."

U.S. Department of Health and Human Services (HHS): This agency has more than 300 major programs related to human health and welfare, the largest of which is Medicare. See www.hhs.gov

Utility Patent: A utility patent may be granted by the U.S. Patent and Trademark Office to anyone who invents or discovers any new, useful, and non-obvious process, machine, article of manufacture, or composition of matter, or any new and useful improvement thereof.

Validation of Data: The procedure carried out to ensure that the data contained in a final clinical trial report match the original observations.

Value Added Tax (VAT): A tax that imposes a levy on businesses at every stage of manufacturing based on the value it adds to a product. Each business in the supply chain pays its own VAT and is subsequently repaid by the next link down the chain; hence, a VAT is ultimately paid by the consumer, being the last link in the supply chain, making it comparable to a sales tax. Generally, VAT only applies to goods bought for consumption within a given country; export goods are exempt from VAT, and purchasers from other countries taking goods back home may apply for a VAT refund.

World Health Organization (WHO): A United Nations agency that assists governments in strengthening health services, furnishing technical assistance and aid in emergencies, working on the prevention and control of epidemics and promoting cooperation among different countries to improve nutrition, housing, sanitation, recreation and other aspects of environmental hygiene. Any country that is a member of the United Nations may become a member of the WHO by accepting its constitution. The WHO currently has 191 member states.

World Trade Organization (WTO): One of the only globally active international organizations dealing with the trade rules between nations. Its goal is to assist the free flow of trade goods, ensuring a smooth, predictable supply of goods to help raise the quality of life of member citizens. Members form consensus decisions that are then ratified by their respective parliaments. The WTO's conflict resolution process generally emphasizes interpreting existing commitments and agreements, and discovers how to ensure trade policies to conform to those agreements, with the ultimate aim of avoiding military or political conflict.

WTO: See "World Trade Organization (WTO)."

Xenotransplantation: The science of transplanting organs such as kidneys, hearts or livers into humans from other mammals, such as pigs or other agricultural animals grown with specific traits for this purpose.

Zinc Fingers: Naturally occurring proteins that bind to DNA to produce desired genetic effects. For example, zinc fingers used in a plant can alter its yield, taste or resistance to drought or insects. They afford very precise changes to DNA which translates into better control when modifying plants and quicker development times compared to typical genetic modification.

Zoonosis: An animal disease that can be transferred to man.

INTRODUCTION

PLUNKETT'S BIOTECH & GENETICS INDUSTRY ALMANAC, the ninth edition of our guide to the biotech and genetics field, is designed to be used as a general source for researchers of all types.

The data and areas of interest covered are intentionally broad, ranging from the ethical questions facing biotechnology, to emerging technology, to an in-depth look at the major for-profit firms (which we call "THE BIOTECH 350") within the many industry sectors that make up the biotechnology and genetics arena.

This reference book is designed to be a general source for researchers. It is especially intended to assist with market research, strategic planning, employment searches, contact or prospect list creation and financial research, and as a data resource for executives and students of all types.

PLUNKETT'S BIOTECH & GENETICS INDUSTRY ALMANAC takes a rounded approach for the general reader. This book presents a complete overview of the entire biotechnology and genetics system (see "How To Use This Book"). For example, you will find trends in the biopharmaceuticals market, along with easy-to-use charts and tables on all facets of biotechnology in general: from the sales and profits of the major drug companies to the amounts of time required in the various stages of drug approval.

THE BIOTECH 350 is our unique grouping of the biggest, most successful corporations in all segments of the global biotechnology and genetics industry. Tens of thousands of pieces of information, gathered from a wide variety of sources, have been researched and are presented in a unique form that can be easily understood. This section includes thorough indexes to THE BIOTECH 350, by geography, industry, sales, brand names, subsidiary names and many other topics. (See Chapter 4.)

Especially helpful is the way in which PLUNKETT'S BIOTECH & GENETICS INDUSTRY ALMANAC enables readers who have no business or scientific background to readily compare the financial records and growth plans of large biotech companies and major industry groups. You'll see the mid-term financial record of each firm, along with the impact of earnings, sales and strategic plans on each company's potential to fuel growth, to serve new markets and to provide investment and employment opportunities.

No other source provides this book's easy-to-understand comparisons of growth, expenditures, technologies, corporations and many other items of great importance to people of all types who may be studying this, one of the most exciting industries in the world today.

By scanning the data groups and the unique indexes, you can find the best information to fit your personal research needs. The major growth companies in biotechnology and genetics are profiled and then ranked using several different groups of specific criteria. Which firms are the biggest employers? Which companies earn the most profits? These things and much more are easy to find.

In addition to individual company profiles, an overview of biotechnology markets and trends is provided. This book's job is to help you sort through easy-to-understand summaries of today's trends in a quick and effective manner.

Whatever your purpose for researching the biotechnology and genetics field, you'll find this book to be a valuable guide. Nonetheless, as is true with all resources, this volume has limitations that the reader should be aware of:

- Financial data and other corporate information can change quickly. A book of this type can be no more current than the data that was available as of the time of editing. Consequently, the financial picture, management and ownership of the firm(s) you are studying may have changed since the date of this book. For example, this almanac includes the most up-to-date sales figures and profits available to the editors as of mid 2010. That means that we have typically used corporate financial data as of late-2009.

- Corporate mergers, acquisitions and downsizing are occurring at a very rapid rate. Such events may have created significant change, subsequent to the publishing of this book, within a company you are studying.

- Some of the companies in THE BIOTECH 350 are so large in scope and in variety of business endeavors conducted within a parent organization, that we have been unable to completely list all subsidiaries, affiliations, divisions and activities within a firm's corporate structure.

- This volume is intended to be a general guide to a vast industry. That means that researchers should look to this book for an overview and, when

conducting in-depth research, should contact the specific corporations or industry associations in question for the very latest changes and data. Where possible, we have listed contact names, toll-free telephone numbers and Internet site addresses for the companies, government agencies and industry associations involved so that the reader may get further details without unnecessary delay.

- Tables of industry data and statistics used in this book include the latest numbers available at the time of printing, generally through late-2009. In a few cases, the only complete data available was for earlier years.

- We have used exhaustive efforts to locate and fairly present accurate and complete data. However, when using this book or any other source for business and industry information, the reader should use caution and diligence by conducting further research where it seems appropriate. We wish you success in your endeavors, and we trust that your experience with this book will be both satisfactory and productive.

Jack W. Plunkett
Houston, Texas
August 2010

HOW TO USE THIS BOOK

The two primary sections of this book are devoted first to the biotechnology and genetics industry as a whole and then to the "Individual Data Listings" for THE BIOTECH 350. If time permits, you should begin your research in the front chapters of this book. Also, you will find lengthy indexes in Chapter 4 and in the back of the book.

THE BIOTECH AND GENETICS INDUSTRY

Glossary: A short list of biotech and genetics industry terms.

Chapter 1: Major Trends Affecting the Biotech & Genetics Industry. This chapter presents an encapsulated view of the major trends and technologies that are creating rapid changes in the biotechnology and genetics industry today.

Chapter 2: Biotech & Genetics Industry Statistics. This chapter presents in-depth statistics on biotechnology markets, spending, research, pharmaceuticals and more.

Chapter 3: Important Biotech & Genetics Industry Contacts – Addresses, Telephone Numbers and Internet Sites. This chapter covers contacts for important government agencies, biotech organizations and trade groups. Included are numerous important Internet sites.

THE BIOTECH 350

Chapter 4: THE BIOTECH 350: Who They Are and How They Were Chosen. The companies compared in this book were carefully selected from the biotech and genetics industry, largely in the United States. 76 of the firms are based outside the U.S. For a complete description, see THE BIOTECH 350 indexes in this chapter.
   Individual Data Listings:
   Look at one of the companies in THE BIOTECH 350's Individual Data Listings. You'll find the following information fields:
   Company Name:
   The company profiles are in alphabetical order by company name. If you don't find the company you are seeking, it may be a subsidiary or division of one of the firms covered in this book. Try looking it up in the Index by Subsidiaries, Brand Names and Selected Affiliations in the back of the book.
   Ranks:
   Industry Group Code: An NAIC code used to group companies within like segments. (See Chapter 4 for a list of codes.)

<u>Ranks Within This Company's Industry Group:</u>
Ranks, within this firm's segment only, for annual sales and annual profits, with 1 being the highest rank.

Business Activities:

A grid arranged into six major industry categories and several sub-categories. A "Y" indicates that the firm operates within the sub-category. A complete Index by Industry is included in the beginning of Chapter 4.

Types of Business:

A listing of the primary types of business specialties conducted by the firm.

Brands/Divisions/Affiliations:

Major brand names, operating divisions or subsidiaries of the firm, as well as major corporate affiliations—such as another firm that owns a significant portion of the company's stock. A complete Index by Subsidiaries, Brand Names and Selected Affiliations is in the back of the book.

Contacts:

The names and titles up to 27 top officers of the company are listed, including human resources contacts.

Address:

The firm's full headquarters address, the headquarters telephone, plus toll-free and fax numbers where available. Also provided is the World Wide Web site address.

Financials:

<u>Annual Sales (2009 or the latest fiscal year available to the editors, plus up to four previous years):</u> These are stated in thousands of dollars (add three zeros if you want the full number). This figure represents consolidated worldwide sales from all operations. 2009 figures may be estimates.

<u>Annual Profits (2009 or the latest fiscal year available to the editors, plus up to four previous years):</u> These are stated in thousands of dollars (add three zeros if you want the full number). This figure represents consolidated, after-tax net profit from all operations. 2009 figures may be estimates.

<u>Stock Ticker, International Exchange, Parent Company:</u> When available, the unique stock market symbol used to identify this firm's common stock for trading and tracking purposes is indicated. Where appropriate, this field may contain "private" or "subsidiary" rather than a ticker symbol. If the firm is a publicly-held company headquartered outside of the U.S., its international ticker and exchange are given. If the firm is a subsidiary, its parent company is listed.

<u>Total Number of Employees:</u> The approximate total number of employees, worldwide, as of the end of 2009 (or the latest data available to the editors).

Apparent Salaries/Benefits:

(The following descriptions generally apply to U.S. employers only.)

A "Y" in appropriate fields indicates "Yes."

Due to wide variations in the manner in which corporations report benefits to the U.S. Government's regulatory bodies, not all plans will have been uncovered or correctly evaluated during our effort to research this data. Also, the availability to employees of such plans will vary according to the qualifications that employees must meet to become eligible. For example, some benefit plans may be available only to salaried workers—others only to employees who work more than 1,000 hours yearly. Benefits that are available to employees of the main or parent company may not be available to employees of the subsidiaries. In addition, employers frequently alter the nature and terms of plans offered.

NOTE: Generally, employees covered by wealth-building benefit plans do not *fully* own ("vest in") funds contributed on their behalf by the employer until as many as five years of service with that employer have passed. All pension plans are voluntary—that is, employers are not obligated to offer pensions.

<u>Pension Plan:</u> The firm offers a pension plan to qualified employees. In this case, in order for a "Y" to appear, the editors believe that the employer offers a defined benefit or cash balance pension plan (see discussions below).The type and generosity of these plans vary widely from firm to firm. Caution: Some employers refer to plans as "pension" or "retirement" plans when they are actually 401(k) savings plans that require a contribution by the employee.

- <u>Defined Benefit Pension Plans:</u> Pension plans that do not require a contribution from the employee are infrequently offered. However, a few companies, particularly larger employers in high-profit-margin industries, offer defined benefit pension plans where the employee is guaranteed to receive a set pension benefit upon retirement. The amount of the benefit is determined by the years of service with the company and the employee's salary during the later years of employment. The longer a person works for the employer, the higher the retirement benefit. These defined benefit plans are funded entirely by the employer. The benefits, up to a reasonable limit, are guaranteed by the Federal Government's Pension Benefit Guaranty

Corporation. These plans are not portable—if you leave the company, you cannot transfer your benefits into a different plan. Instead, upon retirement you will receive the benefits that vested during your service with the company. If your employer offers a pension plan, it must give you a summary plan description within 90 days of the date you join the plan. You can also request a summary annual report of the plan, and once every 12 months you may request an individual benefit statement accounting of your interest in the plan.

- Defined Contribution Plans: These are quite different. They do not guarantee a certain amount of pension benefit. Instead, they set out circumstances under which the employer will make a contribution to a plan on your behalf. The most common example is the 401(k) savings plan. Pension benefits are not guaranteed under these plans.

- Cash Balance Pension Plans: These plans were recently invented. These are hybrid plans—part defined benefit and part defined contribution. Many employers have converted their older defined benefit plans into cash balance plans. The employer makes deposits (or credits a given amount of money) on the employee's behalf, usually based on a percentage of pay. Employee accounts grow based on a predetermined interest benchmark, such as the interest rate on Treasury Bonds. There are some advantages to these plans, particularly for younger workers: a) The benefits, up to a reasonable limit, are guaranteed by the Pension Benefit Guaranty Corporation. b) Benefits are portable—they can be moved to another plan when the employee changes companies. c) Younger workers and those who spend a shorter number of years with an employer may receive higher benefits than they would under a traditional defined benefit plan.

ESOP Stock Plan (Employees' Stock Ownership Plan): This type of plan is in wide use. Typically, the plan borrows money from a bank and uses those funds to purchase a large block of the corporation's stock. The corporation makes contributions to the plan over a period of time, and the stock purchase loan is eventually paid off. The value of the plan grows significantly as long as the market price of the stock holds up. Qualified employees are allocated a share of the plan based on their length of service and their level of salary. Under federal regulations, participants in ESOPs are allowed to diversify their account holdings in set percentages that rise as the

employee ages and gains years of service with the company. In this manner, not all of the employee's assets are tied up in the employer's stock.

Savings Plan, 401(k): Under this type of plan, employees make a tax-deferred deposit into an account. In the best plans, the company makes annual matching donations to the employees' accounts, typically in some proportion to deposits made by the employees themselves. A good plan will match one-half of employee deposits of up to 6% of wages. For example, an employee earning $30,000 yearly might deposit $1,800 (6%) into the plan. The company will match one-half of the employee's deposit, or $900. The plan grows on a tax-deferred basis, similar to an IRA. A very generous plan will match 100% of employee deposits. However, some plans do not call for the employer to make a matching deposit at all. Other plans call for a matching contribution to be made at the discretion of the firm's board of directors. Actual terms of these plans vary widely from firm to firm. Generally, these savings plans allow employees to deposit as much as 15% of salary into the plan on a tax-deferred basis. However, the portion that the company uses to calculate its matching deposit is generally limited to a maximum of 6%. Employees should take care to diversify the holdings in their 401(k) accounts, and most people should seek professional guidance or investment management for their accounts.

Stock Purchase Plan: Qualified employees may purchase the company's common stock at a price below its market value under a specific plan. Typically, the employee is limited to investing a small percentage of wages in this plan. The discount may range from 5 to 15%. Some of these plans allow for deposits to be made through regular monthly payroll deductions. However, new accounting rules for corporations, along with other factors, are leading many companies to curtail these plans—dropping the discount allowed, cutting the maximum yearly stock purchase or otherwise making the plans less generous or appealing.

Profit Sharing: Qualified employees are awarded an annual amount equal to some portion of a company's profits. In a very generous plan, the pool of money awarded to employees would be 15% of profits. Typically, this money is deposited into a long-term retirement account. Caution: Some employers refer to plans as "profit sharing" when they are actually 401(k) savings plans. True profit sharing plans are rarely offered.

Highest Executive Salary: The highest executive salary paid, typically a 2009 amount (or the latest year available to the editors) and typically paid to the Chief Executive Officer.

Highest Executive Bonus: The apparent bonus, if any, paid to the above person.

Second Highest Executive Salary: The next-highest executive salary paid, typically a 2009 amount (or the latest year available to the editors) and typically paid to the President or Chief Operating Officer.

Second Highest Executive Bonus: The apparent bonus, if any, paid to the above person.

Other Thoughts:

Apparent Women Officers or Directors: It is difficult to obtain this information on an exact basis, and employers generally do not disclose the data in a public way. However, we have indicated what our best efforts reveal to be the apparent number of women who either are in the posts of corporate officers or sit on the board of directors. There is a wide variance from company to company.

Hot Spot for Advancement for Women/Minorities: A "Y" in appropriate fields indicates "Yes." These are firms that appear either to have posted a substantial number of women and/or minorities to high posts or that appear to have a good record of going out of their way to recruit, train, promote and retain women or minorities. (See the Index of Hot Spots For Women and Minorities in the back of the book.) This information may change frequently and can be difficult to obtain and verify. Consequently, the reader should use caution and conduct further investigation where appropriate.

Growth Plans/ Special Features:

Listed here are observations regarding the firm's strategy, hiring plans, plans for growth and product development, along with general information regarding a company's business and prospects.

Locations:

A "Y" in the appropriate field indicates "Yes."

Primary locations outside of the headquarters, categorized by regions of the United States and by international locations. A complete index by locations is also in the front of this chapter.

Chapter 1

MAJOR TRENDS AFFECTING THE BIOTECH & GENETICS INDUSTRY

Trends Affecting the Biotechnology & Genetics Industry:

1) The State of the Biotechnology Industry Today
2) A Short History of Biotechnology
3) Ethanol Production Soared, But a Market Glut May Slow Expansion
4) Major Drug Companies Bet on Partnerships With Smaller Biotech Research Firms
5) From India to Singapore to Australia, Nations Compete Fiercely in Biotech Development
6) Medical Trials Conducted Abroad Spark Concerns
7) Gene Therapies and Patients' Genetic Profiles Promise a Personalized Approach to Medicine
8) Breakthrough Drugs for Cancer Treatment— Many More Will Follow
9) Few New Blockbusters: Major Drug Patents Expire While Generic Sales Growth Continues
10) Biotech and Orphan Drugs Pick Up the Slack as Blockbuster Mainstream Drugs Age
11) Biogenerics (Follow-on Biologics) are in Limbo in the U.S.
12) Breakthrough Drug Delivery Systems Evolve
13) Stem Cells—Multiple Sources Stem from New Technologies
14) U.S. Government Stance on Funding for New Stem Cell Research Unclear
15) Stem Cells—Therapeutic Cloning Techniques Advance
16) Stem Cells—A New Era of Regenerative Medicine Takes Shape
17) Nanotechnology Converges with Biotech
18) Agricultural Biotechnology Scores Breakthroughs but Causes Controversy/Selective Breeding Offers a Compromise
19) Focus on Vaccines
20) Ethical Issues Abound
21) Technology Discussion—Genomics
22) Technology Discussion—Proteomics
23) Technology Discussion—Microarrays
24) Technology Discussion—DNA Chips
25) Technology Discussion—SNPs ("Snips")
26) Technology Discussion—Combinatorial Chemistry
27) Technology Discussion—Synthetic Biology
28) Technology Discussion—Recombinant DNA
29) Technology Discussion—Polymerase Chain Reaction (PCR)

1) The State of the Biotechnology Industry Today

Analysts at global accounting firm Ernst & Young estimate global biotech industry revenues for publicly-held companies at $79.1 billion for 2009, an 11.8% decrease from the previous year. The firm also estimates that revenues of publicly-held biotech companies in the U.S. declined 13% to $56.6 billion.

Genetically-engineered drugs, or "biotech" drugs, represent about 9% of the total global prescription drugs market, and about 19% of the U.S. prescription market. The U.S. Centers for Medicare & Medicaid Services (CMS) forecast called for prescription drug purchases in the U.S. to total about $260.1 billion during 2010, representing about $800 per capita. That projected total is up from $246.3 billion in 2009 and a mere $40 billion in 1990. Estimates of the size of this market vary by source. Analysts at the widely respected firm IMS Health placed U.S. domestic prescription drug sales at $300 billion for 2009, representing 5.1% growth over the previous year. By 2019, American drug purchases may reach $457 billion or more, thanks to a rapidly aging U.S. population, increased access to insurance and the continued introduction of expensive new drugs.

Analysts at the noted investment bank Burrill & Company estimate that global research and development (R&D) expenses at all pharmaceutical companies were $65.3 billion in 2009. IMS Health estimates that global drug sales will top $1 trillion for the first time in 2014, a growth of nearly $300 billion over five years. (See www.imshealth.com.)

As of early 2009, there were more than 2,900 medicines in development in the U.S. Advanced generations of drugs developed through biotechnology continue to enter the marketplace. The results may be very promising for patients, as a technology tipping point of medical care is approaching, where drugs that target specific genes and proteins may eventually become widespread. However, it continues to become more difficult and more expensive to introduce a new drug in the U.S. For example, during 2009, the FDA (Food and Drug Administration) approved only seven new biologics (new biotechnology-based drugs, based on living organisms, that have never been marketed in the U.S. in any form) along with 19 new molecular entities or "NMEs" (medications containing chemical compounds that have never before been approved for marketing in the U.S.). This is up from the 18 approved in 2007, but down from the 22 in 2006 (there were 20 approved during 2005 and 36 in 2004).

These NMEs and biologics are novel new active substances that are categorized differently from "NDAs" or New Drug Applications. NDAs may seek approval for drugs based on combinations of substances that have been approved in the past. During 2009, 90 NDAs were approved by the FDA, (compared to 98 NDAs in 2008).

New Drug Application Categories

Applications for drug approval by the FDA fall under the following categories:

BLA (Biologics License Application): An application for approval of a drug synthesized from living organisms. That is, they are drugs created using biotechnology. Such drugs are sometimes referred to as biopharmaceuticals.

NME (New Molecular Entity): A new chemical compound that has never before been approved for marketing in any form in the U.S.

NDA (New Drug Application): An application requesting FDA approval, after completion of the all-important Phase III Clinical Trials, to market a new drug for human use in the U.S. The drug may contain active ingredients that were previously approved by the FDA.

Follow-On Biologics: A term used to describe generic versions of drugs that have been created using biotechnology. Because biotech drugs ("biologics") are made from living cells, a generic version of a drug probably won't be biochemically identical to the original branded version of the drug. Consequently, they are described as "follow-on" drugs to set them apart. Since these drugs won't be exactly the same as the originals, there are concerns that they may not be as safe or effective unless they go through clinical trials for proof of quality. In Europe, these drugs are referred to as "biosimilars."

Priority Reviews: The FDA places some drug applications that appear to promise "significant improvements" over existing drugs for priority approval, with a goal of returning approval within six months.

Accelerated Approval: A process at the FDA for reducing the clinical trial length for drugs designed for certain serious or life-threatening diseases.

Fast Track Development: An enhanced process for rapid approval of drugs that treat certain life-threatening or extremely serious conditions. Fast Track is independent of Priority Review and Accelerated Approval.

Developing a new drug is an excruciatingly slow and expensive endeavor. According to PhRMA, the average time required for the drug discovery, development and clinical trials process is 16 years. The good news is that the median FDA approval time for a "priority" NME is down to about six months, compared to 16.3 months in 2002. As for "standard" NMEs, approval time is about one year, down from nearly two years in 2005.

[handwritten annotation: 16 years drug development. Thus. company history better longer then 10 years.]

The promising era of personalized medicine is slowly, slowly moving closer to fruition. Dozens of exciting new drugs for the treatment of dire diseases such as cancer, AIDS, Parkinson's and Alzheimer's are either on the market or are very close to regulatory approval. In a few instances, doctors are now beginning to make treatment decisions based on a patient's genetic makeup.

In what is potentially one of the most important biotech legal decisions ever to emerge from the courts, a U.S. district judge in Manhattan, in early 2010, ruled as invalid parts of patents claimed by Myriad Genetics on two important breast cancer-related genes, BRCA1 and BRCA2. If this ruling is upheld on appeals, it could lead to the conclusion that vast numbers of gene patents currently claimed by a large number of biotech companies could be invalid. This could hurt the business models of several firms, while opening up the biotech sector in general to a new era of innovation and widespread use of genetic diagnostics, as knowledge of specific genes could be used freely on an industry-wide basis.

Stem cell research is moving ahead briskly on a global basis. The Obama administration relaxed limitations on federal funding of stem cell research that were established by the preceeding administration. In 2009, the National Institutes of Health set new guidelines for funding that will dramatically expand the number of stem cell lines that qualify for research funds from a previous 21 to as many as 700. However, research into certain extremely controversial stem cells, such as those developed via cloning, will not be funded with federal dollars.

Stem cell breakthroughs are occurring rapidly. There is truly exciting evidence of the potential for stem cells to treat many problems, from cardiovascular disease to neurological disorders. Menlo Park, California-based Geron Corporation, for example, has published the results of its experiments that show that when certain cells (called OPCs) derived from stem cells were injected in rats that had spinal cord injuries, the rats quickly recovered. According to the company, "Rats transplanted seven days after injury showed improved walking ability compared to animals receiving a control transplant. The OPC-treated animals showed improved hind limb-forelimb coordination and weight bearing capacity, increased stride length, and better paw placement compared to control-treated animals."

Despite exponential advances in biopharmaceutical knowledge and technology, biotech companies enduring the task of getting new drugs to market continue to face long timeframes, daunting costs and immense risks. Although the number of NDAs submitted to the FDA has grown dramatically since 1996, the number of new drugs receiving final approval remains relatively small. On average, of every 1,000 experimental drug compounds in some form of pre-clinical testing, only one actually makes it to clinical trials. Then, only one in five of those drugs make it to market. Of the drugs that get to market, only one in three bring in enough revenues to recover their costs. Meanwhile, the patent expiration clock is ticking—soon enough, manufacturers of generic alternatives steal market share from the firms that invested all that time and money in the development of the original drug.

In fact, many major drugs have recently gone off patent, or will do so in the near future, which will be a significant boost to generic manufacturers that will quickly issue their own low-priced versions. Drug industry association PhRMA recently estimated that 72% of its members' sales by volume are generic drugs. IMS Health estimated that generic drug sales accounted for nearly two-thirds of all prescriptions sold in America in 2009, by volume, but prices are so low that they accounted for only one-tenth of revenues.

Global Factors Boosting Biotech Today:

1) A rapid aging of the population base of nations in the E.U., as well as Japan and the U.S., including the 76 million surviving Baby Boomers in America who are entering senior years in rising numbers and needing a growing level of health care
2) A renewed, global focus on developing effective vaccines
3) Vast research investments by major pharmaceuticals firms
4) A growing global dependence on genetically-engineered agricultural seeds ("Agribio"), with farmers in 25 nations planting at least some genetically modified seeds as of 2010
5) Aggressive investment in biotechnology research in Singapore, China and India, often with government sponsorship—for example, Singapore's massive Biopolis project
6) A government-subsidized emphasis on renewable energy such as bioethanol and other biofuels as substitutes for petroleum
7) Promising research into synthetic biology
8) Continuing computer-related progress in biotech areas such as gene sequencing

Source: Plunkett Research, Ltd.

According to a study released in 2001 by the Tufts Center for the Study of Drug Development, the cost of developing a new drug and getting it to market averaged $802 million, up from about $500 million in 1996. (Averaged into these figures are the costs of developing and testing drugs that never reach the market.) Expanding on the study to include post-approval research (Phase IV clinical studies), Tufts increased the cost estimate to $897 million. Tufts estimated the average cost to develop a new biotech drug at $1.2 billion in 2006. Even more pessimistic is research released in 2003 by Bain & Co., a consulting firm, which states that the cost is more on the order of $1.7 billion, including such factors as marketing and advertising expenses.

The typical time elapsed from the synthesis of a new chemical drug compound to its introduction to the market remains 12 to 20 years. Considering that the patent for a new compound only lasts about 20 years, a limited amount of time is available to reclaim the considerable investments in research, development, trials and marketing. As a result of these costs and the lengthy time-to-market, young biotech companies encounter a harsh financial reality: commercial profits take years and years to emerge from promising beginnings in the laboratory.

Since national governments pay for a significant part of prescription drug costs in major markets worldwide, the current need for many government agencies to control costs will have a dampening effect on total drug revenues in the U.K., U.S., Japan, France and elsewhere.

However, advances in systems biology (the use of a combination of state-of-the-art technologies, such as molecular diagnostics, advanced computers and extremely deep, efficient genetic databases) may eventually lead to more efficient, faster drug development at reduced costs. Much of this advance will stem from the use of technology to efficiently target the genetic causes of, and develop novel cures for, niche diseases.

The FDA is attempting to help the drug industry bring the most vital drugs to market in shorter time with three programs: Fast Track, Priority Review and Accelerated Approval. The benefits of Fast Track include scheduled meetings to seek FDA input into development as well as the option of submitting a New Drug Application in sections rather than submitting all components at once. The Fast Track designation is intended for drugs that address an unmet medical need, but is independent of Priority Review and Accelerated Approval. Priority drugs are those considered by the FDA to offer improvements over existing drugs or to offer high therapeutic value. The priority program, along with increased budget and staffing at the FDA, are having a positive effect on total approval times for new drugs.

For example, the FDA quickly approved Novartis' new drug Gleevec (a revolutionary and highly effective treatment for patients suffering from chronic myeloid leukemia). After priority review and Fast Track status, it required only two and one-half months in the approval process. This rapid approval, which enabled the drug to promptly begin saving lives, was possible because of two factors aside from the FDA's cooperation. One, Novartis mounted a targeted approach to this niche disease. Its research determined that a specific genetic malfunction causes the disease, and its drug specifically blocks the protein that causes the genetic malfunction. Two, thanks to its use of advanced genetic research techniques, Novartis was so convinced of the effectiveness of this drug that it invested heavily and quickly in its development.

Key Food & Drug Administration (FDA) terms relating to human clinical trials:
Phase I—Small-scale human trials to determine safety. Typically include 20 to 60 patients and are six months to one year in length.
Phase II—Preliminary trials on a drug's safety/efficacy. Typically include 100 to 500 patients and are one and a half to two years in length.
Phase III—Large-scale controlled trials for efficacy/safety; also the last stage before a request for approval for commercial distribution is made to the FDA. Typically include 1,000 to 7,500 patients and are three to five years in length.
Phase IV—Follow-up trials after a drug is released to the public.

Generally, Fast Track approval is reserved for life-threatening diseases such as rare forms of cancer, but new policies are setting the stage for accelerated approval for less deadly but more pervasive conditions such as diabetes and obesity. Approval is also being made easier by the use of genetic testing to determine a drug's efficacy, as well as the practice of drug companies working closely with federal organizations. Examples of these new policies are exemplified in the approval of Iressa, which helps fight certain types of cancer in only 10% of patients but is associated with a genetic marker that can help predict a patient's receptivity; and VELCADE, a cancer drug that received initial approval in only four months because the company that makes it worked closely with the National Cancer Institute to review trials.

Internet Research Tip:
For extensive commentary and analysis on the development and approval of new drugs see:

Tufts Center for the Study of Drug Development
csdd.tufts.edu
Note: This web site gives you the opportunity to download the latest annual edition of the "Outlook", an excellent summary review of trends in drug development.

Small- to mid-size biotech firms continue to look to mature, global pharmaceutical companies for cash, marketing muscle, distribution channels and regulatory expertise.

Personal genetic codes are becoming less expensive and more widely attainable. Stanford University engineer Stephen R. Quake announced in mid 2009 that his new technology for decoding DNA enabled him to make an analysis of his own genome for less than $50,000 using a unique sequencer that is about the size of a household refrigerator. The cost of decoding the most important sections of the human genome for an individual patient has dropped dramatically.

With progress comes setbacks, including a massive award for damages (more than $250 million) that occurred in a small-town Texas court in August 2005. The award was made to the widow of a patient who allegedly had a fatal reaction to Merck & Co.'s Vioxx pain medication (which had previously been removed from the market due to safety concerns). Texas laws capping medical case awards reduced the damages significantly. Nonetheless, recent drug

safety issues and a proliferation of lawsuits such as this may accelerate changes in the business models of drug development firms, discouraging them from risking funds on long-shot drugs intended to benefit the mass market. Meanwhile, drug makers will continue to alter marketing methods and greatly reduce consumer advertising. Virtually all drugs have significant side effect risks for certain types of patients. While drug makers have long practiced a high level of disclosure, those risks will be more clearly communicated in the future.

Global trends are affecting the biotech industry in a big way. Post 9/11, an emphasis was placed by government agencies on the prevention of bioterror risks, such as attacks by the spread of anthrax. This factor, combined with global concern about the possible spread of flu, has been a significant boost to vaccine research and production. At the same time, the rapid rise of offshoring and globalization is contributing to the movement of research, development and clinical trials away from the U.S., Japan and Europe into lower cost technology centers in India and elsewhere. In fact, biotech firms are rising rapidly in India, China, Singapore and South Korea that will provide serious future competition to older companies in the West.

Likewise, retail drug markets have tremendous potential in emerging nations over the mid term. For example, consultants at McKinsey estimated that the drug market in India will grow from $6.3 billion in 2005 to $20 billion in 2015. China offers similar opportunities, while Russia, Brazil and Turkey are also likely to be significant growth markets. This means that major international drug makers will be expanding their presence in these nations. However, it also means that local drug manufacturers have tremendous incentive to expand their research, product lines and marketing within their own nations.

Some of the most exciting developments in the world of technology are occurring in the biotech sector today. These include advances in agricultural biotechnology, the convergence of nanotechnology and information technology with biotech, and breakthroughs in synthetic biotechnology.

Global panic over quickly rising food prices in 2007 and part of 2008 finally gave the genetically modified (GM) seed industry the boost it needed. Agribio (agricultural biotechnology) is spreading rapidly, with genetically modified seeds now planted in at least 25 nations worldwide. This is biotechnology in one of its most productive arenas, the modification of the genetic makeup of seeds in order to make plants resistant to insects, capable of fighting off diseases, loaded with nutrients, able to grow with less water and/or much more productive per acre of planting. This is a science that has evolved through the years to the point that, in a good year, a densely populated nation like India can be capable of growing enough grain to feed its hordes of people. Partly because of rising incomes—leading to more demand for foodstuffs—and a growing global population, a forecast made by analysts at the UN's Food and Agriculture Organization in February 2010 is that agricultural output worldwide needs to increase by 70% by the year 2050. Consumer acceptance of GM food products will increase quickly, along with steady growth in the global population and an expanding global middle class.

2) A Short History of Biotechnology

While the 1900s will be remembered by industrial historians as the Information Technology Era and the Physics Era, the 2000s may be marked by many as the Biotechnology Era because rapid advances in biotechnology will completely revolutionize many aspects of life in coming decades. However, the field of biotechnology can trace its true birth back to the dawn of civilization, when early man discovered the ability to ferment grains to make alcoholic beverages, and learned of the usefulness of cross-pollinating crops in order to create new hybrid strains—the earliest form of genetic engineering. In ancient China, people are thought to have harvested mold from soybean curd to use as an antibiotic as early as 500 B.C.

Robert Hooke first described cells as a concept in 1663 A.D., and in the late 1800s, Gregor Mendel conducted experiments that became the basis of modern theories about heredity. Alexander Fleming discovered the first commercial antibiotic, penicillin, in 1928.

The modern, more common concept of "biotech" could reasonably be said to have its beginnings shortly after World War II. In 1953, scientists James Watson and Francis Crick conceived the "double helix" model of DNA, and thus encouraged a spate of scientists to consider the further implications of human DNA. The Watson/Crick three-dimensional model began to unlock the mysteries of heredity and the methods by which replication of genetic material takes place within cells.

Significant steps toward biotech drugs occurred in the early 1970s. In 1973, Dr. Stanley N. Cohen, a Stanford University genetics professor, and Dr. Herb Boyer, a biochemist, genetic engineer and educator at UC-San Francisco, introduced the concept of gene-splicing and created the first form of recombinant DNA. In 1974, Cesar Milstein and Georges Kohler created monoclonal antibodies, cells that clone over and over again to create large quantities of a specific antibody. Many of today's top biotech drugs are monoclonal antibodies. These two discoveries (recombinant DNA and monoclonal antibodies) created the building blocks of the first modern commercial biotech drugs.

Boyer and Cohen's gene-splicing technique enabled scientists to cut genetic material from the cells of one organism and paste it into another organism. This was an important discovery because the genetic material they moved from one place to another instructs a cell as to how to make a particular protein. The organism on the receiving end of the gene-splicing technique is then able to make that protein. Over time, scientists have perfected the technique of splicing material that enables cells to create proteins that control the creation of insulin, the level of blood pressure and many other human functions. Such genetic engineering enabled, for the first time, the creation of massive vats of isolated proteins grown in bacteria or in cells harvested from mammals—in quantities large enough for the commercial production of new drugs. (In fact, Boyer and Cohen's early experiments involved inserting a gene from an African clawed toad into bacterial DNA for duplication.)

In 1975, the first human gene was isolated, opening the door to gene therapy and creating the interest that led to the beginning of the massive, publicly funded Human Genome Project in 1990. A working draft of the Human Genome was released in 2000, and a complete genome was released in 2003.

In 1976, Bob Swanson of the now-famous Silicon Valley venture capital firm Kleiner Perkins formed a new business, Genentech, in conjunction with Herb Boyer (see above). Other early biotech firms arrived soon after, generally funded by venture capital firms, angel investors and corporate venture partners. These early biotech startups included many companies that grew into today's super-successful biopharma corporations: Amgen, Novartis Diagnostics (formerly Chiron), Biogen Idec and Genzyme. The creation of these startups, focused on the development of new drugs, was particularly noteworthy because it was the first time in decades that new drug companies were launched in significant numbers. In fact, most major drug companies in existence at the beginning of the 1970s could trace their histories back to the early 1900s or before.

The commercial introduction of genetically modified (GM) seeds is a relatively new branch of biotech. By 1987, researchers were gaining enough progress with GM seeds that applications for approval for field testing and certification began to pour into the USDA. The first commercialized food to result was the Flavr Savr Tomato, which was the result of a gene splicing. The added gene prevented the breakdown of cell walls as the fruit ripened, which meant that the tomatoes remained firm for an extended period of time in the truck or on the shelf. In 1995-1996, GM corn with gene modification that enables the plant to produce its own pesticide, received regulatory approval and became commercially available. Today, millions of acres of GM plants, from cotton to soy to corn, are grown worldwide with tremendous efficiency. Significant new advancements in biotech crops are on the horizon. Many researchers are experimenting with GM seeds that grow plant-based pharmaceuticals.

In 2010, one of the most significant biotech developments in years was announced when genetic engineering was used to create an entirely new organism. This is a field known as "synthetic biology."

3) Ethanol Production Soared, But a Market Glut May Slow Expansion

High gasoline prices, effective lobbying by agricultural and industrial interests, and a growing interest in cutting reliance on imported oil put a high national focus on bioethanol in America in recent years. Corn and other organic materials, including agricultural waste, can be converted into ethanol through the use of engineered bacteria and superenzymes manufactured by biotechnology firms. This trend has given a boost to the biotech, agriculture and alternative energy sectors. At present, corn is almost the exclusive source for bioethanol in America. This is a shift of a crop from use in the food chain to use in the energy chain that is unprecedented in all of agricultural history—a shift that is having profound effects on prices for consumers, livestock growers (where corn has long been a traditional animal feed) and food processors.

In addition to the use of ethanol as fuel for cars and trucks, the chemicals industry, faced with high petrochemicals costs, has a new appetite for bioethanol. In fact, bioethanol can be used to create polyethylene—an area that consumes vast quantities of petroleum and natural gas in America and around the globe. Archer Daniels Midland is constructing a plant in Clinton, Iowa that will produce 50,000 tons

of plastic per year through the use of biotechnology to convert corn into polymers. Dow Chemical announced plans in 2008 to construct a sugarcane-to-ethanol-to polyethylene plant in Brazil.

Ethanol is an alcohol produced by a distilling process similar to that used to produce liquors. A small amount of ethanol is added to much of the gasoline sold in America, and most U.S. autos are capable of burning "E10," a gasoline blend that contains 10% ethanol. E85 is an 85% ethanol blend that may grow in popularity due to a shift in automotive manufacturing. Although only about 2,200 or so of the 170,000 U.S. service stations sold E85 as of the middle of 2009, there may be an increase in demand for ethanol in the U.S. due to mandates by the U.S. government calling for reduced dependence on oil.

Yet, despite the millions of vehicles on the road that can run on E85 and billions of dollars in federal subsidies to participating refiners, many oil companies seem unenthusiastic about the adoption of the higher ethanol mix. E85 requires separate gasoline pumps, trucks and storage tanks, as well as substantial cost to the oil companies (the pumps alone cost about $200,000 per gas station to install). The plants needed to create ethanol cost $500 million or more to build. Many drivers who have tried filling up with E85 once revert to regular unleaded when they find as much as a 25% loss in fuel economy when burning the blend.

Ethanol is a very popular fuel source in Brazil. In fact, Brazil is one of the world's largest producers of ethanol, which provides a significant amount of the fuel used in Brazil's cars. This is due to a concerted effort by the government to reduce dependency on petroleum product imports. Brazil has several advantages that make the production of ethanol more efficient there than it is in the United States. After getting an initial boost due to government subsidies and fuel tax strategies beginning in 1975, Brazilian producers developed methods (typically using sugar cane) that enable them to produce ethanol at moderate cost. The fact that Brazil's climate is ideally suited for low-cost sugarcane farming is a great asset. Also, sugar cane can be converted with one less step than corn, which is the primary source for American ethanol, and ethanol plants are often adjacent to sugarcane fields, eliminating costly transportation needs. Brazilian automobiles are typically equipped with engines that can burn pure ethanol or a blend of gasoline and ethanol. Brazilian car manufacturing plants operated by Ford, GM and Volkswagen all make such cars.

In America, partly in response to the energy crisis of the 1970s, Congress instituted federal ethanol production subsidies in 1979. Corn-based grain ethanol production picked up quickly, and federal subsidies have amounted to several billion dollars. The size of these subsidies (about $6 billion yearly as of 2010) and environmental concerns about the production of grain ethanol produced a steady howl of protest from observers through the years. Nonetheless, the Clean Air Act of 1990 further boosted ethanol production by increasing the use of ethanol as an additive to gasoline. Meanwhile, the largest producers of ethanol, such as Archer Daniels Midland (ADM), have reaped massive subsidies from Washington for their output.

The U.S. Energy Act of 2005 specifically requires that oil refiners mix 7.5 billion gallons of renewable fuels such as ethanol in the nation's gasoline supply by 2012. Ethanol production in the U.S. reached 10.75 billion gallons in 2009, compared to 3.8 billion in 2005, according to the U.S. Energy Information Administration (EIA). Iowa, Illinois, Nebraska, Minnesota and South Dakota are the biggest producers, in that order. Although grain farmers and ethanol producers enjoyed high prices at the onset, a glut of ethanol supply eventually caused market prices to plummet. Next, the Energy Independence and Security Act of 2007 called for even more ethanol production, with a goal of 36 billion gallons per year by 2022 including 21 billion gallons to come from cellulosic and other advanced biofuel sources. However, environmental concerns, the sizeable investments needed to construct ethanol refineries and questions about the advisability of using a food grain as a source for fuel may make these goals very unattainable using existing technologies. In addition, the automobile industry expects a significant amount of market share to slowly shift to electric or hybrid-electric vehicles over the long term, which will reduce dependency on liquid fuels, such as gasoline and ethanol.

More recently, some of the largest ethanol production companies have suffered severe financial problems. Notably, VeraSun filed for bankruptcy protection in late 2008, citing high corn prices and difficulty in obtaining trade credit. The Iowa-based company operated 14 ethanol plants in the Midwestern U.S. (Valero, a leading petroleum refiner, purchased some of VeraSun's plants.) The 2008-2009 plummet in the price of crude oil made ethanol look much less attractive from a cost point-of-view. Meanwhile, ethanol factories have generally encountered great difficulty when seeking

profitability in the U.S., despite immense federal government support. New plant construction projects have been cancelled or put on hold, and it is looking very unlikely that the industry can meet the production goals set by congressional mandates.

Traditional grain ethanol is typically made from corn or sugarcane. In contrast to grain ethanol, "cellulosic" ethanol is made from agricultural waste like corncobs, wheat husks, stems, stalks and leaves, which are treated with specially engineered enzymes to break the waste down into its component sugars. The sugars (or sucrose) are then used to make ethanol. Since agricultural waste is plentiful, turning it into energy seems to some people like a good strategy. Cellulosic ethanol can also be made from certain types of plants and grasses.

The trick to cellulosic ethanol production is the creation of efficient enzymes to treat the agricultural waste. The U.S. Department of Energy is investing heavily in research, along with major companies such as Dow Chemical, DuPont and Cargill. Another challenge lies in efficient collection and delivery of cellulosic material to the refinery. It may be considerably more costly to make cellulosic ethanol than to make it from corn. In any event, the U.S. remains far behind Brazil in cost-efficiency, as Brazil's use of sugar cane refined in smaller, nearby biorefineries creates ethanol at much lower costs per gallon.

Iogen, a Canadian biotechnology company, makes enzymes for the breakdown of cellulose for ethanol purposes, and it has been operating a test plant to determine how economical the process may be. The company operates a demonstration biorefinery with a potential output of 5,000 liters per day. This pilot plant in Ottawa utilizes wheat straw. In mid-2009, a Shell gasoline station in Ottawa became the first retail outlet in that nation to sell a blend of gasoline that features 10% cellulosic ethanol.

Internet Research Tip: Cellulosic Ethanol
For an excellent explanation of the steps involved in refining ethanol from agricultural waste, and tour of Iogen's demonstration refinery, see: www.iogen.ca/company/demo_plant/index.html .

In the U.S., the Department of Energy has selected six proposed new cellulosic ethanol refineries to receive a total of $385 million in federal funding. If completed, these six refineries are expected to produce 130 million gallons of ethanol yearly. Iogen's technology may be used in one of the refineries, to be located in Shelley, Idaho. Partners in the refinery include Royal Dutch Shell.

Meanwhile, the Canadian government may support the Canadian biofuel industry with up to 500 million Canadian dollars for construction of next-generation plants. Iogen is expected to receive part of those funds for construction of a commercial scale cellulosic ethanol plant.

In the U.S., BP and Verenium announced plans in February 2009 to form a joint venture to build, on a commercial scale, a cellulosic ethanol plant in Highlands County, Florida. In July 2010, BP Biofuels North America acquired Verenium's cellulosic biofuels business and became the sole investor in the new plant. The plant is expected to cost $300 million and have the capacity to produce 36 million gallons of ethanol yearly from agricultural waste.

Other companies, such as Syngenta, DuPont and Ceres, are genetically engineering crops so that they can be more easily converted to ethanol or other energy producing products. Syngenta, for example, is testing a bio-engineered corn that contains the enzyme amylase. Amylase breaks down the corn's starch into sugar, which is then fermented into ethanol. The refining methods currently used with traditional corn crops add amylase to begin the process.

Environmentalists are concerned that genetically engineering crops for use in fuels will endanger the food supply through cross-pollination with traditional plants. Monsanto is focusing on conventional breeding of plants with naturally higher fermentable starch content as an alternative to genetic engineering.

Another concern relating to ethanol refining is that its production is not as energy efficient as biodiesel made from soybeans. According to a study at the University of Minnesota, the farming and processing of corn grain for ethanol yields only 25% more energy than it consumes, compared to 93% for biodiesel. Likewise, greenhouse gas emissions savings are greater for biodiesel. According to one estimate, producing and burning ethanol results in 12% fewer greenhouse gas emissions than burning gasoline, while producing and burning biodiesel results in a 41% reduction compared to making and burning regular diesel fuel. (While biodiesel cannot be used by gasoline-powered vehicles, ethanol can be added to gasoline and thus have a wider impact. Nonetheless, this comparison with biodiesel is useful as a reminder that corn-based ethanol is not an efficient fuel based on a variety of factors.) A 2009

vote by Illinois' Air Resources Board requires the use of "lower carbon intensity" fuels starting in 2011, which may have a negative long term effect on the use of ethanol.

Global warming concerns were heightened in 2009 by a report by the International Council for Science (ICSU) that concluded that the production of biofuels, including ethanol, has hurt rather than helped the fight against climate change. The report cites findings by a scientist at the Max Planck Institute for Chemistry in Germany that biofuels expand the harmful effects of a gas called nitrous oxide, which may be 300 times worse for global warming than carbon dioxide. The amounts of nitrous oxide released when farming biofuel crops such as corn may negate any advantage gained by reduced carbon dioxide emissions.

In addition, ethanol production requires enormous amounts of water. To produce one gallon of ethanol, up to four gallons of water are consumed by ethanol refineries. Add in the water needed to grow the corn in the first place, and the number grows to as much as 1,700 gallons of water for each gallon of ethanol.

Other concerns regarding the use of corn to manufacture ethanol include the fact that a great deal of energy is consumed in planting and reaping, and then transporting the corn in trucks. Also, high demand for corn for use in biorefineries has, from time-to-time, dramatically driven up the cost per bushel, creating burdens on consumers who rely on corn-based food products such as corn meal and tortillas.

As of mid-2009, new technology was being tested that would produce ethanol from corn cobs that have been stripped of edible kernels. Poet, a South Dakota-based producer of ethanol that operates 26 plants in seven U.S. states, is constructing a plant in Emmetsburg, Iowa that will be one of the first in the U.S. to produce ethanol on a large scale using a non-food source. The plant was slated for completion in 2011.

Novozymes, a Danish bioindustrial product manufacturer, has developed an enzyme blend containing an agent called GH-61 that has the potential to speed chemical reactions. Enzymes containing GH-61 may reduce production costs to the extent that producing ethanol might be competitive on a price basis with fossil fuels. Novozymes says that the cost of the enzyme, called Cellic, is about 50 cents per gallon, or less than a third of the projected $1.90 per gallon total cost of the ethanol (naturally, the retail price would be higher). Poet's Emmetsburg plant will utilize the substance in its ethanol production.

Another potential for ethanol production plants is to retool them to produce other kinds of biochemicals. For example, Houston, Texas startup Glycos Biotechnologies, Inc. is developing an add-on process to use glycerin, a by-product of ethanol, to make chemicals suitable for use in fabrics, insulation and food stuffs. Other firms pursuing similar avenues include Genomatica, Inc., Gevo, Inc. and Myriant Technologies LLC.

SPOTLIGHT: Biofuels

Corn is far from the only source of cellulose for creating biofuels.

Municipal/Agricultural Waste: Cheaply produced, but in limited supply compared to the billions of gallons of fuel needed in the market place.

Wood: Easily harvested and in somewhat healthy supply, however cellulose is more difficult to extract from wood than from other biosources.

Algae: The slimy green stuff does have the potential for high yields per acre, but the process for distilling its cellulose is complex, requiring a source of carbon dioxide to permeate the algae.

Grasses/Wheat: Including switchgrass, miscanthus, sugar cane and wheat straw, the supply would be almost limitless. The challenge here is creating efficient methods for harvesting and infrastructure for delivering it to biorefineries.

Vegetable Oils: Including oil from soybean, canola, sunflower, rapeseed, palm or hemp. It is difficult to keep production costs of these oils low.

SPOTLIGHT: Algae Draws Major Investment

Algae's potential as a source of biofuel got a big boost from an unlikely source in 2009. ExxonMobil announced plans to invest $300 million or more in San Diego, California-based Synthetic Genomics, a company headed by genome pioneer Craig Venter. Dr. Venter is studying ways in which an ideal species of algae can be developed for a unique culturing process. This process induces algae to release their oil (naturally stored as a foodstuff for the organisms), which can then be manipulated so that the oxygen molecules in the oil are disposed of, leaving a pure hydrocarbon suitable for use as biofuel. Another plus to Venter's process is that carbon dioxide claimed from industrial plant exhaust is used in the culturing process and then released in the atmosphere. This does not make algae biofuel production carbon neutral, but it does utilize carbon dioxide twice before it's released. Should the study go well, ExxonMobil has pledged an additional $300 million in funding to further develop the process to an industrial scale.

---

4) Major Drug Companies Bet on Partnerships With Smaller Biotech Research Firms

At pharmaceutical firms both large and small, profits are under constant pressure because blockbuster drugs that have made immense profits for many years eventually lose their patent protection and face vast competition from generic versions. In the U.S., generic drugs now hold between a sixty and seventy percent market share by volume. This puts pressure on large research-based drug firms to develop new avenues for profits. One such avenue is partnerships with and investments in young biotech companies, but profits from such ventures will, in most cases, be slow to appear. Meanwhile, the major, global drug firms are investing billions in-house on biotech research and development projects, but new blockbusters are elusive.

For example, Pfizer historically invested about $7.8 billion yearly on R&D. That money is invested in carefully designed research programs with specific goals. As of early 2010, Pfizer had about 500 projects in development, with 133 of those in Phase I trials or beyond. Biologic drugs accounted for 27 projects under development, and they were part of the firm's "invest to win" areas that focus on potential blockbuster drugs. However, Pfizer's recent acquisition of Wyeth left it with duplicated research labs and personnel, and cutbacks were announced. Today's health care environment is requiring major drug makers to attempt to increase research results while reducing costs.

Much of the future success for the world's major drug companies will lie in harnessing their immense financial power along with their legions of salespeople and marketing specialists to license and sell innovative new drugs that are developed by smaller companies. There are dozens of exciting, smaller biotech companies that are focused on state-of-the-art research that lack the marketing muscle needed to effectively distribute new drugs in the global marketplace. To a large degree, these companies rely on contracts and partnerships with the world's largest drug manufacturers. In addition to money to finance research and salespeople to promote new drugs to doctors, the major drug makers can offer expertise in guiding new drugs through the intricacies of the regulatory process. While these arrangements may not lead to blockbuster drugs that will sell billions of pills yearly to treat mass market diseases, they can and often do lead to very exciting targeted drugs that can produce $300 million to $1 billion in yearly revenues once they are commercialized. A string of these mid-level revenue drugs can add up to a significant amount of yearly income.

A good example of small biotech firms partnering with big pharma is Ligand Pharmaceuticals, a La Jolla, California-based company that focuses on biotech drug development. It does so through partnerships with a veritable who's who of major pharmaceutical firms including Bristol-Myers Squibb, GlaxoSmithKline, Schering-Plough and Pfizer. In April 2010, Ligand received a $6.5 million payment from Roche for a drug that had reached Phase I clinical trials for the treatment of hepatitis C viral infection. The drug, called RG7348, is the result of a collaboration and license agreement between Roche and Metabasis Therapeutics, which was acquired by Ligand in January 2010.

5) From India to Singapore to Australia, Nations Compete Fiercely in Biotech Development

While pharmaceutical companies based in the nations with the largest economies, such as the U.S., U.K. and Japan, struggle to discover the next important drug, companies and government agencies in many other countries are enhancing their positions on the biotech playing field, building their own educational and technological infrastructures. Not surprisingly, countries such as India, Singapore and China, which have already made deep inroads into other technology-based industries, are making major efforts in biotechnology, which is very much an information-based science. Firms that manufacture generics and provide contract research, development and clinical trials services are already common in such nations. In most cases, this is just a beginning, with original drug and technology development the ultimate goal.

The government of Singapore, for example, has made biotechnology one of its top priorities for development, vowing to make it one of the staples of its economy. Its "Biopolis" (www.one-north.sg/hubs_biopolis.aspx), a research and development center, opened in 2004. Biopolis is part of a master planned science and technology park called One-North. The complex is recognized as a center for stem cell and cell therapy research. It is a melting pot of scientists and corporations from all over the world. This status is fostered by the fact that Singapore is centrally-located, has direct airline access to all of the world's major cities and has a well-educated, largely English-speaking population. For example, the Novartis Institute for Tropical Diseases has more than 100 researchers from 18 different nations. Biopolis, is already a large success, with more than 1,000 scientists working in seven public research institutes as of 2009. By the end of 2010, new buildings at Biopolis will expand the development to 4.5 million square feet. (Another unit of the One-North development is called Fusionopolis, a 24-story building housing researchers, designers and entrepreneurs in media, software, communications and entertainment.)

Outsourcing of biotech tasks to India is growing at a very high annual rate. India's total pharmaceuticals industry revenues, largely from the manufacturing of generics, are projected to reach $24 billion in 2010, up about 13% from 2009.

India already has hundreds of firms involved in biotechnology and related support services. In 2005, the nation tightened its intellectual property laws in order to provide strong patent protection to the drug industry. As a result, drug development activity by pharma firms from around the world has increased in Indian locations in recent years, although at least one foreign firm has been disappointed when it attempted to enforce its patents in India. The FDA has approved roughly 900 plants in India for drug manufacturing and raw material production for use in the U.S. (compared to more than 300 approvals in China).

Indian drug companies to watch:
- Piramal Healthcare Limited (formerly Nicholas Piramal, India Ltd.)
- Ranbaxy Laboratories Ltd.
- Dr. Reddy's Laboratories Ltd.

Outsourced services firms and research labs in India will continue to have a growing business in the biotech sector. The costs of developing a new drug in India can be a small fraction of those in the U.S., although drugs developed in India still are required to go through the lengthy and expensive U.S. FDA approval process before they can be sold to American patients. India has its own robust biotech parks, including the well-established S. P. Biotech Park covering 300 acres in Hyderabad.

Stem cell (and cloning) research activity has been brisk in a number of nations outside the U.S. as well. To begin with, certain institutions around the world have stem cell lines in place, and some make them available for purchase. Groups that own existing lines include the National University of Singapore, Monash University in Australia and Hadassah Medical Centre in Israel. Sweden has also stepped onto the stage as a major player in stem cell research, with 40 companies focused on the field, including firms such as Cellartis AB, which has one of the largest lines of stem cells in the world, and NeuroNova AB, which is focusing on regenerating nerve tissue.

More importantly, several Asian nations, including Singapore, South Korea, Japan and China, are investing intensely in biotech research centered on cloning and the development of stem cell therapies. The global lead in the development of stem cell therapies may eventually pass to China, where the Chinese Ministry of Science and Technology readily sees the commercial potential and is enthusiastically funding research. On top of funding from the Chinese government, investments in labs and research are being backed by Chinese universities, private companies, venture capitalists and Hong Kong-based investors. China is the world's largest producer of raw materials for drugs, although it falls far behind India in the export of finished generic drugs.

Meanwhile, leading biotech firms, including Roche, Pfizer and Eli Lilly, are taking advantage of China's high-quality education systems and low operating costs to establish R&D centers there. In this manner, offshore research can be complemented by offshore clinical trials.

Taiwan has planned four biotech research parks. In April 2009, The Taiwanese government passed a biotech development action plan which included a $2 billion venture capital fund, a super-incubator and plans for expansion of the country's existing Development Center for Biotechnology. Meanwhile, Vietnam has plans to open six biotech research labs. Australia also has a rapidly developing biotechnology industry.

South Korea is a world leader in research and development in a wide variety of technical sectors, and it is pushing ahead boldly into biotechnology. Initiatives include the Korea Research Institute of Bioscience and Biotechnology. The combination of government backing and extensive private capital in Korea could make this nation a biotech powerhouse. One area of emphasis there is stem cell research.

In addition to fewer restrictions, many countries outside of the U.S. have lower labor costs, even for highly educated professionals such as doctors and scientists.

Quick Tips: Genetics and Biotech

DNA is a nucleic acid containing genetic instructions needed to produce everything within the body. DNA stands for deoxyribonucleic acid.

A gene is a small segment of DNA. Genes carry "instructions" for the construction and operation of proteins within an organism.

A protein is a large, complex molecule made up of amino acids. Proteins are essential functional or structural components of living cells, and they may include vital enzymes, antibodies or hormones. Genes provide the codes for synthesis of specific proteins.

Genetics is the study of traits inherited via specific components of genes passed from one generation to the next. The field may include the study of mutations to specific genes.

The genome is the way in which genes and DNA are arranged in an organism. A complete genome for an organism represents its full DNA sequence.

Genomics is the study of a complex set of genes in order to identify their sequence and how they are expressed.

> Genetic engineering is an attempt by science to alter an organism by changing, adding or deleting genes. Such engineering lies at the heart of biotechnology.

6) Medical Trials Conducted Abroad Spark Concerns

Another growing trend in health care offshoring is conducting clinical trials in countries outside the U.S. Researchers at Duke University reported in *The New England Journal of Medicine* that in November 2007, 20 of the largest U.S. drug makers conducted trials in 13,521 non-U.S. sites (compared to 10,685 sites within America). The study also reported that the number of countries conducting clinical drug tests has doubled in the last decade. Another study conducted by the Tufts Center for the Study of Drug Development reported that in 2007, only 54% of the roughly 26,000 FDA-regulated chief scientists conducting trials were based in the U.S. (compared to 86% in 1997). The U.S. National Institutes of Health (NIH) reported that in 2008, about 700 clinical trials were conducted in India, China and Russia, up from fewer than 50 in 2003. In 2009, the NIH received 5,454 offshore trial document submissions from 172 countries.

While drug companies save significantly on trial costs overseas, there are a number of ethical concerns related to this trend. On one hand, patients participating in these trials tend to be poorer than those in the U.S. and have less access to health care. These participants may view participation in trials as a form of free medical care, when in fact a therapy offered in a trial may be risky or completely ineffective. Another concern is that clinical tests performed outside the U.S. (and away from the scrutiny of the FDA) may not be properly monitored. In late 2008, a 350-participant study in India of a vaccine made by U.S. manufacturer Wyeth was stopped by Indian authorities after an infant died. The Drug Controller General of India's findings on the matter concluded that supervisory shortcomings were at fault. In 2007, two elderly patients in Poland died in a study of a bird-flu vaccine developed by Novartis. It was revealed that the patients should have been excluded from the trial due to their ages.

7) Gene Therapies and Patients' Genetic Profiles Promise a Personalized Approach to Medicine

Scientists now believe that almost all diseases have some genetic component. For example, some people have a genetic predisposition for breast cancer or heart disease. The understanding of human genetics is hoped to lead to breakthroughs in gene therapy for many ailments. Organizations worldwide are experimenting with personalized drugs that are designed to provide appropriate therapies based on a patient's personal genetic makeup or their lack of specific genes. Genetic therapy involves splicing desired DNA, sometimes taken from a healthy person, into a patient's cells in order to compensate for that patient's inherited genetic defect, or to enable that patient's body to better fight a specific disease.

For example, drugs that target the genetic origins of tumors may eventually offer more effective, longer-lasting and far less toxic alternatives to conventional chemotherapy and radiation. In other cases, genetic therapies, used in combination with surgery or chemotherapy, can reduce the chance of a cancer recurrence. One of the most noted drugs that target specific genetic action is Herceptin, a monoclonal antibody that was developed by Genentech. Approved by the FDA in 1998, Herceptin, when used in conjunction with chemotherapy, shows great promise in significantly reducing breast cancer for certain patients who are known to "overexpress" the HER2 protein (that is, there is an excess of HER2-related protein on tumor cell surfaces, or there is an excess of the HER2 gene itself). A simple test is used to determine if this gene is present in the patient. Herceptin, which works by blocking genetic signals, thus preventing the growth of cancerous cells, may show potential in treating other types of cancer, such as ovarian, pancreatic or prostate cancer.

Another genetic test is marketed by Genomic Health, based in Redwood City, California, (www.genomichealth.com). Its Oncotype DX test provides breast cancer patients with an assessment of the likelihood of the recurrence of their cancer based on the expression of 21 different genes in a tumor. The test enables patients to evaluate the results they may expect from post-operative therapies such as Tamoxifen or chemotherapy. By 2010, more than 90,000 patients had used the test. The test costs about $3,500, and is now covered by Medicare and a number of private health care plans. The firm also offers an Oncotype DX test for colon cancer. Such tests will be standard preventive treatment in coming decades.

By 2009, as many as 1,400 genetic tests were available from various suppliers, including 23andMe's use of microarrays (a lab on a chip used to analyze genetic samples) that test human saliva for up to 1 million genetic variations linked to disease (about $429 per test) and Navigenics' similar test that detects predisposition to diseases such as

Alzheimer's and type 2 diabetes ($999). On the top end is Knome's $99,000 profile (recently reduced from $350,000), which analyzes an individual's entire genome and provides genetic variation data that can't be detected by microarrays. This is a huge drop in price from the $3 billion spent in 2003 by the Human Genome Project to sequence the first human genome. Prices will fall even more thanks to a $10 million award offered by the X Prize Foundation (www.xprize.org) to the first scientific team to sequence 100 human genomes in 10 days for less than $10,000 each. Likely contenders for the prize include Illumina, Inc., Life Technologies, Inc., 454 Life Sciences (owned by Roche Holding AG) and Helicos BioSciences Corp. A dark horse may be emerging in the race for the prize: a startup called Pacific Biosciences hopes to sequence DNA in 15 minutes for less than $1,000 by 2013.

The states of California and New York issued "cease and desist" letters in mid-2008 to dozens of companies that offer genetic tests, responding to concerns about test validity, and potential patient harm. The companies providing the tests argue that the tests are personal genetic information services as opposed to medical tests. Meanwhile, the Federal Trade Commission has begun investigations into possibly deceptive advertising. Genetic testing is posing a number of issues that are unlikely to be resolved in the near term.

In 2009, a number of prominent scientists expressed doubts about the benefits of genome wide association studies, which compare the genomes of sick patients and healthy individuals in order to pinpoint DNA changes that are likely to cause certain diseases. The genetic variants are proving to be far more numerous, and more complex than first thought. Many geneticists are questioning the efficacy of genome wide studies (which cost millions of dollars to perform) and suggesting alternatives such as decoding entire genomes for individual patients.

Despite these concerns, genetic testing continues to evolve. In addition to significant strides in cutting costs, the time necessary to sequence DNA is shrinking as well thanks to improved chemicals and faster computers. New technology is simplifying the process which formerly compared multiple copies of DNA sequences to single-molecule sequencing. The new single-molecule version skips the need to compare multiple copies entirely while providing a more complete picture of the genome.

The scientific community's improving knowledge of genes and the role they play in disease is leading to several different tracks for improved treatment results. One track is to profile a patient's genetic makeup for a better understanding of a) which drugs a patient may respond to effectively, and b) whether certain defective genes reside in a patient and are causing a patient's disease or illness. Yet another application of gene testing is to study how a patient's liver is able to metabolize medication, which could help significantly when deciding upon proper dosage. Since today's widely used drugs often produce desired results in only about 50% of patients who receive them, the use of specific medications based on a patient's genetic profile could greatly boost treatment results while cutting costs. Each year, 2.2 million Americans suffer side effects from prescription drugs. Of those, more than 100,000 die, making adverse drug reaction a leading cause of death in the U.S. A Journal of the American Medical Association study states that the annual cost of treating these drug reactions totals $4 billion each year.

Pharmacy benefit managers (PBMs) are organizations that provide administrative services in processing and analyzing prescription claims for pharmacy benefit and coverage programs. They are getting involved in gene therapy by beginning to test patients for genetic variations that might indicate which drugs would be more effective for individual patients. Medco Health Solutions, Inc., for example, is a PBM and pharmacy mail order business that is testing patients who take drugs such as the blood thinner warfarin and breast cancer treatment tamoxifen. CVS Caremark partnered with Generation Health, Inc. in 2010 to offer a similar testing service. PBMs are selling their services to employers who are willing to invest in them for improved health outcomes and lower prescription costs. If personally-tailored prescriptions become a reality, billions of dollars each year could be saved as patients take only those drugs which will do them some good and avoid those which could do them harm.

The second track for use of genetic knowledge is to attack, and attempt to alter, specific defective genes. Generally, pure gene therapy attempts to target defective genes within a patient by introducing new copies of normal genes. These new genes may be introduced through the use of viruses or proteins that carry them into the patient's body.

Since 1989, more than 1,400 clinical trials of some form of gene therapy have been conducted with only 47 reaching phase III and none making it to regulatory approval in the U.S. China-based Shenzhen Sibiono GeneTech achieved the world's first pure gene therapy to be approved (in China) for wide commercial use, in October 2003. The drug is sold under the brand name Gendicine as a treatment for a head and neck cancer known as squamous cell carcinoma (HNSC). The company that developed the drug claims that it has proven effective in trials, however observers outside of China are highly skeptical and want to see more extensive trials. The drug is generally not available outside of China.

Other applications of gene therapy are already in research in the U.S. and elsewhere for treatment of a wide variety of diseases. For example, gene therapy may be highly effective in the treatment of rare immune system disorders, melanoma (a malignant skin cancer for which there is currently no effective cure once the disease has spread to other organs) and cystic fibrosis. In 2006, two male patients suffering from a rare malady called chronic granulomatous disease (CGD), which makes patients terribly vulnerable to infections, were able to cease taking daily antibiotics due to a gene therapy that introduced healthy genes to replace defective ones in their bloodstreams.

In 2008, a team of scientists led by a group at the University of Pennsylvania successfully introduced a healthy gene in six patients suffering from Leber's congenital amaurosis, a condition that leads to blindness. Four of the six patients' vision improved. In 2010, a gene therapy for a rare immunodeficiency disease commonly known as the "bubble boy syndrome" went into trial, seven years after similar trials in Paris and London resulted in five of the 20 participating patients contracting leukemia. This prompted the FDA to restrict selected gene therapy studies. The new trial, which is being conducted at five sites around the world and involves a therapy stripped of the feature believed to have caused the cancer, is evidence of a rekindled interest in gene therapy's potential for treating rare diseases.

Roche Pharmaceuticals and Affymetrix, Inc. have developed a lab-on-a-chip (the AmpliChip CYP450) that can detect more than two dozen variations of two different genes. These genes are important to a patient's reaction to and use of drug therapies because the genes regulate the way in which the liver metabolizes a large number of common pharmaceuticals, such as beta blockers and antidepressants. A quick analysis of a tiny bit of a patient's blood can lead to much more effective use of prescriptions. Additional chips for specific types of genetic analysis will follow.

Another gene discovered by scientists appears to play a major role in widespread forms of the more

than 30 different types of congenital heart defects, the most common of all human birth defects. Genes that are used by bacteria to trigger the infection process have also been identified, which could lead to powerful vaccines and antibiotics against life-threatening bacteria such as salmonella.

In the area of heart disease, about 300,000 patients yearly undergo heart bypass surgery in an effort to deliver increased blood flow to and from the heart. Clogged arteries are bypassed with arteries moved from the leg or elsewhere in the patient's body. Genetic experts have now developed the biobypass. That is, they have determined which genes and human proteins create a condition known as angiogenesis, which is the growth of new blood vessels that can increase blood flow without traumatic bypass surgery. This technique may become widespread in the near future.

Gene therapy is still in its infancy, and is not without its failures. In 2010, results of a study conducted by Nina Paynter of Brigham and Women's Hospital in Boston collected 101 variants genetically linked to heart disease. However, the variants were shown to have no value in forecasting disease among thousands of study participants over 12 years. Many genetic experts believe that the genetic testings will not progress to the extent where it can make a clinical difference in treating, diagnosing or predicting illness for at least another decade. However, the potential for gene therapy and genetic testing used for the choice and dosage of medications is almost limitless. Watch for major investment by pharmaceutical companies over the mid-term in further study and test development.

8) Breakthrough Drugs for Cancer Treatment— Many More Will Follow

Multi-kinase inhibitors are now on the leading edge of new drug development. This exciting class of drugs has the potential to target a wide variety of defective proteins. Kinases are enzymes that influence certain basic functions within cells, such as cell division. (This enzyme catalyzes the transfer of phosphates from Adenosine triphosphate (ATP) to proteins, thus causing changes in protein function.) Defective kinases can lead to diseases such as cancer. Kinase-blocking drugs are able to inhibit cell activity that can cause both tumor growth and blood vessel creation—thus they are able to shut down the blood vessels that enable tumors to grow and thrive. A good example is PTK/ZK, a drug developed by Novartis and Bayer Schering AG for the treatment of cancer of the colon, brain and other organs.

Earlier, in 2001, Novartis launched the drug Gleevec for the treatment of a blood cell cancer known as chronic myeloid leukemia. Gleevec operates by destroying diseased blood cells without harming normal cells. Patient results have been excellent. Additional projects include Genentech's highly effective colorectal and lung cancer drug Avastin, and the lung and pancreatic cancer drug Tarceva, developed by OSI Pharmaceuticals, Genentech and Roche Group.

Dozens of new kinase-blocking drugs are in various stages of development and testing. This could become a tremendous breakthrough in treatment of cancer patients and other groups. From a business standpoint, the market for such drugs is growing rapidly. By 2010, the global market may top $11.8 billion.

Another promising drug for cancer treatment is Iressa, made by AstraZeneca PLC, which was first regarded as a failure for the treatment of lung cancer. However, oncologists noted that the medication did improve outcomes for patients who were Asian or nonsmokers, or those with a particular genetic mutation involving malignant cell growth. A recent study of 400 Asian patients taking Iressa slowed or arrested the progression of the disease for a median 9.5 months (compared to 6.3 months for chemotherapy).

In 2009, a drug called Cerepro was approved for use in France and the Netherlands, making it the first approved gene therapy to hit the market in Europe. Cerepro targets fatal brain tumors called malignant glioma, and is manufactured by Ark Therapeutics in London. The drug, administered by multiple injections following the surgical removal of a malignant tumor of the brain, helps healthy tissue avoid the growth of a new tumor.

In 2010, the FDA approved the first vaccine to treat advanced prostate cancer. The vaccine, called Provenge, is a series of three injections which include some of the patient's own cells. The drug is designed to "train" the patient's immune system to destroy malignant cells.

Scientists are working on ways to deliver gene therapies so that they will be accepted by healthy cells rather than fought by the body's immune system. Gene therapies are often introduced as viruses, which can be neutralized in this way. A company called Hybrid Systems Limited (www.hybridsystems.co.uk) has a portfolio of patents relating to the development of polymer coatings which protect these viruses from attack by immune systems.

9) Few New Blockbusters: Major Drug Patents Expire While Generic Sales Growth Continues

Historically, the drug industry has been one of the world's most profitable business sectors. IMS Health reported that $300.3 billion was spent on prescription drugs in America during 2009, up 5.1% from 2008. Nearly 20% of pharmaceutical company revenues are invested in discovering new medicines, well above the average of 4% reinvested in research and development in most industries. Advanced technology has allowed drug companies to saturate their development programs with smarter, more promising drugs. R&D (research and development) budgets are staggering.

Once the patent on an existing blockbuster drug expires, competing drug companies are allowed to market cheaper generic brands which are identical chemical compounds. This takes huge revenues away from the maker of the original drug. According to a Kaiser Family Foundation study, the average branded drug costs three times as much as the average generic drug. When a generic equivalent of a drug hits the market, it is frequently 50% to 75% cheaper than the branded drug.

Drug makers are challenged to price their branded drugs in a way that will earn a good return on investment prior to the expiration of patent protection. The current U.S. patent policy grants drug manufacturers the normal 20 years protection from the date of the original patent (which is most likely filed very early in the research process), plus a period of 14 years after FDA approval. Since typical drugs take 10 to 15 years to research, prove in trials and bring to market, this effectively gives the patent holder only 19 to 24 years before low-price competition from generic manufacturers begins. Branded drugs tend to promptly lose 15% to 30% of market share when generic versions come on the market, eventually losing as much as 75% to 90% after a few years.

Another boost to generic drug sales was Wal-Mart's landmark move in 2006 when it began offering a large number of generic drugs for a flat $4 per month. Other discount retailers including Target, Walgreens and Kmart jumped on the generic bandwagon with their own deep discounts. In late 2007, Publix, a grocery store chain in the Southeast U.S. with more than 700 pharmacies, launched a free generic antibiotic program for prescriptions of several common drugs, for up to 14 days. Since then Wal-Mart further expanded its generic program. It offers 90-day supplies of hundreds of generic drugs for $10 in an effort to undercut mail-order pharmacy businesses as of 2010. Wal-Mart is also targeting women's health by adding $9 generic prescriptions for up to a 30-day supply of generics that treat osteoporosis, breast cancer, menopause and hormone deficiencies.

Then there is the issue of "follow-on" drugs. These are drugs that hold their own patent for therapies similar to or competing with the original breakthrough patent. Examples include Zantac and Prilosec, drugs which were introduced by major firms to compete with the huge success of a pioneering drug, Tagamet. Follow-on drugs tend to get through regulatory approval much faster, and often are brought to market at much lower prices. Both factors create a significant competitive advantage for follow-on patent owners.

Major drug companies are trying to get on the generic gravy train by quietly creating their own generic drug subsidiaries. Pfizer, for example, has a division called Greenstone, Ltd., which produces generic versions of its blockbuster drugs including Zoloft, an antidepressant that brought in upwards of $2 billion in 2006 sales, at which time its patent expired. Greenstone, as of late 2010, offered 39 generic compounds. Likewise, Schering-Plough created generic subsidiary Warrick Pharmaceuticals. Naturally, independent generic drug companies oppose the practice.

IMS Health estimated that generic drug sales accounted for nearly two-thirds of all prescriptions sold in America in 2009, by volume, but prices are so low that they accounted for only one-tenth of revenues. The total number of generic prescriptions in the U.S. rose 5.9%, while branded drug prescriptions fell by 7.6% in 2009. This is likely due to consumers and health insurance providers seeking ways to reduce spending.

From 2007 through 2012, patents will expire on branded drugs in the U.S. with about $60 billion in combined annual sales. For example, by 2011 a generic version of cholesterol-controlling Lipitor is expected to be on the market. Lipitor has long been the world's best selling branded drug (selling as much as $13 billion yearly at one time), but sales have been declining thanks to a generic substitute for a competing drug named Zocor; the generic of Zocor costs about 60% less than Lipitor.

Pharmaceutical company profits and research budgets are under pressure due to a wide range of additional challenges. Blockbuster drugs that might provide the high returns needed to bring a drug to market are increasingly difficult to develop. While research and development remains extremely expensive, blockbuster results are simply not emerging at the rate that they did in the past.

Meanwhile, health coverage payors are fighting to reduce the immense amounts that they spend on prescription drugs, and future revenues for drug makers will be impacted by cost-control measures at the individual level. In addition, the U.S. Government's Medicare drug benefit is putting even greater pressure on drug companies to sell drugs at lower prices. Drug makers are reacting to these pressures as best they can.

Highly effective new drug therapies can be incredibly expensive for patients. Take Iressa, a breakthrough treatment for lung cancer. It costs as much as $1,800 monthly per patient. Likewise, the groundbreaking colon cancer treatment Erbitux can cost as much as $30,000 for a seven-week treatment. A standard treatment of the latest drug regimens for some diseases can run up a $250,000 or higher bill over a couple of years.

The good news is that as the demand for new and improved treatments intensifies, so do the abilities of modern technology. In addition to expediting the process and lowering the costs of drug discovery and development, advanced pharmaceutical technology promises to increase the number of diseases that are treatable with drugs, enhance the effectiveness of those drugs and increase the ability to predict disease, not just the ability to react to it.

Consumers' voracious need for drugs will continue, thanks in part to the rapidly aging populations of such nations as the U.S. (where Baby Boomers made up a market of about 76 million in 2010), Japan, Italy and the U.K. Insurance companies may raise co-payments for drugs, strike deals with drug companies and employ pharmacy benefit management tactics in an effort to fend off rising pharmaceutical costs.

One trend is creating a new category for blockbuster drugs, based on vanity, convenience or personal choices. Until recently, pharmaceutical research was focused primarily on curing life-threatening or severely debilitating illnesses. But a current generation of drugs, commonly referred to as "lifestyle" drugs, is transforming the pharmaceutical industry. Lifestyle drugs target a variety of human conditions, ranging from the painful to the inconvenient, including obesity, impotence, memory loss, urinary urgency and depression. Drug companies also continue to develop lifestyle treatments for hair loss and skin wrinkles in an effort to capture their share of the huge anti-aging market aimed at the Baby Boomer generation. The use of lifestyle drugs dramatically increases the total annual consumer intake of pharmaceuticals, and creates a great deal of controversy over which drugs should be covered by managed care and which should be paid for by the consumer alone.

In coming years, taming pharmaceutical costs will be one of the biggest challenges facing the health care system. Prescription drug costs already account for more than 10% of all health care expenditures in the U.S. Managed care must be able to determine which promising new drugs can deliver meaningful clinical benefits proportionate to their monetary costs.

Several developments are fueling the fire under the controversy over drug costs in the U.S. To begin with, it has become common knowledge that pharmaceutical firms tend to price their drugs at vastly lower prices outside the U.S. market. The result is that U.S. consumers and their managed care payors are bearing a disproportionate share of the costs of developing new drugs. In 2009, the Obama administration signed a sweeping health care reform bill into law which may greatly impact health care costs and the way in which health care overall is managed. The legislation is likely to have very large effects on the pharmaceuticals industry. For example, extending coverage to a large number of people under some sort of universal care will significantly boost total drug sales in terms of unit volume. However, the heavy hand of government as an ever-larger buyer of health care could lead to demands for reduced drug pricing. If profits are restrained, it may discourage research or investment in biotechnology projects.

Factors leading to high expenditures in the American health care system:
- 76 million Baby Boomers are beginning to enter their senior years. The lifespan of Americans is increasing, and chronic illnesses are increasing as the population ages.
- Obesity-related illnesses are estimated to cost more than $147 billion yearly.
- Fraud, abuse and billing errors in the Medicare and Medicaid system cost an estimated $68 billion in 2007 alone.
- Malpractice insurance, lawsuits and "defensive" treatment practices intended to limit exposure to lawsuits add an estimated $75 billion to overall health care costs.

- Breakthroughs in research and development are creating significant new drug therapies, allowing a wide range of popular, but expensive, treatments that were not previously available. An excellent example is the rampant use of antidepressants such as Prozac and Zoloft.
- "Lifestyle" drug use is high, as shown by the popularity of such drugs as Viagra (for the treatment of sexual dysfunction), Propecia (for the treatment of male baldness) and Botox (for the treatment of facial wrinkles). These drugs are generally extremely expensive.

Source: Plunkett Research, Ltd.

10) Biotech and Orphan Drugs Pick Up the Slack as Blockbuster Mainstream Drugs Age

While Big Pharma traditionally concentrated on mass-market drugs that treat a broad spectrum of ailments, biotech companies have often developed treatments for rare disorders or maladies such as certain cancers that only affect a small portion of the population. For example, biotech pioneers Genentech and Biogen Idec developed Rituxan for the treatment of non-Hodgkin's lymphoma, an important but relatively small market.

Drugs such as Rituxan are commonly referred to as "orphan drugs," which means that they treat illnesses that no other drug on the market addresses. Technically, a drug designated by the FDA with orphan status provides therapeutic benefit for a disease or condition that affects less than 200,000 people in the U.S. Almost half of all drugs produced by biotech companies are for orphan diseases. These drugs enjoy a unique status due to the Orphan Drug Act of 1983, which gives pharmaceutical companies a seven-year monopoly on the drug without having to file for patent protection, plus a 50% tax credit for research and development costs. As of 2009 there were 2,113 drugs designated "orphan" by the FDA, including a few hundred on the market and the balance pending approval.

Analysts at global accounting firm Ernst & Young estimate global biotech industry revenues for publicly-held companies at $79.1 billion for 2009, an 11.8% decrease from the previous year. The firm also estimates that revenues of publicly-held biotech companies in the U.S. declined 13% to $56.6 billion.

Genetically-engineered drugs, or "biotech" drugs, represent about 9% of the total global prescription drugs market, and about 19% of the U.S. prescription market. As the wellspring of mainstream Big Pharma blockbusters begins to dry, many big pharmaceutical companies are rushing to develop their own orphan drugs. Pfizer released Sutent, a drug used to treat cancerous kidney and stomach-lining tumors in 2006, long after biotech firm Genentech had four drugs used for cancer on the market. Genentech's colon cancer treatment, Avastin, had about $5.9 billion in sales in 2009, while Pfizer's Sutent sold only $964 million that same year.

Commentary: The Challenges Facing the Biopharmaceuticals Industry

- Working with governments to develop methods to safely and effectively speed approval of new drugs.
- Working with the investment community to build confidence and foster patience in the investors for the lengthy timeframe required for commercialization of promising new compounds.
- Working with civic, government, religious and academic leaders to deal with ethical questions centered on stem cells and other new technologies in a manner that will enable research and development to move forward.
- Overcoming, through research, the technical obstacles to therapeutic cloning.
- Emphasizing fair and appropriate pricing models that will enable payors (both private and public) to afford new drugs and diagnostics while providing ample profit incentives to the industry.
- Developing appropriate standards that fully realize the potential of systems biology (that is, the use of advanced information technology and the resources of genetic databases) in a manner that will create the synergies necessary to accelerate and lower the total cost of new drug development.
- Fostering payor acceptance, diagnostic practices and physician practices that will harness the full potential of genetically targeted, personalized medicine when a large base of new biopharma drugs becomes available.

Source: Plunkett Research, Ltd.

---

11) Biogenerics (Follow-on Biologics) are in Limbo in the U.S.

Patents on the first biotech-based drugs began expiring in 2001. Nonetheless, the FDA has no formal path for enabling drug makers to obtain approval of a generic version of biotechnology-based drugs without extensive clinical trials. Consequently, biotech drug makers as of yet haven't faced the type of generic competition that constantly challenges makers of traditional drugs, despite the fact that

generic drug makers report that the biologics industry could reach $100 billion in sales as early as 2011. However, this situation is likely to change in the near future as more and more biotech drugs go off-patent.

Because biotech drugs ("biologics") are made from living cells, a generic version of a drug probably won't be biochemically identical to the original branded version of the drug. Therefore, they are described as "follow-on biologics" to set them apart. There are concerns that follow-on biologics may not be as safe or effective as the originals unless they go through clinical trials for proof of quality.

In addition, the development of generic versions of biologics is very costly. Industry analysts estimate that initial generic development for one compound runs between $100 million and $150 million, compared to $5 million to $10 million for non-biotech generic drugs. However, the economic potential for the marketing of these generic compounds is staggering, which makes the cost easier to swallow. Merck & Co. set up a business unit in 2009 to develop generic biologics, and is planning to invest $1.5 billion in research and development in this area by 2015.

In Europe, these drugs are referred to as "biosimilars." In the European Union, the first biosimilars were approved in April 2006 by the EMA (European Medicines Agency), the regulatory body responsible for new drugs. The first approved biosimilar was Omnitrope, a generic substitute for growth hormone Genotropin. The second was Valtropin.

FDA regulations for approval of generics, written in 1984, allow companies to achieve approval of generic versions of chemically-based, traditionally-manufactured pharmaceuticals with expired patents in a relatively short period of time. However, the law doesn't address drugs developed through biotechnology. The question remains whether the FDA will approve such generic biologics without lengthy and expensive clinical trials. It's a whole new game in the generics sector. Would-be generics makers also face the fact that biotech drugs can be vastly more complicated to manufacture than chemicals-based drugs.

Like the Europeans (EMA), the FDA did approve a simple generic biotech drug, a human growth hormone, in May 2006. However, there are no published guidelines for getting more complex biogenerics to market. The FDA is under mounting pressure from the U.S. House of Representatives and from a number of state governors who are petitioning for these drugs.

In order to obtain approval of a biogeneric under existing regulations, a manufacturer would be forced to prove that the new drug contains the same ingredient as the original drug and that the two drugs are bioequivalent. This may be impossible without clinical trials.

India-based Dr. Reddy's Laboratories, one of that nation's leading drug firms, plans to create at least one follow-on biologic yearly over the mid-term. It already sells two such biogenerics in India, versions of Roche's Rituxan and Amgen's Neupogen. Other Indian biogeneric manufacturers, including Reliance Life Sciences and Ranbaxy, are also marketing products. Ranbaxy began marketing biogenerics in India as early as 2003. In 2008, Reliance Life Sciences launched three biosimilars (ReliPoietin, ReliFeron and ReliGrast) for sale in South Asia, South East Asia and several countries in Latin America. This was followed by a fourth (TPA Reteplase) in 2009.

In Europe, the EMA has issued guidelines for approval of biosimilars. These include:

1) Comparability studies are required. The proof or lack of proof of comparability to the original drug will dictate how many new clinical studies may be required.
2) Clinical studies may be required to prove the biosimilar's safety and efficacy.
3) Nonclinical studies may be required as well.
4) After the biosimilar is approved and brought to market, continuing safety and efficacy study commitments will be required.

In 2008, the Biologics Price Competition and Innovation Act of 2007 regarding FDA regulation of follow-on biologics stalled in Congress and never became law. If the act had passed, FDA guidelines for biogenerics would have closely mirrored those of the EMA. In 2009, the Obama administration's sweeping health care reform initiative included no legislation regarding guidelines for the approval of biosimilars. The reform plan does include a new bill that would grant 13.5 years of intellectual property protection to biologic drugs, which would significantly extend the life of branded biotech-based drugs. This would protect the original maker of a biologic, because a generic competitor would not be able to utilize data from the clinical trials conducted by the original maker until the 13.5 years had passed and the original patent had also expired.

12) Breakthrough Drug Delivery Systems Evolve

Controlling how drugs are delivered is a huge business. Sales of drugs using new drug delivery systems are expected to balloon.

Until the biotech age, drugs were generally comprised of small chemical molecules capable of being absorbed by the stomach and passed into the blood stream—drugs that were swallowed as pills or liquids. However, many new biotech drugs require injection (or some other form of delivery) directly into the bloodstream, because they are based on large molecules that cannot be absorbed by the stomach. A number of new drug delivery techniques that provide an alternative to needles are in development.

In the near future, there may be an implantable device, controlled by a small chip, capable of releasing variable doses of multiple potent medications over an extended period, potentially up to one year. The miniscule implants may bear a series of tiny wells, sealed with membranes that dissolve and release the contents when a command is received. Chips that can receive commands beamed through the skin are also theoretically possible. This technology would help treatment of conditions such as Parkinson's disease or cancer, where doctors need to vary medications and dosages. A firm named MicroCHIPS is a leader in this field.

Other potential needle-free drug delivery systems include synthetic molecules attached to a drug, making it harder for the stomach to render the medicine useless before it reaches the blood. High-tech inhalers, which force medicine through the lungs, are also in the works. For the patient, this means less pain and the promise of better outcomes. Needle-free systems may also make toxic drugs safer and give older drugs new life. For example, the painkiller Fentanyl is available in a lozenge form.

Some firms are developing techniques to encapsulate or rearrange drug molecules into more sturdy compounds that release steady, even doses over a prolonged period. A patch developed by Alza Corp., now a subsidiary of Johnson & Johnson, has a network of microscopic needles that penetrate painlessly into the first layer of the skin. Vyteris announced an agreement granting Laboratory Corporation of America Holdings rights to represent LidoSite, the first FDA-approved active transdermal patch using electric current for pain relief from blood draws and venipunctures. In 2010, Vyteris reported positive Phase II trials for delivering a fertility hormone using its smart patch technology. Another device, made by Echo Therapeutics (formerly Sontra Medical Corp.), uses ultrasound and gel in its Prelude

SkinPrep System to agitate and open temporary pores in the skin that can then receive a drug. A similar technology, called the Symphony tCGM System, can provide continuous, transdermal glucose monitoring. Yet another potential drug delivery system is edible film; quickly dissolving films treated with medication that melt on the tongue. Already in use as a breath freshener (Listerine brand PocketPaks, made by Pfizer), edible film is now the delivery method of choice for Novartis' Triaminic, Theraflu Thin Strips and Gas-X Thin Strips, as well as Enlyten's PediaStrips that replenish electrolytes in children and Electrolyte SportStrips for adults.

13) Stem Cells—Multiple Sources Stem from New Technologies

During the 1980s, a biologist at Stanford University, Irving L. Weissman, was the first to isolate the stem cell that builds human blood (the mammalian hematopoietic cell). Later, Weissman isolated a stem cell in a laboratory mouse and went on to co-found SysTemix, Inc. (now part of drug giant Novartis) and StemCells, Inc. to continue this work in a commercial manner.

In November 1998, two different university-based groups of researchers announced that they had accomplished the first isolation and characterization of the human embryonic stem cell (HESC). One group was led by James A. Thomson at the University of Wisconsin at Madison. The second was led by John D. Gearhart at the Johns Hopkins University School of Medicine at Baltimore. The HESC is among the most versatile basic building blocks in the human body. Embryos, when first conceived, begin creating small numbers of HESCs, and these cells eventually differentiate and develop into the more than 200 cell types that make up the distinct tissues and organs of the human body. If scientists can reproduce and then guide the development of these basic HESCs, then they could theoretically grow replacement organs and tissues in the laboratory—even such complicated tissue as brain cells or heart cells.

Ethical and regulatory difficulties have arisen from the fact that the only source for human "embryonic" stem cells is, logically enough, human embryos. A laboratory can obtain these cells in one of three ways: 1) by inserting a patient's DNA into an egg, thus producing a blastocyst that is a clone of the patient—which is then destroyed after only a few days of development; 2) by harvesting stem cells from aborted fetuses; or 3) by harvesting stem cells from embryos that are left over and unused after an in vitro fertilization of a hopeful mother. (Artificial in vitro fertilization requires the creation of a large number of test tube embryos per instance, but only one of these embryos is used in the final process.)

A rich source of similar but "non-embryonic" stem cells is bone marrow. Doctors have been performing bone marrow transplants in humans for years. This procedure essentially harnesses the healing power of stem cells, which proliferate to create healthy new blood cells in the recipient. Several other non-embryonic stem cell sources have great promise (see "Methods of developing 'post-embryonic' stem cells without the use of human embryos" below).

In the fall of 2001, a small biotech company called Advanced Cell Technology, Inc. announced the first cloning of a human embryo. The announcement set off yet another firestorm of rhetoric and debate on the scientific and ethical questions that cloning and related stem cell technology inspire. While medical researchers laud the seemingly infinite possibilities stem cells promise for fighting disease and the aging process, conservative theologians, many government policy makers, certain ethics organizations and pro-life groups decry the harvest of cells from aborted fetuses and the possibility of cloning as an ethical and moral abomination.

Stem cell research has been underway for years at biotech companies including Stem Cells, Inc., Geron and ViaCord (formerly ViaCell which was acquired by PerkinElmer in 2007). One company, Osiris Therapeutics, has been at work long enough to have two clinical trial programs in progress. Osiris derives its stem cells from the bone marrow of healthy adults between the ages of 18 and 30 who are volunteers. Prior to harvesting the stem cells, Osiris screens blood samples of the donors to make sure that they are free of diseases such as HIV and hepatitis. Osiris believes that this approach to gathering stem cells places them outside of the embryonic source controversy. In 2010, Geron announced that it had received FDA clearance to proceed with Phase I trials of a stem cell-based therapy for acute spinal cord injuries.

The potential benefits of stem cell-based therapies are staggering. Neurological disorders might be aided with the growth of healthy cells in the brain. Injured cells in the spinal column might be regenerated. Damaged organs such as hearts, livers and kidneys might be infused with healthy cells. In China, physicians in more than 100 hospitals are already injecting stem cells into damaged spinal cords with varying levels of success.

Methods of developing "post-embryonic" stem cells without the use of human embryos:

- Adult Skin Cells—Exposure of harvested adult skin cells to viruses that carry specific genes, capable of reprogramming the skin cells so that they act as stem cells.
- Parthenogenesis—manipulation of unfertilized eggs.
- Other Adult Cells—Harvesting adult stem cells from bone marrow or brain tissue.
- Other Cells—harvesting of stem cells from human umbilical cords, placentas or other cells.
- De-Differentiation—use of the nucleus of an existing cell, such as a skin cell, that is altered by an egg that has had its own nucleus removed.
- Transdifferentiation—making a skin cell de-differentiate back to its primordial state so that it can then morph into a useable organ cell, such as heart tissue.
- Pluripotent state cells (iPSCs). Adult cells are drawn from a skin biopsy and treated with reprogramming factors.
- Most recently, researchers have found it possible to harvest stem cells from a wide variety of tissue. The looming question, however, is whether these cells can be successfully reprogrammed.

HESCs (typically harvested from five-day-old human embryos which are destroyed during the process) are used because of their ability to evolve into any cell or tissue in the body. Their versatility is undeniable, yet the implications of the death of the embryo are at the heart of the ethical and moral concerns.

Meanwhile, scientists have discovered that there are stem cells in existence in many diverse places in the adult human body, and they are thus succeeding in creating stem cells without embryos, by utilizing "post-embryonic" cells, such as cells from marrow. Such cells are already showing the ability to differentiate and function in animal and human recipients. Best of all, these types of stem cells may not be plagued by problems found in the use of HESCs, such as the tendency for HESCs to form tumors when they develop into differentiated cells.

Studies were published in 2006 and 2007 regarding the reprogramming of adult mouse skin cells into stem cells. Initially, Shinya Yamanaka of Kyoto (Japan) University announced that he and coworkers had exposed skin cells harvested from adult mice to viruses carrying four specific genes. Yamanaka had determined that these four genes are apparently responsible for a stem cell's ability to develop into virtually any type of tissue. The technique is easy to replicate, and it was confirmed by additional studies in the U.S. published in 2007. This may be a tremendous breakthrough in stem cell research. However, scientists are still a long way from overcoming daunting technical challenges and adapting this technique for use in humans.

In late 2007, two scientific papers were published regarding induced pluripotent state cells (iPSCs). These are human stem cells produced without human cloning or the use of human embryos or eggs. Adult cells are drawn from a skin biopsy and treated with four reprogramming factors, rendering cells that can produce all human cell types and grow indefinitely. iPSCs are patient-specific, making them superior to HESCs because cell rejection by a patient's immune system is avoided. However, the risk of tumor formation appears to be higher for iPSCs due to the reprogramming genes which remain in the cell once the process is complete. Lawrence Goldstein, a researcher at the University of California at San Diego is already using iPSCs to study cells from patients with Alzheimer's in order to develop new drug therapies. Ipierian, a small biopharmaceutical company based in San Francisco, California and formerly known as iZumi Bio, is working on using iPSCs to test potential drugs against selected neuro-degenerative diseases such as ALS (Lou Gehrig's Disease) and Parkinson's. The technology attempts to create cells in the laboratory to show signs of these diseases and then expose them to potential drugs. Although further development is necessary in order to make iPSCs viable for general clinical use, they are a profound breakthrough in stem cell research, and they do not use embryo-derived cells. Stem cell pioneer James A. Thomson is now deeply involved in iPSC research and is cofounder of a research firm called Cellular Dynamics.

Additional stem cell sources continue to be identified. In 2008, researchers at Children's Hospital of Pittsburgh found that stem cells can be harvested from the walls of small blood vessels such as capillaries. In 2009, the *Journal of Translational Medicine* reported that stem cells taken from human fallopian tubes have the potential to become a number of different cell types.

14) U.S. Government Stance on Funding for New Stem Cell Research Unclear

Shortly after taking office, U.S. President Barack Obama reversed an eight-year ban on the use of federal funding for embryonic stem cell research. Specifically, Obama issued Executive Order 13505, entitled "Removing Barriers to Responsible Scientific Research Involving Human Stem Cells." This executive order, dated March 9, 2009, charged the National Institutes of Health (NIH) with issuing new guidelines for stem cell research which became effective in July 2009. The order further authorized the NIH to "support and conduct responsible, scientifically worthy human stem cell research, including human embryonic stem cell research, to the extent permitted by law." The important words here are "human embryonic," since the harvesting of stem cells from discarded human embryos is what started the stem cell funding controversy in the first place. Further, this wording clearly eliminates the possibility of funding research projects involving stem cells that result from cloning. Under previous U.S. regulations, federal research funds were granted only for work with 21 specific lines of stem cells that existed in 2001. Harvesting and developing new embryonic lines did not qualify.

By mid-April 2009, the NIH had issued a new policy statement. The issue of funding remains politically charged. The NIH is taking a middle road. Its guidelines state that embryos donated for such research must be given voluntarily and without financial inducement. (Such embryos typically are donated by couples who have completed fertility treatments and have no need for remaining, redundant embryos. This is a common practice in seeking laboratory-aided pregnancies.)

However, all bets may be off due to an August 2010 ruling by a federal district judge that blocks the Obama administration's executive order. The ruling says that the order violates a ban on federal money being used to destroy embryos. This is a stunning blow to NIH scientists and researchers working with embryonic cells. Further clarification as to what exactly will be legal with regard to funding as well as actual research is expected in the near-term.

Once a stem cell starts to replicate, a large colony, or line, of self-replenishing cells can theoretically continue to reproduce forever. Unfortunately, only about a dozen of the stem cell lines existing in 2001 were considered to be useful, and some scientists believe that these lines were getting tired.

The use of non-federal funding, however, was not restricted during the eight-year ban, although many groups did want to see further state or federal level restrictions on stem cell research or usage. A major confrontation continues between American groups that advocate the potential health benefits of stem cell therapies and groups that decry the use of stem cells on ethical or religious terms. Meanwhile, stem cell development forged ahead in other technologically advanced nations.

In November 2004, voters in California approved a unique measure that provides $3 billion in state funding for stem cell research. Connecticut, Massachusetts and New Jersey also passed legislation that permits embryonic-stem cell research. California already has a massive biotech industry, spread about San Diego and San Francisco in particular. As approved, California's Proposition 71 created an oversight committee that determines how and where grants will be made, and an organization, the California Institute for Regenerative Medicine (www.cirm.ca.gov), to issue bonds for funding and to manage the entire program. The money is being invested in research over 10 years.

In June 2007, the California Institute for Regenerative Medicine (CIRM) approved grants totaling more than $50 million to finance construction of shared research laboratories at 17 academic and non-profit institutions. These facilities are scheduled to be complete and available to researchers within six months to two years of the grant awards. By March 2010, CIRM had approved 328 grants totaling more than $1 billion. The grants are funding dedicated laboratory space for the culture of human embryonic stem cells (HESCs), particularly those that fall outside federal guidelines.

In the private sector, funding for stem cell research was generous up until the global economic crisis began in 2008. For example, the Juvenile Diabetes Research Foundation has an $8 million stem cell research program underway and Stanford University has used a $12 million donation to create a research initiative. Likewise, major, privately funded efforts have been launched at Harvard and at the University of California at San Francisco.

Corporate investment in stem cells has also been strong. AstraZeneca Pharmaceuticals invested $77 million in a startup firm in San Diego called BrainCells, Inc. to study how antidepressants might be used to spur brain cell growth. GlaxoSmithKline agreed to pay OncoMed Pharmaceuticals up to $1.4 billion in late 2007 for four radical new drugs that target cancer stem cells. In 2009, Geron Corporation began an FDA-approved trial of embryonic stem cells in recent spinal cord injuries in eight to ten patients.

President Obama's reversal of the stem cell funding ban came at a time when funding from private and state sources was drying up. This is good news for further research, but it may be a bit late due to breakthroughs in recent years in stem cell study. In 2007, Shinya Yamanaka, a research scientist in Japan, discovered that adult cells can be reprogrammed to an embryonic state using a relatively easy process. This discovery may open the door to an almost unlimited supply of stem cells in the future and lifts most ethical arguments against using the cells for research. It is hoped that now that the global economy is recovering from the recession of 2008-2009, further funding will support new research.

15) Stem Cells—Therapeutic Cloning Techniques Advance

For scientists, the biggest challenge at present may be to discover the exact process by which stem cells are signaled to differentiate. Another big challenge lies in the fact that broad use of therapeutic cloning may require immense numbers of human eggs in which to grow blastocysts.

The clearest path to "therapeutic" cloning may lie in "autologous transplantation." In this method, a tiny amount of a patient's muscle or other tissue would be harvested. This sample's genetic material would then be de-differentiated; that is, reduced to a simple, unprogrammed state. The patient's DNA sample would then be inserted into an egg to grow a blastocyst. The blastocyst would be manipulated so that its stem cells would differentiate into the desired type of tissue, such as heart tissue. That newly grown tissue would then be transplanted to the patient's body. Many obstacles must be overcome before such a transplant can become commonplace, but the potential is definitely there to completely revolutionize healing through such regenerative, stem cell-based processes. One type of bone marrow stem cell, recently discovered by scientists at the University of Minnesota, appears to have a wide range of differentiation capability.

It is instructive to note that there are two distinct types of embryonic cloning: "reproductive" cloning and "therapeutic" cloning. While they have similar beginnings, the desired end results are vastly different.

"Reproductive" cloning is a method of reproducing an exact copy of an animal—or potentially an exact copy of a human being. A

scientist would remove the nucleus from a donor's unfertilized egg, insert a nucleus from the animal, or human, to be copied, and then stimulate the nucleus to begin dividing to form an embryo. In the case of a mammal, such as a human, the embryo would then be implanted in the uterus of a host female for gestation and birth. The successful birth of a cloned human baby doesn't necessarily mean that a healthy adult human will result. To date, cloned animals have tended to develop severe health problems. For example, a U.S. firm, Advanced Cell Technology, reports that it has engineered the birth of cloned cows that appeared healthy at first but developed severe health problems after a few years. Nonetheless, successful cloning of animals is progressing at labs in many nations.

On the other hand, "therapeutic" cloning is a method of reproducing exact copies of cells needed for research or for the development of replacement tissue. In this case, once again a scientist removes the nucleus from a donor's unfertilized egg, inserts a nucleus from the animal, or human, whose cells are to be copied, and then stimulates the nucleus to begin dividing to form an embryo. However, in therapeutic use, the embryo would never be allowed to grow to any significant stage of development. Instead, it would be allowed to grow for a few hours or days, and stem cells would then be removed from it for use in regenerating tissue.

Because it can provide a source of stem cells, cloning has uses in regenerative medicine that can be vital in treating many types of disease. The main differences between stem cells derived from clones and those derived from aborted fetuses or fertility specimens is that a) they are made from only one source of genes, rather than by mixing sperm and eggs; and b) they are made specifically for scientific purposes, rather than being existing specimens, putting them up to more intense ethical discussions. Cloned stem cells have the added advantage of being 100% compatible with their donors, because they share the same genes, and so would provide the best possible source for replacement organs and tissues. Although the use of cloning for regeneration has stirred heated debate as well, it has not resulted in universal rejection. Most of the industrialized countries, including Canada, Russia, most of Western Europe and most of Asia, have made some government-sanctioned allowances for research into this area.

As a result of government sanction of research and development, some countries have already made progress in the field of regenerative cloning. In an important development in August 2004, scientists at Newcastle University in the U.K. announced that they were granted permission by the Human Fertilisation and Embryology Authority (HFEA), a unit of the British Government, to create human embryos as a source of stem cells for certain therapeutic purposes. Specifically, researchers are cloning early-stage embryos in search of new treatments for such degenerative diseases as Parkinson's disease, Alzheimer's and diabetes. The embryos are destroyed before they are two weeks old and are therefore not develop beyond a tiny cluster of cells.

The good news is that several other cloning methods are on the horizon. Markus Grompe, director of the Oregon Stem Cell Center at Oregon Health and Science University in Portland is working on research similar to that at Newcastle University. Adult donor cells are forced to create a protein called nanog, which is only found in stem cells, yet the process is altered in a way that keeps the cells from forming into embryos. In 2008, researchers at MIT's Picower Institute for Learning and Memory pinpointed stem cells within the spinal cord that may lead to a new non-surgical treatment for spinal cord injuries.

In 2009, researchers at the Chinese Academy of Sciences in Beijing made great strides in stem cell cloning or in this case "reprogramming" when they took mature mouse skin cells and returned them to an embryonic-like state. The reprogrammed cells were injected into early-stage mouse embryos. The result of the study was that out of 37 stem cell lines created by reprogramming, three lines were the genesis of 27 live mouse births, one of which was able to sire offspring of his own. Similar experiments were successfully performed at the National Institute of Biological Sciences in Beijing.

Reprogramming has evolved recently to use proteins to manipulate cells to return to embryonic states. In the past, reprogramming was accomplished by using a virus to carry genes into a mature cell. This is problematic since the virus can cause cancer or spark changes in the target cell that are undesirable. New technology uses a wash made of four proteins that are associated with the genes. The wash is absorbed in the target cell where the proteins trigger additional protein changes, causing the reversion to a primitive state. The research that led to this process was done at Scripps Research Institute in La Jolla, California. Two biotech startups in California, Fate Therapeutics and Stemgent, Inc. are partnering to produce reprogrammed tissue for drug

discovery firms. Another startup, iPierian, Inc. (formerly iZumi Bio, Inc.), which is also in California, is teaming up with Kyoto University to conduct further research in reprogramming.

16) Stem Cells—A New Era of Regenerative Medicine Takes Shape

Many firms are conducting product development and research in the areas of skin replacement, vascular tissue replacement and bone grafting or regeneration. Stem cells, as well as transgenic organs harvested from pigs, are under study for use in humans. At its highest and most promising level, regenerative medicine may eventually utilize human stem cells to create virtually any type of replacement organ or tissue.

In one recent, exciting experiment, doctors took stem cells from bone marrow and injected them into the hearts of patients undergoing bypass surgery. The study showed that the bypass patients who received the stem cells were pumping blood 24% better than patients who had not received them.

In another experiment, conducted by Dr. Mark Keating at Harvard, the first evidence was shown that stem cells may be used for regenerating lost limbs and organs. The regenerative abilities of amphibians have long been known, but exactly how they do it, or how it could be applied to mammals, has been little understood. Much of the regenerative challenge lies in differentiation, or the development of stem cells into different types of adult tissue such as muscle and bone. Creatures such as amphibians have the ability to turn their complex cells back into stem cells in order to regenerate lost parts. In the experiment, Dr. Keating made a serum from the regenerating nub (stem cells) of a newt's leg and applied it to adult mouse cells in a petri dish. He observed the mouse cells "de-differentiate," or turn into stem cells. In a later experiment, de-differentiated cells were turned back into muscle, bone and fat. These experiments could be the first steps to true human regeneration. Keating is continuing to make exciting breakthroughs in regenerative research.

The potential of the relatively young science of tissue engineering appears to be unlimited. Transgenics (the use of organs and tissues grown in laboratory animals for transplantation to humans) is considered by many to have great future potential, and improvements in immune system suppression will eventually make it possible for the human body to tolerate foreign tissue instead of rejecting it. There is also increasing theoretical evidence that malfunctioning or defective vital organs such as livers, bladders and kidneys could be replaced with perfectly functioning "neo-organs" (like spare parts) grown in the laboratory from the patient's own stem cells, with minimal risk of rejection.

The ability of most human tissue to repair itself is a result of the activity of these cells. The potential that cultured stem cells have for transplant medicine and basic developmental biology is enormous.

Diabetics who are forced to cope with daily insulin injection treatments could also benefit from engineered tissues. If they could receive a fully functioning replacement pancreas, diabetics might be able to throw away their hypodermic needles once and for all. This could also save the health care system immense sums, since diabetics tend to suffer from many ailments that require hospitalization and intensive treatment, including blindness, organ failure, diabetic coma and circulatory diseases.

Elsewhere, the harvesting of replacement cartilage, which does not require the growth of new blood vessels, is being used to repair damaged joints and treat urological disorders. Genzyme Corp. won FDA approval for its replacement cartilage product Carticel, the first biologic cell therapy to become licensed. Genzyme's process involves harvesting the patient's own cartilage-forming cells, and, from those cells, re-growing new cartilage in the laboratory. The physician then injects the new cartilage into the damaged area. Full regeneration of the replacement cartilage is expected to take up to 18 months. The Genzyme process can cost up to $30,000, compared to $10,000 for typical cartilage surgery. Other companies are exploring alternative methods that may be less expensive and therefore more attractive to payors.

In 2007, U.S. Army medical doctors were treating five soldiers with an extracellular matrix derived from pig bladders. The material is found in all animals and shows promise for healing and regenerating tissue (it is already in use by veterinarians to help repair torn ligaments in horses). The five soldiers in the test lost fingers in the war in Iraq and the experiment is to see how the matrix affects the wounds and if regeneration of any kind occurs. In 2010, the U.S. Pentagon authorized a $12 million grant from the Department of Defense to fund projects such as the development of bone-fusing cement and muscle-growing cell scaffolds.

Companies to Watch: StemCells, Inc., in Palo Alto, California (www.stemcellsinc.com), is focusing on the use of stem cells to treat damage to major organs such as the liver, pancreas and central nervous system. ViaCord (formerly ViaCell, Inc. before its acquisition by PerkinElmer), in Boston, Massachusetts (www.viacord.com), develops therapies using umbilical cord stems. Also, their ViaCord product enables families to preserve their baby's umbilical cord at the time of birth for possible future use in treating over 40 diseases and genetic disorders. As of mid-2010, more than 600,000 families worldwide had banked their cord blood.

Internet Research Tip:
For an excellent primer on genetics and basic biotechnology techniques, see:

National Center for Biotechnology Information
www.ncbi.nlm.nih.gov

17) Nanotechnology Converges with Biotech

Because of their small size, nanoscale devices can readily interact with biomolecules on both the surface and the inside of cells. By gaining access to so many areas of the body, they have the potential to detect disease and deliver treatment in unique ways. Nanotechnology will create "smart drugs" that are more targeted and have fewer side effects than traditional drugs.

Current applications of nanotechnology in health care include immunosuppressants, hormone therapies, drugs for cholesterol control, and drugs for appetite enhancement, as well as advances in imaging, diagnostics and bone replacement. For example, the NanoCrystal technology developed by Elan, a major biotechnology company, enhances drug delivery in the form of tiny particles, typically less than 2,000 nanometers in diameter. The technology can be used to provide more effective delivery of drugs in tablet form, capsules, powders and liquid dispersions. Abbot Laboratories uses Elan's technology to improve results in its cholesterol drug Tricor. Par Pharmaceutical Companies uses NanoCrystal in its Megace ES drug for the improvement of appetite in people with anorexia.

Since biological processes, including events that lead to cancer, occur at the nanoscale at and inside cells, nanotechnology offers a wealth of tools that are providing cancer researchers with new and innovative ways to diagnose and treat cancer. In America, the National Cancer Institute has established the Alliance for Nanotechnology in

Cancer (http://nano.cancer.gov) in order to foster breakthrough research.

Nanoscale devices have the potential to radically change cancer therapy for the better and to dramatically increase the number of effective therapeutic agents. These devices can serve as customizable, targeted drug delivery vehicles capable of ferrying large doses of chemotherapeutic agents or therapeutic genes into malignant cells while sparing healthy cells, greatly reducing or eliminating the often unpalatable side effects that accompany many current cancer therapies.

At the University of Michigan at Ann Arbor, Dr. James Baker is working with molecules known as dendrimers to create new cancer diagnostics and therapies, thanks to grants from the National Institutes of Health and other funds. This is part of a major effort named the Michigan Nanotechnology Institute for Medicine and Biological Sciences.

A dendrimer is a spherical molecule of uniform size (five to 100 nanometers) and well-defined chemical structure. Dr. Baker's lab is able to build a nanodevice with four or five attached dendrimers. To deliver cancer-fighting drugs directly to cancer cells, Dr. Baker loads some dendrimers on the device with folic acid, while loading others with drugs that fight cancer. Since folic acid is a vitamin, many proteins in the body will bind with it, including proteins on cancer cells. When a cancer cell binds to and absorbs the folic acid on the nanodevice, it also absorbs the anticancer drug. For use in diagnostics, Dr. Baker is able to load a dendrimer with molecules that are visible to an MRI. When the dendrimer, due to its folic acid, binds with a cancer cell, the location of that cancer cell is shown on the MRI. Each of these nanodevices may be developed to the point that they are able to perform several advanced functions at once, including cancer cell recognition, drug delivery, diagnosis of the cause of a cancer cell, cancer cell location information and reporting of cancer cell death. Universities that are working on the leading edge of cancer drug delivery and diagnostics using nanotechnology include MIT and Harvard, as well as Rice University and the University of Michigan.

Meanwhile, at the University of Washington, a research group led by Babak A. Parviz is investigating manufacturing methods that resemble that of plants and other natural organisms by "self-assembly." If man-made machines could be designed to assemble themselves, it could revolutionize manufacturing, especially on the nanoscale level. Researchers are studying ways to program the assembly process by sparking chemical synthesis of nanoscale parts such as quantum dots or molecules which then bind to other parts through DNA hybridization or protein interactions. The group led by Professor Parviz is attempting to produce self-assembled high performance silicon circuits on plastic. It is conceivable that integrated circuits, biomedical sensors or displays could be "grown" at rates exponentially faster than current processes.

18) Agricultural Biotechnology Scores Breakthroughs but Causes Controversy/Selective Breeding Offers a Compromise

Global panic over quickly rising food prices in 2008, coupled with the global economic crisis, gave the genetically modified seed industry the boost it needed. Momentum continues to increase with the United Nations' Food and Agriculture Organisation's 2009 report that world food output must grow by 70% to feed a projected additional 2.3 billion people by 2050. Agribio (agricultural biotechnology) has become a top agenda item in government and corporate research budgets, and consumer acceptance of genetically modified food products will grow quickly.

For the pharmaceuticals industry, the biotech-era technology of "molecular farming" will soon lead to broad commercialization of human drug therapies that are grown via agricultural methods. For example, by inserting human genes into plants, scientists can manipulate them so they grow certain human proteins instead of natural plant proteins. The growth in plants of transgenic protein therapies for humans may become widespread. Such drug development methods may prove to be extremely cost-effective. At the same time, hundreds of antibodies produced in farm animals for use in human drug therapies are currently under development or in clinical trials.

Meanwhile, genetically modified foods (frequently referred to as "GM" for genetically modified, or "GMO" for genetically modified organisms) offer tremendous promise in agriculture.

In November 2009, the Chinese Ministry of Agricultural Biosafety Committee issued biosafety certificates to a pest-resistant GM rice. This rice is of the "Bt" variety. Bt stands for *Bacillus thuringiensis*, which is a naturally-occurring, pest-killing toxin found in soil. Organic farmers often spray a mixture containing Bt on their crops. Bioengineers have developed very successful ways to introduce Bt into plant seeds. As a result, the bacteria become part of the plant itself, with tremendous results. According to the International Rice Research Institute, the Chinese have already more than tripled their rice crop over the past 50 years, largely by improving yield per acre, which is now two-thirds higher than the world's average. The institute estimates that China will need to further boost output by 20% by the year 2030. China recently budgeted $3.5 billion toward GM research on rice, corn and wheat. Positive results from China's massive GM effort may make it more acceptable to other governments to follow suit. This particular strain of Bt rice was created locally at the Huazhong Agricultural University, and is reported to enable an 80% reduction in the use of pesticide while upping yield by as much as eight percent.

Agricultural biotechnology became a significant commercial industry during the 1990s. It was fostered both by startups and by large chemical or seed companies. All of these players were focused on developing genetically modified seeds and plants that had higher yields, better nutritional qualities and/or resistance to diseases or insects. Additional traits of GM plants may include resistance to temperature and moisture extremes. According to the International Service for the Acquisition of Agri-biotech Applications (ISAAA), global acreage of GM crops rose 9% to reach 333 million acres in 2009. This amounts to more than 8% of the world's agricultural acreage. This is mostly in the U.S., but large amounts were also planted in Argentina, Canada, Mexico, India, Romania, Uruguay and South Africa. ISAAA reports that 85% of America's maize crop was GM in 2009, as were 90% of the cotton crops in America, Australia and South Africa. In total, 14 million farms in 25 nations, ranging from simple family establishments to giant commercial operations, were growing GM crops ranging from soy to corn to cotton. Meanwhile, GM seeds have the potential to create vast benefits in low-income nations where reliance on small farms or gardens is high and food is scarce.

At the same time, researchers are modifying the structural makeup of some plants in order to alter leaves, stems, branches, roots or seed structures. The ability to modify the nutritional makeup of plants can have highly desirable effects. For example, Mycogen, an affiliate of Dow AgroSciences (www.dowagro.com/mycogen), has developed sunflower seeds with higher levels of oleic and linoleic acids—acids with exceptional nutritional value. In 2009, Mycogen added 45 new grain corn hybrids, bringing its total grain corn lineup to 132 hybrids which are high-yield and insect and disease resistant. Seven additional silage hybrids, 29 corn hybrids and six new sunflower hybrids were to be added to Mycogen's portfolio in 2010.

There are currently dozens of agribio food products on the market, including a range of fruits, vegetables and nuts. There is significant potential for rapid development of new products, thanks to the same technologies that are pushing development of human gene therapies in the pharmaceutical industry.

U.S. farmers have enjoyed greatly increased crop yields and crop quality thanks to GM seeds, and by some estimates as much as 85% of U.S. food may contain ingredients that have been grown with GM methods. Biotech crops eventually become ingredients in everything from baked goods to soft drinks to clothing.

Although scientists have been able to engineer highly desirable traits in GM seeds for crops (such as disease-resistance and insect-resistance), and the scientific community has given GM foods a clean bill of health for years, such modified foods have faced stiff resistance among many consumers, particularly in Europe. While many areas of biotechnology are controversial, agricultural biotech has been one of the largest targets for consumer backlash and government intervention in the marketplace. Consumer resistance to food products containing material grown in this manner is sometimes fierce.

However, with food prices a major concern, and recent instances of riots in a number of third world countries due to short supplies and very high prices for staples such as rice, GM foods are becoming more acceptable around the world. For example, in Japan and South Korea, a number of manufacturers have begun using genetically engineered corn in soft drinks, snacks and other foods. This is a first, but the manufacturers cannot afford corn starch and corn syrup made from conventionally grown crops. According to Yoon Chang-gyu, director of the Korean Corn Processing Industry Association, non-engineered corn cost Korean millers about $450 a metric ton in early 2008, up from $143 in 2006. Prices for GM corn were considerably less at about $350 per ton.

Syngenta (www.syngenta.com), the result of the merger between the agricultural divisions of AstraZeneca and Novartis, is focused on seeds, crop protection products, insecticides and other agricultural products. With this focus, Syngenta is in a position to make some of the best research, development and marketing decisions. The firm's annual investment in research and development is substantial, at about 10% of revenues. In 2009, its sales of seeds of all types totaled about $2.6 billion, up from $2.0 billion in 2007.

Meanwhile, Monsanto, a major competitor to Syngenta, has invested heavily in biotech seed research with terrific results. From a 2002 loss of $1.7 billion, Monsanto has evolved to a 2009 operating profit of about $2.1 billion (on sales of about $11.7 billion). The company accomplished the turnaround by continuing to invest in genetic engineering and market its products despite protests and controversy. However, some of Monsanto's most important products are going off-patent, which is creating new challenges for the company.

A particular concern among farmers in many parts of the world is that GM crops may infest neighboring plants when they pollinate, thus triggering unintended modification of plant DNA. In any event, there is a vast distrust of GM foods in certain locales. U.S. food growers and processors face significant difficulty exporting to the European Union (EU) because of the reliance that American farmers place on GM seeds.

The European Union, as well as specific nations in Europe, has kept many regulations in place that make the use of GM seeds or the import of GM food products a difficult to impossible task. These restrictions remain a hot topic of debate at the World Trade Organization and elsewhere. Meanwhile, a handful of localities in the U.S. have banned or restricted the planting of GM seeds, hoping to protect traditional crops that local growers are widely-known for. A typical restriction is to require that GM seeds be planted at least a certain distance away from non-GM crops.

Some anti-GM activists have arguments with big business—particularly with the giant corporations like Monsanto that make GM seeds. Some people have accused Monsanto of persecuting farmers who appear to be using Monsanto-developed seeds without paying for them. The company has also received criticism for its history of manufacturing chemicals that have risen to varying levels of infamy, such as PCBs, DDT and Agent Orange. Unfortunately, protestors are sometimes violent or destructive.

Meanwhile, the U.S. Food and Drug Administration (FDA) declared food derived from cloned cows, pigs and goats to be safe for consumption. The European Food Safety Authority has also declared cloned animal products to be safe. However, a number of food companies, including Smithfield Foods, Inc., Kraft Foods, Inc. and Tyson

Foods, Inc., have pledged not to use milk or meat from cloned livestock.

A landmark GM foods compromise may be a selective breeding technique that introduces no foreign DNA such as that used in GM seeds. The technology uses old-school practices in which plants with desirable characteristics such as longer shelf life or resistance to insects are crossbred to create new, hardier specimens. The new twist to the old technique is the use of genetic markers, which make it much easier to isolate plants with a positive trait and the gene that causes it. New plants can also be quickly tested for the presence of the isolated gene. The technology cuts traditional selective breeding time in half.

A number of companies are utilizing gene markers in their breeding programs. Arcadia Biosciences is hoping to develop seeds for wheat that can be eaten by people with the intestinal disorder called Celiac Disease, which affects 1% of Americans and 4% of Europeans. Arcadia (www.arcadiabio.com) has also developed technologies that enable crops to utilize nitrogen (part of common fertilizers) more efficiently, thus reducing the amount of fertilizer needed overall. It is working to develop plants that use water more efficiently, thus producing high crop yields in low water conditions. It has even developed technology that enables plants to be irrigated with saltwater.

Genetic markers are not new, but the ability to use them in a cost effective manner is relatively recent thanks to falling costs since the year 2000. Where it once took several dollars to conduct a plant scan, the same test can now be conducted for pennies, making testing on a large scale possible. Look for crop biotechnology companies including DuPont, Monsanto and Syngenta to invest millions of dollars in selective breeding assisted by gene markers over the near- to mid-term.

Yet another technology may radically impact the world's food supply. Food scientists at Sangamo Biosciences (www.sangamo.com) have developed naturally occurring proteins that bind to DNA called zinc fingers. The fingers (so called because of their shape) can be used to genetically modify cells to produce desired effects such as crop yield, taste or drought resistance in plants. They afford very precise changes to DNA which translates into better control when modifying plants and quicker development times compared to typical genetic modification. Zinc fingers may also be far more acceptable to groups who are against GM foods, because the elements of the zinc finger do not remain in a plant for more than a few days. Dow AgroSciences (a subsidiary of Dow Chemical that is focused on crop production) has invested $20 million in Sangamo, hoping to compete with Monsanto and Syngenta's agribio success.

Nanotechnology is affecting foods as well. As of mid-2010, there were four nano-engineered foods on the market according to The Project on Emerging Nanotechnologies. They were Canola Active Oil, which contains an additive called nanodrops that carry vitamins, minerals and phytochemicals; Nanotea, which is formulated for better taste and increased its selenium supplement qualities; Maternal Water, which is formulated for gestating mothers and uses colloidal silver ion technology to purify mineral water; and Nanoceuticals Slim Shake Chocolate, a chocolate-flavored diet shake that uses nanoclusters to improve taste and health benefits without the need for added sugar. Another food-related product, Primea Ring, is a silver ion coating used to clean milk particles left in espresso machines made by Primea.

Company to Watch:

Sacramento, California-based Ventria, www.ventria.com, has received approval from the USDA every year since 1999 to produce crops for use in biopharmaceuticals. Ventria plants self-pollinating rice or barley specifically because they produce large quantities of proteins by nature and because they are not pollinated by wind or insect activity. Thus, they theoretically should have no effect on nearby traditional plantings.

---

19) Focus on Vaccines

Vaccines in general are experiencing a renaissance as drug companies are once again investing heavily in technologies to protect against bioterrorism and a growing number of medical diseases. Not since the 1950s, when Jonas Salk introduced his world-changing vaccine that virtually wiped out polio, have so many vaccines made such an impact on general health and the health care market. Starting in 2006, for example, Merck licensed three new vaccines, including Gardasil for cervical cancer and another treatment to prevent shingles. Merck posted revenues in vaccines alone in 2009 of $3.6 billion (up from $2.2 billion in 2006). By mid 2010, Merck's Vaccines unit offered seven vaccines. An improved version of Pfizer's Prevnar vaccine, if approved by the FDA, could generate sales of $1.5 billion per year starting in 2010.

In May 2004, the U.S. Congress passed a bill with great bipartisan support that provided for $5.6 billion in funding over ten years for stockpiling vaccines and other medicines in defense of possible bioterror attacks on U.S. population centers. The BioShield bill further enhances the possibility of fast track research in the event of a national emergency, and allows government officials to distribute certain treatments even if they have not yet been approved by the FDA. The intent is to create effective responses to attacks from chemical or biological weapons. Among the greatest concerns are vaccines against and treatments for anthrax and smallpox. Over 100 biotechnology firms are producing products that could be candidates for BioShield contracts.

The biggest single bioterror threat to U.S. population centers is considered to be anthrax. A small parcel of anthrax sprayed over a major city could kill hundreds of thousands of people, and it is relatively easy to manufacture and hard to combat. If the release is detected or the first cases are rapidly diagnosed, quick action could save many lives. Providing the exposed population with antibiotics followed by vaccination could be lifesaving for persons who would otherwise become ill with untreatable inhalation anthrax in the subsequent few weeks. Prophylactic antibiotics alone will prevent disease in persons exposed to antibiotic-susceptible organisms, but incorporating vaccination into the treatment regime can greatly reduce the length of treatment with antibiotics. Without vaccination, antibiotics must be continued for 60 days. However, if effective vaccination can be provided, antibiotic treatment can be reduced to 30 days.

Smallpox is also considered to be a major threat. The national stockpile (fewer than 7 million doses of vaccinia virus vaccine) is insufficient to meet national and international needs in the event of a major bioterror attack. The stockpile is also deteriorating and has a finite life span. The vaccine was made using the traditional method of scarifying and infecting the flanks and bellies of calves and harvesting the infected lymph. No manufacturer exists today with the capability to manufacture calf lymph vaccine by the traditional method. Replacing the stockpile will require the development and licensure of a new vaccine using modern cell-culture methods. This development program, which will include process development, validation of a new manufacturing process, and extensive clinical testing, will be expensive and may take several years.

Dozens of firms, particularly startups, are working on vaccines and antidotes for biological weapons with hopes of obtaining BioShield funds. Overall, the NIH grants $500 to $600 million per year (approximately one-third of the U.S. annual biodefense budget) into product development. The Department of Defense Joint Vaccine Acquisition Program has several experimental vaccines in development.

A total of seven major new, high-security, infectious disease research centers have been funded in the United States, accomplished largely with the assistance of government grants, including military funds. Much of the focus of these labs will be on biodefense, including vaccines. These centers include a 15,000 square foot lab at the CDC in Atlanta, Georgia. Additional new labs include three facilities in Fort Detrick, Maryland; as well as one each in Galveston, Texas (at the University of Texas Medical Branch-UTMB); Hamilton, Montana (Rocky Mountain Laboratories); and Boston, Massachusetts (Boston University Medical Center).

20) Ethical Issues Abound

Significant ethical issues face the biotech industry as it moves forward. They include, for example, the ability to determine an individual's likelihood to develop a disease in the future, based on his or her genetic makeup today; the potential to harvest replacement organs and tissues from animals or from cloned human genetic material; and the ability to alter genetically the basic foods that we eat. These are only a handful of the powers of biotechnology that must be dealt with by society. Watch for intense, impassioned discussion of such issues and a raft of governmental regulation as new technologies and therapies emerge.

The biggest single issue may be privacy. Who should have access to your personal genetic records? Where should they be stored? How should they be accessed? Can you be denied employment or insurance coverage due to your genetic makeup?

> **Internet Research Tip:**
> For the latest biotech developments check out www.biospace.com, a private sector portal for the biotech community, and www.bio.org, the web site of the highly regarded Biotechnology Industry Organization.

21) Technology Discussion—Genomics

The study of genes as a resource for the commercial development of new drugs received a significant boost from computer technology in the late 1970s. Frederick Sanger, a chemist, and Walter Gilbert, a biochemist, developed what is known as

DNA sequencing technology, receiving a Nobel Prize in 1980 for their effort. In the same way that computer technology enabled the rapid growth of the Internet industry, computerization has been the catalyst for the booming biotech industry. For example, Sanger and Gilbert's computerized DNA sequencing technology enables scientists to collect massive amounts of data on human genes at high speed, analyzing how certain genes are connected with specific diseases. Using this technology, the gene responsible for Parkinson's disease was mapped in only nine days—a stark contrast to the nine years required to determine the gene connected with cystic fibrosis using traditional methods. Genomics, the mapping and analysis of genes and their uses, is the basic building block of the biopharmaceuticals business. Pharmacogenomics is the study of genomics for the purpose of creating new pharmaceuticals.

Genes are made up of DNA and reside on densely packed fibers called chromosomes. The genetic information encoded in a gene is copied into RNA (ribonucleic acid—closely related to DNA) and then used to assemble proteins. Think of DNA as a blueprint and RNA as the builder.

The human body contains about 75 trillion cells. Each cell contains 46 chromosomes, arranged in 23 pairs. Each chromosome is a strand of DNA. Each strand of DNA is composed of thousands of segments representing different genes. The sum total of the DNA contained in all 46 chromosomes is called the human genome. The number of protein-coding genes appears to be a surprisingly small 23,000.

In October 2006, the X Prize Foundation announced a prize of $10 million for the first group to accurately produce complete genomes of 100 humans in 10 days or less and while spending less than $10,000 per genome. One of the genomics industry's goals is to be able to market sub-$1,000 genomic studies to people who want a complete map of their DNA. This prize will be a big boost and as of mid 2010 was still up for grabs.

> **Internet Research Tip:**
> For a superior, easy-to-understand, illustrated resource on genetics and molecular biology, see: www.dnaftb.org/dnaftb/15/concept/.

SPOTLIGHT: BGI

BGI, formerly known as the Beijing Genomics Institute, is a genomics research operation headed by Yang Huangming, who was a major player in the Human Genome Project. Dr. Yang's plans for BGI are global in scope, including labs in Hong Kong, North America and Europe. The institute is working on the 1,000 Genomes Project, a three-year effort begun in 2008 that will sequence the genomes of 1,000 anonymous participants around the world (BGI's part of the project covers 200 full human genomes). Previously, BGI successfully sequenced over 1,000 bacteria and about 40 plant and animal species including rice, soybeans, honeybees, pandas and lizards. In addition, BGI is studying cancer by following the pattern of mutating malignant cell tissue in human patients. It is also focusing on cloning, having been the first to clone pigs, and has developed a novel and more effective way to clone mammals. Yet another project is the search for the genetic underpinning of human intelligence. To that end, BGI plans to sample protein-coding genes from 2,000 Chinese children and correlate them against the children's test scores. Although steeped in controversy over a number of its projects, BGI hopes to position China as a premier power in the field of genomics.

22) Technology Discussion—Proteomics

Proteomics is the study of the proteins that a gene produces. A complete set of genetic information is contained in each cell, and this information provides a specific set of instructions to the body. The body carries out these instructions via proteins. Genes encode the genetic information for proteins.

All living organisms are composed largely of proteins. Each protein is a large, complex molecule composed of amino acids. Proteins have three main cellular functions: 1) they act as enzymes, hormones and antibodies, 2) they provide cell structure, and 3) they are involved in cell signaling and cell communication functions.

Proteins are important to researchers because they are the link between genes and pharmaceutical development. They indicate which genes are expressed or are being used. They are important for understanding gene function. They also have unique shapes or structures. Understanding these structures and how potential pharmaceuticals will bind to them is a key element in drug design. Proteomic researchers seek to determine the unique role of each of the hundreds of thousands of proteins in the human body, as well as the relationships that such proteins

have with each other and with various diseases. Microarrays enable the high-speed analysis of these proteins and the discovery of SNPs (see below).

23) Technology Discussion—Microarrays

In the laboratory, scientists use microarrays to map the DNA of a patient's tissues. For example, a physician may take a small biopsy of a cancer patient's tumor. That biopsy would be placed into microarray equipment. There, the microarray would arrange hundreds or thousands of microscopic dots of tumor material onto glass laboratory slides. Lasers would scan the slides, and computer software would compare the contents of the slides to vast databases, in order to determine the exact genes in the tumor. Such information may be extremely useful in treating the patient. Genetic information scanned from microarrays has been used to create giant bioinformatics databases. Ideally, such databases could analyze the genetic blueprint of a patient's tumor in order to assist a physician in determining the drugs that would best treat the specific genes that have mutated into cancerous material, and thereby have the best opportunity to cure the patient.

24) Technology Discussion—DNA Chips

Several chip platforms have been developed to facilitate molecular biology research. For example, industry leader Affymetrix (www.affymetrix.com) has developed chips programmed to act like living cells. These chips can run thousands of tests in short order to help scientists determine a specific gene's expression.

Scientists know that a mutation—or alteration—in a particular gene's DNA often results in a certain disease. However, it can be very difficult to develop a test to detect these mutations, because most large genes have many regions where mutations can occur. For example, researchers believe that mutations in the genes BRCA1 and BRCA2 cause as many as 60 percent of all cases of hereditary breast and ovarian cancers. But there is not one specific mutation responsible for all of these cases. Researchers have already discovered over 800 different mutations in BRCA1 alone.

The DNA chip is a tool used to identify mutations in genes like BRCA1 and BRCA2. The chip, which consists of a small glass plate encased in plastic, is manufactured somewhat like a computer microchip. On the surface, each chip contains thousands of short, synthetic, single-stranded DNA sequences, which together, add up to the normal gene in question. To determine whether an individual possesses a mutation for BRCA1 or BRCA2, a scientist first obtains a sample of DNA from the patient's blood as well as a control sample—one that does not contain a mutation in either gene.

The researcher then denatures the DNA in the samples, a process that separates the two complementary strands of DNA into single-stranded molecules. The next step is to cut the long strands of DNA into smaller, more manageable fragments and then to label each fragment by attaching a fluorescent dye. The individual's DNA is labeled with green dye and the control, or normal, DNA is labeled with red dye. Both sets of labeled DNA are then inserted into the chip and allowed to hybridize (bind) to the synthetic BRCA1 or BRCA2 DNA on the chip. If the individual does not have a mutation for the gene, both the red and green samples will bind to the sequences on the chip.

If the individual does possess a mutation, the individual's DNA will not bind properly in the region where the mutation is located. The scientist can then examine this area more closely to confirm that a mutation is present. (Explanation provided courtesy of National Institutes of Health.)

Until recently, DNA testing equipment manufactured by companies such as Life Technologies and Roche has carried hefty price tags in the $30,000 to $50,000 range. A new machine, the Pixo made by Helixis, can perform DNA testing for as little at $10,000. Pixo uses a smaller, more energy efficient process to heat and cool DNA samples and also holds fewer samples than the more expensive machines (48 compared to 96).

Another breakthrough in DNA testing is a 2009 prototype for the fastest DNA decoder ever made. Designed by Steve Turner, the unit is capable of sequencing millions of DNA letters in as little as 15 minutes. Turner's company, Pacific Biosciences (PacBio), has trademarked its Single Molecule Real Time (SMRT) technology, which can detect biological events at single molecule resolution. As of 2010, PacBio had orders for its SMRT DNA sequencing system from 10 major U.S. research institutions including Baylor College of Medicine, the Broad Institute of MIT and Harvard and the U.S. Department of Energy Joint Genome Institute.

25) Technology Discussion—SNPs ("Snips")

The holy grail of genomics is the search for single nucleotide polymorphisms (SNPs). These are DNA sequence variations that occur when a single nucleotide (A, T, C or G) in the genome sequence is altered. For example an SNP might change the DNA sequence AAGGCTAA to ATGGCTAA. SNPs occur in every 100 to 1,000 bases along the 3 billion-base human genome. SNPs can occur in both coding (gene) and noncoding regions of the genome. Many SNPs have no effect on cell function, but scientists believe others could predispose people to diseases or influence their response to a drug.

Variations in DNA sequence can have a major impact on how humans react to disease; environmental factors such as bacteria, viruses, toxins and chemicals; and drug therapies. This makes SNPs of great value for biomedical research and for developing pharmaceutical products or medical diagnostics. Scientists believe SNP maps will help them identify the multiple genes associated with such complex diseases as cancer, diabetes, vascular disease and some forms of mental illness.

26) Technology Discussion—Combinatorial Chemistry

After researchers determine which protein is involved in a specific disease, they use combinatorial chemistry to find the molecule that controls the protein. High-throughput screening techniques allow scientists to use automatic fluid handlers to mix chemicals under specific test conditions at extremely high speed. Combinatorial chemistry can generate thousands of chemical compound variations in a few hours. Previously, traditional chemistry methods could have required several weeks or even months to do the same work. High-throughput screening enables automated, computerized machines to screen as many as 100,000 chemical compounds daily, seeking potential molecules as drug candidates.

One company, CombinatoRx (www.combinatorx.com), uses a combinatorial array to mix together drugs already on the market to find combinations that could have new and unforeseen uses. The company found hundreds of potential uses for combinations of drugs whose patents had expired and has patented these precise mixtures for new uses. It has received rapid FDA approval in many cases because all the drugs being combined have proven records of safety. In one particularly bizarre instance, the firm found that by combining a particular sedative with an antibiotic, it had made an effective cancer-fighting agent.

27) Technology Discussion—Synthetic Biology

Scientists have followed up on the task of mapping genomes by attempting to directly alter them. This effort has gone past the point of injecting a single gene into a plant cell in order to provide a single trait, as in many agricultural biotech efforts. There are now several projects underway to create entirely new versions of life forms, such as bacteria, with genetic material inserted in the desired combination in the laboratory.

Synthetic biology can be defined as the design and construction of new entities, including enzymes and cells, or the reformatting of existing biological systems. This science capitalizes on previous advances in molecular biology and systems biology, by applying a focus on the design and construction of unique core components that can be integrated into larger systems in order to solve specific problems.

In June 2004, the first international meeting on synthetic biology was held at MIT. Called Synthetic Biology 1.0, the conference brought together researchers who are working to design and build biological parts, devices and integrated biological systems; develop technologies that enable such work; and place this research within its current and future social context. The National Science Foundation agreed, in mid 2006, to invest $16 million in a five-year grant to fund a new Synthetic Biology Engineering Research Center ("SynBERC) at UC Berkeley. An additional $4 million for the project has been raised from other sources, and the NSF is offering the possibility of a further five-year grant.

Elsewhere, engineers and scientists from MIT, Harvard and other institutions are working on a concept called BioBricks, which are strands of DNA with connectors at each end. They can be assembled into higher-level components.

A leading proponent of synthetic biology is Dr. Craig Venter, well known for his efforts in sequencing the human genome. In 2010, the J. Craig Venter Institute announced a significant breakthrough in this field. Twenty-five researchers formed a team that deciphered the genetic instructions of a simple bacterium. In a demonstration project that cost $40 million, they built a computerized genetic code of the bacterium. They then manufactured a series of chemical DNA genetic sequences. Next, they housed these sequences in yeast and E. coli cells, and tied them together in a single one million base-pair genome. Finally, they transplanted this genome into a host cell, transforming this essentially blank cell into a different species of bacteria. Other organizations actively at work on such projects include Amyris Biotechnologies, LS9, Inc. and Joule Unlimited.

28) Technology Discussion—Recombinant DNA

In today's recombinant DNA therapies, the implantation of selected strands of DNA into bacteria allows large-scale production of hormones, antibodies and other exciting new drugs. Additions to the medical arsenal developed from this technique include interferons, growth factor hormone, anticoagulants, human insulin (as opposed to the cow and pig insulin that diabetics have used for 60 years) and more effective immunosuppressive drugs. An anticoagulant proven highly effective for some heart attack patients, TPA, may dramatically reduce heart damage during an attack and cut hospital stays. These drugs are often expensive, but they may greatly reduce the duration of hospital confinement and, in some cases, the need for any hospitalization at all. Long-term, potential applications of this technology include genetically-modified seeds, vaccines, production of insulin, production of blood clotting factors and the production of recombinant pharmaceuticals.

29) Technology Discussion—Polymerase Chain Reaction (PCR)

PCR, also called "molecular photocopying", is a fast, inexpensive method of replicating genetic material in the form of small segments of DNA. The technique was created by Kary B. Mullis of La Jolla, California. His efforts were awarded one-half of the Nobel Prize in Chemistry in 1993, which he shared with another leading DNA researcher, Michael Smith of the University of British Columbia, Canada.

PCR creates sufficient amounts of DNA to allow detailed analysis, which can be vital to researchers. This may lead to better diagnosis and treatment of diseases such as AIDS, Lyme disease, tuberculosis, viral meningitis, cystic fibrosis, Huntington's chorea, sickle cell anemia and many others. PCR also shows promise in providing cancer susceptibility warnings.

Once amplified, the DNA produced by PCR can be used in many different laboratory procedures. For example, most mapping techniques in the Human Genome Project (HGP) rely on PCR. PCR is also valuable in a number of emerging laboratory and clinical techniques, including DNA fingerprinting, detection of bacteria or viruses (particularly AIDS), and diagnosis of genetic disorders.

Chapter 2

BIOTECH & GENETICS INDUSTRY STATISTICS

Biotech Industry Overview

Global	Amount	Units	Year	Source
Public & Private Biotech Companies (US, Europe and Canada)	4,361	Companies	2009	E&Y
Public Biotech Companies	622	Companies	2009	E&Y
Biotech Revenues	79.1	Bil. US$	2009	E&Y
Total Pharmaceutical Sales, Worldwide	837.0	Bil. US$	2009	IMS
Total R&D Expenses, PhRMA Member Companies	47.4	Bil. US$	2009	PhRMA
Total R&D Expenses, All Pharmaceutical Companies	65.3	Bil. US$	2009	Burrill

U.S.	Amount	Units	Year	Source
Public & Private Biotech Companies	1,699	Companies	2009	E&Y
Public Biotech Companies	313	Companies	2009	E&Y
Revenues	56.6	Bil. US$	2009	E&Y
Area of Biotech Crops	64.0	Mil. Hectares	2009	ISAAA
Prescription Drug Sales, Domestic	300.3	Bil. US$	2009	IMS
Number of FDA Approvals for New Drugs (NDAs) and Biologic License Applications (BLAs)	97	Approvals	2009	FDA
Number of Approvals for New Molecular Entities (NMEs) and BLAs	25	Approvals	2009	FDA
Patents Granted for "Multicellular Living Organisms & Unmodified Parts Thereof & Related Processes"	946	Patents	2009	USPTO
Medical Research Funding, NIH	31.2	Bil. US$	2009	NIH
Total Requested Budget for Biological Science Research, U.S. (NSF)	733.0	Mil. US$	2010	NSF
Mean Annual Salary for Biochemists & Biophysicists	88,550	US$	May-09	BLS
Average Cost of Developing a Biologic Drug	1.2	Bil. US$	2005	Tufts

PhRMA[1] Member Statistics, U.S.	Amount	Units	Year	Source
Pharmaceutical Sales, Domestic	183.0	Bil. US$	2009	PhRMA
% Generic (by volume)	64.0	%	2009	PhRMA
Pharmaceutical Sales, Foreign[2]	103.4	Bil. US$	2009	PhRMA
Pharmaceutical R&D Spending, Domestic	35.6	Bil. US$	2009	PhRMA
as a Percentage of Domestic Sales	19.4	%	2009	PhRMA
Share of R&D Spending by Function:				
Prehuman/Preclinical	28.8	%	2008	PhRMA
Phase I	7.1	%	2008	PhRMA
Phase II	10.5	%	2008	PhRMA
Phase III	23.5	%	2008	PhRMA
Approval	5.5	%	2008	PhRMA
Phase IV	12.9	%	2008	PhRMA
Uncategorized	0.5	%	2008	PhRMA

[1] PhRMA = Pharmaceutical Research and Manufacturers Association, a group of leading pharmaceutical and biotechnology companies in the U.S. [2] Not including foreign divisions of foreign companies.

E&Y = Ernst & Young; IMS = IMS Health; Burrill = Burrill & Company; ISAAA = International Service for the Acquisition of Agri-Biotech Applications; Tufts = Tufts University Center for the Study of Drug Development; FDA = U.S. Food & Drug Administration; USPTO = U.S. Patent & Trademark Office; NIH = U.S. National Institutes of Health; NSF = U.S. National Science Foundation; BLS = U.S. Bureau of Labor Statistics

The U.S. Drug Discovery & Approval Process

Years		Investigational New Drug (IND) Application Submitted ↓		New Drug Application FDA Review →→→			FDA Approves 1 New Drug ↓
2	**4**	**6**	**8**	**10**	**12**	**14**	**16**

↑

Drug Discovery 5,000-10,000 Compounds → Pre-Clinical Trials Begin 250 Compounds

↓

Phase I Trials Begin 20-100 Volunteers

↓

Phase II Trials Begin 100-500 Volunteers 5 Compounds → Phase III Trials Begin 1,000-5,000 Volunteers

↑

↑

↓

Phase IV Large Scale Manufacturing Post-Marketing Surveillance

Note: The actual length of the development process varies. On average, it takes between 10 and 15 years and an estimated $1.2 billion to $1.3 billion to create a successful new medicine, as of 2010.

Source: Pharmaceutical Research and Manufacturers Association (PhRMA)

Plunkett Research, Ltd.

www.plunkettresearch.com

U.S. FDA New Drug and Biologic Approvals, 2009

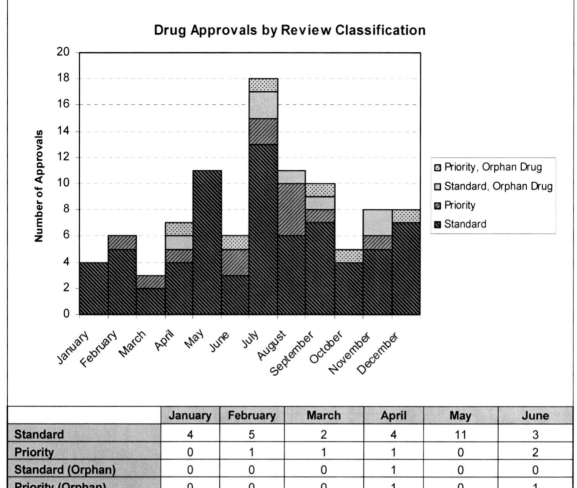

	January	February	March	April	May	June
Standard	4	5	2	4	11	3
Priority	0	1	1	1	0	2
Standard (Orphan)	0	0	0	1	0	0
Priority (Orphan)	0	0	0	1	0	1
	July	**August**	**September**	**October**	**November**	**December**
Standard	13	6	7	4	5	7
Priority	2	4	1	0	1	0
Standard (Orphan)	2	1	1	0	2	0
Priority (Orphan)	1	0	1	1	0	1

Notes: Priority Review classifies drugs that are a significant improvement compared to marketed products, in the treatment, diagnosis, or prevention of a disease. Standard Review classifies drugs that do not qualify for priority review. Orphan Designation is assigned to drugs pursuant to Section 526 of the Orphan Drug Act (Public Law 97-414 as amended). Data refers to CDER's approvals of New Drug Applications (NDAs) and Biologic License Applications (BLAs).

Source: U.S. Food & Drug Administration (FDA)

Plunkett Research, Ltd.

www.plunkettresearch.com

U.S. Pharmaceutical R&D Spending Versus the Number of New Molecular Entity (NME) Approvals: 1993-2009

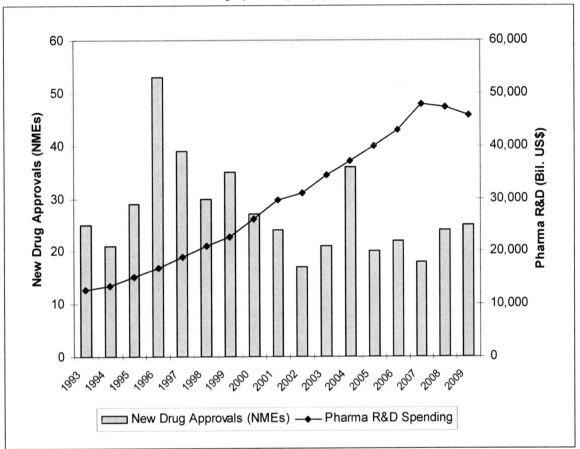

* Beginning in 2004, these figures include new BLAs for therapeutic biologic products transferred from CBER to CDER.

Notes: The FDA defines a New Molecular Entity (NME) as a medication containing an active substance that has never before been approved for marketing in any form in the U.S. Pharmaceutical R&D Spending includes expenditures inside and outside the U.S. by U.S.-owned PhRMA member companies and R&D conducted inside and outside the U.S. by the U.S. divisions of foreign-owned PhRMA member companies. R&D performed by the foreign divisions of foreign-owned PhRMA member companies is excluded.

Source: Pharmaceutical Research and Manufacturers Association (PhRMA); U.S. Food and Drug Administration

Plunkett Research, Ltd.

www.plunkettresearch.com

Employment in Life & Physical Science Occupations by Business Type, U.S.: May 2009

(Wage & Salary in US$)	Employ-ment[1]	Median Hourly Wage	Mean Hourly Wage	Mean Annual Salary[2]	Mean RSE[3] (%)
Animal Scientists	2,190	27.38	31.02	64,510	2.4%
Food Scientists and Technologists	10,790	28.67	30.95	64,370	1.4%
Soil and Plant Scientists	11,830	28.45	31.34	65,180	1.6%
Biochemists and Biophysicists	22,860	39.61	42.57	88,550	1.2%
Microbiologists	16,260	32.01	34.61	71,980	1.7%
Zoologists and Wildlife Biologists	17,460	27.16	29.17	60,670	1.3%
Biological Scientists, All Other	29,630	31.98	33.38	69,430	1.1%
Conservation Scientists	16,810	28.92	29.41	61,180	0.4%
Foresters	10,230	25.89	26.55	55,220	0.5%
Epidemiologists	4,610	29.66	31.22	64,950	0.8%
Medical Scientists, Except Epidemiologists	101,760	35.86	40.75	84,760	1.3%
Life Scientists, All Other	12,320	30.76	34.90	72,590	1.7%
Astronomers	1,240	50.35	49.40	102,740	1.9%
Physicists	13,630	51.15	53.49	111,250	2.1%
Atmospheric and Space Scientists	8,320	40.73	40.94	85,160	2.9%
Chemists	79,910	32.80	34.97	72,740	0.6%
Materials Scientists	8,880	38.61	39.59	82,350	1.7%
Environmental Scientists and Specialists, Including Health	83,530	29.33	32.38	67,360	1.0%
Geoscientists, Except Hydrologists and Geographers	31,860	39.05	44.57	92,710	1.4%
Hydrologists	7,150	35.42	36.91	76,760	1.3%
Physical Scientists, All Other	25,310	45.17	45.62	94,880	1.5%
Agricultural and Food Science Technicians	18,490	16.54	17.72	36,850	1.1%
Biological Technicians	74,560	18.61	19.78	41,140	0.6%
Chemical Technicians	64,420	20.23	21.11	43,900	0.6%
Geological and Petroleum Technicians	14,460	25.60	28.08	58,400	2.3%
Nuclear Technicians	6,290	32.37	32.07	66,700	1.8%
Environmental Science and Protection Technicians, Including Health	30,870	19.61	20.92	43,520	0.7%
Forensic Science Technicians	12,870	24.75	26.47	55,070	0.7%
Forest and Conservation Technicians	31,440	15.80	17.49	36,370	0.3%

[1] Estimates for detailed occupations do not sum to the totals because the totals include occupations not shown separately. Estimates do not include self-employed workers.

[2] Annual wages have been calculated by multiplying the hourly mean wage by a "year-round, full-time" hours figure of 2,080 hours; for those occupations where there is not an hourly mean wage published, the annual wage has been directly calculated from the reported survey data.

[3] The relative standard error (RSE) is a measure of the reliability of a survey statistic. The smaller the relative standard error, the more precise the estimate.

Source: U.S. Bureau of Labor Statistics

Plunkett Research, Ltd.

www.plunkettresearch.com

Federal R&D & R&D Plant Funding for General Science & Basic Research, U.S.: Fiscal Years 2008-2010

(In Millions of US$)

Funding Category and Agency	2008 Actual	2009 Prelim.	2010 Proposed	% Change (09-10)
Total	9,007	9,961	10,584	6.3
National Science Foundation (NSF)	4,506	4,833	5,290	9.5
Biological sciences	616	656	733	11.7
Computer and information science and engineering	535	574	633	10.3
Education and human resources	766	845	858	1.5
Engineering	649	693	765	10.4
Geosciences	758	807	909	12.6
Integrative activities	214	241	271	12.4
Major research equipment and facilities	167	152	117	-23.0
Mathematical and physical sciences	1,171	1,256	1,380	9.9
Office of cyberinfrastructure	185	199	219	10.1
Social, behavioral, and economic sciences	227	240	257	7.1
U.S. polar research programs	447	471	516	9.6
Budget authority adjustment[1]	-1,229	-1,301	-1,368	5.1
Department of Energy (DOE)	3,807	4,326	4,468	3.3
Department of Homeland Security (DHS)[2]	694	802	826	3.0

Notes: Detail may not add to total because of rounding. Percent change derived from unrounded data. Not all federally sponsored basic research is categorized in subfunction 251. Data derived from agencies' submission to Office of Management and Budget, Circular No. A-11, Max Schedule C; budget justification documents; and supplemental data from agencies' budget offices.

[1] Budget authority adjustment subtracts costs for research facilities, major equipment support, and other non-R&D from total NSF budget authority.

[2] In FY 2007 DHS changed its R&D portfolio to reclassify funding in defense and most administration of justice as general science and basic research.

Source: U.S. National Science Foundation

Plunkett Research, Ltd.

www.plunkettresearch.com

National Health Expenditure Amounts by Type of Expenditure, U.S.: Selected Calendar Years, 2004-2019[1]

(In Billions of US$)

Type of Expenditure	2004	2009	2010	2011	2012	2017	2018	2019
National Health Expenditures	1,855.4	2,472.2	2,569.6	2,702.9	2,850.2	3,936.0	4,203.6	4,482.7
Health Services & Supplies	1,733.6	2,306.2	2,395.0	2,518.7	2,655.5	3,660.2	3,908.8	4,169.7
Personal Health Care	1,549.9	2,068.3	2,141.7	2,244.6	2,368.0	3,256.7	3,477.0	3,709.0
Hospital Care	566.5	760.6	788.9	827.3	875.8	1,213.2	1,292.4	1,374.5
Professional Services	581.2	777.3	797.2	832.8	877.9	1,202.1	1,284.4	1,370.7
Physician & Clinical Services	393.6	527.6	535.8	556.1	582.3	776.6	828.1	882.0
Other Professional Services	52.9	69.6	71.4	74.6	79.3	109.0	116.1	123.7
Dental Services	81.5	104.4	107.9	111.8	118.1	162.6	171.4	180.4
Other Personal Health Care	53.3	75.7	82.2	90.3	98.2	154.0	168.9	184.6
Nursing Home & Home Health	53.3	75.7	82.2	90.3	98.2	154.0	168.9	184.6
Home Health Care	42.7	72.2	77.1	82.8	89.1	131.7	142.3	153.8
Nursing Home Care	115.2	144.1	149.3	156.2	163.7	217.8	231.4	245.9
Retail Outlet Sales of Medical Products	244.3	314.1	329.1	345.4	361.4	491.9	526.5	564.1
Prescription Drugs	188.8	246.3	260.1	274.5	287.5	395.4	425.2	457.8
Other Medical Products	55.5	67.8	69.1	70.9	73.9	96.5	101.3	106.4
Durable Medical Equipment	22.8	27.0	27.4	28.1	29.2	38.8	40.9	43.0
Other Non-Durable Medical Products	32.7	40.8	41.6	42.8	44.7	57.7	60.4	63.3
Program Administration & Net Cost of Private Health Insurance	129.8	162.8	172.6	189.5	198.2	280.5	300.6	320.3
Government Public Health Activities	53.8	75.2	80.8	84.6	89.4	123.0	131.3	140.3
Investment	121.8	166.0	174.6	184.2	194.7	275.8	294.8	313.0
Research[2]	38.9	48.0	51.3	54.2	57.7	80.3	85.6	91.2
Structures & Equipment	83.0	117.9	123.3	130.0	137.1	195.5	209.2	221.8

Note: Numbers may not add to totals because of rounding.

[1] The health spending projections were based on the 2008 version of the National Health Expenditures (NHE) released in January 2010.

[2] Research and development expenditures of drug companies and other manufacturers and providers of medical equipment and supplies are excluded from research expenditures. These research expenditures are implicitly included in the expenditure class in which the product falls, in that they are covered by the payment received for that product.

Source: Centers for Medicare & Medicaid Services (CMS), Office of the Actuary

Plunkett Research, Ltd.

www.plunkettresearch.com

U.S. Exports & Imports of Pharmaceutical Products: 2005-1st Quarter 2010

(In Thousands of $US)

Exports

Partner	2005	2006	2007	2008	2009	1Q 2009	1Q 2010
World Total	21,717,180	25,344,854	29,242,543	34,211,517	40,667,260	10,356,666	11,329,983
Germany	801,867	1,357,499	3,396,957	4,666,004	6,112,465	1,563,631	1,882,842
Canada	2,794,759	3,341,662	3,336,081	3,118,117	4,427,850	987,702	1,425,748
United Kingdom	2,771,895	3,604,489	3,683,829	3,936,675	4,264,510	1,280,795	1,276,676
Netherlands	3,496,556	4,212,693	3,626,803	4,235,029	4,817,714	1,327,880	1,094,312
Japan	1,174,992	1,379,970	1,421,708	1,630,790	2,257,441	504,853	899,085
Belgium	1,583,283	1,870,620	2,001,761	2,331,565	2,251,399	518,634	539,101
Switzerland	1,433,356	1,804,761	2,243,737	2,449,143	2,434,773	510,597	509,482
Spain	253,035	231,037	1,004,131	2,136,847	2,345,830	855,928	444,806
Brazil	415,832	526,903	576,222	711,690	745,378	170,888	413,878
Mexico	612,695	697,923	609,542	863,997	1,143,538	225,297	347,959
Italy	889,433	910,333	985,751	1,141,380	1,156,341	260,641	342,463
France	1,223,086	1,266,271	1,206,051	1,386,563	1,469,222	384,101	330,154
Ireland	687,655	488,215	815,909	816,766	843,119	215,341	272,591
Australia	615,396	635,707	808,656	794,229	1,177,809	354,051	266,160
South Korea	213,187	264,299	313,795	357,575	500,630	102,005	135,775

Imports

Partner	2005	2006	2007	2008	2009	1Q 2009	1Q 2010
World Total	35,442,464	42,205,187	48,947,754	52,342,604	55,721,859	12,577,488	14,426,619
Ireland	6,108,618	5,861,418	6,549,044	6,586,462	7,031,826	1,916,006	2,191,338
Germany	4,894,638	6,167,540	7,920,946	6,457,606	7,611,645	1,686,698	1,592,884
United Kingdom	4,097,241	4,894,358	6,451,320	7,576,662	7,145,234	1,656,376	1,260,324
France	3,832,103	3,939,319	4,407,889	5,251,394	4,280,019	954,786	1,225,456
Switzerland	1,498,193	2,203,900	2,305,030	2,985,806	3,909,702	657,163	1,152,214
Canada	2,377,053	3,379,409	4,759,670	4,618,081	4,834,095	1,044,173	1,119,953
Belgium	802,852	1,108,698	1,085,042	1,536,084	2,522,369	292,651	1,001,050
Israel	1,448,237	2,493,233	2,593,992	3,738,582	3,767,229	1,010,479	905,337
Singapore	1,158,317	2,427,199	3,043,012	1,912,644	1,993,478	542,546	562,665
India	283,034	438,401	895,972	1,425,841	1,660,629	293,698	498,772
Italy	1,446,768	1,320,151	1,412,807	2,026,219	1,880,735	389,001	497,522
Japan	1,777,711	1,299,907	1,385,745	1,407,955	1,759,095	364,756	398,496
Denmark	1,087,358	1,005,309	987,529	1,037,855	1,110,809	265,642	353,885
Spain	961,918	694,261	865,519	923,608	811,563	197,310	235,928
Austria	506,581	582,906	838,999	816,397	884,244	184,082	225,162

Note: "Pharmaceutical Products" refers to HS (Harmonized Commodity Description and Coding System) Code 30.

Source: Foreign Trade Division, U.S. Census Bureau

Plunkett Research, Ltd.

www.plunkettresearch.com

U.S. Prescription Drug Expenditures, Aggregate & Per Capita Amounts, Percent Distribution: Selected Calendar Years, 2004-2019

(By Source of Funds)

Year	Total	Out-of-Pocket Payments	Third-Party Payments					Medicare[2]	Medicaid[3]
			Total	Private Health Insurance	Public				
					Total	Federal[1]	State & Local[1]		
Historical Estimates *(In Billions of US$)*									
2004	188.8	46.2	142.6	90.0	52.5	31.4	21.1	3.4	36.3
2005	199.7	48.8	150.9	95.8	55.1	32.6	22.5	3.9	37.2
2006	217.0	46.9	170.1	96.2	73.9	58.7	15.2	39.6	19.1
2007	226.8	48.9	177.8	97.8	80.0	65.4	14.7	46.0	18.8
2008	234.1	48.5	185.5	98.5	87.0	72.5	14.5	52.1	19.4
Projected									
2009	246.3	49.0	197.3	100.8	96.5	82.8	13.7	57.8	21.2
2010	260.1	50.9	209.1	104.5	104.7	89.7	15.0	61.6	23.9
2011	274.5	52.3	222.2	108.3	113.9	95.0	18.9	67.2	25.8
2012	287.5	53.4	234.1	110.9	123.2	103.1	20.2	73.3	27.5
2013	302.9	55.4	247.5	114.9	132.6	111.1	21.5	79.5	29.3
2014	322.1	58.5	263.6	121.5	142.1	119.3	22.8	86.5	31.4
2015	343.8	62.0	281.8	129.1	152.6	128.4	24.2	93.8	33.6
2016	368.4	65.7	302.7	136.9	165.8	139.8	26.0	102.6	35.9
2017	395.4	69.6	325.8	145.5	180.3	152.4	27.9	112.4	38.3
2018	425.2	73.9	351.3	154.9	196.4	166.5	29.9	123.5	40.8
2019	457.8	78.6	379.2	165.0	214.2	182.1	32.1	135.8	43.5
Historical Estimates *(Per Capita Amount in US$)*									
2004	644	158	486	307	179	107	72	N/A	N/A
2005	675	165	510	324	186	110	76	N/A	N/A
2006	726	157	569	322	247	196	51	N/A	N/A
2007	752	162	589	324	265	217	49	N/A	N/A
2008	769	159	609	324	286	238	48	N/A	N/A
Projected									
2009	802	160	642	328	314	269	45	N/A	N/A
2010	839	164	675	337	338	289	48	N/A	N/A
2011	878	167	711	346	364	304	60	N/A	N/A
2012	911	169	742	351	391	327	64	N/A	N/A
2013	952	174	778	361	417	349	68	N/A	N/A
2014	1,003	182	821	378	443	372	71	N/A	N/A
2015	1,062	192	870	399	471	397	75	N/A	N/A
2016	1,128	201	927	419	508	428	80	N/A	N/A
2017	1,201	211	989	442	548	463	85	N/A	N/A
2018	1,280	223	1,058	466	591	501	90	N/A	N/A
2019	1,367	235	1,132	493	640	544	96	N/A	N/A

Notes: Per capita amounts based on July 1 Census resident-based population estimates. Numbers and percents may not add to totals because of rounding. The health spending projections were based on the 2008 version of the National Health Expenditures (NHE) released in January 2010.

[1] Includes Medicaid SCHIP Expansion and SCHIP. [2] Subset of Federal funds. [3] Subset of Federal and State and Local Funds. N/A = Not Applicable.

Source: Centers for Medicare & Medicaid Services (CMS), Office of the Actuary
Plunkett Research, Ltd.
www.plunkettresearch.com

U.S. Prescription Drug Expenditures: 1965-2019

(In Millions of US$)

Year	TOTAL	Private			Public								
		Total Private	Out-of-Pocket	Insurance	Total Public	Federal				State & Local		
						Total Fed	Medicare	Medicaid	Other	Total S&L	Medicaid	Other
1965	3,715	3,571	3,441	130	144	58	0	0	58	85	0	85
1966	3,985	3,776	3,594	182	209	95	0	50	45	115	53	62
1967	4,227	3,950	3,712	238	277	127	0	99	29	150	106	44
1968	4,742	4,437	4,120	317	305	141	0	114	27	164	106	59
1969	5,149	4,761	4,362	399	388	191	0	165	26	197	135	62
1970	5,497	5,015	4,531	484	482	237	0	224	13	245	193	53
1971	5,877	5,309	4,752	558	568	297	0	281	16	271	214	57
1972	6,324	5,678	5,035	644	646	328	0	308	20	318	255	63
1973	6,817	6,083	5,341	742	734	355	0	336	19	379	308	71
1974	7,422	6,567	5,714	854	855	443	0	419	24	412	321	91
1975	8,052	7,032	6,068	963	1,020	508	0	478	30	512	392	120
1976	8,722	7,565	6,476	1,089	1,157	610	0	576	35	547	394	153
1977	9,196	7,946	6,754	1,192	1,250	625	0	587	37	626	449	177
1978	9,891	8,531	7,115	1,416	1,360	655	0	613	41	706	493	212
1979	10,744	9,178	7,471	1,707	1,566	743	0	701	42	823	536	288
1980	12,049	10,249	8,466	1,783	1,800	857	0	813	44	943	595	348
1981	13,398	11,338	8,843	2,494	2,060	979	0	932	47	1,081	679	403
1982	15,029	12,840	10,272	2,568	2,189	990	0	934	56	1,199	735	464
1983	17,323	14,807	11,253	3,554	2,516	1,133	0	1,067	66	1,383	841	542
1984	19,617	16,669	12,502	4,167	2,948	1,311	2	1,238	71	1,638	982	656
1985	21,795	18,564	13,608	4,957	3,230	1,421	21	1,323	78	1,809	1,009	800
1986	24,290	20,196	15,450	4,746	4,094	1,834	44	1,707	83	2,261	1,279	982
1987	26,888	22,259	16,405	5,854	4,630	2,077	75	1,910	92	2,552	1,486	1,067
1988	30,646	25,322	18,333	6,989	5,324	2,398	102	2,184	111	2,926	1,630	1,296
1989	34,757	28,827	20,150	8,677	5,931	2,617	139	2,353	124	3,314	1,729	1,584
1990	40,290	32,997	22,372	10,625	7,293	3,247	185	2,915	147	4,046	2,162	1,884
1991	44,381	35,940	23,043	12,896	8,442	3,837	231	3,437	169	4,604	2,607	1,997
1992	47,573	38,051	23,418	14,633	9,522	4,476	293	3,990	194	5,046	2,894	2,152
1993	50,991	40,446	24,094	16,352	10,545	5,143	372	4,540	231	5,402	3,181	2,221
1994	54,301	42,614	23,384	19,230	11,688	5,768	496	4,990	282	5,919	3,638	2,282
1995	60,876	47,734	23,354	24,380	13,142	6,657	721	5,557	378	6,485	4,143	2,342
1996	68,485	53,751	24,175	29,576	14,735	7,878	1011	6,308	560	6,856	4,517	2,339
1997	77,617	61,124	25,636	35,488	16,494	9,081	1,259	7,025	797	7,413	5,077	2,336
1998	88,541	69,474	27,545	41,929	19,067	10,732	1,561	8,096	1,076	8,334	5,971	2,363
1999	104,606	81,409	30,406	51,003	23,197	13,213	1,919	9,756	1,538	9,984	7,238	2,745
2000	120,580	92,886	33,401	59,485	27,694	15,883	2,096	11,636	2,151	11,812	8,541	3,270

(Continued on next page)

U.S. Prescription Drug Expenditures: 1965-2019 (cont.)

(In Millions of US$)

Year	TOTAL	Private			Public								
		Total Private	Out-of-Pocket	Insurance	Total Public	Federal				State & Local		
						Total Fed	Medicare	Medicaid	Other	Total S&L	Medicaid	Other
2001	138,337	105,244	36,027	69,217	33,093	19,210	2,447	13,807	2,956	13,883	10,038	3,845
2002	157,640	118,300	40,352	77,947	39,341	23,102	2,472	16,222	4,408	16,239	11,589	4,650
2003	174,165	127,532	44,110	83,422	46,633	27,769	2,478	19,393	5,898	18,865	13,128	5,737
2004	188,790	136,243	46,209	90,033	52,548	31,404	3,379	21,282	6,744	21,144	15,007	6,136
2005	199,701	144,556	48,764	95,793	55,144	32,626	3,937	21,262	7,427	22,518	15,911	6,607
2006	217,000	143,096	46,904	96,192	73,904	58,703	39,571	10,881	8,250	15,202	8,208	6,993
2007	226,758	146,728	48,930	97,797	80,030	65,365	46,029	10,717	8,619	14,666	8,043	6,623
2008	234,073	147,042	48,528	98,513	87,031	72,505	52,081	11,514	8,911	14,526	7,898	6,628
2009	246,297	149,791	49,006	100,785	96,506	82,786	57,776	14,497	10,512	13,720	6,677	7,043
2010	260,072	155,395	50,931	104,464	104,677	89,688	61,573	16,507	11,609	14,989	7,435	7,553
2011	274,541	160,670	52,347	108,323	113,871	95,003	67,199	15,048	12,756	18,868	10,727	8,141
2012	287,534	164,286	53,431	110,855	123,248	103,081	73,283	16,060	13,738	20,166	11,433	8,733
2013	302,944	170,313	55,440	114,873	132,632	111,112	79,476	17,133	14,503	21,520	12,210	9,311
2014	322,079	179,988	58,524	121,464	142,092	119,297	86,460	18,340	14,496	22,795	13,076	9,719
2015	343,814	191,174	62,048	129,126	152,640	128,397	93,813	19,611	14,973	24,243	13,988	10,255
2016	368,351	202,591	65,654	136,936	165,761	139,758	102,551	20,947	16,259	26,003	14,952	11,052
2017	395,385	215,065	69,603	145,461	180,321	152,439	112,420	22,349	17,671	27,882	15,960	11,922
2018	425,213	228,794	73,934	154,860	196,419	166,523	123,469	23,826	19,227	29,896	17,022	12,874
2019	457,781	243,615	78,592	165,023	214,166	182,112	135,800	25,376	20,935	32,055	18,143	13,912

Notes: Federal and State and Local "Other" funds include Medicaid SCHIP Expansion and SCHIP. The health spending projections were based on the 2008 version of the NHE released in January 2010. 2009-2019 are projections.

Source: Centers for Medicare & Medicaid Services (CMS), Office of the Actuary

Plunkett Research, Ltd.

www.plunkettresearch.com

Total U.S. Biotechnology Patents Granted per Year by Patent Class: 1977-2009

(Original & Cross-Reference Classifications; Duplicates Eliminated)

Year	Patent Class*						
	47	71	119	426	435	800	930
1977-87	1,853	938	2,271	7,582	7,668	66	1,573
1988	183	61	234	793	1,070	23	317
1989	221	80	265	1,157	1,432	22	255
1990	235	60	311	1,030	1,405	13	6
1991	244	68	391	1,003	1,561	29	12
1992	241	94	399	908	1,969	48	58
1993	226	73	341	899	2,257	38	76
1994	273	78	355	823	2,175	99	66
1995	272	93	313	884	2,250	91	38
1996	345	75	348	902	3,084	253	59
1997	296	100	362	814	4,140	282	85
1998	469	89	553	980	6,130	498	79
1999	325	87	501	1,261	6,215	670	79
2000	351	64	469	1,267	5,604	631	48
2001	305	97	447	1,226	6,275	666	46
2002	306	112	540	1,064	5,733	555	33
2003	283	70	532	1,015	5,297	525	31
2004	186	60	396	874	4,620	614	40
2005	159	46	407	519	4,132	521	29
2006	157	37	442	636	5,277	778	32
2007	124	32	380	524	5,195	904	23
2008	97	33	285	420	4,789	762	18
2009	141	40	308	462	4,915	946	12
Total	7,292	2,487	10,851	27,043	93,197	9,034	3,015

* The patent classes are as follows:
47: Plant Husbandry
71: Chemical Fertilizers
119: Animal Husbandry
426: Food or Edible Material: Processes, Compositions and Products
435: Chemistry: Molecular Biology & Microbiology
800: Multicellular Living Organisms and Unmodified Parts Thereof and Related Processes
930: Peptide or Protein Sequence
The agricultural classes 47, 71, 119 and 426 include only a small portion of patents that are related to biotechnology. All patents in classes 435, 800 and 930 are biotechnology-related.

Source: U.S. Patent & Trademark Office (USPTO)

Plunkett Research, Ltd.

www.plunkettresearch.com

Research Funding for Biological Sciences, U.S. National Science Foundation: Fiscal Year 2009-2011

(In Millions of US$)

	FY 2009 Actual	FY 2010 Current Plan	FY 2011 Requested	Change over 2010 Estimate	
				Amount	Percent
Molecular and Cellular Biosciences (MCB)	$121.28	$125.59	$133.69	$8.10	6.4%
Integrative Organismal Systems (IOS)	212.34	216.25	226.70	10.45	4.8%
Environmental Biology (EB)	120.37	142.55	155.59	13.04	9.1%
Biological Infrastructure (BI)	117.95	126.86	145.63	18.77	14.8%
Emerging Frontiers (EF)	84.68	103.29	106.20	2.91	2.8%
Total Biological Sciences Activity	**$656.62**	**$714.54**	**$767.81**	**$53.27**	**7.5%**
Research	502.57	520.64	577.84	$57.20	11.0%
Education	36.01	45.66	52.45	$6.79	14.9%
Infrastructure	107.20	135.45	123.23	-$12.22	-9.0%
Stewardship	10.84	12.79	14.29	$1.50	11.7%

Note: Totals may not add due to rounding.

Source: U.S. National Science Foundation
Plunkett Research, Ltd.
www.plunkettresearch.com

Global Area of Biotech Crops by Country: 2009

(In Millions of Hectares)

Rank	Country	Area	Biotech Crops
1*	USA	64.0	Soybean, maize, cotton, canola, squash, papaya, alfalfa, sugarbeet
2*	Brazil	21.4	Soybean, maise, cotton
3*	Argentina	21.3	Soybean, maize, cotton
4*	India	8.4	Cotton
5*	Canada	8.2	Canola, maize, soybean, sugarbeet
6*	China	3.7	Cotton, tomato, poplar, papaya, sweet pepper
7*	Paraguay	2.2	Soybean
8*	South Africa	2.1	Maize, soybean, cotton
9*	Uruguay	0.8	Soybean, maize
10*	Bolivia	0.8	Soybean
11*	Philippines	0.5	Maize
12*	Australia	0.2	Cotton, canola
13*	Burkina Faso	0.1	Cotton
14*	Spain	0.1	Maize
15*	Mexico	0.1	Cotton, soybean
16	Chile	<0.1	Maize, soybean, canola
17	Colombia	<0.1	Cotton, carnation
18	Honduras	<0.1	Maize
19	Czech Republic	<0.1	Maize
20	Portugal	<0.1	Maize
21	Romania	<0.1	Maize
22	Poland	<0.1	Maize
22	Costa Rica	<0.1	Maize
24	Egypt	<0.1	Maize
25	Slovakia	<0.1	Maize

* 15 biotech mega-countries growing 50,000 hectares or more of biotech crops.

Source: Clive James, ISAAA

Plunkett Research, Ltd.

www.plunkettresearch.com

R&D as a Percentage of U.S. Biopharmaceutical Sales, PhRMA Member Companies: 1970-2009

Year	Domestic R&D as a % of Domestic Sales	Total R&D as a % of Total Sales
2009*	19.0	16.0
2008	19.4	16.6
2007	19.8	17.5
2006	19.4	17.1
2005	18.6	16.9
2004	18.4	16.1
2003	18.3	16.5
2002	18.4	16.1
2001	18.0	16.7
2000	18.4	16.2
1999	18.2	15.5
1998	21.1	16.8
1997	21.6	17.1
1996	21.0	16.6
1995	20.8	16.7
1994	21.9	17.3
1993	21.6	17.0
1992	19.4	15.5
1991	17.9	14.6
1990	17.7	14.4
1989	18.4	14.8
1988	18.3	14.1
1987	17.4	13.4
1986	16.4	12.9
1985	16.3	12.9
1984	15.7	12.1
1983	15.9	11.8
1982	15.4	10.9
1981	14.8	10.0
1980	13.1	8.9
1979	12.5	8.6
1978	12.2	8.5
1977	12.4	9.0
1976	12.4	8.9
1975	12.7	9.0
1974	11.8	9.1
1973	12.5	9.3
1972	12.6	9.2
1971	12.2	9.0
1970	12.4	9.3

*Estimated

Source: Pharmaceutical Research and Manufacturers of America (PhRMA), *PhRMA Annual Membership Survey*, 2010

Plunkett Research, Ltd.

www.plunkettresearch.com

Biologics & Biotechnology R&D, PhRMA Member Companies: 2008

(In Millions of US$)

	Amount	Share
Biotechnology-Derived Therapeutic Proteins	10,542.3	22.2
Vaccines	1,600.8	3.4
CellorGene Therapy	176.9	0.4
All Other Biologics	1,337.8	2.8
Total Biologics/Biotechnology R&D	13,657.7	28.8
Non-Biologics/Biotechnology R&D	30,057.5	63.4
Uncategorized R&D	3,667.9	7.7
Total R&D	47,383.1	100.0

Note: All figures include company-financed R&D only. Total values may be affected by rounding.

Source: Pharmaceutical Research and Manufacturers of America (PhRMA), *PhRMA Annual Membership Survey*, 2010

Plunkett Research, Ltd.

www.plunkettresearch.com

Domestic & Foreign Pharmaceutical Sales, PhRMA Member Companies: 1975-2009

(In Millions of US$)

Year	Domestic Sales	APC	Sales Abroad[1]	APC	Total Sales	APC
2009[2]	183,026.4	-0.1%	103,370.9	0.5%	286,397.3	0.1%
2008	183,167.2	-1.1%	102,842.4	16.6%	286,009.6	4.6%
2007	185,209.2	4.2%	88,213.4	14.8%	273,422.6	7.4%
2006	177,736.3	7.0%	76,870.2	10.0%	254,606.4	7.9%
2005	166,155.5	3.4%	69,881.0	0.1%	236,036.5	2.4%
2004[3]	160,751.0	8.6%	69,806.9	14.6%	230,557.9	10.3%
2003[3]	148,038.6	6.4%	60,914.4	13.4%	208,953.0	8.4%
2002	139,136.4	6.4%	53,697.4	12.1%	192,833.8	8.0%
2001	130,715.9	12.8%	47,886.9	5.9%	178,602.8	10.9%
2000	115,881.8	14.2%	45,199.5	1.6%	161,081.3	10.4%
1999	101,461.8	24.8%	44,496.6	2.7%	145,958.4	17.1%
1998	81,289.2	13.3%	43,320.1	10.8%	124,609.4	12.4%
1997	71,761.9	10.8%	39,086.2	6.1%	110,848.1	9.1%
1996	64,741.4	13.3%	36,838.7	8.7%	101,580.1	11.6%
1995	57,145.5	12.6%	33,893.5	(4)	91,039.0	(4)
1994	50,740.4	4.4%	26,870.7	1.5%	77,611.1	3.4%
1993	48,590.9	1.0%	26,467.3	2.8%	75,058.2	1.7%
1992	48,095.5	8.6%	25,744.2	15.8%	73,839.7	11.0%
1991	44,304.5	15.1%	22,231.1	12.1%	66,535.6	14.1%
1990	38,486.7	17.7%	19,838.3	18.0%	58,325.0	17.8%
1989	32,706.6	14.4%	16,817.9	-4.7%	49,524.5	7.1%
1988	28,582.6	10.4%	17,649.3	17.1%	46,231.9	12.9%
1987	25,879.1	9.4%	15,068.4	15.6%	40,947.5	11.6%
1986	23,658.8	14.1%	13,030.5	19.9%	36,689.3	16.1%
1985	20,742.5	9.0%	10,872.3	4.0%	31,614.8	7.3%
1984	19,026.1	13.2%	10,450.9	0.4%	29,477.0	8.3%
1983	16,805.0	14.0%	10,411.2	-2.4%	27,216.2	7.1%
1982	14,743.9	16.4%	10,667.4	0.1%	25,411.3	9.0%
1981	12,665.0	7.4%	10,658.3	1.4%	23,323.3	4.6%
1980	11,788.6	10.7%	10,515.4	26.9%	22,304.0	17.8%
1979	10,651.3	11.2%	8,287.8	21.0%	18,939.1	15.3%
1978	9,580.5	12.0%	6,850.4	22.2%	16,430.9	16.1%
1977	8,550.4	7.5%	5,605.0	10.2%	14,155.4	8.6%
1976	7,951.0	11.4%	5,084.3	9.7%	13,035.3	10.8%
1975	7,135.7	10.3%	4,633.3	19.1%	11,769.0	13.6%
Average		10.1%		9.7%		9.8%

Notes: Total values may be affected by rounding. APC = Annual Percent Change.

[1] Sales abroad includes sales generated outside the United States by U.S.-owned PhRMA member companies and sales generated abroad by the U.S. divisions of foreign-owned PhRMA member companies. Sales generated abroad by the foreign divisions of foreign-owned PhRMA member companies are excluded. Domestic sales, however, includes sales generated within the United States by all PhRMA member companies.

[2] Estimated.

[3] Revised in 2007 to reflect updated data.

[4] Sales Abroad affected by merger and acquisition activity.

Source: Pharmaceutical Research and Manufacturers of America (PhRMA), *PhRMA Annual Membership Survey*, 2010
Plunkett Research, Ltd.
www.plunkettresearch.com

Sales By Geographic Area, PhRMA Member Companies: 2008

(In Millions of US$; Latest Year Available)

Geographic Area*	Amount	Share
Africa	1,294.2	0.5%
Americas		
United States	183,167.1	64.0%
Canada	7,002.8	2.4
Mexico	3,140.9	1.1
Brazil	3,120.9	1.1
Other Latin America[1]	4,597.2	0.0
Asia-Pacific		
Japan	10,496.2	3.7%
China	2,570.0	0.9
India	698.3	0.2
Other Asia-Pacific	4,787.4	1.7
Australia & New Zealand	3,687.1	1.3%
Europe		
France	10,342.1	3.6%
Germany	7,780.8	2.7
Italy	7,033.2	2.5
Spain	6,663.8	2.3
United Kingdom	6,297.6	2.2
Other Western European nations	13,232.9	4.6
Turkey	1,767.4	0.6
Russia	1,318.5	0.5
Other Eastern European nations[2]	4,929.2	1.7
Middle East[3]	2,076.0	0.7%
Uncategorized	6.1	0.0%
Total Sales	286,009.6	100.0%

Note: Total values may be affected by rounding.

* Sales abroad includes sales generated outside the United States by U.S.-owned PhRMA member companies and sales generated abroad by the U.S. divisions of foreign-owned PhRMA member companies. Sales generated abroad by the foreign divisions of foreign-owned PhRMA member companies are excluded. Domestic sales, however, includes sales generated within the United States by all PhRMA member companies.

[1] Other South American, Central American, and all Caribbean nations.

[2] Cyprus, Czech Republic, Estonia, Hungary, Poland, Slovenia, Bulgaria, Lithuania, Latvia, Romania, Slovakia, Malta and the Newly Independent States.

[3] Saudi Arabia, Yemen, United Arab Emirates, Iraq, Iran, Kuwait, Israel, Jordan, Syria, Afghanistan and Qatar.

Source: Pharmaceutical Research and Manufacturers of America (PhRMA), *PhRMA Annual Membership Survey*, 2010

Plunkett Research, Ltd.

www.plunkettresearch.com

Domestic U.S. Biopharmaceutical R&D Breakdown, PhRMA Member Companies: 2008

(In Millions of US$; Latest Year Available)

	Dollars ($)	Share (%)
R&D Expenditures for Human-Use Pharmaceuticals		
Domestic	34,936.4	73.7
Abroad*	11,456.0	24.2
Total Human-Use R&D	46,392.4	97.9
R&D Expenditures for Veterinary-Use Pharmaceuticals		
Domestic	634.7	1.3
Abroad*	356.0	0.8
Total Vet-Use R&D	990.7	2.1
Total R&D	**$47,383.1**	**100.0%**

Domestic R&D by Type of Project

Licensed-In	6,567.3	18.5
Self-Originated	27,474.7	77.2
Uncategorized	1,529.0	4.3
Total R&D	**$35,571.1**	**100.0%**

R&D By Function

Prehuman/Preclinical	12,795.6	27.0
Phase I	3,889.6	8.2
Phase II	6,089.7	12.9
Phase III	15,407.4	32.5
Approval	2,225.8	4.7
Phase IV	6,835.8	14.4
Uncategorized	139.1	0.3
Total R&D	**$47,383.1**	**100.0%**

Note: All figures include company-financed R&D only. Total values may be affected by rounding.

* R&D Abroad includes expenditures outside the United States by U.S.-owned PhRMA member companies and R&D conducted abroad by the U.S. divisions of foreign-owned PhRMA member companies. R&D performed abroad by the foreign divisions of foreign-owned PhRMA member companies is excluded. Domestic R&D, however, includes R&D expenditures within the United States by all PhRMA member companies.

Source: Pharmaceutical Research and Manufacturers Association (PhRMA), *PhRMA Annual Membership Survey*, 2010

Plunkett Research, Ltd.

www.plunkettresearch.com

R&D by Global Geographic Area, PhRMA Member Companies: 2008

(In Millions of US$; Latest Year Available)

Geographic Area*	Dollars	Share
Africa	40.7	0.1
Americas		
United States	35,571.1	75.1
Canada	572.2	1.2
Mexico	81.2	0.2
Brazil	96.7	0.2
Other Latin America[1]	210.4	0.4
Asia-Pacific		
Japan	925.3	2.0
China	93.2	0.2
India	94.4	0.2
Other Asia-Pacific	318.1	0.7
Australia & New Zealand	190.3	0.4
Europe		
France	540.8	1.1
Germany	781.2	1.6
Italy	284.0	0.6
Spain	301.7	0.6
United Kingdom	2,732.9	5.8
Other Western European nations	4,046.4	8.5
Turkey	40.6	0.1
Russia	80.4	0.2
Central & Eastern European nations[2]	338.3	0.7
Middle East[3]	43.2	0.1
Total R&D	**$47,383.1**	**100.0%**

Note: All figures include company-financed R&D only. Total values may be affected by rounding.

* R&D Abroad includes expenditures outside the United States by U.S.-owned PhRMA member companies and R&D conducted abroad by the U.S. divisions of foreign-owned PhRMA member companies. R&D performed abroad by the foreign divisions of foreign-owned PhRMA member companies is excluded. Domestic R&D, however, includes R&D expenditures within the United States by all PhRMA member companies.

[1] Other South American, Central American, and all Caribbean nations.

[2] Cyprus, Czech Republic, Estonia, Hungary, Poland, Slovenia, Bulgaria, Lithuania, Latvia, Romania, Slovakia, Malta and the Newly Independent States.

[3] Saudi Arabia, Yemen, United Arab Emirates, Iraq, Iran, Kuwait, Israel, Jordan, Syria, Afghanistan and Qatar.

Source: Pharmaceutical Research and Manufacturers of America (PhRMA), *PhRMA Annual Membership Survey*, 2010

Plunkett Research, Ltd.

www.plunkettresearch.com

Domestic Biopharmaceutical R&D Scientific, Professional & Technical Personnel by Function, PhRMA Member Companies: 2008

(Latest Year Available)

Function	Personnel	Share (%)
Prehuman/Preclinical	26,113	28.8
Phase I	6,409	7.1
Phase II	9,526	10.5
Phase III	21,356	23.5
Approval	5,025	5.5
Phase IV	11,739	12.9
Uncategorized	477	0.5
Total R&D Staff	**80,645**	**88.9**
Supported R&D Nonstaff	10,067	11.1
TOTAL R&D PERSONNEL	**90,712**	**100.0%**

Source: Pharmaceutical Research and Manufacturers of America (PhRMA), *PhRMA Annual Membership Survey*, 2010

Plunkett Research, Ltd.

www.plunkettresearch.com

Domestic U.S. Biopharmaceutical R&D & R&D Abroad, PhRMA Member Companies: 1970-2009

(In Millions of US$)

Year	Domestic R&D	Annual % Chg.	R&D Abroad[1]	Annual % Chg.	Total R&D	Annual % Chg.
2009[2]	34,806.0	-2.2	10,976.1	-7.1	45,782.1	-3.4
2008	35,571.1	-2.8	11,812.0	4.6	47,383.1	-1.1
2007	36,608.4	7.8	11,294.8	25.4	47,903.1	11.5
2006	33,967.9	9.7	9,005.6	1.3	42,973.5	7.8
2005	30,969.0	4.8	8,888.9	19.1	39,857.9	7.7
2004	29,555.5	9.2	7,462.6	1.0	37,018.1	7.4
2003	27,064.9	5.5	7,388.4	37.9	34,453.3	11.1
2002	25,655.1	9.2	5,357.2	-13.9	31,012.2	4.2
2001	23,502.0	10.0	6,220.6	33.3	29,772.7	14.4
2000	21,363.7	15.7	4,667.1	10.6	26,030.8	14.7
1999	18,471.1	7.4	4,219.6	9.9	22,690.7	8.2
1998	17,127.9	11.0	3,839.0	9.9	20,966.9	10.8
1997	15,466.0	13.9	3,492.1	6.5	18,958.1	12.4
1996	13,627.1	14.8	3,278.5	-1.6	16,905.6	11.2
1995	11,874.0	7.0	3,333.5	(3)	15,207.4	(3)
1994	11,101.6	6.0	2,347.8	3.8	13,449.4	5.6
1993	10,477.1	12.5	2,262.9	5.0	12,740.0	11.1
1992	9,312.1	17.4	2,155.8	21.3	11,467.9	18.2
1991	7,928.6	16.5	1,776.8	9.9	9,705.4	15.3
1990	6,802.9	13.0	1,617.4	23.6	8,420.3	14.9
1989	6,021.4	15.0	1,308.6	0.4	7,330.0	12.1
1988	5,233.9	16.2	1,303.6	30.6	6,537.5	18.8
1987	4,504.1	16.2	998.1	15.4	5,502.2	16.1
1986	3,875.0	14.7	865.1	23.8	4,740.1	16.2
1985	3,378.7	13.3	698.9	17.2	4,077.6	13.9
1984	2,982.4	11.6	596.4	9.2	3,578.8	11.2
1983	2,671.3	17.7	546.3	8.2	3,217.6	16.0
1982	2,268.7	21.3	505.0	7.7	2,773.7	18.6
1981	1,870.4	20.7	469.1	9.7	2,339.5	18.4
1980	1,549.2	16.7	427.5	42.8	1,976.7	21.5
1979	1,327.4	13.8	299.4	25.9	1,626.8	15.9
1978	1,166.1	9.7	237.9	11.6	1,404.0	10.0
1977	1,063.0	8.1	213.1	18.2	1,276.1	9.7
1976	983.4	8.8	180.3	14.1	1,163.7	9.6
1975	903.5	13.9	158.0	7.0	1,061.5	12.8
1974	793.1	12.0	147.7	26.3	940.8	14.0
1973	708.1	8.1	116.9	64.0	825.0	13.6
1972	654.8	4.5	71.3	24.9	726.1	6.2
1971	626.7	10.7	57.1	9.2	683.8	10.6
1970	566.2	-----	52.3	-----	618.5	-----
Average		11.6%		15.5%		12.2%

Note: All figures include company-financed R&D only. Total values may be affected by rounding.

[1] Estimated.

[2] R&D Abroad includes expenditures outside the United States by U.S.-owned PhRMA member companies and R&D conducted abroad by the U.S. divisions of foreign-owned PhRMA member companies. R&D performed abroad by the foreign divisions of foreign-owned PhRMA member companies is excluded. Domestic R&D, however, includes R&D expenditures within the United States by all PhRMA member companies.

[3] R&D Abroad affected by merger and acquisition activity.

Source: Pharmaceutical Research and Manufacturers of America (PhRMA), *PhRMA Annual Membership Survey*, 2010
Plunkett Research, Ltd.
www.plunkettresearch.com

Chapter 3

IMPORTANT BIOTECH & GENETICS INDUSTRY CONTACTS

I.   Agricultural Biotechnology Industry Associations

International Service for the Acquisition of Agri-Biotech Applications (ISAAA)
417 Bradfield Hall
Cornell University
Ithaca, NY 14853 US
Phone: 607-255-1724
Fax: 607-255-1215
E-mail Address: *americenter@isaaa.org*
Web Address: www.isaaa.org
The International Service for the Acquisition of Agri-Biotech Applications (ISAAA) is a not-for-profit organization that provides bioengineered seeds to poor and developing countries. In general, such seeds will enhance production per acre due to resistance to drought, insects and disease, and will offer additional crop enhancements.

Plant Biotechnology Institute (PBI)
110 Gymnasium Pl.
Saskatoon, SK S7N 0W9 Canada
Phone: 306-975-5248
Fax: 306-975-4839
E-mail Address: *PBI-Info@nrc-cnrc.gc.ca*
Web Address: pbi-ibp.nrc-cnrc.gc.ca
The Plant Biotechnology Institute (PBI) is a member of Canada's National Research Council and is engaged in research regarding the genomics, metabolic pathways, gene expression, genetic transformation and structured biology of plants and crops.

II.   Agricultural Biotechnology Resources

Ag BioTech Infonet
Phone: 208-263-5236
E-mail Address: *karen@hillnet.com*
Web Address: www.biotech-info.net
Ag BioTech Infonet provides information on the application of biotechnology and genetic engineering in agricultural production, food processing and marketing.

Biotechnology Knowledge Center
Monsanto Company
800 N. Lindbergh Blvd.
St. Louis, MO 63167 US
Phone: 314-694-1000
Web Address: www.biotechknowledge.com

The Biotechnology Knowledge Center, sponsored by Monsanto Company, is an online resource containing news, technical documents, a discussion board and a glossary.

UK Agricultural Biodiversity Coalition (UKabc)
UK Food Group
94 White Lion St.
London, N1 9PF UK
Phone: 44-20-7713-5813
Fax: 44-207-837-1141
E-mail Address: *ukabc@ukabc.org*
Web Address: www.ukabc.org
The UK Agricultural Biodiversity Coalition (UKabc) provides links to life science and seed companies, databases and information resources, publicly funded research bodies and industry associations. UKabc is an activity of the UK Food Group (UKFG).

III.   Agriculture Industry Associations

Agricultural Institute of Canada (AIC)
280 Albert St., Ste. 900
Ottawa, ON K1P 5G8 Canada
Phone: 613-232-9459
Fax: 613-594-5190
Toll Free: 888-277-7980
E-mail Address: *office@aic.ca*
Web Address: www.aic.ca
The Agricultural Institute of Canad (AIC) is one of Canada's largest agricultural trade organizations promoting the professional and scientific development of agriculture nationwide.

U.S. Grains Council
1400 K St. NW, Ste. 1200
Washington, DC 20005 US
Phone: 202-789-0789
Fax: 202-898-0522
Web Address: www.grains.org
The U.S. Grains Council develops export markets for U.S. barley, corn, grain sorghum and related products. Founded in 1960, the Council is a private, nonprofit corporation with nine international offices and programs in more than 50 countries. Membership funds trigger matching market development funds from the U.S. government and support from cooperating groups in foreign countries to produce an annual development program valued at more than $25 million. The organization's website contains extensive grain production statistics and a special resource center on biotechnology.

IV.   Alternative Energy-Ethanol

Renewable Fuels Association (RFA)
1 Massachusetts Ave. NW, Ste. 820
Washington, DC 20001 US
Phone: 202-289-3835
E-mail Address: *info@ethanolrfa.org*
Web Address: www.ethanolrfa.org
The Renewable Fuels Association (RFA) is a trade organization representing the ethanol industry. It publishes a wealth of useful information, including a listing of biorefineries and monthly U.S. fuel ethanol production and demand.

V.   Biology-Synthetic

BioBricks Foundation (BBF)
One Kendall Sq.
PMB 126
Cambridge, MA 02139 US
E-mail Address: *info@biobricks.org*
Web Address: www.biobricks.org
The BioBricks Foundation (BBF) is a not-for-profit organization founded by engineers and scientists from MIT, Harvard, and UCSF with significant experience in both nonprofit and commercial biotechnology research. BBF encourages the development and responsible use of technologies based on BioBrick standard DNA parts that encode basic biological functions.

Synthetic Biology
Web Address: syntheticbiology.org
Synthetic Biology is a consortium of individuals, labs and groups working together to advance the development of biological engineering. Its members include 19 labs from 11 different universities, including MIT, Princeton and Harvard. The site's FAQ includes discussions of synthetic biology, its applications and the difference between synthetic and systems biology. Synthetic Biology does not maintain a headquarters, instead allowing members to update and edit the web site in order to disseminate knowledge.

SyntheticBiology.org
E-mail Address: *synbioadmin@openwetware.org*
Web Address: www.syntheticbiology.org
SyntheyicBiology.org is an open source web site dedicated to disseminating news about synthetic biology. It was originally founded by a group of students, faculty and staff from MIT and Harvard.

VI.   Biotechnology and Biological Industry Associations

All India Biotech Association (AIBA)
Web Address: www.aibaonline.com
All India Biotech Association (AIBA) was established in 1994 as a nonprofit group to represent India's biotechnology industry.

American Peptide Society (APS)
7668 El Camino Real, Ste. 104-140
La Costa, CA 92009-7932 US
Phone: 858-455-4752
Fax: 775-667-5332
E-mail Address:
aps_member@americanpeptidesociety.org
Web Address: www.americanpeptidesociety.org
The American Peptide Society (APS) is a nonprofit organization for the advancement and promotion of knowledge and research in the field of peptide chemistry and biology.

American Society for Microbiology (ASM)
1752 N St. NW
Washington, DC 20036 US
Phone: 202-737-3600
Web Address: www.asm.org
The American Society for Microbiology (ASM) is a life science membership organization that specializes in the advancement of the study of bacteria, viruses, rickettsiae, mycoplasma, fungi, algae and protozoa.

American Society of Gene Therapy (ASGT)
555 East Wells St., Ste. 1100
Milwaukee, WI 53202 US
Phone: 414-278-1341
Fax: 414-276-3349
E-mail Address: *info@asgt.org*
Web Address: www.asgt.org
The American Society of Gene Therapy (ASGT) is a nonprofit medical and professional organization that represents researchers and scientists devoted to the discovery of new gene therapies. ASGT was established in 1996 by Dr. George Stamatoyannopoulos, professor of medicine at the University of Washington 's School of Medicine and a group of the country's leading researchers in gene therapy. With more than 2,000 members in the United States and worldwide, ASGT is the largest association of individuals involved in gene therapeutics.

Association of Biotechnology Led Enterprises (ABLE)
123/C, 16th Main Rd., 5th Cross, 4th Block
Near Sony World showroom / Headstart school
Koramangala, Bangalore 560034 India
Fax: 91-80-2553-3938
E-mail Address: *info@ableindia.org*
Web Address: www.ableindia.org
The Association of Biotechnology Led Enterprises
(ABLE) is an organization focused on accelerating
the pace of Biotechnology in India by enabling
strategic alliances between researchers, the
government and the global Biotech industry.

AusBiotech
322 Glenferrie Rd., Lv. 1
Malvern, VIC 3144 Australia
Phone: 03-9828-1400
Fax: 03-9824-5188
E-mail Address: *admin@ausbiotech.org*
Web Address: www.ausbiotech.org
AusBiotech is a professional organization for the
biotech industry in Australia, with members in the
human health, agricultural, medical device,
bioinformatics, environmental and industrial sectors.

BioAlberta
314 Capital Pl., Ste. 9707-110
Edmonton, AB T5K 2L9 Canada
Phone: 780-425-3804
Fax: 780-409-9263
E-mail Address: *info@bioalberta.com*
Web Address: www.bioalberta.com
BioAlberta is a private, nonprofit industry association
representing Alberta, Canada's biotech industry.

Biochemical Society
16 Procter St.
London, WC1V 6NX UK
Phone: 020-7280-4100
Fax: 020-7280-4170
E-mail Address: *genadmin@biochemistry.org*
Web Address: www.biochemsoc.org
The Biochemical Society is a professional association
that promotes the advancement of the science of
biochemistry, viewing cellular and molecular life
sciences as a seamless continuum.

BIOCOM
4510 Executive Dr., Plz. 1
San Diego, CA 92121 US
Phone: 858-455-0300
Fax: 858-455-0022

E-mail Address: *inquiries@biocom.org*
Web Address: www.biocom.org
BIOCOM is a trade association for the life science
industry in San Diego and Southern California.

BioIndustry Association
14/15 Belgrave Sq.
London, SW1X 8PS UK
Phone: 44-20-7565-7190
Fax: 44-20-7565-7191
E-mail Address: *admin@bioindustry.org*
Web Address: www.bioindustry.org
The BioIndustry Association promotes bioscience
development in the U.K. The organization operates a
public affairs program, a conference and seminar
program, trade missions and publications for internal
and external audiences.

Bioindustry Association of Korea (BAK)
97-3, Guro 6-dong, Guro-gu
DaeLim Office Valley, 6th Fl.
Seoul, 152-841 Korea
Phone: 82-2-855-7993
Fax: 82-2-855-7006
E-mail Address: *bioindus@chol.com*
Web Address: www.bak.or.kr
The Bioindustry Association of Korea (BAK) was
established in 1991 to promote and facilitate growth
and development of the bioindustry within Korea.

Biomedical Engineering Society
8401 Corporate Dr., Ste. 140
Landover, MD 20785-2224 US
Phone: (301) 459-1999
Fax: (301) 459-2444
E-mail Address: *info@bmes.org*
Web Address: www.bmes.org
The Biomedical Engineering Society (BMES)
members are the in biomedical engineering and
bioengineering industry.

Biomedical Engineering Society of India
Sree Chitra Tirunal Institute for Medical Sciences
and Technology
Satelmond Palace Campus, Poojapura
Thiruvananthapuram, 695 012 India
Phone: 91-471-2520-282
Fax: 91-471-2341-814
E-mail Address: *niranjan@sctimst.ac.in*
Web Address: www.bmesi.org.in
The Biomedical Engineering Society of India is an all
India association which seeks to advance
interdisciplinary cooperation among scientists,

engineers, and medical doctors for the growth of teaching, research and practices of biomedical engineering.

BioQuebec

381, rue Notre-Dame Ouest, Bureau 300
Montreal, QB H2Y 1V2 Canada
Phone: 514 733-8411
Fax: 514 733-8272
E-mail Address: *reception@bioquebec.com*
Web Address: www.bioquebec.com
BIOQuebec is a biotechnology and life science industry association representing more than 175 member companies and R&D centers in Quebec. The association works to create a positive influence on the growth of the life science industry, including capital access, public policy, workforce development and international promotion.

BioSingapore

P.O. Box 950
Rocher Post Office
, 911837 Singapore
Phone: 65 6336 5371
Fax: 65 6336 5316
E-mail Address: *enquiries@biosingapore.org.sg*
Web Address: www.biosingapore.org
BioSingapore is an industry association for life sciences business in Singapore, helping local businesses network with financial institutions and other important sectors.

BIOTECanada

130 Albert St., Ste. 420
Ottawa, ON K1P 5G4 Canada
Phone: 613-230-5585
Fax: 613-563-8850
E-mail Address: *info@biotech.ca*
Web Address: www.biotech.ca
BIOTECanada is a trade organization that promotes the Canadian biotech industry.

Biotechnology and Biological Sciences Research Council (BBSRC)

N. Star Ave.
Polaris House
Swindon, SN2 1UH UK
Phone: 44-0-1793-413200
Fax: 44-0-1793-413201
E-mail Address: *webmaster@bbsrc.ac.uk*
Web Address: www.bbsrc.ac.uk

The Biotechnology and Biological Sciences Research Council (BBSRC) provides funding for biotech research in the U.K.

Biotechnology Association of Alabama

500 Beacon Pkwy. W.
Birmingham, AL 35209 US
Phone: 205-290-9593
E-mail Address: *knugentphd@yahoo.com*
Web Address: www.bioalabama.com
The Biotechnology Association of Alabama is a nonprofit association that represents the biotech industry in Alabama.

Biotechnology Industry Organization (BIO)

1201 Maryland Ave. SW, Ste. 900
Washington, DC 20024 US
Phone: 202-962-9200
Fax: 202-488-6301
E-mail Address: *info@bio.org*
Web Address: www.bio.org
The Biotechnology Industry Organization (BIO) is involved in the research and development of health care, agricultural, industrial and environmental biotechnology products. BIO has both small and large member organizations.

Biotechnology Research Institute of the National Research Council Canada (NRC-BRI)

6100 Royalmount Ave.
Montréal, QC H4P 2R2 Canada
Phone: 514-496-6100
E-mail Address: *bri-info@cnrc-nrc.gc.ca*
Web Address: www.bri.nrc-cnrc.gc.ca
The Biotechnology Research Institute of the National Research Council Canada (NRC-BRI) is one of the largest Canadian research facilities that focuses solely on biotechnology.

Brazilian Association of Biotechnology Companies (ABRABI)

Av. José Cândido da Silveira, 2.100
Horto - Belo Horizonte
MG, CEP 31170-000 Brazil
Phone: 55 31 3486-1733
E-mail Address: *abrabi@abrabi.org.br*
Web Address: www.abrabi.org.br/whats-abrabi.htm
Established in 1986, ABRABI is the national entity that represents the Brazilian Biotechnology sector. ABRABI congregates companies, in all areas, that are part of the biotech chain, namely small innovative companies, national and international pharmaceutical and diagnosis companies, agribusinesses and

industrial biotech companies, bioinformatics and nanobiotechnology.

California Healthcare Institute (CHI)
1020 Prospect St., Ste. 310
La Jolla, CA 92037 US
Phone: 858-551-6677
Fax: 858-551-6688
E-mail Address: *chi@chi.org*
Web Address: www.chi.org
California Healthcare Institute (CHI) is an independent organization that represents the biomedical industry in California.

Connecticut BioScience Cluster (CURE)
300 George St., Ste. 561
New Haven, CT 06511 US
Phone: 203-777-8747
Fax: 203-777-8754
E-mail Address: *twallace@curenet.org*
Web Address: www.curenet.org
Connecticut BioScience Cluster (CURE) is a partnership with the Department of Economic and Community Development to promote Connecticut's biotech clusters.

Council for Biotechnology Information (CBI)
1201 Maryland Ave. SW, Ste. 900
Washington, DC 20024 US
Phone: 202-962-9200
Web Address: www.whybiotech.com
The Council for Biotechnology Information (CBI) is a trade organization dedicated to promoting biotech.

Environmental Mutagen Society (EMS)
1821 Michael Faraday Dr., Ste. 300
Reston, VA 20190 US
Phone: 703-438-8220
Fax: 703-438-3113
E-mail Address: *emshq@ems-us.org*
Web Address: www.ems-us.org
Environmental Mutagen Society (EMS) provides information on the process of biological mutagenesis, and the application of this process in the field of genetic toxicology.

EuropaBio
Ave. de l'Armée 6
Brussels, 1040 Belgium
Phone: 32 2 735 03 13
Fax: 32 2 735 49 60
E-mail Address: *info@europabio.org*
Web Address: www.europabio.org

EuropaBio's mission is to promote an innovative and dynamic biotechnology-based industry in Europe. EuropaBio, (the European Association for Bioindustries), has 79 corporate and 5 associate members operating worldwide, 5 Bioregions and 25 national biotechnology associations representing some 1,800 small and medium sized enterprises.

Federation of American Societies for Experimental Biology (FASEB)
9650 Rockville Pike
Bethesda, MD 20814 US
Phone: 301-634-7000
Fax: 301-634-7001
E-mail Address: *webmaster@faseb.org*
Web Address: www.faseb.org
The Federation of American Societies for Experimental Biology (FASEB) works with its 21 member societies to advance biological science through collaborative advocacy for research policies that promote scientific progress leading to improvements in human health.

German Medical Technology Association (BVMed)
Bundesverband Medizintechnologie e.V.
Reinhardtstr. 29b
Berlin, 10117 Germany
Phone: 49 (0) 30 246 255-0
Fax: 49 (0) 30 246 255-99
E-mail Address: *info@bvmed.de*
Web Address: www.bvmed.de
The German Medical Technology Association (BVMed) represents about 200 manufacturers and service providers of medical devices.

International Biometric Society (IBS)
1444 I St. NW, Ste. 700
Washington, DC 20005 US
Phone: 202-712-9049
Fax: 202-216-9646
E-mail Address: *ibs@bostrom.com*
Web Address: www.tibs.org
The International Biometric Society (IBS) is an association that is devoted to the development and application of statistical and mathematical theory and methods in the biosciences.

International Society for Clinical Biostatistics (ISCB)
Teglgaarden 24
Birkerod, DK-3460 Denmark
Phone: 45-4567-2279

Fax: 45-7022-1571
E-mail Address: *office@iscb.info*
Web Address: www.iscb.info
The International Society for Clinical Biostatistics (ISCB) is a professional organization that aims to stimulate research on the biostatistical principles and methodology used in clinical research, to increase the relevance of statistical theory to clinical medicine and to promote better understanding of the use and interpretation of biostatistics by the general public.

International Society for Stem Cell Research (ISSCR)
111 Deer Lake Rd., Ste. 100
Deerfield, IL 60015 US
Phone: 847-509-1944
Fax: 847-480-9282
E-mail Address: *isscr@isscr.org*
Web Address: www.isscr.org
The International Society for Stem Cell Research is an independent, nonprofit organization established to promote the exchange and dissemination of information and ideas relating to stem cells, to encourage the general field of research involving stem cells and to promote professional and public education in all areas of stem cell research and application.

Iowa Biotechnology Association
900 E. Des Moines St.
Des Moines, IA 50309 US
Phone: 515-327-9156
Fax: 515-327-1407
E-mail Address: *dgetter@netins.net*
Web Address: www.iowabiotech.com
The Iowa Biotech Association promotes research and education in biotechnology in Iowa.

Japan Bioindustry Association
2-26-9 Hatchobori, Chuo-ku
Grande Bldg. 8F
Tokyo, 104-0032 Japan
Phone: 03-5541-2731
Fax: 03-5541-2737
Web Address: www.jba.or.jp
JBA is a nonprofit organization dedicated to the promotion of bioscience, biotechnology and bioindustry in both Japan and the rest of the world. Established through the support and cooperation of industry, academia, and government, JBA is the only organization of its kind in Japan.

Korea Bio Venture Association (KOBIOVEN)
F8th 703- 5 Seo-il Plz.
Yoeksam-dong, Kangnam-gu
Seoul, 135-513 Korea
Phone: 82-2-552-4771
Fax: 82-2-552-4840
E-mail Address: *kobioven@kobioven.or.kr*
Web Address: eng.kobioven.or.kr
Korea Bio Venture Association (KOBIOVEN) was created to provide support to the biotech industry in Korea. It is affiliated with the Ministry of Commerce, Industry and Energy.

Massachusetts Biotechnology Council (MBC)
1 Cambridge Ctr., 9th Fl.
Cambridge, MA 02142 US
Phone: 617-674-5100
Fax: 617-674-5101
E-mail Address: *info_request@massbio.org*
Web Address: www.massbio.org
The Massachusetts Biotechnology Council (MBC) is a nonprofit organization that promotes the Massachusetts biotech industry.

MdBio
9713 Key West Ave.
Rockville, MD 20850 US
Phone: 240-243-4026
E-mail Address: *rzakour@techcouncilmd.com*
Web Address: www.mdbio.org
MdBio is a nonprofit organization that promotes biotech in Maryland. Areas of emphasis include corporate and business development, networking and community building, education and workforce development and communications.

Michigan Biotechnology Association
3520 Green Ct., Ste. 450
Ann Arbor, MI 48105-1579 US
Phone: 734-527-9150
Fax: 734-302-4933
E-mail Address: *info@michbio.org*
Web Address: www.michbio.org
The Michigan Biotechnology Association is a nonprofit organization dedicated to promoting the biotech industry in Michigan.

Missouri Biotechnology Association (MBA)
428 E. Capitiol
P.O. Box 148
Jefferson City, MO 65102-0148 US
Phone: 573-761-7600
Fax: 573-761-7601

E-mail Address: *gillespie@mobio.org*
Web Address: www.mobio.org
The Missouri Biotechnology Association (MBA) is an industry organization that promotes the biotech industry in Missouri.

National Association for Biomedical Research (NABR)
818 Connecticut Ave. NW, Ste. 900
Washington, DC 20006 US
Phone: 202-857-0540
Fax: 202-659-1902
E-mail Address: *info@nabr.org*
Web Address: www.nabr.org
The National Association for Biomedical Research (NABR) is an nonprofit group that advocates for sound public policy that recognizes the vital role of humane animal use in biomedical research, higher education, and product safety testing.

New York Biotechnology Association (NYBA)
25 Health Sciences Dr., Ste. 203
Stony Brook, NY 11790 US
Phone: 631-444-8895
Fax: 631-444-8896
E-mail Address: *info@nyba.org*
Web Address: www.nyba.org
The New York Biotechnology Association (NYBA) is a not-for-profit trade association dedicated to the development and growth of biotechnology-related industries and institutions in New York State.

Organibio
28 rue St. Dominique
Paris, F-75007 France
Phone: 33-147 53 09 12
Fax: 33-147 53 73 76
Web Address: www.organibio.org
ORGANIBIO represents all of the segments of the French bioscience industry, particularly biotechnology with applications in pharmaceuticals and agribio.

Pennsylvania Bioscience Association (PBA)
7 Great Valley Pkwy., Ste. 290
Malvern, PA 19355 US
Phone: 610-578-9220
Fax: 610-578-9219
E-mail Address: *info@pennsylvaniabio.org*
Web Address: www.pennsylvaniabio.com
The Pennsylvania Bioscience Association (PBA) is a nonprofit organization promoting bioscience in Pennsylvania.

Scotland Biotechnology
5 Atlantic Quay
150 Broomielaw
Glasgow, G2 8LU UK
Phone: 0141-248-2700
E-mail Address: *enquiries@scotent.co.uk*
Web Address: www.scottish-enterprise.com/sedotcom_home/sig/life-sciences.htm
Scotland Lifesciences promotes the biotech industry of Scotland, as well provides information on medical devices and pharmaceuticals.

Society for Biomaterials
15000 Commerce Pkwy., Ste. C
Mt. Laurel, NJ 08054 US
Phone: 856-439-0826
Fax: 856-439-0525
E-mail Address: *info@biomaterials.org*
Web Address: www.biomaterials.org
The Society for Biomaterials is a professional society that promotes advances in all phases of materials research and development by encouraging cooperative educational programs, clinical applications and professional standards in the biomaterials field.

Society for In Vitro Biology (SIVB)
514 Daniels St., Ste. 411
Raleigh, NC 27605 US
Phone: 919-420-7940
Fax: 919-420-7939
Toll Free: 888-588-1923
E-mail Address: *sivb@sivb.org*
Web Address: www.sivb.org
The Society for In Vitro Biology (SIVB) is an association of scientists working to foster the exchange of knowledge in the field of in vitro biology. The group maintains a variety of publications, national and local conferences, meetings and workshops.

Society for Industrial Microbiology (SIM)
3929 Old Lee Hwy., Ste. 92A
Fairfax, VA 22030-2421 US
Phone: 703-691-3357
Fax: 703-691-7991
E-mail Address: *simhq@simhq.org*
Web Address: www.simhq.org
The Society for Industrial Microbiology (SIM) is a nonprofit professional association that works for the advancement of microbiological sciences as they apply to industrial products, biotechnology, materials and processes.

Tennessee Biotechnology Association (TBA)
111 10th Ave. S., Ste. 110
Nashville, TN 37203 US
Phone: 615-255-6270
Fax: 615-255-0094
E-mail Address: *jrolwing@tnbio.org*
Web Address: www.tnbio.org
The Tennessee Biotechnology Association (TBA) is a statewide clearinghouse that supports the biotech industry in Tennessee.

The German Association of Biotechnology Industries (DIB)
Deutsche Industrievereinigung Biotechnologie
HallerstraBe 6
Berlin, 10587 Germany
Phone: 49 (30) 343816-0
Fax: 49 (30) 343819-28
E-mail Address: *post@lv-no.vci.de*
Web Address: www.nordostchemie.de
The German Association of Biotechnology Industries (DIB) represents biotechnology companies and their interests.

VBU Association of German Biotech Companies
c/o DECHEMA e.V.
Theodor-Heuss-Allee 25
Frankfurt am Main, 60486 Germany
Phone: 49 (0)69 - 7564-124
Fax: 49 (0)69 - 7564-169
E-mail Address: *vbu@dechema.de*
Web Address: www.v-b-u.org
VBU Association of German Biotech Companies was founded in October 1996. It now has over 210 corporate members, making it the largest and oldest industrial biotechnology association in Germany. Since its foundation VBU has devoted itself to the promotion of science and technology and the transfer of research findings into innovations.

Virginia Biotechnology Association
800 E. Leigh St., Ste. 14
Richmond, VA 23219-1534 US
Phone: 804-643-6360
Fax: 804-643-6361
E-mail Address: *questions@vabio.org*
Web Address: www.vabio.org
The Virginia Biotechnology Association is a nonprofit organization that promotes biotechnology in Virginia.

Wisconsin Biotechnology and Medical Devices Association (WBMDA)
455 Science Dr., Ste. 160
Madison, WI 53711 US
Phone: 608-236-4693
Fax: 608-236-4695
E-mail Address: *info@wisbiomed.org*
Web Address: www.wisconsinbiotech.org
The Wisconsin Biotechnology and Medical Devices Association (WBMDA) is a professional organization devoted to the promotion of the biotech industry in Wisconsin.

VII. Biotechnology Investing

BioTech Stock Report
P.O. Box 7274
Beaverton, OR 97007-7274 US
Phone: 503-649-1355
Fax: 503-649-4490
E-mail Address: *info@biotechnav.com*
Web Address: www.biotechnav.com
The BioTech Stock Report is a monthly newsletter that provides analysis, commentary, news and company developments for biotechnology investors.

Burrill & Company
1 Embarcadero Ctr., Ste. 2700
San Francisco, CA 94111 US
Phone: 415-591-5400
Fax: 415-591-5401
E-mail Address: *burrill@b-c.com*
Web Address: www.burrillandco.com
Burrill & Company is a leading private merchant bank concentrated on companies in the life sciences industries: biotechnology, pharmaceuticals, medical technologies, agricultural technologies, animal health and nutraceuticals.

Medical Technology Stock Letter
P.O. Box 40460
Berkeley, CA 94704 US
Phone: 510-843-1857
Fax: 510-843-0901
E-mail Address: *mtsl@bioinvest.com*
Web Address: www.bioinvest.com
The Medical Technology Stock Letter is a newsletter that provides financial advice about investing in biotechnology. It is distributed by mail and electronically.

VIII.   Biotechnology Resources

About Biotech
Web Address: biotech.about.com
About Biotech provides news and information on the
biotech industry, expertly compiled and edited by
Theresa Phillips, PhD.

BioAbility
3200 Chapel Hill/Nelson Blvd., Ste. 201
Research Triangle Park, NC 27709-4569 US
Phone: 919-544-5111
Fax: 919-544-5401
E-mail Address: *info@bioability.com*
Web Address: www.bioability.com
BioAbility provides strategic business information to
the biotechnology, pharmaceutical and life science
industries.

BioBasics
E-mail Address: *info@biotech.gc.ca*
Web Address: biobasics.gc.ca
BioBasics is a Canadian web site that offers
information and links related to gene therapy, genetic
testing and xenotransplantation. It also contains
information on food, health, industrial biotechnology,
natural resources and sustainable development.

Bioengineering Industry Links
University of Pennsylvania
Bioengineering, School of Eng. & Applied Science
210 S. 33rd St., Rm. 240 Skirkanich Hall
Philadelphia, PA 19104 US
Phone: 215-898-8501
Fax: 215-573-2071
E-mail Address: *beoffice@seas.upenn.edu*
Web Address:
www.seas.upenn.edu/be/misc/bmelink/cell.html
Bioengineering Industry Links is a web site provided
by the University of Pennsylvania's Department of
Bioengineering. This site features links to companies
involved in cell and tissue engineering.

Biofind
E-mail Address: *info@biofind.com*
Web Address: www.biofind.com
Biofind offers a biotech news directory, job search,
chat room and event announcements, as well as a
place to post announcements about biotech
innovations.

BioMed Central
34-42 Cleveland St.

Middlesex House
London, W1T 4LB UK
Phone: 44-20-7323-0323
Fax: 44-20-7631-9926
E-mail Address: *info@biomedcentral.com*
Web Address: www.biomedcentral.com
BioMed Central is an independent publishing house
that prints approximately 160 peer-reviewed journals
for the medical industry. Its web site provides free,
open access to all of its research.

Biospace, Inc.
Ramshorn Executive Ctr.
2399 Hwy. 34, Bldg. A-5
Manasquan, NJ 08736-1528 US
Phone: 732-528-3688
Fax: 732-528-3668
Toll Free: 888-246-7722
E-mail Address: *support@biospace.com*
Web Address: www.biospace.com
Biospace.com offers information, news and profiles
on biotech companies. It also provides an outlet for
business and scientific leaders in bioscience to
communicate with each other.

BioTech
The Ellington Lab, University of Texas at Austin
Chem & Biochem Dept., 1 University Station A4800
Austin, TX 78712-0165 US
E-mail Address: *feedback@biotech.icmb.utexas.edu*
Web Address: biotech.icmb.utexas.edu
The BioTech web site offers a comprehensive
dictionary of biotech terms, plus extensive research
data regarding biotechnology. BioTech is located in
The Ellington Lab at the University of Texas at
Austin.

Biotech Rumor Mill
E-mail Address: *info@biofind.com*
Web Address: www.biofind.com/rumor-mill
The Biotech Rumor Mill is an online discussion
forum that attracts participants from many biotech
disciplines. Rumor Mill service is provided by
Biofind Limited.

Biotechnology Information Directory Section
Cato Research Ltd.
4364 S. Alston Ave.
Durham, NC 27713-2280 US
Phone: 919-361-2286
Fax: 919-361-2290
E-mail Address: *info@cato.com*
Web Address: www.cato.com/pub.shtml

The Biotechnology Information Directory Section contains links to companies, research institutes, universities, sources of information and other directories related to biotechnology, pharmaceutical development and similar fields. The directory is a service of Cato Research Ltd.

Biotechterms.org

E-mail Address: *webmaster@biotechterms.org*
Web Address: biotechterms.org
Biotechterms.org is an online version of Technomic Publishing's Glossary of Biotechnology Terms by Kimball R. Nill.

Biotech-U

Web Address: www.biotech-u.com
Biotech-U provides a web site offering Internet-based courses on the basics of biotechnology.

BioWorld Online

3525 Piedmont Rd.
Bldg. 6, Ste. 400
Atlanta, GA 30305 US
Toll Free: 800-688-2421
E-mail Address: *customerservice@bioworld.com*
Web Address: www.bioworld.com
BioWorld Online is a news and information site that offers in-depth resources about the biotech industry and leading companies.

Centre for Cellular and Molecular Biology (CCMB)

Council of Scientific & Industrial Research
Uppal Rd.
Hyderabad, 500 007 India
Phone: 91-40-27160222
Fax: 91-040-27160591
Web Address: www.ccmb.res.in
Centre for Cellular and Molecular Biology (CCMB) is one of the constituent Indian national laboratories of the Council of Scientific and Industrial Research (CSIR), the multidisciplinary research and development organization of the Government of India.

Electronic Journal of Biotechnology

Av. Brasil 2950
P.O. Box 4059
Valparaíso, Chile
E-mail Address: *edbiotec@ecv.cl*
Web Address: www.ejbiotechnology.info

The Electronic Journal of Biotechnology is an online journal that publishes information about the biotech industry.

Genetic Engineering & Biotechnology News

140 Huguenot St., 3rd Fl.
New Rochelle, NY 10801-5215 US
Phone: 914-740-2100
Fax: 914-740-2201
Toll Free: 800-799-9436
E-mail Address: *info@liebertpub.com*
Web Address: www.genengnews.com
Genetic Engineering News is a widely read magazine that offers weekly news on topics in biotechnology, bioregulation, bioprocess, bioresearch and technology transfer. It is published by Mary Ann Liebert, Inc.

GrantsNet

1200 New York Ave. NW
Washington, DC 20005 US
Phone: 202-326-6550
Web Address: www.grantsnet.org
GrantsNet is a free online service to locate funding for training in the biomedical science industry and undergraduate science education, provided through ScienceCareers.org.

Institute for Biological Sciences (IBS)

1200 Montreal Rd.
Ottawa, Ontario K1A 0R6 Canada
Phone: 613-990-6111
Toll Free: 800-267-0441
E-mail Address: *Micha.Hage-Badr@nrc-cnrc.gc.ca*
Web Address: ibs-isb.nrc-cnrc.gc.ca
The Institute for Biological Sciences (IBS) is a branch of Canada's National Research Council and focuses its research and development programs on life sciences operations designed to fight age-related and infectious diseases.

Institute of Bioinformatics and Applied Biotechnology (IBAB)

G-05, Tech Park Mall, Int'l Tech. Pk. Bangalore Whitefield Rd.
Bangalore, 560 066 India
Phone: 91-80-2841-0029
Fax: 91-80-2841-2761
E-mail Address: *info@ibab.ac.in*
Web Address: www.ibab.ac.in
Institute of Bioinformatics and Applied Biotechnology's (IBAB) is a joint venture of the corporate sector and Karnataka State Government in

India. Its serves as the advisory body to Karnataka and provides educational programs in related bioinformatics and biotechnology industries.

Korea Institute of Bioscience & Biotechnology (KRIBB)
111 Gwahangno
Yuseong-gu
Daejeon, 305-806 Korea
Phone: 042-860-4114
Fax: 042-861-1759
Web Address: www.kribb.re.kr
KRIBB is a Korean government research institute dedicated to biotechnology research across a broad span of expertise, from basic studies for the fundamental understanding of life phenomena to applied studies such as new drug discovery, novel biomaterials, integrated biotechnology, and bioinformation.

LifeSciences World
E-mail Address: *info@lifesciencesworld.com*
Web Address: www.biotechfind.com
LifeSciences World is a directory of life science news, jobs, events, articles, reports and links to information on biotechnology, pharmaceuticals and medical devices.

Medical Biochemistry Subject List
E-mail Address: *miking@iupui.edu*
Web Address: themedicalbiochemistrypage.org
The Medical Biochemistry Subject List, produced by Indiana State University, is a text-based introduction to biochemistry.

MedWeb: Biomedical Internet Resources
Web Address: www.medweb.emory.edu/medweb
MedWeb: Biomedical Internet Resources is a web site operated by Emory University, which lists resources by medical field, and allows users to search for articles by topic or date.

Microbiology Network
150 Parkway Dr.
N. Chili, NY 14514 US
Phone: 585-594-8273
Fax: 585-594-3338
E-mail Address: *scott.sutton@microbiol.org*
Web Address: www.microbiol.org
The Microbiology Network is a communication starting point for microbiologists, this site includes a discussion forum, user groups and extensive file libraries.

Recombinant Capital
2033 N. Main St., Ste. 1050
Walnut Creek, CA 94596-3722 US
Phone: 925-952-3870
Fax: 925-952-3871
E-mail Address: *info@recap.com*
Web Address: www.recap.com
Recombinant Capital provides databases of public information, including press releases, clinical trial references and a library of analyzed collaborations. It hopes to facilitate alliances between biotechnology companies.

Signals
2033 N. Main St., Ste. 1050
Walnut Creek, CA 94596-3722 US
Phone: 925-952-3870
Fax: 925-952-3871
E-mail Address: *signals_edit@recap.com*
Web Address: www.signalsmag.com
Signals is an online magazine of analysis for the biotechnology industry. The site provides trends dealing with biotechnology, medical device and pharmaceutical companies involved in the development of medical products.

The Microbiology Network
150 Parkway Dr.
N. Chili, NY 14514 US
Phone: 585-594-8273
Fax: 585-594-3338
E-mail Address: *scott.sutton@microbiol.org*
Web Address: www.microbiol.org
The Microbiology Network is a virtual library containing lists of organizations and associations in the fields of microbiology, biology and general science. Users can also submit new links.

University of California at Davis Biotechnology Program
1 Shields Ave.
301 Life Sciences
Davis, CA 95616 US
Phone: 530-752-3260
Fax: 530-752-4125
E-mail Address: *biotechprogram@ucdavis.edu*
Web Address: www.biotech.ucdavis.edu
The University of California at Davis Biotechnology Program provides useful biotech information and links, as well as the administrative home for UC Davis' Biotechnology Program.

University of Pennsylvania's Center for Bioethics
3401 Market St., Ste. 320
Philadelphia, PA 19104 US
Phone: 215-898-7136
Web Address: www.bioethics.upenn.edu
The University of Pennsylvania's Center for
Bioethics is a world-renowned resource. It
incorporates the work of more than twenty people
from the university's schools of law, medicine,
business, philosophy, public policy and religious
studies, as well as other departments. Resources
include the PennBioethics newsletter.

Windhover Information
10 Hoyt St.
Norwalk, CT 06851 US
Phone: 203-838-4401
Fax: 203-838-3214
E-mail Address:
fdcwindhover.custcare@elsevier.com
Web Address: www.windhoverinfo.com
Windhover Information provides analysis and
commentary on healthcare and biotech business
strategy.

IX. Brazilian Government Agencies-Scientific

National Council for Scientific & Technological Development
SEPN 507, Block B
Ed Sede CNPq
Brasilia, DF 70740-901 Brazil
Phone: 55-61-2108-9000
Fax: 55-61-2108-9394
Web Address: www.cnpq.br
The National Council for Scientific & Technological
Development (Conselho Nacional de
Desenvolvimento Cientifico e Tecnologico, or
CNPq) is a Brazilian government agency affiliated
with the country's Ministry of Science and
Technology. CNPq works to promote scientific and
technological research in Brazil through grants and
other support services. The organization also seeks to
encourage the development of Brazilian scientists
and researchers through the awarding of scholarships
and fellowships to students in the sciences.

X. Canadian Government Agencies-Health Care

Canadian Institutes of Health Research (CIHR)
160 Elgin St., 9th Fl.
Ottawa, ON K1A 0W9 Canada
Phone: 613-941-2672
Fax: 613-954-1800
Toll Free: 888-603-4178
E-mail Address: *info@cihr-irsc.gc.ca*
Web Address: www.cihr-irsc.gc.ca
Canadian Institutes of Health Research (CIHR) is the
Government of Canada's agency for health research.
CIHR's mission is to create new scientific knowledge
and to catalyze its translation into improved health,
more effective health services and products, and a
strengthened Canadian health-care system.
Composed of 13 Institutes, CIHR provides leadership
and support to more than 13,000 health researchers
and trainees across Canada. The agency provides
grants for research in the fields of biomedical,
clinical, health systems and environmental health.

Health Canada (Health Portfolio, Canadian Minister of Health)
Health Canada, Brooke Claxton Building, Tunney's
Pasture
Postal Locator: 0906C
Ottawa, ON K1A 0K9 Canada
E-mail Address: *Info@hc-sc.gc.ca*
Web Address: www.hc-sc.gc.ca
The Minister of Health is responsible for maintaining
and improving the health of Canadians. This is
supported by the Health Portfolio which comprises
Health Canada, the Public Health Agency of Canada,
the Canadian Institutes of Health Research, the
Hazardous Materials Information Review
Commission, the Patented Medicine Prices Review
Board and Assisted Human Reproduction Canada.
The Health Portfolio consists of approximately
12,000 full-time equivalent employees and an annual
budget of over $3.8 billion.

Patented Medicine Prices Review Board (PMPRB)
333 Laurier Avenue West
Suite 1400, Box L40
Ottawa, ON K1P 1C1 Canada
Phone: 613-952-7360

Fax: 613-952-7626
Toll Free: 877-861-2350
E-mail Address: *pmprb@pmprb-cepmb.gc.ca*
Web Address: www.pmprb-cepmb.gc.ca
The Patented Medicine Prices Review Board
(PMPRB) is an independent quasi-judicial body
established by Parliament of Canada in 1987 under
the Patent Act. It publishes a wealth of information
about the Canadian drug industry and drug
development. Its role includes the regulation of drug
prices. Although the PMPRB is part of the Health
Portfolio, it carries out its mandate at arms-length
from the Minister of Health. It also operates
independently of other bodies such as Health Canada,
which approves drugs for safety and efficacy, and
public drug plans, which approve the listing of drugs
on their respective formularies for reimbursement
purposes.

XI.   Careers-Biotech

Biotechemployment.com
Phone: 561-630-5201
E-mail Address: *jobs@Biotechemployment.com*
Web Address: www.biotechemployment.com
Biotechemployment.com is an online resource for job
seekers in biotechnology. The site's features includes
resume posting, job search agents and employer
profiles. It is part of the eJobstores.com, Inc., which
includes the Health Care Job Store sites.

Chase Group (The)
10955 Lowell Ave., Ste. 500
Overland Park, KS 66210 US
Phone: 913-663-3100
Fax: 913-663-3131
E-mail Address: *chase@chasegroup.com*
Web Address: www.chasegroup.com
The Chase Group is an executive search firm
specializing in biomedical and pharmaceutical
placement.

XII.   Careers-General Job Listings

CareerBuilder, Inc.
200 N. LaSalle St., Ste. 1100
Chicago, IL 60601 US
Phone: 773-527-3600
Toll Free: 800-638-4212
Web Address: www.careerbuilder.com
CareerBuilder, Inc. focuses on the needs of
companies and also provides a database of job

openings. The site has 1.5 million jobs posted by
300,000 employers, and receives an average 23
million unique visitors monthly. The company also
operates online career centers for 150 newspapers,
1,000 partners and other online portals such as
America Online. Resumes are sent directly to the
company, and applicants can set up a special e-mail
account for job-seeking purposes. CareerBuilder is
primarily a joint venture between three newspaper
giants: The McClatchy Company (which recently
acquired former partner Knight Ridder), Gannett Co.,
Inc. and Tribune Company. In 2007, Microsoft
acquired a minority interest in CareerBuilder,
allowing the site to ally itself with MSN.

HotJobs
45 W. 18th St., 6th Fl.
New York, NY 10011 US
Phone: 646-351-5300
Web Address: www.hotjobs.yahoo.com
HotJobs, designed for experienced professionals,
employers and job seekers, is a Monster-owned site
that provides company profiles, a resume posting
service and a resume workshop. The site allows
posters to block resumes from being viewed by
certain companies and provides a notification service
of new jobs.

JobCentral
DirectEmployers Association, Inc.
9002 N. Purdue Rd., Quad III, Ste. 100
Indianapolis, IN 46268 US
Phone: 317-874-9000
Fax: 317-874-9100
Toll Free: 866-268-6206
E-mail Address: *info@jobcentral.com*
Web Address: www.jobcentral.com
JobCentral, operated by the nonprofit
DirectEmployers Association, Inc., links users
directly to hundreds of thousands of job opportunities
posted on the sites of participating employers, thus
bypassing the usual job search sites. This saves
employers money and allows job seekers to access
many more job opportunities.

Wall Street Journal - CareerJournal
Wall Street Journal
200 Liberty St.
New York, NY 10281 US
Phone: 212-416-2000
Toll Free: 800-568-7625
E-mail Address: *onelinejournal@wsj.com*
Web Address: cj.careercast.com/careers/jobsearch

The Wall Street Journal's CareerJournal, an executive career site, features a job database with thousands of available positions; career news and employment related articles; and advice regarding resume writing, interviews, networking, office life and job hunting.

XIII. Careers-Health Care

Health Care Source
8 Winchester Pl.
Winchester, MA 01890 US
Phone: 781-368-1033
Fax: 800-829-6600
Toll Free: 800-869-5200
E-mail Address: *info@healthcaresource.com*
Web Address: www.healthcaresource.com
Health Care Source offers career-related information and job finding tools for health care professionals.

Medzilla, Inc.
P.O. Box 1710
Marysville, WA 98270 US
Phone: 360-657-5681
Fax: 775-514-9440
E-mail Address: *mgroutage@medzilla.com*
Web Address: www.medzilla.com
Medzilla, Inc.'s web site offers job searches, salary surveys, a search agent and information on health care employment.

Monster Career Advice-Healthcare
Monster Worldwide, Inc.
622 3rd Ave.
New York, NY 10017 US
Phone: 212-351-7000
Toll Free: 800-666-7837
Web Address: career-advice.monster.com/Healthcare/job-category-3975.aspx
Monster Career Advice-Healthcare, a service of Monster Worldwide, Inc., provides job listings, job searches and search agents for the medical field.

PracticeLink
P.O. Box 100
415 2nd Ave.
Hinton, WV 25951 US
Toll Free: 800-776-8383
Web Address: www.practicelink.com
PracticeLink is one of the largest physician employment web sites. It is a free service used by more than 18,000 practice-seeking physicians annually to quickly search and locate potential physician practice opportunities. PracticeLink is

financially supported by more than 700 hospitals, medical groups, private practices and health care systems that advertise more than 5,000 opportunities.

RPh on the Go USA, Inc.
5510 Howard St.
Skokie, IL 60077-2620 US
Phone: 847-588-7170
Fax: 847-588-7060
Toll Free: 800-553-7359
E-mail Address: *lbalaguer@rphonthego.com*
Web Address: www.rphonthego.com
RPh on the Go USA, Inc. places temporary and permanent qualified professionals in the pharmacy community.

XIV. Careers-Job Reference Tools

NewsVoyager
4401 Wilson Blvd., Ste. 900
Arlington, VA 22203-1867 US
Phone: 571-366-1000
Fax: 571-366-1195
E-mail Address: *sally.clarke@naa.org*
Web Address: www.newsvoyager.com
NewsVoyager, a service of the Newspaper Association of America (NAA), links individuals to local, national and international newspapers. Job seekers can search through thousands of classified sections.

Vault.com, Inc.
75 Varick St., 8th Fl.
New York, NY 10013 US
Phone: 212-366-4212
E-mail Address: *feedback@staff.vault.com*
Web Address: www.vault.com
Vault.com, Inc. is a comprehensive career web site for employers and employees, with job postings and valuable information on a wide variety of industries. Vault gears many of its features toward MBAs. The site has been recognized by Forbes and Fortune Magazines.

XV. Careers-Science

Chem Jobs
ChemIndustry.com
730 E. Cypress Ave.
Monrovia, CA 91016 US
Phone: 626-930-0808
Fax: 626-930-0102

E-mail Address: *info@chemindustry.com*
Web Address: www.chemjobs.net
Chem Jobs is a leading Internet site for job seekers in chemistry and related fields, aimed at chemists, biochemists, pharmaceutical scientists and chemical engineers looking for work. The web site is powered by Chemindustry.com.

New Scientist Jobs
New Scientist, Lacon House
84 Theobald's Rd.
London, WC1X 8NS UK
Phone: 44-20-7611-1200
Fax: 44-20-7611-1250
E-mail Address: *nssales@elsevier.com.*
Web Address: www.newscientistjobs.com
New Scientist Jobs is a web site produced by the publishers of New Scientist Magazine, which helps jobseekers and employers in the bioscience fields find each other. The site includes a job search engine and a free-of-charge e-mail job alert service. New Scientist Jobs is owned by Reed Business Information Ltd.

Science Jobs
E-mail Address: *help@science-jobs.org*
Web Address: www.science-jobs.org/index.htm
Science Jobs is a web site that contains many useful categories of links, including employment newsgroups, scientific journals and placement agencies. It also links to sites containing information regarding internship and fellowship opportunities for high school students and undergrads.

XVI. Chinese Government Agencies-Science & Technology

China Ministry of Science and Technology (MOST)
15B Fuxing Rd.
Beijing, 100862 China
Web Address: www.most.gov.cn
The China Ministry of Science and Technology (MOST) has information and links to its various departments including the Departments of Personnel; Social Development; Rural Science and Technology; Basic Research; and High and New Technology Development and Industrialization.

XVII. Clinical Trials

Clinical Trials
8600 Rockville Pike
Bethesda, MD 20894 US
Phone: 301-594-5983
Fax: 301-402-1384
Toll Free: 888-346-3656
Web Address: www.clinicaltrials.gov
Clinical Trials offers up-to-date information for locating federally and privately supported clinical trials for a wide range of diseases and conditions. It is a service of the National Institutes of Health (NIH).

Institute of Clinical Research (ICR)
Institute House
Boston Dr.
Bourne End, Buckinghamshire SL8 5YS UK
Phone: 44-1628-536960
Fax: 44-1628-530641
E-mail Address: *info@icr-global.org*
Web Address: www.icr-global.org
The Institute of Clinical Research (ICR) is a professional organization for clinical researchers in the pharmaceutical industry in the U.K.

Office of Biotechnology Activities, NIH
6705 Rockledge Dr., Ste. 750, MSC 7985
Bethesda, MD 20892-7985 US
Phone: 301-496-9838
Fax: 301-496-9839
E-mail Address: *oba@od.nih.gov*
Web Address: www4.od.nih.gov/oba/
This unit of the U.S. National Institutes of Health operates a web site with links to clinical research in recombinant DNA and gene transfer, along with information on the National Science Advisory Board for Biosecurity.

XVIII. Communications Professional Associations

Health and Science Communications Association (HeSCA)
39 Wedgewood Dr., Ste. A
Jewett City, CT 06351 US
Phone: 860-376-5915
Fax: 860-376-6621

E-mail Address: *hesca@hesca.org*
Web Address: www.hesca.org
The Health and Science Communications Association (HeSCA) is an association of communications professionals committed to sharing knowledge and resources in the health sciences arena.

XIX.   Corporate Information Resources

bizjournals.com
120 W. Morehead St., Ste. 400
Charlotte, NC 28202 US
Web Address: www.bizjournals.com
Bizjournals.com is the online media division of American City Business Journals, the publisher of dozens of leading city business journals nationwide. It provides access to research into the latest news regarding companies small and large.

Business Wire
44 Montgomery St., 39th Fl.
San Francisco, CA 94104 US
Phone: 415-986-4422
Fax: 415-788-5335
Toll Free: 800-227-0845
Web Address: www.businesswire.com
Business Wire offers news releases, industry- and company-specific news, top headlines, conference calls, IPOs on the Internet, media services and access to tradeshownews.com and BW Connect On-line through its informative and continuously updated web site.

Edgar Online, Inc.
50 Washington St., 11th Fl.
Norwalk, CT 06854 US
Phone: 203-852-5666
Fax: 203-852-5667
Toll Free: 800-416-6651
Web Address: www.edgar-online.com
Edgar Online, Inc. is a gateway and search tool for viewing corporate documents, such as annual reports on Form 10-K, filed with the U.S. Securities and Exchange Commission.

PR Newswire Association LLC
810 7th Ave., 32nd Fl.
New York, NY 10019 US
Phone: 201-360-6700
Toll Free: 800-832-5522
E-mail Address: *information@prnewswire.com*
Web Address: www.prnewswire.com

PR Newswire Association LLC provides comprehensive communications services for public relations and investor relations professionals ranging from information distribution and market intelligence to the creation of online multimedia content and investor relations web sites. Users can also view recent corporate press releases. The Association is owned by United Business Media plc.

Silicon Investor
100 W. Main
P.O. Box 29
Freeman, MO 64746 US
E-mail Address: *admin_dave@techstocks.com*
Web Address: siliconinvestor.advfn.com
Silicon Investor is focused on providing information about technology companies. The company's web site serves as a financial discussion forum and offers quotes, profiles and charts.

XX.   Economic Data & Research

Eurostat
Phone: 32-2-299-9696
Toll Free: 80-0-6789-1011
Web Address: www.epp.eurostat.ec.europa.eu
Eurostat is the European Union's service that publishes a wide variety of comprehensive statistics on European industries, populations, trade, agriculture, technology, environment and other matters.

India Brand Equity Foundation (IBEF)
249-F Sector 18
Udyog Vihar Phase IV
Gurgaon, Haryana 122015 India
Phone: 91-124-4014060
Fax: 91-124-4013873
E-mail Address: *ceo@ibef.org*
Web Address: www.ibef.org
India Brand Equity Foundation (IBEF) is a public-private partnership between the Ministry of Commerce and Industry, Government of India, and the Confederation of Indian Industry. The Foundation's primary objective is to build positive economic perceptions of India globally. It aims to effectively present the India business perspective and leverage business partnerships in a globalizing market-place.

Statistics Canada
150 Tunney's Pasture Driveway
Ottawa, ON K1A 0T6 Canada

Phone: 613-951-8116
Fax: 613-951-0581
Toll Free: 800-263-1136
Web Address: www.statcan.gc.ca
Statistics Canada is a complete portal to Canadian economic data and statistics.

STAT-USA/Internet
STAT-USA, HCHB, U.S. Dept. of Commerce
Rm. 4885
Washington, DC 20230 US
Phone: 202-482-1986
Fax: 202-482-2164
Toll Free: 800-782-8872
E-mail Address: *statmail@esa.doc.gov*
Web Address: www.stat-usa.gov
STAT-USA/Internet offers daily economic news, statistical releases and databases relating to export and trade, as well as the domestic economy. It is provided by STAT-USA, which is an agency in the Economics & Statistics Administration of the U.S. Department of Commerce. The site mainly consists of two main databases, the State of the Nation (SOTN), which focuses on the current state of the U.S. economy; and the Global Business Opportunities (GLOBUS) & the National Trade Data Bank (NTDB), which deals with U.S. export opportunities, global political/socio-economic conditions and other world economic issues.

XXI. Engineering, Research & Scientific Associations

Agency For Science, Technology And Research (A*STAR)
1 Fusionopolis Way
20-10 Connexis N. Twr.
138632 Singapore
Phone: 65-6826-6111
Fax: 65-6777-1711
Web Address: www.a-star.edu.sg
The Agency For Science, Technology And Research (A*STAR) of Singapore comprises the Biomedical Research Council (BMRC), the Science and Engineering Research Council (SERC), Exploit Technologies Pte Ltd (ETPL), the A*STAR Graduate Academy (A*GA) and the Corporate Planning and Administration Division (CPAD). Both Councils fund the A*STAR public research institutes which conducts research in specific niche areas in science, engineering and biomedical science.

American Association for the Advancement of Science (AAAS)
1200 New York Ave. NW
Washington, DC 20005 US
Phone: 202-326-6400
E-mail Address: *webmaster@aaas.org*
Web Address: www.aaas.org
The American Association for the Advancement of Science (AAAS) is the world's largest scientific society and the publisher of Science magazine. It is an international nonprofit organization dedicating to advancing science.

American Society for Healthcare Engineering (ASHE)
1 N. Franklin, 28th Fl.
Chicago, IL 60606 US
Phone: 312-422-3800
Fax: 312-422-4571
E-mail Address: *ashe@aha.org*
Web Address: www.ashe.org
The American Society for Healthcare Engineering (ASHE) is the advocate and resource for continuous improvement in the health care engineering and facilities management professions.

American Society of Agricultural and Biological Engineers (ASABE)
2950 Niles Rd.
St. Joseph, MI 49085 US
Phone: 269-429-0300
Fax: 269-429-3852
E-mail Address: *hq@asabe.org*
Web Address: www.asabe.org
The American Society of Agricultural and Biological Engineers (ASABE) is a nonprofit professional and technical organization interested in engineering knowledge and technology for food and agriculture and associated industries.

Association of Official Analytical Chemists (AOAC)
481 N. Frederick Ave., Ste. 500
Gaithersburg, MD 20877-2417 US
Phone: 301-924-7077
Fax: 301-924-7089
Toll Free: 800-379-2622
E-mail Address: *aoac@aoac.org*
Web Address: www.aoac.org
The Association of Official Analytical Chemists (AOAC) is a nonprofit scientific association committed to worldwide confidence in analytical results.

China Association for Science and Technology (CAST)
3 Fuxing Rd.
Beijing, 100863 China
Phone: 8610-68571898
Fax: 8610-68571897
E-mail Address: *english@cast.org.cn*
Web Address: english.cast.org.cn
The China Association for Science and Technology (CAST) is the largest national non-governmental organization of scientific and technological workers in China. The association has 167 member organizations in the field of engineering, science and technology.

Chinese Academy of Sciences (CAS)
52 Sanlihe Rd.
Beijing, 100864 China
Phone: 86-10-68597289
Fax: 86-10-68512458
E-mail Address: *bulletin@mail.casipm.ac.cn*
Web Address: english.cas.ac.cn
By 2010, the Chinese Academy of Sciences (CAS) plans on maintaining roughly 80 national institutes for science and technological innovation each with distinctive features with roughly thirty planned for internationally acknowledged, high-level research.

DECHEMA (Society for Chemical Engineering and Biotechnology)
Theodor-Heuss-Allee 25
Frankfurt am Main, 60486 Germany
Phone: 0049 69 7564-0
Fax: 0049 69 7564-201
Web Address: dechema.de/
The DECHEMA (Society for Chemical Engineering and Biotechnology) is a nonprofit scientific and technical society based in Frankfurt on Main. It was founded in 1926. Today it has over 5000 private and institutional members. Its aim is to promote research and technical advances in the areas of chemical engineering, biotechnology and environmental protection. The group's work is interdisciplinary, with scientists, engineers, and technologists working together under one roof. Experts from science, business, and government departments cooperate in working parties and subject divisions.

Federation of Technology Industries (FHI)
Federatie Van Technologiebranches
Dodeweg 6B
AK LEUSDEN, 3832 Netherlands
Phone: (033) 465 75 07

Fax: (033) 461 66 38
E-mail Address: *info@fhi.nl*
Web Address: federatie.fhi.nl
The Federation of Technology Industries (FHI) is the Dutch trade organization representing industrial electronics, industry automation, laboratory technology and medical technolgy.

German Association of High-Tech Industries (SPECTARIS)
Verband der Hightech-Industrie
Kirchweg 2
Koln, 50858 Germany
Phone: 0221/948628-0
Fax: 0221/948628-80
E-mail Address: *info@spectaris.de*
Web Address: www.spectaris.de/
German Association of High-Tech Industries (SPECTARIS) is the trade association for technology and research in the consumer optics, photonics, biotech and medical technology.

Institute of Bioengineering and Nanotechnology, Singapore
31 Biopolis Way
The Nanos 04-01
138669 Singapore
Phone: 65-6824-7000
Fax: 65-6478-9080
E-mail Address: *enquiry@ibn.a-star.edu.sg*
Web Address: www.ibn.a-star.edu.sg
As a scientific research institute, Institute of Bioengineering and Nanotechnology focuses its activities on developing the following key areas; developing a critical knowledge base in bioengineering and nanotechnology; generating new biomaterials, devices and processes; and producing and publishiing high-quality scientific research.

Institute of Biological Engineering (IBE)
1020 Monarch St., Ste. 300B
Lexington, KY 40512 US
Phone: 859-977-7450
Fax: 859-977-7441
E-mail Address: *sclements@ibe.org*
Web Address: www.ibeweb.org
The Institute of Biological Engineering (IBE) is a professional organization encouraging inquiry and interest in biological engineering and professional development for its members.

Institute of Electrical and Electronics Engineers (IEEE)
3 Park Ave., 17th Fl.
New York, NY 10016-5997 US
Phone: 212-419-7900
Fax: 212-752-4929
E-mail Address: *ieeeusa@ieee.org*
Web Address: www.ieee.org
The Institute of Electrical and Electronics Engineers (IEEE) is a nonprofit, technical professional association of more than 375,000 individual members in approximately 160 countries. The IEEE sets global technical standards and acts as an authority in technical areas ranging from computer engineering, biomedical technology and telecommunications, to electric power, aerospace and consumer electronics.

International Commission of Agricultural and Biosystems Engineering (CIGR)
Univ. of Tsukuba-Life and Environmental Sciences
1-1-1 Tennodai
Tsukuba, Ibaraki 305-8572 Japan
Phone: 81-29-853-6989
Fax: 81-29-853-7496
E-mail Address: *biopro@sakura.cc.tsukuba.ac.jp*
Web Address: www.cigr.org
International Commission of Agricultural and Biosystems Engineering (CIGR) encourages and facilitates interregional exchange and the development of sciences and technologies in the field of agricultural engineering.

International Federation for Medical and Biological Engineering
Faculty of Electrical Engineering and Computing
Univ. of Zagreb, Unska 3
Zagreb, HR 10000 Croatia
Phone: 385-1-6129-938
Fax: 385-1-6129-652
E-mail Address: *ratko.magjarevic@fer.hr*
Web Address: www.ipem.ac.uk
The International Federation for Medical and Biological Engineering (IFMBE) is a federation of national and transnational organizations that represent national interests in medical and biological engineering. The objectives of the IFMBE are scientific, technological, literary, and educational.

International Society of Pharmaceutical Engineers (ISPE)
3109 W. Dr. Martin Luther King, Jr. Blvd., Ste. 250
Tampa, FL 33607 US
Phone: 813-960-2105

Fax: 813-264-2816
E-mail Address: *ASK@ispe.org*
Web Address: www.ispe.org
The International Society of Pharmaceutical Engineers (ISPE) is a worldwide nonprofit society dedicated to educating and advancing pharmaceutical manufacturing professionals and the pharmaceutical industry.

International Union of Microbiological Societies (IUMS)
P.O. Box 85167
Utrecht, 3508 AD The Netherlands
Phone: 31-30-212-2600
Fax: 31-30-251-2097
E-mail Address: *samson@cbs.knaw.nl*
Web Address: www.iums.org
The International Union of Microbiological Societies (IUMS) works to promote the study of microbiological sciences around the world through its three divisions: Bacteriology & Applied Microbiology (BAM); Mycology; and Virology. The association is one of the 29 Scientific Unions of the International Council of Science (ICSU).

Japan Science and Technology Agency (JST)
Kawaguchi Ctr. Bldg.
4-1-8 Honcho, Kawaguchi-shi
Saitama, 332-0012 Japan
Phone: 81-48-226-5601
Fax: 81-48-226-5651
E-mail Address: *www-admin@tokyo.jst.go.jp*
Web Address: www.jst.go.jp/EN
The Japan Science and Technology Agency (JST) aims to act as a core organization for implementation of the nation's science and technology policies by conducting research and development with particular emphasis on new technological needs.

National Academy of Science (NAS)
500 5th St. NW
Washington, DC 20001 US
Phone: 202-334-2000
E-mail Address: *worldwidewebfeedback@nas.edu*
Web Address: www.nationalacademies.org
The National Academies are private, nonprofit, self-perpetuating societies of scholars engaged in scientific and engineering research dedicated to the furtherance of science and technology and to their use for the general welfare. Four organizations comprise the Academies: The National Academy of Engineering, the National Research Council, the

National Academy of Sciences and the Institute of Medicine.

National Medical Research Council (NMRC)
11 Biopolis Way
Helios 09-10/11
138667 Singapore
Phone: 65-6325-8130
Fax: 65-6324-3735
E-mail Address: *MOH_NMRC@MOH.GOV.SG*
Web Address: www.nmrc.gov.sg
National Medical Research Council (NMRC) oversees the development and advancement of medical research in Singapore.

Netherlands Organization for Applied Scientific Research (TNO)
Schoemakerstraat 97, Bldg. A
Delft, NL-2628 VK The Netherlands
Phone: 31-15-269-69-00
Fax: 31-15-261-24-03
E-mail Address: *infodesk@tno.nl*
Web Address: www.tno.nl
The Netherlands Organization for Applied Scientific Research (TNO) is a contract research organization that provides a link between fundamental research and practical application.

Research in Germany, German Academic Exchange Service (DAAD)
Kennedyallee 50
Bonn, 53175 Germany
Phone: 49(0)228 882 - 0
Fax: 49(0)228 882 - 660
Web Address: www.research-in-germany.de
The Research in Germany portal, German Academic Exchange Service (DAAD),
is an information platform and contact point for those looking to find out more about Germany's research landscape and its latest research achievements. The portal is an initiative of the Federal Ministry of Education and Research.

Royal Society (The)
6-9 Carlton House Ter.
London, SW1Y 5AG UK
Phone: 44-20-7451-2500
Fax: 44-20-7930-2170
E-mail Address: *info@royalsociety.org*
Web Address: www.royalsoc.ac.uk
The Royal Society is the UK's leading scientific organization. It operates as a national academy of science, supporting scientists, engineers,

technologists and research. On its website, you will find a wealth of data about the research and development initiatives of its fellows and foreign members.

Royal Society of Chemistry (RSC)
Burlington House, Piccadilly
London, W1J 0BA UK
Phone: 44-20-7437-8656
Fax: 44-20-7437-8883
Web Address: www.rsc.org
The Royal Society of Chemistry (RSC) is one of Europe's largest organizations for advancing the chemical sciences.

Society for Chemical Engineering and Biotechnology (DECHEMA)
Theodor-Heuss-Allee 25
Frankfurt am Main, 60486 Germany
Phone: (069) 7564-0
Fax: (069) 7564-201
Web Address: www.dechema.de
The Society for Chemical Engineering and Biotechnology (DECHEMA) is an interdisciplinary organization representing private firms and institutions in Germany. DECHEMA cooperates with government, science and business to promote chemical engineering, biotechnology and environmental protection.

XXII. Engineering, Research & Scientific Resources

Steacie Institute for Molecular Sciences (SIMS)
100 Sussex Dr., Rm. 1151
Ottawa, ON K1A 0R6 Canada
Phone: 613-991-5419
Fax: 613-954-5242
E-mail Address: *Helene.Letourneau@nrc-cnrc.gc.ca*
Web Address: steacie.nrc-cnrc.gc.ca
The Steacie Institute for Molecular Sciences (SIMS) was created to facilitate the collaboration of scientific communities researching molecular sciences both within Canada and internationally.

XXIII. Food Industry Resources, General

Consultative Group on International Agricultural Research (CGIAR)
1818 H St. NW
The World Bank, MSN G6-601
Washington, DC 20433 US

Phone: 202-473-8951
Fax: 202-473-8110
E-mail Address: *cgiar@cgiar.org*
Web Address: www.cgiar.org
The Consultative Group on International Agricultural Research (CGIAR), established in 1971, and operating as part of The World Bank, is a strategic partnership of countries, international and regional organizations and private foundations supporting the work of 15 international Centers. In collaboration with national agricultural research systems, civil society and the private sector, the CGIAR fosters sustainable agricultural growth through high-quality science aimed at benefiting the poor through stronger food security, better human nutrition and health, higher incomes and improved management of natural resources.

Food and Agriculture Organization of the United Nations
Viale delle Terme di Caracalla
Rome, 00153 Italy
Phone: 39-06-57051
Fax: 39-06-570-53152
E-mail Address: *fao-hq@fao.org*
Web Address: www.fao.org
The Food and Agriculture Organization of the United Nations leads international efforts to defeat hunger. Serving both developed and developing countries, FAO acts as a forum where nations meet to negotiate agreements and debate policy. FAO is also a source of knowledge and information. It helps nations in their efforts to modernize and improve agriculture, forestry and fisheries practices and ensure good nutrition for all. The FAO's website provides extensive statistical data on food production, costs and world hunger.

XXIV. Genetics & Genomics Industry Associations

American College of Medical Genetics (ACMG)
9650 Rockville Pike
Bethesda, MD 20814-3998 US
Phone: 301-634-7127
Fax: 301-634-7275
E-mail Address: *acmg@acmg.net*
Web Address: www.acmg.net
The American College of Medical Genetics (ACMG) provides education, resources and a voice for the medical genetics profession. The ACMG promotes the development and implementation of methods to diagnose, treat and prevent genetic disease.

European Society of Human Genetics (ESHG)
Vienna Medical Academy
Alser Strasse 4
Vienna, 1090 Austria
Phone: 43-1-405-13-83-20
Fax: 43-1-407-82-74
E-mail Address: *office@eshg.org*
Web Address: www.eshg.org
The European Society of Human Genetics (ESHG) is a organization that promotes the sharing of information among genetics societies in Europe.

Genetics Society of America (GSA)
9650 Rockville Pike
Bethesda, MD 20814-3998 US
Phone: 301-634-7300
Fax: 301-634-7079
Toll Free: 866-486-4343
E-mail Address: *society@genetics-gsa.org*
Web Address: www.genetics-gsa.org
The Genetics Society of America (GSA) includes over 4,000 scientists and educators interested in the field of genetics. The society promotes the communication of advances in genetics through publication of the journal GENETICS, and by sponsoring scientific meetings focused on key organisms widely used in genetic research.

International Cancer Genome Consortium (ICGC)
215 Euston Road
Wellcome Trust
London, NW1 7262 UK
Phone: 44 (0) 20 7611 7262
Web Address: www.icgc.org
The International Cancer Genome Consortium (ICGC) plans to decode the genomes from 25,000 cancer samples and create a resource of freely available data that will help cancer researchers around the world. Funded projects in several participating nations will examine more than 10,000 tumors for cancer types that affect a diversity of organs including blood, brain, breast, colon, kidney, liver, lung, pancreas, stomach, oral cavity and ovary.

Bioinformatics Institute (A*STAR Singapore)
30 Biopolis St.
07-01 Matrix
138671 Singapore
Phone: 65-6478-8298
E-mail Address: *enquiry@bii.a-star.edu.sg*
Web Address: www.bii.a-star.edu.sg

The Bioinformatics Institute has developed and deployed analytical tools and computational techniques for biology research both in-house and through close collaboration with experimental and clinical groups within and outside the Biopolis and Singapore. Its web site includes information on current research as well as links, databases, and events.

XXV.   Genetics & Genomics Industry Resources

Genetic Education Center: Human Genome Project Resources (GEC)
3901 Rainbow Blvd.
Kansas City, KS 66160 US
Phone: 913-588-5000
E-mail Address: *dcollins@kumc.edu*
Web Address: www.kumc.edu/gec/prof/hgc.html
The Genetic Education Center: Human Genome Project Resources (GEC) provides a list of links related to the human genome including human genome centers, information sites and news articles.

Genome Institute of Singapore
60 Biopolis St.
02-01 Genome
138672 Singapore
Phone: 65-6478-8000
E-mail Address: *genome@gis.a-star.edu.sg*
Web Address: www.gis.a-star.edu.sg
The Genome Institute of Singapore (GIS) is a Singaporean national initiative with a global vision. GIS pursues the integration of technology, genetics and biology towards the goal of individualized medicine.

International Communication Forum in Human Genetics (ICFHG)
E-mail Address: *admin@hum-molgen.de*
Web Address: www.hum-molgen.de
The International Communication Forum in Human Genetics (ICFHG) contains news articles, a bulletin board and a variety of other services related to human molecular genetics.

National Human Genome Research Institute (NHGRI)
31 Center Dr.
Bldg. 31, Rm. 4B09
Bethesda, MD 20892-2152 US
Phone: 301-402-0911
Fax: 301-402-2218

Web Address: www.nhgri.nih.gov
The National Human Genome Research Institute (NHGRI) led the human genome project until its completion in April 2003. The agency, a division of the National Institutes of Health, now provides research news and information about the field of human genetics.

XXVI.   Government Agencies-Singapore

Singapore Government Online (SINGOV)
140 Hill St., MICA Bldg., 5th Fl.
179369 Singapore
E-mail Address: *singov_webmaster@mica.gov.sg*
Web Address: www.gov.sg
Singapore Government Online (SINGOV) is the default homepage for the Singapore Government and is a portal for governmenal information. The website lists governmental agencies, news, information, policies and inititives.

XXVII.   Health Care Business & Professional Associations

American Association of Immunologists
9650 Rockville Pike
Bethesda, MD 20814 US
Phone: 301-634-7178
Fax: 301-634-7887
E-mail Address: *infoaai@aai.org*
Web Address: www.aai.org
The American Association of Immunologists is a nonprofit organization that represents professionals in the immunology field.

American Society of Clinical Oncology (ASCO)
2318 Mill Rd., Ste. 800
Alexandria, VA 22314 US
Phone: 571-483-1300
Fax: 703-299-1044
Web Address: www.asco.org
The American Society of Clinical Oncology (ASCO) is a nonprofit organization, founded in 1964, with overarching goals of improving cancer care and prevention and ensuring that all patients with cancer receive care of the highest quality. Nearly 25,000 oncology practitioners belong to ASCO, representing all oncology disciplines.

Association of Clinical Research Professionals (ACRP)
500 Montgomery St., Ste. 800

Alexandria, VA 22314 US
Phone: 703-254-8100
Fax: 703-254-8101
E-mail Address: *office@acrpnet.org*
Web Address: www.acrpnet.org
The Association of Clinical Research Professionals
(ACRP) is an organization for professionals in the
pharmaceutical, biotechnology and medical device
industries, as well as those in hospital, academic
medical centers and physician office settings.

Association of Food and Drug Officials (AFDO)
2550 Kingston Rd., Ste. 311
York, PA 17402 US
Phone: 717-757-2888
Fax: 717-755-8089
E-mail Address: *afdo@afdo.org*
Web Address: www.afdo.org
The Association of Food and Drug Officials (AFDO)
is a trusted resource for building consensus and
promoting uniformity on public health and consumer
protection issues related to the regulation of foods,
drugs, devices, cosmetics and consumer products.

Health Industry Distributors Association (HIDA)
310 Montgomery St.
Alexandria, VA 22314-1516 US
Phone: 703-549-4432
Fax: 703-549-6495
E-mail Address: *sandler@hida.org*
Web Address: www.hida.org
The Health Industry Distributors Association (HIDA)
is the international trade association representing
medical products distributors.

Hong Kong Medical Association
15 Hennessy Rd.
5/F Duke of Windsor Social Service Bldg.
Wanchai, Hong Kong China
Phone: 2527-8285
E-mail Address: *hkma@hkma.org*
Web Address: www.hkma.org
The Hong Kong Medical Association's objective is to
promote the welfare of the medical profession and
the health of the public of Hong Kong.

Society for Pharmaceutical and Medical Device Professionals (ISPE)
3109 W. Dr. Martin Luther King, Jr. Blvd., Ste. 250
Tampa, FL 33607 US
Phone: 813-960-2105
Fax: 813-264-2816
E-mail Address: *ASK@ispe.org*

Web Address: www.ispe.org
The Society for Pharmaceutical and Medical Device
Professionals (ISPE) works with its members by
providing extensive education, training, technical
publications, conferences and networking
opportunities.

Society of Clinical Research Associates (SOCRA)
530 W. Butler Ave., Ste. 109
Chalfont, PA 18914 US
Phone: 215-822-8644
Fax: 215-822-8633
Toll Free: 800-7627292
E-mail Address: *socramail@aol.com*
Web Address: www.socra.org
The Society of Clinical Research Associates
(SOCRA) works to provide training and continuing
education and to establish and maintain an
international certification program for clinical
research professionals.

Society of Toxicology (SOT)
1821 Michael Faraday Dr., Ste. 300
Reston, VA 20190 US
Phone: 703-438-3115
Fax: 703-438-3113
E-mail Address: *sothq@toxicology.org*
Web Address: www.toxicology.org
The Society of Toxicology (SOT) is an association
that works to advance the science of enhancing
human, animal and environmental health through the
understanding of toxicology.

XXVIII.   Health Care Resources

Access Excellence
1350 Connecticut Ave. NW, 5th Fl.
Washington, DC 20036 US
Phone: 650-712-1723
Web Address: www.accessexcellence.org
Access Excellence provides information for high
school biology and life science teachers. It is
produced by the National Health Museum.

Singapore Medical Council (SMC)
16 College Rd., 01-01 College of Medicine Bldg.
169854 Singapore
Phone: 65-6372-3065
Fax: 65-6221-0558
E-mail Address: *moh_smc@moh.gov.sg*
Web Address: www.smc.gov.sg
The Singapore Medical Council (SMC), a statutory
board under the Ministry of Health, maintains the

Register of Medical Practitioners in Singapore, administers the compulsory continuing medical education program and also governs and regulates the professional conduct and ethics of registered medical practitioners.

XXIX. Health Facts-Global

Organisation for Economic Co-Operation and Development (OECD) - Health Statistics
2 rue André Pascal, F-75775
Paris, Cedex 16 France
Phone: 33-145-24-8200
Fax: 33-145-24-8500
E-mail Address: *health.contact@oecd.org*
Web Address: www.oecd.org
The Organisation for Economic Co-Operation and Development (OECD) offers extensive health statistics on a country-by-country basis. Data ranges from health expenditures per capita to health expenditures as percent of GDP for the 30 nations with the world's largest economies.

XXX. Immunization Resources

CDC National Immunization Information Hotline (NIIH)
1600 Clifton Rd.
Atlanta, GA 30333 US
Toll Free: 800-232-4636
Web Address:
The CDC National Immunization Information Hotline (NIIH) offers up-to-date immunization information, including vaccine schedules, side effects, contraindications, recommendations and more.

XXXI. Industry Research/Market Research

Forrester Research
400 Technology Sq.
Cambridge, MA 02139 US
Phone: 617-613-6000
Fax: 617-613-5200
Toll Free: 866-367-7378
Web Address: www.forrester.com
Forrester Research identifies and analyzes emerging trends in technology and their impact on business. Among the firm's specialties are the financial services, retail, health care, entertainment, automotive and information technology industries.

Marketresearch.com
11200 Rockville Pike, Ste. 504
Rockville, MD 20852 US
Phone: 240-747-3000
Fax: 240-747-3004
Toll Free: 800-298-5699
E-mail Address: *customerservice@marketresearch.com*
Web Address: www.marketresearch.com
Marketresearch.com is a leading broker for professional market research and industry analysis. Users are able to search the company's database of research publications including data on global industries, companies, products and trends.

Plunkett Research, Ltd.
P.O. Drawer 541737
Houston, TX 77254-1737 US
Phone: 713-932-0000
Fax: 713-932-7080
E-mail Address: *customersupport@plunkettresearch.com*
Web Address: www.plunkettresearch.com
Plunkett Research, Ltd. is a leading provider of market research, industry trends analysis and business statistics. Since 1985, it has served clients worldwide, including corporations, universities, libraries, consultants and government agencies. At the firm's web site, visitors can view product information and pricing and access a great deal of basic market information on industries such as financial services, infotech, e-commerce, health care and biotech.

XXXII. Libraries-Medical Data

Library of the National Medical Society
Web Address: www.medical-library.org
The Library of the National Medical Society provides a free resource of medical information for both health care consumers and medical professionals.

Weill Cornell Medical Library
Weill Cornell Medical College, C.V. Starr Biomedical Information Center
1300 York Ave.
New York, NY 10021-4896 US
Phone: 212-746-6055
E-mail Address: *infodesk@med.cornell.edu*
Web Address: library.med.cornell.edu
The Weill Cornell Medical Library houses information on the biomedical sciences, as well as

performing data retrieval, management and evaluation.

XXXIII.   Nanotechnology Associations

IndiaNano

Range Hills Rd., Shivaji Nagar
IndiaCo Ctr., 4th Fl., Symphony, S. 210 A/1
Pune, Maharashtra 411 020 India
Phone: 91-20-25513254
Fax: 91-20-25513243
E-mail Address: *info@indianano.com*
Web Address: www.indianano.com
IndiaNano is a nonprofit organization located in India and supported by academic and industry experts. It aims to develop collaboration in order to advance technologies, including nanotechnology.

Nano Science and Technology Institute (NSTI)

1 Kendall Sq., PMB 308
Cambridge, MA 02139 US
Phone: 508-357-2925
Fax: 925-886-8461
E-mail Address: *mlaudon@nsti.org*
Web Address: www.nsti.org
The Nano Science and Technology Institute (NSTI) is engaged in the promotion and integration of nano and other advanced technologies through education, technology and business development. NSTI offers consulting services, continuing education programs, scientific and business publishing and community outreach.

NCI Alliance for Nanotechnology in Cancer

c/o Nat'l Cancer Institute, Office of Tech. & Industrial Rel.
Bldg. 21, Rm. 10A49, 31 Center Dr., MSC 2580
Bethesda, MD 20892-2580 US
Phone: 301-496-1550
Fax: 301-496-7807
E-mail Address: *cancer.nano@mail.nih.gov*
Web Address: nano.cancer.gov
The NCI Alliance for Nanotechnology in Cancer, a service of the National Cancer Institute, is dedicated to using nanotechnology to advance the ways we prevent, treat and diagnose cancer. It especially seeks advances in the near and medium terms and to lower the barriers for those advances to be handed off to the private sector for commercial development. The Alliance focuses on translational research and development work in six major challenge areas, where nanotechnology can have the biggest and fastest impact on cancer treatment.

XXXIV.   Online Health Data

Medscape

76 Ninth Ave., Ste. 719
New York, NY 10011 US
Phone: 212-624-3700
Toll Free: 888-506-6098
E-mail Address:
medscapecustomersupport@medscape.net
Web Address: www.medscape.com
Medscape, an online resource for better patient care, provides links to journal articles, health care-related sites and health care information.

PubMed

E-mail Address: *pubmedcentral@nih.gov*
Web Address: www.ncbi.nlm.nih.gov/entrez/query
PubMed provides access to over 17 million MEDLINE citations dating back to the mid-1960s and additional life science journals. PubMed includes links to open access full text articles.

XXXV.   Patent Organizations

European Patent Office

Av. de Cortenbergh 60
Brussels, 1000 Belgium
Phone: 32-2-274-15-90
Web Address: www.epo.org
The European Patent Office (EPO) provides a uniform application procedure for individual inventors and companies seeking patent protection in up to 38 European countries. It is the executive arm of the European Patent Organization and is supervised by the Administrative Council.

U.S. Patent and Trademark Office (PTO)

U.S. Patent and Trademark Office
Office of Public Affairs, P. O. Box 1450
Alexandria, VA 22313-1450 US
Phone: 571-272-1000
Fax: 571-273-8300
Toll Free: 800-786-9199
E-mail Address: *usptoinfo@uspto.gov*
Web Address: www.uspto.gov
The U.S. Patent and Trademark Office (PTO) administers patent and trademark laws for the U.S. and enables registration of patents and trademarks.

World Intellectual Property Organization (WIPO)

34 chemin des Colombettes
Geneva 20, CH-1211 Switzerland

Phone: 41-22-338-9111
Fax: 41-22-733-54-28
E-mail Address: *publicinf@wipo.int*
Web Address: www.wipo.int
The WIPO has a United Nations mandate to assist organizations and companies in filing patents and other intellectual property data on a global basis. At its website, you can download free copies of its WIPO magazine, and you can search its international patent applications.

XXXVI.   Patent Resources

Patent Docs
E-mail Address: *patentdocs@gmail.com*
Web Address: patentdocs.typepad.com/patent_docs/
Patent Docs is an excellent blog about patent law and patent news in the fields of biotechnology and pharmaceuticals.

Patent Law for Non-Lawyers
E-mail Address: *info@thinkbiotech.com*
Web Address: www.dnapatent.com/law
Patent Law for Non-Lawyers is an informative site detailing the patent process in the fields of biotechnology and engineering. The site assumes a working knowledge of the industry.

XXXVII.   Pharmaceutical Industry Associations (Drug Industry)

Academy of Pharmaceutical Physicians and Investigators (APPI)
500 Montgomery St., Ste. 800
Alexandria, VA 22314 US
Phone: 703-254-8100
Fax: 703-254-8101
Toll Free: 866-225-2779
E-mail Address: *office@acrpnet.org*
Web Address: www.appinet.org
The Academy of Pharmaceutical Physicians and Investigators (APPI) is an association that arose when the American Academy of Pharmaceutical Physicians and the Association of Clinical Research Professionals merged. It is a membership organization that provides scientific and educational activities on issues concerning pharmaceutical medicine.

Accreditation Council for Pharmacy Education (ACPE)
20 N. Clark St., Ste. 2500

Chicago, IL 60602-5109 US
Phone: 312-664-3575
Fax: 312-664-4652
E-mail Address: *info@acpe-accredit.org*
Web Address: www.acpe-accredit.org
The Accreditation Council for Pharmacy Education (ACPE) provides accreditation for pharmaceutical programs.

American Association of Pharmaceutical Sciences (AAPS)
2107 Wilson Blvd., Ste. 700
Arlington, VA 22201-3042 US
Phone: 703-243-2800
Fax: 703-243-9650
Web Address: www.aapspharmaceutica.com
The American Association of Pharmaceutical Scientists (AAPS) represents scientists in the pharmaceutical field. Members are given access to international forum, scientific programs, ongoing education, opportunities for networking and professional development.

American Association of Pharmacy Technicians (AAPT)
P.O. Box 1447
Greensboro, NC 27402 US
Phone: 336-333-9356
Fax: 336-333-9068
Toll Free: 877-368-4771
E-mail Address: *secretary@pharmacytechnician.com*
Web Address: www.pharmacytechnician.com
The American Association of Pharmacy Technicians (AAPT) provides leadership and represents the interests of pharmacy technicians in the United States. The group also offers continuing education programs and services to its members.

American Pharmaceutical Association (APhA)
1100 15th St. NW, Ste. 400
Washington, DC 20005-1707 US
Phone: 202-628-4410
Fax: 202-783-2351
Toll Free: 800-237-2742
E-mail Address: *infocenter@aphanet.org*
Web Address: www.pharmacist.com
American Pharmaceutical Association (APhA) is a national professional society that provides news and information to pharmacists.

Society for Automation in Pharmacy (ASAP)
492 Norristown Rd., Ste. 160
Blue Bell, PA 19422 US

Phone: 610-825-7783
Fax: 610-825-7641
E-mail Address: *will@computertalk.com*
Web Address: www.asapnet.org
The American Society for Automation in Pharmacy (ASAP) is a nonprofit organization that seeks to advance the application of computer technology in the pharmacists role as caregiver and in the efficient operation and management of a pharmacy.

American Society for Clinical Pharmacology and Therapeutics (ASCPT)
528 N. Washington St.
Alexandria, VA 22314 US
Phone: 703-836-6981
Fax: 703-836-5223
E-mail Address: *info@ascpt.org*
Web Address: www.ascpt.org
The American Society for Clinical Pharmacology and Therapeutics (ASCPT) is a nonprofit organization that is devoted to the discovery, development, regulation, and use of safe and effective medications necessary for the prevention and treatment of illness.

American Society for Pharmacology and Experimental Therapeutics (ASPET)
9650 Rockville Pike
Bethesda, MD 20814-3995 US
Phone: 301-634-7060
Fax: 301-634-7061
E-mail Address: *info@aspet.org*
Web Address: www.aspet.org
The American Society for Pharmacology and Experimental Therapeutics (ASPET) is a scientific society whose members conduct basic and clinical pharmacological research in academia, industry and the government.

American Society of Consultant Pharmacists (ASCP)
1321 Duke St.
Alexandria, VA 22314-3563 US
Phone: 703-739-1300
Fax: 800-220-1321
Toll Free: 800-355-2727
E-mail Address: *info@ascp.com*
Web Address: www.ascp.com
The American Society of Consultant Pharmacists (ASCP) is an international professional association that provides leadership, education, advocacy and resources to advance the practice of consultant and senior care pharmacy.

American Society of Pharmacognosy (ASP)
3149 Dundee Rd., Ste. 270
Northbrook, IL 60062 US
Phone: 623-202-3500
Fax: 847-656-2800
E-mail Address: *asphocog@aol.com*
Web Address: www.phcog.org
The American Society of Pharmacognosy (ASP) is a volunteer organization that promotes the growth and development of pharmacognosy, the study of the physical, chemical, biochemical and biological properties of drugs of natural origin and drugs from natural sources.

Association of the British Pharmaceutical Industry (ABPI)
12 Whitehall
London, SW1A 2DY UK
Phone: 44-870-890-4333
Fax: 44-20-7747-1414
Web Address: www.abpi.org.uk
The Association of the British Pharmaceutical Industry (ABPI) is a trade association that provides research and information for the British pharmaceuticals industry.

Canadian Pharmacists Association (CPHA)
1785 Alta Vista Dr.
Ottawa, ON K1G 3Y6 Canada
Phone: 613-523-7877
Fax: 613-523-0445
Toll Free: 800-917-9489
E-mail Address: *info@pharmacists.ca*
Web Address: www.pharmacists.ca
The Canadian Pharmacists Association (CPHA) is a professional organization providing drug information, pharmacy practice support material, patient information and news about the pharmacy industry.

Canadian Research-Based Pharmaceutical Companies Association (Rx&D)
55 Metcalfe St., Ste. 1220
Ottawa, ON K1P 6L5 Canada
Phone: 613-236-0455
Fax: 613-236-6756
E-mail Address: *info@canadapharma.org*
Web Address: www.canadapharma.org
The Canadian Research-Based Pharmaceutical Companies Association (Rx&D) is a trade organization providing news and information to the Canadian biotech industry.

Canadian Society for Pharmaceutical Sciences (CSPS)
3126 Dentistry/Pharmacy Ctr.
University of Alberta Campus
Edmonton, Alberta T6G 2N8 Canada
Phone: 780-492-0950
Fax: 780-492-0951
E-mail Address: csps@cspscanada.org
Web Address: www.cspscanada.org
The Canadian Society for Pharmaceutical Sciences (CSPS) is a nonprofit organization that works to advance pharmaceutical research. CSPS also maintains the Journal of Pharmacy and Pharmaceutical Sciences, an international online publication.

Controlled Release Society (CRS)
3340 Pilot Knob Rd.
St. Paul, MN 55421 US
Phone: 651-454-7250
Fax: 651-454-0766
E-mail Address: crs@scisoc.org
Web Address: www.controlledrelease.org
The Controlled Release Society (CRS) is an organization that promotes the science of the controlled delivery of bioactive substances.

Drug, Chemical & Associated Technologies Association (DCAT)
1 Washington Blvd., Ste. 7
Robbinsville, NJ 08691 US
Phone: 609-448-1000
Fax: 609-448-1944
Toll Free: 800-640-3228
E-mail Address: mtimony@dcat.org
Web Address: www.dcat.org
The Drug, Chemical & Associated Technologies Association (DCAT) is a business development association whose membership is made up of companies that manufacture, distribute or provide services to the pharmaceutical, chemical, nutritional and related industries.

Generic Pharmaceutical Association (GPhA)
2300 Clarendon Blvd., Ste. 400
Arlington, VA 22201 US
Phone: 703-647-2480
Web Address: www.gphaonline.org
The Generic Pharmaceutical Association (GPhA) represents the manufacturers and distributors of finished generic pharmaceutical products, manufacturers and distributors of bulk active pharmaceutical chemicals, and suppliers of other

goods and services to the generic pharmaceutical industry.

International Academy of Compounding Pharmacists (IACP)
4638 Riverstone Blvd.
Missouri City, TX 77459 US
Phone: 281-933-8400
Fax: 281-495-0602
Toll Free: 800-927-4227
E-mail Address: iacpinfo@iacprx.org
Web Address: www.iacprx.org
The International Academy of Compounding Pharmacists (IACP) is a nonprofit association that seeks to protect, promote and advance the art the customizing, compounding pharmacy profession.

International Association for Pharmaceutical Technology (APV)
Kurfurstenstrasse 59
Mainz, 55118 Germany
Phone: 49-6131-9769
Fax: 49-6131-97-6969
E-mail Address: apv@apv-mainz.de
Web Address: www.apv-mainz.de
The International Association for Pharmaceutical Technology (APV) is a nonprofit scientific association that is located in Mainz, Germany, and publishes its own scientific journal.

International Federation of Pharmaceutical Manufacturers Associations (IFPMA)
15 Ch. Louis-Dunant
P.O. Box 195, 1211
Geneva, 20 Switzerland
Phone: 41-22-338-32-00
Fax: 41-22-338-32-99
E-mail Address: admin@ifpma.org
Web Address: www.ifpma.org
The International Federation of Pharmaceutical Manufacturers Associations (IFPMA) is a nonprofit organization that represents the world's research-based pharmaceutical and biotech companies.

International Federation of Pharmaceutical Wholesalers (IFPW)
10569 Crestwood Dr.
Manassas, VA 20109 US
Phone: 703-331-3714
Fax: 703-331-3715
E-mail Address: info@ifpw.com
Web Address: www.ifpw.com

The International Federation of Pharmaceutical Wholesalers (IFPW) collaborates with its six member associations to promote the pharmaceutical distribution industry.

International Pharmaceutical Excipients Council of the Americas (IPEC-Americas)
1655 N. Fort Myer Dr., Ste. 700
Arlington, VA 22209 US
Phone: 703-875-2127
Fax: 703-525-5157
E-mail Address: *info@ipecamericas.org*
Web Address: www.ipecamericas.org
International Pharmaceutical Excipients Council of the Americas (IPEC-Americas) is a trade organization that promotes standardized approval criteria for drug inert ingredients, or excipients, among different nations. The organization also works to promote safe and useful excipients in the U.S.

International Pharmaceutical Federation (FIP)
Andries Bickerweg 5
P.O. Box 84200
The Hague, AE 2508 The Netherlands
Phone: 31-70-3021970
Fax: 31-70-3021999
E-mail Address: *fip@fip.org*
Web Address: www.fip.org
The International Pharmaceutical Federation (FIP) is a global federation of national associations representing 2 million pharmacists and pharmaceutical scientists around the world.

International Pharmaceutical Students Federation (IPSF)
P.O. Box 84200
Den Haag, 2506 AE The Netherlands
Phone: 31-70-302-1992
Fax: 31-70-302-1999
E-mail Address: *president@ipsf.org*
Web Address: www.ipsf.org
The International Pharmaceutical Students Federation (IPSF) is an organization that aims to promote the interests of pharmacy students and encourage international co-operation amongst them.

International Society for Pharmacoepidemiology (ISPE)
5272 River Rd., Ste. 630
Bethesda, MD 20816 US
Phone: 301-718-6500
Fax: 301-656-0989
E-mail Address: *ispe@paimgmt.com*

Web Address: www.pharmacoepi.org
The International Society for Pharmacoepidemiology (ISPE) is an international organization dedicated to the health of the public by advancing the study of the effects and determinants of pharmacology on epidemic diseases and to help provide risk benefit assessments on drugs with large scale distributions.

International Society of Regulatory Toxicology & Pharmacology (ISRTP)
6546 Belleview Dr.
Columbia, MD 21046-1054 US
Phone: 410-992-9083
Fax: 410-740-9181
E-mail Address: *c.carr65@comcast.net*
Web Address: www.isrtp.org
The International Society of Regulatory Toxicology & Pharmacology (ISRTP) is an association of professionals that mediates between policy makers and scientists in order to promote sound toxicologic and pharmacologic science as a basis for regulation affecting the environment and human safety and health.

International Union of Basic and Clinical Pharmacology (IUPHAR)
3901 Rainbow Blvd.
Mail Stop 4016
Kansas City, KS 66160 US
Phone: 913-588-7533
Fax: 913-588-7373
E-mail Address: *iuphar@kumc.edu*
Web Address: www.iuphar.org
The International Union of Basic and Clinical Pharmacology (IUPHAR) is a nonprofit association representing the interests of pharmacologists around the world.

Korean Research-based Pharmaceutical Industry Association (KRPIA)
201-6 Guui-Dong, Kwangjin-Gu
5th Fl., Dae Han Bldg.
Seoul, 143-200 Korea
Phone: 82-2-456-8553
Fax: 82-2-456-8320
E-mail Address: *krpia@krpia*
Web Address: www.krpia.or.kr
The Korean Research-based Pharmaceutical Industry Association (KRPIA) is an association of 24 research-based pharmaceutical companies operating in Korea.

LEEM (French Pharmaceutical Companies Association)
88 Rue de la Faisanderie
Paris, 75016 France
Phone: 01 45 03 88 88
Web Address: www.leem.org
LEEM (Les Entreprises du Médicament) represents the 335 pharmaceutical companies operating in France, engaged in the research and/or development of medicines for human use.

National Association of Boards of Pharmacy (NABP)
1600 Feehanville Dr.
Mount Prospect, IL 60056 US
Phone: 847-391-4406
Fax: 847-391-4502
E-mail Address: *exec-office@nabp.net*
Web Address: www.nabp.net
The National Association of Boards of Pharmacy (NABP) is the association of the member boards and jurisdictions in the field of pharmacy for the development, implementation, and enforcement of uniform standards for the purpose of protecting the public health.

National Association of Pharmaceutical Manufacturers (NAPM)
6 De Veer Ln.
Arcadia, Pretoria 0007 South Africa
Phone: 012-323-7529
Fax: 086-529-4245
E-mail Address: *napm@mweb.co.za*
Web Address: www.napm.co.za
The National Association of Pharmaceutical Manufacturers (NAPM) is a nonprofit trade association consisting of South African, generic-based pharmaceutical manufacturers and distributors.

Parenteral Drug Association (PDA)
4350 E. West Hwy., Ste. 200
Bethesda, MD 20814 US
Phone: 301-656-5900
Fax: 301-986-1093
E-mail Address: *swan@pda.org*
Web Address: www.pda.org
The Parenteral Drug Association (PDA) is a global provider of science, technology and regulatory information and education for the pharmaceutical and biopharmaceutical community.

Pharmaceutical Information and Pharmacovigilance Association (PIPA)
P.O. Box 254
Haslemere, Surrey GU27 9AF UK
Phone: 07531-899537
E-mail Address: *pipa@pipaonline.org*
Web Address: www.aiopi.org.uk
The Pharmaceutical Information and Pharmacovigilance Association (PIPA) is a professional organization that promotes the advancement of information in the pharmaceutical industry in the U.K.

Pharmaceutical Research and Manufacturers of America (PhRMA)
950 F St. NW, Ste. 300
Washington, DC 20004 US
Phone: 202-835-3400
Fax: 202-835-3414
Web Address: www.phrma.org
Pharmaceutical Research and Manufacturers of America (PhRMA) represents the nation's leading research-based pharmaceutical and biotechnology companies.

Pharmacy Council of India
Kotla Rd., Aiwan-E-Ghalib, Marg
Combined Councils' Bldg.
New Delhi, 110-002 India
Phone: 011-23239184
Fax: 011-23239184
E-mail Address: *pci@ndb.vsnl.net.in*
Web Address: www.pci.nic.in
The Pharmacy Council of India provides regulation of pharmacists under the Pharmacy Act and is a statutory body working under the Ministry of Health Family Welfare, Government of India.

PharmaSUG
421 New Parkside Dr.
Chapel Hill, NC 27516 US
E-mail Address: *syamala.ponnapalli@gmail.com*
Web Address: www.pharmasug.org
PharmaSUG is an organization for professional users of SAS software in the pharmaceuticals industry.

Royal Pharmaceutical Society of Great Britain (RPSGB)
1 Lambeth High St.
London, SE1 7JN UK
Phone: 020-7735-9141
Fax: 020-7735-7629
E-mail Address: *enquiries@rpsgb.org*

Web Address: www.rpsgb.org.uk
The Royal Pharmaceutical Society of Great Britain
(RPSGB) is the regulatory agency for pharmacists in
England, Wales and Scotland.

**Singapore Association of Pharmaceutical
Industries (SAPI)**
151 Chin Swee Rd.
02-13A/14 Manhattan House
169876 Singapore
Phone: 65-6738-0966
Fax: 65-6738-0977
E-mail Address: *FokTaiHung@sapi.org.sg*
Web Address: www.sapi.org.sg
Singapore Association of Pharmaceutical Industries
(SAPI) represents a wide spectrum of pharmaceutical
related businesses, namely the trading houses,
manufacturers, representative offices and pharmacies.

Society of Infectious Diseases Pharmacists (SIDP)
823 Congress Ave., Ste. 230
Austin, TX 78701 US
Phone: 512-479-0425
Fax: 512-495-9031
E-mail Address: *sidp@eami.com*
Web Address: www.sidp.org
The Society of Infectious Diseases Pharmacists
(SIDP) is an association of health professionals
dedicated to promoting the appropriate use of
antimicrobials, providing its members with
education, advocacy and leadership in all aspects of
the treatment of infectious diseases.

| XXXVIII.   Pharmaceutical Industry
             Resources (Drug Industry) |

American Institute of the History of Pharmacy
777 Highland Ave.
Madison, WI 53705-2222 US
Phone: 608-262-5378
E-mail Address: *AIHP@AIHP.org*
Web Address: www.pharmacy.wisc.edu/aihp/
The American Institute of the History of Pharmacy is
a nonprofit national organization that works to
advance knowledge of the role of pharmacy in
history through its programs and print publications.

Board of Pharmaceutical Specialties (BPS)
1100 15th St. NW, Ste. 400
Washington, DC 20005-1707 US
Phone: 202-429-7591
Fax: 202-429-6304
E-mail Address: *bps@aphanet.org*

Web Address: www.bpsweb.org
The Board of Pharmaceutical Specialties (BPS)
operates a certification program for specialized
clinical pharmacists.

European Medicines Agency (EMA)
7 Westferry Circus
Canary Wharf
London, E14 4HB UK
Phone: 44-2074188400
Fax: 44-2074188416
E-mail Address: *info@emea.europa.eu*
Web Address: www.emea.europa.eu
The European Medicines Agency (EMA) is the
European agency charged with approving new drugs
and monitoring the efficacy of existing drugs.

Pharmaportal.com
Web Address: www.pharmaportal.com
Pharmaportal.com is a pharmaceutical portal
containing information about the Pharmaceutical
Magazine, as well as links for executives in the
industry to meet each other.

Tufts Center for the Study of Drug Development
75 Kneeland St., Ste. 1100
Boston, MA 02111 US
Phone: 617-636-2170
Fax: 617-636-2425
E-mail Address: *csdd@tufts.edu*
Web Address: csdd.tufts.edu
The Tufts Center for the Study of Drug Development,
an affiliate of Tuft's University, provides analyses
and commentary on pharmaceutical issues. Its
mission is to improve the quality and efficiency of
pharmaceutical development, research and
utilization. It is famous, among other things, for its
analysis of the true total cost of developing and
commercializing a new drug. Tuft's Center conducts
research in areas of drug development, public policy
and regulation, and biotechnology.

United States Pharmacopeia
12601 Twinbrook Pkwy.
Rockville, MD 20852-1790 US
Phone: 800-227-8772
Fax: 301-816-8148
Toll Free: 301-881-0666
E-mail Address: *custsvc@usp.org*
Web Address: www.usp.org
The United States Pharmacopeia is the official public
standards-setting authority for all over-the-counter
and prescription medicines, dietary supplements and

other healthcare products manufactured and sold in the United States.

XXXIX.   Research & Development, Laboratories

Battelle Memorial Institute
505 King Ave.
Columbus, OH 43201-2693 US
Phone: 614-424-5853
Toll Free: 800-201-2011
Web Address: www.battelle.org
Battelle Memorial Institute serves commercial and governmental customers in developing new technologies and products. The institute adds technology to systems and processes for manufacturers; pharmaceutical and agrochemical industries; trade associations; and government agencies supporting energy, the environment, health, national security and transportation.

Commonwealth Scientific and Industrial Research Organization (CSRIO)
CSIRO Enquiries, Bag 10
Clayton South, Victoria 3169 Australia
Phone: 61-3-9545-2176
Fax: 61-3-9545-2175
E-mail Address: *enquiries@csiro.au*
Web Address: www.csiro.au
The Commonwealth Scientific and Industrial Research Organization (CSRIO) is Australia's national science agency and a leading international research agency. CSRIO performs research in Australia over a broad range of areas including agriculture, minerals and energy, manufacturing, communications, construction, health and the environment.

Computational Neurobiology Laboratory
The Salk Institute
10010 N. Torrey Pines Rd.
La Jolla, CA 92037 US
Phone: 858-453-4100
E-mail Address: *sejnowski@salk.edu*
Web Address: www.cnl.salk.edu
The Computational Neurobiology Laboratory at The Salk Institute strives to understand the computational resources of the brain from the biophysical to the systems levels.

Council of Scientific & Industrial Research (CSIR)
2 Rafi Marg

Anusandhan Bhawan
New Delhi, 110 001 India
Phone: 011-23710618
Fax: 011-23713011
E-mail Address: *itweb@csir.res.in*
Web Address: www.csir.res.in
The Council of Scientific & Industrial Research (CSIR) is a government-funded organization that promotes research and development initiatives in India. It operates in the fields of energy, biotechnology, space, science and technology.

German Cancer Research Center
Im Neuenheimer Feld 280
Heidelberg, D-69120 Germany
Phone: 49 (0) 6221 420
Fax: 49 (0) 6221 422995
Web Address: www.dkfz.de
The German Cancer Research Center (Deutsches Krebsforschungszentrum, DKFZ) is the largest biomedical research institute in Germany and is a member of the Helmholtz Association of National Research Centers. More than 2,000 staff members, including 850 scientists, are investigating the mechanisms of cancer and are working to identify cancer risk factors. They provide the foundations for developing novel approaches in the prevention, diagnosis, and treatment of cancer. In addition, the staff of the Cancer Information Service (KID) offers information about the widespread disease of cancer for patients, their families, and the general public.

National Research Council Canada (NRC)
NRC Communications & Corp. Rel.
1200 Montreal Rd., Bldg. M-58
Ottawa, ON K1A 0R6 Canada
Phone: 613-993-9101
Fax: 613-952-9907
Toll Free: 877-672-2672
E-mail Address: *info@nrc-cnrc.gc.ca*
Web Address: www.nrc-cnrc.gc.ca
National Research Council Canada (NRC) is a government organization of 20 research institutes that carry out multidisciplinary research with partners in industries and sectors key to Canada's economic development.

SRI International
333 Ravenswood Ave.
Menlo Park, CA 94025-3493 US
Phone: 650-859-2000
E-mail Address: *ellie.javadi@sri.com*
Web Address: www.sri.com

SRI International is a nonprofit organization offering a wide range of services, including engineering services, information technology, pure and applied physical sciences, product development, pharmaceutical discovery, biopharmaceutical discovery and policy issues. SRI conducts research for commercial and governmental customers.

XL. Robotics Associations

Laboratory Robotics Interest Group (LRIG)
1730 W. Circle Dr.
Martinsville, NJ 08836-2147 US
Phone: 732-302-1038
Fax: 732-875-0270
E-mail Address: *andy.zaayenga@lab-robotics.org*
Web Address: www.lab-robotics.org
Laboratory Robotics Interest Group (LRIG) is a membership group focused on the application of robotics in the laboratory.

XLI. Science & Technology Resources

Technology Review
1 Main St., 7th Fl.
Cambridge, MA 02142 US
Phone: 617-475-8000
Fax: 617-475-8042
Toll Free: 800-877-5230
Web Address: www.technologyreview.com
Technology Review, an MIT enterprise, publishes tech industry news, covers innovation and writes in-depth articles about research, development and cutting-edge technologies.

XLII. Technology Transfer Associations

Association of University Technology Managers (AUTM)
111 Deer Lake Rd., Ste. 100
Deerfield, IL 60015 US
Phone: 847-559-0846
Fax: 847-480-9282
E-mail Address: *info@autm.net*
Web Address: www.autm.net
The Association of University Technology Managers (AUTM) is a nonprofit professional association whose members belong to over 350 research institutions, universityies, teaching hospitals, government agencies and corporations from 45 countries. The association's mission is to advance the field of technology transfer, and enhance members'

ability to bring academic and nonprofit research to people around the world.

Federal Laboratory Consortium for Technology Transfer
950 North Kings Highway, Ste. 208
Cherry Hill, NJ 08304 US
Phone: 856-667-7727
Fax: 856-667-8009
E-mail Address: *flcmso@federallabs.org*
Web Address: www.federallabs.org
In keeping with the aims of the Federal Technology Transfer Act of 1986 and other related legislation, the Federal Laboratory Consortium (FLC) works to facilitate the sharing of research results and technology developments between federal laboratories and the mainstream U.S. economy. FLC affiliates include federal laboratories, large and small businesses, academic and research institutions, state and local governments, and various federal agencies. The group has regional support offices and local contacts throughout the U.S.

Licensing Executives Society (U.S.A. and Canada), Inc.
1800 Diagonal Rd., Ste. 280
Alexandria, VA 22314 US
Phone: 703-836-3106
Fax: 703-836-3107
E-mail Address: *info@les.org*
Web Address: www.lesusacanada.org
Licensing Executives Society (U.S.A. and Canada), Inc., established in 1965, is a professional association composed of about 5,000 members who work in fields related to the development, use, transfer, manufacture and marketing of intellectual property. Members include executives, lawyers, licensing consultants, engineers, academic researchers, scientists and government officials. The society is part of the larger Licensing Executives Society International, Inc. (same headquarters address), with a worldwide membership of some 12,000 members in 30 national societies, representing approximately 80 countries.

The State Science and Technology Institute
5015 Pine Creek Dr.
Westerville, OH 43081 US
Phone: 614-901-1690
Fax: 614-901-1696
Web Address: www.ssti.org
The State Science and Technology Institute (SSTI) is a national nonprofit group that serves as a resource

for technology-based economic development. In addition to the information on its web site, the Institute publishes a free weekly digest of news and issues related to technology-based economic development efforts, as well as a members-only publication listing application information, eligibility criteria and submission deadlines for a variety of funding oportunities, federal and otherwise.

XLIII. Trade Associations-General

BUSINESSEUROPE
168 Ave. de Cortenbergh
Brussels, 1000 Belgium
Phone: 32-0-2-237-65-11
Fax: 32-0-2-231-14-45
E-mail Address: *main@businesseurope.eu*
Web Address: www.businesseurope.eu
BUSINESSEUROPE is a major European trade federation that operates in a manner similar to a chamber of commerce. Its members are the central national business federations of the 34 countries throughout Europe from which they come. Companies cannot become direct members of BUSINESSEUROPE, though there is a support group which offers the opportunity for firms to encourage BUSINESSEUROPE objectives in various ways.

The Associated Chambers of Commerce and Industry of India (ASSOCHAM)
1 Community Ctr. Zamrudpur
Kailash Colony
New Delhi, 110 048 India
Phone: 46550555
Fax: 46536481
E-mail Address: *assocham@nic.in*
Web Address: www.assocham.org
The Associated Chambers of Commerce and Industry of India (ASSOCHAM) has a membership of more than 300 chambers and trade associations and serves members from all over India.

XLIV. Trade Associations-Global

World Trade Organization (WTO)
Centre William Rappard
Rue de Lausanne 154
Geneva 21, CH-1211 Switzerland
Phone: 41-22-739-51-11
Fax: 41-22-731-42-06
E-mail Address: *enquiries@wto.og*
Web Address: www.wto.org

The World Trade Organization (WTO) is a global organization dealing with the rules of trade between nations. To become a member, nations must agree to abide by certain guidelines. Membership increases a nation's ability to import and export efficiently.

XLV. U.S. Government Agencies

Bureau of Economic Analysis (BEA)
1441 L St. NW
Washington, DC 20230 US
Phone: 202-606-9900
E-mail Address: *customerservice@bea.gov*
Web Address: www.bea.gov
The Bureau of Economic Analysis (BEA), an agency of the U.S. Department of Commerce, is the nation's economic accountant, preparing estimates that illuminate key national, international and regional aspects of the U.S. economy.

Bureau of Labor Statistics (BLS)
2 Massachusetts Ave. NE
Washington, DC 20212-0001 US
Phone: 202-691-5200
Web Address: stats.bls.gov
The Bureau of Labor Statistics (BLS) is the principal fact-finding agency for the Federal Government in the field of labor economics and statistics. It is an independent national statistical agency that collects, processes, analyzes and disseminates statistical data to the American public, U.S. Congress, other federal agencies, state and local governments, business and labor. The BLS also serves as a statistical resource to the Department of Labor.

Center for Biologics Evaluation and Research (CBER)
1401 Rockville Pike, Ste. 200N
Rockville, MD 20852-1448 US
Phone: 301-827-1800
Toll Free: 800-835-4709
E-mail Address: *octma@fda.hhs.gov*
Web Address: www.fda.gov/Cber
The Center for Biologics Evaluation and Research (CBER) regulates biologic products for use in humans. It is a source for a broad variety of data on drugs, including blood products, counterfeit drugs, exports, drug shortages, recalls and drug safety.

Center for Devices and Radiological Health (CDRH)
1350 Piccard Dr., HFZ-210
Rockville, MD 20850 US

Phone: 301-276-3103
Fax: 301-443-8818
Toll Free: 800-638-2041
E-mail Address: *dsmica@cdrh.fda.gov*
Web Address: www.fda.gov/cdrh
The Center for Devices and Radiological Health
(CDRH) is a unit of the FDA that regulates medical
devices and radiation-emitting products.

Center for Drug Evaluation and Research (CDER)

5600 Fishers Ln., HFD-240
Rockville, MD 20852-9787 US
Phone: 301-796-3400
Toll Free: 888-463-6332
E-mail Address: *druginfo@cder.fea.gov*
Web Address: www.fda.gov/cder
The Center for Drug Evaluation and Research
(CDER) is a division of the FDA that offers a wealth
of information on new drug approval statistics and
the approval process.

Center for Food Safety and Applied Nutrition-Biotechnology (CFSAN)

5100 Paint Branch Pkwy., HFS-555
College Park, MD 20740-3835 US
Toll Free: 888-723-3366
Web Address: www.cfsan.fda.gov/list.html
The Center for Food Safety and Applied Nutrition-
Biotechnology (CFSAN) is an FDA site that provides
information about genetically engineered food
products.

Centers for Disease Control and Prevention (CDC)

1600 Clifton Rd.
Atlanta, GA 30333 US
Phone: 404-639-3311
Toll Free: 800-232-4636
E-mail Address: *cdcinfo@cdc.gov*
Web Address: www.cdc.gov
The Centers for Disease Control and Prevention
(CDC), headquartered in Atlanta and established as
an operating health agency within the U.S. Public
Health Service, is the federal agency charged with
protecting the public health of the nation by
providing leadership and direction in the prevention
and control of diseases and other preventable
conditions and responding to public health
emergencies.

National Cancer Institute (NCI)

6116 Executive Blvd., Ste. 3036A
Bethesda, MD 20892-8322 US
Toll Free: 800-422-6237
E-mail Address: *ncergovstaff@mail.nih.gov.*
Web Address: www.cancer.gov
The National Cancer Institute (NCI) is the Federal
Government's principal agency for cancer research
and training.

National Center for Biotechnology Information (NCBI)

U.S. National Library of Medicine
National Institues of Health
8600 Rockville Pike, Bldg. 38A
Bethesda, MD 20894 US
Phone: 301-496-2475
Fax: 301-480-9241
E-mail Address: *info@ncbi.nlm.nih.gov*
Web Address: www.ncbi.nlm.nih.gov
The National Center for Biotechnology Information
(NCBI) creates public databases, conducts research in
computational biology, develops software for
analyzing genome data and disseminates biomedical
information. It is part of the U.S. National Library of
Medicine (NLM), which is located on the campus of
the National Institutes of Health (NIH).

National Center for Research Resources (NCRR)

6701 Democracy Blvd.
MSC 4874, 1 Democracy Plz., 9th Fl.
Bethesda, MD 20892-4874 US
Phone: 301-435-0888
Fax: 301-480-3558
E-mail Address: *info@ncrr.nih.gov*
Web Address: www.ncrr.nih.gov
The National Center for Research Resources (NCRR)
supports primary health and life sciences research to
create and develop critical resources, models and
technologies.

National Center for Toxicological Research

3900 NCTR Rd.
Jefferson, AR 72079 US
Phone: 870-543-7130
Toll Free: 800-216-7331
E-mail Address: *rhuber@fda.hhs.gov*
Web Address: www.fda.gov/nctr
The mission of the National Center for Toxicological
Research is to conduct peer-reviewed scientific
research that supports and anticipates the FDA's
current and future regulatory needs.

National Heart, Lung, and Blood Institute (NHLBI)
P.O. Box 30105
Bethesda, MD 20824-0105 US
Phone: 301-592-8573
Fax: 240-629-3246
E-mail Address: *nhlbiinfo@nhlbi.nih.gov*
Web Address: www.nhlbi.nih.gov
The National Heart, Lung, and Blood Institute (NHLBI) provides leadership for a national program in diseases of the heart, blood vessels, lung and blood; blood resources; and sleep disorders.

National Institute of Standards and Technology (NIST)
100 Bureau Dr., Stop 1070
Gaithersburg, MD 20899-1070 US
Phone: 301-975-6478
E-mail Address: *inquiries@nist.gov*
Web Address: www.nist.gov
The National Institute of Standards and Technology (NIST) is an agency of the U.S. Department of Commerce's Technology Administration. It works with various industries to develop and apply technology, measurements and standards.

National Institutes of Health (NIH)
9000 Rockville Pike
Bethesda, MD 20892 US
Phone: 301-496-4000
Toll Free: 800-411-1222
E-mail Address: *nihinfo@od.nih.gov*
Web Address: www.nih.gov
The National Institutes of Health (NIH) is the leader of medical and behavioral research for the nation, with over 15 institutes ranging from the National Cancer Institute to the National Institute of Mental Health.

National Science Foundation (NSF)
4201 Wilson Blvd.
Arlington, VA 22230 US
Phone: 703-292-5111
Toll Free: 800-877-8339
E-mail Address: *info@nsf.gov*
Web Address: www.nsf.gov
The National Science Foundation (NSF) is an independent U.S. government agency responsible for promoting science and engineering. The foundation provides grants and funding for research.

U.S. Census Bureau
4600 Silver Hill Rd.
Washington, DC 20233-8800 US
Phone: 301-763-4636
Fax: 301-457-3670
Toll Free: 800-923-8282
E-mail Address: *pio@census.gov*
Web Address: www.census.gov
The U.S. Census Bureau is the official collector of data about the people and economy of the U.S. Founded in 1790, it provides official social, demographic and economic information.

U.S. Department of Commerce (DOC)
1401 Constitution Ave. NW
Washington, DC 20230 US
Phone: 202-482-2000
E-mail Address: *cgutierrez@doc.gov*
Web Address: www.commerce.gov
The U.S. Department of Commerce (DOC) regulates trade and provides valuable economic analysis of the economy.

U.S. Department of Labor (DOL)
Frances Perkins Bldg.
200 Constitution Ave. NW
Washington, DC 20210 US
Toll Free: 866-487-2365
Web Address: www.dol.gov
The U.S. Department of Labor (DOL) is the government agency responsible for labor regulations. This site provides tools to help citizens find out whether companies are complying with family and medical-leave requirements.

U.S. Food and Drug Administration (FDA)
5600 Fishers Ln.
Rockville, MD 20857 US
Toll Free: 888-463-6332
Web Address: www.fda.gov
The U.S. Food and Drug Administration (FDA) promotes and protects the public health by helping safe and effective products reach the market in a timely way and by monitoring products for continued safety after they are in use. It regulates both prescription and over-the-counter drugs as well as medical devices and food products.

U.S. Securities and Exchange Commission (SEC)
100 F St. NE
Washington, DC 20549 US
Phone: 202-551-6000
Toll Free: 888-732-6585

E-mail Address: *publicinfo@sec.gov*
Web Address: www.sec.gov
The U.S. Securities and Exchange Commission
(SEC) is a nonpartisan, quasi-judicial regulatory
agency responsible for administering federal
securities laws. These laws are designed to protect
investors in securities markets and ensure that they
have access to disclosure of all material information
concerning publicly traded securities. Visitors to the
web site can access the EDGAR database of
corporate financial and business information.

Chapter 4

THE BIOTECH 350:
WHO THEY ARE AND HOW THEY WERE
CHOSEN

Includes Indexes by Company Name, Industry & Location,
And a Complete Table of Sales, Profits and Ranks

The companies chosen to be listed in PLUNKETT'S BIOTECH & GENETICS INDUSTRY ALMANAC comprise a unique list. THE BIOTECH 350 (the actual count is 364 companies) were chosen specifically for their dominance in the many facets of biotechnology and genetics in which they operate. Complete information about each firm can be found in the "Individual Profiles," beginning at the end of this chapter. These profiles are in alphabetical order by company name.

THE BIOTECH 350 includes leading companies from all parts of the United States as well as many other nations, and from all biotech and genetics related industry segments: pharmaceuticals; diagnostics; research and development; and support services.

Simply stated, the list contains 364 of the largest, most successful, fastest growing firms in biotech and related industries in the world. To be included in our list, the firms had to meet the following criteria:

1) Generally, these are corporations based in the U.S., however, the headquarters of 76 firms are located in other nations.

2) Prominence, or a significant presence, in biotech, genetics and supporting fields. (See the following Industry Codes section for a complete list of types of businesses that are covered).

3) The companies in THE BIOTECH 350 do not have to be exclusively in the biotech and genetics field.

4) Financial data and vital statistics must have been available to the editors of this book, either directly from the company being written about or from outside sources deemed reliable and accurate by the editors. A small number of companies that we would like to have included are not listed because of a lack of sufficient, objective data.

INDUSTRY LIST, WITH CODES

**This book refers to the following list of unique
industry codes, based on the 2007 NAIC code
system (NAIC is used by many analysts as a
replacement for older SIC codes because NAIC is
more specific to today's industry sectors, see
www.census.gov/NAICS). Companies profiled in
this book are given a primary NAIC code,
reflecting the main line of business of each firm.**

Agriculture

Agriculture
11511   Agricultural Crop Production Support, Seeds,
             Fertilizers

Energy

Alternative Energy
325193   Ethanol Fuel Manufacturing

Financial Services

Stocks & Investments
523110   Investment Banking

Health Care

Health Products, Manufacturing
325411   Medicinals & Botanicals, Manufacturing
325412   Drugs (Pharmaceuticals), Discovery &
             Manufacturing
325412A   Drug Delivery Systems
325412B   Veterinary Products Manufacturing
325413   Diagnostic Substances Manufacturing
325414   Biological Products, Manufacturing
33911   Medical/Dental/Surgical Equipment &
             Supplies, Manufacturing

Health Products, Wholesale Distribution
423450   Medical/Dental/Surgical Equipment &
             Supplies, Distribution
***Health Care-Clinics, Labs and
Organizations***
6215   Laboratories & Diagnostic Services--Medical

InfoTech

Computers & Electronics Manufacturing
3345   Instrument Manufacturing, including
             Measurement, Control, Test & Navigational
Software
511210D   Computer Software, Healthcare &
             Biotechnology

Manufacturing

Textiles Manufacturing
313   Textiles, Fabrics, Sheets/Towels,
             Manufacturing
Chemicals
325   Chemicals, Manufacturing
325199   Other Basic Organic Chemicals/Biofuels

Nanotechnology

Nanotechnology
541711   Nanotechnology-Biotech/Health

Services

Consulting & Professional Services
541690   Consulting--Scientific & Technical
541712   Research & Development-Physical,
             Engineering & Life Sciences
541910   Market Research

INDEX OF RANKINGS WITHIN INDUSTRY GROUPS

Company	Industry Code	2009 Sales (U.S. $ thousands)	Sales Rank	2009 Profits (U.S. $ thousands)	Profits Rank
Agricultural Crop Production Support, Seeds, Fertilizers					
ARCADIA BIOSCIENCES INC	11511				
DOW AGROSCIENCES LLC	11511				
DU PONT AGRICULTURE & NUTRITION	11511	8,287,000	3	1,224,000	3
MONSANTO CO	11511	11,724,000	1	2,109,000	1
PIONEER HI-BRED INTERNATIONAL INC	11511				
S&W SEED CO	11511				
SYNGENTA AG	11511	10,992,000	2	1,374,000	2
Biological Products, Manufacturing					
ALDAGEN INC	325414				
ALPHARMA ANIMAL HEALTH	325414				
ANIKA THERAPEUTICS INC	325414	40,136	5	3,688	4
BIOHEART INC	325414	400	9	-3,800	6
CELLARTIS AB	325414				
CELLULAR DYNAMICS INTERNATIONAL	325414				
CSL LIMITED	325414	4,387,910	2	997,925	1
EISAI CO LTD	325414	8,244,890	1	502,850	2
GENENCOR INTERNATIONAL	325414				
GTC BIOTHERAPEUTICS INC	325414	2,826	7	-26,953	8
JUBILANT BIOSYS LTD	325414				
LIFECELL CORPORATION	325414				
LIFECORE BIOMEDICAL INC	325414				
NOVOZYMES	325414	1,544,030	3	218,230	3
ORGANOGENESIS INC	325414				
POLYDEX PHARMACEUTICALS	325414	4,825	6	-1,326	5
SERACARE LIFE SCIENCES INC	325414	44,434	4	-15,379	7
STEMCELLS INC	325414	608	8	-27,026	9
VIACORD INC	325414				
Chemicals, Manufacturing					
AKZO NOBEL NV	325	18,566,600	4	1,162,670	4
BASF SE	325	67,557,500	1	1,879,080	1
BAYER AG	325	42,346,100	2	1,846,390	2
BAYER CORP	325				
E I DU PONT DE NEMOURS & CO (DUPONT)	325	26,109,000	3	1,755,000	3
INTERNATIONAL ISOTOPES INC	325	6,123	7	-4,633	7
LONZA GROUP	325	2,536,130	5	152,730	6
SIGMA-ALDRICH CORP	325	2,147,600	6	346,700	5
Computer Software, Healthcare & Biotechnology					
ACCELRYS INC	511210D	80,981	4	94	4
CEGEDIM SA	511210D	1,161,910	1	72,710	1
ERESEARCH TECHNOLOGY INC	511210D	93,823	3	10,687	2
MEDIDATA SOLUTIONS INC	511210D	140,400	2	5,182	3
PHARSIGHT CORP	511210D				
TRIPOS INTERNATIONAL	511210D				

Company	Industry Code	2009 Sales (U.S. $ thousands)	Sales Rank	2009 Profits (U.S. $ thousands)	Profits Rank
Consulting--Scientific & Technical					
I3 CANREG	541690				
Diagnostic Substances Manufacturing					
AFFYMETRIX INC	325413	327,094	5	-23,909	19
AVIVA BIOSCIENCES CORP	325413				
BIOSITE INC	325413				
CALIPER LIFE SCIENCES	325413	130,412	10	-8,225	15
CELSIS INTERNATIONAL PLC	325413	52,500	15	6,800	10
CEPHEID	325413	170,627	8	-22,502	18
DECODE GENETICS LTD	325413				
GENOMIC HEALTH INC	325413	194,548	7	-9,411	16
GEN-PROBE INC	325413	489,302	4	91,783	5
HEMAGEN DIAGNOSTICS INC	325413	5,419	19	-808	12
HYCOR BIOMEDICAL INC	325413				
IDEXX LABORATORIES INC	325413	1,031,633	2	122,225	3
IMMUNOMEDICS INC	325413	30,021	17	2,274	11
LIFE TECHNOLOGIES CORP	325413	3,280,344	1	144,594	1
LUMINEX CORPORATION	325413	120,643	12	17,729	7
MERIDIAN BIOSCIENCE INC	325413	148,274	9	32,759	6
NEOGEN CORPORATION	325413	118,721	13	13,874	8
NEOPROBE CORPORATION	325413	9,518	18	-39,606	20
ORASURE TECHNOLOGIES INC	325413	77,026	14	-7,813	14
PROMEGA CORP	325413				
QIAGEN NV	325413	1,009,825	3	137,767	2
SEQUENOM INC	325413	37,863	16	-71,012	21
SIEMENS HEALTHCARE DIAGNOSTICS	325413				
SPECTRAL DIAGNOSTICS INC	325413	3,283	20	-2,764	13
TECHNE CORP	325413	263,956	6	105,242	4
THERMO SCIENTIFIC	325413				
TRINITY BIOTECH PLC	325413	125,907	11	11,824	9
VERMILLION INC	325413	0		-22,048	17
Drug Delivery Systems					
ALKERMES INC	325412A	326,839	4	130,505	3
ARADIGM CORPORATION	325412A	4,883	14	-13,427	12
BIOVAIL CORPORATION	325412A	820,430	2	176,455	2
DEPOMED INC	325412A	57,728	7	-22,008	15
DURECT CORP	325412A	24,293	10	-30,288	16
EMISPHERE TECHNOLOGIES	325412A	92	19	-21,243	14
FLAMEL TECHNOLOGIES SA	325412A	42,100	8	-11,400	11
GENEREX BIOTECHNOLOGY	325412A	1,119	17	-45,812	19
HOSPIRA INC	325412A	3,879,300	1	403,900	1
IMPAX LABORATORIES INC	325412A	358,409	3	50,061	4
INSITE VISION INC	325412A	9,798	13	-14,155	13
MDRNA INC	325412A	14,732	12	-8,046	10
NANOBIO CORPORATION	325412A				
NEKTAR THERAPEUTICS	325412A	71,931	6	-102,519	20
NEOPHARM INC	325412A	0		-7,461	9
NEXMED INC	325412A	2,974	15	-32,043	17

Company	Industry Code	2009 Sales (U.S. $ thousands)	Sales Rank	2009 Profits (U.S. $ thousands)	Profits Rank
NOVAVAX INC	325412A	325	18	-38,374	18
NOVEN PHARMACEUTICALS	325412A				
ONCOGENEX PHARMACEUTICALS	325412A	25,539	9	-5,476	7
PENWEST PHARMACEUTICALS	325412A	23,812	11	-1,500	5
SKYEPHARMA PLC	325412A	80,470	5	-2,020	6
SOLIGENIX INC	325412A	2,800	16	-6,000	8
Drugs (Pharmaceuticals), Discovery & Manufacturing					
3SBIO INC	325412	46,429	87	12,224	56
4SC AG	325412	2,470	135	-21,380	117
ABBOTT LABORATORIES	325412	30,764,700	8	5,745,800	10
ACCESS PHARMACEUTICALS	325412	352	147	-17,340	110
ACTELION LTD	325412	1,619,800	35	284,470	30
ADAMIS PHARMACEUTICALS	325412	660	144	-1,895	76
ADOLOR CORP	325412	37,361	93	-47,914	145
AEOLUS PHARMACEUTICALS	325412	0		-2,296	80
AETERNA ZENTARIS INC	325412	61,870	82	-24,190	126
AKORN INC	325412	75,891	79	-25,306	129
ALCON INC	325412	6,499,000	20	2,007,000	15
ALEXION PHARMACEUTICALS	325412	386,800	49	295,166	29
ALIMERA SCIENCES INC	325412				
ALLERGAN INC	325412	4,447,600	23	621,300	23
ALLIANCE PHARMACEUTICAL	325412	79	154	-1,203	74
ALLOS THERAPEUTICS INC	325412	3,585	132	-73,553	151
ALSERES PHARMACEUTICALS	325412	0		-10,777	99
ALTANA AG	325412	1,605,500	37	14,130	54
AMARIN CORPORATION PLC	325412				
AMGEN INC	325412	14,642,000	13	4,605,000	11
AMYLIN PHARMACEUTICALS	325412	753,993	44	-186,256	165
ANGIOTECH PHARMACEUTICALS INC	325412	279,678	59	-22,868	121
ANTIGENICS INC	325412	3,334	133	-30,318	134
APP PHARMACEUTICALS INC	325412				
ARENA PHARMACEUTICALS INC	325412	10,387	123	-138,387	163
ARIAD PHARMACEUTICALS INC	325412	8,302	128	-80,008	153
ARQULE INC	325412	25,198	104	-36,136	138
ASTELLAS PHARMA INC	325412	10,439,500	15	1,848,410	18
ASTRAZENECA CANADA	325412				
ASTRAZENECA PLC	325412	32,804,000	7	7,544,000	7
AVANIR PHARMACEUTICALS	325412	4,176	131	-21,996	119
AVAX TECHNOLOGIES INC	325412				
AVI BIOPHARMA INC	325412	17,585	111	-25,159	128
AXCAN PHARMA INC	325412				
BAYER SCHERING PHARMA AG	325412				
BIOCRYST PHARMACEUTICALS	325412	74,589	80	-13,452	105
BIOGEN IDEC INC	325412	4,377,300	24	977,100	19
BIOMARIN PHARMACEUTICAL	325412	324,656	55	-488	73
BIOTIME INC	325412	1,925	138	-5,144	90
BRISTOL-MYERS SQUIBB CO	325412	18,808,000	11	10,612,000	3
CAMBREX CORP	325412	236,277	63	10,392	59

Company	Industry Code	2009 Sales (U.S. $ thousands)	Sales Rank	2009 Profits (U.S. $ thousands)	Profits Rank
CANGENE CORP	325412	224,910	65	56,420	48
CARACO PHARMACEUTICAL LABORATORIES	325412	337,177	53	20,537	51
CARDIOME PHARMA CORP	325412	54,672	83	-1,276	75
CARDIUM THERAPEUTICS INC	325412	445	146	-11,680	102
CELGENE CORP	325412	2,689,893	31	766,747	21
CELL THERAPEUTICS INC	325412	80	152	-95,400	158
CELLDEX THERAPEUTICS INC	325412	15,180	112	-36,525	139
CEL-SCI CORPORATION	325412	80	153	-40,910	141
CEPHALON INC	325412	2,192,308	32	210,727	36
CERUS CORPORATION	325412	17,982	110	-24,135	125
CHIRON CORP	325412				
COLUMBIA LABORATORIES INC	325412	32,196	98	-21,870	118
COMPUGEN LTD	325412	250	148	-3,831	85
CORTEX PHARMACEUTICALS	325412	0		-10,788	100
CRUCELL NV	325412	338,163	52	23,938	50
CUBIST PHARMACEUTICALS	325412	562,144	46	79,600	45
CUMBERLAND PHARMACEUTICALS	325412	43,537	88	5,777	64
CURAGEN CORPORATION	325412				
CURIS INC	325412	8,590	127	-9,823	98
CYPRESS BIOSCIENCE INC	325412	27,335	103	-28,252	131
CYTRX CORPORATION	325412	9,500	125	-4,800	87
DAIICHI SANKYO CO LTD	325412	8,881,960	18	-2,272,830	170
DAINIPPON SUMITOMO PHARMA	325412	3,086,110	28	233,610	34
DENDREON CORPORATION	325412	101	151	-220,161	167
DISCOVERY LABORATORIES	325412	0		-30,240	133
DIVI'S LABORATORIES LIMITED	325412				
DR REDDY'S LABORATORIES	325412	1,365,000	40	-102,000	159
DSM PHARMACEUTICALS INC	325412				
DUSA PHARMACEUTICALS INC	325412	29,808	100	-2,508	81
DYAX CORP	325412	21,643	106	-62,419	146
ELAN CORP PLC	325412	1,113,000	42	-176,200	164
ELI LILLY & COMPANY	325412	21,836,000	10	4,328,800	12
ENDO PHARMACEUTICALS HOLDINGS INC	325412	1,460,841	38	266,336	31
ENTREMED INC	325412	5,284	129	-8,216	95
ENZO BIOCHEM INC	325412	89,572	76	-23,564	123
ENZON PHARMACEUTICALS INC	325412	184,622	66	683	71
EUSA PHARMA (USA) INC	325412				
EXELIXIS INC	325412	151,759	67	-135,220	162
FOREST LABORATORIES INC	325412	3,636,055	27	767,743	20
GALDERMA PHARMA SA	325412				
GENENTECH INC	325412				
GENTA INC	325412	218	149	-86,301	156
GENVEC INC	325412	13,857	114	-18,362	113
GENZYME CORP	325412	4,515,525	22	422,300	27
GENZYME ONCOLOGY	325412				
GERON CORPORATION	325412	1,726	140	-70,184	149
GILEAD SCIENCES INC	325412	7,011,383	19	2,635,755	13

Company	Industry Code	2009 Sales (U.S. $ thousands)	Sales Rank	2009 Profits (U.S. $ thousands)	Profits Rank
GLAXOSMITHKLINE PLC	325412	43,608,100	5	8,502,420	5
HARBOR BIOSCIENCES INC	325412	0		-15,626	106
HEMISPHERX BIOPHARMA INC	325412	1,083	141	-12,446	103
HI-TECH PHARMACAL CO INC	325412	108,651	72	9,817	60
HUMAN GENOME SCIENCES	325412	275,749	60	5,659	66
IDERA PHARMACEUTICALS INC	325412	34,518	96	7,546	62
IMCLONE SYSTEMS INC	325412				
IMMTECH PHARMACEUTICALS	325412	2,383	136	-6,502	92
IMMUNOGEN INC	325412	27,988	102	-31,937	136
INCYTE CORP	325412	9,265	126	-211,870	166
INSMED INCORPORATED	325412	10,373	124	118,350	40
INSPIRE PHARMACEUTICALS	325412	92,159	75	-39,976	140
INTELLIPHARMACEUTICS INTERNATIONAL INC	325412	700	143	-2,000	78
INTERMUNE INC	325412	48,700	86	-116,020	161
IRONWOOD PHARMACEUTICALS INC	325412	36,102	95	-71,185	150
ISIS PHARMACEUTICALS INC	325412	121,600	70	150,672	38
JAZZ PHARMACEUTICALS	325412	128,449	69	-6,836	93
JOHNSON & JOHNSON	325412	61,897,000	1	12,266,000	2
KENDLE INTERNATIONAL INC	325412	551,912	47	15,237	53
KERYX BIOPHARMACEUTICALS	325412	25,194	105	10,485	58
KING PHARMACEUTICALS INC	325412	1,776,500	34	91,953	43
KV PHARMACEUTICAL CO	325412	312,327	57	-313,627	168
LA JOLLA PHARMACEUTICAL	325412	0		-8,634	97
LANNETT COMPANY INC	325412	119,002	71	6,534	63
LEXICON PHARMACEUTICALS	325412	10,700	121	-82,780	155
LIGAND PHARMACEUTICALS	325412	38,940	91	-1,948	77
LORUS THERAPEUTICS INC	325412	170	150	-8,610	96
MANHATTAN PHARMACEUTICALS INC	325412	0		-2,793	82
MAXYGEN INC	325412	36,376	94	-32,157	137
MEDICINES CO (THE)	325412	404,241	48	-76,229	152
MEDICIS PHARMACEUTICAL	325412	571,915	45	75,951	47
MEDIMMUNE INC	325412				
MERCK & CO INC	325412	27,428,300	9	13,024,200	1
MERCK KGAA	325412	10,268,000	16	485,500	26
MERCK SERONO SA	325412				
MILLENNIUM PHARMACEUTICALS INC	325412				
MYLAN INC	325412	5,092,785	21	247,748	33
MYRIAD GENETICS INC	325412	326,527	54	84,615	44
NABI BIOPHARMACEUTICALS	325412	10,489	122	-18,727	114
NOVARTIS AG	325412	45,075,000	4	8,400,000	6
NOVO-NORDISK AS	325412	9,335,490	17	1,968,060	17
NPS PHARMACEUTICALS INC	325412	84,147	78	-17,862	112
NYCOMED	325412	4,278,460	25	308,430	28
ONCOTHYREON INC	325412	2,100	137	-17,200	109
ONYX PHARMACEUTICALS INC	325412	250,390	62	-16,161	107
OSI PHARMACEUTICALS INC	325412				
OXIGENE INC	325412	0		-24,728	127

Company	Industry Code	2009 Sales (U.S. $ thousands)	Sales Rank	2009 Profits (U.S. $ thousands)	Profits Rank
PAIN THERAPEUTICS INC	325412	20,563	108	-3,467	84
PALATIN TECHNOLOGIES INC	325412	11,352	120	-4,802	88
PAR PHARMACEUTICAL COMPANIES INC	325412	1,193,159	41	76,928	46
PDL BIOPHARMA	325412	318,184	56	189,660	37
PEREGRINE PHARMACEUTICALS INC	325412	18,151	109	-16,524	108
PERRIGO CO	325412	2,006,862	33	144,049	39
PFIZER INC	325412	50,009,000	2	8,644,000	4
PHARMACYCLICS INC	325412	0		-23,447	122
PHARMOS CORP	325412	0		-3,100	83
PONIARD PHARMACEUTICALS	325412	0		-45,715	144
POZEN INC	325412	32,187	99	1,959	70
PROGENICS PHARMACEUTICALS	325412	48,947	85	-30,612	135
QLT INC	325412	42,100	89	99,400	42
QUESTCOR PHARMACEUTICALS	325412	88,320	77	26,629	49
RANBAXY LABORATORIES LTD	325412	1,619,700	36	103,328	41
RAPTOR PHARMACEUTICAL	325412				
REGENERON PHARMACEUTICALS INC	325412	379,268	50	-67,830	148
REPLIGEN CORPORATION	325412	29,362	101	5,746	65
REPROS THERAPEUTICS INC	325412	551	145	-27,234	130
RIGEL PHARMACEUTICALS INC	325412	750	142	-111,547	160
ROCHE HOLDING LTD	325412	48,287,100	3	7,348,180	8
SALIX PHARMACEUTICALS	325412	232,890	64	-43,619	143
SANOFI-AVENTIS SA	325412	41,776,800	6	7,153,240	9
SAVIENT PHARMACEUTICALS	325412	2,960	134	-90,853	157
SCICLONE PHARMACEUTICALS	325412	72,411	81	11,945	57
SCIOS INC	325412				
SEATTLE GENETICS	325412	51,965	84	-81,683	154
SENETEK PLC	325412	1,907	139	-5,105	89
SEPRACOR INC	325412				
SHIONOGI INC	325412				
SHIRE CANADA INC	325412				
SHIRE PLC	325412	3,007,700	29	491,600	25
SIGA TECHNOLOGIES INC	325412	13,812	115	-17,618	111
SIMCERE PHARMACEUTICAL GROUP	325412	272,062	61	3,871	69
SPECTRUM PHARMACEUTICALS	325412	38,025	92	-19,046	116
STIEFEL LABORATORIES INC	325412				
SUPERGEN INC	325412	41,253	90	4,737	67
TAKEDA PHARMACEUTICAL COMPANY LTD	325412	16,813,800	12	2,561,790	14
TAMIR BIOTECHNOLOGY INC	325412	0		-4,539	86
TARGETED GENETICS CORP	325412	12,171	119	7,948	61
TARO PHARMACEUTICAL INDUSTRIES	325412				
TELIK INC	325412	0		-23,693	124
TEVA PARENTERAL MEDICINES	325412				
TEVA PHARMACEUTICAL INDUSTRIES	325412	13,899,000	14	2,000,000	16
TITAN PHARMACEUTICALS	325412				
TOLMAR INC	325412				
TRIMERIS INC	325412	15,180	113	12,296	55
TRIUS THERAPUTICS INC	325412	5,000	130	-22,700	120

Company	Industry Code	2009 Sales (U.S. $ thousands)	Sales Rank	2009 Profits (U.S. $ thousands)	Profits Rank
UCB SA	325412	4,233,520	26	696,980	22
UNIGENE LABORATORIES	325412	12,792	117	-13,380	104
UNITED THERAPEUTICS CORP	325412	369,848	51	19,462	52
UNITED-GUARDIAN INC	325412	13,277	116	3,879	68
URIGEN PHARMACEUTICALS	325412	0		-2,257	79
VALEANT PHARMACEUTICALS INTERNATIONAL	325412	830,461	43	263,741	32
VAXGEN INC	325412	0		-6,208	91
VERNALIS PLC	325412	20,720	107	-18,810	115
VERTEX PHARMACEUTICALS	325412	101,889	73	-642,178	169
VICAL INC	325412	12,686	118	-28,558	132
VIROPHARMA INC	325412	310,449	58	-11,077	101
WARNER CHILCOTT PLC	325412	1,435,816	39	514,118	24
WATSON PHARMACEUTICALS	325412	2,793,000	30	222,000	35
XECHEM INTERNATIONAL	325412				
XENOPORT INC	325412	34,273	97	-66,334	147
XOMA LTD	325412	98,430	74	550	72
ZIOPHARM ONCOLOGY INC	325412	0		-7,649	94
ZYMOGENETICS INC	325412	136,972	68	-42,981	142
Ethanol Fuel Manufacturing					
AMYRIS BIOTECHNOLOGIES	325193				
Instrument Manufacturing, including Measurement, Control, Test & Navigational					
AGILENT TECHNOLOGIES INC	3345	4,481,000	1	-31,000	4
HARVARD BIOSCIENCE INC	3345	85,772	4	7,233	3
ILLUMINA INC	3345	666,324	3	72,281	2
MILLIPORE CORP	3345	1,654,410	2	179,190	1
WHATMAN PLC	3345				
Investment Banking					
BURRILL & COMPANY	523110				
Laboratories & Diagnostic Services--Medical					
23ANDME INC	6215				
BIORELIANCE CORP	6215				
CML HEALTHCARE INCOME FUND	6215	504,520	2	54,000	2
MDS INC	6215	231,000	3	-135,000	5
MEDTOX SCIENTIFIC INC	6215	84,108	4	1,299	3
ORCHID CELLMARK INC	6215	59,062	5	-1,542	4
QUEST DIAGNOSTICS INC	6215	7,455,243	1	729,111	1
RULES BASED MEDICINE INC	6215				
SPECIALTY LABORATORIES INC	6215				
Market Research					
IMS HEALTH INC	541910	2,189,745	1	258,455	1
Medical/Dental/Surgical Equipment & Supplies, Distribution					
THERMO FISHER SCIENTIFIC	423450	10,109,700	1	850,300	1
Medical/Dental/Surgical Equipment & Supplies, Manufacturing					
AASTROM BIOSCIENCES INC	33911	182	15	-15,946	13
ABAXIS INC	33911	105,600	8	12,000	7
ADVANSOURCE BIOMATERIALS CORPORATION	33911	1,915	14	-2,517	10

Company	Industry Code	2009 Sales (U.S. $ thousands)	Sales Rank	2009 Profits (U.S. $ thousands)	Profits Rank
BAUSCH & LOMB INC	33911				
BAXTER INTERNATIONAL INC	33911	12,562,000	2	2,215,000	2
BIO RAD LABORATORIES INC	33911	1,784,244	5	144,620	4
BIOCOMPATIBLES INTERNATIONAL PLC	33911	42,400	12	-9,240	12
BIOMET INC	33911	2,504,100	4	-749,200	15
BIOSPHERE MEDICAL INC	33911	31,443	13	-2,674	11
BRACHYTHERAPY SERVICES	33911				
COVIDIEN PLC	33911	10,677,000	3	907,000	3
GE HEALTHCARE	33911	16,015,000	1	2,420,000	1
GENZYME BIOSURGERY	33911				
INSTITUT STRAUMANN AG	33911	705,990	6	140,350	5
INTEGRA LIFESCIENCES HOLDINGS CORP	33911	682,487	7	50,955	6
SYNOVIS LIFE TECHNOLOGIES	33911	58,211	10	2,706	9
TENGION INC	33911				
THERAGENICS CORP	33911	78,326	9	3,075	8
VIVUS INC	33911	50,041	11	-54,291	14
Medicinals & Botanicals, Manufacturing					
PROTEIN POLYMER TECHNOLOGIES	325411				
VENTRIA BIOSCIENCE	325411				
Nanotechnology-Biotech/Health					
ARRYX INC	541711				
BIND BIOSCIENCES	541711				
BIONANOMATRIX	541711				
Other Basic Organic Chemicals/Biofuels Manufacturing					
SYNTHETIC GENOMICS	325199				
Research & Development--Physical, Engineering & Life Sciences					
ADVANCED CELL TECHNOLOGY	541712	1,416	17	-36,760	15
ALBANY MOLECULAR RESEARCH INC	541712	196,417	6	-16,695	11
APPLIED MOLECULAR EVOLUTION INC	541712				
ARRAY BIOPHARMA INC	541712	24,982	12	-127,815	18
BIOANALYTICAL SYSTEMS INC	541712	31,784	11	-5,463	9
CANGENE BIOPHARMA INC	541712				
CELERA CORPORATION	541712	167,100	7	-32,700	14
CHARLES RIVER LABORATORIES INTERNATIONAL INC	541712	1,202,551	3	112,602	3
COMMONWEALTH BIOTECHNOLOGIES INC	541712	3,273	16	-2,360	7
COVANCE INC	541712	1,962,626	1	175,882	1
ENCORIUM GROUP INC	541712	21,167	14	-3,870	8
EVOTEC AG	541712	56,570	9	-60,340	17
ICON PLC	541712	887,612	5	91,558	4
INFINITY PHARMACEUTICALS	541712	49,539	10	-32,505	13
LIFE SCIENCES RESEARCH INC	541712				
ORE PHARMACEUTICALS HOLDINGS INC	541712	175	18	-8,383	10
PACIFIC BIOMARKERS INC	541712	10,881	15	1,236	6
PAREXEL INTERNATIONAL	541712	1,050,755	4	39,307	5
PHARMACEUTICAL PRODUCT DEVELOPMENT INC	541712	1,416,770	2	159,295	2
PHARMANET DEVELOPMENT GROUP INC	541712				

Company	Industry Code	2009 Sales (U.S. $ thousands)	Sales Rank	2009 Profits (U.S. $ thousands)	Profits Rank
PRA INTERNATIONAL	541712				
QUINTILES TRANSNATIONAL	541712				
SANGAMO BIOSCIENCES INC	541712	22,187	13	-18,587	12
VERENIUM CORPORATION	541712	65,911	8	-56,240	16
Textiles, Fabrics, Sheets/Towels, Manufacturing					
TOYOBO CO LTD	313	4,083,660	1	-139,040	1
Veterinary Products Manufacturing					
HESKA CORP	325412B	75,678	1	2,242	1
IMMUCELL CORPORATION	325412B	4,509	2	-216	2
SYNBIOTICS CORP	325412B				
VIRBAC CORP	325412B				

ALPHABETICAL INDEX

CYPRESS BIOSCIENCE INC	ICON PLC
CYTRX CORPORATION	IDERA PHARMACEUTICALS INC
DAIICHI SANKYO CO LTD	IDEXX LABORATORIES INC
DAINIPPON SUMITOMO PHARMA CO LTD	ILLUMINA INC
DECODE GENETICS LTD	IMCLONE SYSTEMS INC
DENDREON CORPORATION	IMMTECH PHARMACEUTICALS INC
DEPOMED INC	IMMUCELL CORPORATION
DISCOVERY LABORATORIES INC	IMMUNOGEN INC
DIVI'S LABORATORIES LIMITED	IMMUNOMEDICS INC
DOW AGROSCIENCES LLC	IMPAX LABORATORIES INC
DR REDDY'S LABORATORIES LIMITED	IMS HEALTH INC
DSM PHARMACEUTICALS INC	INCYTE CORP
DU PONT AGRICULTURE & NUTRITION	INFINITY PHARMACEUTICALS INC
DURECT CORP	INSITE VISION INC
DUSA PHARMACEUTICALS INC	INSMED INCORPORATED
DYAX CORP	INSPIRE PHARMACEUTICALS INC
E I DU PONT DE NEMOURS & CO (DUPONT)	INSTITUT STRAUMANN AG
EISAI CO LTD	INTEGRA LIFESCIENCES HOLDINGS CORP
ELAN CORP PLC	INTELLIPHARMACEUTICS INTERNATIONAL INC
ELI LILLY & COMPANY	INTERMUNE INC
EMISPHERE TECHNOLOGIES INC	INTERNATIONAL ISOTOPES INC
ENCORIUM GROUP INC	IRONWOOD PHARMACEUTICALS INC
ENDO PHARMACEUTICALS HOLDINGS INC	ISIS PHARMACEUTICALS INC
ENTREMED INC	JAZZ PHARMACEUTICALS
ENZO BIOCHEM INC	JOHNSON & JOHNSON
ENZON PHARMACEUTICALS INC	JUBILANT BIOSYS LTD
ERESEARCH TECHNOLOGY INC	KENDLE INTERNATIONAL INC
EUSA PHARMA (USA) INC	KERYX BIOPHARMACEUTICALS INC
EVOTEC AG	KING PHARMACEUTICALS INC
EXELIXIS INC	KV PHARMACEUTICAL CO
FLAMEL TECHNOLOGIES SA	LA JOLLA PHARMACEUTICAL
FOREST LABORATORIES INC	LANNETT COMPANY INC
GALDERMA PHARMA SA	LEXICON PHARMACEUTICALS INC
GE HEALTHCARE	LIFE SCIENCES RESEARCH INC
GENENCOR INTERNATIONAL INC	LIFE TECHNOLOGIES CORP
GENENTECH INC	LIFECELL CORPORATION
GENEREX BIOTECHNOLOGY	LIFECORE BIOMEDICAL INC
GENOMIC HEALTH INC	LIGAND PHARMACEUTICALS INC
GEN-PROBE INC	LONZA GROUP
GENTA INC	LORUS THERAPEUTICS INC
GENVEC INC	LUMINEX CORPORATION
GENZYME BIOSURGERY	MANHATTAN PHARMACEUTICALS INC
GENZYME CORP	MAXYGEN INC
GENZYME ONCOLOGY	MDRNA INC
GERON CORPORATION	MDS INC
GILEAD SCIENCES INC	MEDICINES CO (THE)
GLAXOSMITHKLINE PLC	MEDICIS PHARMACEUTICAL CORP
GTC BIOTHERAPEUTICS INC	MEDIDATA SOLUTIONS INC
HARBOR BIOSCIENCES INC	MEDIMMUNE INC
HARVARD BIOSCIENCE INC	MEDTOX SCIENTIFIC INC
HEMAGEN DIAGNOSTICS INC	MERCK & CO INC
HEMISPHERX BIOPHARMA INC	MERCK KGAA
HESKA CORP	MERCK SERONO SA
HI-TECH PHARMACAL CO INC	MERIDIAN BIOSCIENCE INC
HOSPIRA INC	MILLENNIUM PHARMACEUTICALS INC
HUMAN GENOME SCIENCES INC	MILLIPORE CORP
HYCOR BIOMEDICAL INC	MONSANTO CO
I3 CANREG	MYLAN INC

MYRIAD GENETICS INC
NABI BIOPHARMACEUTICALS
NANOBIO CORPORATION
NEKTAR THERAPEUTICS
NEOGEN CORPORATION
NEOPHARM INC
NEOPROBE CORPORATION
NEXMED INC
NOVARTIS AG
NOVAVAX INC
NOVEN PHARMACEUTICALS
NOVO-NORDISK AS
NOVOZYMES
NPS PHARMACEUTICALS INC
NYCOMED
ONCOGENEX PHARMACEUTICALS
ONCOTHYREON INC
ONYX PHARMACEUTICALS INC
ORASURE TECHNOLOGIES INC
ORCHID CELLMARK INC
ORE PHARMACEUTICALS HOLDINGS INC
ORGANOGENESIS INC
OSI PHARMACEUTICALS INC
OXIGENE INC
PACIFIC BIOMARKERS INC
PAIN THERAPEUTICS INC
PALATIN TECHNOLOGIES INC
PAR PHARMACEUTICAL COMPANIES INC
PAREXEL INTERNATIONAL CORP
PDL BIOPHARMA
PENWEST PHARMACEUTICALS CO
PEREGRINE PHARMACEUTICALS INC
PERRIGO CO
PFIZER INC
PHARMACEUTICAL PRODUCT DEVELOPMENT INC
PHARMACYCLICS INC
PHARMANET DEVELOPMENT GROUP INC
PHARMOS CORP
PHARSIGHT CORP
PIONEER HI-BRED INTERNATIONAL INC
POLYDEX PHARMACEUTICALS
PONIARD PHARMACEUTICALS INC
POZEN INC
PRA INTERNATIONAL
PROGENICS PHARMACEUTICALS
PROMEGA CORP
PROTEIN POLYMER TECHNOLOGIES
QIAGEN NV
QLT INC
QUEST DIAGNOSTICS INC
QUESTCOR PHARMACEUTICALS
QUINTILES TRANSNATIONAL CORP
RANBAXY LABORATORIES LIMITED
RAPTOR PHARMACEUTICAL CORP
REGENERON PHARMACEUTICALS INC
REPLIGEN CORPORATION
REPROS THERAPEUTICS INC
RIGEL PHARMACEUTICALS INC

ROCHE HOLDING LTD
RULES BASED MEDICINE INC
S&W SEED CO
SALIX PHARMACEUTICALS
SANGAMO BIOSCIENCES INC
SANOFI-AVENTIS SA
SAVIENT PHARMACEUTICALS INC
SCICLONE PHARMACEUTICALS
SCIOS INC
SEATTLE GENETICS
SENETEK PLC
SEPRACOR INC
SEQUENOM INC
SERACARE LIFE SCIENCES INC
SHIONOGI INC
SHIRE CANADA INC
SHIRE PLC
SIEMENS HEALTHCARE DIAGNOSTICS
SIGA TECHNOLOGIES INC
SIGMA-ALDRICH CORP
SIMCERE PHARMACEUTICAL GROUP
SKYEPHARMA PLC
SOLIGENIX INC
SPECIALTY LABORATORIES INC
SPECTRAL DIAGNOSTICS INC
SPECTRUM PHARMACEUTICALS INC
STEMCELLS INC
STIEFEL LABORATORIES INC
SUPERGEN INC
SYNBIOTICS CORP
SYNGENTA AG
SYNOVIS LIFE TECHNOLOGIES INC
SYNTHETIC GENOMICS
TAKEDA PHARMACEUTICAL COMPANY LTD
TAMIR BIOTECHNOLOGY INC
TARGETED GENETICS CORP
TARO PHARMACEUTICAL INDUSTRIES
TECHNE CORP
TELIK INC
TENGION INC
TEVA PARENTERAL MEDICINES INC
TEVA PHARMACEUTICAL INDUSTRIES
THERAGENICS CORP
THERMO FISHER SCIENTIFIC INC
THERMO SCIENTIFIC
TITAN PHARMACEUTICALS
TOLMAR INC
TOYOBO CO LTD
TRIMERIS INC
TRINITY BIOTECH PLC
TRIPOS INTERNATIONAL
TRIUS THERAPUTICS INC
UCB SA
UNIGENE LABORATORIES
UNITED THERAPEUTICS CORP
UNITED-GUARDIAN INC
URIGEN PHARMACEUTICALS INC
VALEANT PHARMACEUTICALS INTERNATIONAL

VAXGEN INC
VENTRIA BIOSCIENCE
VERENIUM CORPORATION
VERMILLION INC
VERNALIS PLC
VERTEX PHARMACEUTICALS INC
VIACORD INC
VICAL INC
VIRBAC CORP
VIROPHARMA INC
VIVUS INC
WARNER CHILCOTT PLC
WATSON PHARMACEUTICALS INC
WHATMAN PLC
XECHEM INTERNATIONAL
XENOPORT INC
XOMA LTD
ZIOPHARM ONCOLOGY INC
ZYMOGENETICS INC

INDEX OF U.S. HEADQUARTERS LOCATION BY STATE

To help you locate members of THE BIOTECH 350 geographically, the city and state of the headquarters of each company are in the following index.

ALABAMA
BIOCRYST PHARMACEUTICALS INC; Birmingham

ARIZONA
MEDICIS PHARMACEUTICAL CORP; Scottsdale

CALIFORNIA
23ANDME INC; Mountain View
ABAXIS INC; Union City
ACCELRYS INC; San Diego
ADAMIS PHARMACEUTICALS CORPORATION; Del Mar
ADVANCED CELL TECHNOLOGY INC; Santa Monica
AEOLUS PHARMACEUTICALS INC; Mission Viejo
AFFYMETRIX INC; Santa Clara
AGILENT TECHNOLOGIES INC; Santa Clara
ALLERGAN INC; Irvine
ALLIANCE PHARMACEUTICAL CORP; San Diego
AMGEN INC; Thousand Oaks
AMYLIN PHARMACEUTICALS INC; San Diego
AMYRIS BIOTECHNOLOGIES INC; Emeryville
APPLIED MOLECULAR EVOLUTION INC; San Diego
ARADIGM CORPORATION; Hayward
ARCADIA BIOSCIENCES INC; Davis
ARENA PHARMACEUTICALS INC; San Diego
AVANIR PHARMACEUTICALS; Aliso Viejo
AVIVA BIOSCIENCES CORP; San Diego
BIO RAD LABORATORIES INC; Hercules
BIOMARIN PHARMACEUTICAL INC; Novato
BIOSITE INC; San Diego
BIOTIME INC; Alameda
BURRILL & COMPANY; San Francisco
CARDIUM THERAPEUTICS INC; San Diego
CELERA CORPORATION; Alameda
CEPHEID; Sunnyvale
CERUS CORPORATION; Concord
CHIRON CORP; Emeryville
CORTEX PHARMACEUTICALS INC; Irvine
CYPRESS BIOSCIENCE INC; San Diego
CYTRX CORPORATION; Los Angeles
DEPOMED INC; Menlo Park
DURECT CORP; Cupertino
EXELIXIS INC; South San Francisco
GENENTECH INC; South San Francisco
GENOMIC HEALTH INC; Redwood City
GEN-PROBE INC; San Diego
GERON CORPORATION; Menlo Park
GILEAD SCIENCES INC; Foster City
HARBOR BIOSCIENCES INC; San Diego

HYCOR BIOMEDICAL INC; Garden Grove
ILLUMINA INC; San Diego
IMPAX LABORATORIES INC; Hayward
INSITE VISION INC; Alameda
INTERMUNE INC; Brisbane
ISIS PHARMACEUTICALS INC; Carlsbad
JAZZ PHARMACEUTICALS; Palo Alto
LA JOLLA PHARMACEUTICAL; San Diego
LIFE TECHNOLOGIES CORP; Carlsbad
LIGAND PHARMACEUTICALS INC; La Jolla
MAXYGEN INC; Redwood City
NEKTAR THERAPEUTICS; San Carlos
NEXMED INC; San Diego
ONYX PHARMACEUTICALS INC; Emeryville
OXIGENE INC; San Francisco
PAIN THERAPEUTICS INC; San Mateo
PEREGRINE PHARMACEUTICALS INC; Tustin
PHARMACYCLICS INC; Sunnyvale
PONIARD PHARMACEUTICALS INC; South San Francisco
PROTEIN POLYMER TECHNOLOGIES; La Jolla
QUESTCOR PHARMACEUTICALS; Union City
RAPTOR PHARMACEUTICAL CORP; Novato
RIGEL PHARMACEUTICALS INC; South San Francisco
S&W SEED CO; Five Points
SANGAMO BIOSCIENCES INC; Richmond
SCICLONE PHARMACEUTICALS; Foster City
SENETEK PLC; Napa
SEQUENOM INC; San Diego
SPECIALTY LABORATORIES INC; Valencia
SPECTRUM PHARMACEUTICALS INC; Irvine
STEMCELLS INC; Palo Alto
SUPERGEN INC; Dublin
SYNTHETIC GENOMICS; La Jolla
TELIK INC; Palo Alto
TEVA PARENTERAL MEDICINES INC; Irvine
TITAN PHARMACEUTICALS; South San Francisco
TRIUS THERAPUTICS INC; San Diego
URIGEN PHARMACEUTICALS INC; San Francisco
VALEANT PHARMACEUTICALS INTERNATIONAL; Aliso Viejo
VAXGEN INC; South San Francisco
VERMILLION INC; Fremont
VICAL INC; San Diego
VIVUS INC; Mountain View
WATSON PHARMACEUTICALS INC; Corona
XENOPORT INC; Santa Clara
XOMA LTD; Berkeley

COLORADO
ALLOS THERAPEUTICS INC; Westminster
ARRAY BIOPHARMA INC; Boulder
HESKA CORP; Loveland
TOLMAR INC; Fort Collins
VENTRIA BIOSCIENCE; Fort Collins

CONNECTICUT
ALEXION PHARMACEUTICALS INC; Cheshire
CURAGEN CORPORATION; Branford
IMS HEALTH INC; Norwalk
PENWEST PHARMACEUTICALS CO; Danbury

DELAWARE
DU PONT AGRICULTURE & NUTRITION; Wilmington
E I DU PONT DE NEMOURS & CO (DUPONT); Wilmington
INCYTE CORP; Wilmington

FLORIDA
BIOHEART INC; Sunrise
NOVEN PHARMACEUTICALS; Miami

GEORGIA
ALIMERA SCIENCES INC; Alpharetta
THERAGENICS CORP; Buford

IDAHO
INTERNATIONAL ISOTOPES INC; Idaho Falls

ILLINOIS
ABBOTT LABORATORIES; Abbott Park
AKORN INC; Lake Forest
APP PHARMACEUTICALS INC; Schaumburg
ARRYX INC; Chicago
BAXTER INTERNATIONAL INC; Deerfield
CELSIS INTERNATIONAL PLC; Chicago
HOSPIRA INC; Lake Forest
NEOPHARM INC; Lake Bluff
SIEMENS HEALTHCARE DIAGNOSTICS; Deerfield

INDIANA
BIOANALYTICAL SYSTEMS INC; West Lafayette
BIOMET INC; Warsaw
DOW AGROSCIENCES LLC; Indianapolis
ELI LILLY & COMPANY; Indianapolis

IOWA
PIONEER HI-BRED INTERNATIONAL INC; Johnston

MAINE
IDEXX LABORATORIES INC; Westbrook
IMMUCELL CORPORATION; Portland

MARYLAND
BIORELIANCE CORP; Rockville
CANGENE BIOPHARMA INC; Baltimore
ENTREMED INC; Rockville
GENVEC INC; Gaithersburg
HEMAGEN DIAGNOSTICS INC; Columbia
HUMAN GENOME SCIENCES INC; Rockville
MEDIMMUNE INC; Gaithersburg
NABI BIOPHARMACEUTICALS; Rockville
NOVAVAX INC; Rockville
UNITED THERAPEUTICS CORP; Silver Spring

MASSACHUSETTS
ADVANSOURCE BIOMATERIALS CORPORATION;
Wilmington
ALKERMES INC; Waltham
ALSERES PHARMACEUTICALS INC; Hopkinton
ANIKA THERAPEUTICS INC; Bedford
ANTIGENICS INC; Lexington
ARIAD PHARMACEUTICALS INC; Cambridge
ARQULE INC; Woburn
BIND BIOSCIENCES; Cambridge
BIOGEN IDEC INC; Cambridge
BIOSPHERE MEDICAL INC; Rockland
CALIPER LIFE SCIENCES; Hopkinton
CELLDEX THERAPEUTICS INC; Needham
CHARLES RIVER LABORATORIES
INTERNATIONAL INC; Wilmington
CUBIST PHARMACEUTICALS INC; Lexington
CURIS INC; Cambridge
DUSA PHARMACEUTICALS INC; Wilmington
DYAX CORP; Cambridge
GENZYME BIOSURGERY; Cambridge
GENZYME CORP; Cambridge
GENZYME ONCOLOGY; Cambridge
GTC BIOTHERAPEUTICS INC; Framingham
HARVARD BIOSCIENCE INC; Holliston
IDERA PHARMACEUTICALS INC; Cambridge
IMMUNOGEN INC; Waltham
INFINITY PHARMACEUTICALS INC; Cambridge
IRONWOOD PHARMACEUTICALS INC; Cambridge
MILLENNIUM PHARMACEUTICALS INC; Cambridge
MILLIPORE CORP; Billerica
ORE PHARMACEUTICALS HOLDINGS INC;
Cambridge
ORGANOGENESIS INC; Canton
PAREXEL INTERNATIONAL CORP; Waltham
REPLIGEN CORPORATION; Waltham
SEPRACOR INC; Marlborough
SERACARE LIFE SCIENCES INC; Milford
THERMO FISHER SCIENTIFIC INC; Waltham
VERENIUM CORPORATION; Cambridge
VERTEX PHARMACEUTICALS INC; Cambridge
VIACORD INC; Cambridge

MICHIGAN
AASTROM BIOSCIENCES INC; Ann Arbor
CARACO PHARMACEUTICAL LABORATORIES;
Detroit
NANOBIO CORPORATION; Ann Arbor
NEOGEN CORPORATION; Lansing
PERRIGO CO; Allegan

MINNESOTA
LIFECORE BIOMEDICAL INC; Chaska
MEDTOX SCIENTIFIC INC; St. Paul
SYNOVIS LIFE TECHNOLOGIES INC; St. Paul
TECHNE CORP; Minneapolis

MISSOURI
KV PHARMACEUTICAL CO; St. Louis
MONSANTO CO; St. Louis
PHARSIGHT CORP; St. Louis
SIGMA-ALDRICH CORP; St. Louis
SYNBIOTICS CORP; Kansas City
TRIPOS INTERNATIONAL; St. Louis

NEVADA
PDL BIOPHARMA; Incline Village

NEW JERSEY
CAMBREX CORP; East Rutherford
CELGENE CORP; Summit
COLUMBIA LABORATORIES INC; Livingston
COVANCE INC; Princeton
EMISPHERE TECHNOLOGIES INC; Cedar Knolls
ENZON PHARMACEUTICALS INC; Bridgewater
GENTA INC; Berkeley Heights
IMCLONE SYSTEMS INC; Branchburg
IMMUNOMEDICS INC; Morris Plains
INTEGRA LIFESCIENCES HOLDINGS CORP;
Plainsboro
JOHNSON & JOHNSON; New Brunswick
LIFE SCIENCES RESEARCH INC; East Millstone
LIFECELL CORPORATION; Branchburg
MEDICINES CO (THE); Parsippany
MERCK & CO INC; Whitehouse Station
NPS PHARMACEUTICALS INC; Bedminster
ORCHID CELLMARK INC; Princeton
PALATIN TECHNOLOGIES INC; Cranbury
PAR PHARMACEUTICAL COMPANIES INC;
Woodcliff Lake
PHARMANET DEVELOPMENT GROUP INC;
Princeton
PHARMOS CORP; Iselin
QUEST DIAGNOSTICS INC; Madison
SAVIENT PHARMACEUTICALS INC; E. Brunswick
SCIOS INC; New Brunswick
SHIONOGI INC; Florham Park
SOLIGENIX INC; Princeton
TAMIR BIOTECHNOLOGY INC; Somerset
UNIGENE LABORATORIES; Boonton
WARNER CHILCOTT PLC; Rockaway
XECHEM INTERNATIONAL; Edison

NEW YORK
ALBANY MOLECULAR RESEARCH INC; Albany
BAUSCH & LOMB INC; Rochester
BRISTOL-MYERS SQUIBB CO; New York
ENZO BIOCHEM INC; New York
FOREST LABORATORIES INC; New York
GENENCOR INTERNATIONAL INC; Rochester
HI-TECH PHARMACAL CO INC; Amityville
IMMTECH PHARMACEUTICALS INC; New York
KERYX BIOPHARMACEUTICALS INC; New York
MANHATTAN PHARMACEUTICALS INC; New York
MEDIDATA SOLUTIONS INC; New York

OSI PHARMACEUTICALS INC; Melville
PFIZER INC; New York
PROGENICS PHARMACEUTICALS; Tarrytown
REGENERON PHARMACEUTICALS INC; Tarrytown
SIGA TECHNOLOGIES INC; New York
UNITED-GUARDIAN INC; Hauppauge
ZIOPHARM ONCOLOGY INC; New York

NORTH CAROLINA
ALDAGEN INC; Durham
DSM PHARMACEUTICALS INC; Greenville
INSPIRE PHARMACEUTICALS INC; Durham
PHARMACEUTICAL PRODUCT DEVELOPMENT
INC; Wilmington
POZEN INC; Chapel Hill
PRA INTERNATIONAL; Raleigh
QUINTILES TRANSNATIONAL CORP; Durham
SALIX PHARMACEUTICALS; Morrisville
STIEFEL LABORATORIES INC; Research Triangle Park
TRIMERIS INC; Durham

OHIO
KENDLE INTERNATIONAL INC; Cincinnati
MERIDIAN BIOSCIENCE INC; Cincinnati
NEOPROBE CORPORATION; Dublin

PENNSYLVANIA
ADOLOR CORP; Exton
AVAX TECHNOLOGIES INC; Philadelphia
BAYER CORP; Pittsburgh
BIONANOMATRIX; Philadelphia
CEPHALON INC; Frazer
DISCOVERY LABORATORIES INC; Warrington
ENDO PHARMACEUTICALS HOLDINGS INC; Chadds
Ford
ERESEARCH TECHNOLOGY INC; Philadelphia
EUSA PHARMA (USA) INC; Langhorne
HEMISPHERX BIOPHARMA INC; Philadelphia
LANNETT COMPANY INC; Philadelphia
MYLAN INC; Canonsburg
ORASURE TECHNOLOGIES INC; Bethlehem
TENGION INC; East Norriton
VIROPHARMA INC; Exton

TENNESSEE
ALPHARMA ANIMAL HEALTH; Bristol
CUMBERLAND PHARMACEUTICALS; Nashville
KING PHARMACEUTICALS INC; Bristol

TEXAS
ACCESS PHARMACEUTICALS INC; Dallas
LEXICON PHARMACEUTICALS INC; The Woodlands
LUMINEX CORPORATION; Austin
REPROS THERAPEUTICS INC; The Woodlands
RULES BASED MEDICINE INC; Austin
VIRBAC CORP; Fort Worth

UTAH
MYRIAD GENETICS INC; Salt Lake City

VIRGINIA
BRACHYTHERAPY SERVICES INC; Springfield
CEL-SCI CORPORATION; Vienna
COMMONWEALTH BIOTECHNOLOGIES INC;
Richmond
INSMED INCORPORATED; Richmond

WASHINGTON
AVI BIOPHARMA INC; Bothell
CELL THERAPEUTICS INC; Seattle
DENDREON CORPORATION; Seattle
MDRNA INC; Bothell
ONCOGENEX PHARMACEUTICALS; Bothell
ONCOTHYREON INC; Seattle
PACIFIC BIOMARKERS INC; Seattle
SEATTLE GENETICS; Bothell
TARGETED GENETICS CORP; Seattle
ZYMOGENETICS INC; Seattle

WISCONSIN
CELLULAR DYNAMICS INTERNATIONAL; Madison
PROMEGA CORP; Madison

INDEX OF NON-U.S. HEADQUARTERS LOCATION BY COUNTRY

AUSTRALIA
CSL LIMITED; Parkville

BELGIUM
THERMO SCIENTIFIC; Erembodegem
UCB SA; Brussels

CANADA
AETERNA ZENTARIS INC; Quebec City
ANGIOTECH PHARMACEUTICALS INC; Vancouver
ASTRAZENECA CANADA; Mississauga
AXCAN PHARMA INC; Mont-Saint-Hilaire
BIOVAIL CORPORATION; Mississauga
CANGENE CORP; Winnipeg
CARDIOME PHARMA CORP; Vancouver
CML HEALTHCARE INCOME FUND; Mississauga
GENEREX BIOTECHNOLOGY; Toronto
I3 CANREG; Dundas
INTELLIPHARMACEUTICS INTERNATIONAL INC; Toronto
LORUS THERAPEUTICS INC; Toronto
MDS INC; Ottawa
POLYDEX PHARMACEUTICALS; Toronto
QLT INC; Vancouver
SHIRE CANADA INC; Saint-Laurent
SPECTRAL DIAGNOSTICS INC; Toronto

CHINA
3SBIO INC; Shenyang
SIMCERE PHARMACEUTICAL GROUP; Nanjing

DENMARK
NOVO-NORDISK AS; Bagsvaerd
NOVOZYMES; Bagsvaerd

FINLAND
ENCORIUM GROUP INC; Espoo

FRANCE
CEGEDIM SA; Boulogne-Billancourt
FLAMEL TECHNOLOGIES SA; Venissieux Cedex
SANOFI-AVENTIS SA; Paris

GERMANY
4SC AG; Planegg-Martinsried
ALTANA AG; Wesel
BASF SE; Ludwigshafen
BAYER AG; Leverkusen
BAYER SCHERING PHARMA AG; Berlin
EVOTEC AG; Hamburg
MERCK KGAA; Darmstadt

ICELAND
DECODE GENETICS LTD; Reykjavik

INDIA
DIVI'S LABORATORIES LIMITED; Hyderabad
DR REDDY'S LABORATORIES LIMITED; Hyderabad
JUBILANT BIOSYS LTD; Noida
RANBAXY LABORATORIES LIMITED; Gurgaon

IRELAND
AMARIN CORPORATION PLC; Dublin

ISRAEL
COMPUGEN LTD; Tel Aviv
TARO PHARMACEUTICAL INDUSTRIES; Haifa Bay
TEVA PHARMACEUTICAL INDUSTRIES; Petach Tikva

JAPAN
ASTELLAS PHARMA INC; Tokyo
DAIICHI SANKYO CO LTD; Tokyo
DAINIPPON SUMITOMO PHARMA CO LTD; Osaka
EISAI CO LTD; Tokyo
TAKEDA PHARMACEUTICAL COMPANY LTD; Osaka
TOYOBO CO LTD; Osaka

SWEDEN
CELLARTIS AB; Goteborg

SWITZERLAND
ACTELION LTD; Allschwil
ALCON INC; Hunenberg
GALDERMA PHARMA SA; Lausanne
INSTITUT STRAUMANN AG; Basel
LONZA GROUP; Basel
MERCK SERONO SA; Geneva 20
NOVARTIS AG; Basel
NYCOMED; Zurich
ROCHE HOLDING LTD; Basel
SYNGENTA AG; Basel

THE NETHERLANDS
AKZO NOBEL NV; Amsterdam
CRUCELL NV; Leiden
QIAGEN NV; Venlo

UNITED KINGDOM
ASTRAZENECA PLC; London
BIOCOMPATIBLES INTERNATIONAL PLC; Farnham
COVIDIEN PLC; Dublin
ELAN CORP PLC; Dublin
GE HEALTHCARE; Chalfont St. Giles
GLAXOSMITHKLINE PLC; Middlesex
ICON PLC; Dublin
SHIRE PLC; Dublin 24
SKYEPHARMA PLC; London
TRINITY BIOTECH PLC; Bray
VERNALIS PLC; Winnersh
WHATMAN PLC; Maidstone

INDEX BY REGIONS OF THE U.S.
WHERE THE FIRMS HAVE LOCATIONS

WEST
23ANDME INC
ABAXIS INC
ABBOTT LABORATORIES
ACCELRYS INC
ACTELION LTD
ADAMIS PHARMACEUTICALS CORPORATION
ADVANCED CELL TECHNOLOGY INC
AEOLUS PHARMACEUTICALS INC
AFFYMETRIX INC
AGILENT TECHNOLOGIES INC
ALCON INC
ALLERGAN INC
ALLIANCE PHARMACEUTICAL CORP
ALLOS THERAPEUTICS INC
ALPHARMA ANIMAL HEALTH
AMGEN INC
AMYLIN PHARMACEUTICALS INC
AMYRIS BIOTECHNOLOGIES INC
ANGIOTECH PHARMACEUTICALS INC
APPLIED MOLECULAR EVOLUTION INC
ARADIGM CORPORATION
ARCADIA BIOSCIENCES INC
ARENA PHARMACEUTICALS INC
ARRAY BIOPHARMA INC
ASTELLAS PHARMA INC
ASTRAZENECA PLC
AVANIR PHARMACEUTICALS
AVI BIOPHARMA INC
AVIVA BIOSCIENCES CORP
BASF SE
BAUSCH & LOMB INC
BAXTER INTERNATIONAL INC
BAYER AG
BAYER CORP
BAYER SCHERING PHARMA AG
BIO RAD LABORATORIES INC
BIOANALYTICAL SYSTEMS INC
BIOGEN IDEC INC
BIOMARIN PHARMACEUTICAL INC
BIOMET INC
BIOSITE INC
BIOTIME INC
BRISTOL-MYERS SQUIBB CO
BURRILL & COMPANY
CALIPER LIFE SCIENCES
CANGENE CORP
CARDIUM THERAPEUTICS INC
CELERA CORPORATION
CELL THERAPEUTICS INC
CEPHALON INC
CEPHEID
CERUS CORPORATION
CHARLES RIVER LABORATORIES
INTERNATIONAL INC

CHIRON CORP
CORTEX PHARMACEUTICALS INC
COVANCE INC
COVIDIEN PLC
CUBIST PHARMACEUTICALS INC
CYPRESS BIOSCIENCE INC
CYTRX CORPORATION
DENDREON CORPORATION
DEPOMED INC
DISCOVERY LABORATORIES INC
DU PONT AGRICULTURE & NUTRITION
DURECT CORP
E I DU PONT DE NEMOURS & CO (DUPONT)
ELAN CORP PLC
ELI LILLY & COMPANY
EXELIXIS INC
FOREST LABORATORIES INC
GE HEALTHCARE
GENENCOR INTERNATIONAL INC
GENENTECH INC
GENOMIC HEALTH INC
GEN-PROBE INC
GENZYME CORP
GERON CORPORATION
GILEAD SCIENCES INC
GLAXOSMITHKLINE PLC
HARBOR BIOSCIENCES INC
HARVARD BIOSCIENCE INC
HESKA CORP
HOSPIRA INC
HYCOR BIOMEDICAL INC
ICON PLC
IDEXX LABORATORIES INC
ILLUMINA INC
IMPAX LABORATORIES INC
IMS HEALTH INC
INSITE VISION INC
INTEGRA LIFESCIENCES HOLDINGS CORP
INTERMUNE INC
INTERNATIONAL ISOTOPES INC
ISIS PHARMACEUTICALS INC
JAZZ PHARMACEUTICALS
JOHNSON & JOHNSON
KENDLE INTERNATIONAL INC
KING PHARMACEUTICALS INC
LA JOLLA PHARMACEUTICAL
LANNETT COMPANY INC
LIFE TECHNOLOGIES CORP
LIGAND PHARMACEUTICALS INC
MAXYGEN INC
MDRNA INC
MEDICIS PHARMACEUTICAL CORP
MEDIDATA SOLUTIONS INC
MEDIMMUNE INC
MERCK & CO INC
MERCK KGAA
MILLIPORE CORP
MONSANTO CO

MYRIAD GENETICS INC
NEKTAR THERAPEUTICS
NEXMED INC
NOVARTIS AG
NOVO-NORDISK AS
NOVOZYMES
ONCOGENEX PHARMACEUTICALS
ONCOTHYREON INC
ONYX PHARMACEUTICALS INC
OSI PHARMACEUTICALS INC
OXIGENE INC
PACIFIC BIOMARKERS INC
PAIN THERAPEUTICS INC
PAREXEL INTERNATIONAL CORP
PDL BIOPHARMA
PEREGRINE PHARMACEUTICALS INC
PFIZER INC
PHARMACEUTICAL PRODUCT DEVELOPMENT INC
PHARMACYCLICS INC
PHARMANET DEVELOPMENT GROUP INC
PHARSIGHT CORP
PIONEER HI-BRED INTERNATIONAL INC
PONIARD PHARMACEUTICALS INC
PRA INTERNATIONAL
PROMEGA CORP
PROTEIN POLYMER TECHNOLOGIES
QIAGEN NV
QLT INC
QUEST DIAGNOSTICS INC
QUESTCOR PHARMACEUTICALS
QUINTILES TRANSNATIONAL CORP
RAPTOR PHARMACEUTICAL CORP
RIGEL PHARMACEUTICALS INC
ROCHE HOLDING LTD
S&W SEED CO
SANGAMO BIOSCIENCES INC
SANOFI-AVENTIS SA
SCICLONE PHARMACEUTICALS
SEATTLE GENETICS
SENETEK PLC
SEQUENOM INC
SIEMENS HEALTHCARE DIAGNOSTICS
SIGA TECHNOLOGIES INC
SIGMA-ALDRICH CORP
SPECIALTY LABORATORIES INC
SPECTRUM PHARMACEUTICALS INC
STEMCELLS INC
STIEFEL LABORATORIES INC
SUPERGEN INC
SYNBIOTICS CORP
SYNGENTA AG
SYNOVIS LIFE TECHNOLOGIES INC
SYNTHETIC GENOMICS
TAKEDA PHARMACEUTICAL COMPANY LTD
TARGETED GENETICS CORP
TECHNE CORP
TELIK INC
TEVA PARENTERAL MEDICINES INC

TEVA PHARMACEUTICAL INDUSTRIES
THERAGENICS CORP
THERMO FISHER SCIENTIFIC INC
THERMO SCIENTIFIC
TITAN PHARMACEUTICALS
TOLMAR INC
TRINITY BIOTECH PLC
TRIUS THERAPUTICS INC
URIGEN PHARMACEUTICALS INC
VALEANT PHARMACEUTICALS INTERNATIONAL
VAXGEN INC
VENTRIA BIOSCIENCE
VERENIUM CORPORATION
VERMILLION INC
VERTEX PHARMACEUTICALS INC
VICAL INC
VIVUS INC
WATSON PHARMACEUTICALS INC
XENOPORT INC
XOMA LTD
ZYMOGENETICS INC

SOUTHWEST
ABBOTT LABORATORIES
ACCESS PHARMACEUTICALS INC
AGILENT TECHNOLOGIES INC
ALCON INC
ARCADIA BIOSCIENCES INC
ASTRAZENECA PLC
AVANIR PHARMACEUTICALS
BASF SE
BAXTER INTERNATIONAL INC
BAYER AG
BAYER CORP
BIO RAD LABORATORIES INC
BIOHEART INC
BRISTOL-MYERS SQUIBB CO
CELERA CORPORATION
CHARLES RIVER LABORATORIES
INTERNATIONAL INC
COVANCE INC
COVIDIEN PLC
DU PONT AGRICULTURE & NUTRITION
E I DU PONT DE NEMOURS & CO (DUPONT)
ELI LILLY & COMPANY
GALDERMA PHARMA SA
GENZYME CORP
GLAXOSMITHKLINE PLC
HOSPIRA INC
ICON PLC
IDEXX LABORATORIES INC
IMS HEALTH INC
INTERNATIONAL ISOTOPES INC
JOHNSON & JOHNSON
LEXICON PHARMACEUTICALS INC
LIFE TECHNOLOGIES CORP
LUMINEX CORPORATION
MEDICIS PHARMACEUTICAL CORP

MEDIDATA SOLUTIONS INC
MERCK & CO INC
MILLIPORE CORP
MONSANTO CO
MYLAN INC
ONCOTHYREON INC
ORCHID CELLMARK INC
PFIZER INC
PHARMACEUTICAL PRODUCT DEVELOPMENT INC
PIONEER HI-BRED INTERNATIONAL INC
PRA INTERNATIONAL
QUEST DIAGNOSTICS INC
QUINTILES TRANSNATIONAL CORP
REPROS THERAPEUTICS INC
ROCHE HOLDING LTD
RULES BASED MEDICINE INC
SANOFI-AVENTIS SA
SIGMA-ALDRICH CORP
TEVA PHARMACEUTICAL INDUSTRIES
THERAGENICS CORP
THERMO FISHER SCIENTIFIC INC
VIRBAC CORP

MIDWEST
AASTROM BIOSCIENCES INC
ABBOTT LABORATORIES
AFFYMETRIX INC
AGILENT TECHNOLOGIES INC
AKORN INC
AKZO NOBEL NV
ALKERMES INC
ALPHARMA ANIMAL HEALTH
ALTANA AG
ANGIOTECH PHARMACEUTICALS INC
APP PHARMACEUTICALS INC
ARRYX INC
ASTELLAS PHARMA INC
ASTRAZENECA PLC
AVANIR PHARMACEUTICALS
BASF SE
BAXTER INTERNATIONAL INC
BAYER AG
BAYER CORP
BIOANALYTICAL SYSTEMS INC
BIOMET INC
BRISTOL-MYERS SQUIBB CO
CAMBREX CORP
CARACO PHARMACEUTICAL LABORATORIES
CELLULAR DYNAMICS INTERNATIONAL
CELSIS INTERNATIONAL PLC
CEPHALON INC
CHARLES RIVER LABORATORIES
INTERNATIONAL INC
CHIRON CORP
COVANCE INC
COVIDIEN PLC
CSL LIMITED
CUBIST PHARMACEUTICALS INC

DOW AGROSCIENCES LLC
DU PONT AGRICULTURE & NUTRITION
E I DU PONT DE NEMOURS & CO (DUPONT)
EISAI CO LTD
ELI LILLY & COMPANY
FOREST LABORATORIES INC
GE HEALTHCARE
GENENCOR INTERNATIONAL INC
GENZYME CORP
GLAXOSMITHKLINE PLC
HESKA CORP
HOSPIRA INC
ICON PLC
IDEXX LABORATORIES INC
IMMTECH PHARMACEUTICALS INC
INTEGRA LIFESCIENCES HOLDINGS CORP
JOHNSON & JOHNSON
KENDLE INTERNATIONAL INC
KING PHARMACEUTICALS INC
KV PHARMACEUTICAL CO
LIFE TECHNOLOGIES CORP
LIFECORE BIOMEDICAL INC
LONZA GROUP
MEDICIS PHARMACEUTICAL CORP
MEDTOX SCIENTIFIC INC
MERCK & CO INC
MERCK KGAA
MERIDIAN BIOSCIENCE INC
MILLIPORE CORP
MONSANTO CO
MYLAN INC
NANOBIO CORPORATION
NEOGEN CORPORATION
NEOPHARM INC
NEOPROBE CORPORATION
NOVARTIS AG
ORCHID CELLMARK INC
PFIZER INC
PHARMACEUTICAL PRODUCT DEVELOPMENT INC
PHARMANET DEVELOPMENT GROUP INC
PHARSIGHT CORP
PIONEER HI-BRED INTERNATIONAL INC
PRA INTERNATIONAL
PROMEGA CORP
QUEST DIAGNOSTICS INC
QUINTILES TRANSNATIONAL CORP
ROCHE HOLDING LTD
SANOFI-AVENTIS SA
SHIRE PLC
SIGMA-ALDRICH CORP
SYNGENTA AG
SYNOVIS LIFE TECHNOLOGIES INC
TAKEDA PHARMACEUTICAL COMPANY LTD
TECHNE CORP
TEVA PHARMACEUTICAL INDUSTRIES
THERMO FISHER SCIENTIFIC INC
THERMO SCIENTIFIC
TOLMAR INC

TRINITY BIOTECH PLC
TRIPOS INTERNATIONAL
UCB SA
VENTRIA BIOSCIENCE
VIACORD INC
VIRBAC CORP
WATSON PHARMACEUTICALS INC

SOUTHEAST
AKZO NOBEL NV
ALIMERA SCIENCES INC
ALPHARMA ANIMAL HEALTH
AMGEN INC
ANGIOTECH PHARMACEUTICALS INC
ASTRAZENECA PLC
AVANIR PHARMACEUTICALS
AXCAN PHARMA INC
BASF SE
BAUSCH & LOMB INC
BAXTER INTERNATIONAL INC
BAYER AG
BAYER CORP
BIOCRYST PHARMACEUTICALS INC
BIOMET INC
BRISTOL-MYERS SQUIBB CO
CANGENE CORP
CELERA CORPORATION
CHARLES RIVER LABORATORIES
INTERNATIONAL INC
COVANCE INC
COVIDIEN PLC
CRUCELL NV
CSL LIMITED
CUBIST PHARMACEUTICALS INC
CUMBERLAND PHARMACEUTICALS
DENDREON CORPORATION
DU PONT AGRICULTURE & NUTRITION
DURECT CORP
E I DU PONT DE NEMOURS & CO (DUPONT)
ELAN CORP PLC
ELI LILLY & COMPANY
GENZYME CORP
GLAXOSMITHKLINE PLC
ICON PLC
IDEXX LABORATORIES INC
JOHNSON & JOHNSON
KING PHARMACEUTICALS INC
LIFE TECHNOLOGIES CORP
MEDICIS PHARMACEUTICAL CORP
MERCK & CO INC
MERCK KGAA
MERIDIAN BIOSCIENCE INC
MILLIPORE CORP
MONSANTO CO
MYLAN INC
NEKTAR THERAPEUTICS
NOVARTIS AG
NOVEN PHARMACEUTICALS

ORCHID CELLMARK INC
PFIZER INC
PIONEER HI-BRED INTERNATIONAL INC
QUEST DIAGNOSTICS INC
QUINTILES TRANSNATIONAL CORP
ROCHE HOLDING LTD
SANOFI-AVENTIS SA
SHIONOGI INC
SIGMA-ALDRICH CORP
STIEFEL LABORATORIES INC
SYNGENTA AG
SYNOVIS LIFE TECHNOLOGIES INC
TEVA PHARMACEUTICAL INDUSTRIES
THERAGENICS CORP
THERMO FISHER SCIENTIFIC INC
THERMO SCIENTIFIC
UCB SA
UNITED THERAPEUTICS CORP
VERENIUM CORPORATION
WATSON PHARMACEUTICALS INC

NORTHEAST
ABBOTT LABORATORIES
ACCELRYS INC
ACTELION LTD
ADOLOR CORP
ADVANCED CELL TECHNOLOGY INC
ADVANSOURCE BIOMATERIALS CORPORATION
AETERNA ZENTARIS INC
AGILENT TECHNOLOGIES INC
AKORN INC
AKZO NOBEL NV
ALBANY MOLECULAR RESEARCH INC
ALCON INC
ALDAGEN INC
ALEXION PHARMACEUTICALS INC
ALKERMES INC
ALLOS THERAPEUTICS INC
ALPHARMA ANIMAL HEALTH
ALSERES PHARMACEUTICALS INC
ALTANA AG
AMARIN CORPORATION PLC
AMGEN INC
ANGIOTECH PHARMACEUTICALS INC
ANIKA THERAPEUTICS INC
ANTIGENICS INC
ARIAD PHARMACEUTICALS INC
ARQULE INC
ARRAY BIOPHARMA INC
ASTELLAS PHARMA INC
ASTRAZENECA PLC
AVANIR PHARMACEUTICALS
AVAX TECHNOLOGIES INC
AXCAN PHARMA INC
BASF SE
BAUSCH & LOMB INC
BAXTER INTERNATIONAL INC
BAYER AG

BAYER CORP
BIND BIOSCIENCES
BIO RAD LABORATORIES INC
BIOCRYST PHARMACEUTICALS INC
BIOGEN IDEC INC
BIOMET INC
BIONANOMATRIX
BIORELIANCE CORP
BIOSPHERE MEDICAL INC
BIOVAIL CORPORATION
BRACHYTHERAPY SERVICES INC
BRISTOL-MYERS SQUIBB CO
BURRILL & COMPANY
CALIPER LIFE SCIENCES
CAMBREX CORP
CANGENE BIOPHARMA INC
CANGENE CORP
CEGEDIM SA
CELERA CORPORATION
CELGENE CORP
CELLDEX THERAPEUTICS INC
CEL-SCI CORPORATION
CELSIS INTERNATIONAL PLC
CEPHALON INC
CHARLES RIVER LABORATORIES
INTERNATIONAL INC
CHIRON CORP
CML HEALTHCARE INCOME FUND
COLUMBIA LABORATORIES INC
COMMONWEALTH BIOTECHNOLOGIES INC
COVANCE INC
COVIDIEN PLC
CRUCELL NV
CSL LIMITED
CUBIST PHARMACEUTICALS INC
CURAGEN CORPORATION
CURIS INC
DAIICHI SANKYO CO LTD
DAINIPPON SUMITOMO PHARMA CO LTD
DENDREON CORPORATION
DISCOVERY LABORATORIES INC
DIVI'S LABORATORIES LIMITED
DR REDDY'S LABORATORIES LIMITED
DSM PHARMACEUTICALS INC
DU PONT AGRICULTURE & NUTRITION
DUSA PHARMACEUTICALS INC
DYAX CORP
E I DU PONT DE NEMOURS & CO (DUPONT)
EISAI CO LTD
ELAN CORP PLC
ELI LILLY & COMPANY
EMISPHERE TECHNOLOGIES INC
ENCORIUM GROUP INC
ENDO PHARMACEUTICALS HOLDINGS INC
ENTREMED INC
ENZO BIOCHEM INC
ENZON PHARMACEUTICALS INC
ERESEARCH TECHNOLOGY INC

EUSA PHARMA (USA) INC
FOREST LABORATORIES INC
GALDERMA PHARMA SA
GE HEALTHCARE
GENENCOR INTERNATIONAL INC
GENEREX BIOTECHNOLOGY
GEN-PROBE INC
GENTA INC
GENVEC INC
GENZYME BIOSURGERY
GENZYME CORP
GENZYME ONCOLOGY
GILEAD SCIENCES INC
GLAXOSMITHKLINE PLC
GTC BIOTHERAPEUTICS INC
HARVARD BIOSCIENCE INC
HEMAGEN DIAGNOSTICS INC
HEMISPHERX BIOPHARMA INC
HI-TECH PHARMACAL CO INC
HOSPIRA INC
HUMAN GENOME SCIENCES INC
I3 CANREG
ICON PLC
IDERA PHARMACEUTICALS INC
IDEXX LABORATORIES INC
IMCLONE SYSTEMS INC
IMMTECH PHARMACEUTICALS INC
IMMUCELL CORPORATION
IMMUNOGEN INC
IMMUNOMEDICS INC
IMPAX LABORATORIES INC
IMS HEALTH INC
INCYTE CORP
INFINITY PHARMACEUTICALS INC
INSMED INCORPORATED
INSPIRE PHARMACEUTICALS INC
INSTITUT STRAUMANN AG
INTEGRA LIFESCIENCES HOLDINGS CORP
IRONWOOD PHARMACEUTICALS INC
JOHNSON & JOHNSON
JUBILANT BIOSYS LTD
KENDLE INTERNATIONAL INC
KERYX BIOPHARMACEUTICALS INC
KING PHARMACEUTICALS INC
LANNETT COMPANY INC
LEXICON PHARMACEUTICALS INC
LIFE SCIENCES RESEARCH INC
LIFE TECHNOLOGIES CORP
LIFECELL CORPORATION
LONZA GROUP
MANHATTAN PHARMACEUTICALS INC
MEDICINES CO (THE)
MEDICIS PHARMACEUTICAL CORP
MEDIDATA SOLUTIONS INC
MEDIMMUNE INC
MEDTOX SCIENTIFIC INC
MERCK & CO INC
MERCK KGAA

MERCK SERONO SA
MERIDIAN BIOSCIENCE INC
MILLENNIUM PHARMACEUTICALS INC
MILLIPORE CORP
MONSANTO CO
MYLAN INC
NABI BIOPHARMACEUTICALS
NOVARTIS AG
NOVAVAX INC
NOVEN PHARMACEUTICALS
NOVO-NORDISK AS
NOVOZYMES
NPS PHARMACEUTICALS INC
NYCOMED
ORASURE TECHNOLOGIES INC
ORCHID CELLMARK INC
ORE PHARMACEUTICALS HOLDINGS INC
ORGANOGENESIS INC
OSI PHARMACEUTICALS INC
OXIGENE INC
PALATIN TECHNOLOGIES INC
PAR PHARMACEUTICAL COMPANIES INC
PAREXEL INTERNATIONAL CORP
PENWEST PHARMACEUTICALS CO
PERRIGO CO
PFIZER INC
PHARMACEUTICAL PRODUCT DEVELOPMENT INC
PHARMANET DEVELOPMENT GROUP INC
PHARMOS CORP
PHARSIGHT CORP
PIONEER HI-BRED INTERNATIONAL INC
PONIARD PHARMACEUTICALS INC
POZEN INC
PRA INTERNATIONAL
PROGENICS PHARMACEUTICALS
QIAGEN NV
QUEST DIAGNOSTICS INC
QUINTILES TRANSNATIONAL CORP
RANBAXY LABORATORIES LIMITED
REGENERON PHARMACEUTICALS INC
REPLIGEN CORPORATION
ROCHE HOLDING LTD
RULES BASED MEDICINE INC
SALIX PHARMACEUTICALS
SANOFI-AVENTIS SA
SAVIENT PHARMACEUTICALS INC
SCIOS INC
SEPRACOR INC
SEQUENOM INC
SERACARE LIFE SCIENCES INC
SHIONOGI INC
SHIRE PLC
SIEMENS HEALTHCARE DIAGNOSTICS
SIGA TECHNOLOGIES INC
SIGMA-ALDRICH CORP
SOLIGENIX INC
STEMCELLS INC
STIEFEL LABORATORIES INC

SYNBIOTICS CORP
SYNGENTA AG
SYNTHETIC GENOMICS
TAKEDA PHARMACEUTICAL COMPANY LTD
TAMIR BIOTECHNOLOGY INC
TARO PHARMACEUTICAL INDUSTRIES
TENGION INC
TEVA PHARMACEUTICAL INDUSTRIES
THERAGENICS CORP
THERMO FISHER SCIENTIFIC INC
THERMO SCIENTIFIC
TOLMAR INC
TOYOBO CO LTD
TRIMERIS INC
TRINITY BIOTECH PLC
UCB SA
UNIGENE LABORATORIES
UNITED THERAPEUTICS CORP
UNITED-GUARDIAN INC
VERENIUM CORPORATION
VERTEX PHARMACEUTICALS INC
VIACORD INC
VIROPHARMA INC
VIVUS INC
WARNER CHILCOTT PLC
WATSON PHARMACEUTICALS INC
WHATMAN PLC
XECHEM INTERNATIONAL
ZIOPHARM ONCOLOGY INC

INDEX OF FIRMS WITH
INTERNATIONAL OPERATIONS

JOHNSON & JOHNSON
JUBILANT BIOSYS LTD
KENDLE INTERNATIONAL INC
KING PHARMACEUTICALS INC
LIFE SCIENCES RESEARCH INC
LIFE TECHNOLOGIES CORP
LONZA GROUP
LORUS THERAPEUTICS INC
LUMINEX CORPORATION
MDS INC
MEDICINES CO (THE)
MEDIDATA SOLUTIONS INC
MEDIMMUNE INC
MERCK & CO INC
MERCK KGAA
MERCK SERONO SA
MERIDIAN BIOSCIENCE INC
MILLIPORE CORP
MONSANTO CO
MYLAN INC
NEKTAR THERAPEUTICS
NEOGEN CORPORATION
NEXMED INC
NOVARTIS AG
NOVO-NORDISK AS
NOVOZYMES
NYCOMED
ONCOGENEX PHARMACEUTICALS
ORCHID CELLMARK INC
ORGANOGENESIS INC
OSI PHARMACEUTICALS INC
OXIGENE INC
PAREXEL INTERNATIONAL CORP
PERRIGO CO
PFIZER INC
PHARMACEUTICAL PRODUCT DEVELOPMENT INC
PHARMANET DEVELOPMENT GROUP INC
PHARSIGHT CORP
PIONEER HI-BRED INTERNATIONAL INC
POLYDEX PHARMACEUTICALS
PRA INTERNATIONAL
PROMEGA CORP
QIAGEN NV
QLT INC
QUEST DIAGNOSTICS INC

QUINTILES TRANSNATIONAL CORP
RANBAXY LABORATORIES LIMITED
ROCHE HOLDING LTD
RULES BASED MEDICINE INC
SANOFI-AVENTIS SA
SCICLONE PHARMACEUTICALS
SENETEK PLC
SEPRACOR INC
SEQUENOM INC
SERACARE LIFE SCIENCES INC
SHIONOGI INC
SHIRE CANADA INC
SHIRE PLC
SIEMENS HEALTHCARE DIAGNOSTICS
SIGMA-ALDRICH CORP
SIMCERE PHARMACEUTICAL GROUP
SKYEPHARMA PLC
SOLIGENIX INC
SPECTRAL DIAGNOSTICS INC
SPECTRUM PHARMACEUTICALS INC
STEMCELLS INC
STIEFEL LABORATORIES INC
SYNBIOTICS CORP
SYNGENTA AG
TAKEDA PHARMACEUTICAL COMPANY LTD
TARO PHARMACEUTICAL INDUSTRIES
TECHNE CORP
TEVA PARENTERAL MEDICINES INC
TEVA PHARMACEUTICAL INDUSTRIES
THERMO FISHER SCIENTIFIC INC
THERMO SCIENTIFIC
TOYOBO CO LTD
TRINITY BIOTECH PLC
TRIPOS INTERNATIONAL
UCB SA
UNITED THERAPEUTICS CORP
VALEANT PHARMACEUTICALS INTERNATIONAL
VERNALIS PLC
VERTEX PHARMACEUTICALS INC
VIRBAC CORP
WARNER CHILCOTT PLC
WHATMAN PLC
XECHEM INTERNATIONAL

Individual Profiles
On Each Of
THE BIOTECH 350

23ANDME INC

www.23andme.com

Industry Group Code: 6215 Ranks within this company's industry group: Sales: Profits:

Drugs:	Other:		Clinical:		Computers:		Services:	
Discovery:	AgriBio:		Trials/Services:	Y	Hardware:		Specialty Services:	Y
Licensing:	Genetic Data:	Y	Labs:	Y	Software:	Y	Consulting:	
Manufacturing:	Tissue Replacement:		Equipment/Supplies:		Arrays:		Blood Collection:	
Generics:			Research/Development Svcs.:	Y	Database Management:		Drug Delivery:	
			Diagnostics:	Y			Drug Distribution:	

TYPES OF BUSINESS:

Genetic Testing Services
Genetic Research

BRANDS/DIVISIONS/AFFILIATES:

Genentech Inc
Google Inc
New Enterprise Associates
Personal Genome Service

CONTACTS: Note: Officers with more than one job title may be intentionally listed here more than once.

Anne Wojcicki, Pres./Co-Founder
Linda Avey, Co-Founder

Phone: 650-938-6300	Fax:
Toll-Free:	
Address: 1390 Shorebird Way, Mountain View, CA 94043 US	

GROWTH PLANS/SPECIAL FEATURES:

23AndMe, Inc. is a start-up company based in California that specializes in personalized genetic analysis. The firm offers its customers a detailed genetic profile using DNA analysis technologies and web-based interactive tools. To utilize 23AndMe's Personal Genome Service, a customer receives a sample collection kit by mail, along with instructions on how to collect the saliva necessary for analysis. The kit includes a pre-paid, pre-addressed envelope, along with a barcoded specimen tube. The firm's technicians extract DNA from cells in the saliva, which is then processed on an Illumina BeadChip, which reads more than 550,000 SNPs (single nucleotide polymorphisms), as well as a custom designed set of more than 25,000 additional SNPs. Using this analysis, the firm creates a genetic profile of the client, and in 2-4 weeks the customer can review results through an interactive web site. Personal Genome Service provides customers insight into ancestry, genetic predisposition to disease and genetic sensitivity to certain drugs. Personal Genome Service can also determine customers' carrier status for 24 inheritable diseases, including cystic fibrosis and sickle cell anemia. 23AndMe utilizes regular security audits and multiple encryption layers to protect customers' privacy. The compiled genetic information is also used for scientific research. The firm ships products to over 50 countries. 23andMe's investors include Genentech, Inc.; Google, Inc.; and New Enterprise Associates, a venture capital firm. In April 2009, the company entered a partnership with Palomar Pomerado Health (PPH) in California to offer Personal Genome Service at PPH's outpatient health centers. In June 2010, 23AndMe selected Informed Medical Decisions, Inc. to provide independent genetic counseling services to Personal Genome Service customers.

23AndMe offers its employees medical, dental and vision coverage, a 401(k) plan, stock options, catered lunches, on-site yoga, public transit passes, subsidized gym memberships and free genotyping for employees and a family member.

FINANCIALS: Sales and profits are in thousands of dollars—add 000 to get the full amount. 2009 Note: Financial information for 2009 was not available for all companies at press time.

2009 Sales: $	2009 Profits: $	U.S. Stock Ticker: Private
2008 Sales: $	2008 Profits: $	Int'l Ticker: Int'l Exchange:
2007 Sales: $	2007 Profits: $	Employees:
2006 Sales: $	2006 Profits: $	Fiscal Year Ends:
2005 Sales: $	2005 Profits: $	Parent Company:

SALARIES/BENEFITS:

Pension Plan:	ESOP Stock Plan:	Profit Sharing:	Top Exec. Salary: $	Bonus: $
Savings Plan: Y	Stock Purch. Plan:		Second Exec. Salary: $	Bonus: $

OTHER THOUGHTS:

Apparent Women Officers or Directors: 3
Hot Spot for Advancement for Women/Minorities: Y

LOCATIONS: ("Y" = Yes)

West:	Southwest:	Midwest:	Southeast:	Northeast:	International:
Y					

3SBIO INC

www.3sbio.com

Industry Group Code: 325412 Ranks within this company's industry group: Sales: 87 Profits: 56

Drugs:		Other:		Clinical:	Computers:	Services:	
Discovery:	Y	AgriBio:	Y	Trials/Services:	Hardware:	Specialty Services:	
Licensing:	Y	Genetic Data:		Labs:	Software:	Consulting:	
Manufacturing:	Y	Tissue Replacement:		Equipment/Supplies:	Arrays:	Blood Collection:	
Generics:				Research/Development Svcs.:	Database Management:	Drug Delivery:	
				Diagnostics:		Drug Distribution:	Y

TYPES OF BUSINESS:

Biopharmaceutical Manufacturing & Design
Anemia Treatments
Cancer Treatments
Exporting Biopharmaceuticals

BRANDS/DIVISIONS/AFFILIATES:

Shenyang Sunshine Pharmaceutical Company Limited
EPIAO
TPIAO
INTEFEN
INLEUSIN
Tietai Iron Sucrose Supplement
NuPIAO
NuLeusin

CONTACTS: Note: Officers with more than one job title may be intentionally listed here more than once.

Jing Lou, CEO
David Chen, COO/Exec. VP
Bo Tan, CFO
Bin Huang, VP-Human Resources
Yingfei Wei, Chief Scientific Officer
Dongmei Su, CTO
Ke Li, Corp. Sec.
Yingfei Wei, VP-Bus. Dev.
Dan Lou, Chmn.

Phone: 86-24-2581-1820	Fax:
Toll-Free:	
Address: No. 3 A1, Rd. 10, Shenyang Development Zone, Shenyang, 110027 China	

GROWTH PLANS/SPECIAL FEATURES:

3SBio, Inc. is a leading biotechnology company that researches, develops, manufactures and markets biopharmaceutical products, primarily in China. Substantially operating through wholly-owned subsidiary Shenyang Sunshine Pharmaceutical Company Ltd., 3SBio currently markets five drugs. EPIAO, an injectable drug, stimulates the production of red blood cells for anemic patients, reducing the need for transfusions. TPIAO is a protein-based treatment for chemotherapy-induced thrombocytopenia, a deficiency of platelets. INTEFEN is a treatment for lymphatic or hematopoietic system carcinomas, such as lymphoma and leukemia, as well as for viral infections such as hepatitis C. INLEUSIN has several applications, including the treatment of renal cell carcinoma, a common type of kidney cancer; metastatic melanoma, a skin cancer; and thoracic fluid build-up caused by cancer or tuberculosis. Lastly, Tietai Iron Sucrose Supplement is an injectable anemia treatment indicated for end-stage renal disease requiring iron replacement therapy. The company's development products, mostly related to its marketed products, include the following. NuPIAO is a second-generation EPIAO product for treating anemia following renal failure or chemotherapy. NuLeusin is a next generation INLEUSIN product, designed for the treatment of certain skin and kidney cancers. The company is adapting TPIAO for the treatment of idiopathic thrombocytopenic purpura (ITP), an immune disorder that causes the body to destroy its own platelets. A human papilloma virus (HPV) vaccine is also being developed. Finally, an anti-TNF (Tumor necrosis factor alpha) antibody is underway; TNF is responsible for regulating the inflammatory process and plays an underlying role in rheumatoid arthritis, psoriasis and other inflammatory disorders. EPIAO sales generate approximately 61.9% of the firm's revenue; TPIAO, 28.3%; INTEFEN and INLEUSIN, 2.2%; and Tietai, 3.4%, while export sales account for 4.2%. 3SBio's primary export markets include Egypt, Korea, Brazil, Pakistan, Thailand and Bangladesh. In February 2010, 3SBio announced a partnership with Panacor Bioscience to commercialize a treatment for elevated phosphate levels.

FINANCIALS: Sales and profits are in thousands of dollars—add 000 to get the full amount. 2009 Note: Financial information for 2009 was not available for all companies at press time.

2009 Sales: $46,429	2009 Profits: $12,224	U.S. Stock Ticker: SSRX
2008 Sales: $35,653	2008 Profits: $5,795	Int'l Ticker: Int'l Exchange:
2007 Sales: $24,700	2007 Profits: $11,174	Employees: 550
2006 Sales: $16,373	2006 Profits: $3,907	Fiscal Year Ends: 12/31
2005 Sales: $	2005 Profits: $	Parent Company:

SALARIES/BENEFITS:

Pension Plan:	ESOP Stock Plan:	Profit Sharing:	Top Exec. Salary: $	Bonus: $
Savings Plan:	Stock Purch. Plan:		Second Exec. Salary: $	Bonus: $

OTHER THOUGHTS:

Apparent Women Officers or Directors: 1
Hot Spot for Advancement for Women/Minorities: Y

LOCATIONS: ("Y" = Yes)

West:	Southwest:	Midwest:	Southeast:	Northeast:	International:
					Y

Note: Financial information, benefits and other data can change quickly and may vary from those stated here.

4SC AG

www.4sc.de

Industry Group Code: 325412 Ranks within this company's industry group: Sales: 135   Profits: 117

Drugs:		Other:	Clinical:	Computers:	Services:
Discovery:	Y	AgriBio:	Trials/Services:	Hardware:	Specialty Services:
Licensing:	Y	Genetic Data:	Labs:	Software:	Consulting:
Manufacturing:		Tissue Replacement:	Equipment/Supplies:	Arrays:	Blood Collection:
Generics:			Research/Development Svcs.:	Database Management:	Drug Delivery:
			Diagnostics:		Drug Distribution:

TYPES OF BUSINESS:

Drug Discovery & Pre-Clinical Development
Drugs-Oncology
Drugs-Inflammation
Drugs-Autoimmune Diseases
High Throughput Screening

BRANDS/DIVISIONS/AFFILIATES:

4SCan Technology
4SC-101
4SC-201
4SC-203
4SC-205
4SC-202
4SC-207

CONTACTS: *Note: Officers with more than one job title may be intentionally listed here more than once.*

Ulrich Dauer, CEO
Enno Spillner, CFO
Marion Fischbach, Mgr.-Human Resources
Daniel Vitt, Chief Science Officer
Bernd Hentsch, Chief Dev. Officer
Manfred Groppel, Dir.-Bus. Dev.
Yvonne Alexander, Mgr.-Public Rel.
Yvonne Alexander, Mgr.-Investor Rel.
Roland Baumgartner, Mgr.-Customer Rel.
Jorg Neermann, Chmn.
Tanja Wieber, Mgr.-Purchasing

Phone: 49-89-700-763-0	Fax: 49-89-700-763-29
Toll-Free:	
Address: Am Klopferspitz 19a, Planegg-Martinsried, 82152 Germany	

GROWTH PLANS/SPECIAL FEATURES:

4SC AG is a German-based drug discovery and development company that focuses on possible treatments for autoimmune diseases and cancer. The company develops its compounds to the clinical phase, after which they are licensed to the biopharmaceutical industry for further testing, development and commercialization. 4SC's technology platform uses a virtual high throughput screening technology known as 4SCan, which allows the high volume analysis of molecules based on protein structures, biological activity or homology modeling on a computer, rather than in a laboratory. Using this technology, the company has compiled a virtual library of small molecule drug candidates. The firm's product pipeline comprises six drug candidates in different development stages. The six drug candidates include 4SC-101; 4SC-201; 4SC-203; 4SC-205; 4SC-202; and 4SC-207. 4SC-101, also known as Vidofludimus, is the company's most advanced drug candidate in the field of autoimmune diseases. It is a small molecule drug that can be orally administered and is being developed for the treatment of rheumatoid arthritis and inflammatory bowel diseases. There are two clinical trials in progress for 4SC-101. 4SC-201, also known as Resminostat, is an orally administered pan histone deacetylase (HDAC) inhibitor. It is the company's most advanced cancer candidate and is in two Phase II studies for liver cancer and Hodgkin's lymphoma. 4SC-203 is a novel multi-target kinase inhibitor and is in preclinical testing. 4SC-205, a small molecule inhibitor of the human kinesin spindle protein Eg5, is in Phase I trials. 4SC-202 is an orally administered selective HDAC inhibitor that has completed the preclinical phase. Lastly, 4SC-207 is a novel, orally available cell-cycle blocker and is in the preclinical phase. The company has collaborations with AiCuris, ProQinase and ViroLogik.

FINANCIALS: Sales and profits are in thousands of dollars—add 000 to get the full amount. 2009 Note: Financial information for 2009 was not available for all companies at press time.

2009 Sales: $2,470	2009 Profits: $-21,380	**U.S. Stock Ticker:** VSC
2008 Sales: $4,220	2008 Profits: $-16,830	**Int'l Ticker:** Int'l Exchange:
2007 Sales: $2,100	2007 Profits: $-12,700	Employees: 94
2006 Sales: $5,700	2006 Profits: $-8,600	Fiscal Year Ends:
2005 Sales: $3,200	2005 Profits: $-9,800	Parent Company:

SALARIES/BENEFITS:

Pension Plan:	ESOP Stock Plan:	Profit Sharing:	Top Exec. Salary: $	Bonus: $
Savings Plan:	Stock Purch. Plan:		Second Exec. Salary: $	Bonus: $

OTHER THOUGHTS:

Apparent Women Officers or Directors: 2
Hot Spot for Advancement for Women/Minorities: Y

LOCATIONS: ("Y" = Yes)

West:	Southwest:	Midwest:	Southeast:	Northeast:	International: Y

AASTROM BIOSCIENCES INC

www.aastrom.com

Industry Group Code: 33911 Ranks within this company's industry group: Sales: 15 Profits: 13

Drugs:	Other:	Clinical:		Computers:	Services:
Discovery:	AgriBio:	Trials/Services:	Y	Hardware:	Specialty Services:
Licensing:	Genetic Data:	Labs:		Software:	Consulting:
Manufacturing:	Tissue Replacement: Y	Equipment/Supplies:		Arrays:	Blood Collection:
Generics:		Research/Development Svcs.:		Database Management:	Drug Delivery:
		Diagnostics:			Drug Distribution:

TYPES OF BUSINESS:

Cell Products Development
Regenerative medicine products
Clinical & Pre-clinical Development

BRANDS/DIVISIONS/AFFILIATES:

Single-Pass Perfusion (SPP)

CONTACTS: Note: Officers with more than one job title may be intentionally listed here more than once.

Timothy M. Mayleben, CEO
Timothy M. Mayleben, Pres.
Scott Durbin, CFO
Ronnda L. Bartel, Chief Scientific Officer
Julie A. Caudill, Corp. Sec.
Sheldon A. Schaffer, VP-Corp. Dev.
Kimberli O'Meara, Contact-Investor
Sharon Watling, VP-Clinical & Regulatory
George Dunbar, Chmn.

Phone: 734-930-5555	Fax: 734-665-0485
Toll-Free:	
Address: 24 Frank Lloyd Wright Dr., Ann Arbor, MI 48105 US	

GROWTH PLANS/SPECIAL FEATURES:

Aastrom Biosciences, Inc. is a development stage company focused on the development of autologous cell products for use in regenerative medicine. The company's pre-clinical and clinical products development programs utilize patient-derived bone marrow stem and progenitor cell populations, which are being investigated for their ability to aid in the regeneration of vascular, bone, cardiac and neural tissues. The firm's primary business is to develop Tissue Repair Cell (TRC) based products using Single-Pass Perfusion (SPP) to replicate early-stage stem and progenitor cells. The TRC platform technology is based on its cell products (a unique cell mixture containing large numbers of stromal, stem and progenitor cells, produced outside the body from a small amount of bone marrow taken from a patient) and the means to produce these products in an automated process. TRC-based products have been used in over 350 patients. The pre-clinical data for the TRCs showed a substantial increase in the stem and progenitor cells that can develop into tissues such as hematopoietic (i.e., blood forming) or mesenchymal (i.e., developing into tissues characteristic of certain internal organs), as well as stromal progenitor cells that produce various growth factors. The company demonstrated in the laboratory that TRCs can progress into bone cell and blood vessel cell lineages. Based on these pre-clinical observations, the TRCs are currently in active clinical trials for bone regeneration and vascular regeneration applications. The firm developed a patented manufacturing system to produce human cells for clinical use. Aastrom currently owns 26 U.S. patents and 41 foreign patents. TRC therapies are currently undergoing three late-stage U.S. clinical trials: two in dilated cardiomyopathy (inflammation of the heart muscles) and one in critical limb ischemia (severe obstruction of the arteries in the lower extremities).

FINANCIALS: Sales and profits are in thousands of dollars—add 000 to get the full amount. 2009 Note: Financial information for 2009 was not available for all companies at press time.

2009 Sales: $ 182	2009 Profits: $-15,946	U.S. Stock Ticker: ASTM
2008 Sales: $ 522	2008 Profits: $-20,133	Int'l Ticker: Int'l Exchange:
2007 Sales: $ 685	2007 Profits: $-17,594	Employees: 48
2006 Sales: $ 863	2006 Profits: $-16,475	Fiscal Year Ends: 6/30
2005 Sales: $ 909	2005 Profits: $-11,811	Parent Company:

SALARIES/BENEFITS:

Pension Plan:	ESOP Stock Plan:	Profit Sharing:	Top Exec. Salary: $375,000	Bonus: $
Savings Plan:	Stock Purch. Plan:		Second Exec. Salary: $239,583	Bonus: $19,721

OTHER THOUGHTS:

Apparent Women Officers or Directors: 5
Hot Spot for Advancement for Women/Minorities: Y

LOCATIONS: ("Y" = Yes)

West:	Southwest:	Midwest:	Southeast:	Northeast:	International:
		Y			

ABAXIS INC

www.abaxis.com

Industry Group Code: 33911 Ranks within this company's industry group: Sales: 8 Profits: 7

Drugs:	Other:	Clinical:		Computers:	Services:
Discovery:	AgriBio:	Trials/Services:		Hardware:	Specialty Services:
Licensing:	Genetic Data:	Labs:		Software:	Consulting:
Manufacturing:	Tissue Replacement:	Equipment/Supplies:	Y	Arrays:	Blood Collection:
Generics:		Research/Development Svcs.:		Database Management:	Drug Delivery:
		Diagnostics:	Y		Drug Distribution:

TYPES OF BUSINESS:

Point-of-Care Blood Analyzer Systems Equipment
Veterinary Blood Analyzer Systems
Reagents & Supplies

BRANDS/DIVISIONS/AFFILIATES:

VetScan VS2
VetScan Classic
VetScan Vspro
Piccolo Xpress
VetScan HM2
Orbos Discrete Lyophilization Process
VetScan HM5
Piccolo xpress

CONTACTS: Note: Officers with more than one job title may be intentionally listed here more than once.

Clint Severson, CEO
Donald Wood, COO
Clint Severson, Pres.
Al Santa Ines, CFO
Kenneth Aron, CTO
Martin Mulroy, VP-Veterinary Sales & Mktg., North America
Brenton G.A. Hanlon, VP-Medical Sales & Mktg., North America
Clint Severson, Chmn.
Vladimir E. Ostoich, VP-Gov't Affairs & Mktg., Pacific Rim

Phone: 510-675-6500	Fax: 510-441-6150
Toll-Free:	
Address: 3240 Whipple Rd., Union City, CA 94587 US	

GROWTH PLANS/SPECIAL FEATURES:

Abaxis, Inc. develops, manufactures and markets portable blood analysis systems for use in veterinary or human patient-care settings to provide clinicians with rapid blood constituent measurements. The company provides over 90% of the chemical tests in animal and medical diagnoses. The medical market accounted for 23% of total revenue in 2009, and the veterinary market accounted for 70%. The firm's analysis systems includes the VetScan VS2, VetScan HM2, VetScan VSpro, VetScan Profiles, VetScan i-STAT 1 Handheld Analyzer and VetScan Classic in the veterinary market; and the Piccolo xpress and the Piccolo xpress Panels in the human medical market. Abaxis' blood analysis systems consist of a compact analyzer and a series of single-use plastic discs, called reagent discs, containing all the chemicals required to perform a panel of up to 27 diagnostic tests. In addition to blood analysis systems, Abaxis sells the VetScan HM5 hematology analyzer that provides a 22-parameter blood count analysis, including a five-part white blood cell differential. To produce the dry reagents used in the reagent disks, Abaxis uses its Orbos Discrete Lyophilization Process (Orbos process), a process which freeze dries reagents in small quantities, enabling efficient manufacturing of reagents in a convenient and stable format. Abaxis licenses its Orbos process to bioMerieux, Cepheid and GE Healthcare. In October 2009, the company received USDA approval for its Canine Heartworm Antigen Test that is performed on its VetScan VS2 system. In December of the same year, the firm entered the production animal diagnostics market with the development of an Avian Influenza Antigen test with the ability to show results in 15 minutes.

Employees are offered medical, dental and vision insurance; life insurance; disability coverage; a 401(k) plan; flexible spending accounts; and memberships to financial institutions, retail stores and clubs.

FINANCIALS: Sales and profits are in thousands of dollars—add 000 to get the full amount. 2009 Note: Financial information for 2009 was not available for all companies at press time.

2009 Sales: $105,600	2009 Profits: $12,000	**U.S. Stock Ticker: ABAX**
2008 Sales: $100,551	2008 Profits: $12,503	**Int'l Ticker:** Int'l Exchange:
2007 Sales: $86,221	2007 Profits: $10,073	Employees: 339
2006 Sales: $68,928	2006 Profits: $7,475	Fiscal Year Ends: 3/31
2005 Sales: $52,758	2005 Profits: $4,851	Parent Company:

SALARIES/BENEFITS:

Pension Plan:	ESOP Stock Plan:	Profit Sharing:	Top Exec. Salary: $355,770	Bonus: $226,406
Savings Plan: Y	Stock Purch. Plan:		Second Exec. Salary: $208,654	Bonus: $129,375

OTHER THOUGHTS:

Apparent Women Officers or Directors: 1
Hot Spot for Advancement for Women/Minorities:

LOCATIONS: ("Y" = Yes)

West:	Southwest:	Midwest:	Southeast:	Northeast:	International:
Y					Y

ABBOTT LABORATORIES

www.abbott.com

Industry Group Code: 325412 Ranks within this company's industry group: Sales: 8   Profits: 10

Drugs:		Other:	Clinical:		Computers:		Services:	
Discovery:	Y	AgriBio:	Trials/Services:		Hardware:		Specialty Services:	
Licensing:		Genetic Data:	Labs:		Software:		Consulting:	
Manufacturing:	Y	Tissue Replacement:	Equipment/Supplies:	Y	Arrays:		Blood Collection:	
Generics:			Research/Development Svcs.:		Database Management:		Drug Delivery:	
			Diagnostics:				Drug Distribution:	

TYPES OF BUSINESS:

Pharmaceuticals Manufacturing
Nutritional Products
Diagnostics
Consumer Health Products
Medical & Surgical Devices
Pharmaceutical Products
Animal Health
Eye Care Products

BRANDS/DIVISIONS/AFFILIATES:

Solvay Pharmaceuticals
Abbott Medical Optics Inc
Experimental & Applied Sciences Inc
Vysis Inc
Humira
Multi-Link Vision
PediaSure
Ensure

CONTACTS: Note: Officers with more than one job title may be intentionally listed here more than once.

Miles D. White, CEO
Thomas C. Freyman, CFO
Stephen R. Fussell, Sr. VP-Human Resources
John C. Landgraf, Sr. VP-Mfg. & Pharmaceuticals
Laura J. Schumacher, General Counsel/Exec. VP/Sec.
Richard W. Ashley, Exec. VP-Corp. Dev.
Melissa Brotz, VP-External Comm.
Thomas C. Freyman, Exec. VP-Finance
Olivier Bohoun, Exec. VP-Pharmaceutical Prod.
Holger Liepmann, Exec. VP-Nutritional Prod.
John M. Capek, Exec. VP-Medical Devices
Edward L. Michael, Exec. VP-Diagnostics Prod.
Miles D. White, Chmn.
John C. Landgraf, Sr. VP-Supply

Phone: 847-937-6100	Fax: 847-937-9555
Toll-Free:	
Address: 100 Abbott Park Rd., Abbott Park, IL 60064 US	

GROWTH PLANS/SPECIAL FEATURES:

Abbott Laboratories develops, manufactures and sells health care products and technologies ranging from pharmaceuticals to medical devices. The firm markets its products in more than 130 countries. The company operates in four segments: pharmaceutical products, diagnostics products, nutritional products and vascular products. The pharmaceutical segment deals with adult and pediatric conditions such as rheumatoid arthritis, HIV, epilepsy and manic depression. The diagnostic instruments and test segment deals with a range of medical tests to diagnose infectious diseases, cancer, diabetes and genetic conditions. Products include Humira for the treatment of arthritis, psoriasis and Crohn's disease; Aluvia, a protease inhibitor for HIV infection; and Synthroid for hypothyroidism The diagnostic products segment includes diagnostic systems and tests such as the Commander immunoassay system, ARCHITECT chemistry system and i-STAT hematology systems, which are manufactured, marketed and sold to blood banks, hospitals, commercial laboratories, physicians' offices and plasma protein therapeutic companies. The nutritional products segment offers consumer products such as Similac, Ensure, PediaSure and Zone Perfect, as well as feeding devices in healthcare institutions. The vascular products segment consists of coronary, endovascular and vessel closure devices, used in the treatment of vascular disease. Products include the Multi-Link Vision coronary metallic stents, Voyager balloon dilatation systems and StarClose vessel closure devices. These products are generally marketed and sold directly to hospitals from Abbot-owned distribution centers and public warehouses. The company's other products include the FreeStyle line of diabetes products, as well as medical devices for the eye and a line of animal products for the veterinary market. The company operates internationally in Europe, Asia Pacific, Africa, Latin America and the Middle East. In February 2010, the firm acquired Solvay Pharmaceuticals from Belgium-based Solvay SA for $6.2 billion.

Employees are offered medical, dental and vision insurance; flexible spending accounts; adoption assistance; an employee assistance program; legal services; tuition assistance and life insurance.

FINANCIALS: Sales and profits are in thousands of dollars—add 000 to get the full amount. 2009 Note: Financial information for 2009 was not available for all companies at press time.

2009 Sales: $30,764,700	2009 Profits: $5,745,800	U.S. Stock Ticker: ABT
2008 Sales: $29,527,600	2008 Profits: $4,880,700	Int'l Ticker: Int'l Exchange:
2007 Sales: $25,914,200	2007 Profits: $3,606,300	Employees: 73,000
2006 Sales: $22,476,322	2006 Profits: $1,716,755	Fiscal Year Ends: 12/31
2005 Sales: $22,337,808	2005 Profits: $3,372,065	Parent Company:

SALARIES/BENEFITS:

Pension Plan: Y	ESOP Stock Plan:	Profit Sharing: Y	Top Exec. Salary: $1,852,319	Bonus: $3,900,000
Savings Plan: Y	Stock Purch. Plan:		Second Exec. Salary: $914,461	Bonus: $1,286,000

OTHER THOUGHTS:

Apparent Women Officers or Directors: 10
Hot Spot for Advancement for Women/Minorities: Y

LOCATIONS: ("Y" = Yes)

West:	Southwest:	Midwest:	Southeast:	Northeast:	International:
Y	Y	Y		Y	Y

Note: Financial information, benefits and other data can change quickly and may vary from those stated here.

ACCELRYS INC

www.accelrys.com

Industry Group Code: 511210D Ranks within this company's industry group: Sales: 4   Profits: 4

Drugs:	Other:	Clinical:	Computers:		Services:
Discovery:	AgriBio:	Trials/Services:	Hardware:		Specialty Services:
Licensing:	Genetic Data:	Labs:	Software:	Y	Consulting:
Manufacturing:	Tissue Replacement:	Equipment/Supplies:	Arrays:		Blood Collection:
Generics:		Research/Development Svcs.:	Database Management:		Drug Delivery:
		Diagnostics:			Drug Distribution:

TYPES OF BUSINESS:

Software - Simulation & Informatics
Computational Nanotechnology Tools
Informatics Software
Modeling & Simulation Software

BRANDS/DIVISIONS/AFFILIATES:

SciTegic Pipeline Pilot
Pipeline Pilot Enterprise Server 7.5

CONTACTS: *Note: Officers with more than one job title may be intentionally listed here more than once.*

Max Carnecchia, CEO
Max Carnecchia, Pres.
Michael A. Piraino, CFO/Sr. VP
Todd Johnson, Exec. VP-Sales, Mktg. & Svcs.
Judith Ohrn Hicks, VP-Human Resources
Matt Hahn, Sr. VP-R&D
David Mersten, General Counsel/Corp. Sec./Sr. VP
Maria Krinsky, Contact-Press
Paul Burrin, Chief Mktg. Officer/Sr. VP
Kenneth L. Coleman, Chmn.

Phone: 858-799-5000	Fax: 858-799-5100
Toll-Free:	
Address: 10188 Telesis Ct., Ste. 100, San Diego, CA 92121-3752 US	

GROWTH PLANS/SPECIAL FEATURES:

Accelrys, Inc. develops and commercializes scientific business intelligence software and services designed to accelerate the discovery and development of new drugs and materials. Its customers include pharmaceutical, biotechnology and life science companies as well as companies operating in the energy, aerospace and consumer packaged goods markets. Biologists, chemists, other scientists and information technology professionals at these firms use Accelrys products to aggregate, mine, integrate, analyze, simulate, manage and interactively report scientific data. The firm offers products in four primary categories: Pipeline Pilot, Informatics, Discovery Studio and Materials Studio. Accelrys' Pipeline Pilot platform allows users to aggregate, integrate and mine vast quantities of both structured and unstructured scientific data such as chemical structures, biological sequences and complex digital images. It also filters, normalizes and performs statistical analysis on this scientific data and provides interactive visual reports to both scientists and scientific managers. Its Informatics data management and analysis product lines are designed to work with Pipeline Pilot and include Accord cheminformatics database tools geared towards pharmaceutical development; bioinformatics collection, mining, analysis and sharing tools; and the Accelrys Biological Registration System, a multi-entity, fully-integrated system designed to facilitate registration, relationship-tracking and reporting on biological entities. Accelrys' Discovery Studio life science modeling and simulation products are designed to aid in the drug discovery process with tools for examining molecular properties, study systems and more. Materials Studio modeling and simulation products are designed for chemical, materials and pharmaceutical development. The firm also offers consulting, training, support and contract research services. In April 2010, Accelrys and Symyx Technologies, Inc. agreed to merge.

Employees are offered medical, dental and vision insurance; flexible health and dependent care spending accounts; an employee assistance program; life insurance; short-and long-term disability coverage; a retirement savings plan; educational savings programs; employee stock equity programs; tuition reimbursement; a computer loan program; and subsidized gym memberships.

FINANCIALS: Sales and profits are in thousands of dollars—add 000 to get the full amount. 2009 Note: Financial information for 2009 was not available for all companies at press time.

2009 Sales: $80,981	2009 Profits: $ 94	U.S. Stock Ticker: ACCL
2008 Sales: $79,739	2008 Profits: $1,321	Int'l Ticker: Int'l Exchange:
2007 Sales: $80,955	2007 Profits: $-1,525	Employees: 364
2006 Sales: $82,001	2006 Profits: $-7,739	Fiscal Year Ends: 3/31
2005 Sales: $79,030	2005 Profits: $-16,578	Parent Company:

SALARIES/BENEFITS:

Pension Plan:	ESOP Stock Plan:	Profit Sharing:	Top Exec. Salary: $350,000	Bonus: $315,000
Savings Plan: Y	Stock Purch. Plan: Y		Second Exec. Salary: $275,000	Bonus: $104,000

OTHER THOUGHTS:

Apparent Women Officers or Directors: 2
Hot Spot for Advancement for Women/Minorities: Y

LOCATIONS: ("Y" = Yes)

West:	Southwest:	Midwest:	Southeast:	Northeast:	International:
Y				Y	Y

Note: Financial information, benefits and other data can change quickly and may vary from those stated here.

ACCESS PHARMACEUTICALS INC

www.accesspharma.com

Industry Group Code: 325412 Ranks within this company's industry group: Sales: 147 Profits: 110

Drugs:		Other:		Clinical:	Computers:		Services:	
Discovery:	Y	AgriBio:		Trials/Services:	Hardware:		Specialty Services:	
Licensing:	Y	Genetic Data:		Labs:	Software:		Consulting:	
Manufacturing:	Y	Tissue Replacement:		Equipment/Supplies:	Arrays:		Blood Collection:	
Generics:				Research/Development Svcs.:	Database Management:		Drug Delivery:	Y
				Diagnostics:			Drug Distribution:	

TYPES OF BUSINESS:

Pharmaceutical Development
Drug Delivery Systems
Polymer Technology
Oncology Products

BRANDS/DIVISIONS/AFFILIATES:

MuGard
ProLindac
MacroChem Corp.
Somanta Pharmaceuticals
Thiarabine
Cobalamin

CONTACTS: *Note: Officers with more than one job title may be intentionally listed here more than once.*

Jeffrey B. Davis, CEO
Jeffrey B. Davis, Pres.
Stephen B. Thompson, CFO/VP
Frank Jacobucci, VP-Mktg. & Sales
David P. Nowotnik, Sr. VP-R&D
Phillip Wise, VP-Bus. Dev. & Strategy
Esteban Cvitkovic, Vice Chmn.
Esteban Cvitkovic, Sr. Dir.-Clinical Oncology R&D
Steven H. Rouhandeh, Chmn.

Phone: 214-905-5100	Fax: 214-905-5101
Toll-Free:	
Address: 2600 Stemmons Freeway, Ste. 176, Dallas, TX 75207 US	

GROWTH PLANS/SPECIAL FEATURES:

Access Pharmaceuticals, Inc. is a biopharmaceutical company developing products for use in the treatment of cancer, the supportive care of cancer and other disease states. Beyond its own research, Access also incorporates the work of its recent acquisitions, Somanta Pharmaceuticals and MacroChem Corp. The company has one technology approved by the FDA, MuGard, which is a viscous polymer solution that coats the oral cavity and is used in the treatment of mucositis, an issue that is faced by many chemotherapy patients. The firm also has three drug delivery technology platforms: synthetic polymer targeted delivery, which is designed to exploit enhanced permeability and retention at tumor sites to selectively accumulate drug and control drug release; Cobalamin-medicated oral delivery, which utilizes vitamin B12 to increase absorption of orally consumed medicines; and Cobalamin-medicated targeted delivery. Other drugs in development include ProLindac and Thiarabine. ProLindac, a nanopolymer DACH-platinum prodrug, is the company's lead development candidate for the treatment of cancer. The firm recently completed a Phase II clinical trial, which had positive results; it also initiated a study of ProLindac combined with Paclitaxel in second line treatment of advanced ovarian cancer patients. Thiarabine, or 4-thio Ara-C, is a next generation nucleoside analog licensed from Southern Research Institute. It is a potential treatment for leukemia and lymphoma. Previously named SR9025 and OSI-7836, the compound has been in two Phase I/II solid tumor human clinical trials and was shown to have anti-tumor activity. The firm intends to initiate additional Phase II clinical trials with Thiarabine. In October 2009, the company signed an agreement with iMedicor for the North American commercial launch of MuGard.

FINANCIALS: Sales and profits are in thousands of dollars—add 000 to get the full amount. 2009 Note: Financial information for 2009 was not available for all companies at press time.

2009 Sales: $ 352	2009 Profits: $-17,340	U.S. Stock Ticker: ACCP
2008 Sales: $ 291	2008 Profits: $-31,431	Int'l Ticker: Int'l Exchange:
2007 Sales: $ 57	2007 Profits: $-36,652	Employees: 10
2006 Sales: $	2006 Profits: $-12,874	Fiscal Year Ends: 12/31
2005 Sales: $	2005 Profits: $-1,700	Parent Company:

SALARIES/BENEFITS:

Pension Plan:	ESOP Stock Plan:	Profit Sharing:	Top Exec. Salary: $200,000	Bonus: $
Savings Plan: Y	Stock Purch. Plan:		Second Exec. Salary: $175,675	Bonus: $

OTHER THOUGHTS:

Apparent Women Officers or Directors:
Hot Spot for Advancement for Women/Minorities:

LOCATIONS: ("Y" = Yes)

West:	Southwest:	Midwest:	Southeast:	Northeast:	International:
	Y				Y

Note: Financial information, benefits and other data can change quickly and may vary from those stated here.

ACTELION LTD

www.actelion.com

Industry Group Code: 325412 Ranks within this company's industry group: Sales: 35 Profits: 30

Drugs:		Other:	Clinical:	Computers:	Services:
Discovery:	Y	AgriBio:	Trials/Services:	Hardware:	Specialty Services:
Licensing:		Genetic Data:	Labs:	Software:	Consulting:
Manufacturing:		Tissue Replacement:	Equipment/Supplies:	Arrays:	Blood Collection:
Generics:			Research/Development Svcs.:	Database Management:	Drug Delivery:
			Diagnostics:		Drug Distribution:

TYPES OF BUSINESS:
Drugs, Discovery & Development
Pharmaceutical Research
Cardiovascular Treatment
Genetic Disorder Treatment

BRANDS/DIVISIONS/AFFILIATES:
G-Protein Coupled Receptors (GPCRs)
Tracleer
Zavesca
Ventavis
Bosentan
Miglustat
Macitentan
Almorexant

CONTACTS:
Note: Officers with more than one job title may be intentionally listed here more than once.
Jean-Paul Clozel, CEO
Andrew J. Oakley, CFO/VP
Guy Braunstein, Sr. VP/Head-Clinical Dev.
Marian Borovsky, General Counsel/VP/Corp. Sec.
Otto Schwarz, Pres., Bus. Oper.
Simon Buckingham, Pres., Corp. & Bus. Dev.
Roland Haefeli, Head-Public Affairs
Roland Haefeli, VP/Head-Investor Rel.
Thomas Weller, VP/Head-Drug Discovery, Chemistry
Isaac Kobrin, Chief Medical Officer/Sr. VP
Martine Clozel, Sr. VP/Head-Pharmacology & Pre-Clinical Dev.
Robert E. Cawthorn, Chmn.

Phone: 41-61-565-65-65	Fax: 41-61-565-65-00
Toll-Free:	
Address: Gewerbestrasse 16, Allschwil, CH-4123 Switzerland	

GROWTH PLANS/SPECIAL FEATURES:
Actelion, Ltd. is a biopharmaceutical company that focuses on the discovery, development and marketing of drugs for unaddressed medical needs. Actelion focuses its drug discovery efforts on the design and synthesis of novel low molecular weight, drug-like molecules. Additional drug discovery platforms include G-Protein Coupled Receptors (GPCRs), aspartic proteinases, anti-infectives and ion channels. Actelion's most recognized product is Tracleer, a dual endothelin receptor antagonist used for pulmonary arterial hypertension (PAH). Another popular Actelion drug, Zavesca, is one of the first approved oral drug therapy treatments for a genetic lipid metabolic disorder called Gaucher disease. The company's Ventavis is an inhaled formulation of iloprost for the treatment of PAH. Additional drugs that are currently in the developmental stage include Bosentan, Miglustat, Almorexant, Clazosentan and Macitentan. Bosentan is an orally active dual endothelin receptor antagonist for the treatment of PAH. Miglustat is a low molecular weight inhibitor of glucosylceramide synthase and glucosidase for the treatment of Gaucher disease. Almorexant is a first-in-class receptor antagonist for the treatment of sleep disorders. Clazosentan is an intravenous endothelin receptor antagonist intended for the prevention and treatment of vasospasms, a life-threatening condition that leads to neurological deficits after a patient suffers an aneurysm. Macitentan is a tissue-targeting endothelin receptor antagonist intended for treatment of the cardiovascular system. Actelion is based in Switzerland and has subsidiaries in more than 25 countries.

Actelion offers its employees performance based bonuses and employee stock ownership programs.

FINANCIALS:
Sales and profits are in thousands of dollars—add 000 to get the full amount. 2009 Note: Financial information for 2009 was not available for all companies at press time.

2009 Sales: $1,619,800	2009 Profits: $284,470	**U.S. Stock Ticker:**
2008 Sales: $1,254,960	2008 Profits: $273,810	**Int'l Ticker: ATLN** Int'l Exchange: Zurich-SWX
2007 Sales: $1,291,000	2007 Profits: $122,100	Employees: 2,200
2006 Sales: $926,800	2006 Profits: $236,300	Fiscal Year Ends: 12/31
2005 Sales: $545,168	2005 Profits: $103,135	Parent Company:

SALARIES/BENEFITS:
Pension Plan:	ESOP Stock Plan: Y	Profit Sharing:	Top Exec. Salary: $	Bonus: $
Savings Plan:	Stock Purch. Plan:		Second Exec. Salary: $	Bonus: $

OTHER THOUGHTS:
Apparent Women Officers or Directors: 1
Hot Spot for Advancement for Women/Minorities:

LOCATIONS: ("Y" = Yes)
West:	Southwest:	Midwest:	Southeast:	Northeast:	International:
Y				Y	Y

ADAMIS PHARMACEUTICALS CORPORATION

www.adamispharmaceuticals.com
Industry Group Code: 325412 Ranks within this company's industry group: Sales: 144 Profits: 76

Drugs:		Other:	Clinical:	Computers:	Services:
Discovery:	Y	AgriBio:	Trials/Services:	Hardware:	Specialty Services:
Licensing:	Y	Genetic Data:	Labs:	Software:	Consulting:
Manufacturing:		Tissue Replacement:	Equipment/Supplies:	Arrays:	Blood Collection:
Generics:			Research/Development Svcs.:	Database Management:	Drug Delivery:
			Diagnostics:		Drug Distribution:

TYPES OF BUSINESS:

Drugs
Allergy Medications
Respiratory Diseases
Pediatric Conditions
Vaccines

BRANDS/DIVISIONS/AFFILIATES:

Adamis Labs
Adamis Viral Therapies
Cellegy Pharmaceuticals
Colby Pharmaceuticals Corporation

CONTACTS: Note: Officers with more than one job title may be intentionally listed here more than once.

Dennis J. Carlo, CEO
Dennis J. Carlo, Pres.
Robert O. Hopkins, CFO/VP
Thomas Moll, VP-Research
Karen K. Daniels, VP-Oper.
David J. Marguglio, VP-Corp. Dev.
Maurizio Zanetti, Chief Scientific Advisor
Richard C. Williams, Chmn.

Phone: 858-401-3984	Fax:
Toll-Free:	
Address: 2658 Del Mar Heights Rd., Ste. 555, Del Mar, CA 92014 US	

GROWTH PLANS/SPECIAL FEATURES:

Adamis Pharmaceuticals Corporation is a commercial-stage pharmaceutical company specializing in the development of medications for allergy, respiratory disease and pediatric conditions. Operations of Adamis are split between the Adamis Labs division and Adamis Viral Therapies (Adamis' research and development division). Work in the Adamis Labs division has focused on developing and launching its PFS Syringe, a pre-filled epinephrine syringe used for treatment of anaphylactic shock. The syringe was launched and made commercially available in July 2009. A second product being developed is a generic inhaled nasal steroid for treating seasonal and perennial allergic rhinitis. Adamis Viral Therapies focuses its work on developing products for prevention and treatment of certain infectious viral diseases. Its primary focus is now on developing a vaccine for the avian influenza virus. In February 2009, Cellegy Pharmaceuticals was acquired through merger by Adamis Pharmaceuticals. In February 2010, Adamis signed an agreement with Colby Pharmaceuticals Corporation to acquire exclusive licensing rights for its three small molecule compounds designed for the treatment of prostate cancer. APC-100, one of the acquired compounds, is an orally bioavailable drug that is hoped to have increased efficacy over standard prostate cancer treatments.

FINANCIALS: Sales and profits are in thousands of dollars—add 000 to get the full amount. 2009 Note: Financial information for 2009 was not available for all companies at press time.

2009 Sales: $ 660	2009 Profits: $-1,895	U.S. Stock Ticker: ADMP
2008 Sales: $	2008 Profits: $-1,534	Int'l Ticker: Int'l Exchange:
2007 Sales: $	2007 Profits: $-1,927	Employees: 14
2006 Sales: $2,660	2006 Profits: $9,672	Fiscal Year Ends: 12/31
2005 Sales: $12,199	2005 Profits: $-5,008	Parent Company:

SALARIES/BENEFITS:

Pension Plan:	ESOP Stock Plan:	Profit Sharing:	Top Exec. Salary: $273,770	Bonus: $
Savings Plan:	Stock Purch. Plan:		Second Exec. Salary: $182,103	Bonus: $

OTHER THOUGHTS:

Apparent Women Officers or Directors: 1
Hot Spot for Advancement for Women/Minorities:

LOCATIONS: ("Y" = Yes)

West:	Southwest:	Midwest:	Southeast:	Northeast:	International:
Y					

ADOLOR CORP

www.adolor.com

Industry Group Code: 325412 Ranks within this company's industry group: Sales: 93   Profits: 145

Drugs:		Other:	Clinical:	Computers:	Services:
Discovery:	Y	AgriBio:	Trials/Services:	Hardware:	Specialty Services:
Licensing:	Y	Genetic Data:	Labs:	Software:	Consulting:
Manufacturing:		Tissue Replacement:	Equipment/Supplies:	Arrays:	Blood Collection:
Generics:			Research/Development Svcs.:	Database Management:	Drug Delivery:
			Diagnostics:		Drug Distribution:

TYPES OF BUSINESS:

Drugs, Discovery & Development
Pain Management Products
Gastrointestinal Products

BRANDS/DIVISIONS/AFFILIATES:

Entereg
ADL5945
ADL7445

CONTACTS: Note: Officers with more than one job title may be intentionally listed here more than once.

Michael R. Dougherty, CEO
Michael R. Dougherty, Pres.
Stephen W. Webster, CFO
Michaed D. Adelman, VP-Mktg. & Sales
George R. Maurer, Sr. VP-Mfg. & Pharmaceutical Tech.
John M. Limongelli, General Counsel/Sr. VP/Sec.
Stephen W. Webster, Sr. VP-Finance
David P. Geoghegan, VP-Mfg.
Kevin Darryl White, VP-Regulatory Affairs
David M. Madden, Chmn.

Phone: 484-595-1500	Fax: 484-595-1520
Toll-Free:	
Address: 700 Pennsylvania Dr., Exton, PA 19341 US	

GROWTH PLANS/SPECIAL FEATURES:

Adolor Corporation is a development stage biopharmaceutical corporation specializing in the discovery and development of prescription pain management products. The company's leading product is FDA approved Entereg (alvimopan), which was developed in collaboration with Glaxo Group Limited to treat gastrointestinal (GI) complications caused by post-operative ileus (POI). The firm is also exploring treatments for opioid bowel dysfunction (OBD). These complications are typically caused by opioids such as morphine, which provide several beneficial pain-management results with the central nervous system (CNS) but can lead to serious side effects. The company has identified a series of novel, orally active delta agonists that selectively stimulate the delta opioid receptor, whereas all marketed opioid drugs currently interact with the mu receptors in the brain and spinal cord. The company is collaborating with Pfizer, Inc. to develop new alternatives for pain management based on this research. Adolor is also exploring methods designed to alleviate pain resulting from inflammation and nerve damage. In September 2009, the firm acquired the rights to the opioid receptor antagonist and clinical stage-product candidate, OpRA III, from Eli Lilly and Company, and subsequently identified the product candidate as ADL5945. In November of the same year, the company began Phase I clinical trials for ADL7445, an oral mu opioid receptor antagonist for OBD.

FINANCIALS: Sales and profits are in thousands of dollars—add 000 to get the full amount. 2009 Note: Financial information for 2009 was not available for all companies at press time.

2009 Sales: $37,361	2009 Profits: $-47,914	U.S. Stock Ticker: ADLR
2008 Sales: $49,456	2008 Profits: $-30,122	Int'l Ticker: Int'l Exchange:
2007 Sales: $9,120	2007 Profits: $-48,443	Employees: 114
2006 Sales: $15,087	2006 Profits: $-69,738	Fiscal Year Ends: 12/31
2005 Sales: $15,719	2005 Profits: $-56,797	Parent Company:

SALARIES/BENEFITS:

Pension Plan:	ESOP Stock Plan:	Profit Sharing:	Top Exec. Salary: $453,949	Bonus: $120,381
Savings Plan: Y	Stock Purch. Plan:		Second Exec. Salary: $453,949	Bonus: $109,438

OTHER THOUGHTS:

Apparent Women Officers or Directors:
Hot Spot for Advancement for Women/Minorities:

LOCATIONS: ("Y" = Yes)

West:	Southwest:	Midwest:	Southeast:	Northeast:	International:
				Y	

ADVANCED CELL TECHNOLOGY INC

www.advancedcell.com

Industry Group Code: 541712 Ranks within this company's industry group: Sales: 17   Profits: 15

Drugs:		Other:		Clinical:	Computers:		Services:	
Discovery:	Y	AgriBio:		Trials/Services:	Hardware:		Specialty Services:	
Licensing:		Genetic Data:	Y	Labs:	Software:		Consulting:	
Manufacturing:		Tissue Replacement:	Y	Equipment/Supplies:	Arrays:		Blood Collection:	
Generics:				Research/Development Svcs.:	Database Management:		Drug Delivery:	
				Diagnostics:			Drug Distribution:	

TYPES OF BUSINESS:

Human Stem Cell Research
Patent Licensing

BRANDS/DIVISIONS/AFFILIATES:

Mytogen, Inc.

CONTACTS: Note: Officers with more than one job title may be intentionally listed here more than once.

William M. Caldwell, IV, CEO
William M. Caldwell, IV, Principle Financial Officer
Robert Lanza, Chief Scientific Officer
Roger Gay, Sr. Dir.-Mfg.
Rita Parker, Dir.-Oper.
Edward Mickunas, VP-Regulatory
William M. Caldwell, IV, Chmn.

Phone: 310-576-0611	Fax: 310-576-0662
Toll-Free:	
Address: P.O. Box 1700, Santa Monica, CA 90406 US	

GROWTH PLANS/SPECIAL FEATURES:

Advanced Cell Technology, Inc. (ACT) is a biotechnology company focusing on developing and commercializing embryonic and adult stem cell technology in the emerging field of regenerative medicine. Regenerative medicine treats chronic degenerative diseases and facilitates regenerative repair of acute disease, such as trauma, infarction and burns. ACT owns or licenses over 150 patents and patent applications related to the field of stem cell therapy, including nuclear transfer, which allows the production of stem cells genetically matched to the patient, and a reduced complexity library of stem cells for acute clinical applications. ACT divides its research into three categories. Cellular reprogramming involves turning stem cells into one of over 200 different human cell types that may be therapeutically relevant in treating diseased or destroyed tissue, and which are tailored to each patient's needs. The blastomere program is dedicated to further development of its technology for generating human embryonic stem cells (hESCs) without destroying the developmental capabilities of the embryo. ACT is one of the first companies to develop such a technique. Lastly, its stem cell differentiation segment controls the differentiation, culture and growth of the company's stem cells. Based on its stem cell technology, ACT has three identified cellular product platforms: Retinal pigment epithelial (RPE) therapy, Myoblast stem cell (cells derived from the patient's skeletal leg muscle) therapy and its Hemangioblast (HG) platform. The company's RPE platform is developing treatments for degenerative retinal diseases such as, Stargardt's Macular Dystrophy (for which no therapy currently exists) and Age-related macular degeneration. Additionally, the firm is developing a Myoblast Program to treat heart failure, currently undergoing Phase II clinical trials. The HG platform is developing preclinical treatments for cardiovascular disease, stroke and cancer based on hemangioblast cells. In June 2010, ACT was granted a patent covering its methods for producing RPE cells from hESCs.

FINANCIALS: Sales and profits are in thousands of dollars—add 000 to get the full amount. 2009 Note: Financial information for 2009 was not available for all companies at press time.

2009 Sales: $1,416	2009 Profits: $-36,760	U.S. Stock Ticker: ACTC
2008 Sales: $ 787	2008 Profits: $-33,904	Int'l Ticker: Int'l Exchange:
2007 Sales: $ 647	2007 Profits: $-15,899	Employees: 14
2006 Sales: $ 441	2006 Profits: $-18,720	Fiscal Year Ends: 12/31
2005 Sales: $ 395	2005 Profits: $-9,394	Parent Company:

SALARIES/BENEFITS:

Pension Plan:	ESOP Stock Plan:	Profit Sharing:	Top Exec. Salary: $417,500	Bonus: $140,000
Savings Plan:	Stock Purch. Plan:		Second Exec. Salary: $311,250	Bonus: $81,250

OTHER THOUGHTS:

Apparent Women Officers or Directors:
Hot Spot for Advancement for Women/Minorities:

LOCATIONS: ("Y" = Yes)

West:	Southwest:	Midwest:	Southeast:	Northeast:	International:
Y				Y	

Note: Financial information, benefits and other data can change quickly and may vary from those stated here.

ADVANSOURCE BIOMATERIALS CORPORATION

www.advbiomaterials.com
Industry Group Code: 33911 Ranks within this company's industry group: Sales: 14 Profits: 10

Drugs:	Other:	Clinical:	Computers:	Services:	
Discovery:	AgriBio:	Trials/Services:	Hardware:	Specialty Services:	Y
Licensing:	Genetic Data:	Labs:	Software:	Consulting:	
Manufacturing:	Tissue Replacement:	Equipment/Supplies: Y	Arrays:	Blood Collection:	
Generics:		Research/Development Svcs.:	Database Management:	Drug Delivery:	
		Diagnostics:		Drug Distribution:	

TYPES OF BUSINESS:
Polymer Material Manufacturing
Medical Device Design
Polyurethane-Based Biomaterials

BRANDS/DIVISIONS/AFFILIATES:
AdvanSource Biomaterials Corp.
HydroMed
CardioTech International, Ltd.
PolyBlend
ChronoFlex
HydroThane
ChronoSil
CardioPass

CONTACTS: *Note: Officers with more than one job title may be intentionally listed here more than once.*
Michael F. Adams, CEO
Michael F. Adams, Pres.
David Volpe, Interim CFO
Andrew M. Reed, VP-Science
Andrew M. Reed, VP-Tech.
William J. O'Neill, Jr., Chmn.

Phone: 978-657-0075	Fax: 978-657-0074
Toll-Free:	
Address: 229 Andover St., Wilmington, MA 01887 US	

GROWTH PLANS/SPECIAL FEATURES:
AdvanSource Biomaterials Corp., formerly CardioTech International, Inc., develops advanced polymer materials used in the design and development of medical devices for a range of anatomical site and disease state treatment. The firm provides products with both short and long-term implant applications, including stents, artificial hearts, catheters and vascular access ports. The company sells its products under the ChronoFlex, ChronoThane, ChronoPrene, HydroThane, HydroMed and PolyBlend brand names. ChronoFlex is a family of medical grade polyurethanes for the prevention of in-vivo formation of environmental stress cracking (ESC). ChronoThane is an aromatic ether-based product line of thermoplastic polymers. ChronoPrene thermoplastic elastomeric materials are made of easy molding, high flow rubber with high chemical resistance to alcohols, acids and bases. HydroThane is a family of highly absorbent thermoplastic, hydrophilic, elastomers allowing for targeted water absorption and tissue like elasticity. HydroMed is a line of hydrophilic polyurethanes used as lubricious coatings for medical devices. The PolyBlend product line includes extremely soft, extrudable and injection moldable polyurethane elastomeric alloys. AdvanSource's services include polymer customization, including the addition of radiopaque polymers; manufacturing; university collaboration; feasibility advising; and product concept support through its Concept Center. AdvanSource serves the interventional radiology, diabetes management, peripheral vascular, gastroenterology, ear, nose and throat, cardiovascular, drug delivery, orthopedic, endoscopy, neurology, oncology, urology and spine market segments. In October 2009, the company filed for a U.S. patent on ChronoSil, a silicone-urethane copolymer product, which has properties typical of polyurethanes, but in addition, the feel and characteristics of silicones. In early 2010, the firm announced it would cease development of its CardioPass synthetic coronary artery bypass graft upon advice that suggested sizes smaller than its 4 and 5mm grafts were needed.

FINANCIALS: Sales and profits are in thousands of dollars—add 000 to get the full amount. 2009 Note: Financial information for 2009 was not available for all companies at press time.

2009 Sales: $1,915	2009 Profits: $-2,517	**U.S. Stock Ticker: ASB**
2008 Sales: $1,950	2008 Profits: $-6,090	**Int'l Ticker:** Int'l Exchange:
2007 Sales: $2,275	2007 Profits: $-2,962	Employees: 20
2006 Sales: $22,381	2006 Profits: $-5,069	Fiscal Year Ends: 3/31
2005 Sales: $21,841	2005 Profits: $-1,595	Parent Company:

SALARIES/BENEFITS:

Pension Plan:	ESOP Stock Plan:	Profit Sharing:	Top Exec. Salary: $292,000	Bonus: $36,000
Savings Plan:	Stock Purch. Plan:		Second Exec. Salary: $196,000	Bonus: $

OTHER THOUGHTS:
Apparent Women Officers or Directors:
Hot Spot for Advancement for Women/Minorities:

LOCATIONS: ("Y" = Yes)

West:	Southwest:	Midwest:	Southeast:	Northeast:	International:
				Y	

AEOLUS PHARMACEUTICALS INC

www.aeoluspharma.com

Industry Group Code: 325412 **Ranks within this company's industry group:** Sales: Profits: 80

Drugs:		Other:	Clinical:	Computers:	Services:
Discovery:	Y	AgriBio:	Trials/Services:	Hardware:	Specialty Services:
Licensing:		Genetic Data:	Labs:	Software:	Consulting:
Manufacturing:		Tissue Replacement:	Equipment/Supplies:	Arrays:	Blood Collection:
Generics:			Research/Development Svcs.:	Database Management:	Drug Delivery:
			Diagnostics:		Drug Distribution:

TYPES OF BUSINESS:

Pharmaceutical Development
Catalytic Antioxidants
Biological Weapons Countermeasures

BRANDS/DIVISIONS/AFFILIATES:

AEOL 10150

CONTACTS: *Note: Officers with more than one job title may be intentionally listed here more than once.*

John L. McManus, CEO
John L. McManus, Pres.
Michael McManus, CFO
Brian J. Day, Chief Scientific Officer
Michael McManus, Sec.
Michael McManus, Treas.
David C. Cavalier, Chmn.

Phone: 949-481-9825	Fax: 949-481-9829
Toll-Free:	
Address: 26361 Crown Valley Pkwy., Ste. 150, Mission Viejo, CA 92691 US	

GROWTH PLANS/SPECIAL FEATURES:

Aeolus Pharmaceuticals, Inc. is a Southern California-based biopharmaceutical company. It is developing a new class of catalytic antioxidant compounds as countermeasures against biological, chemical and radiological weapons as well as for diseases and disorders of the central nervous system and oncology. The company's lead drug candidate is AEOL 10150, the first drug in its class to enter human clinical evaluation. AEOL 10150 is a small molecule catalytic antioxidant with the ability to scavenge a broad range of reactive oxygen species, or free radicals. As a catalytic antioxidant, it mimics and amplifies the body's natural enzymatic systems for eliminating these damaging compounds. Early indications suggest its usefulness against the effects of ARS (acute radiation syndrome) in the lungs and gastro-intestinal (GI) tract. It is also hoped to be an effective countermeasure against exposure to chlorine and sulfur mustard gas. Tests on mice demonstrate that AEOL 10150 protects healthy lung tissue from radiation injury delivered in a single dose or by fractionated radiation therapy doses, and that the drug does not negatively affect tumor radiotherapy. It is also thought to be useful as a treatment for amyotrophic lateral sclerosis (ALS), also known as Lou Gehrig's disease. Aeolus boasts positive safety results after two Phase I single dose studies were completed without serious adverse effects reported. Aeolus has also selected AEOL 11207 as its second development candidate. Collected data suggests the compound may be useful as a potential once-every-other-day oral therapeutic treatment option for central nervous system disorders including, epilepsy and Parkinson's disease.

FINANCIALS: Sales and profits are in thousands of dollars—add 000 to get the full amount. 2009 Note: Financial information for 2009 was not available for all companies at press time.

2009 Sales: $	2009 Profits: $-2,296	U.S. Stock Ticker: AOLS	
2008 Sales: $	2008 Profits: $-2,973	Int'l Ticker: Int'l Exchange:	
2007 Sales: $	2007 Profits: $-3,024	Employees: 2	
2006 Sales: $ 92	2006 Profits: $-5,728	Fiscal Year Ends: 9/30	
2005 Sales: $ 252	2005 Profits: $-6,905	Parent Company:	

SALARIES/BENEFITS:

Pension Plan:	ESOP Stock Plan:	Profit Sharing:	Top Exec. Salary: $250,200	Bonus: $
Savings Plan:	Stock Purch. Plan: Y		Second Exec. Salary: $	Bonus: $

OTHER THOUGHTS:

Apparent Women Officers or Directors:
Hot Spot for Advancement for Women/Minorities:

LOCATIONS: ("Y" = Yes)

West:	Southwest:	Midwest:	Southeast:	Northeast:	International:
Y					

AETERNA ZENTARIS INC

www.aezsinc.com

Industry Group Code: 325412 Ranks within this company's industry group: Sales: 82 Profits: 126

Drugs:		Other:		Clinical:	Computers:	Services:
Discovery:	Y	AgriBio:	Y	Trials/Services:	Hardware:	Specialty Services:
Licensing:	Y	Genetic Data:	Y	Labs:	Software:	Consulting:
Manufacturing:	Y	Tissue Replacement:		Equipment/Supplies:	Arrays:	Blood Collection:
Generics:				Research/Development Svcs.:	Database Management:	Drug Delivery:
				Diagnostics:		Drug Distribution:

TYPES OF BUSINESS:

Drug Development
Oncology Products
Endocrine Therapy Products

BRANDS/DIVISIONS/AFFILIATES:

Cetrotide
Perifosine
AEZS-108
AEZS-130
AEZS-120
Zentaris GmbH

CONTACTS: Note: Officers with more than one job title may be intentionally listed here more than once.

Jurgen Engel, CEO
Jurgen Engel, Pres.
Dennis Turpin, CFO/Sr. VP
Matthias Seeber, Sr. VP-Admin.
Matthias Seeber, Sr. VP-Legal Affairs
Chantal Gravel, Dir.-Bus. Oper.
Nicholas J. Pelliccione, Sr. VP-Regulatory Affairs & Quality Assurance
Paul Blake, Chief Medical Officer/Sr. VP
Juergen Ernst, Chmn.

Phone: 418-652-8525	Fax: 418-652-0881
Toll-Free:	
Address: 1405 du Parc-Technologique Blvd., Quebec City, QC G1P 4P5 Canada	

GROWTH PLANS/SPECIAL FEATURES:

AEterna Zentaris Inc. is a Canadian biopharmaceutical company focused on endocrine therapy and oncology. The company is devoted to discovering and developing drugs for the treatment of certain forms of cancer, endocrine disorders and infectious diseases. The firm has one product on the commercial market, Cetrotide. Cetrotide was the first hormone antagonist treatment approved for in vitro fertilization. The drug is administered to women in order to prevent premature ovulation in order to increase fertility success rates. Cetrotide is approved in over 80 countries. The drug is marketed worldwide by Merck Serono, except in Japan where it is marketed by Shionogi and Nippon Kayaku. Currently, the company has two drugs in Phase I clinical trials, two in Phase II and two in Phase III. The firm's Phase II drugs include Perifosine, for multiple cancers, and AEZS-108, for ovarian and endometrial cancer. The drugs in Phase III clinical trials include Perifosine, for multiple myeloma and advanced metastatic cancer, and Solorel (AEZS-130), a diagnostic for use in adult growth hormone deficiency. In addition, the firm has two products in preclinical in vivo testing and three in preclinical development, including AEZS-120, a potential prostate cancer vaccine. In addition to its North American operations, the company owns 100% of Zentaris GmbH, an integrated clinical research company based in Germany. In September 2009, the U.S. Food and Drug Administration (FDA) granted orphan drug status to the firm's Perifosine formulation for the treatment of myeloma, thereby easing some of the regulatory burdens involved in bringing the drug to market.

FINANCIALS: Sales and profits are in thousands of dollars—add 000 to get the full amount. 2009 Note: Financial information for 2009 was not available for all companies at press time.

2009 Sales: $61,870	2009 Profits: $-24,190	**U.S. Stock Ticker: AEZS**	
2008 Sales: $38,478	2008 Profits: $-59,817	**Int'l Ticker: AEZ** Int'l Exchange: Toronto-TSX	
2007 Sales: $42,100	2007 Profits: $-32,300	Employees: 109	
2006 Sales: $38,800	2006 Profits: $33,400	Fiscal Year Ends: 12/31	
2005 Sales: $47,204	2005 Profits: $10,571	Parent Company:	

SALARIES/BENEFITS:

Pension Plan:	ESOP Stock Plan:	Profit Sharing:	Top Exec. Salary: $458,040	Bonus: $
Savings Plan:	Stock Purch. Plan:		Second Exec. Salary: $366,000	Bonus: $

OTHER THOUGHTS:

Apparent Women Officers or Directors: 1
Hot Spot for Advancement for Women/Minorities:

LOCATIONS: ("Y" = Yes)

West:	Southwest:	Midwest:	Southeast:	Northeast:	International:
				Y	Y

AFFYMETRIX INC

www.affymetrix.com

Industry Group Code: 325413 Ranks within this company's industry group: Sales: 5 Profits: 19

Drugs:	Other:		Clinical:	Computers:		Services:	
Discovery:	AgriBio:		Trials/Services:	Hardware:	Y	Specialty Services:	
Licensing:	Genetic Data:	Y	Labs:	Software:		Consulting:	
Manufacturing:	Tissue Replacement:		Equipment/Supplies:	Arrays:	Y	Blood Collection:	
Generics:			Research/Development Svcs.:	Database Management:		Drug Delivery:	
			Diagnostics:			Drug Distribution:	

TYPES OF BUSINESS:

Chips-Genetics
DNA Array Technology
Genomics

BRANDS/DIVISIONS/AFFILIATES:

GeneChip
CustomExpress
CustomSeq
USB Corporation
Panomics, Inc.

CONTACTS: Note: Officers with more than one job title may be intentionally listed here more than once.

Kevin M. King, CEO
Kevin M. King, Pres.
John C. Batty, CFO
Rick Runkel, General Counsel
Doug Farrell, VP-Investor Rel.
John C. Batty, Exec. VP-Finance
Andrew Last, Chief Commercial Officer
Stephen P. A. Fodor, Chmn.

Phone: 408-731-5000	**Fax:** 408-731-5380
Toll-Free: 888-362-2447	
Address: 3420 Central Expwy., Santa Clara, CA 95051 US	

GROWTH PLANS/SPECIAL FEATURES:

Affymetrix, Inc. develops, manufactures, sells and services consumables and systems for genetic analysis in the life sciences and clinical healthcare markets. The firm sells its products directly to pharmaceutical, biotechnology, agrichemical, diagnostics and consumer products companies, as well as academic research centers, government research laboratories, private foundation laboratories and clinical reference laboratories, in North America and Europe. The company also sells its products through life science supply specialists acting as authorized distributors in Latin America, India, the Middle East and Asia Pacific regions. Affymetrix markets products for two principal applications: monitoring of gene or exon expression levels and investigation of genetic variation. Its catalogue GeneChip expression arrays are available for the study of human, rat, mouse and a range of other mammalian and model organisms. Human, mouse and rat exon analysis arrays are also available. The firm's integrated GeneChip microarray platform includes disposable DNA probe arrays consisting of nucleic acid sequences set out in an ordered, high density pattern; certain reagents for use with the probe arrays; a scanner and other instruments used to process the probe arrays; and software to analyze and manage genomic or genetic information obtained from the probe arrays. Additionally, the company markets CustomExpress and CustomSeq products, which enable its customers to design their own custom GeneChip expression arrays or a sequence of arrays for organisms of interest to them. Subsidiaries include Panomics, Inc. and USB Corporation.

Affymetrix offers its employees medical, dental and vision coverage; a 401(k) plan; life and disability insurance; a tuition assistance plan; health fitness membership discounts; a lunch program; a family resources program; domestic partner benefits; and reimbursement accounts.

FINANCIALS: Sales and profits are in thousands of dollars—add 000 to get the full amount. 2009 Note: Financial information for 2009 was not available for all companies at press time.

2009 Sales: $327,094	2009 Profits: $-23,909	**U.S. Stock Ticker:** AFFX
2008 Sales: $410,249	2008 Profits: $-307,919	**Int'l Ticker:** Int'l Exchange:
2007 Sales: $371,320	2007 Profits: $12,593	Employees: 1,128
2006 Sales: $355,317	2006 Profits: $-13,704	Fiscal Year Ends: 12/31
2005 Sales: $367,602	2005 Profits: $65,787	Parent Company:

SALARIES/BENEFITS:

Pension Plan:	ESOP Stock Plan:	Profit Sharing:	Top Exec. Salary: $655,673	Bonus: $
Savings Plan: Y	Stock Purch. Plan:		Second Exec. Salary: $491,346	Bonus: $

OTHER THOUGHTS:

Apparent Women Officers or Directors: 1
Hot Spot for Advancement for Women/Minorities:

LOCATIONS: ("Y" = Yes)

West:	Southwest:	Midwest:	Southeast:	Northeast:	International:
Y		Y			Y

AGILENT TECHNOLOGIES INC

www.agilent.com

Industry Group Code: 3345 Ranks within this company's industry group: Sales: 1   Profits: 4

Drugs:	Other:	Clinical:		Computers:	Services:
Discovery:	AgriBio:	Trials/Services:		Hardware:	Specialty Services:
Licensing:	Genetic Data:	Labs:		Software:	Consulting:
Manufacturing:	Tissue Replacement:	Equipment/Supplies:	Y	Arrays:	Blood Collection:
Generics:		Research/Development Svcs.:		Database Management:	Drug Delivery:
		Diagnostics:			Drug Distribution:

TYPES OF BUSINESS:

Test Equipment
Communications Test Equipment
Integrated Circuits Test Equipment
Optoelectronics Test Equipment
Image Sensors
Bioinstrumentation
Software Products
Informatics Products

BRANDS/DIVISIONS/AFFILIATES:

Varian
Stratagene Corp

CONTACTS: Note: Officers with more than one job title may be intentionally listed here more than once.

William P. Sullivan, CEO
William P. Sullivan, Pres.
Didier Hirsch, CFO
Jean M. Halloran, Sr. VP-Human Resources
Darlene J. S. Solomon, CTO
Marie Oh Huber, General Counsel/Sec./Sr. VP
Sheila Barr Robertson, VP-Corp. Dev. & Strategy
Amy Flores, Mgr.-Public Rel.
Alicia Rodriguez, VP-Investor Rel.
Ron Nersesian, Sr. VP/Pres., Electronic Measurement Group
Mike McMullen, Sr. VP/Pres., Chemical Analysis Group
Nick Roelofs, VP/Gen. Mgr.-Life Sciences Solutions
James G. Cullen, Chmn.

Phone: 408-345-8886	Fax: 408-345-8474
Toll-Free: 877-424-4536	
Address: 5301 Stevens Creek Blvd., Santa Clara, CA 95051 US	

GROWTH PLANS/SPECIAL FEATURES:

Agilent Technologies, Inc. is a measurement technology company with two main business segments: electronic measurement and bio-analytical measurement. The electronic measurement business provides products for the communications testing market and the general purpose test market. These products include testing equipment for fiber optic networks; broadband and data networks; wireless communications and microwave networks; general purpose instruments; modular instruments and test software; digital test design products; high-frequency electronic design tools; and electronic manufacturing test equipment. It also assists in installing, activating and maintaining optical, wireless, wireline and large-company networks. The company handles the final assembly and testing of its electronic measurement products and relies on a direct sales force as well as channel partners including manufacturer's representatives to sell, implement and support the company's products. Customers in this segment include manufacturers of electronic products, handsets and network equipment and service providers who implement, maintain and manage communication networks and services. The bio-analytical measurement business serves two main markets: life sciences, including pharma, biotech, CRO, CMO, academic and government customers; and chemical analysis, comprised of customers from the petrochemical, chemical, environmental, forensics, homeland security, bioagriculture an food safety sectors. Its main product categories are gas chromatography, liquid chromatography, mass spectrometry, microarrays, lab automation and robotics; electrophoresis; PCR instrumentation; software and informatics; bioreagents; and other consumables. Stratagene Corp., a subsidiary, specializes in life science research tools. In January 2010, the European Commission granted Agilent antitrust clearance for its acquisition of Varian, a scientific instrument and vacuum technologies provider. In February 2010, Agilent Technologies agreed to sell its network solutions business to JDSU, a communications test and measurement products company.

Agilent offers its employees medical, dental, vision, life and disability plans; an employee and family assistance plan; onsite fitness centers; recreational sports leagues; charitable gift matching; on-site vaccinations, screenings and massages; and discounts on certain expenses.

FINANCIALS: Sales and profits are in thousands of dollars—add 000 to get the full amount. 2009 Note: Financial information for 2009 was not available for all companies at press time.

2009 Sales: $4,481,000	2009 Profits: $-31,000	**U.S. Stock Ticker: A**
2008 Sales: $5,774,000	2008 Profits: $693,000	**Int'l Ticker:** Int'l Exchange:
2007 Sales: $5,420,000	2007 Profits: $638,000	Employees: 19,600
2006 Sales: $4,973,000	2006 Profits: $3,307,000	Fiscal Year Ends: 10/31
2005 Sales: $4,685,000	2005 Profits: $327,000	Parent Company:

SALARIES/BENEFITS:

Pension Plan: Y	ESOP Stock Plan:	Profit Sharing:	Top Exec. Salary: $986,667	Bonus: $1,305,563
Savings Plan: Y	Stock Purch. Plan: Y		Second Exec. Salary: $699,996	Bonus: $627,775

OTHER THOUGHTS:

Apparent Women Officers or Directors: 6
Hot Spot for Advancement for Women/Minorities: Y

LOCATIONS: ("Y" = Yes)

West:	Southwest:	Midwest:	Southeast:	Northeast:	International:
Y	Y	Y		Y	Y

Note: Financial information, benefits and other data can change quickly and may vary from those stated here.

AKORN INC

www.akorn.com

Industry Group Code: 325412 Ranks within this company's industry group: Sales: 79 Profits: 129

Drugs:		Other:	Clinical:	Computers:	Services:
Discovery:	Y	AgriBio:	Trials/Services:	Hardware:	Specialty Services:
Licensing:		Genetic Data:	Labs:	Software:	Consulting:
Manufacturing:	Y	Tissue Replacement:	Equipment/Supplies:	Arrays:	Blood Collection:
Generics:			Research/Development Svcs.:	Database Management:	Drug Delivery:
			Diagnostics:		Drug Distribution:

TYPES OF BUSINESS:

Ophthalmic & Hospital Drugs & Injectables
Contract Services
Vaccines

BRANDS/DIVISIONS/AFFILIATES:

Akorn (New Jersey), Inc.
Akorn-Strides, LLC

CONTACTS: Note: Officers with more than one job title may be intentionally listed here more than once.

Raj Rai, CEO
Timothy Dick, CFO
John R. Sabat, Sr. VP-Global Sales & Mktg.
Sean Brynjelsen, VP-R&D
Mark M. Silverberg, Exec. VP-Tech. Svcs.
Joseph Bonaccorsi, General Counsel/Sr. VP/Sec.
Mark M. Silverberg, Exec. VP-Oper.
Sean Brynjelsen, VP-New Bus. Dev.
John R. Sabat, Sr. VP-National Accounts
Mark M. Silverberg, Exec. VP-Global Quality Assurance
Sam Boddapati, Sr. VP-Regulatory Affairs
Michael P. Stehn, Sr. VP-Oper.
John N. Kapoor, Chmn.

Phone: 847-279-6100	Fax: 800-943-3694
Toll-Free: 800-932-5676	
Address: 1925 W. Field Ct., Ste. 300, Lake Forest, IL 60045 US	

GROWTH PLANS/SPECIAL FEATURES:

Akorn, Inc. provides diagnostic and therapeutic pharmaceuticals for specialty areas such as ophthalmology, rheumatology, anesthesia and antidotal medicine. The company operates in four segments: ophthalmic; hospital drugs and injectables; biologics and vaccines; and contract services. The ophthalmic division markets a line of diagnostic and therapeutic products for eye health. Diagnostic products, primarily used in the office setting, include mydriatics, cycloplegics, anesthetics, topical stains, gonioscopic solutions and angiography dyes. Therapeutic products, sold primarily to wholesalers and other national account customers, include antibiotics, anti-infectives, steroids, glaucoma medications, decongestants /antihistamines and anti-edema medications. Non-pharmaceutical products include various artificial tear solutions, lubricating ointments, eyelid cleansers and contact lens accessories. Through the hospital drugs and injectables segment, Akorn markets a line of specialty injectable pharmaceutical products, including antidotes, anesthesia and treatments for rheumatoid arthritis and pain management. The biologics and vaccines segment focuses on adult Tetanus-Diphtheria vaccines as well as the production of flu vaccines. The contract services segment manufactures products for third-party pharmaceutical and biotechnology customers based on their specifications. Akorn (New Jersey), Inc., a wholly-owned subsidiary, is involved in manufacturing, product development and administrative activities related to the ophthalmic and hospital drugs and injectables segments. Customers include physicians, optometrists, wholesalers, group purchasing organizations and other pharmaceutical companies. Akorn currently holds seven U.S. patents, two pending U.S. patents and one pending international patent. The firm's has a joint venture with Strides Arcolab Limited called Akorn-Strides, LLC, which focuses on developing liquid, lyophilized and dry powder formulations of generic injectable drugs. In April 2010, the company received FDA approval for the generic version of Dilaudid-HP Hyromorphone Hydrochloride 10mg/ml injection, used for pain relief.

Employees are offered medical, dental and vision insurance; flexible spending accounts; life insurance; adoption assistance; an employee assistance program; and educational assistance.

FINANCIALS: Sales and profits are in thousands of dollars—add 000 to get the full amount. 2009 Note: Financial information for 2009 was not available for all companies at press time.

2009 Sales: $75,891	2009 Profits: $-25,306	U.S. Stock Ticker: AKRX
2008 Sales: $93,598	2008 Profits: $-7,939	Int'l Ticker: Int'l Exchange:
2007 Sales: $52,895	2007 Profits: $-19,168	Employees: 329
2006 Sales: $71,250	2006 Profits: $-5,960	Fiscal Year Ends: 12/31
2005 Sales: $44,484	2005 Profits: $-8,609	Parent Company:

SALARIES/BENEFITS:

Pension Plan:	ESOP Stock Plan:	Profit Sharing:	Top Exec. Salary: $256,370	Bonus: $
Savings Plan: Y	Stock Purch. Plan: Y		Second Exec. Salary: $255,458	Bonus: $

OTHER THOUGHTS:

Apparent Women Officers or Directors:
Hot Spot for Advancement for Women/Minorities:

LOCATIONS: ("Y" = Yes)

West:	Southwest:	Midwest:	Southeast:	Northeast:	International:
		Y		Y	

Note: Financial information, benefits and other data can change quickly and may vary from those stated here.

AKZO NOBEL NV

www.akzonobel.com

Industry Group Code: 325 Ranks within this company's industry group: Sales: 4   Profits: 4

Drugs:		Other:	Clinical:	Computers:	Services:
Discovery:	Y	AgriBio:	Trials/Services:	Hardware:	Specialty Services:
Licensing:		Genetic Data:	Labs:	Software:	Consulting:
Manufacturing:		Tissue Replacement:	Equipment/Supplies:	Arrays:	Blood Collection:
Generics:			Research/Development Svcs.:	Database Management:	Drug Delivery:
			Diagnostics:		Drug Distribution:

TYPES OF BUSINESS:

Specialty Chemicals
Coatings
Decorative Paints

BRANDS/DIVISIONS/AFFILIATES:

Axko Nobel India Limited
Eka Chemicals AB
Devoe
Sikkens
Dulux
Hammerite
Liquid Nails
Ralph Lauren Paint

CONTACTS: *Note: Officers with more than one job title may be intentionally listed here more than once.*

Hanz Wijers, CEO
Keith Nichols, CFO
Tim Van der Zanden, Head-Corp. Media Rel.
Huib Wurfbain, Dir.-Investor Rel.
Hans de Vriese, Dir.-Control
Leif Darner, Dir.-Performance Coatings
Rob Frohn, Dir.-Specialty Chemicals
Tex Gunning, Dir.-Decorative Paints
Karel Vuursteen, Chmn.-Supervisory Board

Phone: 31-20-502-7555	Fax: 31-20-502-7666
Toll-Free:	
Address: Strawinskylaan 2555, Amsterdam, 1077 ZZ The Netherlands	

GROWTH PLANS/SPECIAL FEATURES:

Akzo Nobel N.V. produces paints, coatings and chemicals, and operates in over 80 countries. The company's operations are divided into three segments: Decorative Paints, Performance Coatings and Specialty Chemicals. The Decorative Paints division includes paint, lacquer and varnish products, as well as adhesives, floor leveling compounds, mixing machines and training courses. Brands in this division consist of Sikkens, Dulux, Ralph Lauren Paint, Devoe Paint, Mulco, Hammerite and Liquid Nails Adhesive. The segment has offices in the U.K., Continental Europe, the Americas and Asia. Akzo Nobel's Performance Coatings division makes a variety of chemical products including powder, industrial and marine coatings; wood finishes and adhesives; and a line of car refinishes. The company's Specialty Chemicals division produces pulp and paper chemicals; polymer chemicals such as metal alkyls and suspending agents; surfactants used in hair and skincare products; base chemicals such as salt and chlor-alkali products used in the manufacture of glass and plastics; and functional chemicals used in toothpaste, ice cream and flame retardants. Subsidiary Eka Chemicals AB, a provider of colloidal silica products, operates in this division. In November 2009, Azko Nobel agreed to acquire the powder coatings business of The Dow Chemical Company. In December 2009, the company introduced a new line of paints under the Martha Stewart Living brand. In April 2010, the firm renamed its Indian decorative paints business, formerly ICI India, Azko Nobel India Limited.

FINANCIALS: Sales and profits are in thousands of dollars—add 000 to get the full amount. 2009 Note: Financial information for 2009 was not available for all companies at press time.

2009 Sales: $18,566,600	2009 Profits: $1,162,670	U.S. Stock Ticker:
2008 Sales: $20,387,300	2008 Profits: $981,340	Int'l Ticker: AKZA Int'l Exchange: Amsterdam-Euronext
2007 Sales: $18,481,000	2007 Profits: $12,770,600	Employees: 58,600
2006 Sales: $15,836,300	2006 Profits: $1,821,700	Fiscal Year Ends: 12/31
2005 Sales: $20,540,000	2005 Profits: $1,518,400	Parent Company:

SALARIES/BENEFITS:

Pension Plan:	ESOP Stock Plan:	Profit Sharing:	Top Exec. Salary: $	Bonus: $
Savings Plan:	Stock Purch. Plan:		Second Exec. Salary: $	Bonus: $

OTHER THOUGHTS:

Apparent Women Officers or Directors: 1
Hot Spot for Advancement for Women/Minorities:

LOCATIONS: ("Y" = Yes)

West:	Southwest:	Midwest:	Southeast:	Northeast:	International:
		Y	Y	Y	Y

ALBANY MOLECULAR RESEARCH INC
www.amriglobal.com

Industry Group Code: 541712 Ranks within this company's industry group: Sales: 6 Profits: 11

Drugs:	Other:	Clinical:	Computers:	Services:
Discovery:	AgriBio:	Trials/Services:	Hardware:	Specialty Services:
Licensing:	Genetic Data:	Labs:	Software:	Consulting:
Manufacturing:	Tissue Replacement:	Equipment/Supplies:	Arrays:	Blood Collection:
Generics:		Research/Development Svcs.: Y	Database Management:	Drug Delivery:
		Diagnostics:		Drug Distribution:

TYPES OF BUSINESS:
Contract Drug Discovery & Development
Custom Biotech & Genomic Research
Chemistry Research
Manufacturing Services
Consulting Services
Analytical Chemistry Services

BRANDS/DIVISIONS/AFFILIATES:
Excelsyn Ltd.

CONTACTS: Note: Officers with more than one job title may be intentionally listed here more than once.
Thomas E. D'Ambra, CEO
Thomas E. D'Ambra, Pres.
Mark T. Frost, CFO
Steve Jennings, Sr. VP-Sales & Mktg.
Brian D. Russell, VP-Human Resources
Bruce J. Sargent, VP-Discovery R&D
Harold Meckler, VP-Science & Tech.
Steven R. Hagen, VP-Pharmaceutical Dev. & Mfg.
Mark T. Frost, VP-Admin.
Peter Jerome, Dir.-Investor Rel.
Mark T. Frost, Treas.
Michael P. Trova, Sr. VP-Chemistry
Thomas E. D'Ambra, Chmn.

Phone: 518-464-0279	Fax: 518-512-2020
Toll-Free:	
Address: 26 Corporate Cir., Albany, NY 12212-5098 US	

GROWTH PLANS/SPECIAL FEATURES:

Albany Molecular Research, Inc. (AMRI) provides scientific services, technologies and products. Its fee-for-service contract services platform encompasses drug discovery, development and manufacturing services and a separate technology division consisting of proprietary technology investments, internal drug discovery and niche generic active pharmaceutical ingredient product development. It derives revenue from discovering then licensing new compounds with commercial potential for service fees, milestone and royalty payments. Some of the products of this research led to the development of the active ingredient (fexofenadine HCl) for a non-sedating antihistamine marketed by Sanofi-Aventis S.A. as Allegra in the U.S. and as Telfast outside the U.S. Since its launch in 1995, AMRI has earned more than $367.3 million in royalty and milestone revenue from this product. In addition to developing its own drugs, AMRI has increasingly acted as a custom research and development source for the pharmaceutical, genomic and biotechnology industries. It provides contract services across the entire product development cycle, from lead discovery to commercial manufacturing. The company's services allow pharmaceutical companies to outsource their chemistry departments in order to pursue a greater number of drug discovery and development opportunities. An integral part of these contract operations consists of several facilities in India, Hungary and Singapore, which were launched as part of a strategic move to globalize its services. The firm also has several domestic research facilities in New York and Washington. In February 2009, AMRI acquired Excelsyn Ltd., a U.K.-based chemical development and manufacturing services firm.

The firm offers its employees a 401(k) plan.

FINANCIALS: Sales and profits are in thousands of dollars—add 000 to get the full amount. 2009 Note: Financial information for 2009 was not available for all companies at press time.

2009 Sales: $196,417	2009 Profits: $-16,695	U.S. Stock Ticker: AMRI
2008 Sales: $229,260	2008 Profits: $20,560	Int'l Ticker: Int'l Exchange:
2007 Sales: $192,511	2007 Profits: $8,936	Employees: 1,266
2006 Sales: $179,807	2006 Profits: $2,183	Fiscal Year Ends: 12/31
2005 Sales: $183,906	2005 Profits: $16,321	Parent Company:

SALARIES/BENEFITS:
Pension Plan:	ESOP Stock Plan:	Profit Sharing:	Top Exec. Salary: $484,615	Bonus: $202,750
Savings Plan: Y	Stock Purch. Plan:		Second Exec. Salary: $309,692	Bonus: $95,550

OTHER THOUGHTS:
Apparent Women Officers or Directors: 2
Hot Spot for Advancement for Women/Minorities: Y

LOCATIONS: ("Y" = Yes)
West:	Southwest:	Midwest:	Southeast:	Northeast:	International:
				Y	Y

Note: Financial information, benefits and other data can change quickly and may vary from those stated here.

ALCON INC

www.alcon.com

Industry Group Code: 325412 Ranks within this company's industry group: Sales: 20 Profits: 15

Drugs:		Other:	Clinical:		Computers:		Services:	
Discovery:	Y	AgriBio:	Trials/Services:		Hardware:		Specialty Services:	
Licensing:		Genetic Data:	Labs:		Software:		Consulting:	
Manufacturing:	Y	Tissue Replacement:	Equipment/Supplies:	Y	Arrays:		Blood Collection:	
Generics:			Research/Development Svcs.:		Database Management:		Drug Delivery:	
			Diagnostics:				Drug Distribution:	

TYPES OF BUSINESS:

Eye Care Products
Ophthalmic Products & Equipment
Contact Lens Care Products
Surgical Instruments

BRANDS/DIVISIONS/AFFILIATES:

Opti-Free
Patanol
AcrySof
Systane
WaveLight AG
Alcon Surgical
Nestle Corporation
Novartis

CONTACTS: *Note: Officers with more than one job title may be intentionally listed here more than once.*

Kevin J. Buehler, CEO
Kevin J. Buehler, Pres.
Richard Croarkin, CFO/Sr. VP
Sabri Markabi, Sr. VP-R&D/Chief Medical Officer
Ed McGough, Sr. VP-Tech.
Ed McGough, Sr. VP-Global Mfg.
Elaine E. Whitbeck, General Counsel/Chief Legal Officer
Doug MacHatton, VP-Public Rel.
John Selzer, Dir.-Investor Rel.
Richard Croarkin, Sr. VP-Finance
Doug MacHatton, VP-Treasury
Cary Rayment, Chmn.
Bill Barton, Sr. VP-Int'l Markets

Phone: 41-41-785-8888	Fax:
Toll-Free:	
Address: Bosch 69, Hunenberg, CH-6331 Switzerland	

GROWTH PLANS/SPECIAL FEATURES:

Alcon, Inc. is a leading eye care products company. The firm manages its business through two business segments: Alcon United States and Alcon International. Its portfolio spans three key ophthalmic categories: pharmaceutical, surgical and consumer eye care products. The divisions develop, manufacture and market ophthalmic pharmaceuticals, surgical equipment and devices, contact lens care products and other consumer eye care products that treat diseases and conditions of the eye. Alcon maintains manufacturing plants, laboratories and offices in 75 countries and offers its products and services in over 180 countries. The firm holds approximately 5,750 global patents and 3,950 pending patent applications. The company makes more than 10,000 unique products, including prescription and over-the-counter drugs, contact lens solutions, surgical instruments, intraocular lenses and office systems for ophthalmologists. Its brand names include Patanol solution for eye allergies, AcrySof intraocular lenses, Systane lubricant drops for dry eye and the Opti-Free system for contact lens care. The firm also has research and development laboratories in Texas, California, Florida, Switzerland and Spain. Alcon Surgical creates implantable lenses, viscoelastics and medical tools specifically made for ocular surgeons, including instruments for cataract removal and absorbable sutures. Sales of glaucoma products account for 41.9% of total pharmaceutical sales. In September 2009, the firm agreed to acquire ESBATech AG, a Swiss Biotechnology firm. Novartis AG, which owns a 25% stake in the company, agreed to acquire the remaining 52% from Nestle SA in January 2010, as part of its plan to ultimately takeover 100% of the firm. In March 2010, Alcon acquired the rights to two drugs, Durezol and Zyclorin, from Sirion Therapeutics.

Alcon offers its employees 401(k) and retirement plans; medical, dental, vision, life and AD&D insurance; paid time off; a wellness program including on-site or discounted fitness centers, flu shots and weight watchers discounts; and an employee assistance program.

FINANCIALS: Sales and profits are in thousands of dollars—add 000 to get the full amount. 2009 Note: Financial information for 2009 was not available for all companies at press time.

2009 Sales: $6,499,000	2009 Profits: $2,007,000	U.S. Stock Ticker: ACL
2008 Sales: $6,294,000	2008 Profits: $2,046,000	Int'l Ticker: Int'l Exchange:
2007 Sales: $5,599,600	2007 Profits: $1,586,400	Employees: 15,700
2006 Sales: $4,896,600	2006 Profits: $1,348,100	Fiscal Year Ends: 12/31
2005 Sales: $4,368,500	2005 Profits: $931,000	Parent Company:

SALARIES/BENEFITS:

Pension Plan:	ESOP Stock Plan:	Profit Sharing:	Top Exec. Salary: $1,250,000	Bonus: $1,375,000
Savings Plan: Y	Stock Purch. Plan:		Second Exec. Salary: $570,833	Bonus: $390,000

OTHER THOUGHTS:

Apparent Women Officers or Directors: 1
Hot Spot for Advancement for Women/Minorities: Y

LOCATIONS: ("Y" = Yes)

West:	Southwest:	Midwest:	Southeast:	Northeast:	International:
Y	Y			Y	Y

Note: Financial information, benefits and other data can change quickly and may vary from those stated here.

ALDAGEN INC

www.aldagen.com

Industry Group Code: 325414 Ranks within this company's industry group: Sales: Profits:

Drugs:		Other:		Clinical:	Computers:	Services:
Discovery:	Y	AgriBio:		Trials/Services:	Hardware:	Specialty Services:
Licensing:		Genetic Data:		Labs:	Software:	Consulting:
Manufacturing:	Y	Tissue Replacement:	Y	Equipment/Supplies:	Arrays:	Blood Collection:
Generics:				Research/Development Svcs.:	Database Management:	Drug Delivery:
				Diagnostics:		Drug Distribution:

TYPES OF BUSINESS:

Regenerative Cell Therapy Development

BRANDS/DIVISIONS/AFFILIATES:

ALD-201
ALD-301
ALD-401
Intersouth Partners

CONTACTS: Note: Officers with more than one job title may be intentionally listed here more than once.

W. Thomas Amick, CEO
Edward L. Field, COO
Edward L. Field, Pres.
W. Thomas Amick, Chmn.

Phone: 919-484-2571	Fax: 919-484-8792
Toll-Free:	
Address: 2810 Meridian Pkwy., Ste. 148, Durham, NC 27713 US	

GROWTH PLANS/SPECIAL FEATURES:

Aldagen, Inc. is a biopharmaceutical firm. The company is developing proprietary regenerative cell therapies for cardiovascular applications. The firm has three product candidates currently in clinical trials: ALD-201, a treatment for ischemic heart failure (also known as myocardial ischaemia); ALD-301, a treatment for critical limb ischemia (a severe obstruction of the arteries); and ALD-401, which could potentially be used for the post-acute treatment of ischemic stroke. These product candidates consist of a certain amount of stem cells from a patient's own bone marrow. Aldagen isolates these stem cells (usually representing less than 1% of all cells in bone marrow) using its proprietary technology. Based on the firm's preclinical studies and clinical trials, these cell populations appear to have the potential to promote the regeneration of multiple types of cells and tissues. An example of the type of this regeneration includes angiogenesis, which is the growth of new blood vessels. Aldagen is partially owned by venture capital firm Intersouth Partners.

FINANCIALS: Sales and profits are in thousands of dollars—add 000 to get the full amount. 2009 Note: Financial information for 2009 was not available for all companies at press time.

2009 Sales: $	2009 Profits: $	**U.S. Stock Ticker:** Private	
2008 Sales: $	2008 Profits: $	**Int'l Ticker:** Int'l Exchange:	
2007 Sales: $	2007 Profits: $	Employees:	
2006 Sales: $	2006 Profits: $	Fiscal Year Ends: 12/31	
2005 Sales: $	2005 Profits: $	Parent Company:	

SALARIES/BENEFITS:

Pension Plan:	ESOP Stock Plan:	Profit Sharing:	Top Exec. Salary: $	Bonus: $
Savings Plan:	Stock Purch. Plan:		Second Exec. Salary: $	Bonus: $

OTHER THOUGHTS:

Apparent Women Officers or Directors:
Hot Spot for Advancement for Women/Minorities:

LOCATIONS: ("Y" = Yes)

West:	Southwest:	Midwest:	Southeast:	Northeast: Y	International:

ALEXION PHARMACEUTICALS INC

www.alexionpharm.com

Industry Group Code: 325412 Ranks within this company's industry group: Sales: 49   Profits: 29

Drugs:		Other:	Clinical:	Computers:	Services:
Discovery:	Y	AgriBio:	Trials/Services:	Hardware:	Specialty Services:
Licensing:		Genetic Data:	Labs:	Software:	Consulting:
Manufacturing:	Y	Tissue Replacement:	Equipment/Supplies:	Arrays:	Blood Collection:
Generics:			Research/Development Svcs.:	Database Management:	Drug Delivery:
			Diagnostics:		Drug Distribution:

TYPES OF BUSINESS:

Therapeutic Products
Hematologic Diseases
Neurological Diseases
Cancer
Autoimmune Disorders

BRANDS/DIVISIONS/AFFILIATES:

Soliris
Alexion Europe SAS
ALXN6000
Eculizumab

CONTACTS: *Note: Officers with more than one job title may be intentionally listed here more than once.*

Leonard Bell, CEO/Treas./Sec.
Vikas Sinha, CFO/Sr. VP
Glenn Melrose, VP-Human Resources
Stephen P. Squinto, Exec. VP/Head-R&D
James Bilotta, CIO/VP
M. Stacy Hooks, Sr. VP-Tech. Oper.
Claude Nicaise, Sr. VP-Strategic Product Dev.
Daniel N. Caron, Exec. Dir.-Eng.
M. Stacy Hooks, Sr. VP-Mfg.
Thomas Dubin, General Counsel/Sr. VP
Daniel N. Caron, Exec. Dir.-Site Oper.
Jeremy Springhorn, VP-Corp. Strategy & Bus. Dev.
Camille Bedrosian, Chief Medical Officer/Sr. VP
David Hallal, Sr. VP-Commercial Oper., Americas
Max Link, Chmn.
Patrice Coissac, Gen. Mgr./Pres., Alexion Int'l

Phone: 203-272-2596	Fax: 203-271-8190
Toll-Free:	
Address: 352 Knotter Dr., Cheshire, CT 06410 US	

GROWTH PLANS/SPECIAL FEATURES:

Alexion Pharmaceuticals, Inc. engages in the discovery, development and commercialization of therapeutic products designed to treat patients with severe diseases such as hematologic diseases, neurologic disease, cancer and autoimmune disorders. The company devotes substantially all of its resources to drug discovery, research, and product and clinical development. The firm has one marketed product, Soliris (eculizumab), the first therapy approved for the treatment of paroxysmal nocturnal hemoglobinuria (PNH). PNH is a rare genetic deficiency blood disorder, in which a patient's own complement system attacks and destroys blood cells. Soliris, a genetically altered antibody known as C5 complement inhibitor, treats PNH by selectively blocking the production of inflammation-causing proteins in the complement cascade. The company's other products are currently in preclinical or phase trial stages. These products include intravenous eculizumab for the treatment of a variety of conditions including Atypical Hemolytic Uremic Syndrome, Dense Deposit Disease, Acute Humoral Rejection and Myasthenia Gravis; and anti-CD200 MAb (monoclonal antibody) for the treatment of multiple myeloma and leukemia. The firm's products focus on anti-inflammatory therapeutics for disease in which the complement cascade is activated. Alexion also operates several subsidiaries in Europe, including Alexion Europe SAS, which supports commercial and regulatory international operations. The firm's primary sources of revenue are product sales and contract research revenue.

The firm offers its employees a 401(k) plan; medical, dental, vision, life and disability insurance; a college savings plan; and automobile and homeowners insurance discounts.

FINANCIALS: Sales and profits are in thousands of dollars—add 000 to get the full amount. 2009 Note: Financial information for 2009 was not available for all companies at press time.

2009 Sales: $386,800	2009 Profits: $295,166	U.S. Stock Ticker: ALXN
2008 Sales: $259,099	2008 Profits: $33,149	Int'l Ticker: Int'l Exchange:
2007 Sales: $72,041	2007 Profits: $-92,290	Employees: 673
2006 Sales: $1,558	2006 Profits: $-131,514	Fiscal Year Ends: 12/31
2005 Sales: $1,064	2005 Profits: $-108,750	Parent Company:

SALARIES/BENEFITS:

Pension Plan:	ESOP Stock Plan:	Profit Sharing:	Top Exec. Salary: $622,752	Bonus: $750,000
Savings Plan:	Stock Purch. Plan:		Second Exec. Salary: $399,538	Bonus: $184,000

OTHER THOUGHTS:

Apparent Women Officers or Directors: 1
Hot Spot for Advancement for Women/Minorities:

LOCATIONS: ("Y" = Yes)

West:	Southwest:	Midwest:	Southeast:	Northeast:	International:
				Y	Y

Note: Financial information, benefits and other data can change quickly and may vary from those stated here.

ALIMERA SCIENCES INC
www.alimerasciences.com

Industry Group Code: 325412 Ranks within this company's industry group: Sales: Profits:

Drugs:	Other:	Clinical:	Computers:	Services:
Discovery: Y	AgriBio:	Trials/Services:	Hardware:	Specialty Services:
Licensing:	Genetic Data:	Labs:	Software:	Consulting:
Manufacturing:	Tissue Replacement:	Equipment/Supplies:	Arrays:	Blood Collection:
Generics:		Research/Development Svcs.:	Database Management:	Drug Delivery:
		Diagnostics:		Drug Distribution:

TYPES OF BUSINESS:
Pharmaceuticals Research & Development
Retinal Disease Medication

BRANDS/DIVISIONS/AFFILIATES:
Iluvien

CONTACTS: Note: Officers with more than one job title may be intentionally listed here more than once.
Dan Myers, CEO
Dan Myers, Pres.
Rick Eiswirth, CFO
Dave Holland, VP-Mktg.
Ken Green, Sr. VP/Chief Scientific Officer
Susan Caballa, Sr. VP-Regulatory & Medical Affairs

Phone: 678-990-5740	Fax: 678-990-5744

Toll-Free:

Address: 6120 Windward Pkwy., Ste. 290, Alpharetta, GA 30005 US

GROWTH PLANS/SPECIAL FEATURES:
Alimera Sciences, Inc. is a biopharmaceutical company researching, developing and commercializing prescription ophthalmic pharmaceuticals. It is primarily focused on diseases affecting the retina. Iluvien, the firm's most advanced product candidate, is currently being developed in order to treat diabetic macular edema (DME), a disease affecting patients with diabetes that can lead to severe vision loss or blindness. It is composed of fluocinolone acetonide (FA), a compound known for its history of treating ocular diseases, contained inside a small, cylindrical ploymide tube and administered as an intravitreal insert into the patient's eye. Its design provides for a sustained-release that delivers a low, steady daily dose of FA over an anticipated 24 to 36 month period. The drug is currently undergoing two Phase III clinical trials in 956 patients across the U.S., Canada, Europe and India to test its safety and usefulness in treating DME. Additionally, three Phase II studies are underway to determine Iluvien's efficacy for the treatment of dry and wet age-related macular degeneration and retinal vein occlusion. The firm is also at work testing two classes of nicotinamide adenine dinucleotide phosphate (NADPH) oxidase inhibitors for the treatment of dry age-related macular degeneration. Alimera has acquired an exclusive, worldwide license from Emory University to continue its testing of these compunds. In June 2010, the firm submitted a New Drug Application (NDA) to the FDA for Iluvien.

FINANCIALS: Sales and profits are in thousands of dollars—add 000 to get the full amount. 2009 Note: Financial information for 2009 was not available for all companies at press time.
2009 Sales: $	2009 Profits: $	U.S. Stock Ticker: ALIM
2008 Sales: $	2008 Profits: $	Int'l Ticker: Int'l Exchange:
2007 Sales: $	2007 Profits: $	Employees:
2006 Sales: $	2006 Profits: $	Fiscal Year Ends: 12/31
2005 Sales: $	2005 Profits: $	Parent Company:

SALARIES/BENEFITS:
Pension Plan:	ESOP Stock Plan:	Profit Sharing:	Top Exec. Salary: $	Bonus: $
Savings Plan:	Stock Purch. Plan:		Second Exec. Salary: $	Bonus: $

OTHER THOUGHTS:
Apparent Women Officers or Directors: 1
Hot Spot for Advancement for Women/Minorities:

LOCATIONS: ("Y" = Yes)
West:	Southwest:	Midwest:	Southeast: Y	Northeast:	International:

ALKERMES INC

www.alkermes.com

Industry Group Code: 325412A Ranks within this company's industry group: Sales: 4   Profits: 3

Drugs:	Other:	Clinical:	Computers:	Services:
Discovery:	AgriBio:	Trials/Services:	Hardware:	Specialty Services:
Licensing:	Genetic Data:	Labs:	Software:	Consulting:
Manufacturing: Y	Tissue Replacement:	Equipment/Supplies:	Arrays:	Blood Collection:
Generics:		Research/Development Svcs.:	Database Management:	Drug Delivery:
		Diagnostics:		Drug Distribution:

TYPES OF BUSINESS:
Drug Delivery Systems
Pulmonary Drug Delivery Systems
Sustained Release Injection Delivery Systems

BRANDS/DIVISIONS/AFFILIATES:
Vivitrol
Risperdal Consta
Medisorb
Janssen Pharmaceutica, Inc
Amylin Pharmaceuticals Inc.
ALKS 9070
ALKS 37
ALKS 33

CONTACTS: *Note: Officers with more than one job title may be intentionally listed here more than once.*
Richard F. Pops, CEO
Gordon G. Pugh, COO/Sr. VP
Richard F. Pops, Pres.
James M. Frates, CFO/Sr. VP
Madeline D. Coffin, VP-Human Resources
Elliot W. Ehrich, Sr. VP-R&D/Chief Medical Officer
Kathryn L. Biberstein, General Counsel/Sr. VP/Sec.
Blair Jackson, VP-Bus. Dev. & Program Leadership
Rebecca J. Peterson, VP-Corp. Comm.
James M. Frates, Treas.
Kathryn L. Biberstein, Chief Compliance Officer/Sr. VP-Gov't Rel.
Dennis J. Bucceri, VP-Regulatory Affairs
Stephen King, VP-Commercial Oper.
Michael J. Landine, Sr. VP-Corp. Dev.
Richard F. Pops, Chmn.

Phone: 781-609-6000	Fax: 781-890-0524
Toll-Free:	
Address: 852 Winter St., Waltham, MA 02451 US	

GROWTH PLANS/SPECIAL FEATURES:

Alkermes, Inc. is a biotechnology company that specializes in the development of sophisticated drug delivery technologies. The company currently markets two commercial products: Risperdal Consta, a long-acting atypical antipsychotic medication for schizophrenia and bipolar I disorder; and Vivitrol, an injectable medication for the treatment of alcohol dependence. Risperdal Consta is a long-acting formulation of risperidone, a product of Janssen Pharmaceutica, Inc. Risperdal Consta is the first long-acting, atypical antipsychotic to be approved by the U.S. Food and Drug Administration (FDA). The medication uses the firm's proprietary Medisorb technology to maintain therapeutic medication levels in the body through just one injection every two weeks. Risperdal Consta is approved in 85 countries and marketed in 60 countries; Janssen continues to launch the product around the world. Revenues related to Risperdal Consta account for about 83% of Alkermes' total revenue. Vivitrol is an extended-release formulation for the treatment of alcohol dependence. Each injection of Vivitrol provides medication for one month and alleviates the need for patients to make daily medication dosing decisions. Vivitrol is also being developed for the treatment of opioid dependence. In addition to its marketed products, Alkermes also develops extended-release injectable, pulmonary and oral products for the treatment of central nervous system disorders, addiction, diabetes and autoimmune disorders. Additional products in development include an injectable form of Amylin's Byetta, used to treat type 2 diabetes; ALKS 37 for the treatment of opioid-induced constipation; ALKS 9070, a sustained-release version of aripiprazole for the treatment of schizophrenia; and ALKS 33, an oral opioid modulator being developed for the treatment of addiction. In January 2010, the firm relocated its headquarters from Cambridge, Massachusetts to Waltham, Massachusetts.

Alkermes offers its employees medical, dental and vision insurance; a 401(k) plan; tuition reimbursement; flexible spending accounts; an employee assistance program; and transportation benefits.

FINANCIALS: Sales and profits are in thousands of dollars—add 000 to get the full amount. 2009 Note: Financial information for 2009 was not available for all companies at press time.

2009 Sales: $326,839	2009 Profits: $130,505	**U.S. Stock Ticker: ALKS**
2008 Sales: $240,717	2008 Profits: $166,979	Int'l Ticker: Int'l Exchange:
2007 Sales: $239,965	2007 Profits: $9,445	Employees: 570
2006 Sales: $166,601	2006 Profits: $3,818	Fiscal Year Ends: 3/31
2005 Sales: $76,126	2005 Profits: $-73,916	Parent Company:

SALARIES/BENEFITS:

Pension Plan:	ESOP Stock Plan:	Profit Sharing:	Top Exec. Salary: $639,567	Bonus: $395,325
Savings Plan: Y	Stock Purch. Plan:		Second Exec. Salary: $509,615	Bonus: $315,000

OTHER THOUGHTS:
Apparent Women Officers or Directors: 3
Hot Spot for Advancement for Women/Minorities: Y

LOCATIONS: ("Y" = Yes)

West:	Southwest:	Midwest:	Southeast:	Northeast:	International:
		Y		Y	

ALLERGAN INC

www.allergan.com

Industry Group Code: 325412 Ranks within this company's industry group: Sales: 23 Profits: 23

Drugs:		Other:	Clinical:	Computers:	Services:
Discovery:	Y	AgriBio:	Trials/Services:	Hardware:	Specialty Services:
Licensing:		Genetic Data:	Labs:	Software:	Consulting:
Manufacturing:	Y	Tissue Replacement:	Equipment/Supplies:	Arrays:	Blood Collection:
Generics:			Research/Development Svcs.:	Database Management:	Drug Delivery:
			Diagnostics:		Drug Distribution:

TYPES OF BUSINESS:

Pharmaceutical Development
Eye Care Supplies
Dermatological Products
Neuromodulator Products
Obesity Intervention Products
Urologic Products
Medical Aesthetics

BRANDS/DIVISIONS/AFFILIATES:

Restasis
Lumigan
Optive
Latisse
Botox
Juvederm
Ozurdex
Serica Technologies, Inc.

CONTACTS: Note: Officers with more than one job title may be intentionally listed here more than once.

David E.I. Pyott, CEO
F. Michael Ball, Pres.
Jeffrey L. Edwards, CFO/Exec. VP-Finance
Dianne Dyer-Bruggeman, Exec. VP-Human Resources
Scott M. Whitcup, Exec. VP-R&D
Raymond H. Diradoorian, Exec. VP-Global Tech. Oper.
Douglas S. Ingram, Chief Admin. Officer/Chief Ethics Officer
Douglas S. Ingram, Corp. Sec./Exec. VP
Jeffrey L. Edwards, Exec. VP-Bus. Dev.
James F. Barlow, Principal Acct. Officer/Sr. VP
David E.I. Pyott, Chmn.

Phone: 714-246-4500	Fax: 714-246-6987
Toll-Free: 800-347-4500	
Address: 2525 Dupont Dr., Irvine, CA 92612 US	

GROWTH PLANS/SPECIAL FEATURES:

Allergan, Inc. is a technology-driven global health care company that develops and commercializes specialty pharmaceutical products, biologics and medical devices for the ophthalmic, neurological, medical aesthetics, medical dermatology, breast aesthetics, obesity intervention, urological and other specialty markets in more than 100 countries. The company focuses on treatments for chronic dry eye, glaucoma, retinal disease, psoriasis, acne, movement disorders, neuropathic pain and genitourinary diseases. The company operates in two segments: specialty pharmaceuticals and medical devices. The specialty pharmaceuticals segment includes eye care products, such as Restasis ophthalmic emulsion; Lumigan ophthalmic solution; Optive lubricant eye drops and the Refresh line of artificial tears; Botox, used in the treatment of neuromuscular disorders, pain management, the temporary improvement of wrinkles and for certain other therapeutic and aesthetic indications; skin care products, principally tazarotene products in cream and gel formulations for the treatment of acne, facial wrinkles and psoriasis, marketed under the name Tazorac; eyelash growth solution sold under the name Latisse; and urologics products, including Sanctura XR, a medication for overactive bladder. The medical devices segment includes breast implants for augmentation, revision and reconstructive surgery; obesity intervention products, including the Lap-Band, an adjustable gastric banding system, and the Orbera intragastric balloon system; and facial aesthetics products, including the Juvederm line of dermal filler products. In June 2009, the FDA approved Allergan's Ozurdex, a steroid implant that is injected, for the treatment of macular edema following branch retinal vein occlusion or central retinal vein occlusion. In 2010, the company acquired Serica Technologies, Inc., a medical device company focused on the development of biodegradable silk-based scaffolds for use in tissue regeneration procedures, for approximately $70 million.

Employees are offered medical, vision and dental insurance; flexible spending accounts; life insurance; disability coverage; adoption assistance; education assistance; an employee assistance program; scholarship awards for employees' children; and group auto and home insurance.

FINANCIALS: Sales and profits are in thousands of dollars—add 000 to get the full amount. 2009 Note: Financial information for 2009 was not available for all companies at press time.

2009 Sales: $4,447,600	2009 Profits: $621,300	U.S. Stock Ticker: AGN
2008 Sales: $4,339,700	2008 Profits: $578,600	Int'l Ticker: Int'l Exchange:
2007 Sales: $3,879,000	2007 Profits: $499,300	Employees: 8,300
2006 Sales: $3,010,100	2006 Profits: $-127,400	Fiscal Year Ends: 12/31
2005 Sales: $2,319,200	2005 Profits: $403,900	Parent Company:

SALARIES/BENEFITS:

Pension Plan: Y	ESOP Stock Plan:	Profit Sharing:	Top Exec. Salary: $1,300,000	Bonus: $1,560,000
Savings Plan: Y	Stock Purch. Plan:		Second Exec. Salary: $661,500	Bonus: $463,100

OTHER THOUGHTS:

Apparent Women Officers or Directors: 3
Hot Spot for Advancement for Women/Minorities: Y

LOCATIONS: ("Y" = Yes)

West:	Southwest:	Midwest:	Southeast:	Northeast:	International:
Y					Y

Note: Financial information, benefits and other data can change quickly and may vary from those stated here.

ALLIANCE PHARMACEUTICAL CORP

www.allp.com

Industry Group Code: 325412 Ranks within this company's industry group: Sales: 154   Profits: 74

Drugs:		Other:	Clinical:	Computers:	Services:
Discovery:	Y	AgriBio:	Trials/Services:	Hardware:	Specialty Services:
Licensing:		Genetic Data:	Labs:	Software:	Consulting:
Manufacturing:		Tissue Replacement:	Equipment/Supplies:	Arrays:	Blood Collection:
Generics:			Research/Development Svcs.:	Database Management:	Drug Delivery:
			Diagnostics:		Drug Distribution:

TYPES OF BUSINESS:
Drugs-Cardiovascular & Respiratory
Blood Substitutes

BRANDS/DIVISIONS/AFFILIATES:
Oxygent

CONTACTS: *Note: Officers with more than one job title may be intentionally listed here more than once.*
Duane J. Roth, CEO
B. Jack Defranco, COO
B. Jack DeFranco, Pres.
B. Jack DeFranco, CFO
Duane J. Roth, Chmn.

Phone: 858-779-1458	Fax: 858-427-0646
Toll-Free:	
Address: 4660 La Jolla Village Dr., Ste. 825, San Diego, CA 92122 US	

GROWTH PLANS/SPECIAL FEATURES:

Alliance Pharmaceutical Corp. develops therapeutic and diagnostic products that utilize perfluorochemicals, which are chemical substances with high oxygen-carrying capacity. Artificial oxygen carriers are intended to help ease blood shortages and avoid transfusion-related safety issues, such as the presence of blood transmissible diseases like Chagas disease in a blood supply. The company's leading product candidate, Oxygent, is an intravascular oxygen carrier designed to augment oxygen delivery in surgical patients. Oxygent is also intended to be an emulsion-based red blood cell substitute. The blood substitute is sterile and universally compatible with all blood types. Oxygent's primary indication is to provide oxygen to tissues during surgeries. In clinical studies, Oxygent has been shown to correct or mean arterial pressure, heart rate and mixed venous oxygen tension. Alliance has formed many collaborative relationships with other companies to test, manufacture and eventually distribute Oxygent in other countries, including Beijing Double-Crane Pharmaceutical Co., Ltd. in China; Leo Pharma A/S in Europe, including EU member and applicant countries, and Canada; and Il Yang Pharm. Co., Ltd. in South Korea.

FINANCIALS: Sales and profits are in thousands of dollars—add 000 to get the full amount. 2009 Note: Financial information for 2009 was not available for all companies at press time.

2009 Sales: $ 79	2009 Profits: $-1,203	U.S. Stock Ticker: ALLP.OB
2008 Sales: $ 106	2008 Profits: $-1,780	Int'l Ticker: Int'l Exchange:
2007 Sales: $ 573	2007 Profits: $-4,093	Employees:
2006 Sales: $ 129	2006 Profits: $-9,575	Fiscal Year Ends: 6/30
2005 Sales: $1,477	2005 Profits: $-5,743	Parent Company:

SALARIES/BENEFITS:

Pension Plan:	ESOP Stock Plan:	Profit Sharing:	Top Exec. Salary: $235,000	Bonus: $
Savings Plan:	Stock Purch. Plan:		Second Exec. Salary: $120,000	Bonus: $

OTHER THOUGHTS:
Apparent Women Officers or Directors:
Hot Spot for Advancement for Women/Minorities:

LOCATIONS: ("Y" = Yes)

West:	Southwest:	Midwest:	Southeast:	Northeast:	International:
Y					

ALLOS THERAPEUTICS INC

www.allos.com

Industry Group Code: 325412 Ranks within this company's industry group: Sales: 132 Profits: 151

Drugs:		Other:		Clinical:	Computers:		Services:	
Discovery:	Y	AgriBio:	Y	Trials/Services:	Hardware:		Specialty Services:	
Licensing:	Y	Genetic Data:		Labs:	Software:		Consulting:	
Manufacturing:	Y	Tissue Replacement:		Equipment/Supplies:	Arrays:		Blood Collection:	
Generics:				Research/Development Svcs.:	Database Management:		Drug Delivery:	
				Diagnostics:			Drug Distribution:	

TYPES OF BUSINESS:

Cancer Treatment Drugs
Small-Molecule Therapies

BRANDS/DIVISIONS/AFFILIATES:

Folotyn

CONTACTS: Note: Officers with more than one job title may be intentionally listed here more than once.

Paul L. Berns, CEO
Paul L. Berns, Pres.
James V. Caruso, Chief Commercial Officer/Exec. VP
Marc H. Graboyes, General Counsel/Sr. VP
Bruce K. Bennett, VP-Pharmaceutical Oper.
Monique Greer, Contact-Corp. Comm.
Monique Greer, Contact-Investor Rel.
David C. Clark, VP-Finance/Treas.
Charles Q. Morris, Exec. VP/Chief Medical Officer
Stephen J. Hoffman, Chmn.

Phone: 303-426-6262	Fax: 303-426-4731
Toll-Free:	
Address: 11080 CirclePoint Rd., Ste. 200, Westminster, CO 80020 US	

GROWTH PLANS/SPECIAL FEATURES:

Allos Therapeutics, Inc. focuses on developing and commercializing small molecule drugs for cancer treatment. The company is currently focused on the development and commercialization of one drug: Folotyn (pralatrexate injection). Folotyn is a targeted antifolate inhibitor designed to accumulate preferentially in cancer cells. Folotyn targets the inhibition of dihydrofolate reductase, or DHFR, an enzyme critical in the folate pathway, thereby interfering with DNA and RNA synthesis and triggering cancer cell death. January 2010 marked the firm's commercial launch of Folotyn in the U.S., after receiving U.S. Food and Drug Administration (FDA) approval for the drug when used as a single agent in the treatment of patients with relapsed or refractory T-cell lymphoma, or PTCL. Trials are currently underway relating to the use of Folotyn in combination therapy regimens, and in other potential applications across a variety of hematological malignancies and solid tumor indications. Ongoing trials and development activities are focused on indications such as cutaneous T-cell lymphoma, non-small cell lung cancer and bladder cancer. In addition to organizing a nationwide group of drug sales representatives in early 2010 to promote Folotyn in the U.S. medical community, the company also established Allos Support for Assisting Patients (ASAP), to help make the drug more widely available by providing reimbursement resources for uninsured and underinsured patients. Allos Therapeutics maintains exclusive worldwide commercial rights to the drug, and has announced plans to submit applications in late 2010 for Folotyn to be approved for sale as a single agent treatment within the European Union.

The company offers its employees a 401(k) plan; health, dental and vision insurance; group life insurance; short and long-term disability; educational reimbursement; an employee stock purchase plan; and flexible spending accounts.

FINANCIALS: Sales and profits are in thousands of dollars—add 000 to get the full amount. 2009 Note: Financial information for 2009 was not available for all companies at press time.

2009 Sales: $3,585	2009 Profits: $-73,553	U.S. Stock Ticker: ALTH
2008 Sales: $	2008 Profits: $-51,730	Int'l Ticker: Int'l Exchange:
2007 Sales: $	2007 Profits: $-39,370	Employees: 170
2006 Sales: $	2006 Profits: $-30,212	Fiscal Year Ends: 12/31
2005 Sales: $	2005 Profits: $-20,137	Parent Company:

SALARIES/BENEFITS:

Pension Plan:	ESOP Stock Plan:	Profit Sharing:	Top Exec. Salary: $500,800	Bonus: $340,700
Savings Plan: Y	Stock Purch. Plan: Y		Second Exec. Salary: $407,000	Bonus: $183,500

OTHER THOUGHTS:

Apparent Women Officers or Directors: 1
Hot Spot for Advancement for Women/Minorities:

LOCATIONS: ("Y" = Yes)

West:	Southwest:	Midwest:	Southeast:	Northeast:	International:
Y				Y	

Note: Financial information, benefits and other data can change quickly and may vary from those stated here.

ALPHARMA ANIMAL HEALTH

www.alpharmaah.com

Industry Group Code: 325414 Ranks within this company's industry group: Sales: Profits:

Drugs:		Other:		Clinical:	Computers:	Services:
Discovery:	Y	AgriBio:	Y	Trials/Services:	Hardware:	Specialty Services:
Licensing:		Genetic Data:		Labs:	Software:	Consulting:
Manufacturing:	Y	Tissue Replacement:		Equipment/Supplies:	Arrays:	Blood Collection:
Generics:				Research/Development Svcs.:	Database Management:	Drug Delivery:
				Diagnostics:		Drug Distribution:

TYPES OF BUSINESS:

Drugs-Animal Health
Human Pharmaceuticals
Animal Feed Additives

BRANDS/DIVISIONS/AFFILIATES:

BIO-COX
Histostat
BMD
Albac
3-Nitro
Deccox
Proflora
King Pharmaceuticals Inc

CONTACTS: Note: Officers with more than one job title may be intentionally listed here more than once.

Eric J. Bruce, Pres.
Debbie Thompson, Global Mktg. Associate
Jack Howarth, VP-Investor Rel.

Phone:	Fax:
Toll-Free: 800-834-6470	
Address: 501 5th St., Bristol, TN 37620 US	

GROWTH PLANS/SPECIAL FEATURES:

Alpharma Animal Health, formerly Alpharma, Inc., is a multinational pharmaceutical company that manufactures and markets pharmaceutical products for farm animals. The company has operations in the U.S., Canada, Asia Pacific, Europe, the Middle East and Africa. The firm divides its operations based on these geographic locations; in addition, the U.S. group is divided into poultry and livestock. Alpharma markets over 100 animal feed additives and water soluble therapeutic products, including BMD, a feed additive that promotes growth and feed efficiency, and prevents/treats diseases in poultry and swine; Albac, a feed additive for poultry, swine and calves; BIO-COX, which prevents coccidiosis in poultry; and 3-Nitro, Zoamix and Histostat feed grade anticoccidials. Additional product names include Aureomycin, Beta Mos and Deccox. Alpharma Animal Health manufactures its products at facilities in Maryland, Colorado, West Virginia, Illinois, Arkansas and China. The company is a wholly-owned subsidiary of King Pharmaceuticals, Inc. In January 2010, Alpharma introduced its new poultry feed additive, Proflora, which is a non-drug, non-antibiotic chemical that improves poultry production.

The company offers its employees medical, dental and vision; flexible spending accounts; prescription drug coverage; life insurance; short and long term disability; a 401(k) matching plan; and employee assistance program; tuition assistance; adoption reimbursement; and travel assistance.

FINANCIALS: Sales and profits are in thousands of dollars—add 000 to get the full amount. 2009 Note: Financial information for 2009 was not available for all companies at press time.

2009 Sales: $	2009 Profits: $	U.S. Stock Ticker: Subsidiary
2008 Sales: $	2008 Profits: $	Int'l Ticker: Int'l Exchange:
2007 Sales: $722,425	2007 Profits: $-13,581	Employees:
2006 Sales: $653,828	2006 Profits: $82,544	Fiscal Year Ends: 12/31
2005 Sales: $553,617	2005 Profits: $133,769	Parent Company: KING PHARMACEUTICALS INC

SALARIES/BENEFITS:

Pension Plan:	ESOP Stock Plan:	Profit Sharing:	Top Exec. Salary: $291,054	Bonus: $
Savings Plan: Y	Stock Purch. Plan: Y		Second Exec. Salary: $	Bonus: $

OTHER THOUGHTS:

Apparent Women Officers or Directors: 1
Hot Spot for Advancement for Women/Minorities: Y

LOCATIONS: ("Y" = Yes)

West:	Southwest:	Midwest:	Southeast:	Northeast:	International:
Y		Y	Y	Y	Y

ALSERES PHARMACEUTICALS INC

www.alseres.com

Industry Group Code: 325412 Ranks within this company's industry group: Sales: Profits: 99

Drugs:		Other:	Clinical:	Computers:	Services:
Discovery:		AgriBio:	Trials/Services:	Hardware:	Specialty Services:
Licensing:	Y	Genetic Data:	Labs:	Software:	Consulting:
Manufacturing:		Tissue Replacement:	Equipment/Supplies:	Arrays:	Blood Collection:
Generics:			Research/Development Svcs.:	Database Management:	Drug Delivery:
			Diagnostics:		Drug Distribution:

TYPES OF BUSINESS:

Pharmaceutical Discovery & Development
Patent & Development Rights
CNS Disorder Treatments

BRANDS/DIVISIONS/AFFILIATES:

Oncomodulin
Inosine
Cethrin
Altropane

CONTACTS: Note: Officers with more than one job title may be intentionally listed here more than once.

Peter G. Savas, CEO
Mark J. Pykett, COO
Mark J. Pykett, Pres.
Kenneth L. Rice, Jr., CFO
Noel J. Cusack, Sr. VP-Preclinical Dev.
Kenneth L. Rice, Jr., Exec. VP-Admin.
Kenneth L. Rice, Jr., In-House Counsel
Richard M. Thorn, Sr. VP-Program Oper.
Kenneth L. Rice, Jr., Exec. VP-Finance
James R. Weston, Sr. VP-Regulatory Affairs & Quality
Mark Hurtt, Chief Medical Officer
Susan M. Flint, Sr. VP-Drug Dev.
Peter G. Savas, Chmn.

Phone: 508-497-2360	Fax: 508-497-9964
Toll-Free:	
Address: 239 South St., Hopkinton, MA 01748 US	

GROWTH PLANS/SPECIAL FEATURES:

Alseres Pharmaceuticals, Inc. is a biotechnology company engaged in the research and development of diagnostic and therapeutic products primarily for central nervous system (CNS) disorders. The company's research and development is based on three technology platforms: the regenerative therapeutics program, which focuses on nerve repair, restoring movement and returning sensory activity in patients with loss of CNS function from traumas or degenerative diseases; the molecular imaging program, which focuses on the diagnosis of Parkinsonian Syndromes, including Parkinson's Disease (PD), Dementia and Attention Deficit Hyperactivity Disorder (ADHD); and the neurodegenerative disease program, which focuses on treating the symptoms of PD and slowing or stopping the progression of PD. The regenerative therapeutics program has three drugs under development: Cethrin for the treatment of acute spinal cord injuries (SCIs); Inosine for the treatment of SCIs, TBIs (traumatic brain injuries) and stroke; and Oncomodulin for the treatment of optic nerve injuries and Glaucoma. It also has research programs dedicated to other regenerative therapies such as bone repair. Due to budgets shortages however, the firm has suspended development on all drugs save for Cethrin, which is its leading candidate for market approval. Within the molecular imaging program, the company has two products in the pre-clinical stages of development, as well as Altropane for PD in Phase III clinical trials and for Dementia with Lewy Bodies (DLB) in Phase II of clinical trials. Based on budget shortcomings Alseres has been forced to greatly reduce research and development related to its neurodegenerative disease program. The firm owns or has licensed 41 issued U.S. patents and has 29 pending U.S. patent applications.

FINANCIALS: Sales and profits are in thousands of dollars—add 000 to get the full amount. 2009 Note: Financial information for 2009 was not available for all companies at press time.

2009 Sales: $	2009 Profits: $-10,777	U.S. Stock Ticker: ALSE
2008 Sales: $	2008 Profits: $-20,847	Int'l Ticker: Int'l Exchange:
2007 Sales: $	2007 Profits: $-19,548	Employees: 5
2006 Sales: $	2006 Profits: $-26,355	Fiscal Year Ends: 12/31
2005 Sales: $	2005 Profits: $-11,501	Parent Company:

SALARIES/BENEFITS:

Pension Plan:	ESOP Stock Plan:	Profit Sharing:	Top Exec. Salary: $408,085	Bonus: $
Savings Plan:	Stock Purch. Plan:		Second Exec. Salary: $304,260	Bonus: $

OTHER THOUGHTS:

Apparent Women Officers or Directors: 1
Hot Spot for Advancement for Women/Minorities:

LOCATIONS: ("Y" = Yes)

West:	Southwest:	Midwest:	Southeast:	Northeast:	International:
				Y	

ALTANA AG

www.altana.com

Industry Group Code: 325412 Ranks within this company's industry group: Sales: 37 Profits: 54

Drugs:	Other:	Clinical:	Computers:	Services:
Discovery:	AgriBio:	Trials/Services:	Hardware:	Specialty Services:
Licensing:	Genetic Data:	Labs:	Software:	Consulting:
Manufacturing: Y	Tissue Replacement:	Equipment/Supplies: Y	Arrays:	Blood Collection:
Generics:		Research/Development Svcs.:	Database Management:	Drug Delivery:
		Diagnostics:		Drug Distribution:

TYPES OF BUSINESS:

Specialty Chemical Manufacturing
Imaging Products
Electrical Insulation
Coatings
Inks

BRANDS/DIVISIONS/AFFILIATES:

BYK Additives & Instruments
ECKART Effect Pigments
ELANTAS Electrical Insulation
ACTEGA Coatings & Sealants
BYK-Chemie
DyStar group
Water Ink Technologies Inc
Aquaprint GmbH

CONTACTS: Note: Officers with more than one job title may be intentionally listed here more than once.

Matthias L. Wolfgruber, CEO
Martin Babilas, CFO
Roland Peter, Pres., Additives & Instruments Div.
Christoph Schlunken, Pres., Effect Pigments Div.
Wolfgang Schutt, Pres., Electrical Insulation Div.
Guido Forstbach, Pres., Coatings & Sealants Div.
Fritz Frohlich, Chmn.

Phone: 49-281-670-8	Fax: 49-281-670-376
Toll-Free:	
Address: Abelstrasse 43, Wesel, 46483 Germany	

GROWTH PLANS/SPECIAL FEATURES:

Altana AG is an international chemicals company that develops, manufactures and markets products for a range of targeted, highly specialized applications. The company serves customers in the coatings, paint and plastics, printing, cosmetics, electrical and electronics industries. Altana has 35 production facilities and 47 service and research labs worldwide. The firm operates through four divisions: BYK Additives & Instruments; ECKART Effect Pigments; ELANTAS Electrical Insulation; and ACTEGA Coatings & Sealants. BYK Additives & Instruments offers a range of chemical additives, produced by subsidiary BYK-Chemie, that help to improve and regulate the quality and processability of coatings and plastics. This division also offers testing and measuring equipment, produced by subsidiary BYK-Gardner, allowing manufacturers to predetermine the color, gloss and other physical properties of paints and plastics products. ECKART Effect Pigments develops and produces metallic effect and pearlescent pigments, as well as gold-bronze and zinc pigments, used to produce certain optical effects in paints, inks, cosmetics and coatings. ELANTAS Electrical Insulation produces insulating materials used in electrical and electronics applications, including electric motors, household appliances, cars, generators, transformers, capacitors, televisions, computers, wind mills, circuit boards and sensors. ACTEGA Coatings & Sealants develops and produces specialty coatings and sealants used primarily by the graphic arts and packaging industries. These products, which include sealants for glass bottles and metal cans as well as coatings for flexible packaging, help to regulate the physical properties of packaging, preserve the freshness of contents and contribute to the overall appearance of packaged goods. Altana acquired the high-performance additives business of DyStar group in March 2009. In October 2009, the firm completed the acquisition of Water Ink Technologies, Inc., a U.S. ink and coatings company. In March 2010, Altana acquired German coating manufacturer Aquaprint GmbH.

FINANCIALS: Sales and profits are in thousands of dollars—add 000 to get the full amount. 2009 Note: Financial information for 2009 was not available for all companies at press time.

2009 Sales: $1,605,500	2009 Profits: $14,130	U.S. Stock Ticker:
2008 Sales: $1,774,430	2008 Profits: $136,760	Int'l Ticker: ALT Int'l Exchange: Frankfurt-Euronext
2007 Sales: $1,825,660	2007 Profits: $171,790	Employees:
2006 Sales: $2,039,470	2006 Profits: $89,500	Fiscal Year Ends: 12/31
2005 Sales: $1,405,400	2005 Profits: $679,100	Parent Company:

SALARIES/BENEFITS:

Pension Plan:	ESOP Stock Plan:	Profit Sharing:	Top Exec. Salary: $	Bonus: $
Savings Plan:	Stock Purch. Plan:		Second Exec. Salary: $	Bonus: $

OTHER THOUGHTS:

Apparent Women Officers or Directors: 1
Hot Spot for Advancement for Women/Minorities:

LOCATIONS: ("Y" = Yes)

West:	Southwest:	Midwest:	Southeast:	Northeast:	International:
		Y		Y	Y

AMARIN CORPORATION PLC

www.amarincorp.com

Industry Group Code: 325412 Ranks within this company's industry group: Sales: Profits:

Drugs:		Other:	Clinical:	Computers:	Services:
Discovery:	Y	AgriBio:	Trials/Services:	Hardware:	Specialty Services:
Licensing:	Y	Genetic Data:	Labs:	Software:	Consulting:
Manufacturing:	Y	Tissue Replacement:	Equipment/Supplies:	Arrays:	Blood Collection:
Generics:			Research/Development Svcs.:	Database Management:	Drug Delivery:
			Diagnostics:		Drug Distribution:

TYPES OF BUSINESS:

Drugs, Cardiovascular
Drugs, Cardiovascular Health
Drugs, Triglyceride Reduction

BRANDS/DIVISIONS/AFFILIATES:

AMR101
Amarin Pharma, Inc.

CONTACTS: Note: Officers with more than one job title may be intentionally listed here more than once.

Declan Doogan, Interim CEO
John Thero, CFO
Tom Maher, General Counsel/Corp. Sec.
Paresh Soni, Sr. VP-Corp. Dev.
Conor Dalton, VP-Finance
Thomas G. Lynch, Chmn.

Phone: 353-1-6699-020	Fax: 353-1-6699-028
Toll-Free:	
Address: 1st Fl., Block 3, Shellbourne Rd., Ballsbridge, Dublin, 4 Ireland	

GROWTH PLANS/SPECIAL FEATURES:

Amarin Corporation plc is a specialty pharmaceutical company focused on cardiovascular disorders. The company's platforms utilize the cardiovascular system's innate quality of being affected by polyunsaturated fatty acids. Amarin's lipophilic drugs are fat-soluble, allowing easy transportation across the blood-brain barrier. The company is also utilizing a combinational lipid platform, where bioactive lipids are attached either to other drugs or other lipids. Amarin's current leading development project is its prescription grade Omega-3 fatty acid AMR101 for hypertriglyceridemia. Hypertriglyceridemia is a condition in which patients exhibit high blood levels of triglycerides, a known risk indicator for cardiovascular disease, which AMR101 may help to combat. In December 2009, Amarin began enrolling patients in two separate AMR101 Phase III clinical trials: the Marine trial and the Anchor trial. The first, the Marine trial, is the primary stand-alone trial, and is focused on patients with very high triglyceride levels (more than 500 milligrams per deciliter of blood, or 500 mg/dl). The Anchor trial, meanwhile, is focused on patients with high triglyceride levels (greater than 200 mg/dl) who are already undergoing enzyme-inhibitor therapy to reduce their cholesterol; the results of this trial, in conjunction with the Marine trial, may allow Amarin to expand the indications of AMR101 when the drug is brought to market. The company's cardiovascular pipeline also includes additional potential product candidates, including potential combination product candidates. These additional candidates are currently in the development and pre-clinical stages, with no clinical trials commenced as of mid-2010. Subsidiary Amarin Pharma, Inc. oversees the company's U.S. operations in Mystic, Connecticut. In October 2009, the company announced that it would discontinue its development activities related to treating disorders of the central nervous systems; this work had previously involved AMR101 and several other drug formulations aimed at addressing diseases such as Huntington's, Parkinson's and Myasthenia Gravis.

FINANCIALS: Sales and profits are in thousands of dollars—add 000 to get the full amount. 2009 Note: Financial information for 2009 was not available for all companies at press time.

2009 Sales: $	2009 Profits: $	U.S. Stock Ticker: AMRN
2008 Sales: $	2008 Profits: $-20,021	Int'l Ticker: Int'l Exchange:
2007 Sales: $	2007 Profits: $-38,197	Employees: 27
2006 Sales: $ 111	2006 Profits: $-23,707	Fiscal Year Ends: 8/31
2005 Sales: $ 500	2005 Profits: $-19,630	Parent Company:

SALARIES/BENEFITS:

Pension Plan: Y	ESOP Stock Plan:	Profit Sharing:	Top Exec. Salary: $	Bonus: $
Savings Plan:	Stock Purch. Plan:		Second Exec. Salary: $	Bonus: $

OTHER THOUGHTS:

Apparent Women Officers or Directors:
Hot Spot for Advancement for Women/Minorities:

LOCATIONS: ("Y" = Yes)

West:	Southwest:	Midwest:	Southeast:	Northeast:	International:
				Y	Y

AMGEN INC

www.amgen.com

Industry Group Code: 325412 Ranks within this company's industry group: Sales: 13   Profits: 11

Drugs:		Other:	Clinical:	Computers:	Services:
Discovery:	Y	AgriBio:	Trials/Services:	Hardware:	Specialty Services:
Licensing:		Genetic Data:	Labs:	Software:	Consulting:
Manufacturing:	Y	Tissue Replacement:	Equipment/Supplies:	Arrays:	Blood Collection:
Generics:			Research/Development Svcs.:	Database Management:	Drug Delivery:
			Diagnostics:		Drug Distribution:

TYPES OF BUSINESS:

Drugs-Diversified
Oncology Drugs
Nephrology Drugs
Inflammation Drugs
Neurology Drugs

BRANDS/DIVISIONS/AFFILIATES:

Aranesp
EPOGEN
Neulasta
NEUPOGEN
Enbrel
Sensipar
Vectibix
Nplate

CONTACTS: Note: Officers with more than one job title may be intentionally listed here more than once.

Kevin W. Sharer, CEO
Kevin W. Sharer, Pres.
Robert A. Bradway, CFO/Exec. VP
Brian McNamee, Sr. VP-Human Resources
Roger M. Perlmutter, Exec. VP-R&D
Thomas J. (Tom) Flanagan, CIO/Sr. VP
David J. Scott, General Counsel/Sr. VP/Corp. Sec.
Fabrizio Bonanni, Exec. VP-Oper.
David Beier, Sr. VP-Corp. Affairs & Global Gov't
Anna Richo, Sr. VP/Chief Compliance Officer
George J. Morrow, Exec. VP-Global Commercial Oper.
Kevin W. Sharer, Chmn.

Phone: 805-447-1000	Fax: 805-447-1010
Toll-Free: 800-772-6436	
Address: 1 Amgen Center Dr., Thousand Oaks, CA 91320-1799 US	

GROWTH PLANS/SPECIAL FEATURES:

Amgen, Inc. is a global biotechnology medicines company that discovers, develops, manufactures and markets human therapeutics based on cellular and molecular biology. Its products are used for treatment in the fields of supportive cancer care, nephrology and inflammation. Amgen's primary products include Aranesp, EPOGEN, Neulasta, NEUPOGEN and ENBREL, which together represent 93% of the company's sales. Aranesp and EPOGEN stimulate the production of red blood cells to treat anemia and belong to a class of drugs referred to as erythropoiesis-stimulating agents. Aranesp is used for the treatment of anemia both in chronic kidney failure and in concomitant chemotherapy. EPOGEN is used to treat anemia associated with end-stage renal disease. Neulasta and NEUPOGEN selectively stimulate the production of neutrophils, one type of white blood cell that helps the body fight infections. ENBREL inhibits tumor necrosis factor (TNF), a substance induced in response to inflammatory and immunological responses, such as rheumatoid arthritis and psoriasis. Other marketed products include Sensipar/Mimpara, which lowers serum calcium levels; Vectibix, used to treat specific progressions of metastatic colorectal cancer; and Nplate, used to treat low platelet count. Amgen maintains sales and marketing forces primarily in the U.S., Europe and Canada, and markets its products to healthcare providers including physicians or their clinics, dialysis centers, hospitals and pharmacies. Amgen focuses its research and development efforts in the core areas of oncology, hematology, inflammation, bone, nephrology, cardiology and neurology. In 2009, Nplate was approved in the European Union for the treatment of idiopathic thrombocytopenic purpura (ITP).

Amgen offers its employees health coverage, disability coverage, life insurance, a tuition reimbursement plan, dependent care spending accounts, a 401(k) plan and a stock purchase plan, childcare services, voluntary group home and auto insurance, telecommuting and remote work options and recreation and fitness classes.

FINANCIALS: Sales and profits are in thousands of dollars—add 000 to get the full amount. 2009 Note: Financial information for 2009 was not available for all companies at press time.

2009 Sales: $14,642,000	2009 Profits: $4,605,000	**U.S. Stock Ticker: AMGN**
2008 Sales: $15,003,000	2008 Profits: $4,052,000	**Int'l Ticker:** Int'l Exchange:
2007 Sales: $14,771,000	2007 Profits: $3,166,000	Employees: 16,900
2006 Sales: $14,268,000	2006 Profits: $2,950,000	Fiscal Year Ends: 12/31
2005 Sales: $12,430,000	2005 Profits: $3,674,000	Parent Company:

SALARIES/BENEFITS:

Pension Plan:	ESOP Stock Plan:	Profit Sharing:	Top Exec. Salary: $1,682,308	Bonus: $3,790,000
Savings Plan: Y	Stock Purch. Plan: Y		Second Exec. Salary: $991,131	Bonus: $1,195,000

OTHER THOUGHTS:

Apparent Women Officers or Directors: 3
Hot Spot for Advancement for Women/Minorities: Y

LOCATIONS: ("Y" = Yes)

West:	Southwest:	Midwest:	Southeast:	Northeast:	International:
Y			Y	Y	Y

AMYLIN PHARMACEUTICALS INC

www.amylin.com

Industry Group Code: 325412 Ranks within this company's industry group: Sales: 44 Profits: 165

Drugs:		Other:		Clinical:	Computers:	Services:
Discovery:	Y	AgriBio:		Trials/Services:	Hardware:	Specialty Services:
Licensing:	Y	Genetic Data:		Labs:	Software:	Consulting:
Manufacturing:		Tissue Replacement:		Equipment/Supplies:	Arrays:	Blood Collection:
Generics:				Research/Development Svcs.:	Database Management:	Drug Delivery:
				Diagnostics:		Drug Distribution:

TYPES OF BUSINESS:

Pharmaceutical Discovery & Development
Drugs, Obesity
Drugs, Diabetes

BRANDS/DIVISIONS/AFFILIATES:

SYMLIN
BYETTA
Exenatide LAR
Integrated Neurohormonal Therapy for Obesity
Pramlintide/Metreleptin

CONTACTS: Note: Officers with more than one job title may be intentionally listed here more than once.

Daniel M. Bradbury, CEO
Daniel M. Bradbury, Pres.
Mark G. Foletta, CFO
Vincent P. Mihalik, Sr. VP-Sales & Mktg.
Roger Marchetti, Sr. VP-Human Resources
Christian Weyer, Sr. VP-R&D
Roger Marchetti, Sr. VP-Info. Mgmt.
Marcea B. Lloyd, General Counsel/Sr. VP-Gov't Affairs
Paul Marshall, Sr. VP-Oper.
Mark J. Gergen, Sr. VP-Corp. Dev.
Marcea B. Lloyd, Sr. VP-Corp. Affairs
Mark G. Foletta, Sr. VP-Finance
Vincent P. Mihalik, Chief Commercial Officer
Orville G. Kolterman, Chief Medical Officer/Sr. VP
Harry J. Leonhardt, VP-Legal & Governance/Corp. Sec.
Paulo F. Costa, Chmn.

Phone: 858-552-2200	Fax: 858-552-2212
Toll-Free:	
Address: 9360 Towne Centre Dr., San Diego, CA 92121 US	

GROWTH PLANS/SPECIAL FEATURES:

Amylin Pharmaceuticals, Inc. is engaged in the discovery, development and commercialization of drug candidates for the treatment of diabetes, obesity and other diseases. Amylin's research process is focused on identifying potentially useful peptide hormones, experimenting to discover their therapeutic applications and developing new treatments. The firm has amassed a polypeptide hormone library encompassing over 1,000 potentially useful biologics. Amylin is currently marketing two medicines to treat diabetes: BYETTA (exenatide) injection and SYMLIN (pramlintide acetate) injection. Additionally, the company is in Phase III of developing exenatide once weekly and in Phase I for the development of exenatide nasal, both drugs targeted for the treatment of diabetes. Amylin's Integrated Neurohormonal Therapy for Obesity (INTO) program studies the safety and efficacy of multiple neurohormones when used in combination with pramlintide to treat obesity. The company's obesity drugs, which are primarily a part of Amylin's INTO program, include Pramlintide/Metreleptin, currently in Phase II. Amylin has strategic alliance partnerships with Alkermes, Inc. and Eli Lilly for the development of the exenatide once-weekly diabetes treatment. In October 2009, the U.S. FDA approved an expanded indication to include BYETTA as a first-line, stand-alone medication (along with diet and exercise) to improve glycemic control in adults with type-2 diabetes.

Amylin provides its employees with medical, dental and voluntary vision plans with domestic partner coverage; life and disability insurance; flexible spending accounts; a 401(k) plan; an employee stock purchase plan; an employee assistance program; education assistance programs; and discounted gym memberships.

FINANCIALS: Sales and profits are in thousands of dollars—add 000 to get the full amount. 2009 Note: Financial information for 2009 was not available for all companies at press time.

2009 Sales: $753,993	2009 Profits: $-186,256	U.S. Stock Ticker: AMLN
2008 Sales: $765,342	2008 Profits: $-321,941	Int'l Ticker: Int'l Exchange:
2007 Sales: $780,997	2007 Profits: $-211,136	Employees: 1,800
2006 Sales: $510,875	2006 Profits: $-218,856	Fiscal Year Ends: 12/31
2005 Sales: $140,474	2005 Profits: $-206,832	Parent Company:

SALARIES/BENEFITS:

Pension Plan:	ESOP Stock Plan:	Profit Sharing:	Top Exec. Salary: $662,019	Bonus: $814,280
Savings Plan: Y	Stock Purch. Plan: Y		Second Exec. Salary: $431,539	Bonus: $265,400

OTHER THOUGHTS:

Apparent Women Officers or Directors: 4
Hot Spot for Advancement for Women/Minorities: Y

LOCATIONS: ("Y" = Yes)

West:	Southwest:	Midwest:	Southeast:	Northeast:	International:
Y					Y

Note: Financial information, benefits and other data can change quickly and may vary from those stated here.

AMYRIS BIOTECHNOLOGIES INC

www.amyrisbiotech.com

Industry Group Code: 325193 Ranks within this company's industry group: Sales: Profits:

Drugs:		Other:		Clinical:	Computers:	Services:
Discovery:	Y	AgriBio:	Y	Trials/Services:	Hardware:	Specialty Services:
Licensing:	Y	Genetic Data:		Labs:	Software:	Consulting:
Manufacturing:		Tissue Replacement:		Equipment/Supplies:	Arrays:	Blood Collection:
Generics:				Research/Development Svcs.:	Database Management:	Drug Delivery:
				Diagnostics:		Drug Distribution:

TYPES OF BUSINESS:

Ethanol Production
Artemisinin
Renewable Fuels

BRANDS/DIVISIONS/AFFILIATES:

Amyris Fuels, LLC
Amyris Brasil S.A.
Biofene
No Compromise Fuels
SMA Industria Quimica S.A.
Soliance

CONTACTS: Note: Officers with more than one job title may be intentionally listed here more than once.

John Melo, CEO
Mario Portela, COO
Jeryl L. Hilleman, CFO
Jack D. Newman, Sr. VP-Research
Neil Renninger, CTO
Jeff Lievense, Sr. VP-Process Dev. & Mfg.
Tamara Tompkins, General Counsel/Sr. VP
Joel Cherry, Sr. VP-Research Programs & Oper.
Kinkead Reiling, Sr. VP-Corp. Dev.
Peter Boynton, Chief Commercial Officer

Phone: 510-450-0761	Fax: 510-225-2645
Toll-Free:	
Address: 5885 Hollis Street, Ste. 100, Emeryville, CA 94608 US	

GROWTH PLANS/SPECIAL FEATURES:

Amyris Biotechnologies, Inc. specializes in developing renewable products for both the healthcare and energy sectors. Founded in 2003, Amyris began with the desire to find and provide a reliable, affordable source of artemisinin, an extract from the Chinese Sweet Wormwood plant used in the treatment of Malaria. Treatment of patients with Artemisinin-based Combination Therapies (ACTs) is currently expensive, due to the difficulty of extracting the compound. Amyris likewise develops renewable, cost-effective chemicals with a wide scope of applicability. The renewable farnesene molecule, Biofene, is the basis for a number of applications including; emollients, flavors and fragrances, surfactants, isoprene, industrial and automotive oils and lubricants. The firm also runs a wholly-owned subsidiary, Amyris Fuels, LLC, with the goal of developing and distributing renewable, No Compromise fuels in the U.S. No Compromise fuels are designed to be drop-in fuels, offering environmental benefits while meeting engine and distribution infrastructure requirements. The company generates revenue to support its development enterprise by selling third-party ethanol to wholesale customers in the southeastern U.S. In addition, Amyris Brasil S.A. is a majority-owned subsidiary investigating renewable applications for Brazilian sugarcane, such as a supplement or replacement for petroleum-based diesel fuel. The firm is also exploring a sustainable alternative to current, petroleum-based jet fuel. In April 2010, the company established a joint venture, SMA Industria Quimica S.A., with Grupo Sao Martinho to build a production facility for Amyris renewable products. In June 2010, Amyris agreed to supply its No Compromise diesel fuel to Shell Oil. Also in June, the company partnered with cosmetics company Soliance to provide its technology to produce squalane for use in cosmetics.

Employee benefits include; comprehensive medical, dental and vision insurance; flexible spending account plans; life and AD&D insurance; 401(k) plans; and educational reimbursement. Additional perks include an on-site gym and yoga classes.

FINANCIALS: Sales and profits are in thousands of dollars—add 000 to get the full amount. 2009 Note: Financial information for 2009 was not available for all companies at press time.

2009 Sales: $	2009 Profits: $	U.S. Stock Ticker: Private
2008 Sales: $	2008 Profits: $	Int'l Ticker: Int'l Exchange:
2007 Sales: $	2007 Profits: $	Employees:
2006 Sales: $	2006 Profits: $	Fiscal Year Ends:
2005 Sales: $	2005 Profits: $	Parent Company:

SALARIES/BENEFITS:

Pension Plan:	ESOP Stock Plan:	Profit Sharing:	Top Exec. Salary: $	Bonus: $
Savings Plan: Y	Stock Purch. Plan:		Second Exec. Salary: $	Bonus: $

OTHER THOUGHTS:

Apparent Women Officers or Directors: 3
Hot Spot for Advancement for Women/Minorities: Y

LOCATIONS: ("Y" = Yes)

West:	Southwest:	Midwest:	Southeast:	Northeast:	International:
Y					Y

ANGIOTECH PHARMACEUTICALS INC

www.angiotech.com

Industry Group Code: 325412 Ranks within this company's industry group: Sales: 59 Profits: 121

Drugs:		Other:	Clinical:		Computers:	Services:
Discovery:	Y	AgriBio:	Trials/Services:	Y	Hardware:	Specialty Services:
Licensing:	Y	Genetic Data:	Labs:		Software:	Consulting:
Manufacturing:	Y	Tissue Replacement:	Equipment/Supplies:	Y	Arrays:	Blood Collection:
Generics:			Research/Development Svcs.:		Database Management:	Drug Delivery:
			Diagnostics:			Drug Distribution:

TYPES OF BUSINESS:

Medical Device Coatings
Drugs, Inflammatory Disease
Surgical Equipment & Technology
Dialysis Catheters

BRANDS/DIVISIONS/AFFILIATES:

TAXUS
Vascular Wrap
CoSeal
Vitagel Surgical Hemostat
BioSeal
MultiStem
Quill SRS
Haemacure Corp

CONTACTS: Note: Officers with more than one job title may be intentionally listed here more than once.

William L. Hunter, CEO
William L. Hunter, Pres.
K. Thomas Bailey, CFO
Sean Cunliffe, Sr. VP-Sales & Mktg.
Tammy Neske, Sr. VP-Human Resources
Jeffery M. Gross, Sr. VP-R&D
Victor Diaz, Sr. VP-Global Mfg. & Supply Chain Mgmt.
David D. McMasters, General Counsel/Sr. VP-Legal
Jonathan W. Chen, Sr. VP-Bus. Dev. & Financial Strategy
Rick Smith, Corp. Comm.
Rick Smith, Investor Rel.
Jay Dent, Sr. VP-Finance
Rui Avelar, Chief Medical Officer
Steven Bryant, Sr. VP-Sales & Mktg., Specialty Medical Devices
David T. Howard, Chmn.

Phone: 604-221-7676 Fax: 604-221-2330
Toll-Free:
Address: 1618 Station St., Vancouver, BC V6A 1B6 Canada

GROWTH PLANS/SPECIAL FEATURES:

Angiotech Pharmaceuticals, Inc. is a Canadian pharmaceutical company that develops medical products for complications associated with medical device implants, wound closure, surgical interventions and acute injury. The company consists of two segments: pharmaceutical technologies and medical products. The pharmaceutical technologies segment develops, licenses and sells technologies that improve the performances of medical devices and the outcomes of surgical procedures. The medical products segment establishes development and marketing partnerships with medical device, pharmaceutical and biomaterials companies. The firm's products address restenosis treatment, surgical adhesions, surgical sealants, surgical hemostats, systemic programs and clinical programs. Angiotech's flagship products are the Quill SRS (Self-Retaining System) suture, which does not require knots, and TAXUS, a polymeric formulation that delivers paclitaxel as a stent coating for restenosis. CoSeal is the firm's surgical sealant, which aids in the healing of suture lines and synthetic grafts during surgery. Hemostatic devices, like Vitagel Surgical Hemostat, control bleeding during surgery and help prevent complications including blood loss and ineffective closure. The firm also has various products in clinical trials, including the HemoStream central venous catheters to administer fluid and nourishment to critically ill patients; Vascular Wrap, a paclitaxel-eluting mesh surgical implant to treat complications with vascular graft implants; BioSeal, a biopsy track plug; the Option Vena Cava filter; and the MultiStem stem cell therapy program. In February 2009, Canada approved Angiotech's license to market the Quill SRS. In May 2009, the company announced U.S. FDA approval of its TAXUS Liberte Atom and Element stents. In April 2010, Angiotech acquired Haemacure Corp. Included in the transaction were all of Haemacure's research and development activities, manufacturing operations, key personnel and intellectual property rights to pursue development of Haemacure's human biomaterial products.

FINANCIALS: Sales and profits are in thousands of dollars—add 000 to get the full amount. 2009 Note: Financial information for 2009 was not available for all companies at press time.

2009 Sales: $279,678	2009 Profits: $-22,868	U.S. Stock Ticker: ANPI
2008 Sales: $283,272	2008 Profits: $-741,176	Int'l Ticker: ANP Int'l Exchange: Toronto-TSX
2007 Sales: $287,694	2007 Profits: $-65,940	Employees: 1,350
2006 Sales: $315,075	2006 Profits: $10,714	Fiscal Year Ends: 12/31
2005 Sales: $199,600	2005 Profits: $-1,200	Parent Company:

SALARIES/BENEFITS:

Pension Plan:	ESOP Stock Plan:	Profit Sharing:	Top Exec. Salary: $691,769	Bonus: $363,200
Savings Plan:	Stock Purch. Plan:		Second Exec. Salary: $487,190	Bonus: $156,875

OTHER THOUGHTS:

Apparent Women Officers or Directors: 1
Hot Spot for Advancement for Women/Minorities:

LOCATIONS: ("Y" = Yes)

West:	Southwest:	Midwest:	Southeast:	Northeast:	International:
Y		Y	Y	Y	Y

ANIKA THERAPEUTICS INC   www.anikatherapeutics.com

Industry Group Code: 325414 Ranks within this company's industry group: Sales: 5   Profits: 4

Drugs:		Other:		Clinical:	Computers:		Services:	
Discovery:	Y	AgriBio:		Trials/Services:	Hardware:		Specialty Services:	
Licensing:	Y	Genetic Data:		Labs:	Software:		Consulting:	
Manufacturing:	Y	Tissue Replacement:		Equipment/Supplies:	Arrays:		Blood Collection:	
Generics:				Research/Development Svcs.:	Database Management:		Drug Delivery:	
				Diagnostics:			Drug Distribution:	

TYPES OF BUSINESS:

Tissue Protection, Healing & Repair Drugs
Hyaluronic Acid Based Drugs
Tissue Augmentation Products
Osteoarthritis Pain Relief
Post-Surgical Anti-Adhesion

BRANDS/DIVISIONS/AFFILIATES:

Orthovisc
Elevess
Amvisc
Staarvisc-II
ShellGel
Incert
Hyvisc
Fidia Farmaceutici SpA

CONTACTS: Note: Officers with more than one job title may be intentionally listed here more than once.

Charles H. Sherwood, CEO
Frank J. Luppino, COO
Charles H. Sherwood, Pres.
Kevin W. Quinlan, CFO
Thomas J. Chambers, VP-Mktg. & Sales
William J. Mrachek, VP-Human Resources
Irina B. Kulinets, VP-Regulatory & Clinical Affairs
Andrew J. Carter, CTO
Randall W. Wilhoite, VP-Oper.
Gregory T. Fulton, Chief Commercial Officer

Phone: 781-457-9000	Fax: 781-305-9720
Toll-Free:	
Address: 32 Wiggins Ave., Bedford, MA 01730 US	

GROWTH PLANS/SPECIAL FEATURES:

Anika Therapeutics, Inc. develops, manufactures and markets therapeutic products for tissue protection, healing and repair. These products are based on hyaluronic acid (HA), a naturally occurring, biocompatible polymer found throughout the body that is essential to several physiological functions such as the protection and lubrication of soft tissues and joints; the maintenance of the structural integrity of tissues; and the transport of molecules to and within cells. The firm's currently marketed products consist of Orthovisc, an HA product which provides lubrication for the knee and helps cushion the knee joint; Monovisc, a single-injection drug designed to alleviate arthritis pain; Staarvisc-II, Amvisc, Amvisc Plus and ShellGel, each an injectable ophthalmic viscoelastic HA product; Hyvisc, an HA product used for treatment of joint dysfunction in horses; Incert-S, an HA-based anti-adhesive for surgical applications; and Elevess, an HA-based injectable dermal filler used for cosmetic tissue applications and products associated with joint health. Another osteoarthritis solution, Cingal, is currently in pre-clinical development. Anika is working to produce more Elevess products in the future, as there is a growing demand for soft tissue fillers for facial wrinkles, scar remediation and lip augmentation. Orthovisc is marketed in the U.S. by DePuy Mitek, a subsidiary of Johnson & Johnson, and outside the U.S. by distributors in roughly 20 countries. Hyvisc is marketed in the U.S. through Boehringer Ingelheim Vetmedica, Inc. Through the December 2009 acquisition of Fidia Farmaceutici S.p.A. (FAB), Anika also offers over 20 FAB products, including Hyalofast, a biodegradable support for human bone marrow mesenchymal stem cells; Hyalograft 3D, for the regeneration of skin; Hyalograft C Autograft for cartilage regeneration; Hyalomatrix, for treatment of burns and ulcers; Hyalonect, a woven gauze used as a graft wrap; and Hyaloss, which are fibers used to mix blood/bone grafts to form a paste for bone regeneration.

FINANCIALS: Sales and profits are in thousands of dollars—add 000 to get the full amount. 2009 Note: Financial information for 2009 was not available for all companies at press time.

2009 Sales: $40,136	2009 Profits: $3,688	U.S. Stock Ticker: ANIK	
2008 Sales: $35,780	2008 Profits: $3,629	Int'l Ticker: Int'l Exchange:	
2007 Sales: $30,830	2007 Profits: $6,035	Employees: 133	
2006 Sales: $26,841	2006 Profits: $4,604	Fiscal Year Ends: 12/31	
2005 Sales: $29,835	2005 Profits: $5,893	Parent Company:	

SALARIES/BENEFITS:

Pension Plan:	ESOP Stock Plan:	Profit Sharing:	Top Exec. Salary: $428,480	Bonus: $192,816
Savings Plan: Y	Stock Purch. Plan:		Second Exec. Salary: $261,888	Bonus: $62,583

OTHER THOUGHTS:

Apparent Women Officers or Directors: 1
Hot Spot for Advancement for Women/Minorities:

LOCATIONS: ("Y" = Yes)

West:	Southwest:	Midwest:	Southeast:	Northeast:	International:
				Y	

Plunkett Research, Ltd. 169

ANTIGENICS INC

www.antigenics.com

Industry Group Code: 325412 Ranks within this company's industry group: Sales: 133   Profits: 134

Drugs:		Other:	Clinical:	Computers:	Services:
Discovery:	Y	AgriBio:	Trials/Services:	Hardware:	Specialty Services:
Licensing:		Genetic Data:	Labs:	Software:	Consulting:
Manufacturing:	Y	Tissue Replacement:	Equipment/Supplies:	Arrays:	Blood Collection:
Generics:			Research/Development Svcs.:	Database Management:	Drug Delivery:
			Diagnostics:		Drug Distribution:

TYPES OF BUSINESS:
Cancers & Infectious Diseases Products & Technologies

BRANDS/DIVISIONS/AFFILIATES:
Oncophage
Aroplatin
AG-707
QS-21 Stimulon

CONTACTS: Note: Officers with more than one job title may be intentionally listed here more than once.
Garo H. Armen, CEO
Shalini Sharp, CFO/VP
John Cerio, VP-Human Resources
Marcel Rozencweig, Interim Chief Medical Officer
Stephen Monks, VP-Mfg. & Process Dev.
Karen Higgins Valentine, General Counsel/VP/Sec.
Daniel Levey, VP-Bus. Dev. & Scientific Affairs
Brad Miles, Contact-Media
Christine M. Klaskin, VP-Finance
Kerry A. Wentworth, VP-Clinical, Regulatory & Quality
Lori Baranauskas, Interim Head-Quality & Regulatory Affairs
Cristina Musselli, Sr. Dir.-Preclinical Dev. & Transitional Medicine
Leah Isakov, Sr. Dir.-Biostatistics & Data Mgmt.
Garo H. Armen, Chmn.

Phone: 212-994-8200	Fax: 212-994-8299
Toll-Free:	
Address: 3 Forbes Rd., Lexington, MA 02421-7305 US	

GROWTH PLANS/SPECIAL FEATURES:
Antigenics, Inc. develops technologies and products to treat cancers and infectious diseases, primarily based on immunological approaches. The company's most advanced product candidate is Oncophage (vitespen), a personalized therapeutic cancer vaccine candidate based on a heat shock protein called gp96. It has been tested in several cancer indications, including in Phase III clinical trials for the treatment of renal cell carcinoma (the most common type of kidney cancer) and for metastatic melanoma. Oncophage has also been tested in Phase I and Phase II clinical trials in a range of indications, and is in Phase II clinical trials for glioma (a type of brain cancer). The firm's product candidate portfolio also includes AG-707, a therapeutic vaccine program in Phase I clinical trial for the treatment of genital herpes; QS-21 Stimulon, an adjuvant used in numerous vaccines under development for diseases including hepatitis, human immunodeficiency virus (HIV), influenza, cancer, Alzheimer's disease, malaria, and tuberculosis; and Aroplatin, a liposomal chemotherapeutic in Phase I clinical trial for the treatment of solid tumors and B-cell lymphomas. Antigenics is based in New York, New York and has research, development and manufacturing facilities in Lexington, Massachusetts. The firm has approximately 183 issued patents and 64 patents pending worldwide.

FINANCIALS: Sales and profits are in thousands of dollars—add 000 to get the full amount. 2009 Note: Financial information for 2009 was not available for all companies at press time.
2009 Sales: $3,334	2009 Profits: $-30,318	U.S. Stock Ticker: AGEN
2008 Sales: $2,651	2008 Profits: $-28,488	Int'l Ticker: Int'l Exchange:
2007 Sales: $5,552	2007 Profits: $-36,795	Employees: 54
2006 Sales: $ 692	2006 Profits: $-51,881	Fiscal Year Ends: 12/31
2005 Sales: $ 630	2005 Profits: $-74,104	Parent Company:

SALARIES/BENEFITS:
Pension Plan:	ESOP Stock Plan:	Profit Sharing:	Top Exec. Salary: $440,000	Bonus: $
Savings Plan: Y	Stock Purch. Plan: Y		Second Exec. Salary: $240,000	Bonus: $

OTHER THOUGHTS:
Apparent Women Officers or Directors: 7
Hot Spot for Advancement for Women/Minorities: Y

LOCATIONS: ("Y" = Yes)
West:	Southwest:	Midwest:	Southeast:	Northeast:	International:
				Y	

Note: Financial information, benefits and other data can change quickly and may vary from those stated here.

APP PHARMACEUTICALS INC

www.apppharma.com

Industry Group Code: 325412 Ranks within this company's industry group: Sales: Profits:

Drugs:		Other:	Clinical:	Computers:	Services:
Discovery:	Y	AgriBio:	Trials/Services:	Hardware:	Specialty Services:
Licensing:	Y	Genetic Data:	Labs:	Software:	Consulting:
Manufacturing:	Y	Tissue Replacement:	Equipment/Supplies:	Arrays:	Blood Collection:
Generics:			Research/Development Svcs.:	Database Management:	Drug Delivery:
			Diagnostics:		Drug Distribution:

TYPES OF BUSINESS:

Pharmaceuticals Manufacturing
Injectable Drugs
Anti-Infective Drugs
Critical Care Drugs

BRANDS/DIVISIONS/AFFILIATES:

Fresenius Kabi Pharmaceuticals Holding Inc
APP Pharmaceuticals LLC

CONTACTS: Note: Officers with more than one job title may be intentionally listed here more than once.

John Ducker, CEO
J. Frank Harmon, COO/Exec. VP
John Ducker, Pres.
Richard J. Tajak, CFO/Exec. VP
James W. Callanan, VP-Human Resources
Christopher Bryant, Chief Scientific Officer/Exec. VP
Richard E. Maroun, Chief Admin. Officer
Richard E. Maroun, General Counsel/Corp. Sec.
Scott W. Meacham, Exec. VP-Commercial Oper.
Debra L. Ross, Dir.-Comm.

Phone: 847-413-2075	Fax: 800-743-7082
Toll-Free: 888-386-1300	
Address: 1501 E. Woodfield Rd., Ste. 300 E., Schaumburg, IL 60173-5837 US	

GROWTH PLANS/SPECIAL FEATURES:

APP Pharmaceuticals, Inc., a subsidiary of Fresenius Kabi Pharmaceuticals Holding, Inc., is a fully integrated biopharmaceutical company that develops, manufactures and markets multi-source and branded injectable pharmaceutical products for use in both patient and ambulatory settings, focusing on the oncology, anti-infective, anesthetic/analgesic and critical care markets. The firm offers injectable products in each of the three basic forms: liquid, powder and lyophilized (freeze-dried). APP manufactures and markets more than 160 products, including about 130 injectable products, as well as jellies, creams and topical solutions, oral solutions, empty sterilized vials, sterile water for injections, universal syringe tip adaptors, transfer pins and transfer sets, tamper-evident seals, sterile mineral oil, IV transfer spikes, pumping chamber sets and vented needles. The firm's newer injectable products include chlorothiazide sodium, a diuretic and antihypertensive; recuronium bromide, a neuromuscular blocking agent; cefepime hydrochloride, an antibiotic; colistimethate sodium and polymyxin B sulfate, anti-infectives; irinotecan hydrochloride, an antineoplastic; and caffeine citrate, a central nervous system stimulant, also offered as an oral solution. APP manufactures a comprehensive range of dosage formulations, used in hospitals, long-term care facilities, alternate care sites and clinics. In January 2009, the firm launched a rocuronium bromide injection, the generic version of the medication Zemuron, which is used as a muscle relaxant. During 2009, APP received approval from the U.S. Food and Drug Administration (FDA) to market its chlorothiazide sodium, sumatriptan succinate, deferoxamine mesylate, penicillin G potassium and idarubicin hydrochloride injections.

Parent company Fresenius Kabi Pharmaceuticals offers its employees profit sharing, corporate pensions, health insurance, subsidized childcare and accident, injury and life insurance.

FINANCIALS: Sales and profits are in thousands of dollars—add 000 to get the full amount. 2009 Note: Financial information for 2009 was not available for all companies at press time.

2009 Sales: $	2009 Profits: $	U.S. Stock Ticker: Subsidiary
2008 Sales: $	2008 Profits: $	Int'l Ticker: Int'l Exchange:
2007 Sales: $647,374	2007 Profits: $34,358	Employees:
2006 Sales: $583,201	2006 Profits: $-46,897	Fiscal Year Ends: 12/31
2005 Sales: $385,082	2005 Profits: $17,657	Parent Company: FRESENIUS KABI PHARMACEUTICALS HOLDING INC

SALARIES/BENEFITS:

Pension Plan: Y	ESOP Stock Plan:	Profit Sharing: Y	Top Exec. Salary: $	Bonus: $
Savings Plan:	Stock Purch. Plan:		Second Exec. Salary: $	Bonus: $

OTHER THOUGHTS:

Apparent Women Officers or Directors: 1
Hot Spot for Advancement for Women/Minorities:

LOCATIONS: ("Y" = Yes)

West:	Southwest:	Midwest:	Southeast:	Northeast:	International:
		Y			Y

APPLIED MOLECULAR EVOLUTION INC

www.amevolution.com

Industry Group Code: 541712 Ranks within this company's industry group: Sales: Profits:

Drugs:		Other:		Clinical:	Computers:	Services:
Discovery:	Y	AgriBio:		Trials/Services:	Hardware:	Specialty Services:
Licensing:		Genetic Data:	Y	Labs:	Software:	Consulting:
Manufacturing:		Tissue Replacement:		Equipment/Supplies:	Arrays:	Blood Collection:
Generics:				Research/Development Svcs.:	Database Management:	Drug Delivery:
				Diagnostics:		Drug Distribution:

TYPES OF BUSINESS:

Research-Directed Molecular Evolution
Human Biotherapeutics
Small-Molecule Drugs

BRANDS/DIVISIONS/AFFILIATES:

AME System
DirectAME
ExpressAME
ScreenAME

CONTACTS: Note: Officers with more than one job title may be intentionally listed here more than once.

Thomas F. Bumol, Pres.
Cheryl C. Gabele, Dir.-Admin.
Cheryl C. Gabele, Dir.-Finance

Phone: 858-597-4990	Fax: 858-597-4950
Toll-Free:	
Address: 3520 Dunhill St., San Diego, CA 92121 US	

GROWTH PLANS/SPECIAL FEATURES:

Applied Molecular Evolution, Inc. (AME), a subsidiary of Eli Lilly, is a drug development company and a leader in the application of directed molecular evolution to the development of biotherapeutics. The company uses its proprietary AMEsystem technology to develop improved versions of currently marketed, U.S. FDA-approved biopharmaceuticals as well as novel human biotherapeutics. It does so by adjusting amino acids in individual proteins to create positive characteristics. There are three components to the AMEsystem technology. DirectAME is a gene synthesis process that enables the rapid production of a library of variant genes based on an initial gene. ExpressAME consists of gene expression systems that produce proteins from the genes generated using DirectAME. ScreenAME is a series of tests that facilitate the selection and identification of proteins with the desired commercial properties from the protein libraries produced using ExpressAME. While AME had collaborative projects in the past, all of them were completed before its acquisition by Eli Lilly, and its employees now devote themselves entirely to working on projects for Eli Lilly. The AME scientists have assisted in developing eight new molecules in Lilly's pipeline, and four additional molecules it developed are in clinical trials. In October 2009, Lilly opened a new biotech facility in San Diego, allowing AME scientists to focus on discovery, engineering and conducting clinical trials for potential biologic medicines, with an emphasis on oncology, diabetes and autoimmune diseases.

Employee benefits include: Health and dental insurance; a prescription drug plan; a 401(k) plan; a retirement plan; flexible spending accounts; financial planning services; educational assistance; domestic partner benefits; on-site employee health services; an employee assistance program; on-site food services; banking, postal and dry cleaning services; free gym membership; maternity and paternity leave; nursing mothers program; adoption assistance; flextime and flex week; on-site child centers; telecommuting; and education leave.

FINANCIALS: Sales and profits are in thousands of dollars—add 000 to get the full amount. 2009 Note: Financial information for 2009 was not available for all companies at press time.

2009 Sales: $	2009 Profits: $	U.S. Stock Ticker: Subsidiary
2008 Sales: $	2008 Profits: $	Int'l Ticker: Int'l Exchange:
2007 Sales: $10,200	2007 Profits: $	Employees:
2006 Sales: $	2006 Profits: $	Fiscal Year Ends: 12/31
2005 Sales: $	2005 Profits: $	Parent Company: ELI LILLY AND COMPANY

SALARIES/BENEFITS:

Pension Plan: Y	ESOP Stock Plan:	Profit Sharing:	Top Exec. Salary: $	Bonus: $
Savings Plan: Y	Stock Purch. Plan:		Second Exec. Salary: $	Bonus: $

OTHER THOUGHTS:

Apparent Women Officers or Directors: 1
Hot Spot for Advancement for Women/Minorities:

LOCATIONS: ("Y" = Yes)

West:	Southwest:	Midwest:	Southeast:	Northeast:	International:
Y					

ARADIGM CORPORATION

www.aradigm.com

Industry Group Code: 325412A Ranks within this company's industry group: Sales: 14   Profits: 12

Drugs:		Other:		Clinical:		Computers:		Services:	
Discovery:		AgriBio:		Trials/Services:		Hardware:		Specialty Services:	
Licensing:		Genetic Data:		Labs:		Software:		Consulting:	
Manufacturing:	Y	Tissue Replacement:		Equipment/Supplies:	Y	Arrays:		Blood Collection:	
Generics:				Research/Development Svcs.:		Database Management:		Drug Delivery:	Y
				Diagnostics:				Drug Distribution:	

TYPES OF BUSINESS:

Drug Delivery Systems
Pulmonary Drug Delivery Systems

BRANDS/DIVISIONS/AFFILIATES:

AERx
AERx iDMS
AERx Essence

CONTACTS: Note: Officers with more than one job title may be intentionally listed here more than once.

Igor Gonda, CEO
Igor Gonda, Pres.
Nancy Pecota, CFO
D. Jeffery Grimes, General Counsel/VP-Legal Affairs/Sec.
Nancy Pecota, Contact-Investor Rel.
Virgil D. Thompson, Chmn.

Phone: 510-265-9000	Fax: 510-265-0277
Toll-Free:	
Address: 3929 Point Eden Way, Hayward, CA 94545 US	

GROWTH PLANS/SPECIAL FEATURES:

Aradigm Corporation is a specialty pharmaceutical company focused on the development and commercialization of a selection of drugs, delivered by inhalation, for the treatment of severe respiratory disease by pulmonologists. Aradigm's operations are centered on its hand-held AERx pulmonary drug delivery system, which creates aerosols from liquid drug formulations. This system is marketed as a replacement for medical devices such as nebulizers, metered-dose inhalers and dry powder inhalers. It is particularly suitable for drugs where highly efficient and precise delivery to the respiratory tract is advantageous or essential. Aradigm is focusing on product development for the treatment of respiratory disease, including identifying opportunities that it can develop and commercialize in the U.S. without a partner. In selecting its development programs, the company primarily seeks drugs approved by the U.S. FDA that can be reformulated for existing and new indications in respiratory disease. Aradigm is developing liposomal ciprofloxacin for the treatment of cystic fibrosis (Phase 2), bronchiectasis (Phase 2); inhalation of anthrax (pre-clinical) and smoking cessation (Phase 1). In collaboration with other companies, the firm is additionally developing treatments for pulmonary arterial hypertension, asthma and other chronic obstructive diseases of airways. The company has also produced AERx handheld devices for use in diabetes and pain management, including the AERx iDMS and AERx Essence, a small palm-sized purely mechanical device that offers the same pulmonary delivery of biologics as the electromechanical models.

Aradigm offers its employees medical, dental, life, and disability insurance; a 529 college savings plan; tuition reimbursement and an educational rewards program.

FINANCIALS: Sales and profits are in thousands of dollars—add 000 to get the full amount. 2009 Note: Financial information for 2009 was not available for all companies at press time.

2009 Sales: $4,883	2009 Profits: $-13,427	**U.S. Stock Ticker: ARDM**
2008 Sales: $ 251	2008 Profits: $-22,608	**Int'l Ticker:** Int'l Exchange:
2007 Sales: $ 961	2007 Profits: $-24,201	Employees: 15
2006 Sales: $4,814	2006 Profits: $-13,027	Fiscal Year Ends: 12/31
2005 Sales: $10,507	2005 Profits: $-29,215	Parent Company:

SALARIES/BENEFITS:

Pension Plan:	ESOP Stock Plan:	Profit Sharing:	Top Exec. Salary: $380,000	Bonus: $
Savings Plan: Y	Stock Purch. Plan: Y		Second Exec. Salary: $238,000	Bonus: $

OTHER THOUGHTS:

Apparent Women Officers or Directors: 1
Hot Spot for Advancement for Women/Minorities:

LOCATIONS: ("Y" = Yes)

West:	Southwest:	Midwest:	Southeast:	Northeast:	International:
Y					

ARCADIA BIOSCIENCES INC

www.arcadiabio.com

Industry Group Code: 11511 Ranks within this company's industry group: Sales: Profits:

Drugs:	Other:		Clinical:	Computers:	Services:
Discovery:	AgriBio:	Y	Trials/Services:	Hardware:	Specialty Services:
Licensing:	Genetic Data:		Labs:	Software:	Consulting:
Manufacturing:	Tissue Replacement:		Equipment/Supplies:	Arrays:	Blood Collection:
Generics:			Research/Development Svcs.:	Database Management:	Drug Delivery:
			Diagnostics:		Drug Distribution:

TYPES OF BUSINESS:

Agricultural-Based Technologies
Environment Health Technologies
Human Health Technologies

BRANDS/DIVISIONS/AFFILIATES:

TILLING

CONTACTS: Note: Officers with more than one job title may be intentionally listed here more than once.

Eric Rey, CEO
Eric Rey, Pres.
Vic Knauf, Chief Scientific Officer
Steve Brandwein, VP-Admin.
Wendy Neal, Chief Legal Officer/VP
Roger Salameh, VP-Bus. Dev., Agriculture
Steve Brandwein, VP-Finance
Frank Flider, VP-Bus. Dev., Nutrition
Don Emlay, Dir.-Regulatory Affairs & Compliance

Phone: 530-756-7077	Fax: 530-756-7027
Toll-Free:	
Address: 202 Cousteau Place, Ste. 200, Davis, CA 95618 US	

GROWTH PLANS/SPECIAL FEATURES:

Arcadia Biosciences, Inc. specializes in developing and adapting agricultural technologies that are designed to benefit the environment as well as human health. The company uses advanced breeding techniques, genetic screening and genetic engineering to develop its product portfolio. It has engineered plants that are able to produce crops in places and conditions that are not suitable for traditional crops. The crop portfolio includes Nitrogen Use Efficient (NUE) crops, Water Efficient Crops and salt tolerant plants. Its NUE project seeks to minimize the amount of nitrogen fertilizer required to produce crops. NUE crops utilize 50% to 60% less nitrogen than conventional fertilizers to produce an equivalent yield. Its salt tolerance project aims to develop plants able to produce normal quality and yields in high saline conditions, with a variety of crop applications including corn, rice, alfalfa, soybeans, turf, wheat and vegetables. These plants not only produce normal results in high saline areas, they are also engineered to bind excess salt into the plant, thus reducing an area's saline levels over time. The company's human health project has two current areas of focus: GLA safflower oil and extended shelf-life produce. The GLA safflower oil project aims to breed new varieties of safflower whose seeds will have as much as 40% GLA (gamma linolenic acid), an omega-6 fatty acid believed to have therapeutic benefits, which could be utilized to manufacture supplements, functional foods and nutraceuticals. Its extended shelf-life produce project also uses the firm's TILLING breeding technology to seek new genetic varieties of tomatoes, lettuce, melons and strawberries. The firm is also currently conducting research, in partnership with Washington State University, focused on the production of wheat for individuals with Celiac disease (an autoimmune disorder triggered by glutens found in wheat), with grant funding from the National Institutes of Health.

FINANCIALS: Sales and profits are in thousands of dollars—add 000 to get the full amount. 2009 Note: Financial information for 2009 was not available for all companies at press time.

2009 Sales: $	2009 Profits: $	U.S. Stock Ticker: Private
2008 Sales: $	2008 Profits: $	Int'l Ticker: Int'l Exchange:
2007 Sales: $4,000	2007 Profits: $	Employees:
2006 Sales: $	2006 Profits: $	Fiscal Year Ends:
2005 Sales: $	2005 Profits: $	Parent Company:

SALARIES/BENEFITS:

Pension Plan:	ESOP Stock Plan:	Profit Sharing:	Top Exec. Salary: $	Bonus: $
Savings Plan: Y	Stock Purch. Plan:		Second Exec. Salary: $	Bonus: $

OTHER THOUGHTS:

Apparent Women Officers or Directors: 1
Hot Spot for Advancement for Women/Minorities:

LOCATIONS: ("Y" = Yes)

West:	Southwest:	Midwest:	Southeast:	Northeast:	International:
Y	Y				

Note: Financial information, benefits and other data can change quickly and may vary from those stated here.

ARENA PHARMACEUTICALS INC

www.arenapharm.com

Industry Group Code: 325412 Ranks within this company's industry group: Sales: 123 Profits: 163

Drugs:		Other:	Clinical:		Computers:		Services:	
Discovery:	Y	AgriBio:	Trials/Services:		Hardware:		Specialty Services:	
Licensing:	Y	Genetic Data:	Labs:		Software:		Consulting:	
Manufacturing:		Tissue Replacement:	Equipment/Supplies:		Arrays:		Blood Collection:	
Generics:			Research/Development Svcs.:	Y	Database Management:		Drug Delivery:	
			Diagnostics:				Drug Distribution:	

TYPES OF BUSINESS:

Oral Drugs Discovery, Development & Commercialization
Obesity Medication
Diabetes Medication

BRANDS/DIVISIONS/AFFILIATES:

Lorcaserin Hydrochloride

CONTACTS: Note: Officers with more than one job title may be intentionally listed here more than once.

Jack Lief, CEO
Jack Lief, Pres.
Robert E. Hoffman, CFO
Dominic P. Behan, Chief Scientific Officer/Sr. VP
Steven W. Spector, General Counsel/Sr. VP/Sec.
Robert E. Hoffman, VP-Finance
K.A. Ajit-Simh, VP-Quality Systems
William R. Shanahan, Chief Medical Officer/VP
Jack Lief, Chmn.

Phone: 858-453-7200	Fax: 858-453-7210
Toll-Free:	
Address: 6166 Nancy Ridge Dr., San Diego, CA 92121 US	

GROWTH PLANS/SPECIAL FEATURES:

Arena Pharmaceuticals, Inc. is a clinical-stage biopharmaceutical company that focuses on the discovery, development and commercialization of oral drugs in four major therapeutic areas: cardiovascular, central nervous system, inflammatory and metabolic diseases. The company's most advanced drug candidate, lorcaserin hydrochloride, has completed Phase III clinical trial programs for the treatment of obesity. Arena initiated three Phase III trials evaluating the efficacy and safety of lorcaserin. The first trial is known as BLOOM (Behavioral modification and Lorcaserin for Overweight and Obesity Management), the second as BLOSSOM (Behavioral modification and Lorcaserin Second Study for Obesity Management) and the third as BLOOM-DM (Behavioral modification and Lorcaserin for Overweight and Obesity Management in Diabetes Mellitus). In addition, the company's internal development programs include APD916 for narcolepsy and cataplexy; APD811for pulmonary arterial hypertension and APD791 for arterial thrombosis. Arena completed a Phase I clinical trial of APD791, an orally available drug candidate used for the treatment and prevention of diseases such as acute coronary syndrome. In addition to internal programs, the firm is developing drugs with two pharmaceutical companies: an oral medication for the treatment of Type II diabetes, APD597, with Ortho-McNeil-Janssen and a Niacin Receptor Agonist with Merck, focusing on treatments for atherosclerosis and other disorders. Arena owns issued patents covering compositions of matter for lorcaserin and related compounds and methods of treatment utilizing lorcaserin and related compounds in 62 countries, with another eight pending applications. In July 2010, Arena announced Eisai, Inc. would have exclusive licensing rights of lorcaserin in the U.S. following the FDA's approval of its New Drug Application.

Arena offers employees medical, dental and vision insurance; life, AD&D and disability insurance; flexible spending accounts; a 401(k) plan; an employee assistance program; stock options and a stock purchase plan.

FINANCIALS: Sales and profits are in thousands of dollars—add 000 to get the full amount. 2009 Note: Financial information for 2009 was not available for all companies at press time.

2009 Sales: $10,387	2009 Profits: $-138,387	U.S. Stock Ticker: ARNA
2008 Sales: $9,809	2008 Profits: $-237,573	Int'l Ticker: Int'l Exchange:
2007 Sales: $19,332	2007 Profits: $-143,166	Employees: 358
2006 Sales: $30,569	2006 Profits: $-86,248	Fiscal Year Ends: 12/31
2005 Sales: $23,233	2005 Profits: $-67,901	Parent Company:

SALARIES/BENEFITS:

Pension Plan:	ESOP Stock Plan:	Profit Sharing:	Top Exec. Salary: $688,788	Bonus: $343,922
Savings Plan: Y	Stock Purch. Plan: Y		Second Exec. Salary: $399,333	Bonus: $128,800

OTHER THOUGHTS:

Apparent Women Officers or Directors:
Hot Spot for Advancement for Women/Minorities:

LOCATIONS: ("Y" = Yes)

West:	Southwest:	Midwest:	Southeast:	Northeast:	International:
Y					

ARIAD PHARMACEUTICALS INC

www.ariad.com

Industry Group Code: 325412 Ranks within this company's industry group: Sales: 128   Profits: 153

Drugs:		Other:		Clinical:		Computers:		Services:	
Discovery:	Y	AgriBio:		Trials/Services:		Hardware:		Specialty Services:	
Licensing:	Y	Genetic Data:		Labs:		Software:		Consulting:	
Manufacturing:		Tissue Replacement:		Equipment/Supplies:		Arrays:		Blood Collection:	
Generics:				Research/Development Svcs.:	Y	Database Management:		Drug Delivery:	
				Diagnostics:				Drug Distribution:	

TYPES OF BUSINESS:
Signaling Inhibitors Drugs

BRANDS/DIVISIONS/AFFILIATES:
AP26113
AP24534
Ridaforolimus
ARGENT Technology

CONTACTS: Note: Officers with more than one job title may be intentionally listed here more than once.
Harvey J. Berger, CEO
Harvey J Berger, Pres.
Edward M. Fitzgerald, CFO/Exec. VP/Treas.
Timothy P. Clackson, Pres., R&D/Chief Scientific Officer
Kelly M. Schmitz, VP-IT
David C. Dalgarno, VP-Research Tech.
Andreas Woppmann, VP-Mfg. Oper.
Raymond T. Keane, General Counsel/Sr. VP/Sec.
Kelly M. Schmitz, VP-Oper.
Pierre F. Dodion, Sr. VP-Corp. Dev.
Maria E. Cantor, VP-Corp. Comm.
Maria E. Cantor, VP-Investor Rel.
Joseph Bratica, VP-Finance/Controller
Frank G. Haluska, Chief Medical Officer/VP-Clinical Research
John D. Luliucci, Sr. VP-Dev.
David L. Berstein, Sr. VP/Chief Intellectual Property Officer
Daniel M. Bollag, Sr. VP-Regulatory Affairs & Quality
Harvey J. Berger, Chmn.

Phone: 617-494-0400	Fax: 617-494-8144
Toll-Free:	
Address: 26 Landsdowne St., Cambridge, MA 02139-4234 US	

GROWTH PLANS/SPECIAL FEATURES:

Ariad Pharmaceuticals, Inc. is an oncology company that is engaged in the discovery and development of its ARGENT technology, which revolves around medicines used to treat cancers by regulating cell signaling with small molecules. The company's lead product candidate is ridaforolimus, an inhibitor of the protein mTOR, which is a master switch in cancers. Blocking mTOR creates a starvation-like effect in cancer cells by interfering with cell growth, division, metabolism and angiogenesis. It is currently being developed by Ariad in partnership with Merck & Co., Inc. The firm has tested ridaforolimus in Phase III trials for patients with metastatic soft-tissue and bone sarcomas and is planning to initiate multiple clinical trials of ridaforolimus, including Phase II clinical trials in endometrial, breast and prostate cancers. Ariad's second product candidate, AP24534 is an orally active kinase inhibitor, which is in development for the treatment of various forms of leukemia and blood-based cancers. AP24534 has shown to be a potent inhibition of Bcr-Abl, a kinase that causes chronic myeloid leukemia, and of Flt3, a kinase involved in acute myeloid leukemia and the progression of multiple solid tumors. The firms final product candidate is AP26113, a potential anaplastic lymphoma kinase inhibitor for use in patients with various cancers, including lymphoma, non-small cell lung cancer and neuroblastoma. The firm licenses both the intellectual property related to the ARGENT technology and the product candidates resulting from the application of this technology, including mTOR inhibitors.

Ariad offers employees medical, dental, life and short and long-term disability insurance; a 401(k) plan; stock options; an employee stock purchase plan; tuition reimbursement and a parking or public transportation pass.

FINANCIALS: Sales and profits are in thousands of dollars—add 000 to get the full amount. 2009 Note: Financial information for 2009 was not available for all companies at press time.

2009 Sales: $8,302	2009 Profits: $-80,008	U.S. Stock Ticker: ARIA
2008 Sales: $7,082	2008 Profits: $-71,052	Int'l Ticker: Int'l Exchange:
2007 Sales: $3,583	2007 Profits: $-58,522	Employees: 149
2006 Sales: $ 896	2006 Profits: $-61,928	Fiscal Year Ends: 12/31
2005 Sales: $1,217	2005 Profits: $-55,482	Parent Company:

SALARIES/BENEFITS:

Pension Plan:	ESOP Stock Plan:	Profit Sharing:	Top Exec. Salary: $607,500	Bonus: $
Savings Plan: Y	Stock Purch. Plan: Y		Second Exec. Salary: $350,000	Bonus: $168,000

OTHER THOUGHTS:
Apparent Women Officers or Directors: 2
Hot Spot for Advancement for Women/Minorities: Y

LOCATIONS: ("Y" = Yes)

West:	Southwest:	Midwest:	Southeast:	Northeast:	International:
				Y	

ARQULE INC

www.arqule.com

Industry Group Code: 325412 Ranks within this company's industry group: Sales: 104   Profits: 138

Drugs:		Other:		Clinical:		Computers:		Services:	
Discovery:	Y	AgriBio:		Trials/Services:		Hardware:	Y	Specialty Services:	Y
Licensing:		Genetic Data:	Y	Labs:		Software:	Y	Consulting:	
Manufacturing:		Tissue Replacement:		Equipment/Supplies:		Arrays:		Blood Collection:	
Generics:				Research/Development Svcs.:	Y	Database Management:	Y	Drug Delivery:	
				Diagnostics:				Drug Distribution:	

TYPES OF BUSINESS:

Research-Drug Discovery
Small-Molecule Compounds
Systems & Software
Predictive Modeling

BRANDS/DIVISIONS/AFFILIATES:

Activated Checkpoint Therapy (ACT)
ARQ 197
ARQ 501
ARQ 621
ARQ 761

CONTACTS: Note: Officers with more than one job title may be intentionally listed here more than once.

Paolo Pucci, CEO
Peter S. Lawrence, COO
Peter S. Lawrence, Pres.
Anthony S. Messina, VP-Human Dev.
Thomas C.K. Chan, Chief Scientific Officer
Brian Schwartz, Chief Medical Officer
Patrick J. Zenner, Chmn.

Phone: 781-994-0300	Fax: 781-376-6019
Toll-Free:	
Address: 19 Presidential Way, Woburn, MA 01801-5140 US	

GROWTH PLANS/SPECIAL FEATURES:

ArQule, Inc. is a clinical-stage biotechnology company engaged in the research and development of cancer drugs that act selectively against cancer cells, target multiple tumor types and are better tolerated by patients than conventional cancer treatments. The company's lead approach to cancer therapy development is based on its Activated Checkpoint Therapy (ACT) platform, which restores the cells' internal damage response mechanism, or checkpoint, which triggers programmed cell death, or apoptosis, in cells with damaged DNA. Because cancer cells invariably contain both damaged DNA and de-activated damage response checkpoints, ACT is designed to restore damaged checkpoint functions, allowing cells to trigger apoptosis, selectively destroying cancer cells instead of both cancerous and healthy cells as traditional chemotherapy does. ArQule's lead products consist of ARQ 197, an orally-administered inhibitor of the c-Met receptor tyrosine kinase, a molecule that plays a key role in cancer growth, survival and metastasis (spreading from one part of the body to another); ARQ 501, an intravenously administered novel activator of the cell's damage response mechanism mediated by the E2F-1 transcription factor; and ARQ 761, an intravenously administered second generation activator of E2F-1. Other products include ARQ 621, a new drug focused on inhibition of the Eg5 kinesin spindle protein and compounds designed to inhibit B-RAF kinase. Early-stage clinical trial results, which are available for ARQ 197 and ARQ 501, have demonstrated anti-cancer activity across multiple types of tumors. ArQule's current research efforts are concentrated on the design of kinase inhibitors that are powerful, selective and do not interfere with ATP (adenosine triphosphate, a crucial element in normal cell metabolism), a technology the firm calls AKIP (ArQule Kinase Inhibitor Platform).

ArQule offers its employees tuition reimbursement; a 401(k) and employee stock purchase plan; credit union membership; an employee assistance program; flexible spending accounts; and medical, dental, vision and disability insurance.

FINANCIALS: Sales and profits are in thousands of dollars—add 000 to get the full amount. 2009 Note: Financial information for 2009 was not available for all companies at press time.

2009 Sales: $25,198	2009 Profits: $-36,136	**U.S. Stock Ticker: ARQL**
2008 Sales: $14,141	2008 Profits: $-50,864	Int'l Ticker: Int'l Exchange:
2007 Sales: $9,165	2007 Profits: $-53,372	Employees: 111
2006 Sales: $6,626	2006 Profits: $-31,440	Fiscal Year Ends: 12/31
2005 Sales: $6,628	2005 Profits: $-7,520	Parent Company:

SALARIES/BENEFITS:

Pension Plan:	ESOP Stock Plan:	Profit Sharing:	Top Exec. Salary: $450,000	Bonus: $202,500
Savings Plan: Y	Stock Purch. Plan: Y		Second Exec. Salary: $383,654	Bonus: $138,375

OTHER THOUGHTS:

Apparent Women Officers or Directors: 1
Hot Spot for Advancement for Women/Minorities:

LOCATIONS: ("Y" = Yes)

West:	Southwest:	Midwest:	Southeast:	Northeast: Y	International:

ARRAY BIOPHARMA INC

www.arraybiopharma.com

Industry Group Code: 541712 **Ranks within this company's industry group:** Sales: 12 Profits: 18

Drugs:		Other:	Clinical:	Computers:		Services:
Discovery:	Y	AgriBio:	Trials/Services:	Hardware:		Specialty Services:
Licensing:	Y	Genetic Data:	Labs:	Software:		Consulting:
Manufacturing:		Tissue Replacement:	Equipment/Supplies:	Arrays:	Y	Blood Collection:
Generics:			Research/Development Svcs.:	Y	Database Management:	Y	Drug Delivery:
			Diagnostics:			Drug Distribution:

TYPES OF BUSINESS:

Drug Development & Research Services
Small-Molecule Drugs
Arrays

BRANDS/DIVISIONS/AFFILIATES:

Array Discovery Platform
MEK
ARRY-614
ARRY-162

CONTACTS: *Note: Officers with more than one job title may be intentionally listed here more than once.*

Robert E. Conway, CEO
David L. Snitman, COO
Kevin Koch, Pres.
R. Michael Carruthers, CFO
Kevin Koch, Chief Scientific Officer
John R. Moore, General Counsel/VP
David L. Snitman, VP-Bus. Dev.
John A. Josey, VP-Discovery Chemistry
James D. Winkler, VP-Discovery & Translational Biology
Gary M. Clark, VP-Biostatistics & Data Mgmt.
Bengt Bergstrom, VP-Clinical Dev.
Kyle A. Lefkoff, Chmn.

Phone: 303-381-6600	Fax: 303-449-5376
Toll-Free: 877-633-2436	
Address: 3200 Walnut St., Boulder, CO 80301 US	

GROWTH PLANS/SPECIAL FEATURES:

Array BioPharma, Inc. is a biopharmaceutical company focused on discovering, developing and commercializing targeted small molecule drugs for the treatment of cancer and inflammatory diseases. The firm's scientists use its Array Discovery Platform, an integrated set of drug discovery technologies including predictive informatics, high throughput screening and information databases of existent drugs, to invent novel small-molecule drugs in collaboration with leading pharmaceutical and biotechnology companies, as well as for its own pipeline of proprietary drugs. The company holds collaborative partnerships with AstraZeneca; Genentech; InterMune; Cellgene Corp.; Amgen Inc.; Eli Lilly and Company; and Novartis. Array's prime research focuses are cancer and inflammatory diseases. The company has four cancer drugs and one inflammation drugs in its advanced pipeline. The clinical candidate ARRY-162, or MEK for Cancer, is an orally active drug that is currently in Phase I clinical trials for the treatment of cancer. Other cancer products in development consist of inhibitors targeting breast cancer, acute myeloid leukemia, multiple myeloma and solid tumors. The inflammatory and metabolic drugs aim to regulate biological targets that control inflammation. Its primary inflammatory drug candidate is ARRY-614, an inhibitor for myelodysplastic syndrome (MDS). In collaboration with its partner companies, Array is also developing a glucokinase activation drug for the treatment of Type 2 diabetes; and a hepatitis C virus (HCV) protease inhibitor. In January 2009, the company announced several cost-cutting measures, including layoffs of 10% of its workforce. In April 2010, Array entered a strategic agreement with Novartis for the development of certain MEK inhibitors.

Array offers its employees medical, dental, vision and prescription coverage; life insurance and disability coverage; an employee assistance program; flexible spending accounts; a 401(k) plan; a stock purchase plan; and educational assistance.

FINANCIALS: Sales and profits are in thousands of dollars—add 000 to get the full amount. 2009 Note: Financial information for 2009 was not available for all companies at press time.

2009 Sales: $24,982	2009 Profits: $-127,815	U.S. Stock Ticker: ARRY
2008 Sales: $28,808	2008 Profits: $-96,288	Int'l Ticker: Int'l Exchange:
2007 Sales: $36,970	2007 Profits: $-55,443	Employees: 355
2006 Sales: $45,003	2006 Profits: $39,614	Fiscal Year Ends: 6/30
2005 Sales: $45,505	2005 Profits: $-23,244	Parent Company:

SALARIES/BENEFITS:

Pension Plan:	ESOP Stock Plan:	Profit Sharing:	Top Exec. Salary: $513,750	Bonus: $
Savings Plan: Y	Stock Purch. Plan: Y		Second Exec. Salary: $405,000	Bonus: $

OTHER THOUGHTS:

Apparent Women Officers or Directors:
Hot Spot for Advancement for Women/Minorities:

LOCATIONS: ("Y" = Yes)

West:	Southwest:	Midwest:	Southeast:	Northeast:	International:
Y				Y	

Note: Financial information, benefits and other data can change quickly and may vary from those stated here.

ARRYX INC
www.arryx.com

Industry Group Code: 541711 Ranks within this company's industry group: Sales: Profits:

Drugs:	Other:	Clinical:	Computers:		Services:
Discovery:	AgriBio:	Trials/Services:	Hardware:	Y	Specialty Services:
Licensing:	Genetic Data:	Labs:	Software:		Consulting:
Manufacturing:	Tissue Replacement:	Equipment/Supplies:	Arrays:		Blood Collection:
Generics:		Research/Development Svcs.:	Database Management:		Drug Delivery:
		Diagnostics:			Drug Distribution:

TYPES OF BUSINESS:
Holographic Laser Steering
Optics
Nonmaterial Manipulation Technology

BRANDS/DIVISIONS/AFFILIATES:
BioRyx 200
University of Chicago
Laser Tweezers
HOTkit Complete Holo-Tweezers Solution

CONTACTS: Note: Officers with more than one job title may be intentionally listed here more than once.
Michael Reese, CFO
Daniel Mueth, Lead Scientist
Daniel Mueth, CTO
Jan Conneely, VP/Gen. Mgr.
Ken Bradley, VP
Nicole S. Williams, Chmn.

Phone: 312-726-6675	Fax: 312-726-6652
Toll-Free:	
Address: 316 N. Michigan Ave., Ste. 400, Chicago, IL 60601 US	

GROWTH PLANS/SPECIAL FEATURES:

Arryx, Inc., a subsidiary of Haemonetics Corporation, develops advanced systems and products for the optoelectronics and biotechnology industries. The company created the BioRyx 200 system, which employs holographic laser steering to improve productivity and profitability for manufacturing and processing in industries ranging from pharmaceuticals to integrated circuit manufacturing. The system integrates proprietary laser steering technology with easy-to-use software to give researchers the ability to manipulate multiple microscopic objects independently and simultaneously in three dimensions. The firm is the exclusive licensee of holographic optical trapping (HOT) technology, originally developed at the University of Chicago. HOT uses a holographic device, such as a spatial light modulator, to operate beams of light to capture and manipulate microscopic and nanoscopic objects such as carbon nanotubes in optical traps. The optical traps are also called Laser Tweezers. The system can grab, move, spin, assemble, stretch, join, separate and otherwise control materials ranging in size from 1/1000th the diameter of a human hair to the size of human cells. HOT uses nonstandard beam profiles such as Bessel beams, which are non-diffracting, and optical vortices, which are optical fields with phase singularities. Applications include isolating valuable cells or molecules from other cells, tissues and contaminants; detecting and measuring the presence of materials to increase test sophistication and sensitivity; and manufacturing sensors to detect biohazards, chemical hazards and other contaminants on a universal basis for the rapidly growing homeland security industry. Clients of the firm have included universities and government institutions, such as the National Institute of Standards and Technology. It also offers the HOTkit Holo-Tweezers Solution, a holographic optical trapping add-on allowing scientists to incorporate event manipulation into existing microscope or imaging technologies.

Arryx offers its employees medical, dental and vision plans; life and disability insurance; a 401(k) plan; flex time; and a stock purchase plan.

FINANCIALS: Sales and profits are in thousands of dollars—add 000 to get the full amount. 2009 Note: Financial information for 2009 was not available for all companies at press time.

2009 Sales: $	2009 Profits: $	U.S. Stock Ticker: Subsidiary
2008 Sales: $	2008 Profits: $	Int'l Ticker: Int'l Exchange:
2007 Sales: $	2007 Profits: $	Employees:
2006 Sales: $	2006 Profits: $	Fiscal Year Ends: 12/31
2005 Sales: $	2005 Profits: $	Parent Company: HAEMONETICS CORPORATION

SALARIES/BENEFITS:

Pension Plan:	ESOP Stock Plan:	Profit Sharing:	Top Exec. Salary: $	Bonus: $
Savings Plan: Y	Stock Purch. Plan: Y		Second Exec. Salary: $	Bonus: $

OTHER THOUGHTS:
Apparent Women Officers or Directors: 2
Hot Spot for Advancement for Women/Minorities:

LOCATIONS: ("Y" = Yes)

West:	Southwest:	Midwest:	Southeast:	Northeast:	International:
		Y			

ASTELLAS PHARMA INC

www.astellas.com

Industry Group Code: 325412 **Ranks within this company's industry group:** Sales: 15 Profits: 18

Drugs:		Other:		Clinical:		Computers:		Services:	
Discovery:	Y	AgriBio:		Trials/Services:		Hardware:		Specialty Services:	
Licensing:		Genetic Data:	Y	Labs:		Software:		Consulting:	
Manufacturing:	Y	Tissue Replacement:		Equipment/Supplies:		Arrays:	Y	Blood Collection:	
Generics:				Research/Development Svcs.:	Y	Database Management:		Drug Delivery:	
				Diagnostics:				Drug Distribution:	

TYPES OF BUSINESS:

Drugs, Manufacturing
Immunological Pharmaceuticals
Over-the-Counter Products
Reagents
Genomic Research
Venture Capital
Drug Licensing

BRANDS/DIVISIONS/AFFILIATES:

Yamanouchi Pharmaceutical Co., Ltd.
Fujisawa Pharmaceutical Co., Ltd.
Prograf
Lipitor
Harnal
Flomax
Micardis
Astellas Venture Management, LLC

CONTACTS: Note: Officers with more than one job title may be intentionally listed here more than once.

Masafumi Nogimori, CEO
Masafumi Nogimori, Pres.
Yasuo Ishii, Chief Sales & Mktg. Officer/Exec. VP
Hirofumi Onosaka, Sr. Corp. Officer
Hitoshi Ohta, Sr. Corp. Officer
Iwaki Miyazaki, Sr. Corp. Officer
Katsuro Yamada, Sr. Corp. Officer
Toichi Takenaka, Chmn.
Yoshihiko Hatanaka, CEO/Pres., Astellas Pharma U.S., Inc.

Phone: 81-3-3244-3000	Fax: 81-3-3244-3272
Toll-Free:	
Address: 2-3-11 Nihonbashi-Honcho, Chuo-ku, Tokyo, 103-8411 Japan	

GROWTH PLANS/SPECIAL FEATURES:

Astellas Pharma, Inc. is one of the largest pharmaceuticals manufacturers in Japan. It was formed from the recent merger of Yamanouchi Pharmaceutical Co., Ltd. and Fujisawa Pharmaceutical Co., Ltd. In addition to developing its own pharmaceuticals, Astellas pursues in-licensing and co-promotion agreements with biotechnology firms and a host of other pharmaceutical companies such as Pfizer, Inc. Nearly all of Astellas's sales relate to pharmaceuticals, led by Prograf, which is used as an immunosuppressant in conjunction with organ transplantation. Other products target needs in dermatology, urology, gastrointestinal disorders, immunology, infectious diseases, psychiatry and cardiology. Some of the firm's main products include Lipitor (developed by Pfizer) for high cholesterol; Micardis (co-promoted with Nippon Boehringer Ingelheim), a hypertension treatment; Myslee, an insomnia treatment co-promoted with sanofi-aventis S.A.; Seroquel, an antipsychotic; Gaster, for peptic ulcers and gastritis; fungal infection treatments AmBisome and Mycamine; overactive bladder treatment VESIcare; and Harnal for symptoms caused by enlarged prostates. Harnal is marketed by Boehringer Ingelheim Pharmaceuticals, Inc. in the U.S. under the name Flomax. Besides developing its own drugs or marketing drugs developed by others, the firm maintains Los Altos, California based subsidiary, Astellas Venture Management, LLC, which is engaged in investing in biotechnology companies, starting with $30 million in initial capitalization. Astellas has 33 subsidiaries and affiliates located throughout North America, Europe and Asia. In February 2010, the company formed a partnership with Basilea Pharmaceutica, Ltd. to co-develop new drugs. Also in February 2010, the company began a hostile takeover bid for OSI Pharmaceuticals, Inc. In April 2010, the Astellas entered into an exclusive distribution contract with Teijin Pharma to market the TMX-67 drug throughout China and Hong Kong.

FINANCIALS: Sales and profits are in thousands of dollars—add 000 to get the full amount. 2009 Note: Financial information for 2009 was not available for all companies at press time.

2009 Sales: $10,439,500	2009 Profits: $1,848,410	**U.S. Stock Ticker:** ALPMF.PK
2008 Sales: $10,513,900	2008 Profits: $1,918,140	**Int'l Ticker:** 4503 **Int'l Exchange:** Tokyo-TSE
2007 Sales: $9,114,200	2007 Profits: $1,299,700	**Employees:** 7,453
2006 Sales: $8,705,700	2006 Profits: $1,026,200	**Fiscal Year Ends:** 3/31
2005 Sales: $4,425,800	2005 Profits: $333,800	**Parent Company:**

SALARIES/BENEFITS:

Pension Plan: Y	ESOP Stock Plan:	Profit Sharing:	Top Exec. Salary: $	Bonus: $
Savings Plan:	Stock Purch. Plan:		Second Exec. Salary: $	Bonus: $

OTHER THOUGHTS:

Apparent Women Officers or Directors: 1
Hot Spot for Advancement for Women/Minorities:

LOCATIONS: ("Y" = Yes)

West:	Southwest:	Midwest:	Southeast:	Northeast:	International:
Y		Y		Y	Y

ASTRAZENECA CANADA

www.astrazeneca.ca

Industry Group Code: 325412 Ranks within this company's industry group: Sales: Profits:

Drugs:		Other:	Clinical:	Computers:	Services:
Discovery:	Y	AgriBio:	Trials/Services:	Hardware:	Specialty Services:
Licensing:		Genetic Data:	Labs:	Software:	Consulting:
Manufacturing:	Y	Tissue Replacement:	Equipment/Supplies:	Arrays:	Blood Collection:
Generics:			Research/Development Svcs.:	Database Management:	Drug Delivery:
			Diagnostics:		Drug Distribution:

TYPES OF BUSINESS:

Pharmaceutical Research & Development
Oncology Research
Cardiovascular Research
Neuroscience Research

BRANDS/DIVISIONS/AFFILIATES:

AstraZeneca PLC

CONTACTS: Note: Officers with more than one job title may be intentionally listed here more than once.

Mark Jones, CEO
Mark Jones, Pres.
Vince Rizzi, CFO
Toni Garro, Dir.-Human Resources
Vince Rizzi, VP-Info. Svcs.
Cyndy DeGuisti, Contact-Corp. Comm.
Vince Rizzi, VP-Finance

Phone: 905-277-7111	Fax: 905-270-3248
Toll-Free: 800-565-5877	
Address: 1004 Middlegate Rd., Mississauga, ON L4Y 1M4 Canada	

GROWTH PLANS/SPECIAL FEATURES:

AstraZeneca Canada, a subsidiary of pharmaceutical giant AstraZeneca PLC, is a pharmaceutical research and development company based in Ontario, Canada. AstraZeneca Canada focuses on creating therapeutic interventions in the key areas of cardiovascular, gastrointestinal, oncology, respiratory, neuroscience and infection. The company spends nearly $2 million every week on research initiatives. A significant portion of this funding is invested in the company's Montreal research facility, where the firm employs more than 125 scientists to conduct research in pain treatment. The majority of the company's employees work at its high-tech headquarters in Mississauga, Ontario, where the company engages in the packaging, clinical research, sales, marketing and distribution portion of its business. The firm also sponsors several research chairs and fellowships at many Canadian universities, including the Universities of Toronto, Alberta, Montreal, and Saskatchewan.

AstraZeneca Canada offers its employees pension plans and contributions to those plans; health care expense and personal spending accounts; a savings plan; and paid leave for childcare, elder care, and partner care.

FINANCIALS: Sales and profits are in thousands of dollars—add 000 to get the full amount. 2009 Note: Financial information for 2009 was not available for all companies at press time.

2009 Sales: $	2009 Profits: $	U.S. Stock Ticker: Subsidiary
2008 Sales: $	2008 Profits: $	Int'l Ticker: Int'l Exchange:
2007 Sales: $	2007 Profits: $	Employees:
2006 Sales: $	2006 Profits: $	Fiscal Year Ends:
2005 Sales: $	2005 Profits: $	Parent Company: ASTRAZENECA PLC

SALARIES/BENEFITS:

Pension Plan: Y	ESOP Stock Plan:	Profit Sharing:	Top Exec. Salary: $	Bonus: $
Savings Plan:	Stock Purch. Plan:		Second Exec. Salary: $	Bonus: $

OTHER THOUGHTS:

Apparent Women Officers or Directors:
Hot Spot for Advancement for Women/Minorities:

LOCATIONS: ("Y" = Yes)

West:	Southwest:	Midwest:	Southeast:	Northeast:	International: Y

ASTRAZENECA PLC

www.astrazeneca.com

Industry Group Code: 325412 **Ranks within this company's industry group: Sales: 7   Profits: 7**

Drugs:		Other:	Clinical:	Computers:	Services:
Discovery:	Y	AgriBio:	Trials/Services:	Hardware:	Specialty Services:
Licensing:		Genetic Data:	Labs:	Software:	Consulting:
Manufacturing:	Y	Tissue Replacement:	Equipment/Supplies:	Arrays:	Blood Collection:
Generics:			Research/Development Svcs.:	Database Management:	Drug Delivery:
			Diagnostics:		Drug Distribution:

TYPES OF BUSINESS:

Drugs-Diversified
Pharmaceutical Research & Development

BRANDS/DIVISIONS/AFFILIATES:

Nexium
Seroquel
Crestor
Arimidex
Symbicort
Pulmicort
AstraTech
Novexel

CONTACTS: *Note: Officers with more than one job title may be intentionally listed here more than once.*

David R. Brennan, CEO
Simon Lowth, CFO
Lynn Tetrault, Exec. VP-Human Resources & Corp. Affairs
Jan Lundberg, Exec. VP-Discovery Research
Jeff Pott, General Counsel
David Smith, Exec. VP-Oper.
Anders Ekblom, Exec. VP-Dev.
Tony Zook, Exec. VP-Commercial
Louis Schweitzer, Chmn.

Phone: 44-20-7304-5000	Fax: 44-20-7304-5151
Toll-Free:	
Address: 15 Stanhope Gate, London, W1Y 6LN UK	

GROWTH PLANS/SPECIAL FEATURES:

AstraZeneca plc is a leading global pharmaceutical company that discovers, develops, manufactures and markets prescription pharmaceuticals, biologics and vaccines for the treatment or prevention of diseases in such areas of healthcare as cardiovascular, gastrointestinal, infection, neuroscience, oncology and respiratory and inflammation. The company is the result of the merger of the Zeneca Group with Astra. The firm invests over $4 billion annually in research and development (R&D) and enjoys sales in over 100 countries. It operates 26 manufacturing sites in 18 countries and 17 major research centers in eight countries. AstraZeneca's cardiovascular products include Seloken ZOK, Crestor, Atacand, Plendil, Zestril and Tenormin. The firm's gastrointestinal products include Nexium, Entocort and Prilosec. Merrem, its primary infection product, is an antibiotic for serious hospital-acquired infections. AstraZeneca's neuroscience offering includes Zomig, a migraine treatment; anesthetics Diprivan and Xylocaine; Naropin, a long-acting anesthetic; and Seroquel for the treatment of schizophrenia and bipolar mania. AstraZeneca's cancer treatments include Casodex for prostate cancer; Zoladex; Armidex and Faslodex for breast cancer; Iressa for lung cancer; and Nolvadex. The firm's respiratory and inflammation brands include Pulmicort, Symbicort, Rhinocort, Accolate and Oxis. Nexium, Seroquel, Crestor, Arimidex, Symbicort, Pulmicort, Casodex, Atacand, Synagis, Prilosec and Zoladex all have sales in excess of $1 billion. Subsidiary AstraTech is engaged in the research, development, manufacture and marketing of medical devices and implants. Another subsidiary, Aptium Oncology, is a leading provider of outpatient oncology management and consulting services in the U.S., with full-service outpatient comprehensive cancer centers in California, Florida and New York. The company recently announced plans to implement certain restructuring initiatives in its R&D organization as part of its ongoing restructuring programs. In December 2009, the company announced plans to acquire infection research company Novexel.

FINANCIALS: Sales and profits are in thousands of dollars—add 000 to get the full amount. 2009 Note: Financial information for 2009 was not available for all companies at press time.

2009 Sales: $32,804,000	2009 Profits: $7,544,000	**U.S. Stock Ticker: AZN**
2008 Sales: $31,601,000	2008 Profits: $6,130,000	**Int'l Ticker: AZN** Int'l Exchange: London-LSE
2007 Sales: $29,559,000	2007 Profits: $5,627,000	Employees: 65,000
2006 Sales: $26,475,000	2006 Profits: $6,063,000	Fiscal Year Ends: 12/31
2005 Sales: $23,950,000	2005 Profits: $3,881,000	Parent Company:

SALARIES/BENEFITS:

Pension Plan:	ESOP Stock Plan:	Profit Sharing:	Top Exec. Salary: $	Bonus: $
Savings Plan:	Stock Purch. Plan:		Second Exec. Salary: $	Bonus: $

OTHER THOUGHTS:

Apparent Women Officers or Directors: 4
Hot Spot for Advancement for Women/Minorities: Y

LOCATIONS: ("Y" = Yes)

West:	Southwest:	Midwest:	Southeast:	Northeast:	International:
Y	Y	Y	Y	Y	Y

Note: Financial information, benefits and other data can change quickly and may vary from those stated here.

AVANIR PHARMACEUTICALS
www.avanir.com

Industry Group Code: 325412 Ranks within this company's industry group: Sales: 131 Profits: 119

Drugs:		Other:	Clinical:	Computers:	Services:
Discovery:	Y	AgriBio:	Trials/Services:	Hardware:	Specialty Services:
Licensing:	Y	Genetic Data:	Labs:	Software:	Consulting:
Manufacturing:		Tissue Replacement:	Equipment/Supplies:	Arrays:	Blood Collection:
Generics:			Research/Development Svcs.:	Database Management:	Drug Delivery:
			Diagnostics:		Drug Distribution:

TYPES OF BUSINESS:
Pharmaceutical Discovery & Development
Human Antibody Technology Research
Central Nervous System Research
Allergy & Asthma Drugs
Antibody Generation
Drugs - HSV1 Treatment

BRANDS/DIVISIONS/AFFILIATES:
Xenerex
Abreva
Zenvia

CONTACTS: *Note: Officers with more than one job title may be intentionally listed here more than once.*
Keith A. Katkin, CEO
Keith A. Katkin, Pres.
Michael McFadden, VP-U.S. Sales
Gregory J. Flesher, VP-Bus. Dev.
Eric S. Benevich, VP-Comm.
Christine G. Ocampo, VP-Finance
Randall E. Kaye, Sr. VP-Clinical Research & Medical Affairs
Randall E. Kaye, Chief Medical Officer
Michael McFadden, VP-Managed Markets
Craig A. Wheeler, Chmn.

Phone: 949-389-6700	Fax: 949-643-6800
Toll-Free:	
Address: 101 Enterprise, Ste. 300, Aliso Viejo, CA 92656 US	

GROWTH PLANS/SPECIAL FEATURES:
Avanir Pharmaceuticals is a biopharmaceutical company engaged in acquiring, developing and commercializing therapeutic products for the treatment of central nervous system disorders. The firm developed Abreva (docosonal 10% cream), an over-the-counter, FDA-approved treatment for Type-1 Herpes Simplex (HSV1, more commonly known as cold sores or fever blisters). GlaxoSmithKline Consumer Healthcare is the company's marketing partner for Abreva in the U.S. and Canada. Avanir's lead drug candidate, Zenvia, successfully completed a Phase III trial for the treatment of patients with diabetic peripheral neuropathic pain (DPN pain). It also successfully completed three Phase III clinical trials for the treatment of pseudobulbar affect (PBA). PBA is characterized by unprovoked and uncontrollable episodes of crying or laughing, and afflicts patients with neurological disorders such as Lou Gehrig's disease, Alzheimer's disease, MS, stroke and traumatic brain injury. Another area of development includes testing the company's patented Xenerex antibody technology for discovery of fully human monoclonal antibodies in treating inhaled anthrax. The firm owns or has the rights to 207 issued patents (59 in the U.S. and 148 foreign) and 213 pending applications (25 in the U.S. and 188 foreign). In February 2010, Avanir was issued a new patent for Zenvia from the U.S. Patent and Trademark Office. This patent extends the period of patent protection in the U.S. into late 2025.

Avanir offers major medical, dental, vision and disability insurance; life and AD&D insurance; and a 401(k) plan.

FINANCIALS: Sales and profits are in thousands of dollars—add 000 to get the full amount. 2009 Note: Financial information for 2009 was not available for all companies at press time.

2009 Sales: $4,176	2009 Profits: $-21,996	**U.S. Stock Ticker: AVNR**
2008 Sales: $6,829	2008 Profits: $-17,496	**Int'l Ticker:** Int'l Exchange:
2007 Sales: $9,153	2007 Profits: $-20,933	Employees: 20
2006 Sales: $15,186	2006 Profits: $-62,553	Fiscal Year Ends: 9/30
2005 Sales: $16,691	2005 Profits: $-30,607	Parent Company:

SALARIES/BENEFITS:
Pension Plan:	ESOP Stock Plan:	Profit Sharing:	Top Exec. Salary: $373,248	Bonus: $235,146
Savings Plan: Y	Stock Purch. Plan:		Second Exec. Salary: $326,667	Bonus: $165,000

OTHER THOUGHTS:
Apparent Women Officers or Directors: 1
Hot Spot for Advancement for Women/Minorities:

LOCATIONS: ("Y" = Yes)
West:	Southwest:	Midwest:	Southeast:	Northeast:	International:
Y	Y	Y	Y	Y	

AVAX TECHNOLOGIES INC

www.avax-tech.com

Industry Group Code: 325412 Ranks within this company's industry group: Sales: Profits:

Drugs:		Other:		Clinical:	Computers:		Services:	
Discovery:	Y	AgriBio:	Y	Trials/Services:	Hardware:		Specialty Services:	
Licensing:	Y	Genetic Data:		Labs:	Software:		Consulting:	
Manufacturing:	Y	Tissue Replacement:		Equipment/Supplies:	Arrays:		Blood Collection:	
Generics:				Research/Development Svcs.:	Database Management:		Drug Delivery:	
				Diagnostics:			Drug Distribution:	

TYPES OF BUSINESS:

Drugs-Cancer
Melanoma Treatment
Non-Small Cell Lung Cancer Treatment
Ovarian Cancer Treatment
Vaccine Therapies

BRANDS/DIVISIONS/AFFILIATES:

AC Vaccine
Mvax
Ovax
LungVax
Genopoietic, S.A.

CONTACTS: Note: Officers with more than one job title may be intentionally listed here more than once.

John K. A. Prendergast, CEO
Jean-Louis Misset, Chmn.-Scientific & Advisory Board
Henry E. Schea, III, Dir.-Global Quality & Regulatory Affairs
Isabelle Fourthin, Gen. Mgr.-Genopoietic
John K. A. Prendergast, Chmn.
Isabelle Fourthin, Chief Medical Officer-EMEA

Phone: 215-241-9760	Fax: 215-241-9684
Toll-Free:	
Address: 2000 Hamilton St., Ste. 204, Philadelphia, PA 19130 US	

GROWTH PLANS/SPECIAL FEATURES:

AVAX Technologies, Inc. is a development stage biotechnology company that specializes in the development and commercialization of individualized vaccine therapies and other technologies for the treatment of cancer. The company's vaccine consists of autologous (the patient's own) cancer cells that have been treated with a chemical (or, haptenized) to make them more visible to the patient's immune system. AVAX refers to its cancer vaccine technology as autologous cell vaccine immunotherapy and to the vaccine as AC Vaccine. The firm's previous clinical trials have focused on melanoma, ovarian carcinoma and non-small cell lung cancer. AVAX's AC Vaccine candidates are MVax, currently in Phase III trials for the treatment of melanoma; LungVax, in Phase II trials for the treatment of non-small cell lung cancer; and OVax, in Phase I-II trials for the treatment of ovarian cancer. The company's leading AC Vaccine is MVax, which is designed as an immunotherapy for the post-surgical treatment of late stage (stages three and four) melanoma. AVAX believes that MVax is the first immunotherapy to show a substantial increase in the survival rate for patients with this type of melanoma. Of 214 stage-three melanoma patients treated with MVax, mature studies (in which all surviving patients completed the five-year follow-up) evidenced a five-year overall survival rate of 44%, as opposed to the historical post-surgical survival rates of approximately 22-32%. In total studies of over 600 patients, no serious adverse side effects have yet been reported. Subsidiary Genopoietic, based in Lyons, France, oversees the company's European activities, including a drug manufacturing facility in France. Clinical funding proved elusive during 2009, and the company suspended its Phase III MVax trials for much of the year, but financial headway was made in December 2009 that AVAX hoped would allow trials to resume in 2010.

FINANCIALS: Sales and profits are in thousands of dollars—add 000 to get the full amount. 2009 Note: Financial information for 2009 was not available for all companies at press time.

2009 Sales: $	2009 Profits: $	**U.S. Stock Ticker:** AVXT.PK	
2008 Sales: $	2008 Profits: $	**Int'l Ticker:** Int'l Exchange:	
2007 Sales: $ 617	2007 Profits: $-6,414	Employees:	
2006 Sales: $ 735	2006 Profits: $-5,356	Fiscal Year Ends: 12/31	
2005 Sales: $1,624	2005 Profits: $-3,704	Parent Company:	

SALARIES/BENEFITS:

Pension Plan:	ESOP Stock Plan:	Profit Sharing:	Top Exec. Salary: $1,339,339	Bonus: $2,708,940
Savings Plan:	Stock Purch. Plan:		Second Exec. Salary: $554,769	Bonus: $866,320

OTHER THOUGHTS:

Apparent Women Officers or Directors: 1
Hot Spot for Advancement for Women/Minorities:

LOCATIONS: ("Y" = Yes)

West:	Southwest:	Midwest:	Southeast:	Northeast:	International:
				Y	Y

AVI BIOPHARMA INC

www.avibio.com

Industry Group Code: 325412 Ranks within this company's industry group: Sales: 111 Profits: 128

Drugs:		Other:	Clinical:	Computers:	Services:
Discovery:	Y	AgriBio:	Trials/Services:	Hardware:	Specialty Services:
Licensing:		Genetic Data:	Labs:	Software:	Consulting:
Manufacturing:		Tissue Replacement:	Equipment/Supplies:	Arrays:	Blood Collection:
Generics:			Research/Development Svcs.:	Database Management:	Drug Delivery:
			Diagnostics:		Drug Distribution:

TYPES OF BUSINESS:

Gene-Targeted Pharmaceuticals
Drugs - Cardiovascular Disease
Drugs - Cancer
Drugs - Infectious Disease

BRANDS/DIVISIONS/AFFILIATES:

Translation Suppressing Oligomers (TSO)
Splice Switching Oligomers (SSO)
AVI-4658
Ercole Biotechnology Inc
AVI-6002
AVI-6003

CONTACTS: Note: Officers with more than one job title may be intentionally listed here more than once.

J. David Boyle II, Interim CEO
J. David Boyle II, Interim Pres.
J. David Boyle II, CFO/Sr. VP
Ryszard Kole, Sr. VP-Discovery Research
Dwight D. Weller, Sr. VP-Mfg. & Chemistry
Shirley J. Leow, VP-Clinical Oper. & Project Mgmt.
Paul Medeiros, Sr. VP-Bus. Dev./Chief Business Officer
Steve Shrewsbury, Chief Medical Officer
Steve Shrewsbury, Sr. VP-Preclinical, Clinical & Reg. Affairs
Patrick L. Iversen, Sr. VP-Strategic Alliances
Jacqueline A. Dombroski, VP-Regulatory Affairs & Quality Assurance
William A. Goolsbee, Chmn.

Phone: 425-354-5038	Fax:
Toll-Free:	
Address: 3450 Monte Villa Pkwy., Ste. 101, Bothell, WA 98021 US	

GROWTH PLANS/SPECIAL FEATURES:

AVI BioPharma, Inc. is a biopharmaceutical company that develops drug treatments principally based on RNA therapeutics. Proprietary technologies include Translation Suppressing Oligomers (TSO), which are antisense compounds that are designed to suppress protein translation; and Splice Switching Oligomers (SSO), which block disease-related pathways in protein production. The company's principal products in development target life-threatening diseases, including cardiovascular, genetic and infectious diseases. These products have also been tested in preclinical trials and some clinical studies for the treatment of Duchenne muscular dystrophy (DMD), prevention of Restenosis and the Ebola and Marburg viruses. AVI's leading drug candidate, AVI-4658, currently in Phase I clinical trials, aims to reverse protein deletions that contribute to DMD. The firm's biodefense program is developing two antisense drugs, AVI-6002 and AVI-6003, which treat the Ebola and Margburg hemorrhagic fever viruses, respectively. This research is being supported by the US Department of Defense. The company is also working with Global Therapeutics to develop a cardiovascular restenosis drug for use in bare metal stents. AVI is also affiliated with a number of other institutions, including the US Army Medical Research Institute of Infectious Diseases (AMRIID) to find solutions for infections such as the avian influenza virus; Eleos, Inc. for cancer treatment and the Imperial College of London for muscular dystrophy. The firm owns 191 issued or licensed patent worldwide and 190 pending patent applications. In May 2009, the company announced it had received a $2.5 million grant from the Department of Defense to accelerate the development of its AVI-4658 drug for DMD.

AVI offers employees life, disability, medical, vision and dental insurance; a 401(k) savings plan; paid time off; and employee stock option plans.

FINANCIALS: Sales and profits are in thousands of dollars—add 000 to get the full amount. 2009 Note: Financial information for 2009 was not available for all companies at press time.

2009 Sales: $17,585	2009 Profits: $-25,159	**U.S. Stock Ticker: AVII**
2008 Sales: $21,258	2008 Profits: $-23,953	**Int'l Ticker:** Int'l Exchange:
2007 Sales: $10,985	2007 Profits: $-27,168	Employees: 63
2006 Sales: $ 115	2006 Profits: $-31,073	Fiscal Year Ends: 12/31
2005 Sales: $4,783	2005 Profits: $-16,676	Parent Company:

SALARIES/BENEFITS:

Pension Plan:	ESOP Stock Plan:	Profit Sharing:	Top Exec. Salary: $480,000	Bonus: $181,440
Savings Plan: Y	Stock Purch. Plan: Y		Second Exec. Salary: $324,000	Bonus: $62,111

OTHER THOUGHTS:

Apparent Women Officers or Directors: 2
Hot Spot for Advancement for Women/Minorities:

LOCATIONS: ("Y" = Yes)

West:	Southwest:	Midwest:	Southeast:	Northeast:	International:
Y					

Note: Financial information, benefits and other data can change quickly and may vary from those stated here.

AVIVA BIOSCIENCES CORP

www.avivabio.com

Industry Group Code: 325413 Ranks within this company's industry group: Sales: Profits:

Drugs:	Other:	Clinical:	Computers:		Services:	
Discovery:	AgriBio:	Trials/Services:	Hardware:	Y	Specialty Services:	Y
Licensing:	Genetic Data:	Labs:	Software:		Consulting:	
Manufacturing:	Tissue Replacement:	Equipment/Supplies:	Arrays:		Blood Collection:	
Generics:		Research/Development Svcs.:	Database Management:		Drug Delivery:	
		Diagnostics:			Drug Distribution:	

TYPES OF BUSINESS:
Cellular Biology Equipment
Cancer Cell Isolation Technology
Drug Candidate Screening Technology
Automated Patch Clamp Electrophysiology
Rare Cell Enrichment
Multiple Force Biochips

BRANDS/DIVISIONS/AFFILIATES:
Sealchip
Electrophysiology on Demand (EPOD)
Nav1.5
hERGexpress
China Development Industrial Bank
CapitalBio Corporation
Pac-Link
WI Harper Group

CONTACTS: *Note: Officers with more than one job title may be intentionally listed here more than once.*
David G. Wang, Managing Dir.
Vytas P. Ambutas, Chmn.

Phone: 858-552-0888	Fax: 858-552-9040
Toll-Free: 888-284-8224	
Address: 11045 Roselle St., Ste. 100, San Diego, CA 92121-1230 US	

GROWTH PLANS/SPECIAL FEATURES:

AVIVA Biosciences Corp. develops technologies that enable drug research and development. The firm's key development areas include the integrating of biochips for electrophysiology research, ion channel drug screening and rare-cell isolation. In addition, the company has developed proprietary surface chemistries, microbeads and reagent compositions as solutions to cell biology and bioassay applications in drug development and diagnostic fields. The firm offers Electrophysiology on Demand (EPOD), a drug discovery service that utilizes ion channel drug screening in order to provide clients with a full line of automated electrophysiology instruments, experienced personnel and customer service. The main product line within the EPOD service is Sealchip, a single-use disposable cartridge designed for the study of ion channels on a cell membrane, used in assay development, target validation, late secondary screening and safety screening. Optimized and validated ion channel cell lines are also provided for patch clamp electrophysiology experiments. As well as previously available cell lines, the company can also assist clients in generating or validating new cell lines. AVIVA also develops cancer cell isolation systems that reliably detect targeted tumor cells from blood samples. Through its rare cell enrichment technology platform, which utilizes a proprietary depletion approach, the system removes non-relevant cells and exposes target cells that can then be analyzed and quantified. In addition to its product lines, the firm offers hERGexpress, a screening service for medicinal chemists and toxicologists that provides high-quality data as guidance in assessing the cardiac safety of certain pharmaceutical compounds; and Nav1.5, a similar screening process used to determine if pharmaceutical compounds are blocking the inward sodium current, which can cause arrhythmias. AVIVA is owned by four investors: CapitalBio Corporation; China Development Industrial Bank (CDIB); WI Harper Group; and Pac-Link.

FINANCIALS: Sales and profits are in thousands of dollars—add 000 to get the full amount. 2009 Note: Financial information for 2009 was not available for all companies at press time.

2009 Sales: $	2009 Profits: $	U.S. Stock Ticker: Private
2008 Sales: $	2008 Profits: $	Int'l Ticker: Int'l Exchange:
2007 Sales: $	2007 Profits: $	Employees:
2006 Sales: $	2006 Profits: $	Fiscal Year Ends:
2005 Sales: $	2005 Profits: $	Parent Company:

SALARIES/BENEFITS:

Pension Plan:	ESOP Stock Plan:	Profit Sharing:	Top Exec. Salary: $	Bonus: $
Savings Plan:	Stock Purch. Plan:		Second Exec. Salary: $	Bonus: $

OTHER THOUGHTS:
Apparent Women Officers or Directors:
Hot Spot for Advancement for Women/Minorities:

LOCATIONS: ("Y" = Yes)

West:	Southwest:	Midwest:	Southeast:	Northeast:	International:
Y					

AXCAN PHARMA INC

www.axcan.com

Industry Group Code: 325412 Ranks within this company's industry group: Sales: Profits:

Drugs:	Other:	Clinical:	Computers:	Services:
Discovery:	AgriBio:	Trials/Services:	Hardware:	Specialty Services:
Licensing:	Genetic Data:	Labs:	Software:	Consulting:
Manufacturing: Y	Tissue Replacement:	Equipment/Supplies:	Arrays:	Blood Collection:
Generics:		Research/Development Svcs.:	Database Management:	Drug Delivery:
		Diagnostics:		Drug Distribution:

TYPES OF BUSINESS:

Pharmaceutical Manufacturing
Gastroenterology Treatment Products

BRANDS/DIVISIONS/AFFILIATES:

TPG (Texas Pacific Group)
Salofalk
Canasa
Sudca
Urso
Delursan
Ultrase
Viokase

CONTACTS: *Note: Officers with more than one job title may be intentionally listed here more than once.*

Frank A. G. M. Verwiel, CEO
David W. Mims, COO/Exec. VP
Frank A. G. M. Verwiel, Pres.
Steve Gannon, CFO/Sr. VP
Richard DeVleeschouwer, Sr. VP-Human Resources
Martha Donze, VP-Corp. Admin.
Richard Tarte, General Counsel
Richard Tarte, VP-Corp. Dev.
Isabelle Adjahi, Sr. Dir.-Comm.
Isabelle Adjahi, Sr. Dir.-Investor Rel.
Theresa Stevens, VP-Bus. Dev.
Nicholas Franco, Sr. VP- Int'l Commercial Oper.

Phone: 450-467-5138	Fax: 450-464-9979
Toll-Free: 800-565-3255	
Address: 597 Laurier Blvd., Mont-Saint-Hilaire, QC J3H 6CA8 Canada	

GROWTH PLANS/SPECIAL FEATURES:

Axcan Pharma, Inc. is a leading specialty pharmaceutical company concentrating in the field of gastroenterology, with operations in North America and Europe. Axcan markets and sells pharmaceutical products used in the treatment of a variety of gastrointestinal diseases and disorders, including inflammatory bowel disease, cholestatic liver diseases, irritable bowel syndrome and complications related to pancreatic insufficiency. For the treatment of inflammatory bowel diseases, such as ulcerative colitis and ulcerative proctitis, Axcan markets mesalamine-based products Salofalk and Canasa. Axcan is currently developing products for the prevention and treatment of colorectal cancer, including SUDCA, which is currently in Phase I trial for the prevention of the recurrence of colorectal polyps. For the treatment of the cholestatic liver disease Primary Biliary Cirrhosis (PBC), a condition that causes the slow destruction of bile ducts in the liver, Axcan markets URSO 250 and URSO Forte. For the treatment of both PBC and the cholestatic liver disease Primary Sclerosing Cholangitis (PSC), a condition that narrows the bile ducts inside and outside of the liver through inflammation and scarring, Axcan markets Urso, Urso DS and Delursan. For the treatment of pancreatic insufficiency, Axcan markets Ultrase, Panzytrat and Viokase. For the treatment of duodenal ulcers, Axcan markets Carafate and Sulcrate. Recently, Axcan released Pylera, a therapy for the eradication of Helicobacter pylori, a bacterium recognized as the main cause of gastric and duodenal ulcers. The firm partners with other third party companies to distribute its products in South America, Asia, Oceania and Africa. Private investment firm TPG (Texas Pacific Group) owns Axcan.

FINANCIALS: Sales and profits are in thousands of dollars—add 000 to get the full amount. 2009 Note: Financial information for 2009 was not available for all companies at press time.

2009 Sales: $	2009 Profits: $	U.S. Stock Ticker: Private
2008 Sales: $	2008 Profits: $	Int'l Ticker: Int'l Exchange:
2007 Sales: $349,000	2007 Profits: $	Employees:
2006 Sales: $292,320	2006 Profits: $39,120	Fiscal Year Ends: 9/30
2005 Sales: $251,300	2005 Profits: $26,400	Parent Company: TPG (TEXAS PACIFIC GROUP)

SALARIES/BENEFITS:

Pension Plan:	ESOP Stock Plan:	Profit Sharing:	Top Exec. Salary: $	Bonus: $
Savings Plan:	Stock Purch. Plan:		Second Exec. Salary: $	Bonus: $

OTHER THOUGHTS:

Apparent Women Officers or Directors: 2
Hot Spot for Advancement for Women/Minorities: Y

LOCATIONS: ("Y" = Yes)

West:	Southwest:	Midwest:	Southeast:	Northeast:	International:
			Y	Y	Y

BASF SE

www.basf.com

Industry Group Code: 325 Ranks within this company's industry group: Sales: 1 Profits: 1

Drugs:	Other:	Clinical:		Computers:	Services:	
Discovery:	AgriBio:	Trials/Services:		Hardware:	Specialty Services:	Y
Licensing:	Genetic Data:	Labs:		Software:	Consulting:	
Manufacturing:	Tissue Replacement:	Equipment/Supplies:	Y	Arrays:	Blood Collection:	
Generics:		Research/Development Svcs.:	Y	Database Management:	Drug Delivery:	
		Diagnostics:			Drug Distribution:	

TYPES OF BUSINESS:

Chemicals Manufacturing
Agricultural Products
Oil & Gas Production
Plastics
Coatings
Nanotechnology Research
Nutritional Products
Agricultural Biotechnology

BRANDS/DIVISIONS/AFFILIATES:

Wintershall AG
BASF Canada
BASF Future Business GMBH
Sorex Holdings Ltd
Ciba Holding AG
BASF AG

CONTACTS: Note: Officers with more than one job title may be intentionally listed here more than once.

Jurgen Hambrect, CEO
Kurt W. Bock, CFO
Harald Schwager, Exec. Dir.-Human Resources
Stefan Marcinowski, Exec. Dir.-Specialty Chemicals Research Div.
Kurt W. Bock, Exec. Dir.-Info. Svcs.
Magdalena Moll, Sr. VP-Investor Rel.
Kurt W. Bock, Exec. Dir.-Finance
Andreas Kreimeyer, Exec. Dir.-Inorganics & Petrochemicals
John Feldmann, Exec. Dir.-Construction & Performance Chemicals
Stefan Marcinowski, Exec. Dir.-Crop Protection
Juergen Hambrecht, Chmn.
Martin Brudermueller, Exec. Dir.-Asia Pacific Div.
Hans-Ulrich Engel, Exec. Dir.-Procurement & Logistics

Phone: 49-621-60-0	Fax: 49-621-60-42525
Toll-Free:	
Address: 38 Carl-Bosch St., Ludwigshafen, 67056 Germany	

GROWTH PLANS/SPECIAL FEATURES:

BASF SE, formerly BASF AG, is a chemical manufacturing company that serves customers in more than 170 countries. Around 19% of BASF sales are made to North American industries. The firm operates in six business segments: chemicals; plastics; performance products; agricultural solutions; functional solutions; and oil and gas. The chemicals segment manufactures inorganic, petrochemical and intermediate chemicals for the pharmaceutical, construction, textile and automotive industries. The plastics segment primarily manufactures polystyrene, styrenics and performance polymers for the manufacturing and packaging industries. The performance polymers segment produces pigments, inks, printing supplies, coatings and polymers for the automotive, oil, packaging, textile, detergent, sanitary care, construction and chemical industries. The firm's agricultural solutions segment produces and markets genetically engineered plants, nutritional supplements, herbicides, fungicides and insecticides for use in agriculture, public health and pest control. The functional solutions segment develops automotive and industrial catalysts; construction chemicals; and coatings and refinishes for automotive and construction markets. The oil and gas segment is operated through BASF subsidiary Wintershall AG, which focuses on petroleum and natural gas exploration and production in North America, Asia, Europe, the Middle East and Africa. BASF also employs chemical nanotechnology in pigments that are used to color coatings, paints and plastics; and sunscreen. The company is one of the world's leading R&D firms, with 8,000 employees working in research in 70 sites worldwide, employing a research budget of $1.3 billion Euros yearly. In April 2009, the firm acquired specialty chemicals maker Ciba Holding AG. Also in 2009, BASF announced plans to eliminate 2,000 jobs and shorten hours for another 3,000 workers. In June 2010, the company agreed to purchase Cognis Deutschland GmbH & Co KG, a specialty chemicals company, for approximately $4.1 billion.

U.S. employees are offered medical, dental and vision insurance; life insurance; disability coverage; an employee savings plan; tuition reimbursement; adoption assistance; and a supplier discount.

FINANCIALS: Sales and profits are in thousands of dollars—add 000 to get the full amount. 2009 Note: Financial information for 2009 was not available for all companies at press time.

2009 Sales: $67,557,500	2009 Profits: $1,879,080	**U.S. Stock Ticker:** BASFY
2008 Sales: $83,990,800	2008 Profits: $3,925,610	**Int'l Ticker:** BAS **Int'l Exchange:** Frankfurt-Euronext
2007 Sales: $78,122,600	2007 Profits: $5,479,950	**Employees:** 96,924
2006 Sales: $69,448,400	2006 Profits: $4,575,330	**Fiscal Year Ends:** 12/31
2005 Sales: $52,080,500	2005 Profits: $3,663,700	**Parent Company:**

SALARIES/BENEFITS:

Pension Plan:	ESOP Stock Plan:	Profit Sharing:	Top Exec. Salary: $	Bonus: $
Savings Plan:	Stock Purch. Plan:		Second Exec. Salary: $	Bonus: $

OTHER THOUGHTS:

Apparent Women Officers or Directors: 2
Hot Spot for Advancement for Women/Minorities: Y

LOCATIONS: ("Y" = Yes)

West:	Southwest:	Midwest:	Southeast:	Northeast:	International:
Y	Y	Y	Y	Y	Y

Note: Financial information, benefits and other data can change quickly and may vary from those stated here.

BAUSCH & LOMB INC

www.bausch.com

Industry Group Code: 33911 **Ranks within this company's industry group:** Sales: Profits:

Drugs:	Other:	Clinical:	Computers:	Services:
Discovery:	AgriBio:	Trials/Services:	Hardware:	Specialty Services:
Licensing:	Genetic Data:	Labs:	Software:	Consulting:
Manufacturing: Y	Tissue Replacement:	Equipment/Supplies: Y	Arrays:	Blood Collection:
Generics:		Research/Development Svcs.:	Database Management:	Drug Delivery:
		Diagnostics:		Drug Distribution:

TYPES OF BUSINESS:
Supplies-Eye Care
Contact Lens Products
Ophthalmic Pharmaceuticals
Surgical Products

BRANDS/DIVISIONS/AFFILIATES:
Alrex
Warburg Pincus LLC
Eyeonics Inc
Ocuvite
Lotemax
Alrex
PreserVision
Zyoptic

CONTACTS: *Note: Officers with more than one job title may be intentionally listed here more than once.*
Brent L. Saunders, CEO
Brian J. Harris, CFO/Corp. VP
Paul H. Sartori, Corp. VP-Human Resources
Alan H. Farnsworth, CIO/Sr. VP-IT & Customer Service/Corp. VP
John W. Sheets, Jr., CTO/Corp. VP
A Robert D. Bailey, General Counsel/Corp. VP
Michael Gowen, Exec. VP-Global Bus. Oper. & Process Excellence
Paul H. Sartori, Corp. VP-Public Affairs
Daniel L. Ritz, Dir.-Capital Markets Analysis & Reporting
J. Andy Corley, Corp. VP/Global Pres., Surgical Prod.
David N. Edwards, Corp. VP/Pres., Asia Pacific Region
Peter Valenti III, Corp. VP/Global Pres., Vision Care
Flemming Ornskov, Corp. VP/Global Pres., Pharmaceuticals
Fred Hassan, Chmn.
John H. Brown, Corp. VP/Pres., EMEA

Phone: 585-338-6000	Fax: 585-338-6007
Toll-Free: 800-344-8815	
Address: 1 Bausch & Lomb Pl., Rochester, NY 14604-2701 US	

GROWTH PLANS/SPECIAL FEATURES:

Bausch & Lomb, Inc. (B&L), owned by private equity firm Warburg Pincus LLC, is a world leader in the development, marketing and manufacturing of eye care products. The firm's products are marketed in more than 100 countries and in five categories: contact lenses; lens care; pharmaceuticals; cataract and vitreoretinal surgery; and refractive surgery. In its contact lens category, B&L's product portfolio includes traditional, planned replacement disposable, daily disposable, continuous wear, toric soft contact lenses and rigid gas-permeable materials. The firm's lens care products include multi-purpose solutions, enzyme cleaners and saline solutions. The firm's pharmaceuticals include generic and branded prescription ophthalmic pharmaceuticals, ocular vitamins, over-the-counter medications and vision accessories. Key pharmaceutical trademarks of the firm are Bausch & Lomb, Alrex, Liposic, Lotemax, Ocuvite, PreserVision and Zylet. B&L's cataract and vitreoretinal division offers a broad line of intraocular lenses and delivery systems, as well as the Millennium and Stellaris lines of phacoemulsification equipment used in the extraction of the patient's natural lens during cataract surgery. The company's refractive surgery products include lasers and diagnostic equipment used in the LASIK surgical procedure under the brand Zyoptic. Eyeonics, Inc., is a subsidiary of the company, specializing in ophthalmic medical devices based in Aliso Viejo, California. Bausch & Lomb Pharmaceuticals specializes in the development and manufacture of products for pharmaceutical companies. B&L's global operations include research and development units on six continents and operating offices in over 43 countries. In October 2009, the firm acquired the commercial assets of Tubilux Pharma S.p.A., which develops and markets ophthalmic pharmaceuticals. In December 2009, the company announced plans to close its manufacturing plant in Livingston, Scotland, with the anticipated closure in 2011.

Employees are offered medical and dental coverage; various work/life programs; a vacation buy/sell program; domestic partner benefits; flexible spending accounts; and education reimbursement.

FINANCIALS: Sales and profits are in thousands of dollars—add 000 to get the full amount. 2009 Note: Financial information for 2009 was not available for all companies at press time.

2009 Sales: $	2009 Profits: $	**U.S. Stock Ticker:** Private
2008 Sales: $2,500,000	2008 Profits: $	**Int'l Ticker:** Int'l Exchange:
2007 Sales: $	2007 Profits: $	Employees: 2,500
2006 Sales: $2,293,400	2006 Profits: $	Fiscal Year Ends: 12/31
2005 Sales: $2,353,800	2005 Profits: $19,200	Parent Company: WARBURG PINCUS LLC

SALARIES/BENEFITS:
| Pension Plan: | ESOP Stock Plan: | Profit Sharing: | Top Exec. Salary: $ | Bonus: $ |
| Savings Plan: | Stock Purch. Plan: | | Second Exec. Salary: $ | Bonus: $ |

OTHER THOUGHTS:
Apparent Women Officers or Directors: 2
Hot Spot for Advancement for Women/Minorities:

LOCATIONS: ("Y" = Yes)
West:	Southwest:	Midwest:	Southeast:	Northeast:	International:
Y			Y	Y	Y

Note: Financial information, benefits and other data can change quickly and may vary from those stated here.

BAXTER INTERNATIONAL INC

www.baxter.com

Industry Group Code: 33911 Ranks within this company's industry group: Sales: 2   Profits: 2

Drugs:	Other:	Clinical:		Computers:		Services:	
Discovery:	AgriBio:	Trials/Services:		Hardware:		Specialty Services:	
Licensing:	Genetic Data:	Labs:		Software:		Consulting:	
Manufacturing:	Tissue Replacement:	Equipment/Supplies:	Y	Arrays:		Blood Collection:	
Generics:		Research/Development Svcs.:		Database Management:		Drug Delivery:	
		Diagnostics:				Drug Distribution:	

TYPES OF BUSINESS:

Medical Equipment Manufacturing
Supplies-Intravenous & Renal Dialysis Systems
Medication Delivery Products & IV Fluids
Biopharmaceutical Products
Plasma Collection & Processing
Vaccines
Software
Contract Research

BRANDS/DIVISIONS/AFFILIATES:

Gammagard
BioScience
Ceprotin
Suprane
Tisseel
Clinimix

CONTACTS: Note: Officers with more than one job title may be intentionally listed here more than once.

Robert L. Parkinson, Jr., CEO
Robert L. Parkinson, Jr., Pres.
Robert M. Davis, CFO/VP
Jeanne K. Mason, VP-Human Resources
Norbert G. Riedel, Chief Scientific Officer/VP
Karenann Terrell, CIO/VP
J. Michael Gatling, VP-Mfg.
David P. Scharf, General Counsel/VP
Mary Kay Ladone, VP-Investor Rel.
Robert J. Hombach, Treas./VP
Joy A. Amundson, VP/Pres., Bioscience
Bruce McGillivray, VP/Pres., Renal
Peter J. Arduini, VP/Pres., Medication Delivery
Philip L. Batchelor, VP-Quality
Robert L. Parkinson, Jr., Chmn.
Gerald Lema, VP/Pres., Asia Pacific

Phone: 847-948-4770	Fax: 847-948-3642
Toll-Free: 800-422-9837	
Address: 1 Baxter Pkwy., Deerfield, IL 60015-4625 US	

GROWTH PLANS/SPECIAL FEATURES:

Baxter International, Inc. manufactures and markets products for the treatment of hemophilia, immune disorders, infectious diseases, kidney disease, trauma and other chronic and acute medical conditions, offering expertise in medical devices, pharmaceuticals and biotechnology.   Baxter markets its offerings to hospitals; clinical and medical research labs; blood and blood dialysis centers; rehab facilities; nursing homes; doctor's offices; and patients undergoing supervised home care.   The firm has manufacturing facilities in 27 countries and offers products and services in 100 countries.   Baxter operates in three segments: Medication Delivery, its largest sector, which provides a range of intravenous solutions and specialty products that are used in combination for fluid replenishment, nutrition therapy, pain management and antibiotic therapy; BioScience, which develops biopharmaceuticals, biosurgery products, vaccines, blood collection, processing and storage products and technologies; and Renal, which develops products and provides services to treat end-stage kidney disease.   Products include Suprane, a general inhalation anesthetic; Tisseel, a hemostatic agent; Cryocyte freezing containers; Clinimix injectable nutrition solutions; Ceprotin, for the treatment of venous thrombosis; Gammagard immunoglobulins; V-Link antimicrobial IV connectors; and the Aquarius renal replacement therapy.   In addition, the company provides the following services: Biolife plasma donation, IV pump rental, renal pump servicing, product training and biomedicinal delivery training.   In March 2010, Baxter announced plans to acquire Apatech's outstanding equity, including Actifuse, a sythnetic bone graft material.   In April 2010, the firm and Nycomed announced FDA approval for their collaborative product, TachoSil, a hemostatic patch.

Baxter offers its employees medical and dental coverage; life and AD&D insurance; short- and long-term disability; an employee stock purchase plan; a 401(k) plan; flexible spending accounts; tuition assistance; and access to a credit union.

FINANCIALS: Sales and profits are in thousands of dollars—add 000 to get the full amount. 2009 Note: Financial information for 2009 was not available for all companies at press time.

2009 Sales: $12,562,000	2009 Profits: $2,215,000	**U.S. Stock Ticker: BAX**
2008 Sales: $12,348,000	2008 Profits: $2,014,000	Int'l Ticker: Int'l Exchange:
2007 Sales: $11,263,000	2007 Profits: $1,707,000	Employees: 49,700
2006 Sales: $10,378,000	2006 Profits: $1,397,000	Fiscal Year Ends: 12/31
2005 Sales: $9,849,000	2005 Profits: $956,000	Parent Company:

SALARIES/BENEFITS:

Pension Plan:	ESOP Stock Plan:	Profit Sharing:	Top Exec. Salary: $1,342,000	Bonus: $2,500,560
Savings Plan: Y	Stock Purch. Plan: Y		Second Exec. Salary: $576,923	Bonus: $814,320

OTHER THOUGHTS:

Apparent Women Officers or Directors: 6
Hot Spot for Advancement for Women/Minorities: Y

LOCATIONS: ("Y" = Yes)

West:	Southwest:	Midwest:	Southeast:	Northeast:	International:
Y	Y	Y	Y	Y	Y

Note: Financial information, benefits and other data can change quickly and may vary from those stated here.

BAYER AG

www.bayer.com

Industry Group Code: 325 Ranks within this company's industry group: Sales: 2   Profits: 2

Drugs:	Other:		Clinical:		Computers:	Services:	
Discovery:	AgriBio:	Y	Trials/Services:		Hardware:	Specialty Services:	Y
Licensing:	Genetic Data:		Labs:		Software:	Consulting:	
Manufacturing: Y	Tissue Replacement:		Equipment/Supplies:	Y	Arrays:	Blood Collection:	
Generics:			Research/Development Svcs.:		Database Management:	Drug Delivery:	
			Diagnostics:			Drug Distribution:	

TYPES OF BUSINESS:
Chemicals Manufacturing
Pharmaceuticals
Animal Health Products
Synthetic Materials
Crop Science
Plant Biotechnology
Health Care Products

BRANDS/DIVISIONS/AFFILIATES:
Bayer Corp
Bayer Schering Pharma AG
Bayer HealthCare
Bayer MaterialScience
Bayer Business Services
Bayer Technology Services
Currenta GmbH & Co.
Athenix Corp

CONTACTS: *Note: Officers with more than one job title may be intentionally listed here more than once.*
Werner Wenning, Chmn.-Mgmt. Board
Klaus Kuhn, CFO
Richard Pott, Dir.-Human Resources
Wolfgang Plischke, Dir.-Innovation
Wolfgang Plischke, Dir.-Tech.
Richard Pott, Dir.-Strategy
Michael Schade, Head-Comm.
Alexander Rosar, Head-Investor Rel.
Wolfgang Plischke, Dir.-Environment
A. J. Higgins, Chmn.-Bayer Health Care
F. Berschauer, Chmn.-Bayer Crop Sciences
P. Thomas, Chmn.-Bayer Material Science
Manfred Schneider, Chmn.-Supervisory Board
Klaus Kuhn, Dir.-EMEA

Phone: 49-214-30-1	Fax:
Toll-Free: 800-269-2377	
Address: Bayerwerk Gebaeude W11, Leverkusen, D-51368 Germany	

GROWTH PLANS/SPECIAL FEATURES:

Bayer AG is a German holding company encompassing over 300 consolidated subsidiaries on five continents. The company has six business segments split under either the business division or the service division. Bayers business division contains the Bayer HealthCare, Bayer CropScience, and Bayer MaterialScience segments. The segments that are part of the service division are Bayer Business Services, Bayer Technology Services and Currenta. The Bayer HealthCare segment develops, produces and markets products for the prevention, diagnosis and treatment of human and animal diseases. Bayer CropScience is active in the areas of chemical crop protection and seed treatment; non-agricultural pest and weed control; and plant biotechnology. Bayer MaterialScience develops, manufactures and markets polyurethane, polycarbonate, cellulose derivatives and special metals products. Bayer Business Services offers IT infrastructure and applications, procurement and logistics, human resources and management services. Bayer Technology Services offers process development; process and plant engineering; construction; and optimization services. The Currenta GmbH & Co. segment is a joint venture with Lanxess AG. Currenta offers utility supply, waste management, infrastructure, safety, security, analytics and vocational training services to the chemical industry. The company's pharmaceutical portfolio includes Alka-Seltzer, Asprin, Levitra, Mirena, Aleve, Yaz and the Breeze 2 blood glucose meter. In July 2009, the company's CropScience division opened a new research facility in Canada dedicated to researching and breeding canola oil plants. In November 2009, Bayer acquired Athenix Corp., a biotech company located in North Carolina, for $365 million. It is now a part of Bayer's CropScience division. Athenix specializes in creating pest resistant crops and the study of crop genetics. In January 2010, Bayer acquired the biofungicide Shemer from AgroGreen. The product is used to prevent crops from being destroyed by fungi before and after harvest.

Employees receive sports amenities, flexible work schedules and a varied program of cultural events.

FINANCIALS: Sales and profits are in thousands of dollars—add 000 to get the full amount. 2009 Note: Financial information for 2009 was not available for all companies at press time.

2009 Sales: $42,346,100	2009 Profits: $1,846,390	U.S. Stock Ticker:
2008 Sales: $43,536,000	2008 Profits: $2,273,480	Int'l Ticker: BAYN Int'l Exchange: Frankfurt-Euronext
2007 Sales: $42,831,100	2007 Profits: $6,230,580	Employees: 108,400
2006 Sales: $38,710,400	2006 Profits: $2,249,950	Fiscal Year Ends: 12/31
2005 Sales: $32,662,374	2005 Profits: $1,902,517	Parent Company:

SALARIES/BENEFITS:
Pension Plan:	ESOP Stock Plan:	Profit Sharing:	Top Exec. Salary: $1,700,721	Bonus: $2,985,500
Savings Plan: Y	Stock Purch. Plan: Y		Second Exec. Salary: $1,147,526	Bonus: $1,851,035

OTHER THOUGHTS:
Apparent Women Officers or Directors:
Hot Spot for Advancement for Women/Minorities:

LOCATIONS: ("Y" = Yes)
West:	Southwest:	Midwest:	Southeast:	Northeast:	International:
Y	Y	Y	Y	Y	Y

Note: Financial information, benefits and other data can change quickly and may vary from those stated here.

BAYER CORP

www.bayerus.com

Industry Group Code: 325 Ranks within this company's industry group: Sales: Profits:

Drugs:		Other:		Clinical:		Computers:		Services:	
Discovery:	Y	AgriBio:	Y	Trials/Services:		Hardware:		Specialty Services:	
Licensing:		Genetic Data:		Labs:		Software:		Consulting:	
Manufacturing:	Y	Tissue Replacement:		Equipment/Supplies:	Y	Arrays:		Blood Collection:	
Generics:				Research/Development Svcs.:		Database Management:		Drug Delivery:	
				Diagnostics:	Y			Drug Distribution:	

TYPES OF BUSINESS:

Chemicals Manufacturing
Animal Health Products
Over-the-Counter Drugs
Diagnostic Products
Coatings, Adhesives & Sealants
Polyurethanes & Plastics
Herbicides, Fungicides & Insecticides

BRANDS/DIVISIONS/AFFILIATES:

Bayer
Bayer HealthCare AG
Bayer MaterialSciences LLC
Bayer CropScience LP
Aleve
BREEZE
BaySystems
Alka-Seltzer Plus

CONTACTS: *Note: Officers with more than one job title may be intentionally listed here more than once.*

Gregory S. Babe, CEO
Gregory S. Babe, Pres.
Willy Scherf, CFO
Joyce Burgess, Dir.-Human Resources
Claudio Abreu, CIO
George J. Lykos, Chief Legal Officer
Mark Ryan, Chief Comm. Officer
Arthur Higgins, Chmn.-Bayer HealthCare AG
William Buckner, CEO/Pres., Bayer CropScience, LP
Timothy Roseberry, Chief Procurement Officer/VP-Corp. Materials Mgmt.

Phone: 412-777-2000	Fax: 412-777-2034
Toll-Free:	
Address: 100 Bayer Rd., Pittsburgh, PA 15205-9741 US	

GROWTH PLANS/SPECIAL FEATURES:

Bayer Corporation is the U.S. subsidiary of chemical and pharmaceutical giant Bayer AG. The company operates through four divisions: Bayer HealthCare; Bayer MaterialScience; Bayer Corporate and Business Services; and Bayer CropScience. Bayer HealthCare operates through four units: Bayer Schering Pharma, consumer care, medical care and animal health. Bayer Schering Pharma sells pharmaceuticals including YAZ, Yasmin, Levitra and Mirena. Its consumer care products include analgesics (Aleve and Bayer); cold and cough treatments (Alka-Seltzer Plus and Talcio); digestive relief products (Alka-Mints and Phillips' Milk of Magnesia); topical skin preparations (Domeboro and Bactine); and vitamins (One-A-Day and Flintstones). The medical care division is a leader in self-test blood glucose diagnostic systems; it offers the Breeze and Contour product families that offers alternate site testing and automatic coding and requires smaller blood samples. Its animal health products include vaccines and other preventative measures for farm and domestic animals. Bayer's MaterialScience segment produces coatings, adhesives and sealant raw materials; polyurethanes; and plastics. Bayer CropScience makes products directed toward crop protection, environmental science and bioscience, which include herbicides, fungicides and insecticides. Bayer Corporate and Business Services provides business services to the aforementioned Bayer subsidiaries, such as administration, technology services, mergers/acquisitions and internal auditing. In May 2009, Bayer MaterialScience began construction on a new carbon nanotubes production plant in Germany.

The company offers benefits to its employees including life, disability, medical, dental and vision coverage; prescription drug reimbursement; a 401(k); and adoption assistance.

FINANCIALS: Sales and profits are in thousands of dollars—add 000 to get the full amount. 2009 Note: Financial information for 2009 was not available for all companies at press time.

2009 Sales: $	2009 Profits: $	U.S. Stock Ticker: Subsidiary
2008 Sales: $	2008 Profits: $	Int'l Ticker: Int'l Exchange:
2007 Sales: $	2007 Profits: $	Employees:
2006 Sales: $10,262,800	2006 Profits: $	Fiscal Year Ends: 12/31
2005 Sales: $8,747,200	2005 Profits: $	Parent Company: BAYER AG

SALARIES/BENEFITS:

Pension Plan:	ESOP Stock Plan:	Profit Sharing:	Top Exec. Salary: $	Bonus: $
Savings Plan: Y	Stock Purch. Plan: Y		Second Exec. Salary: $	Bonus: $

OTHER THOUGHTS:

Apparent Women Officers or Directors: 1
Hot Spot for Advancement for Women/Minorities:

LOCATIONS: ("Y" = Yes)

West:	Southwest:	Midwest:	Southeast:	Northeast:	International:
Y	Y	Y	Y	Y	

Note: Financial information, benefits and other data can change quickly and may vary from those stated here.

BAYER SCHERING PHARMA AG   www.bayerscheringpharma.de

Industry Group Code: 325412 Ranks within this company's industry group: Sales: Profits:

Drugs:	Other:		Clinical:		Computers:		Services:	
Discovery:	AgriBio:		Trials/Services:		Hardware:		Specialty Services:	
Licensing:	Genetic Data:	Y	Labs:		Software:		Consulting:	
Manufacturing: Y	Tissue Replacement:		Equipment/Supplies:		Arrays:		Blood Collection:	
Generics:			Research/Development Svcs.:		Database Management:		Drug Delivery:	
			Diagnostics:	Y		Y	Drug Distribution:	

TYPES OF BUSINESS:

Pharmaceuticals Discovery, Development & Manufacturing
Gynecology & Andrology Treatments
Contraceptives
Cancer Treatments
Multiple Sclerosis Treatments
Circulatory Disorder Treatments
Diagnostic & Radiopharmaceutical Agents
Proteomics

BRANDS/DIVISIONS/AFFILIATES:

Bayer AG
Angeliq
Yasmin
Menostar
Betaferon
Gadovist
Magnevist
SciLin

CONTACTS: *Note: Officers with more than one job title may be intentionally listed here more than once.*

Andreas Fibig, Chmn.-Exec. Board
Andreas Busch, Dir.-Labor
Andreas Busch, Dir.-Global Drug Discovery
Bernd Metzner, Dir.-Prod.
Bernd Metzner, Dir.-Admin. & Organization
Ulrich Koestlin, Dir.-Oper.
Kemal Malik, Dir.-Global Dev.
Kemal Malik, Chief Medical Officer
Richard Pott, Chmn.-Supervisory Board

Phone: 49-30-468-1111	Fax: 49-30-468-15305
Toll-Free:	
Address: Mullerstrasse 178, Berlin, 13353 Germany	

GROWTH PLANS/SPECIAL FEATURES:

Bayer Schering Pharma AG, is a major global research-based pharmaceutical company that operates through subsidiaries in more than 100 countries. The firm concentrates its activities on four business areas: women's healthcare, specialty medicine, general medicine and diagnostic imaging. Schering's women's health products include birth control pills (Yasmin), hormone therapy (Angeliq and Menostar) and other contraceptives for women (Mirena). The firm's specialty medicine unit focuses on cancer, central nervous system disease and age related eye disease treatments. Schering has contributed to the body of research on multiple sclerosis (MS) through its Beyond and Benefit studies. Its Betaferon drug reduces the frequency of MS episodes significantly. The general medicine segment focuses on anti-infective treatments, men's health care and cardiovascular, metabolic and thromboembolic diseases. Schering's diagnostics imaging products include a range of contrast media, such as Magnevist, a general MRI contrast agent and Gadovist, a central nervous system MRI contrast agent. The company is a subsidiary of Bayer A.G. In February 2009, the company announced plans to invest in the formation of a research and development facility in Beijing. In July 2009, the company acquired an exclusive supply and distribution agreement with the Polish insulin producer Bioton SA and its subsidiary SciGen to make SciLin insulin.

FINANCIALS: Sales and profits are in thousands of dollars—add 000 to get the full amount. 2009 Note: Financial information for 2009 was not available for all companies at press time.

2009 Sales: $	2009 Profits: $	U.S. Stock Ticker: Subsidiary
2008 Sales: $14,243,900	2008 Profits: $	Int'l Ticker: Int'l Exchange:
2007 Sales: $5,806,500	2007 Profits: $1,980,100	Employees: 25,000
2006 Sales: $8,471,600	2006 Profits: $3,495,700	Fiscal Year Ends: 12/31
2005 Sales: $7,802,800	2005 Profits: $909,900	Parent Company: BAYER AG

SALARIES/BENEFITS:

Pension Plan: Y	ESOP Stock Plan:	Profit Sharing:	Top Exec. Salary: $	Bonus: $
Savings Plan:	Stock Purch. Plan:		Second Exec. Salary: $	Bonus: $

OTHER THOUGHTS:

Apparent Women Officers or Directors: 1
Hot Spot for Advancement for Women/Minorities:

LOCATIONS: ("Y" = Yes)

West:	Southwest:	Midwest:	Southeast:	Northeast:	International:
Y					Y

BIND BIOSCIENCES

www.bindbio.com

Industry Group Code: 541711 Ranks within this company's industry group: Sales: Profits:

Drugs:	Other:	Clinical:	Computers:	Services:
Discovery:	AgriBio:	Trials/Services:	Hardware:	Specialty Services:
Licensing:	Genetic Data:	Labs:	Software:	Consulting:
Manufacturing: Y	Tissue Replacement:	Equipment/Supplies:	Arrays:	Blood Collection:
Generics:		Research/Development Svcs.:	Database Management:	Drug Delivery:
		Diagnostics:		Drug Distribution:

TYPES OF BUSINESS:
Biopharmaceuticals & Nanoparticles

BRANDS/DIVISIONS/AFFILIATES:
Medicinal Nanoengineering Platform
BIND-014

CONTACTS: Note: Officers with more than one job title may be intentionally listed here more than once.
Scott Minick, CEO
Scott Minick, Pres.
Andrea Franz, CFO
Jim Wright, Chief Scientific Officer
Stephen Zale, VP-Prod. Dev.
Ed Schnipper, Chief Medical Officer
Jeff Hrkach, Sr. VP-Pharmaceutical Sciences

Phone: 617-491-3400	Fax: 617-491-0351
Toll-Free:	
Address: 64 Sidney St., Cambridge, MA 02139 US	

GROWTH PLANS/SPECIAL FEATURES:

BIND Biosciences, Inc. is a biopharmaceutical company which uses its proprietary Medicinal Nanoengineering platform to design, engineer and manufacture targeted therapeutics. The company's initial product development efforts are in the areas of oncology, inflammatory disease, cardiovascular disorders and RNAi therapeutics. BIND develops therapeutics that deliver high drug concentrations to target cells and tissues with controlled pharmacodynamic and pharmacokinetic properties, which results in reduced toxicity and increased efficacy. Its Medicinal Nanoengineering platform engineers libraries of drug encapsulated targeted nanoparticles; screens for targeted nanoparticles with biodistribution, drug exposure kinetics and cell- or tissue-specific targeting; and manufactures candidate drug encapsulated targeted nanoparticles using scalable unit operations. BIND-014, the company's lead program, targets a surface protein upregulated in solid tumors and is planned to enter clinical development. The company's drug pipeline is also focused on improving the efficacy and expanding the applicability of existing drugs. The firm's intellectual property portfolio consists of over 50 patent families of issued and pending patent applications covering targeted nanoparticles, controlled release polymer systems, long-circulating nanoparticles and nanoparticles synthesis. BIND's major investors include ARCH Venture Partners; DHK Investment; Endeavour Vision; Flagship Ventures; NanoDimension; and Polaris Venture Partners.

FINANCIALS: Sales and profits are in thousands of dollars—add 000 to get the full amount. 2009 Note: Financial information for 2009 was not available for all companies at press time.

2009 Sales: $	2009 Profits: $	U.S. Stock Ticker: Private
2008 Sales: $	2008 Profits: $	Int'l Ticker: Int'l Exchange:
2007 Sales: $	2007 Profits: $	Employees:
2006 Sales: $	2006 Profits: $	Fiscal Year Ends:
2005 Sales: $	2005 Profits: $	Parent Company:

SALARIES/BENEFITS:

Pension Plan:	ESOP Stock Plan:	Profit Sharing:	Top Exec. Salary: $	Bonus: $
Savings Plan:	Stock Purch. Plan:		Second Exec. Salary: $	Bonus: $

OTHER THOUGHTS:
Apparent Women Officers or Directors: 1
Hot Spot for Advancement for Women/Minorities:

LOCATIONS: ("Y" = Yes)

West:	Southwest:	Midwest:	Southeast:	Northeast: Y	International:

BIO RAD LABORATORIES INC

www.bio-rad.com

Industry Group Code: 33911 Ranks within this company's industry group: Sales: 5   Profits: 4

Drugs:	Other:		Clinical:		Computers:		Services:	
Discovery:	AgriBio:	Y	Trials/Services:		Hardware:		Specialty Services:	
Licensing:	Genetic Data:	Y	Labs:	Y	Software:	Y	Consulting:	
Manufacturing:	Tissue Replacement:		Equipment/Supplies:	Y	Arrays:		Blood Collection:	
Generics:			Research/Development Svcs.:		Database Management:		Drug Delivery:	
			Diagnostics:	Y			Drug Distribution:	

TYPES OF BUSINESS:

Equipment-Life Sciences Research
Clinical Diagnostics Products
Analytical Instruments
Laboratory Devices
Biomaterials
Imaging Products
Assays
Software

BRANDS/DIVISIONS/AFFILIATES:

ProteOn
BioPlex 2200
iScript

CONTACTS: *Note: Officers with more than one job title may be intentionally listed here more than once.*

Norman Schwartz, CEO
Norman Schwartz, Pres.
Christine Tsingos, CFO/VP
Sanford S. Wadler, General Counsel/VP/Sec.
Tina Cuccia, Mgr.-Corp. Comm.
James R. Stark, Corp. Controller
Ronald W. Hutton, Treas.
Brad Crutchfield, VP/Group Mgr.-Life Science
John Goetz, VP/Group Mgr.-Clinical Diagnostics
David Schwartz, Chmn.
Giovanni Magni, VP/Mgr.-Int'l Sales

Phone: 510-724-7000	Fax: 510-741-5815
Toll-Free: 800-424-6723	
Address: 1000 Alfred Nobel Dr., Hercules, CA 94547 US	

GROWTH PLANS/SPECIAL FEATURES:

Bio-Rad Laboratories, Inc., supplies the life science research, health care and analytical chemistry markets with a broad range of products and systems. These are used to separate complex chemical and biological materials and to identify, analyze and purify components. The company operates in two industry segments: life science and clinical diagnostics. The firm's life science division develops products for applications including electrophoresis, image analysis, molecular detection, chromatography, gene transfer, sample preparation and amplification. Products include a range of laboratory instruments, apparatus and consumables used for research in genomics, proteomics and food safety. The Bio-Rad life science division provides its services to universities; medical schools; pharmaceutical manufacturers; industrial research organizations; food testing laboratories; government agencies; and biotechnology researchers. The clinical diagnostics division encompasses an array of technologies incorporated into a variety of tests used to detect, identify and quantify substances in blood or other body fluids and tissues. The test results are used as aids for medical diagnosis, detection, evaluation, monitoring and treatment of diseases and other conditions. This division is known for diabetes monitoring products, quality control systems, blood virus testing, blood typing, toxicology, genetic disorders products, molecular pathology and Internet-based software. Bio-Rad is also an international provider of bovine spongiform encephalopathy (mad cow disease) tests. Bio-Rad's brand name systems include the BioPlex 2200 multiplex testing platform; iScript, reverse transcription reagent kits; and ProteOn, a protein interaction analysis system. In January 2010, the firm acquired the blood transfusion testing, transplantation and infectious diseases diagnostics business operations of Biotest AG.

FINANCIALS: Sales and profits are in thousands of dollars—add 000 to get the full amount. 2009 Note: Financial information for 2009 was not available for all companies at press time.

2009 Sales: $1,784,244	2009 Profits: $144,620	**U.S. Stock Ticker:** BIO
2008 Sales: $1,764,365	2008 Profits: $89,510	**Int'l Ticker:** Int'l Exchange:
2007 Sales: $1,461,052	2007 Profits: $92,994	Employees: 6,600
2006 Sales: $1,273,930	2006 Profits: $103,263	Fiscal Year Ends: 12/31
2005 Sales: $1,180,985	2005 Profits: $81,553	Parent Company:

SALARIES/BENEFITS:

Pension Plan:	ESOP Stock Plan:	Profit Sharing:	Top Exec. Salary: $726,988	Bonus: $788,712
Savings Plan: Y	Stock Purch. Plan:		Second Exec. Salary: $540,065	Bonus: $208,000

OTHER THOUGHTS:

Apparent Women Officers or Directors: 3
Hot Spot for Advancement for Women/Minorities: Y

LOCATIONS: ("Y" = Yes)

West:	Southwest:	Midwest:	Southeast:	Northeast:	International:
Y	Y			Y	Y

BIOANALYTICAL SYSTEMS INC

www.basinc.com

Industry Group Code: 541712 Ranks within this company's industry group: Sales: 11 Profits: 9

Drugs:	Other:	Clinical:		Computers:		Services:	
Discovery:	AgriBio:	Trials/Services:	Y	Hardware:		Specialty Services:	Y
Licensing:	Genetic Data:	Labs:	Y	Software:		Consulting:	Y
Manufacturing:	Tissue Replacement:	Equipment/Supplies:	Y	Arrays:		Blood Collection:	
Generics:		Research/Development Svcs.:	Y	Database Management:		Drug Delivery:	
		Diagnostics:				Drug Distribution:	

TYPES OF BUSINESS:

Contract Research Services
Screening & Testing Services
Bioanalytical Instruments
Immunochemistry
Toxicology Testing
Formulation Development
Testing Equipment

BRANDS/DIVISIONS/AFFILIATES:

Culex
Vetronics
Epsilon
Culex ACB
Culex-L ABC

CONTACTS: Note: Officers with more than one job title may be intentionally listed here more than once.

Anthony S. Chilton, COO
Anthony S. Chilton, Interim Pres.
Michael Cox, CFO
James F. Gitzen, Dir.-Sales
Lina Reeves-Kerner, Sr. VP-Human Resources
John Devine, Gen. Mgr.-Toxicology Research
Ed Burrow, Dir.-Mfg.
Michael Cox, VP-Admin.
Michael Cox, VP-Finance
Craig Bruntlett, Sr. VP-Instruments Div.
Lori Payne, Gen. Mgr.-Northwest Laboratories
Jason Plassard, Gen. Mgr.-West Lafayette Contract Research
John Maltas, Managing Dir.-U.K. & Europe

Phone: 765-463-4527	Fax: 765-497-1102
Toll-Free: 800-845-4246	
Address: 2701 Kent Ave., Purdue Research Park, West Lafayette, IN 47906 US	

GROWTH PLANS/SPECIAL FEATURES:

Bioanalytical Systems, Inc. (BASi) is a contract laboratory research firm and research equipment manufacturer that serves leading pharmaceutical, medical device and biotechnology companies and institutions worldwide. Its principal clients consist of scientists engaged in analytical chemistry, drug safety evaluation, clinical trials, drug metabolism studies, pharmacokinetics and basic neuroscience research at major pharmaceutical organizations. BASi operates two business segments, both of which address bioanalytical, preclinical and clinical research needs of drug developers. The Contract Research Services segment provides product characterization, method development and validation services, including determining the potency, purity, chemical composition, structure and physical properties of a drug compound; bioanalytical testing, mainly targeting drug and metabolite concentrations in complex biological matrices; and stability testing, mainly analyzing product purity, potency and shelf life. It also offers in vivo pharmacology, that is, analyzing drug reactions in living specimens (often animals), as opposed to in vitro (or in the glass) laboratory testing, which takes place outside an organism; and preclinical and pathology services, including pharmacokinetic and safety testing. The Research Products segment focuses on expediting preclinical screening of developmental drugs. This segment designs, develops and manufactures liquid chromatographic and electrochemical instruments, analytical products that utilize the Epsilon electrochemical platform; in vivo sampling products, including the Culex family of automated sampling and dosing instruments, which are used during pharmaceutical research to dose animals and collect biological samples; and physiology monitoring products, consisting of Vetronics, a small animal electro-cardiogram and vital signs monitor mainly used by veterinarians. The firm operates three labs in the U.S. and one in the U.K. In October 2009, BASi launched two new automated blood collection products, Culex ABC and Culex-L ABC.

FINANCIALS: Sales and profits are in thousands of dollars—add 000 to get the full amount. 2009 Note: Financial information for 2009 was not available for all companies at press time.

2009 Sales: $31,784	2009 Profits: $-5,463	U.S. Stock Ticker: BASI
2008 Sales: $41,697	2008 Profits: $-1,489	Int'l Ticker: Int'l Exchange:
2007 Sales: $39,753	2007 Profits: $ 926	Employees: 274
2006 Sales: $43,048	2006 Profits: $-2,670	Fiscal Year Ends: 9/30
2005 Sales: $42,395	2005 Profits: $- 80	Parent Company:

SALARIES/BENEFITS:

Pension Plan:	ESOP Stock Plan:	Profit Sharing:	Top Exec. Salary: $285,000	Bonus: $
Savings Plan: Y	Stock Purch. Plan:		Second Exec. Salary: $195,000	Bonus: $10,000

OTHER THOUGHTS:

Apparent Women Officers or Directors: 4
Hot Spot for Advancement for Women/Minorities: Y

LOCATIONS: ("Y" = Yes)

West:	Southwest:	Midwest:	Southeast:	Northeast:	International:
Y		Y			Y

BIOCOMPATIBLES INTERNATIONAL PLC   www.biocompatibles.com

Industry Group Code: 33911 Ranks within this company's industry group: Sales: 12   Profits: 12

Drugs:		Other:	Clinical:	Computers:		Services:	
Discovery:		AgriBio:	Trials/Services:	Hardware:	Y	Specialty Services:	
Licensing:		Genetic Data:	Labs:	Software:	Y	Consulting:	
Manufacturing:	Y	Tissue Replacement:	Equipment/Supplies:	Y	Arrays:		Blood Collection:	
Generics:			Research/Development Svcs.:	Database Management:		Drug Delivery:	Y
			Diagnostics:			Drug Distribution:	

TYPES OF BUSINESS:

Medical Implant Technology
Supplies-Phosphorycholine Coatings
Embolisation Microspheres
Biomaterials
Cancer Treatment
Drug Delivery Platforms

BRANDS/DIVISIONS/AFFILIATES:

DC Bead
Brachy Sciences
Anchor Seed, Anchor Mark
Biocompatibles Siascopy
CellMed AG
CellBead Neuro
Novabel
Merz Pharmaceuticals GmbH

CONTACTS: *Note: Officers with more than one job title may be intentionally listed here more than once.*

Crispin Simon, CEO
Mike Motion, Dir.-Sales & Mktg., Oncology Prod. Div.
Geoff Tompsett, Dir.-Human Resources
Andy Lewis, Dir.-Research
Geoff Tompsett, Dir.-IT
Ian Ardill, Sec.
Ian Ardill, Dir.-Finance
Tim Maloney, Dir.-Commercial
Alistair Taylor, Dir.-Quality & Regulatory Affairs
Paul Baxter, Dir.-Intellectual Property
John Sylvestor, Managing Dir.-Oncology Prod. Div.
Gerry Brown, Chmn.

Phone: 44-1252-732-732	Fax: 44-1252-732-777
Toll-Free:	
Address: Chapman House, Farnham Business Park, Weydon Ln., Farnham, Surrey GU9 8QL UK	

GROWTH PLANS/SPECIAL FEATURES:

Biocompatibles International plc is an international provider of medical products that combine medical devices with ancillary therapeutic drugs. The company operates through its Oncology Products division and a Licensing division. The Oncology Products division is divided into three groups: Biocompatibles, BrachySciences and Biocompatibles Siascopy. The Biocompatibles division markets drug-eluting beads DC Bead and Precision Bead, which use bead technology to deliver therapeutic agents to tumor location sites. It also offers Bead Block and LC Bead, two bead products that contain no drug, but are used for embolising blood vessels. The BrachySciences division markets brachytherapy products. Employing Anchor Seed and Anchor Marker Technologies radiation-delivering seeds are used to treat prostate cancer. Other brachytherapy products include VraiStrand, which uses synthetic, bioabsorbable polymers to deliver low dose brachytherapy sources to treat prostate cancer. The Siascopy division assists in the diagnosis and management of skin cancer. The Siascope, a handheld medical device, is produced in conjunction with MoleMate and MoleView software for non-invasive skin cancer diagnoses. The Licensing division operates in two separate bodies, CellMed and PC Licensing. CellMed is currently developing several products: CellBead Neuro is used in the treatment of neurological trauma and degeneration, Dermal Beads, cosmetic dermal fillers used for skin augmentation treatments, a drug-eluting bead for stroke treatment and, in agreement with AstraZeneca, CM3, a GLP-1 analog used for the treatment of obesity and diabetes that entered phase I studies in January 2010. PC Licensing's principal relationship is with Medtronic, Inc. in cardiovascular medicine. In March 2009, approval was granted to launch Novabel, a cosmetic dermal filler bead manufactured by its partner Merz Pharmaceuticals GmbH. In June 2010, however, shipments were suspended due to reports of adverse reactions in several patients. In July 2010, Biocompatibles acquired the patents for DC Bead, LC Bead and Bead Block.

FINANCIALS: Sales and profits are in thousands of dollars—add 000 to get the full amount. 2009 Note: Financial information for 2009 was not available for all companies at press time.

2009 Sales: $42,400	2009 Profits: $-9,240	**U.S. Stock Ticker:**
2008 Sales: $29,050	2008 Profits: $- 750	**Int'l Ticker: BII** Int'l Exchange: London-LSE
2007 Sales: $18,112	2007 Profits: $-3,890	Employees: 164
2006 Sales: $11,864	2006 Profits: $9,990	Fiscal Year Ends: 12/31
2005 Sales: $6,880	2005 Profits: $-9,490	Parent Company:

SALARIES/BENEFITS:

Pension Plan:	ESOP Stock Plan:	Profit Sharing:	Top Exec. Salary: $441,517	Bonus: $353,819
Savings Plan:	Stock Purch. Plan:		Second Exec. Salary: $281,222	Bonus: $169,338

OTHER THOUGHTS:

Apparent Women Officers or Directors: 1
Hot Spot for Advancement for Women/Minorities:

LOCATIONS: ("Y" = Yes)

West:	Southwest:	Midwest:	Southeast:	Northeast:	International:
					Y

BIOCRYST PHARMACEUTICALS INC

www.biocryst.com

Industry Group Code: 325412 **Ranks within this company's industry group:** Sales: 80   Profits: 105

Drugs:		Other:	Clinical:	Computers:	Services:
Discovery:	Y	AgriBio:	Trials/Services:	Hardware:	Specialty Services:
Licensing:	Y	Genetic Data:	Labs:	Software:	Consulting:
Manufacturing:		Tissue Replacement:	Equipment/Supplies:	Arrays:	Blood Collection:
Generics:			Research/Development Svcs.:	Database Management:	Drug Delivery:
			Diagnostics:		Drug Distribution:

TYPES OF BUSINESS:

Small-Molecule Pharmaceutical Products
Drugs-Immunological, Infectious & Inflammatory Disease
Drugs-Cancer
Research & Development

BRANDS/DIVISIONS/AFFILIATES:

Fodosine
BCX-4208
Peramivir
Mundipharma
Shionogi & Co., Ltd.
Green Cross Corp.
RAPIACTA

CONTACTS: Note: Officers with more than one job title may be intentionally listed here more than once.

Jon P. Stonehouse, CEO
Jon P. Stonehouse, Pres.
Stuart Grant, CFO/Sr. VP
Robert Stoner, VP-Human Resources
William P. Sheridan, Chief Medical Officer
Alane Barnes, General Counsel/Sec.
Peter L. McCullough, VP-Oper.
David S. McCullough, Sr. VP-Strategic Planning & Corp. Dev.
Robert Bennett, Exec. VP-Investor Rel. & Bus. Dev.
J. Micheal Mills, Controller/ Principal Acct. Officer
Yarlagadda S. Babu, VP-Drug Discovery
Elliott Berger, Sr. VP-Regulatory Affairs
Walter G. Gowan, VP-Pharmaceutical Dev.
David S. McCullough, Sr. VP-Commercialization
Zola P. Horovitz, Chmn.

Phone: 205-444-4600	Fax: 205-444-4640
Toll-Free:	
Address: 2190 Parkway Lake Dr., Birmingham, AL 35244 US	

GROWTH PLANS/SPECIAL FEATURES:

BioCryst Pharmaceuticals, Inc. is a biotechnology company that designs and develops drugs to block key enzymes involved in cancer, viral infections and autoimmune disease. BioCryst has three main drugs in development: forodesine, peramivir and BCX-4208. Forodesine is in Phase II/III trials for the treatment of cutaneous T-cell lymphoma (CTCL); Phase II trials for chronic lymphocytic leukemia (CLL); and Phase I trials for the treatment of acute lymphoblastic leukemia (ALL). The FDA has granted orphan drug status for Forodesine in three indications: T-cell non-Hodgkin lymphoma, including CTCL; CLL and related leukemias, including T-cell prolymphocytic leukemia, adult T-cell leukemia and hairy cell leukemia; and B-cell ALL. Peramivir, a neuraminidase inhibitor, has been approved by the FDA for the intravenous treatment of outpatient seasonal influenza; the company is continuing Phase II/III trials for the treatment of hospitalized acute influenza. BioCryst is developing peramivir for possible global influenza pandemics, including avian and swine flus, based on a four-year $102.6 million award from the U.S. Department of Health and Human Services. BCX-4208 is a second-generation PNP in Phase II development for the treatment of gout. In addition, the firm also owns the rights to inhibitors of parainfluenza, neuraminidase and hepatitis C, which it continues to evaluate in preclinical trials. The firm has corporate alliances with Mundipharma; Green Cross; and Shionogi & Co., Ltd. In October 2009, the FDA issued emergency use authorization (EUA) for peramivir for the treatment of 2009 H1N1 influenza in hospitals. In January 2010, Shionogi & Co. launched peramivir in Japan as RAPIACTA. Also in January 2010, peramivir was approved in Mexico for the treatment of 2009 H1N1 influenza A.

BioCryst offers employees medical, dental, disability and life insurance; a 401(k) retirement plan; and equity participation through a stock option plan and an employee stock purchase plan.

FINANCIALS: Sales and profits are in thousands of dollars—add 000 to get the full amount. 2009 Note: Financial information for 2009 was not available for all companies at press time.

2009 Sales: $74,589	2009 Profits: $-13,452	**U.S. Stock Ticker:** BCRX
2008 Sales: $56,561	2008 Profits: $-24,732	**Int'l Ticker:** Int'l Exchange:
2007 Sales: $71,238	2007 Profits: $-29,056	Employees: 79
2006 Sales: $6,212	2006 Profits: $-43,618	Fiscal Year Ends: 12/31
2005 Sales: $ 152	2005 Profits: $-26,099	Parent Company:

SALARIES/BENEFITS:

Pension Plan:	ESOP Stock Plan:	Profit Sharing:	Top Exec. Salary: $444,500	Bonus: $208,506
Savings Plan: Y	Stock Purch. Plan: Y		Second Exec. Salary: $403,958	Bonus: $122,250

OTHER THOUGHTS:

Apparent Women Officers or Directors: 2
Hot Spot for Advancement for Women/Minorities:

LOCATIONS: ("Y" = Yes)

West:	Southwest:	Midwest:	Southeast:	Northeast:	International:
			Y	Y	

BIOGEN IDEC INC

www.biogenidec.com

Industry Group Code: 325412 **Ranks within this company's industry group:** Sales: 24   Profits: 19

Drugs:		Other:		Clinical:		Computers:		Services:	
Discovery:	Y	AgriBio:		Trials/Services:		Hardware:		Specialty Services:	
Licensing:	Y	Genetic Data:		Labs:	Y	Software:		Consulting:	
Manufacturing:	Y	Tissue Replacement:		Equipment/Supplies:		Arrays:		Blood Collection:	
Generics:				Research/Development Svcs.:		Database Management:		Drug Delivery:	
				Diagnostics:				Drug Distribution:	

TYPES OF BUSINESS:

Drugs-Immunology, Neurology & Oncology
Autoimmune & Inflammatory Disease Treatments
Drugs-Multiple Sclerosis
Drugs-Cancer

BRANDS/DIVISIONS/AFFILIATES:

AVONEX
TYSABRI
RITUXAN
FUMADERM
Fampridine-SR

CONTACTS: Note: Officers with more than one job title may be intentionally listed here more than once.

James C. Mullen, CEO
Robert A. Hamm, COO
James C. Mullen, Pres.
Paul J. Clancy, CFO/Exec. VP
Craig Eric Schneier, Exec. VP-Human Resources
Cecil Pickett, Pres., R&D
Susan H. Alexander, General Counsel/Exec. VP/Corp. Sec.
Tony Kingsley, Sr. VP-U.S. Commercial Oper.
Michael Lytton, Exec. VP-Corp. & Bus. Dev.
Craig Eric Schneier, Exec. VP-Corp. Comm. & Public Affairs
Michael F. MacLean, Chief Acct. Officer/Controller/Sr. VP
John M. Dunn, Exec. VP-New Ventures
Frederick Munschauer, VP-U.S. Medical Affairs
William D. Young, Chmn.
Francesco Granata, Exec. VP-Global Commercial Oper.

Phone: 617-679-2000	Fax: 617-679-2617
Toll-Free:	
Address: 14 Cambridge Ctr., Cambridge, MA 02142 US	

GROWTH PLANS/SPECIAL FEATURES:

Biogen IDEC, Inc. is a biotechnology company that develops, manufactures and markets therapeutic pharmaceuticals in the fields of immunology, neurology and oncology. Biogen currently has four approved products: AVONEX, which is designed to treat relapsing forms of multiple sclerosis (MS) and is used by more than 135,000 patients in 70 countries globally; TYSABRI, which is approved for the treatment of relapsing forms of MS and in the U.S. is approved to treat moderate to severe active Crohn's disease; RITUXAN, which is globally approved for the treatment of relapsed or refractory low-grade or follicular, CD20-positive, B-cell non-Hodgkin's lymphomas (NHLs) and has had approximately 2.1 million patient exposures worldwide; and FUMADERM, which acts as an immunomodulator and is approved in Germany for the treatment of severe psoriasis. RITUXAN, in combination with methotrexate, is also approved to reduce signs and symptoms in adult patients with moderately-to-severely active rheumatoid arthritis who have had an inadequate response to one or more tumor necrosis factor antagonist therapies. The company is working with Genentech and Roche on the development of RITUXAN in additional oncology and other indications. In addition to its approved drugs, the firm currently has 29 drugs under development, including drugs in the company's core areas of focus as well as in other therapeutic areas such as cardiovascular disease and hemophilia. Biogen also generates revenue by licensing drugs it has developed to other companies, including Schering-Plough, Merck and GlaxoSmithKline. In July 2009, Biogen IDEC and Acorda Therapeutics, Inc. agreed to jointly develop and commercialize MS therapy Fampridine-SR in markets outside of the U.S. Biogen offers employees medical, dental and vision insurance; tuition reimbursement; commuter benefits; discounts on health clubs and local events; flexible spending accounts; and an employee assistance program.

Biogen offers employees medical, dental and vision insurance; tuition reimbursement; commuter benefits; discounts on health clubs and local events; flexible spending accounts; and an employee assistance program.

FINANCIALS: Sales and profits are in thousands of dollars—add 000 to get the full amount. 2009 Note: Financial information for 2009 was not available for all companies at press time.

2009 Sales: $4,377,300	2009 Profits: $977,100	U.S. Stock Ticker: BIIB
2008 Sales: $4,097,500	2008 Profits: $783,200	Int'l Ticker: Int'l Exchange:
2007 Sales: $3,171,600	2007 Profits: $638,172	Employees: 4,750
2006 Sales: $2,683,049	2006 Profits: $217,511	Fiscal Year Ends: 12/31
2005 Sales: $2,422,500	2005 Profits: $160,711	Parent Company:

SALARIES/BENEFITS:

Pension Plan:	ESOP Stock Plan:	Profit Sharing:	Top Exec. Salary: $1,192,308	Bonus: $2,400,000
Savings Plan: Y	Stock Purch. Plan: Y		Second Exec. Salary: $816,923	Bonus: $879,450

OTHER THOUGHTS:

Apparent Women Officers or Directors: 4
Hot Spot for Advancement for Women/Minorities: Y

LOCATIONS: ("Y" = Yes)

West:	Southwest:	Midwest:	Southeast:	Northeast:	International:
Y				Y	Y

Note: Financial information, benefits and other data can change quickly and may vary from those stated here.

BIOHEART INC

www.bioheartinc.com

Industry Group Code: 325414 **Ranks within this company's industry group:** Sales: 9 Profits: 6

Drugs:	Other:	Clinical:	Computers:	Services:	
Discovery:	AgriBio:	Trials/Services:	Hardware:	Specialty Services:	Y
Licensing:	Genetic Data:	Labs:	Software:	Consulting:	
Manufacturing:	Tissue Replacement: Y	Equipment/Supplies:	Arrays:	Blood Collection:	
Generics:		Research/Development Svcs.:	Database Management:	Drug Delivery:	
		Diagnostics:		Drug Distribution:	

TYPES OF BUSINESS:

Research & Development-Heart-Related Therapies
Distribution-Medical Devices

BRANDS/DIVISIONS/AFFILIATES:

MyoCell
MyoCell SDF-1
Bioheart Acute Cell Therapy
TGI 1200

CONTACTS: Note: Officers with more than one job title may be intentionally listed here more than once.

Mike Tomas, CEO
Catherine Sulawske-Guck, COO
Mike Tomas, Pres.
Alain Hernandez, VP-Mktg. & Sales
Howard J. Leonhardt, Chief Science Officer
Howard J. Leonhardt, CTO
Linda Palermo, Mgr.-Oper.
Kristin Comella, VP-Corp. Dev.
Angel Rodriguez, Controller
Kristin Comella, VP-Research
Doug Owens, Dir.-Clinical Affairs
Karl E. Groth, Chmn.

Phone: 954-835-1500	Fax: 954-845-9976
Toll-Free:	
Address: 13794 NW 4th St., Ste. 212, Sunrise, FL 33325 US	

GROWTH PLANS/SPECIAL FEATURES:

Bioheart, Inc. discovers, develops and commercializes autologous cell therapies (that is, those derived from the patient themselves), intelligent devices and biologics for the treatment of congestive heart failure, lower limb ischemia, chronic heart ischemia, acute myocardial infarctions and other conditions. Its primary product, MyoCell, is a muscle-derived cell therapy designed to repopulate scar tissue in the patient's heart to improve cardiac function in patients who have experienced heart failure. MyoCell has completed a 40-patient Phase IIa test in Europe and a 20-patient Phase I test in the U.S. The U.S. Food and Drug Administration (FDA) has approved a 330-patient, Phase II/III multicenter test in North America and Europe (the MARVEL Trial). The MARVEL trial has treated 20 patients to date, including six control patients. Bioheart is seeking an amendment with the FDA to consider the MARVEL trial a pivotal trial (from Phase II to Phase III) and to reduce the number of trial patients to 150. MyoCell SDF-1, a related product candidate, utilizes autologous cells that have been genetically modified to increase the amount of SDF-1 protein produced, which the firm maintains will further improve cardiac function. Bioheart is developing Bioheart Acute Cell Therapy, a product that offers an autologous adipose (fat) cell treatment for heart damage. It is designed to be used with Tissue Genesis, Inc.'s TGI 1200 tissue processing system. In January 2010, Bioheart signed a distribution agreement with Alamo Scientific, Inc. to sell the company's home heart failure monitoring system in Texas, Louisiana and Arkansas. In February 2010, the firm commenced the REGEN Phase I clinical trial for MyoCell SDF-1 for the treatment of congestive heart failure (CHF). In May 2010, the company announced plans to establish two centers in the Middle East through which to provide cell therapy procedures to CHF and peripheral arterial disease patients.

FINANCIALS: Sales and profits are in thousands of dollars—add 000 to get the full amount. 2009 Note: Financial information for 2009 was not available for all companies at press time.

2009 Sales: $ 400	2009 Profits: $-3,800	**U.S. Stock Ticker:** BHRT
2008 Sales: $ 200	2008 Profits: $-14,200	**Int'l Ticker:** Int'l Exchange:
2007 Sales: $ 300	2007 Profits: $-18,100	**Employees:**
2006 Sales: $	2006 Profits: $	**Fiscal Year Ends:** 12/31
2005 Sales: $	2005 Profits: $	**Parent Company:**

SALARIES/BENEFITS:

Pension Plan:	ESOP Stock Plan:	Profit Sharing:	Top Exec. Salary: $164,223	Bonus: $162,022
Savings Plan:	Stock Purch. Plan:		Second Exec. Salary: $150,000	Bonus: $

OTHER THOUGHTS:

Apparent Women Officers or Directors: 4
Hot Spot for Advancement for Women/Minorities: Y

LOCATIONS: ("Y" = Yes)

West:	Southwest:	Midwest:	Southeast:	Northeast:	International:
	Y				

BIOMARIN PHARMACEUTICAL INC

www.biomarinpharm.com

Industry Group Code: 325412   Ranks within this company's industry group: Sales: 55   Profits: 73

Drugs:		Other:		Clinical:	Computers:		Services:	
Discovery:	Y	AgriBio:	Y	Trials/Services:	Hardware:		Specialty Services:	
Licensing:	Y	Genetic Data:		Labs:	Software:		Consulting:	
Manufacturing:	Y	Tissue Replacement:		Equipment/Supplies:	Arrays:		Blood Collection:	
Generics:				Research/Development Svcs.:	Database Management:		Drug Delivery:	Y
				Diagnostics:			Drug Distribution:	

TYPES OF BUSINESS:

Biopharmaceutical Product Development
Drugs-Severe Conditions
Pediatric Disease Treatments
Asthma Treatments
Drug Delivery Technologies

BRANDS/DIVISIONS/AFFILIATES:

Naglazyme
Kuvan
Aldurazyme
Genzyme Corporation
PEG-PAL
BH4

CONTACTS: Note: Officers with more than one job title may be intentionally listed here more than once.

Jean-Jacques Bienaime, CEO
Jeffrey H. Cooper, CFO/Sr. VP
Lewis Chapman, VP-Global Mktg.
Mark Wood, VP-Human Resources
Henry J. Fuchs, Chief Medical Officer/Sr. VP
Ed Von Pervieux, CIO/VP
Robert A. Baffi, Sr. VP-Tech. Oper.
Daniel P. Maher, VP-Prod. Dev.
R. Andrew Ramelmeier, VP-Mfg. & Process Dev.
G. Eric Davis, General Counsel/VP/Corp. Sec.
Joshua A. Grass, VP-Bus. Dev./VP-Corp. Dev.
Brian Mueller, Controller/VP
Charles A. O'Neill, VP-Pharmacological Sciences
Dan Oppenheimer, VP-Portfolio Strategy
Luisa Bigornia, VP-Intellectual Property
Stephen Aselage, Chief Bus. Oper./Sr. VP
Jean-Jacques Bienaime, Chmn.
Eduardo O. Thompson, VP/Gen. Mgr.-Latin America
Steven Jungles, VP-Supply Chain

Phone: 415-506-6700	Fax: 415-382-7889
Toll-Free:	
Address: 105 Digital Dr., Novato, CA 94949 US	

GROWTH PLANS/SPECIAL FEATURES:

BioMarin Pharmaceutical, Inc. develops and commercializes biopharmaceutical products for serious diseases and medical conditions. BioMarin has three approved products: Naglazyme, for the treatment of mucopolysaccharidosis VI (MPS-VI); Kuvan, for which the firm was recently granted marketing approval in the U.S. for the treatment of phenylketonuria (PKU); and Aldurazyme, for the treatment of mucopolysaccharidosis I (MPS-I). MPS-VI is a debilitating life-threatening genetic disease for which no other drug treatment currently exists. Naglazyme has been granted orphan drug status in the U.S. and E.U., which confers market exclusivity for the treatment of MPS VI expiring in 2012 and 2016, respectively. BioMarin has been granted orphan drug status in the U.S. for Kuvan, the first drug treatment for PKU, an inherited metabolic disease that affects at least 50,000 diagnosed patients globally under the age of 40. Aldurazyme, developed through a 50/50 joint-venture with Genzyme Corporation, has been approved for marketing in the U.S., European Union (E.U.) and other countries for patients with MPS I. BioMarin's PEG-PAL (formerly referred to as Phenylase) is an investigational enzyme substitution therapy being developed as a subcutaneous injection for those who do not respond to Kuvan. The firm is developing BH4 for the treatment of indications associated with endothelial dysfunction. BioMarin initiated Phase II clinical trials of 6R-BH4 for peripheral arterial disease and sickle cell disease as well as preclinical studies for Duchenne Muscular Dystrophy and IgA Nephropathy. In February 2010, the firm agreed to acquire LEAD Therapeutics, Inc., a drug discovery and early stage development company.

BioMarin offers its employees an education assistance program; a flexible spending plan; life, medical, dental and vision insurance; an employee assistance program; AD&D coverage; a 401(k) plan; short and long-term disability; and bi-weekly chair massages.

FINANCIALS: Sales and profits are in thousands of dollars—add 000 to get the full amount. 2009 Note: Financial information for 2009 was not available for all companies at press time.

2009 Sales: $324,656	2009 Profits: $- 488	**U.S. Stock Ticker:** BMRN	
2008 Sales: $296,493	2008 Profits: $30,831	**Int'l Ticker:** Int'l Exchange:	
2007 Sales: $121,581	2007 Profits: $-15,803	Employees: 649	
2006 Sales: $84,209	2006 Profits: $-28,533	Fiscal Year Ends: 12/31	
2005 Sales: $25,669	2005 Profits: $-74,270	Parent Company:	

SALARIES/BENEFITS:

Pension Plan:	ESOP Stock Plan:	Profit Sharing:	Top Exec. Salary: $739,482	Bonus: $794,962
Savings Plan: Y	Stock Purch. Plan: Y		Second Exec. Salary: $361,525	Bonus: $155,459

OTHER THOUGHTS:

Apparent Women Officers or Directors: 3
Hot Spot for Advancement for Women/Minorities: Y

LOCATIONS: ("Y" = Yes)

West:	Southwest:	Midwest:	Southeast:	Northeast:	International:
Y					Y

Note: Financial information, benefits and other data can change quickly and may vary from those stated here.

BIOMET INC

www.biomet.com

Industry Group Code: 33911 **Ranks within this company's industry group:** Sales: 4 Profits: 15

Drugs:	Other:	Clinical:	Computers:	Services:
Discovery:	AgriBio:	Trials/Services:	Hardware:	Specialty Services:
Licensing:	Genetic Data:	Labs:	Software:	Consulting:
Manufacturing:	Tissue Replacement: Y	Equipment/Supplies: Y	Arrays:	Blood Collection:
Generics:		Research/Development Svcs.:	Database Management:	Drug Delivery:
		Diagnostics:		Drug Distribution:

TYPES OF BUSINESS:

Orthopedic Supplies
Electrical Bone Growth Stimulators
Orthopedic Support Devices
Operating Room Supplies
Powered Surgical Instruments
Dental Implants
Imaging Equipment
Human Bone Joint Replacement Systems

BRANDS/DIVISIONS/AFFILIATES:

Biomet 3i
Cartilix
Cytosol Laboratories, Inc.
Blackstone Group LP (The)

CONTACTS: *Note: Officers with more than one job title may be intentionally listed here more than once.*

Jeffrey R. Binder, CEO
Jeffrey R. Binder, Pres.
Daniel P. Florin, CFO/Sr. VP
Peggy Taylor, Sr. VP-Human Resources
Bradley J. Tandy, General Counsel/Sr. VP/Corp. Sec.
Robin T. Barney, Sr. VP-Worldwide Oper.
Glen A. Kashuba, Sr. VP/Pres., Biomet Trauma & Biomet Spine
Gregory W. Sasso, Sr. VP/Pres., Biomet SBU Oper.
Jon C. Serbousek, Sr. VP/Pres., Biomet Orthopedics LLC
Margaret L. Anderson, Sr. VP/Pres., Biomet 3i
Jeffrey R. Binder, Chmn.
Renaat Vermeulen, Sr. VP/Pres., Biomet Europe

Phone: 574-267-6639	Fax: 574-267-8137
Toll-Free:	
Address: 56 E. Bell Dr., P.O. Box 587, Warsaw, IN 46581-0587 US	

GROWTH PLANS/SPECIAL FEATURES:

Biomet, Inc. is an orthopedic medical device company that designs, manufactures and markets surgical and non-surgical products used by orthopedic surgeons and other musculoskeletal medical specialists. Its products are distributed in more than 90 countries. The company's portfolio encompasses reconstructive products, fixation devices, spinal products and other products. Reconstructive products include knee, hip and shoulder joint replacement systems, as well as dental reconstructive implants, bone cements and cement delivery systems. Subsidiary Biomet 3i is manufactures dental implants, abutments and related products. Fixation devices include electrical stimulation systems used in trauma indications; internal fixation devices, such as nails, plates, screws, pins and wires; external fixation devices; craniomaxillofacial fixation systems; and bone substitution materials. Spinal products include electrical stimulation devices for spinal applications; spinal fixation systems for cervical, thoracolumbar, deformity correction and spacer applications; and bone substitute materials, as well as allograft services for spinal applications. The other product segment includes sports medicine products, orthopedic support products, operating room supplies, casting materials, general surgical instruments and wound care products. Biomet is owned by a private-equity group that includes affiliates of the Blackstone Group, Goldman, Sachs & Co., Kohlberg Kravis Roberts & Co. and Texas Pacific Group, who together paid roughly $11.4 billion for the firm. In October 2009, the company acquired the assets of Cartilix, a developer of cartilage repair technology. In July 2010, the firm acquired the assets of Cytosol Laboratories, Inc., a producer of small volume anti-coagulants for blood component processing.

FINANCIALS: Sales and profits are in thousands of dollars—add 000 to get the full amount. 2009 Note: Financial information for 2009 was not available for all companies at press time.

2009 Sales: $2,504,100	2009 Profits: $-749,200	U.S. Stock Ticker: Private
2008 Sales: $2,383,300	2008 Profits: $-1,018,800	Int'l Ticker: Int'l Exchange:
2007 Sales: $2,107,428	2007 Profits: $335,892	Employees: 7,107
2006 Sales: $2,025,739	2006 Profits: $405,908	Fiscal Year Ends: 5/31
2005 Sales: $1,879,950	2005 Profits: $349,373	Parent Company: BLACKSTONE GROUP LP (THE)

SALARIES/BENEFITS:

Pension Plan:	ESOP Stock Plan:	Profit Sharing:	Top Exec. Salary: $682,500	Bonus: $636,090
Savings Plan:	Stock Purch. Plan:		Second Exec. Salary: $443,378	Bonus: $390,485

OTHER THOUGHTS:

Apparent Women Officers or Directors: 2
Hot Spot for Advancement for Women/Minorities:

LOCATIONS: ("Y" = Yes)

West:	Southwest:	Midwest:	Southeast:	Northeast:	International:
Y		Y	Y	Y	Y

BIONANOMATRIX

www.bionanomatrix.com

Industry Group Code: 541711 Ranks within this company's industry group: Sales: Profits:

Drugs:	Other:		Clinical:		Computers:	Services:
Discovery:	AgriBio:		Trials/Services:		Hardware:	Specialty Services:
Licensing:	Genetic Data:	Y	Labs:		Software:	Consulting:
Manufacturing:	Tissue Replacement:		Equipment/Supplies:	Y	Arrays:	Blood Collection:
Generics:			Research/Development Svcs.:		Database Management:	Drug Delivery:
			Diagnostics:			Drug Distribution:

TYPES OF BUSINESS:

Genome Imaging & Analysis

BRANDS/DIVISIONS/AFFILIATES:

National Institutes of Health
Battelle Ventures
KT Venture Group
Ben Franklin Technology Partners
21Ventures
Complete Genomics, Inc.
National Human Genome Research Institute

CONTACTS: *Note: Officers with more than one job title may be intentionally listed here more than once.*

Edward L. Erickson, CEO
Edward L. Erickson, Pres.
Lorraine LoPresti, CFO
Han Cao, Chief Scientific Officer
Michael Kochersperger, VP-Eng.
Lorraine LoPresti, VP-Admin.
Gary Zweiger, VP-Bus. Dev.
Lorraine LoPresti, VP-Finance
Tracy Warren, Chmn.

Phone: 267-499-2014	Fax: 267-499-2015
Toll-Free:	
Address: 3701 Market St., 4th Fl., Philadelphia, PA 19104 US	

GROWTH PLANS/SPECIAL FEATURES:

BioNanomatrix develops nanoscale whole genome imaging and analytic platforms for applications in biomedical research, genetic diagnostics and personalized medicine. The company's technology is designed to dramatically reduce the time and cost required to analyze genomic DNA. Its key technology, the NanoAnalyzer, is an integrated system that works by detecting, identifying and analyzing long strands of DNA in sequence and in a massively parallel format, which allows for the survey of the DNA of an entire genome without amplification, as opposed to current methods of analyzing short fragments of DNA. Nanochannel arrays are incorporated into a nanochip fluidics device, designed to direct long strands of DNA through the chip into the nanochannels in a linear fashion, making them easy to assess for specific information. The nanochip device's design makes it possible to conduct millions of these analyses simultaneously. BioNanomatrix's development programs are supported in part by grants from the National Institutes of Health (NIH) and an $8.8 million government award to develop, in conjunction with Complete Genomics, Inc., a platform capable of sequencing the entire human genome at a cost of $100. The company's other major investors include Battelle Ventures, KT Venture Group, Ben Franklin Technology Partners and 21Ventures. Recently, the company received a Phase II grant from NHGRI for the continued development of its whole genome imaging and analysis platform. In March 2010, the firm received a U.S. patent for its nanochannel arrays, a key element in its NanoAnalyzer system.

FINANCIALS: Sales and profits are in thousands of dollars—add 000 to get the full amount. 2009 Note: Financial information for 2009 was not available for all companies at press time.

2009 Sales: $	2009 Profits: $	U.S. Stock Ticker: Private
2008 Sales: $	2008 Profits: $	Int'l Ticker:　Int'l Exchange:
2007 Sales: $	2007 Profits: $	Employees:
2006 Sales: $	2006 Profits: $	Fiscal Year Ends:
2005 Sales: $	2005 Profits: $	Parent Company:

SALARIES/BENEFITS:

Pension Plan:	ESOP Stock Plan:	Profit Sharing:	Top Exec. Salary: $	Bonus: $
Savings Plan:	Stock Purch. Plan:		Second Exec. Salary: $	Bonus: $

OTHER THOUGHTS:

Apparent Women Officers or Directors: 2
Hot Spot for Advancement for Women/Minorities: Y

LOCATIONS: ("Y" = Yes)

West:	Southwest:	Midwest:	Southeast:	Northeast:	International:
				Y	

BIORELIANCE CORP

www.bioreliance.com

Industry Group Code: 6215 Ranks within this company's industry group: Sales: Profits:

Drugs:		Other:	Clinical:		Computers:	Services:
Discovery:		AgriBio:	Trials/Services:		Hardware:	Specialty Services:
Licensing:	Y	Genetic Data:	Labs:	Y	Software:	Consulting:
Manufacturing:		Tissue Replacement:	Equipment/Supplies:		Arrays:	Blood Collection:
Generics:			Research/Development Svcs.:	Y	Database Management:	Drug Delivery:
			Diagnostics:			Drug Distribution:

TYPES OF BUSINESS:

Research-Nonclinical Product Testing
Contract Biologics Manufacturing
Biologics Pharmaceutical Services

BRANDS/DIVISIONS/AFFILIATES:

Microbiological Associates
Avista Capital Partners
Invitrogen Corporation
Analytical Services
Laboratory Animal Diagnostic Services (LADS)
Clinical Trial Support Services
CynerGene LLC

CONTACTS: Note: Officers with more than one job title may be intentionally listed here more than once.

David A. Dodd, CEO
David A. Dodd, Pres.
David S. Walker, CFO
David E. Onions, Chief Scientific Officer
David L. Bellitt, VP-Global Comm. Oper., Biologics
Darryl L. Goss, VP-Global Process Excellence
James J. Kramer, VP-Oper., U.S. Biologics
David A. Dodd, Chmn.

Phone: 301-738-1000	Fax: 301-610-2590
Toll-Free: 800-553-5372	
Address: 14920 Broschart Rd., Rockville, MD 20850 US	

GROWTH PLANS/SPECIAL FEATURES:

BioReliance Corp., founded in 1947 as Microbiological Associates, is a global contract services organization providing biological safety testing, toxicology, viral manufacturing and laboratory animal diagnostic services to the pharmaceutical and biopharmaceutical industries. BioReliance provides services to over 600 clients annually, including most of the largest pharmaceutical and biopharmaceutical companies in the world. The company also supports early stage companies lacking the staff, expertise and financial resources to conduct many aspects of product development in-house. BioReliance provides its services throughout the product cycle, starting from early preclinical development through licensed production. Services include biologics safety testing, viral clearance studies, manufacturing, toxicology testing, Analytical Services, Laboratory Animal Diagnostic Services (LADS) and Clinical Trial Support Services. The firm provides GLP compliant genetic, in vivo and molecular toxicology testing on pharmaceutical, biopharmaceutical, medical device and pesticide products. BioReliance's Analytical Services department physiochemically characterizes biotechnology products to determine whether their molecular structures meet predefined criteria and whether their macro-molecules remain safe, stable and efficacious in the final product formulations. The company's LADS department offers a spectrum of animal diagnostic and analytical testing programs, including rodent and simian diagnostics, which are used by laboratories, breeding colonies and animal facilities worldwide for both routine monitoring and emergency situations. BioReliance's Clinical Trial Support Services department has significant expertise in manufacturing and testing viral vectors and vaccines, which may require tests beyond the standard clinical chemistry and hematology required for other clinical trials, including an analysis of patient samples for shed virus, expression of the delivered gene, presence of anti-virus neutralizing antibodies and cytokines. Recently, the company was acquired by private equity firm Avista Capital Partners from Invitrogen Corporation. In April 2010, the company acquired the phramacogenomic testing firm CynerGene LLC.

The company offers its employees medical insurance and a 401(k) among its other benefits.

FINANCIALS: Sales and profits are in thousands of dollars—add 000 to get the full amount. 2009 Note: Financial information for 2009 was not available for all companies at press time.

2009 Sales: $	2009 Profits: $	U.S. Stock Ticker: Private
2008 Sales: $	2008 Profits: $	Int'l Ticker: Int'l Exchange:
2007 Sales: $	2007 Profits: $	Employees:
2006 Sales: $	2006 Profits: $	Fiscal Year Ends: 12/31
2005 Sales: $	2005 Profits: $	Parent Company: AVISTA CAPITAL PARTNERS

SALARIES/BENEFITS:

Pension Plan:	ESOP Stock Plan:	Profit Sharing:	Top Exec. Salary: $	Bonus: $
Savings Plan:	Stock Purch. Plan:		Second Exec. Salary: $	Bonus: $

OTHER THOUGHTS:

Apparent Women Officers or Directors: 1
Hot Spot for Advancement for Women/Minorities:

LOCATIONS: ("Y" = Yes)

West:	Southwest:	Midwest:	Southeast:	Northeast:	International:
				Y	Y

Note: Financial information, benefits and other data can change quickly and may vary from those stated here.

BIOSITE INC

www.biosite.com

Industry Group Code: 325413 Ranks within this company's industry group: Sales: Profits:

Drugs:	Other:	Clinical:		Computers:	Services:	
Discovery:	AgriBio:	Trials/Services:		Hardware:	Specialty Services:	Y
Licensing:	Genetic Data:	Labs:		Software:	Consulting:	Y
Manufacturing:	Tissue Replacement:	Equipment/Supplies:	Y	Arrays:	Blood Collection:	
Generics:		Research/Development Svcs.:	Y	Database Management:	Drug Delivery:	
		Diagnostics:	Y		Drug Distribution:	

TYPES OF BUSINESS:

Medical Diagnostics Products
Rapid Immunoassays
Antibody Development Services
Drug Testing

BRANDS/DIVISIONS/AFFILIATES:

Biosite Discovery
Triage Drugs of Abuse Panel
Triage Cardiac Panel
Triage TOX Drug Screen
Triage BNP Test
Triage Profiler Panels
Triage Parasite Panel
Biosite Encompass

CONTACTS: *Note: Officers with more than one job title may be intentionally listed here more than once.*

Doug Guarino, Contact-Media
David Scott, Chief Scientific Officer-Inverness
Ron Zwanziger, Chmn./CEO-Inverness Medical Innovations

Phone: 858-805-8378	Fax: 858-455-4815
Toll-Free: 888-246-7483	
Address: 9975 Summers Ridge Rd., San Diego, CA 92121 US	

GROWTH PLANS/SPECIAL FEATURES:

Biosite, Inc. is a global diagnostics company dedicated to utilizing biotechnology in the development of diagnostic products. The firm, a wholly-owned subsidiary of Inverness Medical Innovations, validates and patents novel protein biomarkers and panels of biomarkers; develops and markets products; conducts strategic research on its products; and educates healthcare providers about its products. Biosite markets immunoassay diagnostics in the areas of cardiovascular disease, drug overdose and infectious disease. Cardiovascular products include the Triage BNP Test, Triage Cardiac Panel, Triage Profiler Panels and Triage Stroke Panel. The Triage BNP test is used in thousands of hospitals and doctor's offices, and helps in the diagnosis, and severity assessment, of heart failure. The Triage Drugs of Abuse Panel and Triage TOX Drug Screen are rapid, qualitative urine screens that test for up to nine different illicit and prescription drugs, or drug classes, and provide results in less than 15 minutes. The firm's Biosite Discovery research business seeks to identify new protein markers of diseases that lack effective diagnostic tests. Additionally, with Biosite Discovery, the company has the capacity to offer antibody development services to companies seeking high-affinity antibodies for use in drug research. In return, Biosite seeks diagnostic licenses. The firm's Encompass program provides comprehensive education and consultation programs in support of Biosite customers. These programs include training and education; evaluation support; product training; clinical training; outcomes tracking; POC (purchase of care) reimbursement; CLIA (Clinical Laboratory Improvement Amendments) audit reports, audits on request; and consulting services. In all, Biosite's products are used in about half of all U.S. hospitals, as well as in approximately 50 international markets.

Biosite offers employees a benefits package including medical, dental, vision and life insurance; flexible spending accounts; an employee assistance program; educational assistance; a 401(k) plan; and an employee stock purchase plan.

FINANCIALS: Sales and profits are in thousands of dollars—add 000 to get the full amount. 2009 Note: Financial information for 2009 was not available for all companies at press time.

2009 Sales: $	2009 Profits: $	U.S. Stock Ticker: Subsidiary
2008 Sales: $	2008 Profits: $	Int'l Ticker: Int'l Exchange:
2007 Sales: $	2007 Profits: $	Employees:
2006 Sales: $308,592	2006 Profits: $39,994	Fiscal Year Ends: 12/31
2005 Sales: $287,699	2005 Profits: $54,029	Parent Company: INVERNESS MEDICAL INNOVATIONS INC

SALARIES/BENEFITS:

Pension Plan:	ESOP Stock Plan:	Profit Sharing:	Top Exec. Salary: $	Bonus: $
Savings Plan: Y	Stock Purch. Plan:		Second Exec. Salary: $	Bonus: $

OTHER THOUGHTS:

Apparent Women Officers or Directors:
Hot Spot for Advancement for Women/Minorities:

LOCATIONS: ("Y" = Yes)

West:	Southwest:	Midwest:	Southeast:	Northeast:	International:
Y					Y

Note: Financial information, benefits and other data can change quickly and may vary from those stated here.

BIOSPHERE MEDICAL INC

www.biospheremed.com

Industry Group Code: 33911 Ranks within this company's industry group: Sales: 13 Profits: 11

Drugs:		Other:	Clinical:	Computers:		Services:	
Discovery:	Y	AgriBio:	Trials/Services:	Hardware:		Specialty Services:	
Licensing:		Genetic Data:	Labs:	Software:		Consulting:	
Manufacturing:	Y	Tissue Replacement:	Equipment/Supplies:	Arrays:	Y	Blood Collection:	
Generics:			Research/Development Svcs.:	Database Management:		Drug Delivery:	
			Diagnostics:			Drug Distribution:	

TYPES OF BUSINESS:

Drugs-Bioengineered Microspheres
Cancer Treatments
Microsphere Delivery Systems

BRANDS/DIVISIONS/AFFILIATES:

Embosphere
EmboGold
HepaSphere
EmboCath Plus
Sequitor Guidewire
QuadraSphere

CONTACTS: *Note: Officers with more than one job title may be intentionally listed here more than once.*

Richard J. Faleschini, CEO
Richard J. Faleschini, Pres.
Martin J. Joyce, CFO
Melodie R. Domurad, VP-Regulatory, Medical Affairs & Quality Systems
Peter C. Sutcliffe, VP-Mfg.
Martin J. Joyce, Exec. VP-Admin.
Martin J. Joyce, Exec. VP-Finance
David P. Southwell, Chmn.

Phone: 781-681-7900	Fax: 781-792-2745
Toll-Free: 800-394-0295	
Address: 1050 Hingham St., Rockland, MA 02370 US	

GROWTH PLANS/SPECIAL FEATURES:

Biosphere Medical, Inc. develops, manufactures and markets products for medical procedures that use embolotherapy techniques. Embolotherapy, the introduction of biocompatible substances into the circulatory system, works to occlude blood vessels in order to arrest hemorrhaging or to devitalize a structure. Biosphere's core technology consists of patented bioengineered polymers, which help produce miniature spherical embolic particles called microspheres. Embosphere Microspheres (symptomatic uterine fibroids, hypervascularized tumors and arteriovenous malformations) and EmboGold Microspheres (for hypervascularized tumors and arteriovenous malformations), the company's embolic products, are made of an acrylic co-polymer that is cross-linked with gelatin. Due to their uniform, spherical shape and soft, slippery surface, microspheres are easy to inject through microcatheters, resulting in an even distribution within the vessel network. The microspheres selectively block the target tissue's blood supply, which destroys or devitalizes its target. Biosphere's principle focus is the treatment of symptomatic uterine fibroids, which are non-cancerous tumors growing in the uterus, using a procedure called uterine fibroid embolization (UFE). UFE is an alternative to hysterectomy and myomectomy that has been shown to allow faster recovery time and provide equivalent quality of life benefit. In addition to its microspheres, Biosphere produces two microcatheters: the Embocath Plus for general application and the Sequitor Guidewire for increased access to pelvic and visceral anatomy. Biosphere also produces the QuadraSphere and HepaSphere product lines for a number of other medical treatments, including the use of microspheres in the treatment of peripheral arteriovenous malformations and hypervascularized tumors like primary cancer of the liver. QuadraSphere is marketed in the U.S., while HepaSphere is marketed in Russia, Brazil and Europe. HepaSphere is also approved for transarterial chemoembolization of liver cancer, when used in conjunction with doxorubicin, an anticancer drug. In May 2010, the firm agreed to be acquired by Merit Medical Systems, Inc. and merged into Merit subsidiary Merit BioAcquisition, Co.

FINANCIALS: Sales and profits are in thousands of dollars—add 000 to get the full amount. 2009 Note: Financial information for 2009 was not available for all companies at press time.

2009 Sales: $31,443	2009 Profits: $-2,674	U.S. Stock Ticker: BSMD
2008 Sales: $29,258	2008 Profits: $-5,492	Int'l Ticker: Int'l Exchange:
2007 Sales: $26,900	2007 Profits: $-1,854	Employees: 88
2006 Sales: $22,891	2006 Profits: $-2,324	Fiscal Year Ends: 12/31
2005 Sales: $18,484	2005 Profits: $-2,801	Parent Company:

SALARIES/BENEFITS:

Pension Plan:	ESOP Stock Plan:	Profit Sharing:	Top Exec. Salary: $397,800	Bonus: $198,104
Savings Plan: Y	Stock Purch. Plan:		Second Exec. Salary: $250,000	Bonus: $77,150

OTHER THOUGHTS:

Apparent Women Officers or Directors: 2
Hot Spot for Advancement for Women/Minorities:

LOCATIONS: ("Y" = Yes)

West:	Southwest:	Midwest:	Southeast:	Northeast:	International:
				Y	Y

Note: Financial information, benefits and other data can change quickly and may vary from those stated here.

BIOTIME INC

www.biotimeinc.com

Industry Group Code: 325412 Ranks within this company's industry group: Sales: 138   Profits: 90

Drugs:		Other:		Clinical:		Computers:		Services:	
Discovery:	Y	AgriBio:		Trials/Services:		Hardware:		Specialty Services:	
Licensing:	Y	Genetic Data:		Labs:		Software:		Consulting:	
Manufacturing:		Tissue Replacement:		Equipment/Supplies:	Y	Arrays:		Blood Collection:	
Generics:				Research/Development Svcs.:		Database Management:		Drug Delivery:	
				Diagnostics:				Drug Distribution:	

TYPES OF BUSINESS:

Drugs-Surgical
Blood Plasma Expanders
Blood Replacement Solutions
Regenerative Medicine

BRANDS/DIVISIONS/AFFILIATES:

Hextend
PentaLyte
ACTCellerate
Enbryome Sciences, Inc.
ES Cell International Pte Ltd.
OncoCyte Corporation
BioTime Asia, Ltd.
OrthoCyte Corporation

CONTACTS: *Note: Officers with more than one job title may be intentionally listed here more than once.*

Michael D. West, CEO
Robert W. Peabody, COO/Sr. VP
Steven A. Seinberg, CFO
Hal Sternberg, VP-Research
Judith Segall, VP-Admin.
Judith Segall, Corp. Sec.
Judith Segall, Contact-Press
Steven A. Seinberg, Treas.
Walter Funk, VP-Stem Cell Research
Harold Waitz, VP-Regulatory & Quality Control
Alfred D. Kingsley, Chmn.

Phone: 510-521-3390	Fax: 510-521-3389
Toll-Free:	
Address: 1301 Harbor Bay Pkwy., Alameda, CA 94502 US	

GROWTH PLANS/SPECIAL FEATURES:

BioTime, Inc. is a development-stage company engaged in two areas of biomedical research and development: blood plasma volume expanders; and regenerative medicine through stem cell-related products. The firm's blood plasma volume expanders can be used as blood replacement solutions in surgery and emergency trauma treatment. Its lead blood plasma expander product is Hextend, an intravenous solution used in the treatment of hypovolemia, a condition caused by low blood volume, often from blood loss during surgery. Hextend maintains circulatory system fluid volume and blood pressure and keeps vital organs perfused during surgery. Other products include PentaLyte, currently in Phase II development, a pentastarch-based synthetic plasma expander; and HetaCool, in preclinical development, a blood replacement solution formulated for use at low temperatures. The company's regenerative medicine business operates through subsidiary Embryome Sciences, Inc. The regenerative medicine business focuses on the development and sale of advanced human stem cell products and technology for use by university researchers; bioscience and biopharmaceutical companies; and other companies that provide research products to these industries. Embryome's first product is the Embryome.com Database, which provides a detailed map of the embryome. In October 2009, the firm initiated development programs for human therapeutic applications of human embryonic stem (hES) cells for the treatment of cancer, ophthalmologic, skin, musculoskeletal system and hematologic diseases. U.S. subsidiary OncoCyte Corporation conducts the company's cancer research and development programs, while Hong Kong-based subsidiary BioTime Asia, Ltd., conducts hES cell-research related to cancer and other diseases. In July 2009, Embryome Sciences and Millipore Corporation entered a distribution agreement for the firm's ACTCellerate human progenitor cell lines. In September 2009, the company formed subsidiary BioTime Asia. In April 2010, the firm acquired Singapore-based ES Cell International Pte. Ltd. In June 2010, BioTime formed subsidiary OrthoCyte Corporation to develop musculoskeletal treatments based on hES cell technology.

FINANCIALS: Sales and profits are in thousands of dollars—add 000 to get the full amount. 2009 Note: Financial information for 2009 was not available for all companies at press time.

2009 Sales: $1,925	2009 Profits: $-5,144	U.S. Stock Ticker: BTIM
2008 Sales: $1,504	2008 Profits: $-3,781	Int'l Ticker: Int'l Exchange:
2007 Sales: $1,046	2007 Profits: $-1,438	Employees: 17
2006 Sales: $1,162	2006 Profits: $-1,864	Fiscal Year Ends: 12/31
2005 Sales: $ 903	2005 Profits: $-2,074	Parent Company:

SALARIES/BENEFITS:

Pension Plan:	ESOP Stock Plan:	Profit Sharing:	Top Exec. Salary: $258,333	Bonus: $87,917
Savings Plan:	Stock Purch. Plan:		Second Exec. Salary: $165,833	Bonus: $58,500

OTHER THOUGHTS:

Apparent Women Officers or Directors: 1
Hot Spot for Advancement for Women/Minorities:

LOCATIONS: ("Y" = Yes)

West:	Southwest:	Midwest:	Southeast:	Northeast:	International:
Y					Y

Note: Financial information, benefits and other data can change quickly and may vary from those stated here.

BIOVAIL CORPORATION

www.biovail.com

Industry Group Code: 325412A **Ranks within this company's industry group: Sales: 2 Profits: 2**

Drugs:	Other:	Clinical:		Computers:	Services:	
Discovery:	AgriBio:	Trials/Services:	Y	Hardware:	Specialty Services:	
Licensing:	Genetic Data:	Labs:	Y	Software:	Consulting:	
Manufacturing: Y	Tissue Replacement:	Equipment/Supplies:		Arrays:	Blood Collection:	
Generics:		Research/Development Svcs.:	Y	Database Management:	Drug Delivery:	Y
		Diagnostics:			Drug Distribution:	

TYPES OF BUSINESS:

Drug Delivery Systems Technologies
Generic Drugs
Drugs-Hypertension
Drugs-Antidepressants
Contract Research Services
Drug Development

BRANDS/DIVISIONS/AFFILIATES:

BTA Pharmaceuticals, Inc.
Biovail Pharmaceuticals Canada
Biovail Laboratories International SRL
Zovirax
Aplenzin
Xenazine
Cardizem LA
Wellbutrin XL

CONTACTS: *Note: Officers with more than one job title may be intentionally listed here more than once.*

William M. Wells, CEO
Gilbert Godin, COO/Exec. VP
Peggy Mulligan, CFO/Sr. VP
Mark Durham, Sr. VP-Human Resources & Shared Svcs.
John Sebben, VP-Tech. Oper.
Gregory Gubitz, General Counsel
Gregory Gubitz, Sr. VP-Corp. Dev.
Nelson F. Isabel, VP-Corp. Comm.
Nelson F. Isabel, VP-Investor Rel.
Christopher Bovaird, VP-Corp. Finance
Jennifer Tindale, VP/Associate General Counsel/Corp. Sec.
Todd Zator, Controller/VP
Rick Albert, Treas./VP
Christine Mayer, Sr. VP-Bus. Dev., BTA Pharmaceuticals, Inc.
Douglas J.P. Squires, Chmn.
Michel Chouinard, COO-Biovail Laboratories Int'l SRL

Phone: 905-286-3000	Fax: 905-286-3050
Toll-Free:	
Address: 7150 Mississauga Rd., Mississauga, ON L5N 8M5 Canada	

GROWTH PLANS/SPECIAL FEATURES:

Biovail Corporation is a specialty pharmaceutical company focused on the development and commercialization of products that target specialty central nervous system (CNS) disorders without significant existing treatments. The firm's primary area of focus has been on specialty neurology, including epilepsy, Parkinson's disease, Huntington's disease, Lou Gehrig's disease, Alzheimer's disease and multiple sclerosis. The firm's products include Wellbutrin XL, the brand name of bupropion, an anti-depressant; Xenazine, which is used for treatment of chorea associated with Huntington's disease; Aplenzin, an anti-depressant; Cardizem LA and Tiazac, blood pressure medications; Zovirax, a topical form of acyclovir, used in the treatment of the herpes virus; Ralivia and Ultram pain medications; and Glumetza, a diabetes medication. In addition, the company operates a contract research division that provides Biovail and other pharmaceutical companies with a broad range of Phase I/II clinical research services using pharmacokinetic studies and bioanalytical laboratory testing. Biovail Laboratories International SRL (BLS), the firm's primary operating subsidiary, manages the company's intellectual property; develops, manufactures, and sells its pharmaceutical products; and performs strategic planning. Biovail markets its products through its marketing divisions, BTA Pharmaceuticals, Inc. and Biovail Pharmaceuticals Canada, and through other strategic partners to health care professionals. The firm has supply and distribution agreements with GlaxoSmithKline, sanofi-aventis, Ortho-McNeil, Teva Pharmaceuticals, Forest Laboratories and Lundbeck. In February 2010, BLS entered into a collaboration and license agreement with Alexza Pharmaceuticals, Inc., acquiring the U.S. and Canadian rights to commercialize AZ-004, which is used for treatment of agitation in patients with bipolar disorder and schizophrenia. In March 2010, BLS acquired AMPAKINE compounds, including associated intellectual property, from Cortex Pharmaceuticals, Inc. for use with respiratory depression. In June 2010, the company announced plans to merge with Valeant Pharmaceuticals International.

FINANCIALS: Sales and profits are in thousands of dollars—add 000 to get the full amount. 2009 Note: Financial information for 2009 was not available for all companies at press time.

2009 Sales: $820,430	2009 Profits: $176,455	**U.S. Stock Ticker:** BVF
2008 Sales: $757,178	2008 Profits: $199,904	**Int'l Ticker:** BVF **Int'l Exchange:** Toronto-TSX
2007 Sales: $842,818	2007 Profits: $195,539	**Employees:** 1,311
2006 Sales: $1,067,722	2006 Profits: $211,626	**Fiscal Year Ends:** 12/31
2005 Sales: $935,500	2005 Profits: $89,000	**Parent Company:**

SALARIES/BENEFITS:

Pension Plan:	ESOP Stock Plan:	Profit Sharing:	Top Exec. Salary: $	Bonus: $
Savings Plan:	Stock Purch. Plan:		Second Exec. Salary: $	Bonus: $

OTHER THOUGHTS:

Apparent Women Officers or Directors: 3
Hot Spot for Advancement for Women/Minorities: Y

LOCATIONS: ("Y" = Yes)

West:	Southwest:	Midwest:	Southeast:	Northeast:	International:
				Y	Y

BRACHYTHERAPY SERVICES INC

www.brachyservices.com

Industry Group Code: 33911 Ranks within this company's industry group: Sales: Profits:

Drugs:		Other:	Clinical:		Computers:	Services:	
Discovery:	Y	AgriBio:	Trials/Services:		Hardware:	Specialty Services:	
Licensing:		Genetic Data:	Labs:		Software:	Consulting:	
Manufacturing:		Tissue Replacement:	Equipment/Supplies:	Y	Arrays:	Blood Collection:	
Generics:			Research/Development Svcs.:		Database Management:	Drug Delivery:	
			Diagnostics:			Drug Distribution:	

TYPES OF BUSINESS:
Radiation Therapy Products
Radioisotopic Products
Brachytherapy Products
Radiation Therapy Software

BRANDS/DIVISIONS/AFFILIATES:
Prospera
Horizontal Needle Box
STP-110 Precision Stepper
RTP-6000 Precision Stabilizer
Best Theratronics Ltd
Best Medical International
Portola Medical Inc
BrachyPak Surgery Kit

CONTACTS: *Note: Officers with more than one job title may be intentionally listed here more than once.*
Krishnan Suthanthiran, Pres., Best Medical Int'l

Phone: 703-451-2378	Fax: 703-451-5228
Toll-Free: 800-336-4970	
Address: 7643 Fullerton Rd., Springfield, VA 22153 US	

GROWTH PLANS/SPECIAL FEATURES:

BrachyTherapy Services Inc., (BSI) formerly North American Scientific, Inc. (NAS Medical), designs, develops, produces and sells products for radiation therapy. The firm focuses primarily on localized delivery radiation called brachytherapy. Brachytherapy is a minimally invasive medical procedure in which high-dose-rate (HDR) or low-dose-rate (LDR) sealed radioactive sources are temporarily or permanently implanted into cancerous tissue, delivering a therapeutically prescribed dose of radiation that is lethal to the cancerous tissue. BSI manufactures both iodine- and palladium-based brachytherapy seeds, the two most commonly used seeds for the treatment of prostate cancer. The brachytherapy seeds include Prospera I-125 and Prospera Pd-103. The company's brand Prospera is used in the treatment of ocular melanoma and other solid tumor applications. BSI directly markets this product line to ophthalmologists and medical physicists. The company's Prospera Strands includes pre-plugged needles with a synthetic, micro-angled insert and the strand, which, used in conjunction with the needle, is bio-resorbable and designed to hold the seeds at predetermined distances. BSI's products also include the STP-110 Precision Stepper and RTP-6000 Precision Stabilizer, which positions and hold the trans-rectal ultrasound probe during the LDR brachytherapy procedure. The firm also markets the BrachyPak Surgery Kit, a sterile disposable surgical kit for Prospera seed implanting. Other products include needle-loading accessories such as the Horizontal Needle Box and Needle Loading Box. Recently, the company sold its non-therapeutic product line to Eckert & Ziegler Isotope Products, Inc. In March 2009, North American Scientific filed Chapter 11 and sold its ClearPath system to Portola Medical, Inc. In late 2009, Best Medical International acquired North American Scientific and formed BrachyTherapy Services, Inc.

FINANCIALS: Sales and profits are in thousands of dollars—add 000 to get the full amount. 2009 Note: Financial information for 2009 was not available for all companies at press time.

2009 Sales: $	2009 Profits: $	U.S. Stock Ticker: Subsidiary
2008 Sales: $13,927	2008 Profits: $-10,945	Int'l Ticker: Int'l Exchange:
2007 Sales: $15,317	2007 Profits: $-20,998	Employees: 84
2006 Sales: $12,594	2006 Profits: $-17,130	Fiscal Year Ends: 10/31
2005 Sales: $12,032	2005 Profits: $-55,513	Parent Company: BEST MEDICAL INTERNATIONAL

SALARIES/BENEFITS:

Pension Plan:	ESOP Stock Plan:	Profit Sharing:	Top Exec. Salary: $310,708	Bonus: $50,366
Savings Plan: Y	Stock Purch. Plan: Y		Second Exec. Salary: $279,059	Bonus: $

OTHER THOUGHTS:
Apparent Women Officers or Directors:
Hot Spot for Advancement for Women/Minorities:

LOCATIONS: ("Y" = Yes)

West:	Southwest:	Midwest:	Southeast:	Northeast:	International:
				Y	Y

BRISTOL-MYERS SQUIBB CO   www.bms.com

Industry Group Code: 325412 Ranks within this company's industry group: Sales: 11   Profits: 3

Drugs:		Other:	Clinical:		Computers:		Services:
Discovery:	Y	AgriBio:	Trials/Services:		Hardware:		Specialty Services:
Licensing:		Genetic Data:	Labs:		Software:		Consulting:
Manufacturing:	Y	Tissue Replacement:	Equipment/Supplies:	Y	Arrays:		Blood Collection:
Generics:			Research/Development Svcs.:		Database Management:		Drug Delivery:
			Diagnostics:				Drug Distribution:

TYPES OF BUSINESS:

Drugs-Diversified
Medical Imaging Products
Nutritional Products

BRANDS/DIVISIONS/AFFILIATES:

Plavix
Enfamil
Reyataz
Ixempra
Sprycel
Abilify
Medarex
Kosan Biosciences Incorporated

CONTACTS: *Note: Officers with more than one job title may be intentionally listed here more than once.*

Lamberto Andreotti, CEO
Charles Bancroft, CFO/Exec. VP
Anthony McBride, Sr. VP-Human Resources
Elliott Sigal, Exec. VP/Chief Scientific Officer/Pres., R&D
Carlo de Notaristefani, Pres., Tech. Oper. & Global Support Functions
Sandra Leung, General Counsel/Corp. Sec./Sr. VP
Robert T. Zito, Chief Comm. Officer/Sr. VP-Corp. & Bus. Comm.
Anthony C. Hooper, Pres., Americas
John E. Celentano, Pres., Emerging Markets & Asia Pacific
Brian Daniels, Sr. VP-Global Dev. & Medical Affairs
James M. Cornelius, Chmn.
Beatrice Cazala, Pres., Europe & Global Commercialization
Quentin Roach, Chief Procurement Officer/Sr. VP

Phone: 212-546-4000	Fax: 212-546-4020
Toll-Free:	
Address: 345 Park Ave., New York, NY 10154 US	

GROWTH PLANS/SPECIAL FEATURES:

Bristol-Myers Squibb Co. discovers, develops, licenses, manufactures, markets, distributes and sells pharmaceuticals and other health care related products. The company manufactures drugs across multiple therapeutic classes, including cardiovascular; virology, including immunodeficiency virus infection; oncology; affective and other psychiatric disorders; and immunoscience. The firm's pharmaceutical products include chemically-synthesized drugs, or small molecules, and an increasing portion of biological products, or biologics. Small molecule drugs are typically administered orally in the form of a pill, although there are other drug delivery mechanisms that are used as well. Biologics are typically administered to patients through injections. Most of the firm's revenues come from products in the following therapeutic classes: cardiovascular; virology, including human immunodeficiency virus (HIV) infection; oncology; neuroscience; immunoscience; and metabolics. Products include Plavix, Avapro/Avalide, Reyataz, Sprycel and Ixempra. These products are manufactured in the U.S. and Puerto Rico and 11 foreign countries. Bristol-Myers Squibb maintains major research and development (R&D) facilities in Princeton, Hopewell and New Brunswick, New Jersey; Wallingford, Connecticut; and Milpitas, California. Pharmaceutical research and development is also carried out at various other facilities in the U.S., Belgium, Canada, the U.K. and India. Bristol-Myers Squibb invests approximately $3.6 billion in R&D annually. Geographically, the U.S. accounts for roughly 63% of sales; Europe, the Middle East and Asia, 22%; Japan , 3%; and Canada, 3%. In September 2009, the firm acquired Medarex, Inc. In December 2009, as part of the firm's strategy of focusing on its pharmaceutical operations, it divested the Mead Johnson Nutrition Company, which manufactures, markets, distributes and sells infant formulas and other nutritional products, including the entire line of Enfamil products.

Employees are offered medical and dental insurance; health care reimbursement accounts; a pension plan; a 401(k) plan; short-and long-term disability coverage; life insurance; travel accident insurance; an employee assistance plan; and adoption assistance.

FINANCIALS: Sales and profits are in thousands of dollars—add 000 to get the full amount. 2009 Note: Financial information for 2009 was not available for all companies at press time.

2009 Sales: $18,808,000	2009 Profits: $10,612,000	**U.S. Stock Ticker: BMY**
2008 Sales: $17,715,000	2008 Profits: $5,247,000	**Int'l Ticker:** Int'l Exchange:
2007 Sales: $15,617,000	2007 Profits: $2,165,000	Employees: 28,000
2006 Sales: $16,208,000	2006 Profits: $1,585,000	Fiscal Year Ends: 12/31
2005 Sales: $18,605,000	2005 Profits: $3,000,000	Parent Company:

SALARIES/BENEFITS:

Pension Plan: Y	ESOP Stock Plan:	Profit Sharing:	Top Exec. Salary: $1,488,077	Bonus: $4,475,000
Savings Plan: Y	Stock Purch. Plan:		Second Exec. Salary: $1,211,141	Bonus: $2,676,668

OTHER THOUGHTS:

Apparent Women Officers or Directors: 2
Hot Spot for Advancement for Women/Minorities: Y

LOCATIONS: ("Y" = Yes)

West:	Southwest:	Midwest:	Southeast:	Northeast:	International:
Y	Y	Y	Y	Y	Y

BURRILL & COMPANY

www.burrillandco.com

Industry Group Code: 523110 Ranks within this company's industry group: Sales: Profits:

Drugs:	Other:	Clinical:	Computers:	Services:	
Discovery:	AgriBio:	Trials/Services:	Hardware:	Specialty Services:	Y
Licensing:	Genetic Data:	Labs:	Software:	Consulting:	Y
Manufacturing:	Tissue Replacement:	Equipment/Supplies:	Arrays:	Blood Collection:	
Generics:		Research/Development Svcs.:	Database Management:	Drug Delivery:	
		Diagnostics:		Drug Distribution:	

TYPES OF BUSINESS:

Investment Banking-Life Sciences
Strategic Partnership & Spin-Off/Outlicensing Consulting
Life Sciences Publications
Industry Conferences
Venture Capital Funds

BRANDS/DIVISIONS/AFFILIATES:

Burrill Life Sciences Capital Fund
Burrill Biotechnology Capital Fund
Burrill Agbio Capital Fund
Biotech Meeting at Laguna Beach
Burrill Nutraceuticals Capital Fund
Indiana Life Sciences Forum
Burrill Personalized Medicine Meeting
Burrill International Group

CONTACTS: *Note: Officers with more than one job title may be intentionally listed here more than once.*

G. Steven Burrill, CEO
Adriana Petersen, Managing Dir.-Human Resources
Victor Hebert, Chief Admin. Officer
Victor Hebert, Chief Legal Officer
Leslie Errington, Dir.-Corp. Dev.
Peter Winter, Managing Dir.-Comm.
Helena Sen, Controller
Ganesh Kishore, CEO-Malaysian Life Sciences Capital Fund
Tania Fernandez, Dir.-India
Michael Keyoung, Managing Dir.-Burrill Pan-Asia
Hal Gerber, Managing Dir.-Private Equity
John E. Hamer, Managing Dir.-Int'l Group

Phone: 415-591-5400	Fax: 415-591-5401
Toll-Free:	
Address: 1 Embarcadero Ctr., Ste. 2700, San Francisco, CA 94111 US	

GROWTH PLANS/SPECIAL FEATURES:

Burrill & Company is a life-sciences merchant bank focused exclusively on companies involved in biotechnology; pharmaceuticals; drug delivery devices; diagnostics; medical devices; human health care and related medical technologies; nutraceuticals; agricultural biotechnologies; and industrial biomaterials and bioprocesses. The company operates through several business units, including venture capital, merchant banking, private equity, publications and conferences. The venture capital unit manages and offers various funds totaling more than $950 million under management, including the Burrill Life Sciences Capital Fund; Burrill Biotechnology Capital Fund; Burrill Agbio Capital Fund; and Burrill Nutraceuticals Capital Fund. The merchant banking unit assists life science companies in identifying, negotiation and forming strategic partnerships with other companies for access to resources, technologies or collaborations. The unit also works with major life sciences companies to spin off divisions or out-license technologies. The firm's dedicated private equity team focuses on investments in small and mid-cap public companies, along with select spin-offs and buy-outs from larger life sciences corporations. The publications unit publishes monthly indices on biotech industry stock market performance; quarterly reports that highlight important industry developments such as advancements in science, technology breakthroughs and important business transactions and deals; articles and commentary on the biotechnology industry for various publications; and annual biotechnology industry reports. The conference unit annually hosts and sponsors various industry conferences including the Biotech Meeting at Laguna Beach, the Indiana Life Sciences Forum, the Stem Cell Meeting, the Burrill Personalized Medicine Meeting and the Ageing Meeting. Past conferences hosted by the firm have included the Burrill China Life Sciences Meeting, the Burrill India Life Sciences Meeting and the Japan Biotech Meeting. Additionally, the Burrill International Group concentrates on investment opportunities outside the U.S., with current areas of focus including China, India, Japan, Malaysia, Korea, Australia, New Zealand, Canada, Russia, the Middle East and Eastern Europe.

FINANCIALS: Sales and profits are in thousands of dollars—add 000 to get the full amount. 2009 Note: Financial information for 2009 was not available for all companies at press time.

2009 Sales: $	2009 Profits: $	U.S. Stock Ticker: Private
2008 Sales: $	2008 Profits: $	Int'l Ticker: Int'l Exchange:
2007 Sales: $	2007 Profits: $	Employees:
2006 Sales: $	2006 Profits: $	Fiscal Year Ends: 12/31
2005 Sales: $	2005 Profits: $	Parent Company:

SALARIES/BENEFITS:

Pension Plan:	ESOP Stock Plan:	Profit Sharing:	Top Exec. Salary: $	Bonus: $
Savings Plan:	Stock Purch. Plan:		Second Exec. Salary: $	Bonus: $

OTHER THOUGHTS:

Apparent Women Officers or Directors: 16
Hot Spot for Advancement for Women/Minorities: Y

LOCATIONS: ("Y" = Yes)

West:	Southwest:	Midwest:	Southeast:	Northeast:	International:
Y				Y	Y

CALIPER LIFE SCIENCES

www.calipertech.com

Industry Group Code: 325413 Ranks within this company's industry group: Sales: 10 Profits: 15

Drugs:	Other:	Clinical:	Computers:		Services:	
Discovery:	AgriBio:	Trials/Services:	Hardware:	Y	Specialty Services:	
Licensing:	Genetic Data:	Labs:	Software:	Y	Consulting:	
Manufacturing:	Tissue Replacement:	Equipment/Supplies: Y	Arrays:	Y	Blood Collection:	
Generics:		Research/Development Svcs.:	Database Management:		Drug Delivery:	
		Diagnostics:			Drug Distribution:	

TYPES OF BUSINESS:

Bioanalysis Equipment
Microfluidic Systems
High-Throughput Screening Machines
Liquid Handling Systems
Drug Discovery Platforms
Laboratory Automation Solutions
Software

BRANDS/DIVISIONS/AFFILIATES:

Caliper Discovery Alliances & Services
LabChip
NovaScreen Biosciences Corp.

CONTACTS: Note: Officers with more than one job title may be intentionally listed here more than once.

E. Kevin Hrusovsky, CEO
E. Kevin Hrusovsky, Pres.
Peter F. McAree, CFO/Sr. VP
Paula J. Cassidy, VP-Human Resources
Bradley W. Rice, Sr. VP-R&D and Systems
Stephen E. Creager, General Counsel/Corp. Sec./Sr. VP
Bruce J. Bal, Sr. VP-Oper.
William C. Kruka, Sr. VP-Corp. Dev.
David M. Manyak, Exec. VP-Drug Discovery Svcs.
Enrique Bernal, Sr. VP-In Vitro Bus. Dev.
Mark T. Roskey, VP-Applied Biology R&D
Bob Bishop, Chmn.

Phone: 508-435-9500	Fax: 508-435-3439
Toll-Free: 877-522-2447	
Address: 68 Elm St., Hopkinton, MA 01748 US	

GROWTH PLANS/SPECIAL FEATURES:

Caliper Life Sciences, Inc. develops and sells products and services to the life sciences research community, including pharmaceutical and biotechnology companies, government institutions and other not-for-profit research organizations. The company produces integrated systems, consisting of instruments, software, reagents, laboratory automation tools and assay and discovery services. These technologies are designed to enable researchers not only to discover new drugs but to conduct environmental-related, agricultural and forensic testing as well as possible in vitro and in vivo diagnostic applications. The firm's product offerings are organized into three core business areas: optical molecular imaging (imaging), discovery research (research) and caliper discovery alliances and services (CDAS). The imaging business focuses on preclinical imaging and its activities include the expansion of its IVIS imaging instrument product lines, development of new therapeutic area applications and facilitating additional imaging modalities. The research segment is responsible for developing Caliper's core automation and microfluidic technologies, including its flagship product LabChip, for applications such as molecular biology sample preparation for genomics, protomics, cellular screening and forensics. CDAS, operated by subsidiary NovaScreen Biosciences, is responsible for expanding drug discovery collaborations and alliances and increasing sales of drug discovery services by providing outsourcing services and securing contracts with the Environmental Protection Agency under its ToxCast screening program. In December 2009, the company sold subsidiary Xenogen Biosciences Corporation, a provider of in-vivi pre-clinical CRO services, to Taconic Farms, Inc. for approximately $11 million.

The firm offers its employees medical, dental, vision, life and AD&D insurance; short-and long-term disability coverage; a 401(k) plan; a discount stock purchase plan; tuition reimbursement; and an employee assistance plan.

FINANCIALS: Sales and profits are in thousands of dollars—add 000 to get the full amount. 2009 Note: Financial information for 2009 was not available for all companies at press time.

2009 Sales: $130,412	2009 Profits: $-8,225	U.S. Stock Ticker: CALP
2008 Sales: $134,054	2008 Profits: $-68,292	Int'l Ticker: Int'l Exchange:
2007 Sales: $140,707	2007 Profits: $-24,080	Employees: 401
2006 Sales: $107,871	2006 Profits: $-28,934	Fiscal Year Ends: 12/31
2005 Sales: $87,009	2005 Profits: $-14,457	Parent Company:

SALARIES/BENEFITS:

Pension Plan:	ESOP Stock Plan:	Profit Sharing:	Top Exec. Salary: $451,567	Bonus: $
Savings Plan: Y	Stock Purch. Plan: Y		Second Exec. Salary: $261,995	Bonus: $

OTHER THOUGHTS:

Apparent Women Officers or Directors: 2
Hot Spot for Advancement for Women/Minorities: Y

LOCATIONS: ("Y" = Yes)

West:	Southwest:	Midwest:	Southeast:	Northeast:	International:
Y				Y	Y

CAMBREX CORP

Industry Group Code: 325412 Ranks within this company's industry group: Sales: 63 Profits: 59

Drugs:	Other:	Clinical:		Computers:		Services:
Discovery:	AgriBio:	Trials/Services:	Y	Hardware:		Specialty Services:
Licensing:	Genetic Data:	Labs:		Software:		Consulting:
Manufacturing: Y	Tissue Replacement:	Equipment/Supplies:	Y	Arrays:		Blood Collection:
Generics:		Research/Development Svcs.:	Y	Database Management:		Drug Delivery:
		Diagnostics:				Drug Distribution:

TYPES OF BUSINESS:

Contract Pharmaceutical Manufacturing
Contract Research
Pharmaceutical Ingredients
Testing Products & Services

BRANDS/DIVISIONS/AFFILIATES:

Cambrex Charles City, Inc.
Cambrex Kariskoga AB
Cambrex Profarmaco Milano S.r.l.
Cambrex Tallinn AS

CONTACTS: Note: Officers with more than one job title may be intentionally listed here more than once.

Steven M. Klosk, CEO
Steven M. Klosk, Pres.
Greg Sargen, CFO/VP
F. Michael Zachara, General Counsel/VP/Sec.
Paolo Russolo, Pres., Cambrex Profarmaco
John Miller, Chmn.

Phone: 201-804-3000	Fax: 201-804-9852
Toll-Free: 866-286-9133	
Address: 1 Meadowlands Plz., East Rutherford, NJ 07073 US	

GROWTH PLANS/SPECIAL FEATURES:

Cambrex Corp. provides products and services to aid and enhance the discovery and commercialization of therapeutics. The firm manufactures products, which are sold to research organizations, pharmaceutical, biopharmaceutical and generic drug companies. It also provides testing services and safety assessments. Products include Cambrex Camouflage, a polymer resin-based drug delivery technology, which masks the bitter taste of certain drugs, as well as advanced pharmaceutical intermediates, controlled substances, chiral amines and fine chemicals. The firm also manufactures over 70 generic active pharmaceutical ingredients (APIs) for respiratory, pain management, cardiovascular, gastrointestinal, central nervous system and endocrine purposes. These APIs are developed by Italian subsidiary, Cambrex Prodarmaco Milano S.r.l. and serve the pharmaceutical market in Italy, the U.S., Germany, Japan, France, Belgium, England and countries in Eastern Europe. Other subsidiaries of the firm include Cambrex Charles City, Inc. in Iowa; Cambrex Karlskoga AB in Sweden; and Cambrex Tallinn AS in Estonia. The company currently owns 12 issued patents and has eight patent applications pending in the U.S., and owns 26 patents and has 14 patent applications pending in other countries.

Employees are offered medical, dental and vision insurance; health and dependent spending accounts; life insurance; a 401(k) plan; disability coverage; AD&D insurance; tuition reimbursement; scholarship awards; education assistance; an employee assistance program; and business travel accident insurance.

FINANCIALS: Sales and profits are in thousands of dollars—add 000 to get the full amount. 2009 Note: Financial information for 2009 was not available for all companies at press time.

2009 Sales: $236,277	2009 Profits: $10,392	**U.S. Stock Ticker: CBM**
2008 Sales: $249,618	2008 Profits: $7,929	**Int'l Ticker:** Int'l Exchange:
2007 Sales: $252,574	2007 Profits: $209,248	Employees: 854
2006 Sales: $236,659	2006 Profits: $-30,100	Fiscal Year Ends: 12/31
2005 Sales: $223,565	2005 Profits: $-110,458	Parent Company:

SALARIES/BENEFITS:

Pension Plan:	ESOP Stock Plan:	Profit Sharing:	Top Exec. Salary: $431,634	Bonus: $
Savings Plan: Y	Stock Purch. Plan:		Second Exec. Salary: $382,538	Bonus: $

OTHER THOUGHTS:

Apparent Women Officers or Directors: 2
Hot Spot for Advancement for Women/Minorities: Y

LOCATIONS: ("Y" = Yes)

West:	Southwest:	Midwest:	Southeast:	Northeast:	International:
		Y		Y	Y

CANGENE BIOPHARMA INC

www.cblinc.com

Industry Group Code: 541712 Ranks within this company's industry group: Sales: Profits:

Drugs:	Other:	Clinical:		Computers:		Services:	
Discovery:	AgriBio:	Trials/Services:	Y	Hardware:		Specialty Services:	Y
Licensing:	Genetic Data:	Labs:	Y	Software:		Consulting:	Y
Manufacturing: Y	Tissue Replacement:	Equipment/Supplies:		Arrays:		Blood Collection:	
Generics:		Research/Development Svcs.:	Y	Database Management:		Drug Delivery:	
		Diagnostics:				Drug Distribution:	

TYPES OF BUSINESS:

Research & Development
Contract Pharmaceutical Services
Regulatory & Compliance Consulting

BRANDS/DIVISIONS/AFFILIATES:

Cangene Corporation

CONTACTS: *Note: Officers with more than one job title may be intentionally listed here more than once.*

Grant McClarity, VP-R&D
Henry P. Clark, Dir.-New Bus. Dev.
John Langstaff, Pres./CEO-Cangene Corp.
William Labossiere (Bill) Bees, Sr. VP-Oper., Cangene Corp.
Michael Graham, CFO-Cangene Corp.

Phone: 410-843-5000	Fax: 410-843-4414
Toll-Free: 800-441-4225	
Address: 1111 S. Paca St., Baltimore, MD 21230-2591 US	

GROWTH PLANS/SPECIAL FEATURES:

Cangene bioPharma, Inc., a subsidiary of Cangene Corporation, provides contract manufacturing services, such as formulation, filling and packaging of injectable products to the biopharmaceutical industry. Customers typically hire the company to produce developmental stage products for FDA clinical trials or to produce and manufacture FDA approved products for commercial sale. Cangene bioPharma provides services in four major areas: clinical manufacturing, commercial manufacturing, lyophilization and product testing. The company performs clinical manufacturing on a commercial scale and with regulatory approval from agencies in several countries. It has a large selection of pre-qualified vials ranging in size from 3-100cc; stoppers; and syringes in most major sizes. The firm's commercial manufacturing division currently produces 16 commercial products from its facilities. Lyophilization, a dehydration process, is conducted in the company's 240-square-foot lyophilizer, which has a capacity of roughly 60,000 3cc vials, 50,000 5cc vials or 40,000 10cc vials. Products that require lyophilization are filled, partially stoppered and loaded into the pre-chilled lyophilizer throughout the filling period. Its lyophilizer has both low and high speed filling options, temperature control to as low as negative 65 degrees Celsius and vacuum control between 50 and 1,000 millitorr. A second 240-square-foot lyophilizer is expected to be operational early in 2011. The time from product initiation to production in most cases can be reduced to two months by choosing one of the firm's pre-qualified combinations. The company's testing services include identification testing on client supplied raw materials; in-process testing sufficient to ensure bulk formulation is done correctly; and finished product testing as required by the client. Cangene bioPharma's Analytical Services Department can perform Karl Fischer, Dionex IC, Viscosity, Dissolved Oxygen and Osmolality tests, among others. Its Microbiology Department can perform Enhancement & Inhibition, Endotoxin, Bioburden Method Qualification, Bacteriostasis/ Fungistasis, Bulk Sterility, Final Product Sterility and ELISA tests.

Cangene bioPharma offers its employees medical, dental, vision, life and disability insurance.

FINANCIALS: Sales and profits are in thousands of dollars—add 000 to get the full amount. 2009 Note: Financial information for 2009 was not available for all companies at press time.

2009 Sales: $	2009 Profits: $	U.S. Stock Ticker: Subsidiary
2008 Sales: $	2008 Profits: $	Int'l Ticker: Int'l Exchange:
2007 Sales: $	2007 Profits: $	Employees:
2006 Sales: $	2006 Profits: $	Fiscal Year Ends: 7/31
2005 Sales: $	2005 Profits: $	Parent Company: CANGENE CORP

SALARIES/BENEFITS:

Pension Plan:	ESOP Stock Plan:	Profit Sharing:	Top Exec. Salary: $	Bonus: $
Savings Plan: Y	Stock Purch. Plan:		Second Exec. Salary: $	Bonus: $

OTHER THOUGHTS:

Apparent Women Officers or Directors:
Hot Spot for Advancement for Women/Minorities:

LOCATIONS: ("Y" = Yes)

West:	Southwest:	Midwest:	Southeast:	Northeast: Y	International:

Note: Financial information, benefits and other data can change quickly and may vary from those stated here.

CANGENE CORP

www.cangene.com

Industry Group Code: 325412 Ranks within this company's industry group: Sales: 65 Profits: 48

Drugs:		Other:		Clinical:		Computers:		Services:	
Discovery:	Y	AgriBio:		Trials/Services:		Hardware:		Specialty Services:	
Licensing:		Genetic Data:	Y	Labs:		Software:		Consulting:	
Manufacturing:	Y	Tissue Replacement:		Equipment/Supplies:		Arrays:		Blood Collection:	
Generics:				Research/Development Svcs.:	Y	Database Management:		Drug Delivery:	
				Diagnostics:				Drug Distribution:	

TYPES OF BUSINESS:

Drugs-Hyperimmunes
Generic Drugs
Contract Manufacturing
Vaccines
Biodefense-Related Therapeutics
Commercial Therapeutics

BRANDS/DIVISIONS/AFFILIATES:

WinRho SDF
VariZIG
VIG
HepaGam B
Anthrax Immune Globulin (AIG)
Leucotropin
Accretropin
Cangene bioPharma, Inc.

CONTACTS: *Note: Officers with more than one job title may be intentionally listed here more than once.*

John M. Langstaff, CEO
John M. Langstaff, Pres.
Michael Graham, CFO
Paul Brisebois, VP-Mktg. & Sales
Grant McClarty, VP-R&D
Francis St. Hilaire, Corp. Counsel/Sec.
William Bees, Sr. VP-Oper.
Paul Brisebois, VP-Bus. Dev.
Andrew D. Storey, VP-Quality Assurance
Andrew D. Storey, VP-Clinical & Regulatory Affairs
Jack Kay, Chmn.

Phone: 204-275-4200	Fax: 204-269-7003
Toll-Free: 800-768-2304	
Address: 155 Innovation Dr., Winnipeg, MB R3T 5Y3 Canada	

GROWTH PLANS/SPECIAL FEATURES:

Cangene Corp. is a leading global developer and manufacturer of specialty hyperimmune plasma and biotechnology products. The company offers contract research and process development services; bulk products manufacturing; and finished products manufacturing services to biopharmaceutical companies. Cangene's leading product offering is WinRho SDF, a hyperimmune used to prevent hemolytic disease in newborns and to treat immune thrombocytopenic purpura (ITP), a clotting disorder. The company's other approved hyperimmune products include VariZIG, a chicken pox vaccine, developed to prevent chicken pox in pregnant women; VIG (Vaccinia Immune Globulin), an antibody product used to prep recipients for or to treat severe reactions to the smallpox vaccine; Accretropin Injection, a recombinant human growth hormone; and HepaGam B, a specialized antibody for treatment following acute exposure to the hepatitis B virus and preventing its reoccurrence following liver transplant operations. Cangene's products currently under development include Botulism Antitoxin, a product containing neutralizing antibodies to the seven botulinum types; Anthrax Immune Globulin (AIG), an adjunct to antibiotic therapy in critically ill patients with anthrax; and Leucotropin, a recombinant version of a protein that stimulates the production of certain white blood cells, for the treatment of acute radiation syndrome. The company offers contract manufacturing services in the areas of process development, bulk product manufacturing and finished product manufacturing. Services include fermentation and purification process development, optimization and scale-up; formulation development; lyophilization cycle development; cGMP manufacturing; labeling and packaging; quality control testing; stability studies; and regulatory support. In addition, Cangene runs a subsidiary, Cangene bioPharma, located in the U.S., which provides pharmaceutical and biopharmaceutical product development and production services. In addition, it recently acquired Twinstrand Therapeutics, Inc., a company in the early stages of developing a ricin-based prodrug, TST10088, which can target specific activity in a diseased cell while leaving healthy cells unaffected.

FINANCIALS: Sales and profits are in thousands of dollars—add 000 to get the full amount. 2009 Note: Financial information for 2009 was not available for all companies at press time.

2009 Sales: $224,910	2009 Profits: $56,420	**U.S. Stock Ticker:**	
2008 Sales: $156,440	2008 Profits: $27,880	**Int'l Ticker:** CNJ	Int'l Exchange: Toronto-TSX
2007 Sales: $87,030	2007 Profits: $9,510	Employees: 700	
2006 Sales: $100,600	2006 Profits: $12,100	Fiscal Year Ends: 7/31	
2005 Sales: $90,620	2005 Profits: $-13,638	Parent Company:	

SALARIES/BENEFITS:

Pension Plan:	ESOP Stock Plan:	Profit Sharing:	Top Exec. Salary: $	Bonus: $
Savings Plan:	Stock Purch. Plan:		Second Exec. Salary: $	Bonus: $

OTHER THOUGHTS:

Apparent Women Officers or Directors: 1
Hot Spot for Advancement for Women/Minorities:

LOCATIONS: ("Y" = Yes)

West:	Southwest:	Midwest:	Southeast:	Northeast:	International:
Y			Y	Y	Y

CARACO PHARMACEUTICAL LABORATORIES   www.caraco.com

Industry Group Code: 325412 Ranks within this company's industry group: Sales: 53 Profits: 51

Drugs:		Other:	Clinical:	Computers:	Services:
Discovery:	Y	AgriBio:	Trials/Services:	Hardware:	Specialty Services:
Licensing:		Genetic Data:	Labs:	Software:	Consulting:
Manufacturing:	Y	Tissue Replacement:	Equipment/Supplies:	Arrays:	Blood Collection:
Generics:			Research/Development Svcs.:	Database Management:	Drug Delivery:
			Diagnostics:		Drug Distribution:

TYPES OF BUSINESS:
Drugs-Generic

BRANDS/DIVISIONS/AFFILIATES:
Sun Pharmaceutical Industries
Flumadine
Nicardipine HCl
Theophylline
Synalgos-DC

CONTACTS: Note: Officers with more than one job title may be intentionally listed here more than once.
Jitendra N. Doshi, CEO
Mukul Rathi, Interim CFO
Gurpartap Singh Sachdeva, Sr. VP-Bus. Strategies
Robert Kurkiewicz, Sr. VP-Regulatory Affairs
Jayesh Shah, Dir.-Commercial
Akin-Remi Ajac-Ayodele, Dir.-Quality
Thomas Versosky, Dir.-Bus. Strategies
Dilip S. Shanghvi, Chmn.

Phone: 313-871-8400	Fax: 313-871-8314
Toll-Free: 800-818-4555	
Address: 1150 Elijah McCoy Dr., Detroit, MI 48202 US	

GROWTH PLANS/SPECIAL FEATURES:
Caraco Pharmaceutical Laboratories, Ltd., develops, manufactures and markets generic and private-label drugs for prescription and over-the-counter markets. The company's product portfolio includes 42 products in 93 strengths and various package sizes. These drugs relate to a variety of therapeutic segments, including arthritis, pain control, epilepsy, diabetes, antipsychotic and neurological disorders. Pharmaceutical products that the company produces include Flumadine, an influenza A virus medication; Nicardipine HCI, a calcium ion influx inhibitor for the short-term treatment of hypertension; Theophylline, which is used to treat the symptoms of chronic asthma and other chronic lung diseases, and Synalgos-DC, a prescription pain killer. The company also has several drugs awaiting FDA approval. The company has collaborative agreements with several companies, with the most prominent being Sun Pharmaceutical Industries (Sun Pharma), the majority stock holder of Caraco. Under these agreements, Caraco develops generic drugs for each company to market as its own brand. The firm distributes its products through wholesalers, chain drug stores, retail pharmacies, mail-order companies, managed care organizations, hospital groups and nursing homes. Some of the wholesalers that distribute Caraco's products include Amerisource-Bergen Corporation; McKesson Corporation; and Cardinal Health.

Caraco offers employees benefits which include medical, dental, and vision care; paid time off and holiday pay; a 401(k); life and disability insurance; and health and dependent care accounts.

FINANCIALS: Sales and profits are in thousands of dollars—add 000 to get the full amount. 2009 Note: Financial information for 2009 was not available for all companies at press time.

2009 Sales: $337,177	2009 Profits: $20,537	U.S. Stock Ticker: CPD
2008 Sales: $350,367	2008 Profits: $35,388	Int'l Ticker: Int'l Exchange:
2007 Sales: $117,027	2007 Profits: $26,858	Employees: 662
2006 Sales: $82,789	2006 Profits: $-10,423	Fiscal Year Ends: 3/31
2005 Sales: $64,116	2005 Profits: $-2,278	Parent Company: SUN PHARMACEUTICAL INDUSTRIES LTD

SALARIES/BENEFITS:

Pension Plan:	ESOP Stock Plan:	Profit Sharing:	Top Exec. Salary: $429,936	Bonus: $
Savings Plan: Y	Stock Purch. Plan:		Second Exec. Salary: $242,688	Bonus: $

OTHER THOUGHTS:
Apparent Women Officers or Directors:
Hot Spot for Advancement for Women/Minorities:

LOCATIONS: ("Y" = Yes)

West:	Southwest:	Midwest:	Southeast:	Northeast:	International:
		Y			

CARDIOME PHARMA CORP

www.cardiome.com

Industry Group Code: 325412 Ranks within this company's industry group: Sales: 83 Profits: 75

Drugs:		Other:	Clinical:	Computers:	Services:
Discovery:	Y	AgriBio:	Trials/Services:	Hardware:	Specialty Services:
Licensing:		Genetic Data:	Labs:	Software:	Consulting:
Manufacturing:		Tissue Replacement:	Equipment/Supplies:	Arrays:	Blood Collection:
Generics:			Research/Development Svcs.:	Database Management:	Drug Delivery:
			Diagnostics:		Drug Distribution:

TYPES OF BUSINESS:

Drugs-Cardiac

BRANDS/DIVISIONS/AFFILIATES:

Astellas Pharma US, Inc.
Vernakalant (oral)
Vernakalant (iv)
KYNAPID
GED-aPC
Merck & Co. Inc.
Merck Sharp & Dohme (Switzerland) GmbH

CONTACTS: Note: Officers with more than one job title may be intentionally listed here more than once.

Doug Janzen, CEO
Doug Janzen, Pres.
Curtis Sikorsky, CFO
Karim Lalji, Sr. VP-Commercial Affairs
Donald A. McAfee, Chief Scientific Officer
Sheila M. Grant, VP-Prod. Dev.
Taryn Boivin, VP-Mfg. & Pharmaceutical Sciences
Guy F. Cipriani, VP-Bus. Dev.
Gregory N. Beatch, VP-Scientific Affairs
Bob Rieder, Chmn.

Phone: 604-677-6905	Fax: 604-677-6915
Toll-Free: 800-330-9928	
Address: 6190 Agronomy Rd., 6th Fl., Vancouver, BC V6T 1Z3 Canada	

GROWTH PLANS/SPECIAL FEATURES:

Cardiome Pharma Corp. is a drug discovery and development company that is focused on the development of proprietary drugs for the treatment and prevention of cardiovascular diseases. Its primary products are designed to address atrial fibrillation, the most common cardiac arrhythmia (abnormal heart rhythm) condition. The company currently has three late-stage clinical drug programs: a Phase II/post-clinical program for Vernakalant, designed to treat and prevent atrial fibrillation; a Phase I program for GED-aPC, an engineered analog of human activated Protein C; and a preclinical program that specifically focuses on the improvement of cardiovascular function. Cardiome's main product focus is on Vernakalant (formerly known as RSD1235), which is being developed in two forms: the intravenous Vernakalant (iv) and an oral drug formulation, Vernakalant (oral). The drug works by selectively blocking ion channels in the heart that become active during episodes of atrial fibrillation. Vernakalant (iv) is being considered as an intravenous agent to be administered during an atrial fibrillation episode to return the heart rate to normal; Vernakalant (oral) is under consideration for the long-term prevention of atrial fibrillation. Astellas Pharma US, Inc. is the firm's Vernakalant co-development partner. Currently, Vernakalant (iv) is in an ongoing, additional Phase III comparator study for the rapid mediation of atrial fibrillation and return to normal heart rate. KYNAPID is the proposed North American brand name for the intravenous treatment. In May 2009, the firm sold the exclusive global commercialization rights to Vernakalant (oral) to Merck & Co., Inc., and sold rights to the Vernakalant (iv) outside of North America to Merck Sharp & Dohme (Switzerland) GmbH.

Cardiome offers employees medical and dental benefits; an employee incentive stock option program; professional development support; onsite health and wellness resources; parking and transit passes; and gym memberships.

FINANCIALS: Sales and profits are in thousands of dollars—add 000 to get the full amount. 2009 Note: Financial information for 2009 was not available for all companies at press time.

2009 Sales: $54,672	2009 Profits: $-1,276	**U.S. Stock Ticker:** CRME	
2008 Sales: $1,604	2008 Profits: $-60,352	**Int'l Ticker: COM** Int'l Exchange: Toronto-TSX	
2007 Sales: $4,500	2007 Profits: $-78,600	**Employees:** 104	
2006 Sales: $19,000	2006 Profits: $-33,300	**Fiscal Year Ends:** 12/31	
2005 Sales: $15,800	2005 Profits: $-52,300	**Parent Company:**	

SALARIES/BENEFITS:

Pension Plan:	ESOP Stock Plan: Y	Profit Sharing:	Top Exec. Salary: $528,127	Bonus: $725,011
Savings Plan:	Stock Purch. Plan:		Second Exec. Salary: $516,628	Bonus: $715,083

OTHER THOUGHTS:

Apparent Women Officers or Directors: 1
Hot Spot for Advancement for Women/Minorities:

LOCATIONS: ("Y" = Yes)

West:	Southwest:	Midwest:	Southeast:	Northeast:	International:
					Y

CARDIUM THERAPEUTICS INC

www.cardiumthx.com

Industry Group Code: 325412 Ranks within this company's industry group: Sales: 146 Profits: 102

Drugs:	Other:		Clinical:		Computers:		Services:	
Discovery:	AgriBio:		Trials/Services:		Hardware:		Specialty Services:	
Licensing:	Genetic Data:	Y	Labs:		Software:		Consulting:	
Manufacturing:	Tissue Replacement:		Equipment/Supplies:	Y	Arrays:		Blood Collection:	
Generics:			Research/Development Svcs.:	Y	Database Management:		Drug Delivery:	
			Diagnostics:				Drug Distribution:	

TYPES OF BUSINESS:
Biologic Therapeutics & Medical Devices

BRANDS/DIVISIONS/AFFILIATES:
Generx
Corgentin
Tissue Repair Co.
Excellerate
Cardium Biologics

CONTACTS: *Note: Officers with more than one job title may be intentionally listed here more than once.*
Christopher J. Reinhard, CEO
Christopher J. Reinhard, Pres.
Dennis M. Mulroy, CFO
Gabor M. Rubanyi, Chief Scientific Officer
Ted Williams, VP-Tech. Oper.
Ted Williams, VP-Mfg.
Tyler M. Dylan, General Counsel/Exec. VP/Sec./Chief Bus. Officer
Mark McCutchen, VP-Bus. Dev.
Bonnie Ortega, Dir.-Public Rel.
Bonnie Ortega, Dir.-Investor Rel.
Christopher J. Reinhard, Treas.
Robert L. Engler, Chief Medical Advisor
Barabara K. Sosnowski, VP-Biologics Dev./COO-Tissue Repair Co.
Christopher J. Reinhard, Chmn.

Phone: 858-436-1000	Fax: 858-436-1001
Toll-Free:	
Address: 12255 El Camino Real, Ste. 250, San Diego, CA 92130 US	

GROWTH PLANS/SPECIAL FEATURES:
Cardium Therapeutics, Inc. is a medical technology company primarily focused on the development and commercialization of medical devices and biologic therapeutics for the treatment of cardiovascular and ischemic diseases. The company's products are divided between two companies: Cardium Biologics and subsidiary Tissue Repair Company. Product candidates for Cardium Biologics include Generx, a late-stage DNA-based growth factor therapeutic that is being developed to promote and stimulate the growth of collateral circulation in the hearts of patients with ischemic conditions such as recurrent angina; and Corgentin, a DNA-based therapeutic being developed to enhance myocardial healing in and around the infarct zone when used as an adjunct to existing vascular-directed pharmacologic and interventional therapies. The FDA has granted fast track designation to Generx for the potential treatment of myocardial ischemia. The Tissue Repair Company's lead product candidate, Excellarate, is a DNA-activated collagen gel that utilizes the company's Gene Activated Matrix Technology and is formulated with an adenovector delivery carrier; it is being developed for the treatment of chronic diabetic wounds, such as non-healing diabetic foot ulcers. In July 2009, Cardium sold its InnerCool Therapies subsidiary to Royal Philips Electronics for roughly $11.25 million.

FINANCIALS: Sales and profits are in thousands of dollars—add 000 to get the full amount. 2009 Note: Financial information for 2009 was not available for all companies at press time.

2009 Sales: $ 445	2009 Profits: $-11,680	**U.S. Stock Ticker:** CXM
2008 Sales: $ 417	2008 Profits: $-24,598	**Int'l Ticker:** Int'l Exchange:
2007 Sales: $ 446	2007 Profits: $-25,322	Employees: 14
2006 Sales: $ 756	2006 Profits: $-18,593	Fiscal Year Ends: 12/31
2005 Sales: $	2005 Profits: $-5,442	Parent Company:

SALARIES/BENEFITS:
Pension Plan:	ESOP Stock Plan:	Profit Sharing:	Top Exec. Salary: $279,776	Bonus: $
Savings Plan:	Stock Purch. Plan:		Second Exec. Salary: $263,758	Bonus: $

OTHER THOUGHTS:
Apparent Women Officers or Directors: 2
Hot Spot for Advancement for Women/Minorities: Y

LOCATIONS: ("Y" = Yes)
West:	Southwest:	Midwest:	Southeast:	Northeast:	International:
Y					

CEGEDIM SA

www.cegedim.fr

Industry Group Code: 511210D Ranks within this company's industry group: Sales: 1   Profits: 1

Drugs:	Other:	Clinical:		Computers:		Services:	
Discovery:	AgriBio:	Trials/Services:	Y	Hardware:		Specialty Services:	
Licensing:	Genetic Data:	Labs:		Software:	Y	Consulting:	Y
Manufacturing:	Tissue Replacement:	Equipment/Supplies:		Arrays:		Blood Collection:	
Generics:		Research/Development Svcs.:		Database Management:	Y	Drug Delivery:	
		Diagnostics:				Drug Distribution:	

TYPES OF BUSINESS:

Software - Development & Marketing
Databases
Pharmaceuticals Marketing & Tracking Software
Compliance Services

BRANDS/DIVISIONS/AFFILIATES:

Cegedim Dendrite
Dendrite International Inc
Santesurf
Cegedim Strategic Data
Cegers
Infopharm
InfoSante

CONTACTS: *Note: Officers with more than one job title may be intentionally listed here more than once.*

Jean-Claude Labrune, CEO
Karl Guenault, COO
Daniel Flis, Dir.-Comm.
J. E. Umiastowski, Chief Investor Rel. Officer
Pierre Marucchi, Managing Dir.
Laurent Labrune, CRM & Strategic Data
Alain Missoffe, Healthcare Professionals
Bruno Sarfati, Strategic Data
Jean-Claude Labrune, Chmn.

Phone: 33-1-49-09-22-00	Fax: 33-1-46-03-45-95
Toll-Free:	
Address: 127-137, rue d'Aguesseau, Boulogne-Billancourt, 92641 France	

GROWTH PLANS/SPECIAL FEATURES:

Cegedim S.A., formerly Dendrite International, Inc., is a developer of databases and software solutions for pharmaceutical companies, healthcare professionals, health insurance providers and France-based companies in virtually any market sector. The company operated through three divisions: healthcare professionals; customer relationship management (CRM) and strategic Data; and insurance and services. Cegedim's CRM and strategic data services serve pharmaceutical companies in over 80 countries. Through Cegedim Dendrite, the company develops databases that provide pharmaceutical companies with information regarding where their drugs are sold, who prescribes them and why the drugs are prescribed. Additional healthcare products and services include tools for optimizing information resources; report and analysis tools for office and hospital sales forces; strategic marketing, operational marketing and competition monitoring tools and studies; performance measurement and promotional spending auditing tools; and pharmacy order-taking tools. Sales force optimization services are offered through subsidiary Itops; market research studies are offered through subsidiary Cegedim Strategic Data; sales statistics are offered through subsidiaries Cegers, Infopharm and InfoSante; and prescription analysis is offered through Cegedim Customer Information. The healthcare professionals segment provides software products, information systems and consulting services for pharmacists, doctors and other healthcare professionals through subsidiaries Alliadis; Cegedim Rx; AGDF Cedegin RS; and In Practice Systems (INPS) to name a few. Subsidiary Santesurf provides a free and secure French intranet exclusively for healthcare professionals. Cegedim's insurance and services division provides outsourced human resources management, Internet hosting services, electronic bill payment, medical financial leasing, corporate databases, printing services and sample management. Subsidiaries in this sector include Cegedim Activ; Cetip; and iSante.

FINANCIALS: Sales and profits are in thousands of dollars—add 000 to get the full amount. 2009 Note: Financial information for 2009 was not available for all companies at press time.

2009 Sales: $1,161,910	2009 Profits: $72,710	U.S. Stock Ticker:
2008 Sales: $1,121,330	2008 Profits: $44,250	Int'l Ticker: CGM Int'l Exchange: Paris-Euronext
2007 Sales: $995,010	2007 Profits: $58,340	Employees: 8,600
2006 Sales: $868,600	2006 Profits: $60,700	Fiscal Year Ends: 12/31
2005 Sales: $437,240	2005 Profits: $21,447	Parent Company:

SALARIES/BENEFITS:

Pension Plan:	ESOP Stock Plan:	Profit Sharing:	Top Exec. Salary: $	Bonus: $
Savings Plan:	Stock Purch. Plan:		Second Exec. Salary: $	Bonus: $

OTHER THOUGHTS:

Apparent Women Officers or Directors:
Hot Spot for Advancement for Women/Minorities:

LOCATIONS: ("Y" = Yes)

West:	Southwest:	Midwest:	Southeast:	Northeast:	International:
				Y	Y

CELERA CORPORATION

www.celera.com

Industry Group Code: 541712 Ranks within this company's industry group: Sales: 7 Profits: 14

Drugs:	Other:		Clinical:		Computers:		Services:	
Discovery:	AgriBio:		Trials/Services:		Hardware:		Specialty Services:	
Licensing:	Genetic Data:	Y	Labs:		Software:	Y	Consulting:	
Manufacturing:	Tissue Replacement:	Y	Equipment/Supplies:		Arrays:		Blood Collection:	
Generics:			Research/Development Svcs.:	Y	Database Management:	Y	Drug Delivery:	
			Diagnostics:				Drug Distribution:	

TYPES OF BUSINESS:

Research-Human Genome Mapping
Information Management & Analysis Software
Consulting, Research & Development Services

BRANDS/DIVISIONS/AFFILIATES:

4MyHeart.com
Applied Biosystems Group
Abbott Laboratories
ViroSeq HIV-1 Genotyping System
Berkeley HeartLab, Inc.

CONTACTS: Note: Officers with more than one job title may be intentionally listed here more than once.

Kathy Ordonez, CEO
Ugo DeBlasi, CFO
Paul Arata, VP-Human Resources
Thomas White, Chief Scientific Officer
Michael Zoccoli, Sr. VP-Prod. Group
Paul Arata, VP-Admin.
Scott Milsten, General Counsel/VP/Corp. Sec.
Stacey Sias, Sr. VP-Bus. Dev. & Strategic Planning
David P. Speechly, VP-Corp. Affairs
David P. Speechly, Contact-Investor Rel. & Media
Samuel Broder, Chief Medical Officer
H. Robert Superko, VP-Medical Affairs

Phone: 510-749-4200	Fax:
Toll-Free:	
Address: 1401 Harbor Bay Parkway, Alameda, CA 94502 US	

GROWTH PLANS/SPECIAL FEATURES:

Celera Corporation, formerly a subsidiary of Applied Biosystems, Inc. known as the Celera Group, is a diagnostics business with a focus on personalized disease management. The firm operates through three segments: a clinical laboratory testing services business (lab services), a products business (products) and a segment which includes other activities under corporate management (corporate). The lab services business, conducted through Berkely HeartLab, Inc., offers a portfolio of clinical laboratory tests and disease management services designed to help healthcare providers improve cardiovascular disease treatment. The products business group develops, manufactures and oversees the commercialization of molecular diagnostic products, most of which are commercialized by Abbott Molecular, a subsidiary of Abbott Laboratories. The company has five product categories that are sold through its partnership with Abbott: the ViroSeq HIV-1 Genotyping System; products used for the detection of mutations in the CFTR gene, which cause cystic fibrosis; hepatitis C virus ASRs; ASRs for the detection of mutations in the FMR-1 gene, which cause Fragile X Syndrome; and ASRs for the detection of mutations in genes known to be involved in deep vein thrombosis. The corporate segment includes revenues for royalties, licenses, funded collaborations and milestones related to the licensing of certain intellectual property and from the sale of Celera's former small molecule and proteomic programs. Through 4myheart.com, Celera offers cardiovascular patients and physicians a web-based program that makes cardiovascular health records and lab results easily accessible to physicians and patients, among other services.

The firm offers its employees medical, dental, vision, life and disability insurance; and a 401(k) savings plan.

FINANCIALS: Sales and profits are in thousands of dollars—add 000 to get the full amount. 2009 Note: Financial information for 2009 was not available for all companies at press time.

2009 Sales: $167,100	2009 Profits: $-32,700	U.S. Stock Ticker: CRA
2008 Sales: $138,700	2008 Profits: $-104,100	Int'l Ticker: Int'l Exchange:
2007 Sales: $43,400	2007 Profits: $-20,600	Employees: 551
2006 Sales: $46,200	2006 Profits: $-63,600	Fiscal Year Ends: 6/30
2005 Sales: $31,000	2005 Profits: $-77,100	Parent Company:

SALARIES/BENEFITS:

Pension Plan:	ESOP Stock Plan:	Profit Sharing:	Top Exec. Salary: $650,000	Bonus: $338,254
Savings Plan: Y	Stock Purch. Plan:		Second Exec. Salary: $428,000	Bonus: $147,802

OTHER THOUGHTS:

Apparent Women Officers or Directors: 3
Hot Spot for Advancement for Women/Minorities: Y

LOCATIONS: ("Y" = Yes)

West:	Southwest:	Midwest:	Southeast:	Northeast:	International:
Y	Y		Y	Y	

CELGENE CORP

www.celgene.com

Industry Group Code: 325412 Ranks within this company's industry group: Sales: 31 Profits: 21

Drugs:		Other:	Clinical:		Computers:		Services:	
Discovery:	Y	AgriBio:	Trials/Services:		Hardware:		Specialty Services:	
Licensing:	Y	Genetic Data:	Labs:		Software:		Consulting:	
Manufacturing:		Tissue Replacement:	Equipment/Supplies:		Arrays:		Blood Collection:	
Generics:			Research/Development Svcs.:	Y	Database Management:		Drug Delivery:	
			Diagnostics:				Drug Distribution:	

TYPES OF BUSINESS:

Cancer & Immune-Inflammatory Related Diseases Drugs

BRANDS/DIVISIONS/AFFILIATES:

Revlimid
Thalomid
Vidaza
Focalin
Gloucester Pharmaceuticals
Abraxis Bioscience, Inc.
CC-10004
CC-11050

CONTACTS: Note: Officers with more than one job title may be intentionally listed here more than once.

Sol J. Barer, CEO
Robert J. Hugin, COO
Robert J. Hugin, Pres.
David W. Gryska, CFO/Sr. VP
Sol J. Barer, Chmn.

Phone: 908-673-9000	**Fax:** 732-271-4184
Toll-Free:	
Address: 86 Morris Ave., Summit, NJ 07901 US	

GROWTH PLANS/SPECIAL FEATURES:

Celgene Corp. is a global integrated biopharmaceutical company primarily engaged in the discovery, development and commercialization of therapies designed to treat cancer and immune-inflammatory related diseases. The company's commercial stage products are Revlimid, Thalomid, Vidaza and Focalin. Revlimid has been approved by the U.S. FDA and other regulatory agencies in Europe, Latin America, the Middle East and Asia/Pacific for treatment in combination with dexamethasone for multiple myeloma patients who have received at least one prior therapy. In addition, Revlimid has been approved by the FDA and the Canadian Therapeutics Directorate for treatment of patients with transfusion-dependent anemia due to low- or intermediate-1-risk myelodysplastic syndromes (MDS) associated with a deletion 5q cytogenetic abnormality with or without additional cytogenetic abnormalities. Thalomid has been approved by the FDA for treatment in combination with dexamethasone for patients with newly diagnosed multiple myeloma and is also approved for the treatment and suppression of cutaneous manifestations of erythema nodosum leprosum (ENL), an inflammatory complication of leprosy. Vidaza is approved for treatment in patients with various myelodysplastic syndrome subtypes. Focalin, which is used to treat attention deficit hyperactivity disorder (ADHD), is sold exclusively to Novartis Pharma AG. Its portfolio of drug candidates includes IMiDs compounds, which are proprietary to the firm and have demonstrated certain immunomodulatory and other biologically important properties; CC-10004 and CC-11050, oral anti-inflammatory agents; and kinase inhibitors. In January 2010, the firm acquired Gloucester Pharmaceuticals for $340 million. In July 2010, the company agreed to acquire Abraxis Bioscience, Inc., which sells a form of chemotherapy called Abraxane, for approximately $2.9 billion.

Employees are offered medical, dental and vision insurance; health care and dependent care flexible spending accounts; life and AD&D insurance; business travel accident insurance; travel assistance; disability plans; long-term care insurance; a 401(k) plan; an employee assistance program; educational assistance; health services; and fitness benefits.

FINANCIALS: Sales and profits are in thousands of dollars—add 000 to get the full amount. 2009 Note: Financial information for 2009 was not available for all companies at press time.

2009 Sales: $2,689,893	2009 Profits: $766,747	**U.S. Stock Ticker:** CELG
2008 Sales: $2,254,781	2008 Profits: $-1,533,653	**Int'l Ticker:** Int'l Exchange:
2007 Sales: $1,405,820	2007 Profits: $226,433	Employees: 2,550
2006 Sales: $898,873	2006 Profits: $68,981	Fiscal Year Ends: 12/31
2005 Sales: $536,941	2005 Profits: $63,656	Parent Company:

SALARIES/BENEFITS:

Pension Plan:	ESOP Stock Plan:	Profit Sharing:	Top Exec. Salary: $1,057,667	Bonus: $2,685,397
Savings Plan: Y	Stock Purch. Plan:		Second Exec. Salary: $770,000	Bonus: $1,682,455

OTHER THOUGHTS:

Apparent Women Officers or Directors: 2
Hot Spot for Advancement for Women/Minorities:

LOCATIONS: ("Y" = Yes)

West:	Southwest:	Midwest:	Southeast:	Northeast:	International:
				Y	Y

CELL THERAPEUTICS INC
www.celltherapeutics.com

Industry Group Code: 325412 Ranks within this company's industry group: Sales: 152 Profits: 158

Drugs:		Other:		Clinical:	Computers:	Services:
Discovery:	Y	AgriBio:		Trials/Services:	Hardware:	Specialty Services:
Licensing:	Y	Genetic Data:		Labs:	Software:	Consulting:
Manufacturing:		Tissue Replacement:		Equipment/Supplies:	Arrays:	Blood Collection:
Generics:				Research/Development Svcs.:	Database Management:	Drug Delivery:
				Diagnostics:		Drug Distribution:

TYPES OF BUSINESS:
Cancer Treatment Drugs

BRANDS/DIVISIONS/AFFILIATES:
Zevalin
Paclitaxel Poliglumex
Pixantrone
Brostallicin
Systems Medicine LLC
Aequus BioPharma, Inc.
Opaxio

CONTACTS: *Note: Officers with more than one job title may be intentionally listed here more than once.*
James A. Bianco, CEO
Craig W. Philips, Pres.
Jack W. Singer, Chief Medical Officer
Louis A. Bianco, Exec. VP-Admin.
Dan Eramian, Exec. VP-Corp. Comm.
Louis A. Bianco, Exec. VP-Finance
Christina Waters, Pres., Cell Therapuetics Sede Secondaria
Phillip Nudelman, Chmn.

Phone: 206-282-7100	Fax: 206-284-6206
Toll-Free: 800-215-2355	
Address: 501 Elliott Ave. W., Ste. 400, Seattle, WA 98119 US	

GROWTH PLANS/SPECIAL FEATURES:

Cell Therapeutics, Inc. (CTI) develops, acquires and commercializes treatments for cancer. The company's research and in-licensing activities are concentrated on identifying and developing new, less toxic and more effective ways to treat cancer. The firm is currently developing three drugs: Opaxio (paclitaxel poliglumex), Pixantrone and brostallicin, through its wholly-owned subsidiary Systems Medicine LLC. Opaxio, a biologically enhanced chemotherapeutic agent, links a widely used anti-cancer agent, paclitaxel, to a polyglutamate polymer delivery system for the potential treatment of ovarian and other cancers. Pixantrone is an anthracycline derivative for the safer treatment of non-Hodgkin's lymphoma. Finally, brostallicin, which is a small molecule, anti-cancer drug with a unique mechanism of action and composition of matter patent coverage, is being tested for treatment of relapsed/refractory soft tissue sarcoma. CTI is also working on new novel bisplatinum analogues that build on the successes of treatments such as Cisplatin, a platinum-based chemotherapy drug. The company has exclusive rights to 12 issued U.S. patents and 129 U.S. and foreign pending or issued patent applications relating to its polymer drug delivery technology, and has licensed five granted U.S. patents and 394 patents pending for brostallicin to Systems Medicine, LLC. CTI also works with its subsidiaries Systems Medicine, which integrates cellular genomics with medicine, and Aequus BioPharma, Inc., which develops biotherapeutics based on CTI's proprietary Genetic Polymer technology. In September 2008, CTI finalized the closure of its operational facilities in Bresso, Italy, which had been used for pre-clinical research. In April 2010, the company announced a decrease in operating expenses by 21% and of workforce reduction of 36 employees

CTI offers its employees medical, dental, vision, life and disability insurance; an employee assistance program; travel assistance; and an educational assistance program.

FINANCIALS: Sales and profits are in thousands of dollars—add 000 to get the full amount. 2009 Note: Financial information for 2009 was not available for all companies at press time.

		U.S. Stock Ticker: CTIC
2009 Sales: $ 80	2009 Profits: $-95,400	Int'l Ticker: Int'l Exchange:
2008 Sales: $11,432	2008 Profits: $-180,029	Employees: 194
2007 Sales: $ 127	2007 Profits: $-138,108	Fiscal Year Ends: 12/31
2006 Sales: $ 80	2006 Profits: $-135,819	Parent Company:
2005 Sales: $16,092	2005 Profits: $-102,505	

SALARIES/BENEFITS:

Pension Plan:	ESOP Stock Plan:	Profit Sharing:	Top Exec. Salary: $650,000	Bonus: $
Savings Plan: Y	Stock Purch. Plan: Y		Second Exec. Salary: $402,000	Bonus: $

OTHER THOUGHTS:
Apparent Women Officers or Directors: 2
Hot Spot for Advancement for Women/Minorities: Y

LOCATIONS: ("Y" = Yes)

West:	Southwest:	Midwest:	Southeast:	Northeast:	International:
Y					Y

CELLARTIS AB

www.cellartis.com

Industry Group Code: 325414 Ranks within this company's industry group: Sales: Profits:

Drugs:	Other:	Clinical:		Computers:	Services:	
Discovery:	AgriBio:	Trials/Services:	Y	Hardware:	Specialty Services:	Y
Licensing:	Genetic Data:	Labs:		Software:	Consulting:	
Manufacturing:	Tissue Replacement:	Equipment/Supplies:		Arrays:	Blood Collection:	
Generics:		Research/Development Svcs.:		Database Management:	Drug Delivery:	
		Diagnostics:			Drug Distribution:	

TYPES OF BUSINESS:

Cardiac Stemcell Fabrication
Drug Development Tests

BRANDS/DIVISIONS/AFFILIATES:

CONTACTS: *Note: Officers with more than one job title may be intentionally listed here more than once.*

Mats Lundwall, CEO
Johan Hyllner, COO
Lars Pettersson, CFO
Johan Hyllner, Chief Scientific Officer
Kristina Runeberg, Chief Bus. Dev. Officer-Sweden
Barry Middleton, Chief Bus. Dev. Officer-U.K.

Phone: 46-31-758-09-00	Fax: 46-31-758-09-10
Toll-Free:	
Address: Arvid Wallgrens Backe 20, Goteborg, SE-413 46 Sweden	

GROWTH PLANS/SPECIAL FEATURES:

Cellartis AB is a startup biotechnology firm focused on pluripotent stem cells and technology utilized in drug discovery research, toxicity testing and regenerative medicine. Cellartis has developed more than 30 stem cell lines, making it one of the world's largest sources of stem cells. Primarily, the firm is concerned with the development of hepatocytes (cells in the liver) and cardiomyocytes (cells in the heart) for use in drug discovery. Cellartis' products include cardiomyocytes, which are available as clusters and 2D monolayers; hepatocyte-like cells; mesenchymal progenitors; pluripotent stem cells; human stem cell lines; stem cell antibodies; cell culture medium-vitrohes; cell culture tools; and a kit for human embryonic stem cell characterization. Cellartis has partnerships with Novo Nordisk and Lund University to derive insulin-producing cells from stem cells to treat diabetes; Mitsubishi Tanabe Pharma for Parkinson's disease; Invitrogen for multi-lineage reporter stem cell lines; Vitrolife for hES cell culture media; and TATAA Biocenter for specific characterization tools. The company is headquartered in Sweden, with offices in Dundee in the U.K.

FINANCIALS: Sales and profits are in thousands of dollars—add 000 to get the full amount. 2009 Note: Financial information for 2009 was not available for all companies at press time.

2009 Sales: $	2009 Profits: $	**U.S. Stock Ticker: Private**	
2008 Sales: $	2008 Profits: $	**Int'l Ticker:** Int'l Exchange:	
2007 Sales: $	2007 Profits: $	Employees:	
2006 Sales: $	2006 Profits: $	Fiscal Year Ends:	
2005 Sales: $	2005 Profits: $	Parent Company:	

SALARIES/BENEFITS:

Pension Plan:	ESOP Stock Plan:	Profit Sharing:	Top Exec. Salary: $	Bonus: $
Savings Plan:	Stock Purch. Plan:		Second Exec. Salary: $	Bonus: $

OTHER THOUGHTS:

Apparent Women Officers or Directors: 2
Hot Spot for Advancement for Women/Minorities:

LOCATIONS: ("Y" = Yes)

West:	Southwest:	Midwest:	Southeast:	Northeast:	International: Y

CELLDEX THERAPEUTICS INC

www.celldextherapeutics.com

Industry Group Code: 325412 Ranks within this company's industry group: Sales: 112 Profits: 139

Drugs:		Other:	Clinical:	Computers:	Services:
Discovery:	Y	AgriBio:	Trials/Services:	Hardware:	Specialty Services:
Licensing:		Genetic Data:	Labs:	Software:	Consulting:
Manufacturing:	Y	Tissue Replacement:	Equipment/Supplies:	Arrays:	Blood Collection:
Generics:			Research/Development Svcs.:	Database Management:	Drug Delivery:
			Diagnostics:		Drug Distribution:

TYPES OF BUSINESS:

Drugs - Vaccines & Immunotherapeutics
Drugs - Oncology, Inflammatory Disease & Infectious Disease

BRANDS/DIVISIONS/AFFILIATES:

CuraGen Corporation
CholeraGarde
Rotarix
CholeraGarde
CDX-110
CDX-1307
APC Targeting Technology
CDX-1135

CONTACTS: Note: Officers with more than one job title may be intentionally listed here more than once.

Anthony S. Marucci, CEO
Anthony S. Marucci, Pres.
Avery W. (Chip) Catlin, CFO/Sr. VP
Tibor Keler, Chief Scientific Officer/Sr. VP
Thomas Davis, Chief Medical Officer/Sr. VP
Larry Ellberger, Chmn.

Phone: 781-433-0771	Fax: 781-433-0262
Toll-Free:	
Address: 119 4th Ave., Needham, MA 02494 US	

GROWTH PLANS/SPECIAL FEATURES:

Celldex Therapeutics, Inc. is a biopharmaceutical company that applies its Precision Targeted Immunotherapy Platform to generate a pipeline of candidates to treat cancer and other difficult-to-treat diseases. The company's APC Targeting Technology uses human monoclonal antibodies linked to disease associated antigens to efficiently deliver the attached antigens to immune cells known as antigen presenting cells (APC). The firm's immunotherapy platform includes a complementary portfolio of monoclonal antibodies, antibody-targeted vaccines and immunomodulators to create novel disease-specific drug candidates. Celldex currently has one product on the market, five products in clinical development and four products in preclinical development. Its marketed product, Rotarix, is a rotavirus vaccine. In Phase II trials are its CDX-110 vaccine for the tumor specific molecule EGFRvIII and CDX-011, which targets glycoprotein NMB that is expressed in a variety of human cancers, including breast cancer and melanoma. In Phase I is CDX-1307, its primary ACP Targeting Technology product candidate, for the treatment of colorectal, pancreatic, bladder, ovarian and breast cancers. Its Phase I/2 candidates are CDX-1401 for multiple solid tumors; and CDX-1135 for renal disease. In pre-clinical development are its CDX-1127 for immuno-modulation and solid tumors; CDX-1189 for renal disease; CDX-301 for cancer, autoimmune disease and transplants; and CDX-014 for renal and ovarian cancer. The company's bacterial vaccines are the CholeraGarde cholera vaccine, the Ty800 typhoid vaccine and the ETEC E. coli vaccine. In October 2009, the firm acquired and absorbed CuraGen Corporation.

Employees are offered medical, dental and vision coverage; flexible spending accounts; life and AD&D insurance; disability coverage; a 401(k) plan; an employee assistance program; educational assistance; and an employee stock purchase plan.

FINANCIALS: Sales and profits are in thousands of dollars—add 000 to get the full amount. 2009 Note: Financial information for 2009 was not available for all companies at press time.

2009 Sales: $15,180	2009 Profits: $-36,525	**U.S. Stock Ticker:** CLDX
2008 Sales: $7,456	2008 Profits: $-47,501	**Int'l Ticker:** Int'l Exchange:
2007 Sales: $1,406	2007 Profits: $-15,073	Employees: 93
2006 Sales: $ 899	2006 Profits: $-17,835	Fiscal Year Ends: 12/31
2005 Sales: $3,088	2005 Profits: $-18,097	Parent Company:

SALARIES/BENEFITS:

Pension Plan:	ESOP Stock Plan:	Profit Sharing:	Top Exec. Salary: $302,800	Bonus: $137,400
Savings Plan: Y	Stock Purch. Plan: Y		Second Exec. Salary: $300,000	Bonus: $96,600

OTHER THOUGHTS:

Apparent Women Officers or Directors: 1
Hot Spot for Advancement for Women/Minorities:

LOCATIONS: ("Y" = Yes)

West:	Southwest:	Midwest:	Southeast:	Northeast:	International:
				Y	

CELLULAR DYNAMICS INTERNATIONAL   www.cellular-dynamics.com

Industry Group Code: 325414 Ranks within this company's industry group: Sales: Profits:

Drugs:	Other:	Clinical:		Computers:	Services:	
Discovery:	AgriBio:	Trials/Services:	Y	Hardware:	Specialty Services:	Y
Licensing:	Genetic Data:	Labs:		Software:	Consulting:	
Manufacturing:	Tissue Replacement:	Equipment/Supplies:		Arrays:	Blood Collection:	
Generics:		Research/Development Svcs.:		Database Management:	Drug Delivery:	
		Diagnostics:			Drug Distribution:	

TYPES OF BUSINESS:

Cardiac Stemcell Fabrication
Drug Assays

GROWTH PLANS/SPECIAL FEATURES:

Cellular Dynamics International (CDI) is a startup biotechnology firm that utilizes human induced pluripotent stem (iPS) cells from adult tissue. The cells are used to for a variety of applications, such as drug discovery, drug development, toxicity testing and personalized medicine. The company's iCell Cardiomyocytes are human cardiac cells for the purpose of aiding drug discovery, drug predictability efficacy and toxicity screenings. The company has several types of cells in development, mainly hematopoietic (blood) cells, such as Macrophages, a type of white blood cell, which aids in the process of removing foreign material and dying cells; and mast cells, or mastocytes, which are found in both connective tissue and mucous membranes and act as part of the body's immune response system. In December 2009, the firm announced the commercial launch of the iCell Cardiomyocytes for use in drug candidate toxicity screening.

BRANDS/DIVISIONS/AFFILIATES:

iCell

CONTACTS: Note: Officers with more than one job title may be intentionally listed here more than once.

Robert J. Palay, CEO
Emile Nuwaysir, COO/VP
Thomas M. Palay, Pres.
David Snyder, CFO/VP
James A. Thomson, Chief Scientific Officer
Nicholas Seay, CTO/VP
Chris Parker, Chief Commercial Officer/VP

Phone: 608-310-5100	Fax:
Toll-Free: 877-310-5100	
Address: 525 Science Dr., Madison, WI 53711 US	

FINANCIALS: Sales and profits are in thousands of dollars—add 000 to get the full amount. 2009 Note: Financial information for 2009 was not available for all companies at press time.

2009 Sales: $	2009 Profits: $	**U.S. Stock Ticker: Private**
2008 Sales: $	2008 Profits: $	**Int'l Ticker:** Int'l Exchange:
2007 Sales: $	2007 Profits: $	Employees:
2006 Sales: $	2006 Profits: $	Fiscal Year Ends:
2005 Sales: $	2005 Profits: $	Parent Company:

SALARIES/BENEFITS:

Pension Plan:	ESOP Stock Plan:	Profit Sharing:	Top Exec. Salary: $	Bonus: $
Savings Plan:	Stock Purch. Plan:		Second Exec. Salary: $	Bonus: $

OTHER THOUGHTS:

Apparent Women Officers or Directors:
Hot Spot for Advancement for Women/Minorities:

LOCATIONS: ("Y" = Yes)

West:	Southwest:	Midwest:	Southeast:	Northeast:	International:
		Y			

CEL-SCI CORPORATION

www.cel-sci.com

Industry Group Code: 325412 Ranks within this company's industry group: Sales: 153   Profits: 141

Drugs:		Other:	Clinical:	Computers:	Services:
Discovery:	Y	AgriBio:	Trials/Services:	Hardware:	Specialty Services:
Licensing:		Genetic Data:	Labs:	Software:	Consulting:
Manufacturing:	Y	Tissue Replacement:	Equipment/Supplies:	Arrays:	Blood Collection:
Generics:			Research/Development Svcs.:	Database Management:	Drug Delivery:
			Diagnostics:		Drug Distribution:

TYPES OF BUSINESS:

Drugs-Cancer & Infectious Disease
Vaccines
Immunotherapy

BRANDS/DIVISIONS/AFFILIATES:

MultiKine
LEAPS

CONTACTS: *Note: Officers with more than one job title may be intentionally listed here more than once.*

Geert R. Kersten, CEO
Maximilian de Clara, Pres.
Eyal Talor, Chief Scientific Officer
Todd Burkhart, VP-Mfg., Facilities & Commercial Oper.
Patricia B. Prichep, Sr. VP-Oper.
William (Brooke) Jones, VP-Quality Assurance
Daniel Zimmerman, Sr. VP-Research, Cellular Immunology
John Cipriano, Sr. VP-Regulatory Affairs
Maximilian de Clara, Chmn.

Phone: 703-506-9460	Fax: 703-506-9471
Toll-Free:	
Address: 8229 Boone Blvd., Ste. 802, Vienna, VA 22182 US	

GROWTH PLANS/SPECIAL FEATURES:

CEL-SCI Corp. researches and develops immunotherapy products for the treatment of cancer and infectious diseases. Its flagship product, Multikine, is the first of a new class of cancer immunotherapy drugs called Immune SIMULATORs. It simulates the activities of a healthy person's immune system to combat a tumor as well as killing cancer cells in a targeted fashion. The company is currently planning to conduct an international Phase III clinical study of Multikine in the treatment of advanced primary squamous cell carcinoma of the head and neck. Previous trials have shown that Multikine is safe and non-toxic and renders cancer cells much more susceptible to radiation therapy. The company also runs a cold-fill manufacturing and storage facility. Here CEL-SCI, in addition to manufacturing Multikine, can store biologics that require near freezing temperatures to prevent degradation. The company offers fill and finish storage to other pharmaceutical companies. Additionally, CEL-SCI owns a T-cell modulation process technology called Ligand Epitope Antigen Presentations System (LEAPS), through which it has developed the Cel-1000 and 2000 peptides. Cel-1000 has proven affective in protecting animals against herpes, malaria, viral encephalitis and cancer. The company has also tested its effectiveness against avian flu and hepatitis B. LEAPS may also have applications in any disease for which antigenic epitope sequences have been identified, such as infectious diseases, autoimmune diseases, allergic asthma, allergies and select CNS diseases, such as Alzheimer's. CEL-2000 is a peptide vaccine for the prevention or retardation of tissue damage caused by rheumatoid arthritis. In November 2009, the firm began a Phase I clinical trial of the effectiveness of CEL-1000 against the H1N1 virus. In January 2010, CEL-SCI commenced manufacturing of Multikine in its new cold-fill facility.

CEL-SCI offers its employees health, dental, vision, and long term disability insurance; a 401(k) plan; and flexible spending accounts.

FINANCIALS: Sales and profits are in thousands of dollars—add 000 to get the full amount. 2009 Note: Financial information for 2009 was not available for all companies at press time.

2009 Sales: $ 80	2009 Profits: $-40,910	**U.S. Stock Ticker: CVM**
2008 Sales: $ 5	2008 Profits: $-7,703	**Int'l Ticker:** Int'l Exchange:
2007 Sales: $ 57	2007 Profits: $-9,630	Employees: 29
2006 Sales: $ 125	2006 Profits: $-7,939	Fiscal Year Ends: 9/30
2005 Sales: $ 270	2005 Profits: $-3,040	Parent Company:

SALARIES/BENEFITS:

Pension Plan:	ESOP Stock Plan:	Profit Sharing:	Top Exec. Salary: $408,691	Bonus: $
Savings Plan: Y	Stock Purch. Plan:		Second Exec. Salary: $334,720	Bonus: $

OTHER THOUGHTS:

Apparent Women Officers or Directors: 1
Hot Spot for Advancement for Women/Minorities:

LOCATIONS: ("Y" = Yes)

West:	Southwest:	Midwest:	Southeast:	Northeast:	International:
				Y	

CELSIS INTERNATIONAL PLC

www.celsis.com

Industry Group Code: 325413 Ranks within this company's industry group: Sales: 15 Profits: 10

Drugs:	Other:	Clinical:		Computers:		Services:	
Discovery:	AgriBio:	Trials/Services:		Hardware:		Specialty Services:	Y
Licensing:	Genetic Data:	Labs:	Y	Software:	Y	Consulting:	
Manufacturing:	Tissue Replacement:	Equipment/Supplies:	Y	Arrays:	Y	Blood Collection:	
Generics:		Research/Development Svcs.:		Database Management:		Drug Delivery:	
		Diagnostics:	Y			Drug Distribution:	

TYPES OF BUSINESS:

Microbial Contamination Detection Instruments
Laboratory Services Outsourcing
Microbiotics Detection

BRANDS/DIVISIONS/AFFILIATES:

Celsis Rapid Detection
Celsis In Vitro Technologies
Celsis Analytical Services
RapiScreen
AKuScreen
ReACT

CONTACTS: *Note: Officers with more than one job title may be intentionally listed here more than once.*

Jay LeCoque, CEO
Christian Madrolle, Dir.-Finance
Jack Rowell, Chmn.

Phone: 312-476-1200	Fax: 312-476-1201
Toll-Free:	
Address: 600 W. Chicago Ave., Ste. 625, Chicago, IL 60654-2822 US	

GROWTH PLANS/SPECIAL FEATURES:

Celsis International plc is a global provider of life sciences products and laboratory services to the pharmaceutical and consumer products industries. Its core technology involves bioluminescence-based technology, in which ATP (adenosine triphosphate), a nucleotide present in all biological matter, is catalyzed to produce light (as in the abdomen of a firefly). The firm operates through three divisions: Celsis Rapid Detection, Celsis In Vitro Technologies and Celsis Analytical Services. Celsis Rapid Detection uses proprietary enzyme technology to develop and supply screening systems, including instruments, software and reagents, for the rapid detection of microbial contamination in pharmaceutical and consumer products. Celsis's detection product lines include RapiScreen and AKuScreen assays. Celsis In Vitro Technologies uses proprietary hepatocyte (liver cell) and cryopreservation technologies to provide in vitro testing products for pharmaceutical and biotechnological companies, and is a leading provider of ADME-Tox products for drug discovery and development. Celsis Analytical Services provides outsourced laboratory testing services, including analytical chemistry and microbiological laboratory services and stability testing and storage programs, to pharmaceutical, biopharmaceutical, medical device and consumer products companies. The firm has offices in the U.K., the U.S., France, Germany and Belgium. Sales to the Americas generated 65% of the company's 2009 revenue, while sales to Europe generated 24% and sales to other regions generated 11%. Celsis serves the pharmaceuticals, personal care, home care, consumer products and food and beverage industries. During 2009, the pharmaceutical industry generated 59% of the company's revenue, while the remaining 41% came from the consumer market. Products offered by Celsis generated 65% of its 2009 revenue, while services generated 35%. In March 2009, Celsis launched its ReACT RNA-based assay for detecting objectionable organisms without the need for expensive equipment or specialized training. In October 2009, the firm transitioned from being a public company to being a private company.

FINANCIALS: Sales and profits are in thousands of dollars—add 000 to get the full amount. 2009 Note: Financial information for 2009 was not available for all companies at press time.

2009 Sales: $52,500	2009 Profits: $6,800	U.S. Stock Ticker:
2008 Sales: $52,900	2008 Profits: $6,700	Int'l Ticker: CEL Int'l Exchange: London-LSE
2007 Sales: $47,441	2007 Profits: $5,921	Employees: 230
2006 Sales: $33,104	2006 Profits: $4,603	Fiscal Year Ends: 3/31
2005 Sales: $30,397	2005 Profits: $7,999	Parent Company:

SALARIES/BENEFITS:

Pension Plan:	ESOP Stock Plan:	Profit Sharing:	Top Exec. Salary: $451,523	Bonus: $151,235
Savings Plan: Y	Stock Purch. Plan:		Second Exec. Salary: $394,485	Bonus: $142,078

OTHER THOUGHTS:

Apparent Women Officers or Directors:
Hot Spot for Advancement for Women/Minorities:

LOCATIONS: ("Y" = Yes)

West:	Southwest:	Midwest:	Southeast:	Northeast:	International:
		Y		Y	Y

CEPHALON INC

www.cephalon.com

Industry Group Code: 325412 Ranks within this company's industry group: Sales: 32 Profits: 36

Drugs:		Other:	Clinical:		Computers:		Services:	
Discovery:	Y	AgriBio:	Trials/Services:		Hardware:		Specialty Services:	
Licensing:	Y	Genetic Data:	Labs:		Software:		Consulting:	
Manufacturing:		Tissue Replacement:	Equipment/Supplies:		Arrays:		Blood Collection:	
Generics:			Research/Development Svcs.:	Y	Database Management:		Drug Delivery:	
			Diagnostics:				Drug Distribution:	

TYPES OF BUSINESS:

Pharmaceutical Discovery & Development
Neurological Disorder Treatments
Cancer Treatments
Pain Medications
Addiction Treatment

BRANDS/DIVISIONS/AFFILIATES:

Nuvigil
Provigil
Gabitril
Amrix
Fentora
Actiq
Treanda
Trisenox

CONTACTS: Note: Officers with more than one job title may be intentionally listed here more than once.

Frank Baldino, Jr., CEO
J. Kevin Buchi, COO
Wilco Groenhuysen, CFO/Exec. VP
Jeffry L. Vaught, Chief Scientific Officer/Exec. VP
Carl A. Savini, Chief Admin. Officer/Exec. VP
Gerald J. Pappert, General Counsel/Exec. VP
Lesley Russell, Chief Medical Officer/Exec. VP
Valli F. Baldassano, Chief Compliance Officer/Exec. VP
Peter E. Grebow, Exec. VP-Cephalon Ventures
Frank Baldino, Jr., Chmn.
Alain Aragues, Pres., Cephalon Europe/Exec. VP

Phone: 610-344-0200	Fax: 610-738-6590
Toll-Free:	
Address: 41 Moores Rd., Frazer, PA 19355 US	

GROWTH PLANS/SPECIAL FEATURES:

Cephalon, Inc. is a biopharmaceutical company focused on the discovery, development and marketing of products in four core areas: central nervous system (CNS) disorders, pain, oncology and inflammatory disease. The company's CNS products include Nuvigil and Provigil for improving wakefulness in patients with excessive sleepiness associated with narcolepsy, obstructive sleep apnea/hypopnea syndrome (OSA/HS) and shift work sleep disorder (SWSD); and Gabitril for use as adjunctive therapy in the treatment of partial seizures in epileptic patients. The firm's pain therapeutics portfolio includes four marketed products in the U.S.: Amrix, a once-a-day, extended release version of cyclobenzaprine hydrochloride for relief of muscle spasms associated with acute, painful musculoskeletal conditions; Fentora for the management of breakthrough pain in patients with cancer; and Actiq and generic OTFC for the management of breakthrough cancer pain in opioid-tolerant patients. Cephalon's oncology products include two marketed products to treat patients with hematologic cancers: Treanda, a bi-functional hybrid cytotoxic; and Trisenox, an intravenous arsenic-based targeted therapy currently marketed in the U.S. and Europe. The firm also has two commercialized oncology products in Europe: Myocet, a cardio-protective chemotherapy agent used to treat metastatic breast cancer; and Targretin, a treatment for cutaneous T-cell lymphoma. Lastly, the firm's inflammatory disease products include Lupuzor, with which the firm intends to commence a Phase IIb study; CEP-37247, a treatment for patients with sciatica; and Cinquil, an antibody for interleukin-5 (IL-5). The firm markets and sells 30 different branded products in over 50 countries. In April 2010, the firm acquired Mepha AG for $615.4 million and Ception Therapeutics, Inc.

Employees are offered medical, prescription drug, dental and disability coverage; life insurance; healthcare and dependent care reimbursement accounts; a health advocate; a 401(k) plan and profit sharing plan; educational reimbursement; adoption financial assistance; an employee assistance program; employee referral bonuses; and work-life services.

FINANCIALS: Sales and profits are in thousands of dollars—add 000 to get the full amount. 2009 Note: Financial information for 2009 was not available for all companies at press time.

2009 Sales: $2,192,308	2009 Profits: $210,727	**U.S. Stock Ticker: CEPH**
2008 Sales: $1,974,554	2008 Profits: $171,889	**Int'l Ticker:** Int'l Exchange:
2007 Sales: $1,772,638	2007 Profits: $-226,429	Employees: 3,026
2006 Sales: $1,764,069	2006 Profits: $144,816	Fiscal Year Ends: 12/31
2005 Sales: $1,211,892	2005 Profits: $-174,954	Parent Company:

SALARIES/BENEFITS:

Pension Plan:	ESOP Stock Plan:	Profit Sharing: Y	Top Exec. Salary: $1,306,800	Bonus: $1,306,800
Savings Plan: Y	Stock Purch. Plan:		Second Exec. Salary: $590,300	Bonus: $253,800

OTHER THOUGHTS:

Apparent Women Officers or Directors: 3
Hot Spot for Advancement for Women/Minorities: Y

LOCATIONS: ("Y" = Yes)

West:	Southwest:	Midwest:	Southeast:	Northeast:	International:
Y		Y		Y	Y

CEPHEID
www.cepheid.com

Industry Group Code: 325413 Ranks within this company's industry group: Sales: 8   Profits: 18

Drugs:	Other:	Clinical:		Computers:	Services:
Discovery:	AgriBio:	Trials/Services:		Hardware:	Specialty Services:
Licensing:	Genetic Data:	Labs:		Software:	Consulting:
Manufacturing:	Tissue Replacement:	Equipment/Supplies:	Y	Arrays:	Blood Collection:
Generics:		Research/Development Svcs.:		Database Management:	Drug Delivery:
		Diagnostics:	Y		Drug Distribution:

TYPES OF BUSINESS:
Equipment-Biological Testing
Genetic Profiling
DNA Analysis Systems
Molecular Diagnostics

BRANDS/DIVISIONS/AFFILIATES:
Smart Cycler
GeneXpert Infinity
GeneXpert
IVD
ASR
RUO
Xpert C
Xpert vanA

CONTACTS: *Note: Officers with more than one job title may be intentionally listed here more than once.*
John L. Bishop, CEO
Humberto Reyes, COO/Exec. VP
Andrew D. Miller, CFO/Sr. VP
Robert J. Koska, Sr. VP-Sales & Mktg.
Laurie King, Sr. VP-Human Resources
Peter J. Dailey, Sr. VP-R&D
Jan Steuperaert, VP-IT
David H. Persing, CTO/Chief Medical Officer/Exec. VP
Lee Christel, VP-Eng. & System Integration
Joseph H. Smith, Sr. VP-Legal
Humberto Reyes, Exec. VP-Oper.
Joseph H. Smith, Sr. VP-Bus. Dev.
Jared Tipton, Dir.-Corp. Comm.
Jacquie Ross, Contact-Investor Rel.
Michael Myhre, VP/Corp. Controller
Anita Herrstrom-Sjoberg, VP/Managing Dir.-Cepheid AB
Russel K. Enns, Sr. VP-Regulatory, Clinical & Govt. Affairs
Nicolaas Arnold, Sr. VP-Worldwide Commercial Oper.
Sandra Finley, VP-Mktg.
Thomas L. Gutshall, Chmn.
Rika Dutau, VP/Managing Dir.-Cepeid Europe

Phone: 408-541-4191	**Fax:** 408-541-4192
Toll-Free: 888-838-3222	
Address: 904 Caribbean Dr., Sunnyvale, CA 94089 US	

GROWTH PLANS/SPECIAL FEATURES:
Cepheid is a molecular diagnostics company that develops, manufactures and markets fully integrated systems for the clinical genetic assessment, life sciences, industrial and biothreat markets. Cepheid systems enable rapid molecular testing for organisms and genetic-based diseases by implementing automated technology to reduce the complicated and time-intensive steps that are usually involved in molecular testing such as sample preparation, DNA amplification and GeneXpert detection of targeted genes. The company's two principal system platforms are the GeneXpert and the SmartCycler systems. The GeneXpert system integrates sample preparation with DNA amplification and detection, and the SmartCycler integrates DNA amplification and detection in order to rapidly analyze samples. The latest addition to the GeneXpert product line, the GeneXpert Infinity System, is capable of performing up to 1,300 different molecular tests in a 24 hour period with Xpert tests that typically require less than two minutes of hands-on time per test. Both systems are designed for a variety of user types ranging from reference laboratories and hospital central laboratories to satellite testing locations, such as hospital intensive care units and doctors' offices. In addition the GeneXpert is suitable for applications in the biothreat market. Other products include tests designed for use with both the GeneXpert and SmartCycler systems such as IVD medical devices, Analyte Specific Reagents (ASRs) and Research Use Only (RUO) tests with applications in the healthcare industry such as infectious disease, women's health and oncology. In July 2009, Cepheid received clearance from the FDA to market its Xpert C difficile test, an on-demand molecular diagnostic test for the bacterium that causes Clostridium difficile infection (CDI). In January 2010, Cepheid received FDA clearance to market Xpert vanA, its test for vanA, the antimicrobial resistance gene most commonly associated with vancomycin-resistant enterococci (VRE).

Cepheid offers full and part-time employees medical, dental and vision coverage; flexible spending accounts; an employee assistance plan; an income protections plan; an employee stock purchase plan; and a 401(k).

FINANCIALS: Sales and profits are in thousands of dollars—add 000 to get the full amount. 2009 Note: Financial information for 2009 was not available for all companies at press time.

2009 Sales: $170,627	2009 Profits: $-22,502	**U.S. Stock Ticker:** CPHD
2008 Sales: $169,627	2008 Profits: $-22,387	**Int'l Ticker:** Int'l Exchange:
2007 Sales: $129,473	2007 Profits: $-21,423	Employees: 530
2006 Sales: $87,352	2006 Profits: $-25,985	Fiscal Year Ends: 12/31
2005 Sales: $85,010	2005 Profits: $-13,594	Parent Company:

SALARIES/BENEFITS:
Pension Plan:	ESOP Stock Plan:	Profit Sharing:	Top Exec. Salary: $488,462	Bonus: $
Savings Plan: Y	Stock Purch. Plan: Y		Second Exec. Salary: $383,850	Bonus: $

OTHER THOUGHTS:
Apparent Women Officers or Directors: 5
Hot Spot for Advancement for Women/Minorities: Y

LOCATIONS: ("Y" = Yes)
West:	Southwest:	Midwest:	Southeast:	Northeast:	International:
Y					Y

CERUS CORPORATION

www.cerus.com

Industry Group Code: 325412 Ranks within this company's industry group: Sales: 110   Profits: 125

Drugs:		Other:	Clinical:	Computers:	Services:
Discovery:	Y	AgriBio:	Trials/Services:	Hardware:	Specialty Services:
Licensing:	Y	Genetic Data:	Labs:	Software:	Consulting:
Manufacturing:		Tissue Replacement:	Equipment/Supplies:	Arrays:	Blood Collection:
Generics:			Research/Development Svcs.:	Database Management:	Drug Delivery:
			Diagnostics:		Drug Distribution:

TYPES OF BUSINESS:

Pharmaceutical Development
Cancer Vaccines
Blood Treatment Systems

BRANDS/DIVISIONS/AFFILIATES:

INTERCEPT
Helinx
Cerus Europe BV
BioOne Corporation
Grifols SA
Anza Therapeutics Inc

CONTACTS: *Note: Officers with more than one job title may be intentionally listed here more than once.*

Claes Glassell, CEO
Claes Glassell, Pres.
Lainie Corten, Dir.-Mktg.
Laurence M. Corash, Chief Medical Officer/Sr. VP
Lori L. Roll, VP-Admin./Corp. Sec.
William M. Greenman, Chief Bus. Officer
Lainie Corten, Dir.- Global Comm.
Kevin D. Green, VP-Finance/Chief Acct. Officer
B. J. Cassin, Chmn.
Caspar Hogeboom, Managing Dir.-Europe

Phone: 925-288-6000	Fax: 925-288-6001
Toll-Free:	
Address: 2411 Stanwell Dr., Concord, CA 94520 US	

GROWTH PLANS/SPECIAL FEATURES:

Cerus Corp. develops and commercializes novel, proprietary products and technologies within the field of blood safety. Cerus is developing and commercializing the INTERCEPT Blood System for platelets, plasma and red blood cells, which is based on the company's proprietary Helinx technology for controlling biological replication. INTERCEPT is designed to enhance the safety of donated blood components by inactivating viruses, bacteria, parasites and other pathogens, as well as potentially harmful white blood cells. Cerus' Helinx technology prevents the replication of the DNA or RNA of susceptible pathogens, such as hepatitis B, hepatitis C and HIV. INTERCEPT for platelets and plasma is marketed in the European Union and undergoing clinical trials in the US. Trials for INTERCEPT for red blood cells were temporarily halted due to unsatisfactory results in 2003, but a new round of testing has begun. The firm licenses its technology to other companies for international distribution, including subsidiary Cerus Europe B.V., for Europe and BioOne Corporation for select Asian countries. A number of distribution agreements exist for partners in various other countries, including some in Europe, South America and the Middle East. The company owns about 23 US and 47 foreign patents relating to INTERCEPT. In March 2009, Cerus announced it would be laying off about 30% of its workforce, or 31 employees. In May 2009, the company signed an agreement with Grifols S.A. to expand the distribution of INTERCEPT to Italy.

Employees of the firm are offered medical, dental, and vision coverage; life insurance; tuition reimbursement; flexible spending accounts; stock options; and an employee stock purchase plan.

FINANCIALS: Sales and profits are in thousands of dollars—add 000 to get the full amount. 2009 Note: Financial information for 2009 was not available for all companies at press time.

2009 Sales: $17,982	2009 Profits: $-24,135	U.S. Stock Ticker: CERS
2008 Sales: $16,507	2008 Profits: $-29,181	Int'l Ticker: Int'l Exchange:
2007 Sales: $11,044	2007 Profits: $-45,304	Employees: 73
2006 Sales: $30,310	2006 Profits: $-4,779	Fiscal Year Ends: 12/31
2005 Sales: $13,497	2005 Profits: $13,064	Parent Company:

SALARIES/BENEFITS:

Pension Plan:	ESOP Stock Plan:	Profit Sharing:	Top Exec. Salary: $490,000	Bonus: $102,901
Savings Plan: Y	Stock Purch. Plan: Y		Second Exec. Salary: $312,000	Bonus: $32,760

OTHER THOUGHTS:

Apparent Women Officers or Directors: 3
Hot Spot for Advancement for Women/Minorities: Y

LOCATIONS: ("Y" = Yes)

West:	Southwest:	Midwest:	Southeast:	Northeast:	International:
Y					Y

CHARLES RIVER LABORATORIES INTERNATIONAL INC

www.criver.com

Industry Group Code: 541712 Ranks within this company's industry group: Sales: 3   Profits: 3

Drugs:	Other:	Clinical:		Computers:		Services:	
Discovery:	AgriBio:	Trials/Services:		Hardware:		Specialty Services:	Y
Licensing:	Genetic Data:	Labs:	Y	Software:	Y	Consulting:	
Manufacturing:	Tissue Replacement:	Equipment/Supplies:	Y	Arrays:		Blood Collection:	
Generics:		Research/Development Svcs.:		Database Management:		Drug Delivery:	
		Diagnostics:	Y			Drug Distribution:	

TYPES OF BUSINESS:

Animal Research Models
Consulting
Bioactivity Software
Biosafety Testing
Contract Staffing
Laboratory Diagnostics
Intellectual Property Consulting
Analytical Testing

BRANDS/DIVISIONS/AFFILIATES:

NewLab BioQuality AG
MIR Preclinical Services
WuXi PharmaTech
Piedmont Research Center, LLC

CONTACTS: Note: Officers with more than one job title may be intentionally listed here more than once.

James C. Foster, CEO
James C. Foster, Pres.
Thomas F. Ackerman, CFO/Exec. VP
Stephanie B. Wells, Sr. VP-Mktg./Chief Mktg. Officer
David P. Johst, Exec. VP-Human Resources
Real H. Renaud, Pres., Global Research Model Prod. & Svcs.
Nicholas Ventresca, CIO/Sr. VP-IT
David P. Johst, Chief Admin. Officer
David P. Johst, General Counsel/Exec. VP
John C. Ho, Sr. VP-Corp. Strategy
Nancy A. Gillett, Exec. VP/Pres., Global Preclinical Svcs.
Christophe Berthoux, Chief Commercial Officer
Brian Bathgate, Sr. VP/Pres., European Preclinical Svcs.
Christopher Perkin, Sr. VP/Pres., Canadian & Chinese Preclinical Svcs.
James C. Foster, Chmn.
Jorg Geller, Sr. VP-Japanese Oper. & Select Research Model Bus.

Phone: 781-222-6000	Fax:
Toll-Free:	
Address: 251 Ballardvale St., Wilmington, MA 01887 US	

GROWTH PLANS/SPECIAL FEATURES:

Charles River Laboratories International, Inc. (CRL) is a global provider of solutions that accelerate the drug discovery and development process, including research models and outsourced preclinical services. CRL operates 70 facilities in 16 countries. The firm's customer base includes global pharmaceutical companies, biotechnology companies, government agencies, hospitals and academic institutions. CRL operates in two segments: Research Models and Services (RMS); and Preclinical Services (PCS). RMS, which generated 55% of the company's sales in 2009, is focused on the commercial production and sale of research models. This segment includes four services: genetically engineered models and services; consulting and staffing services; research animal diagnostic services; and discovery and imaging services. With approximately 150 different strains, the company primarily provides genetically and virally defined purpose-bred rats and mice as research models. RMS has multiple facilities located throughout North America, Europe and Japan and maintains 180 barrier rooms or isolator facilities. PCS, which generated 45% of CRL's sales in 2009, is engaged in the discovery and development of new drugs, devices and therapies. The company offers particular expertise in the design, execution and reporting of general and specialty toxicology studies, especially those dealing with innovative therapies and biologicals. Additional services include pathology services; bioanalysis, pharmacokinetics and drug metabolism analysis; discovery support; biopharmaceutical services; and Phase I trials in healthy, normal and special populations. The company provides PCS at multiple facilities in the U.S., Canada, Europe and China. In April 2009, the company acquired Piedmont Research Center, LLC for $46 million. In April 2010, the firm agreed to acquire Chinese drug-research contractor WuXi PharmaTech for $1.6 billion.

Employees are offered medical, dental, vision and life insurance; flexible spending accounts; short- and long-term disability; a 401(k) plan; an employee stock purchase plan; an employee assistance plan; auto and homeowner's insurance; a MetLaw legal plan; tuition reimbursement; and fitness reimbursement and wellness programs.

FINANCIALS: Sales and profits are in thousands of dollars—add 000 to get the full amount. 2009 Note: Financial information for 2009 was not available for all companies at press time.

2009 Sales: $1,202,551	2009 Profits: $112,602	**U.S. Stock Ticker: CRL**
2008 Sales: $1,343,493	2008 Profits: $-521,843	Int'l Ticker: Int'l Exchange:
2007 Sales: $1,230,626	2007 Profits: $154,406	Employees: 8,000
2006 Sales: $1,058,385	2006 Profits: $-55,783	Fiscal Year Ends: 12/31
2005 Sales: $993,328	2005 Profits: $141,999	Parent Company:

SALARIES/BENEFITS:

Pension Plan:	ESOP Stock Plan: Y	Profit Sharing:	Top Exec. Salary: $948,500	Bonus: $
Savings Plan: Y	Stock Purch. Plan:		Second Exec. Salary: $454,480	Bonus: $74,296

OTHER THOUGHTS:

Apparent Women Officers or Directors: 4
Hot Spot for Advancement for Women/Minorities: Y

LOCATIONS: ("Y" = Yes)

West:	Southwest:	Midwest:	Southeast:	Northeast:	International:
Y	Y	Y	Y	Y	Y

CHIRON CORP

www.chiron.com

Industry Group Code: 325412 **Ranks within this company's industry group: Sales:** **Profits:**

Drugs:		Other:		Clinical:		Computers:		Services:	
Discovery:	Y	AgriBio:		Trials/Services:		Hardware:		Specialty Services:	
Licensing:		Genetic Data:	Y	Labs:		Software:		Consulting:	
Manufacturing:	Y	Tissue Replacement:		Equipment/Supplies:		Arrays:		Blood Collection:	
Generics:				Research/Development Svcs.:		Database Management:		Drug Delivery:	
				Diagnostics:	Y			Drug Distribution:	

TYPES OF BUSINESS:

Pharmaceuticals Discovery & Development
Biopharmaceuticals
Vaccines
Blood Screening Assays

BRANDS/DIVISIONS/AFFILIATES:

Novartis AG
Vaccines and Diagnostics
Gen-Probe
PROCLEIX
ZymeQuest

CONTACTS: Note: Officers with more than one job title may be intentionally listed here more than once.

Andrin Oswald, Pres.

Phone: 510-655-8730	Fax: 510-655-9910
Toll-Free:	
Address: 4560 Horton St., Emeryville, CA 94608-2916 US	

GROWTH PLANS/SPECIAL FEATURES:

Chiron, the diagnostics business of Novartis AG's Vaccines and Diagnostics division, develops blood screening tools, products and services focused on preventing transfusion-transmitted infection. The company was acquired by Novartis in April 2006. Chiron's transcription-mediated amplification (TMA) products are designed to simplify nucleic acid testing (NAT) by enabling simultaneous detection of multiple viruses within a single tube. TMA technology was developed by Gen-Probe, Chiron's NAT innovation partner. TMA begins by preparing samples for testing by lysing the viruses to release genetic material, a process which involves no pretreatment or handling, thus reducing the risk of contamination. Capture probes hybridize internal control (IC) and viral nucleic acids and bind them to magnetic particles, and then the unbound material is washed away. A DNA copy of the target nucleic acid is created using reverse transcriptase and RNA is synthesized using RNA polymerase. The newly synthesized RNA can then reenter the TMA process and serve as further amplification templates. With this process, billions of copies can potentially be created in under an hour. Simultaneous detection of both IC- and viral-encoded RNA is enabled by dual kinetic assay (DKA) technology, which produces a flash of light for IC-encoded RNA, and a long glow for viral-encoded RNA. Chiron's TMA products include instruments, software and assays, sold under the PROCLEIX brand. Chiron is currently developing an enhanced immunoassay for the detection of abnormal prions (protein particles) in blood and blood products that are associated with variant Creuzfeldt-Jakob Disease, a rare, degenerative and fatal brain disorder believed to be developed by consuming cattle products contaminated with mad-cow disease (bovine spongiform encephalopathy). With its partner ZymeQuest, Chiron is also developing a system to convert A, B and AB red blood cells to enzyme-converted, universally transfusable group O red blood cells. Chiron also offers its clients NAT training and support.

FINANCIALS: Sales and profits are in thousands of dollars—add 000 to get the full amount. 2009 Note: Financial information for 2009 was not available for all companies at press time.

2009 Sales: $	2009 Profits: $	**U.S. Stock Ticker: Subsidiary**	
2008 Sales: $	2008 Profits: $	**Int'l Ticker:** Int'l Exchange:	
2007 Sales: $	2007 Profits: $	Employees:	
2006 Sales: $	2006 Profits: $	Fiscal Year Ends: 12/31	
2005 Sales: $1,921,000	2005 Profits: $187,000	Parent Company: NOVARTIS AG	

SALARIES/BENEFITS:

Pension Plan:	ESOP Stock Plan:	Profit Sharing:	Top Exec. Salary: $	Bonus: $
Savings Plan: Y	Stock Purch. Plan:		Second Exec. Salary: $	Bonus: $

OTHER THOUGHTS:

Apparent Women Officers or Directors:
Hot Spot for Advancement for Women/Minorities:

LOCATIONS: ("Y" = Yes)

West:	Southwest:	Midwest:	Southeast:	Northeast:	International:
Y		Y		Y	Y

CML HEALTHCARE INCOME FUND

www.cmlhealthcare.com

Industry Group Code: 6215 Ranks within this company's industry group: Sales: 2   Profits: 2

Drugs:	Other:	Clinical:		Computers:	Services:
Discovery:	AgriBio:	Trials/Services:		Hardware:	Specialty Services:
Licensing:	Genetic Data:	Labs:	Y	Software:	Consulting:
Manufacturing:	Tissue Replacement:	Equipment/Supplies:		Arrays:	Blood Collection:
Generics:		Research/Development Svcs.:		Database Management:	Drug Delivery:
		Diagnostics:	Y		Drug Distribution:

TYPES OF BUSINESS:

Medical Diagnostic Services
Laboratory Testing Services
Medical Imaging Services

BRANDS/DIVISIONS/AFFILIATES:

CML Healthcare, Inc.
Upper Chesapeake Health System
Imaging Institute (The)
Quarry Lake Imaging Center (The)

CONTACTS: *Note: Officers with more than one job title may be intentionally listed here more than once.*

Paul J. Bristow, CEO
Kent B. Nicholson, COO/Exec. VP
Paul J. Bristow, Pres.
Tom Weber, CFO/Exec. VP
Pam Sharma, Mgr.-Talent & Human Resources
Cameron Duff, Exec. VP-Corp. Dev.
Tom Hudson, Contact-Investor Rel.
Kent Wentzell, Sr. VP-Imaging Svcs.
Stephen R. Wiseman, Chmn.

Phone: 905-565-0043	Fax: 905-565-2844
Toll-Free: 800-263-0801	
Address: 60 Courtneypark Dr. W., Unit 1, Mississauga, ON L5W 0B3 Canada	

GROWTH PLANS/SPECIAL FEATURES:

CML HealthCare Income Fund is an open-ended trust that owns CML Healthcare, Inc., one of Canada's largest diagnostic services businesses, providing laboratory testing services in Ontario and medical imaging services across Canada. The firm operates in two divisions: laboratory services and imaging services. The laboratory services division conducts a wide range of medical tests, including hematology, biochemistry, cytology, microbiology, histology, Holter monitoring, prostate specific antigen (PSA) and HPV (human papillomavirus) testing. These testing services are generally utilized by physicians to assist in diagnosis and treatment. The laboratory services division operates through a network consisting of one central laboratory (a 55,000-square-foot facility located in Mississauga, Ontario), and approximately 121 licensed specimen collection centers. Additional specimens are also sent to the central laboratory directly from physicians' offices. The imaging services division provides medical imaging services, including X-ray and fluoroscopy, ultrasound, mammography, bone densitometry, nuclear medicine, magnetic resonance imaging (MRI) and computed tomography through a network of 110 non-hospital based medical imaging clinics located in Ontario, British Columbia, Alberta, Manitoba and Quebec, in Canada, as well as 23 U.S. clinics located in Maryland, Delaware and Rhode Island. Roughly 95% of the company's annual laboratory testing revenue is paid through the Ontario Ministry of Health and Long Term Care's public health insurance program. In August 2009, the firm entered into a joint venture with Upper Chesapeake Health System to operate a multi-modality imaging facility in Bel Air, Maryland. In October 2009, CML announced the acquisition of two new medical imaging firms, encompassing six imaging centers in the northeastern U.S.: The Imaging Institute, headquartered in Providence, Rhode Island; and The Quarry Lake Imaging Center, based in Baltimore, Maryland.

FINANCIALS: Sales and profits are in thousands of dollars—add 000 to get the full amount. 2009 Note: Financial information for 2009 was not available for all companies at press time.

2009 Sales: $504,520	2009 Profits: $54,000	**U.S. Stock Ticker:**	
2008 Sales: $425,500	2008 Profits: $93,400	**Int'l Ticker: CLC.UN** Int'l Exchange: Toronto-TSX	
2007 Sales: $287,800	2007 Profits: $92,300	Employees: 2,000	
2006 Sales: $265,800	2006 Profits: $90,900	Fiscal Year Ends: 12/31	
2005 Sales: $256,753	2005 Profits: $74,979	Parent Company:	

SALARIES/BENEFITS:

Pension Plan:	ESOP Stock Plan:	Profit Sharing:	Top Exec. Salary: $458,957	Bonus: $176,699
Savings Plan:	Stock Purch. Plan:		Second Exec. Salary: $350,373	Bonus: $40,138

OTHER THOUGHTS:

Apparent Women Officers or Directors: 1
Hot Spot for Advancement for Women/Minorities:

LOCATIONS: ("Y" = Yes)

West:	Southwest:	Midwest:	Southeast:	Northeast:	International:
				Y	Y

COLUMBIA LABORATORIES INC
www.columbialabs.com

Industry Group Code: 325412 Ranks within this company's industry group: Sales: 98 Profits: 118

Drugs:		Other:		Clinical:	Computers:	Services:	
Discovery:	Y	AgriBio:		Trials/Services:	Hardware:	Specialty Services:	
Licensing:	Y	Genetic Data:		Labs:	Software:	Consulting:	
Manufacturing:	Y	Tissue Replacement:		Equipment/Supplies:	Arrays:	Blood Collection:	
Generics:				Research/Development Svcs.:	Database Management:	Drug Delivery:	Y
				Diagnostics:		Drug Distribution:	

TYPES OF BUSINESS:
Men's Healthcare Drugs
Drug Delivery System Technology

BRANDS/DIVISIONS/AFFILIATES:
Prochieve
Crinone
Striant
Bioadhesive Delivery System

CONTACTS: *Note: Officers with more than one job title may be intentionally listed here more than once.*
Frank C. Condella Jr., CEO
Frank C. Condella Jr., Pres.
Lawrence A. Gyenes, CFO/Sr. VP
George W. Creasy, VP-Clinical Research & Dev.
Michael McGrane, General Counsel/Sr. VP/Sec.
Lawrence A. Gyenes, Treas.
Stephen G. Kasnet, Chmn.

Phone: 973-994-3999	Fax: 866-994-3001
Toll-Free: 866-566-5636	
Address: 354 Eisenhower Pkwy., Plz. 1, Fl. 2, Livingston, NJ 07039 US	

GROWTH PLANS/SPECIAL FEATURES:
Columbia Laboratories, Inc. develops, manufactures and markets products that utilize its bioadhesive drug delivery system technology for the optimization of drug delivery. The company's products primarily target men's healthcare disorders. The company's products provide sustained and controlled delivery of active drug ingredients, which are delivered buccally (attached to the mucosal membrane of the gum and cheek). All of the firm's products utilize its patented Bioadhesive Delivery System (BDS) technology, which consists principally of a polymer and an active ingredient. The firm's principal product is STRAINT, used to treat hypogonadism, a deficiency or absence of endogenous testosterone production. The company is also currently conducting a Phase III clinical study evaluating the performance of a former product, PROCHIEVE 8% progesterone gel, in reducing preterm births. Columbia contracts its manufacturing activities to third parties in Europe and maintains a global logistics office in Paris. Merck Serono S.A. sells CRINONE, a progesterone vaginal gel for infertility and pregnancy support, outside of the U.S.; Sandoz S.p.A sells STRAINT in Italy. In July 2010, the company sold the majority of its progesterone-related assets, excepting rights related to its agreement with Merck Serono, to Watson Pharmaceuticals, Inc. for $47 million. Included in the sale are its CRINONE and PROCHIEVE progesterone gel products, designed to deliver progesterone directly to the uterus. Columbia will continue to collaborate with Watson for the development of new progesterone products.

FINANCIALS: Sales and profits are in thousands of dollars—add 000 to get the full amount. 2009 Note: Financial information for 2009 was not available for all companies at press time.

2009 Sales: $32,196	2009 Profits: $-21,870	**U.S. Stock Ticker: CBRX**
2008 Sales: $36,229	2008 Profits: $-14,077	**Int'l Ticker:** Int'l Exchange:
2007 Sales: $29,627	2007 Profits: $-14,292	**Employees:** 62
2006 Sales: $17,393	2006 Profits: $-12,485	**Fiscal Year Ends:** 12/31
2005 Sales: $22,041	2005 Profits: $-10,104	**Parent Company:**

SALARIES/BENEFITS:

Pension Plan:	ESOP Stock Plan:	Profit Sharing:	Top Exec. Salary: $390,000	Bonus: $78,000
Savings Plan:	Stock Purch. Plan:		Second Exec. Salary: $295,700	Bonus: $53,226

OTHER THOUGHTS:
Apparent Women Officers or Directors: 1
Hot Spot for Advancement for Women/Minorities:

LOCATIONS: ("Y" = Yes)

West:	Southwest:	Midwest:	Southeast:	Northeast:	International:
				Y	Y

COMMONWEALTH BIOTECHNOLOGIES INC

www.cbi-biotech.com

Industry Group Code: 541712 Ranks within this company's industry group: Sales: 16   Profits: 7

Drugs:	Other:		Clinical:		Computers:		Services:	
Discovery:	AgriBio:		Trials/Services:		Hardware:		Specialty Services:	
Licensing:	Genetic Data:	Y	Labs:	Y	Software:		Consulting:	
Manufacturing:	Tissue Replacement:	Y	Equipment/Supplies:		Arrays:		Blood Collection:	
Generics:			Research/Development Svcs.:	Y	Database Management:	Y	Drug Delivery:	
			Diagnostics:	Y			Drug Distribution:	

TYPES OF BUSINESS:

Contract Research-Biotech & Genetics
DNA & Genome Sequencing
Genetic Testing Services
Molecular Biology
Biophysical Analysis
Nucleic Acid Synthesis
Peptide Sequencing & Synthesis
Biodefense Services

BRANDS/DIVISIONS/AFFILIATES:

Venturepharm
Mimotopes Pty Ltd
Tripos Discovery Research Limited

CONTACTS: *Note: Officers with more than one job title may be intentionally listed here more than once.*

Bill Guo, Acting CEO
Richard J. Freer, COO
Robert B. Harris, Pres.
Thomas R. Reynolds, Exec. VP-Science
Thomas R. Reynolds, Exec. VP-Tech.
Thomas R. Reynolds, Sec.
James H. Brennan, VP-Financial Oper.
Bill Guo, Chmn.

Phone: 804-648-3820	Fax: 804-648-2641
Toll-Free: 800-735-9224	
Address: 601 Biotech Dr., Richmond, VA 23235 US	

GROWTH PLANS/SPECIAL FEATURES:

Commonwealth Biotechnologies, Inc. (CBI) provides early developmental contract research and services for global biotechnology industries, academic institutions, government agencies and pharmaceutical companies. The company offers a broad array of analytical and synthetic chemistries and biophysical analysis technologies. CBI offers its services in fully-integrated research programs. The firm operates through Mimotopes Pty Ltd, its peptide-based discovery company in Melbourne, Australia; and China-based Venturepharm Laboratories, Ltd., a contract research consortium specializing in drug discovery and development, formulation development, cGMP manufacturing, process scale-up, and clinical trial management. Mimotopes provides research grade peptide synthesis and analysis. It also has several proprietary technologies for peptide preparation, and its patented SynPhase technology provides cost-effective production of large research grade peptide libraries. Mimotopes has a formal peptide alliance with Genzyme Pharmaceuticals, a provider of GMP pharmaceutical grade peptides. It also has an alliance with GL Biochem, a Shanghai-based peptide synthesis and reagent company. Venturepharm provides pharmaceutical development services by integrating research and development, marketing/sales and manufacturing companies. It also provides drug development outsourcing, marketing and sales consulting and clinical research. The company views commercial and government contracts as its most important sources of revenue, and has consequently moved away from piece work for individual investigators. The company attracts customers from its presentations at national trade shows and advertisements in professional journals. In November 2009, the company sold CBI Services and Fairfax Identity Laboratories, two of its divisions, to Bostwick Laboratories, Inc. for $1.1 million. In December 2009, CBI announced plans to acquire the GL Group, a global supplier of research-grade peptide products and peptide reagents.

FINANCIALS: Sales and profits are in thousands of dollars—add 000 to get the full amount. 2009 Note: Financial information for 2009 was not available for all companies at press time.

2009 Sales: $3,273	2009 Profits: $-2,360	U.S. Stock Ticker: CBTE
2008 Sales: $3,716	2008 Profits: $-9,863	Int'l Ticker: Int'l Exchange:
2007 Sales: $3,545	2007 Profits: $-2,758	Employees: 32
2006 Sales: $6,532	2006 Profits: $-1,153	Fiscal Year Ends: 12/31
2005 Sales: $7,803	2005 Profits: $ 79	Parent Company:

SALARIES/BENEFITS:

Pension Plan:	ESOP Stock Plan:	Profit Sharing:	Top Exec. Salary: $191,500	Bonus: $
Savings Plan:	Stock Purch. Plan:		Second Exec. Salary: $167,900	Bonus: $

OTHER THOUGHTS:

Apparent Women Officers or Directors: 1
Hot Spot for Advancement for Women/Minorities:

LOCATIONS: ("Y" = Yes)

West:	Southwest:	Midwest:	Southeast:	Northeast:	International:
				Y	Y

COMPUGEN LTD

www.cgen.com

Industry Group Code: 325412 Ranks within this company's industry group: Sales: 148   Profits: 85

Drugs:		Other:		Clinical:		Computers:		Services:	
Discovery:	Y	AgriBio:	Y	Trials/Services:		Hardware:	Y	Specialty Services:	
Licensing:	Y	Genetic Data:	Y	Labs:		Software:	Y	Consulting:	
Manufacturing:		Tissue Replacement:		Equipment/Supplies:		Arrays:	Y	Blood Collection:	
Generics:				Research/Development Svcs.:	Y	Database Management:	Y	Drug Delivery:	
				Diagnostics:	Y			Drug Distribution:	

TYPES OF BUSINESS:

Drug Discovery
Biotechnology Databases & Diagnostics
Agricultural Biotechnology
Bioinformatics Platform

BRANDS/DIVISIONS/AFFILIATES:

LEADS
MED
Evogene, Ltd.

CONTACTS: Note: Officers with more than one job title may be intentionally listed here more than once.

Anat Cohen-Dayag, CEO
Anat Cohen-Dayag, Pres.
Dikla Czaczkes Axselbrad, CFO
Dorit Amitay, VP-Human Resources
Zurit Levine, VP-R&D
Eli Zangvil, VP-Bus. Dev.
Martin S. Gerstel, Chmn.

Phone: 972-3-765-8585	Fax: 972-3-765-8555
Toll-Free:	
Address: 72 Pinchas Rosen St., Tel Aviv, 69512 Israel	

GROWTH PLANS/SPECIAL FEATURES:

Compugen, Ltd. engages in drug and diagnostic product discovery and commercialization, largely through early stage licensing and co-development agreements. Compugen focuses on using field-focused discovery platforms to predict, select and validate therapeutic drug candidates and diagnostic biomarker candidates. The company's initial discovery platforms have focused mainly on cancer, cardiovascular and immune-related diseases. The firm is structured along the lines of activities including: research and discovery; therapeutics; and diagnostics. Its research and discovery activities are focused on developing field-focused discovery platforms for the prediction and selection of product candidates. Current areas of research include identification of GPCR peptide ligands, Disease Associated Conformation peptide blockers, targets for monoclonal antibodies, drug response genomic markers and pathway kinetic simulation. Therapeutic and diagnostic activities both consist of identification and development of potential therapeutic and diagnostic products. Current therapeutic drug candidates include potential treatments for various types of cancer, inflammatory diseases and cardiovascular conditions. Current diagnostic programs include immunoassay diagnostics, drug-induced toxicity biomarkers, genomic markers and nucleic acid diagnostics. Compugen's core offering is LEADS, a bioinformatics platform for analyzing genomic and protein data; LEADS includes extensive gene information and annotations, such as splice variants, antisense genes, SNPs, novel genes and RNA editing. Another platform utilized by Compugen is MED, a database composed of more than 40,000 public and proprietary microarray experiments representing about 1,400 conditions, including normal tissues, malignant tissues, tissues from drug treated patients, and so forth. The company maintains numerous collaborations for the development and licensing of its product candidates, and currently has therapeutic, diagnostic and research agreements with firms including Pfizer; Biosite; Medarex, Inc.; Merck & Co., Inc.; Ortho-Clinical Diagnostics (a Johnson & Johnson company); Roche; Siemens Healthcare Diagnostics, Inc.; and Teva Pharmaceutical Industries. The company also maintains an investment in Evogene, Limited (formerly a wholly-owned subsidiary), which is focused on agricultural biotechnology, especially plant genomics.

FINANCIALS: Sales and profits are in thousands of dollars—add 000 to get the full amount. 2009 Note: Financial information for 2009 was not available for all companies at press time.

2009 Sales: $ 250	2009 Profits: $-3,831	U.S. Stock Ticker: CGEN
2008 Sales: $ 338	2008 Profits: $-12,527	Int'l Ticker: CGEN Int'l Exchange: Tel Aviv-TASE
2007 Sales: $ 180	2007 Profits: $-12,114	Employees: 37
2006 Sales: $ 215	2006 Profits: $-13,020	Fiscal Year Ends: 12/31
2005 Sales: $ 646	2005 Profits: $-13,978	Parent Company:

SALARIES/BENEFITS:

Pension Plan:	ESOP Stock Plan:	Profit Sharing:	Top Exec. Salary: $	Bonus: $
Savings Plan:	Stock Purch. Plan:		Second Exec. Salary: $	Bonus: $

OTHER THOUGHTS:

Apparent Women Officers or Directors: 5
Hot Spot for Advancement for Women/Minorities: Y

LOCATIONS: ("Y" = Yes)

West:	Southwest:	Midwest:	Southeast:	Northeast:	International:
					Y

Note: Financial information, benefits and other data can change quickly and may vary from those stated here.

CORTEX PHARMACEUTICALS INC

www.cortexpharm.com

Industry Group Code: 325412 Ranks within this company's industry group: Sales: Profits: 100

Drugs:		Other:		Clinical:		Computers:		Services:	
Discovery:	Y	AgriBio:		Trials/Services:		Hardware:		Specialty Services:	
Licensing:	Y	Genetic Data:		Labs:		Software:		Consulting:	
Manufacturing:	Y	Tissue Replacement:		Equipment/Supplies:		Arrays:		Blood Collection:	
Generics:				Research/Development Svcs.:		Database Management:		Drug Delivery:	
				Diagnostics:				Drug Distribution:	

TYPES OF BUSINESS:

Drugs-Neurological
Psychiatric Drugs

BRANDS/DIVISIONS/AFFILIATES:

AMPAKINE
CX1739
CX2007
CX929

CONTACTS: Note: Officers with more than one job title may be intentionally listed here more than once.

Mark A. Varney, CEO
Mark A. Varney, Pres.
Maria S. Messinger, CFO/VP
Pierre V. Tran, Chief Medical Officer/VP-Clinical Dev.
Maria S. Messinger, Corp. Sec.
James H. Coleman, Sr. VP-Bus. Dev.
Janet Vasquez, Contact-Media
Steven A. Johnson, VP-Preclinical Dev.
Roger G. Stoll, Chmn.

Phone: 949-727-3157	Fax: 949-727-3657
Toll-Free:	
Address: 15231 Barranca Pkwy., Irvine, CA 92618 US	

GROWTH PLANS/SPECIAL FEATURES:

Cortex Pharmaceuticals, Inc. focuses on developing novel small-molecule compounds that positively modulate AMPA-type glutamate receptors, a complex of proteins involved in the communication between nerve cells in the brain. The firm's compounds, which are being developed under the brand name AMPAKINE, enhance the activity of the AMPA receptor. AMPAKINE may potentially be used for the treatment of neurological and psychiatric diseases/disorders that involve the depressed functioning of pathways in the brain using glutamate as a neurotransmitter. The drugs have the potential to treat mental illnesses such as Alzheimer's disease; depression; mild cognitive impairment (MCI); schizophrenia; Attention Deficit Hyperactivity Disorder (ADHD); respiratory depression caused by opiate analgesics; and sleep apnea. Ampakine comes in two categories: low impact and high impact compounds, defined by the site to which they bind on the AMPA receptor complex. The company's most advanced low impact Ampakine compound is CX1739, which is currently in Phase I clinical development. The firm has two other compounds in clinical or preclinical development: CX2007 and CX2076. Cortex has several high impact compounds in preclinical testing, including CX929, which has shown the potential to restore depressed levels of the growth factor BDNF in genetic mouse models of Huntington's disease, neurodegenerative genetic disorder. Cortex owns or has exclusive rights to more than 250 issued or allowed U.S. and foreign patents. In March 2010, Cortex sold the AMPAKINE compounds CX717, CX1736, CX1942 and the injectable dosage forms of CX1739 to Biovail Laboratories International SRL for $10 million.

The company offers employees medical, dental, life and vision insurance as well as stock options.

FINANCIALS: Sales and profits are in thousands of dollars—add 000 to get the full amount. 2009 Note: Financial information for 2009 was not available for all companies at press time.

2009 Sales: $	2009 Profits: $-10,788	U.S. Stock Ticker: COR
2008 Sales: $	2008 Profits: $-14,596	Int'l Ticker: Int'l Exchange:
2007 Sales: $	2007 Profits: $-12,969	Employees: 13
2006 Sales: $1,177	2006 Profits: $-16,055	Fiscal Year Ends: 12/31
2005 Sales: $2,577	2005 Profits: $-11,605	Parent Company:

SALARIES/BENEFITS:

Pension Plan:	ESOP Stock Plan:	Profit Sharing:	Top Exec. Salary: $305,250	Bonus: $
Savings Plan: Y	Stock Purch. Plan:		Second Exec. Salary: $298,650	Bonus: $

OTHER THOUGHTS:

Apparent Women Officers or Directors: 2
Hot Spot for Advancement for Women/Minorities:

LOCATIONS: ("Y" = Yes)

West:	Southwest:	Midwest:	Southeast:	Northeast:	International:
Y					

COVANCE INC

www.covance.com

Industry Group Code: 541712 **Ranks within this company's industry group: Sales: 1 Profits: 21**

Drugs:	Other:	Clinical:		Computers:		Services:	
Discovery:	AgriBio:	Trials/Services:	Y	Hardware:		Specialty Services:	Y
Licensing:	Genetic Data:	Labs:		Software:	Y	Consulting:	
Manufacturing:	Tissue Replacement:	Equipment/Supplies:		Arrays:		Blood Collection:	
Generics:		Research/Development Svcs.:	Y	Database Management:		Drug Delivery:	
		Diagnostics:				Drug Distribution:	

TYPES OF BUSINESS:

Pharmaceutical Research & Development
Drug Preclinical/Clinical Trials
Laboratory Testing & Analysis
Approval Assistance
Health Economics & Outcomes Services
Online Tools

BRANDS/DIVISIONS/AFFILIATES:

LabLink
Study Tracker
Trial Tracker

CONTACTS: *Note: Officers with more than one job title may be intentionally listed here more than once.*

Joseph L. Herring, CEO
Wendel Barr, COO/Exec. VP
William E. Klitgaard, CFO/Sr. VP
Donald Kraft, Sr. VP-Human Resources
Steven M. Michael, Chief Scientific Officer-Global Bioanalytical Svcs
James W. Lovett, General Counsel/Sr. VP/Sec.
Richard F. Cimino, Pres., Late Stage Dev.
Richard Cimino, Sr. VP/Pres., Clinical Dev.
Michael Lehmann, Sr. VP/Pres., Nonclinical Safety Assessment
Deborah L. Tanner, Sr. VP/Pres., Global Central Laboratory Svcs.
Joseph L. Herring, Chmn.

Phone: 609-452-4440	Fax:
Toll-Free: 888-268-2623	
Address: 210 Carnegie Ctr., Princeton, NJ 08540 US	

GROWTH PLANS/SPECIAL FEATURES:

Covance, Inc. is a drug development services and contract research firm. It provides a wide range of product development services to pharmaceutical, biotechnology and medical device industries across the globe. The company also provides laboratory testing services for clients in the chemical, agrochemical and food businesses. Covance operates two business segments: early development services and late-stage development services. Covance's early development services include preclinical services (such as toxicology, pharmaceutical development, research products and a bioanalytical testing service) and Phase I clinical services. Its late-stage development services cover clinical development and support; clinical trials; periapproval and market access; and central laboratory operations. Covance has also introduced several Internet-based products: StudyTracker, an Internet-based client access product, which permits customers of toxicology services to review study data and schedules on a near real-time basis; LabLink, a client access program that allows customers of central laboratory services to review and query lab data; and Trial Tracker, a web-enabled clinical trial project management and tracking tool intended to allow both employees and customers of its late-stage clinical business to review and manage all aspects of clinical-trial projects. In April 2009, Covance opened new clinical development offices in Peru and Chile, and expanded one of its offices in Argentina. In July 2009, the firm agreed to sell its Interactive Voice & Web Response Services operations to Phase Forward. Also in July 2009, the company agreed to acquire the Gene Expression Laboratory of Merck & Co., Inc. In September 2009, Covance opened new clinical development offices in Mexico and Brazil. In December 2009, the firm opened new clinical development offices in South Korea and India; and expanded its offices in Japan and Hong Kong.

Covance offers its employees benefits such as medical, dental and vision plans; a range of insurance benefits; employee assistance; a 401(k); financial planning services; and tuition reimbursement.

FINANCIALS: Sales and profits are in thousands of dollars—add 000 to get the full amount. 2009 Note: Financial information for 2009 was not available for all companies at press time.

2009 Sales: $1,962,626	2009 Profits: $175,882	**U.S. Stock Ticker:** CVD
2008 Sales: $1,827,067	2008 Profits: $196,760	**Int'l Ticker:** Int'l Exchange:
2007 Sales: $1,631,516	2007 Profits: $175,929	**Employees:** 10,320
2006 Sales: $1,406,058	2006 Profits: $144,998	**Fiscal Year Ends:** 12/31
2005 Sales: $1,250,400	2005 Profits: $119,600	**Parent Company:**

SALARIES/BENEFITS:

Pension Plan:	ESOP Stock Plan:	Profit Sharing:	Top Exec. Salary: $811,667	Bonus: $730,600
Savings Plan: Y	Stock Purch. Plan: Y		Second Exec. Salary: $471,667	Bonus: $141,600

OTHER THOUGHTS:

Apparent Women Officers or Directors: 3
Hot Spot for Advancement for Women/Minorities: Y

LOCATIONS: ("Y" = Yes)

West:	Southwest:	Midwest:	Southeast:	Northeast:	International:
Y	Y	Y	Y	Y	Y

COVIDIEN PLC

www.covidien.com

Industry Group Code: 33911 Ranks within this company's industry group: Sales: 3 Profits: 3

Drugs:		Other:	Clinical:		Computers:	Services:	
Discovery:		AgriBio:	Trials/Services:		Hardware:	Specialty Services:	
Licensing:		Genetic Data:	Labs:		Software:	Consulting:	
Manufacturing:	Y	Tissue Replacement:	Equipment/Supplies:	Y	Arrays:	Blood Collection:	
Generics:			Research/Development Svcs.:	Y	Database Management:	Drug Delivery:	Y
			Diagnostics:			Drug Distribution:	Y

TYPES OF BUSINESS:

Medical Equipment & Supplies, Manufacturing
Imaging Agents
Pharmaceutical Products
Medical Devices

BRANDS/DIVISIONS/AFFILIATES:

Kendall
Autosuture
AbsorbaTack
Nellcor
Puritan Bennett
Valleylab
Tri-Staple
EEATM Hemorrhoid & Prolapse Stapler Set

CONTACTS: *Note: Officers with more than one job title may be intentionally listed here more than once.*

Richard J. Meelia, CEO
Richard J. Meelia, Pres.
Charles J. Dockendorff, CFO/Exec. VP
Michael P. Dunford, Sr. VP-Human Resources
Steven M. McManama, CIO
John H. Masterson, General Counsel/Sr. VP
Amy A. Wendell, Sr. VP-Bus. Dev. & Strategy
Eric A. Kraus, Sr. VP-Corp. Comm.
Coleman Lannum, VP-Investor Rel.
Richard G. Brown, Jr., Chief Acct. Officer/Controller/VP
Jose E. Almeida, Pres., Medical Devices
Timothy R. Wright, Pres., Pharmaceuticals
James C. Clemmer, Pres., Medical Supplies
Kevin G. DaSilva, Treas./VP
Richard J. Meelia, Chmn.
James M. Muse, Sr. VP-Global Supply Chain

Phone: 353-1-439-3000	Fax:
Toll-Free:	
Address: Cherrywood Business Park, Loughlinstown, Dublin, Ireland UK	

GROWTH PLANS/SPECIAL FEATURES:

Covidien plc, formerly Covidien Ltd., is a global healthcare products company that markets items for both clinical and home settings. The firm manufactures and distributes industry-leading brands such as Kendall, Autosuture, Nellcor and Puritan Bennett. The company's business consists of three segments: medical devices, pharmaceuticals and medical supplies. The medical devices segment offers products such as endomechanical instruments, including surgical staplers and laparoscopic instruments; soft tissue repair products, such as sutures, mesh and biosurgery products; energy devices, including vessel sealing and electrosurgical and ablation products; oximetry and monitoring products, including sensors, monitors and temperature management products; airway and ventilation products, such as ventilator breathing systems and inhalation therapy products; and vascular devices, including vascular therapy and compression products. This division markets these products under the Autosuture, AbsorbaTack and Valleylab brands. The pharmaceuticals segment includes specialty pharmaceuticals, active pharmaceutical ingredients (API), specialty chemicals, contrast products and radiopharmaceuticals. Lastly, the medical supplies segment provides nursing care products, such as incontinence, woundcare, enteral feeding and suction products; medical surgical products, such as operating room supply products and related accessories; SharpSafety products, including needles and syringes; and original equipment manufacturer (OEM) products. The medical devices segment generated 57% of sales in 2009; pharmaceuticals, 27%; and medical supplies, 16%. In September 2009, Covidien agreed to acquire Aspect Medical Systems, Inc. for $210 million. In May 2010, the company announced plans to sell its Specialty Chemicals business to New Mountain Capital, L.L.C. for $280 million. In the same month, it entered the hemorrhoidopexy market by launching its EEATM Hemorrhoid & Prolapse Stapler Set with DST SeriesTM Technology. In June 2010, Covidien launched its Tri-Staple technology platform for endoscopic stapling and agreed to acquire Somanetics Corporation for $250 million and ev3 Inc. for $2.6 billion.

FINANCIALS: Sales and profits are in thousands of dollars—add 000 to get the full amount. 2009 Note: Financial information for 2009 was not available for all companies at press time.

2009 Sales: $10,677,000	2009 Profits: $907,000	U.S. Stock Ticker: COV
2008 Sales: $10,358,000	2008 Profits: $1,361,000	Int'l Ticker: Int'l Exchange:
2007 Sales: $9,317,000	2007 Profits: $-342,000	Employees: 41,800
2006 Sales: $8,313,000	2006 Profits: $1,155,000	Fiscal Year Ends: 9/30
2005 Sales: $	2005 Profits: $	Parent Company:

SALARIES/BENEFITS:

Pension Plan:	ESOP Stock Plan:	Profit Sharing:	Top Exec. Salary: $1,220,808	Bonus: $2,009,735
Savings Plan: Y	Stock Purch. Plan: Y		Second Exec. Salary: $677,177	Bonus: $560,811

OTHER THOUGHTS:

Apparent Women Officers or Directors: 2
Hot Spot for Advancement for Women/Minorities: Y

LOCATIONS: ("Y" = Yes)

West:	Southwest:	Midwest:	Southeast:	Northeast:	International:
Y	Y	Y	Y	Y	Y

CRUCELL NV

www.crucell.com

Industry Group Code: 325412 **Ranks within this company's industry group:** Sales: 52   Profits: 50

Drugs:		Other:		Clinical:		Computers:		Services:	
Discovery:	Y	AgriBio:		Trials/Services:		Hardware:		Specialty Services:	Y
Licensing:	Y	Genetic Data:		Labs:		Software:		Consulting:	
Manufacturing:	Y	Tissue Replacement:		Equipment/Supplies:	Y	Arrays:		Blood Collection:	
Generics:				Research/Development Svcs.:		Database Management:		Drug Delivery:	
				Diagnostics:				Drug Distribution:	Y

TYPES OF BUSINESS:

Biopharmaceuticals
Biopharmaceutical Development Technologies

BRANDS/DIVISIONS/AFFILIATES:

Quinvaxem
Hepavax-Gene
MoRu-Viraten
Epaxal
Dukoral
Inflexal V
PER.C6
AdVac

CONTACTS: Note: Officers with more than one job title may be intentionally listed here more than once.

Ronald H. P. Brus, CEO
Cees de Jong, COO
Ronald H. P. Brus, Pres.
Leonard Kruimer, CFO
Jaap Goudsmit, Chief Scientific Officer
Rene Beukema, General Counsel/Corp. Sec.
Arthur Lahr, Exec. VP-Bus. Dev./Chief Strategy Officer
Jan Pieter Oosterveld, Chmn.

Phone: 31-71-519-91-00	Fax: 31-71-519-98-00
Toll-Free:	
Address: Archimedesweg 4-6, Leiden, 2333 CN The Netherlands	

GROWTH PLANS/SPECIAL FEATURES:

Crucell N.V. focuses on combating infectious diseases. With operations in 10 countries worldwide, Crucell distributed over 115 million vaccine doses in more than 100 countries during 2009, making it one of the largest independent vaccine companies in the world. It has three primary categories of products. Pediatric products comprise the fully liquid Quinvaxem vaccine that protects against five childhood diseases (diphtheria, tetanus, whooping cough, hepatitis B and Haemophilus influenzae type B); aluminum-free hepatitis A vaccine Epaxal Junior (the lack of aluminum reduces pain during administration); recombinant hepatitis B vaccine Hepavax-Gene; and measles and rubella vaccine MoRu-Viraten. Travel and endemic products comprise an adult version of Epaxal, typhoid fever vaccine Vivotif and drinkable cholera vaccine Dukoral. The firm's respiratory product is flu vaccine Inflexal V. Products in development include yellow fever vaccine Flavimun, currently undergoing Phase III trials, and vaccines for tuberculosis, seasonal flu strains, malaria, Ebola and HIV. The tuberculosis and flu vaccines are in Phase II testing, while the others are undergoing Phase I trials. Crucell is also testing human antibodies for the flu and rabies and a blood coagulation factor. The company develops its products through five proprietary technologies. PER.C6, the most important of these technologies, is a human designer cell line for the development and large-scale manufacturing of biopharmaceutical products. AdVac is used with PER.C6 to develop recombinant vaccines. STAR enhances the yield of recombinant human antibodies and proteins on mammalian cell lines. MAbstract is used to discover novel drug targets and identify human antibodies. Lastly, Virosome enables the use of virus antigens in the making of vaccines. In October 2009, Crucell entered into a strategic partnership with Johnson & Johnson to focus on the development of treatments for influenza and other infectious and non-infectious diseases; Johnson & Johnson invested approximately $447.2 million for an 18% stake in Crucell.

FINANCIALS: Sales and profits are in thousands of dollars—add 000 to get the full amount. 2009 Note: Financial information for 2009 was not available for all companies at press time.

2009 Sales: $338,163	2009 Profits: $23,938	**U.S. Stock Ticker:** CRXL
2008 Sales: $267,157	2008 Profits: $14,250	**Int'l Ticker:** CRX **Int'l Exchange:** Zurich-SWX
2007 Sales: $203,786	2007 Profits: $-44,334	Employees: 1,248
2006 Sales: $186,860	2006 Profits: $-123,845	Fiscal Year Ends: 12/31
2005 Sales: $	2005 Profits: $	Parent Company:

SALARIES/BENEFITS:

Pension Plan:	ESOP Stock Plan:	Profit Sharing:	Top Exec. Salary: $	Bonus: $
Savings Plan:	Stock Purch. Plan:		Second Exec. Salary: $	Bonus: $

OTHER THOUGHTS:

Apparent Women Officers or Directors:
Hot Spot for Advancement for Women/Minorities:

LOCATIONS: ("Y" = Yes)

West:	Southwest:	Midwest:	Southeast:	Northeast:	International:
			Y	Y	Y

Note: Financial information, benefits and other data can change quickly and may vary from those stated here.

CSL LIMITED
www.csl.com.au

Industry Group Code: 325414 Ranks within this company's industry group: Sales: 2   Profits: 1

Drugs:		Other:	Clinical:		Computers:	Services:	
Discovery:		AgriBio:	Trials/Services:		Hardware:	Specialty Services:	Y
Licensing:		Genetic Data:	Labs:		Software:	Consulting:	
Manufacturing:	Y	Tissue Replacement:	Equipment/Supplies:	Y	Arrays:	Blood Collection:	Y
Generics:			Research/Development Svcs.:		Database Management:	Drug Delivery:	
			Diagnostics:	Y		Drug Distribution:	

TYPES OF BUSINESS:
Human Blood-Plasma Collection
Plasma Products
Immunohematology Products
Vaccines
Pharmaceutical Marketing
Antivenom
Drugs-Cancer

BRANDS/DIVISIONS/AFFILIATES:
CSL Biotherapies
CSL Bioplasma
CSL Behring LLC
CSL Pharmaceutical

CONTACTS: *Note: Officers with more than one job title may be intentionally listed here more than once.*
Brian McNamee, CEO/Managing Dir.
Peter Turner, COO
Gordon Naylor, CFO
Kenneth J. Roberts, Dir.-Mktg.
Jill Lever, Sr. VP-Human Capital
Andrew Cuthbertson, Chief Scientific Officer/Dir.-R&D
John Akehurst, Dir.-Eng.
Greg Boss, General Counsel/Sr. VP
Paul Walton, Sr. VP-Corp. Dev.
Maria Pikos, Contact-Media Inquiries
Mark Dehring, Dir.-Investor Rel.
Tony M. Cipa, Exec. Dir.-Finance
Peter Turner, Pres., CSL Behring
Jeff Davies, Exec. VP-CSL Biotherapies
Zita Cunningham, VP-Bus. Dev.
Edward Bailey, Australian General Counsel/Corp. Sec.
Elizabeth A. Alexander, Chmn.
Jeff Davies, Pres., Bioplasma Asia Pacific

Phone: 61-3-9389-1911	Fax: 61-3-9389-1434
Toll-Free:	
Address: 45 Poplar Rd., Parkville, VIC 3052 Australia	

GROWTH PLANS/SPECIAL FEATURES:
CSL Limited develops, manufactures and markets pharmaceutical products of biological origin in 27 countries worldwide. The company operates through several subsidiaries that manufacture and distribute pharmaceuticals, vaccines and diagnostics derived from human plasma. The firm's subsidiaries include CSL Behring; CSL Bioplasma; CSL Biotherapies; CSL Plasma; and CSL Bioplasma Immunohaematology. CSL Behring is a world leader in the manufacture of plasma products such as hemophilia treatments, immunoglobulins and wound healing agents. CSL Bioplasma is one of the largest manufacturers of plasma-derived therapeutic products in the southern hemisphere and works with the Red Cross and government entities to supply such products in Australia, New Zealand, Singapore, Malaysia and Hong Kong. It also provides contract plasma fractionation services. CSL Biotherapies, formerly CSL Pharmaceutical, manufactures and markets vaccines and other pharmaceutical products for human use, including children's vaccines, travel vaccines, respiratory vaccines and antivenom. Currently, its primary focus is the manufacturing of flu vaccines. CSL Plasma, based in Florida, is one of the largest collectors of human blood plasma in the world, comprised of over 65 plasma collection centers in the U.S. and Germany as well as plasma testing laboratories and logistics centers in both countries. CSL Bioplasma Immunohaematology is a stand-alone business operating as a subsidiary of CSL Bioplasma. Its core activities include the production of Reagent Red Blood Cells (RRBCs), monoclonal antibodies and the manufacture of reagents and related products for immunohaematology labs. CSL's Research and Development division focuses on plasma replacement therapies, vaccines, immunomodulators and recombinant therapeutic proteins. In October 2010, the firm announced progress in the development of a new vaccine to treat peridontitis, a severe gum disease.

CSL Limited provides flexible work arrangements; ongoing training programs; an employee share purchase program; and tuition reimbursement.

FINANCIALS: Sales and profits are in thousands of dollars—add 000 to get the full amount. 2009 Note: Financial information for 2009 was not available for all companies at press time.

2009 Sales: $4,387,910	2009 Profits: $997,925	U.S. Stock Ticker:
2008 Sales: $2,698,750	2008 Profits: $499,350	Int'l Ticker: CSL Int'l Exchange: Sydney-ASX
2007 Sales: $2,354,470	2007 Profits: $383,400	Employees:
2006 Sales: $2,146,111	2006 Profits: $88,406	Fiscal Year Ends: 6/30
2005 Sales: $1,965,359	2005 Profits: $176,824	Parent Company:

SALARIES/BENEFITS:
Pension Plan:	ESOP Stock Plan:	Profit Sharing:	Top Exec. Salary: $	Bonus: $
Savings Plan:	Stock Purch. Plan: Y		Second Exec. Salary: $	Bonus: $

OTHER THOUGHTS:
Apparent Women Officers or Directors: 3
Hot Spot for Advancement for Women/Minorities: Y

LOCATIONS: ("Y" = Yes)
West:	Southwest:	Midwest:	Southeast:	Northeast:	International:
		Y	Y	Y	Y

Note: Financial information, benefits and other data can change quickly and may vary from those stated here.

CUBIST PHARMACEUTICALS INC

www.cubist.com

Industry Group Code: 325412 Ranks within this company's industry group: Sales: 46 Profits: 45

Drugs:		Other:	Clinical:	Computers:	Services:
Discovery:	Y	AgriBio:	Trials/Services:	Hardware:	Specialty Services:
Licensing:		Genetic Data:	Labs:	Software:	Consulting:
Manufacturing:		Tissue Replacement:	Equipment/Supplies:	Arrays:	Blood Collection:
Generics:			Research/Development Svcs.:	Database Management:	Drug Delivery:
			Diagnostics:		Drug Distribution:

TYPES OF BUSINESS:

Drugs-Infectious Disease
Antimicrobial Drugs
Antiviral Drugs

BRANDS/DIVISIONS/AFFILIATES:

CUBICIN
Calixa Therapeutics Inc.

CONTACTS: Note: Officers with more than one job title may be intentionally listed here more than once.

Michael W. Bonney, CEO
Robert J. (Rob) Perez, COO/Exec. VP
Michael W. Bonney, Pres.
David W.J. McGirr, CFO/Sr. VP
Maureen H. Powers, VP-Human Resources
Steven C. Gilman, Chief Scientific Officer
Anthony S. Murabito, CIO
Steven C. Gilman, Sr. VP-Discovery & Nonclinical Dev.
Tamara L. Joseph, General Counsel/Sr. VP/Corp. Sec.
Ed Campanaro, VP-Clinical Oper.
Praveen Tipirneni, VP-Bus. Dev.
Mary C. Stack, VP-Finance
Gregory Stea, Sr. VP-Commercial Oper.
Santosh J. Vetticaden, Chief Medical Officer/Sr. VP-Clinical Dev.
Mark Battaglini, VP-Gov't Affairs
Dennis D. Keith, VP-Chemistry

Phone: 781-860-8660	Fax: 781-240-0256
Toll-Free:	
Address: 65 Hayden Ave., Lexington, MA 02421 US	

GROWTH PLANS/SPECIAL FEATURES:

Cubist Pharmaceuticals, Inc. is a biopharmaceutical company focused on the research, development and commercialization of pharmaceutical products for the acute care environment. Cubist derives substantially all of its revenues from CUBICIN, a once-daily, bactericidal, intravenous antibiotic with activity against methicillin-resistant Staphylococcus aureus (MRSA). CUBICIN is approved in the U.S. for the treatment of complicated skin and skin structure infections caused by MRSA and certain other Gram-positive bacteria, as well as for MRSA bloodstream infections. Cubist markets CUBICIN to over 2,000 U.S. hospitals and outpatient acute care settings. CUBICIN has received regulatory approval in 66 countries and is marketed in 32 countries. Novartis AG, through a subsidiary, is responsible for regulatory filings, sales, marketing and distribution costs of CUBICIN in Europe, Australia, New Zealand, India and certain Central American, South American and Middle Eastern countries. Merck & Co. subsidiary Banyu Pharmaceutical is responsible for the development and commercialization of CUBICIN in Japan. AstraZeneca develops and commercializes CUBICIN in China and over 100 additional countries. Other international partners for CUBICIN include Medison Pharma for Israel; Sepreacor for Canada; TTY BioPharm for Taiwan; and Kuhnil Pharma for Korea. Under an agreement with AstraZeneca, Cubist promotes MERREM I.V., an antibiotic used for intra-abdominal infections and bacterial meningitis in pediatric patients. In December 2009, Cubist acquired Calixa Therapeutics Inc. for $92.5 million.

Employees are offered tuition reimbursement; skill-based development programs; medical and dental insurance; supplemental vision care; flexible spending accounts; domestic partner coverage; a 401(k) plan; a discounted employee stock purchase plan; short- and long-term disability and life insurance; an adoption assistance program; daily fitness classes; an onsite workout center, massage therapist and meditation room; an annual childcare and eldercare subsidy; an employee assistance program; and free weekly summer ice cream truck visits.

FINANCIALS: Sales and profits are in thousands of dollars—add 000 to get the full amount. 2009 Note: Financial information for 2009 was not available for all companies at press time.

2009 Sales: $562,144	2009 Profits: $79,600	U.S. Stock Ticker: CBST
2008 Sales: $433,641	2008 Profits: $127,892	Int'l Ticker: Int'l Exchange:
2007 Sales: $294,620	2007 Profits: $48,147	Employees: 554
2006 Sales: $194,748	2006 Profits: $- 376	Fiscal Year Ends: 12/31
2005 Sales: $120,645	2005 Profits: $-31,852	Parent Company:

SALARIES/BENEFITS:

Pension Plan:	ESOP Stock Plan:	Profit Sharing:	Top Exec. Salary: $500,000	Bonus: $420,000
Savings Plan: Y	Stock Purch. Plan: Y		Second Exec. Salary: $430,011	Bonus: $207,265

OTHER THOUGHTS:

Apparent Women Officers or Directors: 5
Hot Spot for Advancement for Women/Minorities: Y

LOCATIONS: ("Y" = Yes)

West:	Southwest:	Midwest:	Southeast:	Northeast:	International:
Y		Y	Y	Y	

CUMBERLAND PHARMACEUTICALS   www.cumberlandpharma.com

Industry Group Code: 325412 Ranks within this company's industry group: Sales: 88   Profits: 64

Drugs:		Other:	Clinical:	Computers:	Services:	
Discovery:	Y	AgriBio:	Trials/Services:	Hardware:	Specialty Services:	Y
Licensing:		Genetic Data:	Labs:	Software:	Consulting:	
Manufacturing:		Tissue Replacement:	Equipment/Supplies:	Arrays:	Blood Collection:	
Generics:			Research/Development Svcs.:	Database Management:	Drug Delivery:	
			Diagnostics:		Drug Distribution:	

TYPES OF BUSINESS:
Prescription Drugs Manufacturing

GROWTH PLANS/SPECIAL FEATURES:

Cumberland Pharmaceuticals Inc. is a specialty pharmaceutical company that focuses on acute hospital care and gastroenterology. The firm acquires and markets late-stage and FDA-approved development drugs. Cumberland has three actively marketed products: Acetadote, Kristalose and Caldolor. Acetadote is an infusible drug created to protect against hepatoxic overdoses of acetaminophen, which is a relatively common cause of drug toxicity. Kristalose is a prescription-based powered laxative used to treat acute or chronic constipation. Caldolor is an intravenous ibuprofen variation used to treat pain and fever. Subsidiary, Cumberland Emerging Technologies, Inc. (CET) is a joint venture created with Vanderbilt University and the Tennessee Technology Development Corporation. CET partners with universities and other research organizations to develop early-stage product candidates, in which Cumberland has the opportunity to commercialize. The firm utilizes government grants to fund CET's research and development programs. In addition, the company utilizes a variety of third parties, including third-party manufacturers. Some of the third parties include Cardinal Health Specialty Pharmaceutical Services, that provides warehouses and ship the company's products; Ventiv Commercial Services, LLC, provides a field sales force that is the primary selling team for Kristalose. In June 2009, Cumberland announced the FDA approval of Caldolor, the first non-opioid intravenous pain and fever reliever in the U.S. In August 2009, the company completed its initial public offering (IPO).

BRANDS/DIVISIONS/AFFILIATES:
Cumberland Emerging Technologies, Inc.
Caldolor
Tennessee Technology Development Corporation
Cardinal Health Specialty Pharmaceutical Services
Ventiv Commercial Services LLC
Acetadote
Kristalose

CONTACTS: *Note: Officers with more than one job title may be intentionally listed here more than once.*
A. J. Kazimi, CEO
David L. Lowrance, CFO/VP
Martin E. Cearnal, Chief Commercial Officer/Sr. VP
Amy D. Rock, Sr. Dir.-Scientific & Regulatory Affairs
Barry L. Lee, Dir.-Prod.
Jean W. Marstiller, Corp. Sec./Sr. VP
Leo Pavliv, Sr. VP-Oper.
Angela Novak, Corp. Rel.
Doug Jack, Sr. Mgr.-SEC Reporting
James L. Herman, Corp. Compliance Officer/Sr. Dir.-Nations Acct.
Arther P. Wheeler, Dir.-Medical Affairs
Gordon R. Bernard, Sr. VP/Dir.-Medical
A. J. Kazimi, Chmn.

Phone: 615-255-0068	Fax: 615-255-0094
Toll-Free: 877-484-2700	
Address: 2525 West End Ave., Ste. 950, Nashville, TN 37203 US	

FINANCIALS: Sales and profits are in thousands of dollars—add 000 to get the full amount. 2009 Note: Financial information for 2009 was not available for all companies at press time.

2009 Sales: $43,537	2009 Profits: $5,777	U.S. Stock Ticker: CPIX
2008 Sales: $35,075	2008 Profits: $7,282	Int'l Ticker: Int'l Exchange:
2007 Sales: $28,064	2007 Profits: $6,725	Employees: 108
2006 Sales: $	2006 Profits: $	Fiscal Year Ends: 12/31
2005 Sales: $	2005 Profits: $	Parent Company:

SALARIES/BENEFITS:

Pension Plan:	ESOP Stock Plan:	Profit Sharing:	Top Exec. Salary: $366,000	Bonus: $175,000
Savings Plan:	Stock Purch. Plan:		Second Exec. Salary: $266,000	Bonus: $65,000

OTHER THOUGHTS:
Apparent Women Officers or Directors: 3
Hot Spot for Advancement for Women/Minorities: Y

LOCATIONS: ("Y" = Yes)

West:	Southwest:	Midwest:	Southeast:	Northeast:	International:
			Y		

CURAGEN CORPORATION

www.curagen.com

Industry Group Code: 325412 Ranks within this company's industry group: Sales: Profits:

Drugs:		Other:		Clinical:	Computers:	Services:	
Discovery:	Y	AgriBio:		Trials/Services:	Hardware:	Specialty Services:	
Licensing:		Genetic Data:	Y	Labs:	Software:	Consulting:	
Manufacturing:		Tissue Replacement:		Equipment/Supplies:	Arrays:	Blood Collection:	
Generics:				Research/Development Svcs.:	Database Management:	Drug Delivery:	
				Diagnostics:		Drug Distribution:	

TYPES OF BUSINESS:
Drug Discovery

BRANDS/DIVISIONS/AFFILIATES:
Belinostat
CR011-vcMMAE

CONTACTS: Note: Officers with more than one job title may be intentionally listed here more than once.
Timothy M. Shannon, CEO
Timothy M. Shannon, Pres.
Sean Cassidy, CFO/VP
Ronit Simantov, VP-Medical Dev.
Henri S. Lichenstein, VP-Prod. Dev.
Paul M. Finigan, General Counsel/Exec. VP
Cyrus Karkaria, VP-Oper. & BPS
Elizabeth Crowley, VP-Dev. Oper.
Glenn Schulman, Dir.-Investor Rel.
Hans Scholl, VP-Regulatory Affairs & Quality Assurance
Robert E. Patricelli, Chmn.

Phone: 203-481-1104	Fax: 203-483-2552
Toll-Free: 888-436-6642	
Address: 322 E. Main St., Branford, CT 06405 US	

GROWTH PLANS/SPECIAL FEATURES:

CuraGen Corporation is a biopharmaceutical company dedicated to developing cancer treatments. The firm's two major drug candidates are Belinostat (formerly PXD101) and CR011-vcMMAE, both developed in collaboration with other companies. Belinostat (PXD101), which is an inhibitor for the enzyme histone deactylase (HDAC), was developed with TopoTarget. HDAC inhibitors have been shown to arrest cancer cell growth; induce programmed cell death; and sensitize cancer cells to make them more susceptible to treatments. The drug being tested in both an oral and intravenous format, either as a monotherapy or in conjunction with other drugs, and is in various phases of clinical trials for the treatment of numerous cancers, including Phase II for the treatment of T-cell lymphomas and ovarian cancer; and Phase I for treating solid tumors and soft tissue sarcoma. The firm developed CR011, a fully-human monoclonal antibody drug, with Amgen Fremont (formerly Abgenix), and licensed antibody-drug conjugation (ADC) technology from Seattle Genetics to attach monomethylauristatin E (vcMMAE) to CR011, creating CR011-vcMMAE. CR011 attaches itself to a cancerous cell and transports the ADC inside it, after which the MMAE splits off the antibody and is activated. The drug is currently in Phase I testing for treating metastatic melanoma. Early trials suggest CR011-vcMMAE can cause complete and durable tumor regression, without any notable toxicity or weight loss.

Employees of CuraGen receive medical, dental, prescription, life and AD&D insurance.

FINANCIALS: Sales and profits are in thousands of dollars—add 000 to get the full amount. 2009 Note: Financial information for 2009 was not available for all companies at press time.

2009 Sales: $	2009 Profits: $	U.S. Stock Ticker: CRGN
2008 Sales: $1,174	2008 Profits: $24,781	Int'l Ticker: Int'l Exchange:
2007 Sales: $ 88	2007 Profits: $25,398	Employees: 16
2006 Sales: $2,298	2006 Profits: $-59,839	Fiscal Year Ends: 12/31
2005 Sales: $4,825	2005 Profits: $-73,244	Parent Company:

SALARIES/BENEFITS:
Pension Plan:	ESOP Stock Plan:	Profit Sharing:	Top Exec. Salary: $375,000	Bonus: $136,500
Savings Plan: Y	Stock Purch. Plan:		Second Exec. Salary: $315,000	Bonus: $80,262

OTHER THOUGHTS:
Apparent Women Officers or Directors: 1
Hot Spot for Advancement for Women/Minorities: Y

LOCATIONS: ("Y" = Yes)
West:	Southwest:	Midwest:	Southeast:	Northeast:	International:
				Y	

CURIS INC

www.curis.com

Industry Group Code: 325412 Ranks within this company's industry group: Sales: 127   Profits: 98

Drugs:		Other:		Clinical:	Computers:	Services:
Discovery:	Y	AgriBio:		Trials/Services:	Hardware:	Specialty Services:
Licensing:	Y	Genetic Data:		Labs:	Software:	Consulting:
Manufacturing:		Tissue Replacement:		Equipment/Supplies:	Arrays:	Blood Collection:
Generics:				Research/Development Svcs.:	Database Management:	Drug Delivery:
				Diagnostics:		Drug Distribution:

TYPES OF BUSINESS:

Drugs-Regenerative Medicine
Signaling Pathway Therapeutics

BRANDS/DIVISIONS/AFFILIATES:

GDC-0449

CONTACTS: Note: Officers with more than one job title may be intentionally listed here more than once.

Daniel R. Passeri, CEO
Michael P. Gray, COO
Daniel R. Passeri, Pres.
Michael P. Gray, CFO
Changgeng Qian, VP-Discovery & Preclinical Dev.
Mark Noel, VP-Tech. Mgmt.
Mark Noel, VP-Intellectual Property
Mitchell Keegan, VP-Drug. Dev.
James R. McNab, Jr., Chmn.

Phone: 617-503-6500	Fax: 617-503-6501
Toll-Free:	
Address: 45 Moulton St., Cambridge, MA 02138 US	

GROWTH PLANS/SPECIAL FEATURES:

Curis, Inc. focuses on the discovery and development of products that use signaling pathway drug technologies. Its product development approach involves using small molecules to target components of abnormally regulated signaling pathway networks, the networks used by cells to exchange instructional messages regulating specific biological functions. The firm primarily focuses on targeting signaling pathways in the treatment of cancer, but is also expanding its research to neurological disease and cardiovascular disease. Curis collaborates with other biopharmaceutical companies, including the Hedgehog antagonist program with Genentech and a collaborative partnership with the National Cancer Institute. The Hedgehog program targets the Hedgehog signaling pathway, which controls the development and growth of many tissues in the body. The program's leading drug candidate, GDC-0449, is being tested in three Phase II trials in metastatic colorectal cancer, advanced BCC and advanced ovarian cancer. The company's other main program is its targeted cancer drug development platform, primarily consisting of several proprietary cancer drug programs that target multiple signaling pathways. Curis is also developing a Hedgehog small molecule agonist program, which targets the Hedgehog pathway for the treatment of neurological, cardiovascular and bone diseases. In August 2009, Curis signed an exclusive licensing agreement with Debiopharm Group, under which Debiopharm will have exclusive rights to Curis' Heat Shock Protein Hsp90 technology.

Curis offers employees health, dental and vision benefits; health club membership reimbursement; a Section 125 plan; life and AD&D insurance; short- and long-term disability; business travel accident insurance; a 401(k) plan; an employee assistance program; an educational assistance program; a stock option grant; an employee stock purchase plan; an employee referral bonus; company-paid parking and transportation passes; and a company co-ed softball team.

FINANCIALS: Sales and profits are in thousands of dollars—add 000 to get the full amount. 2009 Note: Financial information for 2009 was not available for all companies at press time.

2009 Sales: $8,590	2009 Profits: $-9,823	U.S. Stock Ticker: CRIS
2008 Sales: $8,367	2008 Profits: $-12,123	Int'l Ticker: Int'l Exchange:
2007 Sales: $16,389	2007 Profits: $-6,964	Employees: 33
2006 Sales: $14,936	2006 Profits: $-8,829	Fiscal Year Ends: 12/31
2005 Sales: $6,002	2005 Profits: $-14,855	Parent Company:

SALARIES/BENEFITS:

Pension Plan:	ESOP Stock Plan:	Profit Sharing:	Top Exec. Salary: $311,538	Bonus: $
Savings Plan: Y	Stock Purch. Plan: Y		Second Exec. Salary: $259,615	Bonus: $

OTHER THOUGHTS:

Apparent Women Officers or Directors: 1
Hot Spot for Advancement for Women/Minorities:

LOCATIONS: ("Y" = Yes)

West:	Southwest:	Midwest:	Southeast:	Northeast:	International:
				Y	

CYPRESS BIOSCIENCE INC

www.cypressbio.com

Industry Group Code: 325412 **Ranks within this company's industry group:** Sales: 103 Profits: 131

Drugs:		Other:		Clinical:		Computers:		Services:	
Discovery:	Y	AgriBio:		Trials/Services:		Hardware:		Specialty Services:	Y
Licensing:	Y	Genetic Data:		Labs:		Software:		Consulting:	
Manufacturing:	Y	Tissue Replacement:		Equipment/Supplies:		Arrays:		Blood Collection:	
Generics:				Research/Development Svcs.:		Database Management:		Drug Delivery:	
				Diagnostics:	Y			Drug Distribution:	

TYPES OF BUSINESS:

Drugs-Manufacturing
Drugs-Fibromyalgia Syndrome
Personalized Medicine Laboratory Services

BRANDS/DIVISIONS/AFFILIATES:

BioLineRx Ltd.
Forest Laboratories Inc
Savella
Avise MCV
Avise PG

CONTACTS: *Note: Officers with more than one job title may be intentionally listed here more than once.*

Jay D. Kranzler, CEO
Sabrina M. Johnson, COO
Sabrina M. Johnson, CFO/Exec. VP
Srinivas G. Rao, Chief Scientific Officer
Mary Gierson, Contact-Investor Rel.
R. Michael Gendreau, Chief Medical Officer/VP-Clinical Dev.
Jay D. Kranzler, Chmn.

Phone: 858-452-2323	Fax: 858-452-1222
Toll-Free:	
Address: 4350 Executive Dr., Ste. 325, San Diego, CA 92121 US	

GROWTH PLANS/SPECIAL FEATURES:

Cypress Bioscience, Inc. develops and markets therapeutics and personalized medicine services that allow physicians to serve unmet medical needs in the areas of pain, rheumatology, physical medicine and rehabilitation. The company's lead product is Savella (milnacipran HCl) for fibromyalgia (FM), a dual reuptake inhibitor that blocks norepinephrine with higher potency than serotonin, two neurotransmitters known to play an essential role in regulating FM symptoms. Savella became available to pharmacies, rheumatologists, pain centers, physicians and physical medicine centers in early 2009. Savella is jointly marketed with Forest Laboratories, Inc. Cypress also provides personalized medicine laboratory services to rheumatologists. These services include Avise PG, a test for measuring the metabolism of methotrexate in rheumatoid arthritis (RA) patients; and Avise MCV, which tests citrullinated vimentin (MCV) to assist in the diagnosis, prognosis and therapeutic monitoring of RA. In January 2009, the FDA approved Savella for use in treating fibromyalgia. In February 2009, Cypress acquired a personalized medical technology platform from Cellatope; the platform uses cell-bound complement activation products (CB-CAP) to diagnose and monitor autoimmune disorders such as lupus. In June 2010, the company entered a license agreement for the development and commercialization of an antipsychotic being developed by BioLineRx, Ltd., for the treatment of schizophrenia.

FINANCIALS: Sales and profits are in thousands of dollars—add 000 to get the full amount. 2009 Note: Financial information for 2009 was not available for all companies at press time.

2009 Sales: $27,335	2009 Profits: $-28,252	**U.S. Stock Ticker:** CYPB
2008 Sales: $16,659	2008 Profits: $-18,226	**Int'l Ticker:** Int'l Exchange:
2007 Sales: $13,441	2007 Profits: $3,488	Employees: 150
2006 Sales: $4,322	2006 Profits: $-8,318	Fiscal Year Ends: 12/31
2005 Sales: $8,384	2005 Profits: $-7,650	Parent Company:

SALARIES/BENEFITS:

Pension Plan: Y	ESOP Stock Plan:	Profit Sharing:	Top Exec. Salary: $601,237	Bonus: $100,207
Savings Plan: Y	Stock Purch. Plan:		Second Exec. Salary: $325,118	Bonus: $20,320

OTHER THOUGHTS:

Apparent Women Officers or Directors: 2
Hot Spot for Advancement for Women/Minorities: Y

LOCATIONS: ("Y" = Yes)

West:	Southwest:	Midwest:	Southeast:	Northeast:	International:
Y					

CYTRX CORPORATION

www.cytrx.com

Industry Group Code: 325412 Ranks within this company's industry group: Sales: 125   Profits: 87

Drugs:		Other:	Clinical:	Computers:	Services:
Discovery:	Y	AgriBio:	Trials/Services:	Hardware:	Specialty Services:
Licensing:	Y	Genetic Data:	Labs:	Software:	Consulting:
Manufacturing:		Tissue Replacement:	Equipment/Supplies:	Arrays:	Blood Collection:
Generics:			Research/Development Svcs.:	Database Management:	Drug Delivery:
			Diagnostics:		Drug Distribution:

TYPES OF BUSINESS:

Pharmaceutical Research & Development
Small-Molecule Drugs

BRANDS/DIVISIONS/AFFILIATES:

Tamibarotene
Iroxanadine
INNO-206
Bafetinib
Arimoclomol
RXi Pharmaceuticals

CONTACTS: *Note: Officers with more than one job title may be intentionally listed here more than once.*

Steven A. Kriegsman, CEO
Steven A. Kriegsman, Pres.
John Y. Caloz, CFO
Daniel Levitt, Chief Medical Officer
Scott Wieland, Sr. VP-Drug Dev.
Scott Geyer, Sr. VP-Mfg.
Benjamin S. Levin, General Counsel/VP-Legal Affairs/Corp. Sec.
David J. Haen, VP-Bus. Dev.
Max Link, Chmn.

Phone: 310-826-5648	Fax: 310-826-6139
Toll-Free:	
Address: 11726 San Vicente Blvd., Ste. 650, Los Angeles, CA 90049 US	

GROWTH PLANS/SPECIAL FEATURES:

CytRx Corporation is a biopharmaceutical company specializing in oncology and the treatment of disorders using its molecular chaperone amplification technology. The company is currently developing treatments for cancer, neurodegenerative disorders and diabetic complications. Cancer treatments include tamibarotene, the company's leading drug candidate. The drug is currently undergoing clinical trial as a treatment for acute promyelocytic leukemia. Additional oncology works include INNO-206, which can be used to combat a variety of different cancers. CytRx plans to initiative three Phase II trials with INNO-206 as a treatment for pancreatic cancer, gastric cancer and soft tissue sarcomas. Bafetinib, another product, is being developed for high-risk B-cell chronic lymphocytic leukemia. The firm's chaperone amplification drugs works by activating molecular chaperones that detect and repair or degrade misfolded proteins that can cause many diseases. These drug candidates include arimoclomol and iroxanadine. Arimoclomol has the potential to treat amyotrophic lateral sclerosis (ALS, or Lou Gherig's Disease) and stroke recovery. Iroxanadine has completed Phase II clinical trials for the treatment of diabetic foot ulcers. In March 2010, CytRx announced that RXi Pharmaceuticals will buy part of the firm's shares in RXi Pharmaceuticals for approximately $3.8 million. Following the transaction, the firm will own a 28% equity interest in RXi Pharmaceuticals. In May 2010, bafetinib entered Phase II clinical trials to treat patients with high-risk B-cell chronic lymphocytic leukemia. The drug will also undergo two additional Phase II clinical trials to test its effect on glioblastoma multiforme and advanced prostate cancer. In addition, the drug was granted Orphan Drug Status for the treatment of Philadelphia chromosome-positive (Ph+) CML by the FDA.

CytRx offers employees medical, dental and vision coverage; a 401(k) plan; life and AD&D insurance; short- and long-term disability insurance; flexible spending accounts; tuition reimbursement; and an incentive stock option plan.

FINANCIALS: Sales and profits are in thousands of dollars—add 000 to get the full amount. 2009 Note: Financial information for 2009 was not available for all companies at press time.

2009 Sales: $9,500	2009 Profits: $-4,800	**U.S. Stock Ticker: CYTR**
2008 Sales: $6,266	2008 Profits: $-27,803	**Int'l Ticker:** Int'l Exchange:
2007 Sales: $7,459	2007 Profits: $-21,890	Employees: 15
2006 Sales: $2,066	2006 Profits: $-16,752	Fiscal Year Ends: 12/31
2005 Sales: $ 184	2005 Profits: $-15,093	Parent Company:

SALARIES/BENEFITS:

Pension Plan:	ESOP Stock Plan:	Profit Sharing:	Top Exec. Salary: $550,000	Bonus: $450,000
Savings Plan: Y	Stock Purch. Plan: Y		Second Exec. Salary: $276,000	Bonus: $75,000

OTHER THOUGHTS:

Apparent Women Officers or Directors:
Hot Spot for Advancement for Women/Minorities:

LOCATIONS: ("Y" = Yes)

West:	Southwest:	Midwest:	Southeast:	Northeast:	International:
Y					

DAIICHI SANKYO CO LTD

www.daiichisankyo.co.jp

Industry Group Code: 325412 Ranks within this company's industry group: Sales: 18 Profits: 170

Drugs:		Other:		Clinical:		Computers:		Services:	
Discovery:	Y	AgriBio:		Trials/Services:		Hardware:		Specialty Services:	
Licensing:	Y	Genetic Data:		Labs:		Software:		Consulting:	
Manufacturing:	Y	Tissue Replacement:		Equipment/Supplies:		Arrays:		Blood Collection:	
Generics:				Research/Development Svcs.:		Database Management:		Drug Delivery:	
				Diagnostics:				Drug Distribution:	

TYPES OF BUSINESS:

Pharmaceuticals
Prescription Drugs
Over-the-Counter Drugs
Functional Foods

BRANDS/DIVISIONS/AFFILIATES:

Sankyo Co., Ltd.
Daiichi Pharmaceutical Co., Ltd.
Daiichi Sankyo Healthcare Co. Ltd.
Daiichi Sankyo, Inc.
Ranbaxy Laboratories Limited
Luitpold Pharmaceuticals, Inc.
PharmaForce, Inc.

CONTACTS: Note: Officers with more than one job title may be intentionally listed here more than once.

Takashi Shoda, CEO
Takashi Shoda, Pres.
Ryouichi Kibushi, Head-Sales & Mktg.
Kyohei Nonose, Head-Group Human Resources
Kazunori Hirokawa, Sr. Exec. Officer/Head-R&D
Takeshi Ogita, Sr. Exec. Officer-IT
Hitoshi Matsuda, Sr. Exec. Officer/Head-Admin.
Tsutomu Une, Sr. Exec. Officer-Global Corp. Strategy
Yoshikazu Takano, Sr. Exec. Officer-External Affairs
Manabu Sakai, Exec. Officer-Global Corp. Finance
Joji Nakayama, Exec. VP/Pres., Japanese Oper.
Akira Nagano, Sr. Exec. Officer-Quality & Safety Mgmt.
Yuki Sato, Head-Pharmaceutical Tech.
Tsutomu Une, Sr. Exec. Officer-Intellectual Property & Hybrids
Kiyoshi Morita, Chmn.

Phone: 81-3-6225-1111	Fax:
Toll-Free:	
Address: 3-5-1, Nihonbashi-honcho, Chuo-ku, Tokyo, 103-8426 Japan	

GROWTH PLANS/SPECIAL FEATURES:

Daiichi Sankyo Co., Ltd., formed from the merger of the century-old Sankyo Co., Ltd. and Daiichi Pharmaceutical Co., Ltd., is primarily a pharmaceutical manufacturing firm. Its products are distributed through roughly 5,000 representatives, about half of which are in Japan with the rest spread throughout the U.S., Europe and Asia. Daiichi Sankyo's research and development is focused mainly on cardiovascular diseases; cancer; glucose metabolic disorders; bone/joint diseases; immunity and allergies; and infectious diseases. The firm has research and development facilities in the U.S., the U.K., Germany, India, China and Japan. Some of the company's major drugs developments include Pravastatin, an antihyperlipidemic agent that is also sold under the name Mevalotin; Levofloxacin, an oral antibacterial agent, also called Cravit; Olmesartan, an antihypertension agent sold in the U.S. as Benicar, and sold in Japan and Europe as Olmetec; and WelChol, an antihyperlipidemic agent and a treatment for Type 2 diabetes. Its OTC (over-the-counter) drug products include Lamisil AT, an athlete's foot and ringworm treatment; and Patecs Felbinac, an external anti-inflammatory analgesic. Besides its domestic network, the firm has some 25 overseas subsidiaries in more than 25 countries. Domestic subsidiaries include Daiichi Sankyo Healthcare Co. Ltd., which offers self-medication options exclusively in Japan, including OTC medicines, skin care products and medical equipment. Overseas subsidiaries include Daiichi Sankyo, Inc. in the U.S., which markets the firm's products as well as conducting clinical trials and overseeing U.S. regulatory affairs and other functions. Geographically, Japan generated around 55% of Daiichi Sankyo's 2009 sales; North America, 24%; Europe, 10%; India, 6%; and other regions, 5%. Daiichi Sankyo also holds an approximate 64% stake in one of India's largest pharmaceutical firms, Ranbaxy Laboratories Limited, with the two companies partnering on a range of drug development and marketing activities worldwide. In January 2010, U.S. subsidiary Luitpold Pharmaceuticals acquired PharmaForce, Inc., a generic drug producer.

FINANCIALS: Sales and profits are in thousands of dollars—add 000 to get the full amount. 2009 Note: Financial information for 2009 was not available for all companies at press time.

2009 Sales: $8,881,960	2009 Profits: $-2,272,830	**U.S. Stock Ticker:**
2008 Sales: $9,282,450	2008 Profits: $1,030,000	**Int'l Ticker: 4568** Int'l Exchange: Tokyo-TSE
2007 Sales: $7,910,000	2007 Profits: $670,000	Employees:
2006 Sales: $8,703,320	2006 Profits: $735,030	Fiscal Year Ends: 3/31
2005 Sales: $8,669,610	2005 Profits: $820,240	Parent Company:

SALARIES/BENEFITS:

Pension Plan:	ESOP Stock Plan:	Profit Sharing:	Top Exec. Salary: $	Bonus: $
Savings Plan:	Stock Purch. Plan:		Second Exec. Salary: $	Bonus: $

OTHER THOUGHTS:

Apparent Women Officers or Directors:
Hot Spot for Advancement for Women/Minorities:

LOCATIONS: ("Y" = Yes)

West:	Southwest:	Midwest:	Southeast:	Northeast:	International:
				Y	Y

DAINIPPON SUMITOMO PHARMA CO LTD   www.ds-pharma.co.jp

Industry Group Code: 325412 Ranks within this company's industry group: Sales: 28   Profits: 34

Drugs:		Other:		Clinical:		Computers:		Services:	
Discovery:	Y	AgriBio:		Trials/Services:		Hardware:		Specialty Services:	Y
Licensing:	Y	Genetic Data:		Labs:		Software:		Consulting:	
Manufacturing:	Y	Tissue Replacement:		Equipment/Supplies:	Y	Arrays:		Blood Collection:	
Generics:				Research/Development Svcs.:		Database Management:		Drug Delivery:	
				Diagnostics:	Y			Drug Distribution:	Y

TYPES OF BUSINESS:

Pharmaceuticals Manufacturing
Pharmaceutical Additives
Animal Feeds and Medications Manufacturing
Sweeteners and Other Food Additives Manufacturing

BRANDS/DIVISIONS/AFFILIATES:

Sepracor Inc
DSP Gokyo Food & Chemical Co., Ltd.
DS Pharma Animal Health Co., Ltd.
DS Pharma Biomedical Co., Ltd.
DSP Distribution Services Co., Ltd.
Sumitomo Pharmaceuticals (Suzhou) Co., Ltd.

CONTACTS: *Note: Officers with more than one job title may be intentionally listed here more than once.*

Masayo Tada, CEO
Masayo Tada, Pres.
Yukio Kitahara, Exec. Dir.-Sales & Mktg.
Masaru Ishidahara, Dir.-Personnel
Yutaka Takeuchi, Exec. Dir.-Tech. R&D
Yutaka Takeuchi, Exec. Dir.-Mfg.
Kazumi Okamura, Sr. Exec. Officer-Admin.
Kazumi Okamura, Sr. Exec. Officer-Legal Affairs
Hiroshi Noguchi, Exec. Dir.-Strategic Planning & Bus. Dev.
Keiichi Ono, Exec. Dir.-Corp. Comm. & Intellectual Property
Hiroshi Nomura, Dir.-Finance & Acct.
Yoshihiro Okada, Exec. Dir.-Drug Dev.
Yasuji Furutani, Exec. Dir.-Regulatory Compliance & Quality
Nobuhiko Tamura, Exec. VP/Chief Science Officer-Sepracor, Inc.
Kenjiro Miyatake, Chmn.
Saburo Hamanaka, Chmn./CEO-Sepracor, Inc.

Phone: 81-6-6203-5321	Fax: 81-6-6202-6028
Toll-Free:	
Address: 2-6-8 Doshomachi, Chuo-ku, Osaka, 541-0045 Japan	

GROWTH PLANS/SPECIAL FEATURES:

Dainippon Sumitomo Pharma Co., Ltd. is a developer and manufacturer of pharmaceuticals and related products, with approximately 22 subsidiaries and operations in Japan, China, the U.K. and the U.S. The company operates in two primary business units, pharmaceuticals and other products. The pharmaceuticals business encompasses the development of drugs to treat diabetes and cardiovascular disease, as well as diseases of the central nervous system such as schizophrenia, epilepsy, anxiety disorders and Parkinson's disease. Products in these areas include Avapro, a therapeutic agent for hypertension; Lonasen, an antipsychotic medication; Prorenal, a vasodilator; Meropen, an ultra-broad spectrum injectable antibiotic; Amlodin, a therapeutic agent for angina pectoris and hypertension; and AmBisome, for systemic fungal infections. The business unit covering other products, meanwhile, develops, manufactures and markets animal health products, including food and medications for domestic animals and livestock; specialty food products, including sweeteners, seasoning agents and pharmaceutical additives; and diagnostic agents and research materials, marketed primarily to the Japanese medical community. Primary Japanese subsidiaries include DSP Gokyo Food & Chemical Co., Ltd.; DS Pharma Animal Health Co., Ltd.; DS Pharma Biomedical Co., Ltd; and DSP Distribution Services Co., Ltd. Sumitomo Pharmaceuticals (Suzhou) Co., Ltd. oversees Chinese drug development operations as well as the sale and marketing of Dainippon Sumitomo's products in China. Massachusetts-based subsidiary Sepracor, Inc., acquired in October 2009 for approximately $2.6 billion, oversees the firm's North American business activities and is focused on the development of viable treatments for respiratory ailments and diseases of the central nervous system. New drug products launched by Dainippon Sumitomo in 2009 and 2010 include Trerief, a therapeutic agent for Parkinson's disease; Miripla, for liver cancer; and Metogluco, an oral drug for the management of hypoglycemia. In July 2010, the company sold its growth hormone business to Japan-based JCR Pharmaceuticals Co., Ltd.

FINANCIALS: Sales and profits are in thousands of dollars—add 000 to get the full amount. 2009 Note: Financial information for 2009 was not available for all companies at press time.

2009 Sales: $3,086,110	2009 Profits: $233,610	U.S. Stock Ticker:	
2008 Sales: $3,085,580	2008 Profits: $299,110	Int'l Ticker: 4506 Int'l Exchange:	
2007 Sales: $3,053,100	2007 Profits: $264,210	Employees:	
2006 Sales: $	2006 Profits: $	Fiscal Year Ends: 3/31	
2005 Sales: $	2005 Profits: $	Parent Company:	

SALARIES/BENEFITS:

Pension Plan:	ESOP Stock Plan:	Profit Sharing:	Top Exec. Salary: $	Bonus: $
Savings Plan: Y	Stock Purch. Plan: Y		Second Exec. Salary: $	Bonus: $

OTHER THOUGHTS:

Apparent Women Officers or Directors:
Hot Spot for Advancement for Women/Minorities:

LOCATIONS: ("Y" = Yes)

West:	Southwest:	Midwest:	Southeast:	Northeast:	International:
				Y	Y

DECODE GENETICS LTD

www.decode.com

Industry Group Code: 325413 Ranks within this company's industry group: Sales: Profits:

Drugs:	Other:	Clinical:	Computers:	Services:
Discovery:	AgriBio:	Trials/Services:	Hardware:	Specialty Services: Y
Licensing:	Genetic Data: Y	Labs:	Software:	Consulting:
Manufacturing:	Tissue Replacement:	Equipment/Supplies:	Arrays:	Blood Collection:
Generics:		Research/Development Svcs.:	Database Management:	Drug Delivery:
		Diagnostics:		Drug Distribution:

TYPES OF BUSINESS:

Genetic Analysis
Genetic Disease Risk Assessment
Genotyping & Sequencing Services
Data Management Services

BRANDS/DIVISIONS/AFFILIATES:

deCODE genetics ehf
Saga Investments LLC
deCODEme
deCODE ProstateCancer
deCODET2
ARUP Laboratories
Illumina
Celera

CONTACTS: *Note: Officers with more than one job title may be intentionally listed here more than once.*

Earl M. Collier, Jr., CEO
Tucker Kelly, CFO
Kari Stafansson, Pres., Research
Hakon Guobjartsson, VP-Informatics
Jeffrey Gulcher, VP-Prod. Dev.
Jakob Sigurosson, Sr. VP-Corp. Dev.
Edward Farmer, Contact-Media & Comm.
Cindy Bayley, VP-Bus. Strategy
Richard Leach, VP-Scientific Svcs.
Unnur Thorsteinsdottir, VP-Research
Hilma Holm, VP-Clinical Research
Kari Stafansson, Chmn.

Phone: 354-570-1900	Fax: 354-570-1903
Toll-Free:	
Address: Sturlugata 8, Reykjavik, 101 Iceland	

GROWTH PLANS/SPECIAL FEATURES:

deCODE genetics, Ltd., owned by Saga Investments LLC, is a human genome analysis and research company based in Iceland. The firm began as a wholly-owned subsidiary of deCODE genetics, Inc., but became a private company after its sale to Saga Investments in January 2010. deCODE's operations include researching key genetic risk factors for certain diseases, including cancer, diabetes and cardiovascular conditions; and the development of DNA-based tests to detect these factors and determine the risk of disease in individuals. Its disease risk assessment tests include deCODET2 for type 2 diabetes; deCODE AF for atrial fibrillation and stroke; deCODE MI for heart attack; deCODE ProstateCancer for prostate cancer; deCODE Glaucoma for a major type of glaucoma; and deCODE BreastCancer for the common forms of breast cancer. The company also offers a personal genome analysis service, deCODEme, a genetic scan that provides individuals with an analysis of their genetic risk for contracting certain diseases. The firm licenses its tests and analytical tools to partner companies in the research, pharmaceuticals and informatics industries, including Celera and Illumina. deCODE also provides value-added scientific services to companies and research institutions; services include genotyping and sequencing, data management and protection, sample storage, sequence imputation and statistical analysis. In January 2009, the company launched its deCODEme Cardio and deCODEme Cancer personal genome tests. In April 2009, deCODE formed a license agreement with Celera for its deCODE risk markers. In January 2010, the firm was acquired by Saga Investments LLC. In July 2010, the company announced a partnership with ARUP Laboratories through which ARUP will offer the company's deCODE ProstateCancer risk assessment test.

FINANCIALS: Sales and profits are in thousands of dollars—add 000 to get the full amount. 2009 Note: Financial information for 2009 was not available for all companies at press time.

2009 Sales: $	2009 Profits: $	U.S. Stock Ticker: Private
2008 Sales: $	2008 Profits: $	Int'l Ticker: Int'l Exchange:
2007 Sales: $	2007 Profits: $	Employees:
2006 Sales: $	2006 Profits: $	Fiscal Year Ends:
2005 Sales: $	2005 Profits: $	Parent Company: SAGA INVESTMENTS LLC

SALARIES/BENEFITS:

Pension Plan:	ESOP Stock Plan:	Profit Sharing:	Top Exec. Salary: $	Bonus: $
Savings Plan:	Stock Purch. Plan:		Second Exec. Salary: $	Bonus: $

OTHER THOUGHTS:

Apparent Women Officers or Directors: 3
Hot Spot for Advancement for Women/Minorities: Y

LOCATIONS: ("Y" = Yes)

West:	Southwest:	Midwest:	Southeast:	Northeast:	International: Y

DENDREON CORPORATION

www.dendreon.com

Industry Group Code: 325412 Ranks within this company's industry group: Sales: 151   Profits: 167

Drugs:		Other:	Clinical:	Computers:	Services:
Discovery:	Y	AgriBio:	Trials/Services:	Hardware:	Specialty Services:
Licensing:		Genetic Data:	Labs:	Software:	Consulting:
Manufacturing:		Tissue Replacement:	Equipment/Supplies:	Arrays:	Blood Collection:
Generics:			Research/Development Svcs.:	Database Management:	Drug Delivery:
			Diagnostics:		Drug Distribution:

TYPES OF BUSINESS:

Cancer Drugs
Antigen Modification Technology

BRANDS/DIVISIONS/AFFILIATES:

Provenge (Sipuleucel-T)
TRPM8
CA-9
CEA
Lapuleucel-T

CONTACTS: *Note: Officers with more than one job title may be intentionally listed here more than once.*

Mitchell H. Gold, CEO
Hans E. Bishop, COO/Exec. VP
Mitchell H. Gold, Pres.
Gregory T. Schiffman, CFO/Sr. VP
Richard J. Ranieri, Sr. VP-Human Resources
David L. Urdal, Chief Scientific Officer/Sr. VP
Rick Hamm, General Counsel/Sec.
Rick Hamm, Sr. VP-Corp. Dev.
Varun Nanda, Sr. VP-Global Comm. Oper.
Mark Frolich, Sr. VP-Clinical Affairs & Chief Medical Officer
Richard B. Brewer, Chmn.

Phone: 206-256-4545	Fax: 206-256-0571
Toll-Free:	
Address: 3005 First Ave., Seattle, WA 98121 US	

GROWTH PLANS/SPECIAL FEATURES:

Dendreon Corporation discovers, develops and commercializes active cellular immunotherapy and small molecule product candidates to treat a wide range of cancers. The company's only FDA-approved product is Provenge (sipuleucel-T), an autologous cellular immunotherapy for the treatment of asymptomatic, metastatic and androgen-independent prostate cancer (AIPC). Provenge is made by introducing the immune system cells from the patient being treated to a protein that works as a prostate cancer antigen. This activates the patient's immune cells to fight prostate cancer. One of the company's products in Phase I clinical development is Lapuleucel-T, which is an investigational active immunotherapy for the treatment of patients with breast, ovarian and colorectal solid tumors expressing the HER2/neu antigen. Dendreon's Antigen Delivery Cassette (ADS) technology is a proprietary tool for antigen modification, which allows for the modification of tumor antigens to supplement the use and processing of the antigen by dendritic cells to strengthen the immune response. The company is currently employing the ADS to develop immunotherapy product candidates targeted to antigens other than PAP and HER2/neu. Two of the antigens targeted in that process are Carbonic AnhydraseIX (CA-9) expressed in renal cell carcinomas and non-small cell lung and breast tumors; and Carcinoembryonic Antigen (CEA) present in a majority of breast, lung and colon cancers. One of the small molecule therapies in research and development is based on TRPM8, an ion channel expressed in 71% of breast cancers, 80% of lung cancers, 93% of colon cancers and 100% of prostate cancers. The lead investigational molecule for TRPM8, D-3263, HCl, is in Phase I clinical trials. In April 2010, the FDA approved Provenge for the treatment of advanced prostate cancer.

The company offers life, medical, dental and vision insurance; AD&D and disability coverage; an employee assistance program; a public transportation subsidy; and a tuition subsidy program.

FINANCIALS: Sales and profits are in thousands of dollars—add 000 to get the full amount. 2009 Note: Financial information for 2009 was not available for all companies at press time.

2009 Sales: $ 101	2009 Profits: $-220,161	U.S. Stock Ticker: DNDN
2008 Sales: $ 111	2008 Profits: $-71,644	Int'l Ticker: Int'l Exchange:
2007 Sales: $ 743	2007 Profits: $-99,264	Employees: 484
2006 Sales: $ 273	2006 Profits: $-91,642	Fiscal Year Ends: 12/31
2005 Sales: $ 210	2005 Profits: $-81,547	Parent Company:

SALARIES/BENEFITS:

Pension Plan:	ESOP Stock Plan:	Profit Sharing:	Top Exec. Salary: $550,000	Bonus: $387,750
Savings Plan: Y	Stock Purch. Plan: Y		Second Exec. Salary: $407,550	Bonus: $191,459

OTHER THOUGHTS:

Apparent Women Officers or Directors: 1
Hot Spot for Advancement for Women/Minorities:

LOCATIONS: ("Y" = Yes)

West:	Southwest:	Midwest:	Southeast:	Northeast:	International:
Y			Y	Y	

DEPOMED INC

www.depomedinc.com

Industry Group Code: 325412A Ranks within this company's industry group: Sales: 7 Profits: 15

Drugs:		Other:		Clinical:	Computers:		Services:	
Discovery:	Y	AgriBio:	Y	Trials/Services:	Hardware:		Specialty Services:	
Licensing:	Y	Genetic Data:		Labs:	Software:		Consulting:	
Manufacturing:	Y	Tissue Replacement:		Equipment/Supplies:	Arrays:		Blood Collection:	
Generics:				Research/Development Svcs.:	Database Management:		Drug Delivery:	Y
				Diagnostics:			Drug Distribution:	

TYPES OF BUSINESS:

Drug Delivery Systems

BRANDS/DIVISIONS/AFFILIATES:

AcuForm
Glumetza
ProQuin XR
Gabapentin GR
Serada

CONTACTS: Note: Officers with more than one job title may be intentionally listed here more than once.

Carl A. Pelzel, CEO
Carl A. Pelzel, Pres.
Tammy L. Cameron, Interim Principal Acct. & Financial Officer
Shay Weisbrich, VP-Mktg. & Sales
Kera Alexander, VP-Human Resources
Michael Sweeney, VP-R&D
Kera Alexander, VP-Admin.
Matthew Gosling, General Counsel/VP
William Callahan, VP-Oper.
Thadd M. Vargas, Sr. VP-Bus. Dev.
Tammy L. Cameron, VP-Finance
Craig R. Smith, Chmn.

Phone: 650-462-5900	Fax: 650-462-9993
Toll-Free:	
Address: 1360 O'Brien Dr., Menlo Park, CA 94025 US	

GROWTH PLANS/SPECIAL FEATURES:

Depomed, Inc. focuses on the development and commercialization of differentiated products that are based on proprietary drug delivery technologies. The company has developed two commercial products: Glumetza, a once-daily treatment for adults with Type 2 diabetes that the firm jointly commercializes in the U.S. with Santarus, Inc.; and ProQuin XR, a once-daily treatment of uncomplicated urinary tract infections. Depomed's most advanced product candidate in development is Gabapentin GR, an extended release form of gabapentin. The company has submitted an investigational new drug application (NDA) for Gabapentin GR, and has completed Phase III studies for its efficacy in the treatment of both menopausal hot flashes, under the name Serada (currently in a third Phase III clinical trial), and postherpetic neuralgia, under the name DM-1796. In addition, the firm has other product candidates in earlier stages of development, including treatments for gastroesophageal reflux disease (DM-3458) and Parkinson's Disease (DM-1992). Depomed's drugs are based on its proprietary drug delivery system, AcuForm. The AcuForm technology is a polymer-based drug delivery platform that provides targeted drug delivery solutions for a wide range of compounds. The technology embraces diffusional, erosional, bilayer and multi-drug systems that can optimize oral drug delivery for both soluble and insoluble drugs. One application of the technology allows standard-sized tablets to be retained in the stomach for 6-8 hours after administration, extending the time of drug delivery to the small intestine. In June 2010, Depomed's NDA for DM-1796 was accepted by the FDA and is currently awaiting market approval.

The company offers medical, dental, life and long-term disability insurance; flexible spending accounts; a 401(k) plan; and an employee stock purchase plan.

FINANCIALS: Sales and profits are in thousands of dollars—add 000 to get the full amount. 2009 Note: Financial information for 2009 was not available for all companies at press time.

2009 Sales: $57,728	2009 Profits: $-22,008	U.S. Stock Ticker: DEPO
2008 Sales: $34,842	2008 Profits: $-15,843	Int'l Ticker: Int'l Exchange:
2007 Sales: $65,582	2007 Profits: $49,219	Employees: 73
2006 Sales: $9,551	2006 Profits: $-39,659	Fiscal Year Ends: 12/31
2005 Sales: $4,405	2005 Profits: $-24,467	Parent Company:

SALARIES/BENEFITS:

Pension Plan:	ESOP Stock Plan:	Profit Sharing:	Top Exec. Salary: $450,000	Bonus: $250,000
Savings Plan: Y	Stock Purch. Plan: Y		Second Exec. Salary: $340,000	Bonus: $110,000

OTHER THOUGHTS:

Apparent Women Officers or Directors: 4
Hot Spot for Advancement for Women/Minorities: Y

LOCATIONS: ("Y" = Yes)

West:	Southwest:	Midwest:	Southeast:	Northeast:	International:
Y					

DISCOVERY LABORATORIES INC   www.discoverylabs.com

Industry Group Code: 325412 Ranks within this company's industry group: Sales: Profits: 133

Drugs:		Other:	Clinical:	Computers:	Services:
Discovery:	Y	AgriBio:	Trials/Services:	Hardware:	Specialty Services:
Licensing:		Genetic Data:	Labs:	Software:	Consulting:
Manufacturing:	Y	Tissue Replacement:	Equipment/Supplies:	Arrays:	Blood Collection:
Generics:			Research/Development Svcs.:	Database Management:	Drug Delivery:
			Diagnostics:		Drug Distribution:

TYPES OF BUSINESS:
Respiratory Disease Treatments
Pulmonary Drug Delivery Products

BRANDS/DIVISIONS/AFFILIATES:
Surfaxin
Aerosurf
KL-4

CONTACTS: *Note: Officers with more than one job title may be intentionally listed here more than once.*
W. Thomas Amick, Interim CEO
Robert J. Capetola, Pres.
John G. Cooper, CFO/Exec. VP
Kathryn Cole, Sr. VP-Human Resources
Russel Clayton, VP-R&D, Regulatory Affairs, Pre-Clinical
Charles F. Katzer, Sr. VP-Mfg. Oper.
David L. Lopez, General Counsel/Exec. VP
Gerald J. Orehostky, Sr. VP-Quality Oper.
Thomas F. Miller, Sr. VP-Corp. Dev. & Commercialization
Robert Segal, Chief Medical Officer/Sr. VP-Medical & Scientific
Mary B. Templeton, Sr. VP/Deputy General Counsel
W. Thomas Amick, Chmn.

Phone: 215-488-9300	Fax: 215-488-9301
Toll-Free:	
Address: 2600 Kelly Road, Ste. 100, Warrington, PA 18976 US	

GROWTH PLANS/SPECIAL FEATURES:

Discovery Laboratories, Inc. (DLI) is a biotechnology company that develops proprietary surfactant technology as surfactant replacement therapies for respiratory disorders and diseases. Surfactants are produced naturally by the lungs and are essential for breathing. Discovery's technology produces a precision-engineered surfactant designed to closely mimic the essential properties of natural human lung surfactant. DLI's technology also utilizes KL-4, a peptide also known as sinapultide, which mimics the essential properties of SP-B, a surfactant protein that lowers surface tension and promotes oxygen exchange. The firm's products, Surfaxin and Surfaxin LS, are used in the prevention of respiratory distress syndrome (RDS) in premature infants. DLI also develops Surfaxin for the prevention and treatment of bronchopulmonary dysplasia in premature infants, and acute respiratory failure in small children. Aerosurf, the company's proprietary SRT in aerosolized form administered through nasal continuous positive airway pressure, is being developed for the prevention and treatment of infants at risk for RDS. DLI has granted development and marketing rights of its SRT products to Dr. Esteve, S.A., one of the largest pharmaceutical companies in Southern Europe. The firm is also working to develop and commercialize aerosol SRT to address a broad range of serious respiratory conditions such as neonatal respiratory failure, cystic fibrosis, chronic obstructive respiratory disorder and asthma. In April 2010, the company announced plans to meet the final requirement for Surfaxin's approval for marketing in the United States: a fetal rabbit Biological Activity Test (BAT) which is used to test quality control release and stability.

Discovery offers its employees medical, vision, dental and life insurance; flexible spending accounts; short and long-term disability; access to a credit union; employee referral bonuses; a 529 college savings plan; and an employee assistance program.

FINANCIALS: Sales and profits are in thousands of dollars—add 000 to get the full amount. 2009 Note: Financial information for 2009 was not available for all companies at press time.

2009 Sales: $	2009 Profits: $-30,240	**U.S. Stock Ticker: DSCO**
2008 Sales: $4,600	2008 Profits: $-39,106	**Int'l Ticker:** Int'l Exchange:
2007 Sales: $	2007 Profits: $-40,005	Employees: 77
2006 Sales: $	2006 Profits: $-46,333	Fiscal Year Ends: 12/31
2005 Sales: $ 134	2005 Profits: $-58,904	Parent Company:

SALARIES/BENEFITS:

Pension Plan:	ESOP Stock Plan:	Profit Sharing:	Top Exec. Salary: $490,000	Bonus: $
Savings Plan: Y	Stock Purch. Plan: Y		Second Exec. Salary: $307,000	Bonus: $

OTHER THOUGHTS:
Apparent Women Officers or Directors: 2
Hot Spot for Advancement for Women/Minorities: Y

LOCATIONS: ("Y" = Yes)

West:	Southwest:	Midwest:	Southeast:	Northeast:	International:
Y				Y	

DIVI'S LABORATORIES LIMITED

www.divislabs.com

Industry Group Code: 325412 **Ranks within this company's industry group:** Sales: Profits:

Drugs:		Other:	Clinical:		Computers:	Services:	
Discovery:	Y	AgriBio:	Trials/Services:		Hardware:	Specialty Services:	Y
Licensing:		Genetic Data:	Labs:		Software:	Consulting:	
Manufacturing:	Y	Tissue Replacement:	Equipment/Supplies:		Arrays:	Blood Collection:	
Generics:			Research/Development Svcs.:	Y	Database Management:	Drug Delivery:	
			Diagnostics:			Drug Distribution:	

TYPES OF BUSINESS:

Pharmaceutical Manufacturing
Pharmaceutical Research & Development

BRANDS/DIVISIONS/AFFILIATES:

Divis Laboratories (USA), Inc.
Divis Laboratories (Europe) AG
Vivital

CONTACTS: *Note: Officers with more than one job title may be intentionally listed here more than once.*

Murali K. Divi, Managing Dir.
L. Kishorebabu, CFO
P. Gundu Rao, Dir.-R&D
P.V. Lakshmi Rajani, Sec.
Kiran S. Divi, Dir.-Bus. Dev.
N. V. Ramana, Exec. Dir.
Madhusudana Rao Divi, Exec. Dir.
Murali K. Divi, Chmn.

Phone: 91-40-2373-1318	Fax: 91-40-2373-3242
Toll-Free:	
Address: Divi Towers, 7-1-77/E/1/303 Dharam Karan Rd., Hyderabad, 500016 India	

GROWTH PLANS/SPECIAL FEATURES:

Divi's Laboratories Limited is an Indian company engaged in pharmaceutical research and development, as well as pharmaceutical manufacturing. The company produces active pharmaceutical ingredients (APIs) and intermediates for generics; peptide components; nucleotide components; carotenoids; and chiral ligands. Some of the APIs manufactured by the company include naproxen, naproxen sodium, niacin, telmisartan, gabapentin and carbidopa. The firm's services include custom synthesis and contract research, specializing in process design for new drug candidates, development, structural elucidation, impurity profile studies, process validation, process justification, process optimization, analytical methods development and validation, environment impact analysis, safety studies and time cycle studies. Divi's operates three manufacturing facilities and four R&D centers. The firm's main manufacturing and research and development facilities are located in Andhra Pradesh, India. Divi's operates two regional subsidiaries, Divi's Laboratories (USA), Inc., located in New Jersey, and Divi's Laboratories (Europe) AG, located in Switzerland. These locations market the firm's nutraceutical products under the brand Vivital. Products offered in this line include, apocarotenal, astaxanithin, beta carotene, canthaxanthin, lutein and lycopene. The company's sales percentages by area breakdown as such: 50.81%, from North America; 31.05% from Europe; India, 7.07%; Far East, 4.24%; South America, 2.83%; and Asia, 1.71%. The rest of the world accounts for the small remainder.

FINANCIALS: Sales and profits are in thousands of dollars—add 000 to get the full amount. 2009 Note: Financial information for 2009 was not available for all companies at press time.

2009 Sales: $	2009 Profits: $	**U.S. Stock Ticker: DIVISLAB**
2008 Sales: $	2008 Profits: $	**Int'l Ticker: 532488** Int'l Exchange: Bombay-BSE
2007 Sales: $	2007 Profits: $	Employees:
2006 Sales: $	2006 Profits: $	Fiscal Year Ends:
2005 Sales: $	2005 Profits: $	Parent Company:

SALARIES/BENEFITS:

Pension Plan:	ESOP Stock Plan: Y	Profit Sharing:	Top Exec. Salary: $	Bonus: $
Savings Plan:	Stock Purch. Plan:		Second Exec. Salary: $	Bonus: $

OTHER THOUGHTS:

Apparent Women Officers or Directors:
Hot Spot for Advancement for Women/Minorities:

LOCATIONS: ("Y" = Yes)

West:	Southwest:	Midwest:	Southeast:	Northeast:	International:
				Y	Y

DOW AGROSCIENCES LLC

www.dowagro.com

Industry Group Code: 11511 Ranks within this company's industry group: Sales: Profits:

Drugs:	Other:		Clinical:	Computers:	Services:
Discovery:	AgriBio:	Y	Trials/Services:	Hardware:	Specialty Services:
Licensing:	Genetic Data:	Y	Labs:	Software:	Consulting:
Manufacturing:	Tissue Replacement:		Equipment/Supplies:	Arrays:	Blood Collection:
Generics:			Research/Development Svcs.:	Database Management:	Drug Delivery:
			Diagnostics:		Drug Distribution:

TYPES OF BUSINESS:

Agricultural Chemicals
Agricultural Biotechnology Products
Herbicides, Pesticides & Fungicides
Plant Genetics

BRANDS/DIVISIONS/AFFILIATES:

Dow Chemical Company (The)
Mycogen
Herculex
WideStrike
Nexera
PhytoGen

CONTACTS: Note: Officers with more than one job title may be intentionally listed here more than once.

Antonio Galindez, CEO
Antonio Galindez, Pres.
Bill Wales, General Counsel/VP-Legal Office/Sec.
Robyn Heine, Leader-Global Public Affairs

Phone: 317-337-3000	Fax:
Toll-Free:	
Address: 9330 Zionsville Rd., Indianapolis, IN 46268 US	

GROWTH PLANS/SPECIAL FEATURES:

Dow AgroSciences, LLC, a wholly-owned subsidiary of Dow Chemical Co., is a global provider of pest management and biotechnology products for agricultural and specialty markets. The company, formerly DowElanco, was formed by a joint venture between Dow Chemical and Eli Lilly; Dow Chemical purchased Eli Lilly's share in 1997. The firm's products are broken into two categories: agricultural chemicals and plant genetics/biotechnology. The agricultural chemicals unit produces herbicides, insecticides, fungicides, pest management solutions such as gas fumigants and termite detection tools and more. The plant genetics and biotechnology business develops agricultural products that protect crops against insects, boost nutritional value and increase crop yields. Products are broken into three segments: traits, which encompasses the brands Herculex and WideStrike; seeds, which includes Mycogen brand seeds and PhytoGen cottonseed; and oils. The subsidiary has operations in over 140 countries worldwide. The firm maintains many collaboration ventures with companies such as Monsanto, a global provider of agricultural products and technology-based solutions; KeyGene, a Research and Development company focused on developing and applying DNA expertise in the molecular genetics of crop plants; Chromatin Inc., which develops and markets proprietary technology that allows whole chromosomes to be designed and integrated into plant cells; and NemGenix, an agricultural biotechnology company based in Perth, Australia. In July 2009, the company agreed to acquire the majority assets of Pfister Hybrids, a corn company based in Illinois. In January 2010, the firm acquired Hyland Seeds, a Canadian agricultural firm. Also in January 2010, the company sold its thifluzamide fungicide business to Nissan Chemical Industries, Ltd., a Japan based chemical firm.

FINANCIALS: Sales and profits are in thousands of dollars—add 000 to get the full amount. 2009 Note: Financial information for 2009 was not available for all companies at press time.

2009 Sales: $	2009 Profits: $	U.S. Stock Ticker: Subsidiary
2008 Sales: $	2008 Profits: $	Int'l Ticker: Int'l Exchange:
2007 Sales: $	2007 Profits: $	Employees:
2006 Sales: $	2006 Profits: $	Fiscal Year Ends: 12/31
2005 Sales: $3,364,000	2005 Profits: $	Parent Company: DOW CHEMICAL COMPANY (THE)

SALARIES/BENEFITS:

Pension Plan: Y	ESOP Stock Plan:	Profit Sharing:	Top Exec. Salary: $	Bonus: $
Savings Plan: Y	Stock Purch. Plan: Y		Second Exec. Salary: $	Bonus: $

OTHER THOUGHTS:

Apparent Women Officers or Directors: 1
Hot Spot for Advancement for Women/Minorities:

LOCATIONS: ("Y" = Yes)

West:	Southwest:	Midwest:	Southeast:	Northeast:	International:
		Y			Y

DR REDDY'S LABORATORIES LIMITED
www.drreddys.com

Industry Group Code: 325412 Ranks within this company's industry group: Sales: 40   Profits: 159

Drugs:		Other:	Clinical:	Computers:	Services:	
Discovery:	Y	AgriBio:	Trials/Services:	Hardware:	Specialty Services:	
Licensing:		Genetic Data:	Labs:	Software:	Consulting:	
Manufacturing:	Y	Tissue Replacement:	Equipment/Supplies:	Arrays:	Blood Collection:	
Generics:			Research/Development Svcs.:	Database Management:	Drug Delivery:	
			Diagnostics:		Drug Distribution:	Y

TYPES OF BUSINESS:
Pharmaceuticals
Generic Drugs
Active Ingredients
Drug Discovery & Development

BRANDS/DIVISIONS/AFFILIATES:
Omez
Reditux
Betapharm
Razo
Mintop
Stamlo
Promius Pharma

CONTACTS: Note: Officers with more than one job title may be intentionally listed here more than once.
G. V. Prasad, CEO/Vice Chmn.
Satish Reddy, COO
Satish Reddy, Managing Dir.
Umang Vohra, CFO/Sr. VP
Prabir Jha, Chief Human Resources Officer/Sr. VP
K. B. Sankara Rao, Exec. VP-Integrated Prod. Dev.
Vilas Dholye, Exec. VP/Head-Formulations Mfg.
V. S. Suresh, Corp. Sec.
Saumen Chakraborty, Pres., Corp. Oper.
Sanjeev Verma, Head-Corp. Comm.
Abijit Mukherjee, Pres., Global Generics
Raghav Chari, Sr. VP-Proprietary Prod.
Amit Patel, Sr. VP/Head-North America Generics
Cartikeya Reddy, Sr. VP/Head-Biologics
Anji Reddy, Chmn.

Phone: 91-40-23731946	Fax: 91-40-23731955
Toll-Free:	
Address: 7-1-27, Ameerpet, Hyderabad, 500 016 India	

GROWTH PLANS/SPECIAL FEATURES:

Dr. Reddy's Laboratories Limited is a global pharmaceutical manufacturer. It operates in three segments: Global Generics, which generated approximately 72% of 2009 revenue; Pharmaceutical Services & Active Ingredients (PSAI), 27%; and Proprietary Products and other, 1%. The Global Generics segment includes branded and unbranded prescription and over-the-counter (OTC) drug products. The firm focuses on cardiovascular, diabetes management, gastro-intestinal and pain management products in India; anti-infective and pain management products in Russia; a broad generics portfolio in North America; and a broad generics portfolio in Europe under the betapharm brand. Some of this segment's products include Omez, the firm's brand of omeprazole; Reditux, the firm's brand of rituximab; Razo, the company's brand of rabeprazole; Mintop, its brand of minoxidil; and Stamlo, its brand of amlodipine. The PSAI segment comprises raw active pharmaceutical ingredients transformed by other manufacturers into pharmaceutical products ready for human consumption such as a tablets, capsules or liquids using additional inactive ingredients. Dr. Reddy's manufactures and markets over 100 bulk active ingredients, the principal ingredients in the finished dosages of drugs. This segment also includes contract research services and contract manufacturing. Lastly, the Proprietary Products segment consists of new chemical entity (NCE) research and the company's Differentiated Formulations business. NCE research is focused on building a NCE pipeline in the metabolic diseases, cardiovascular diseases and antibacterials markets. The Differentiated Formulations business involves applying novel formulation and drug delivery technologies to improve products that have a sizeable track record of human clinical use. The firm's most advanced Differentiated Formulations efforts are in dermatology, where it has launched a broad portfolio of products through Promius Pharma, its wholly-owned subsidiary. In April 2010, the firm launched amlodipine benazepril capsules, a generic version of Lotrel Capsules, in the U.S. In May 2010, the company launched tacrolimus capsules, a generic version of Prograf Capsules, in the U.S.

FINANCIALS: Sales and profits are in thousands of dollars—add 000 to get the full amount. 2009 Note: Financial information for 2009 was not available for all companies at press time.

2009 Sales: $1,365,000	2009 Profits: $-102,000	**U.S. Stock Ticker:** RDY
2008 Sales: $1,230,100	2008 Profits: $116,900	**Int'l Ticker: 500124** Int'l Exchange: Bombay-BSE
2007 Sales: $1,488,010	2007 Profits: $215,790	Employees: 9,575
2006 Sales: $541,304	2006 Profits: $36,620	Fiscal Year Ends: 3/31
2005 Sales: $438,000	2005 Profits: $4,800	Parent Company:

SALARIES/BENEFITS:

Pension Plan:	ESOP Stock Plan:	Profit Sharing:	Top Exec. Salary: $	Bonus: $
Savings Plan: Y	Stock Purch. Plan: Y		Second Exec. Salary: $	Bonus: $

OTHER THOUGHTS:
Apparent Women Officers or Directors: 1
Hot Spot for Advancement for Women/Minorities:

LOCATIONS: ("Y" = Yes)

West:	Southwest:	Midwest:	Southeast:	Northeast:	International:
				Y	Y

Note: Financial information, benefits and other data can change quickly and may vary from those stated here.

DSM PHARMACEUTICALS INC

www.dsm.com

Industry Group Code: 325412 Ranks within this company's industry group: Sales: Profits:

Drugs:		Other:		Clinical:		Computers:		Services:	
Discovery:		AgriBio:		Trials/Services:	Y	Hardware:		Specialty Services:	Y
Licensing:		Genetic Data:		Labs:		Software:		Consulting:	
Manufacturing:	Y	Tissue Replacement:		Equipment/Supplies:	Y	Arrays:		Blood Collection:	
Generics:				Research/Development Svcs.:	Y	Database Management:		Drug Delivery:	
				Diagnostics:				Drug Distribution:	

TYPES OF BUSINESS:

Drugs-Custom Manufacturing
Oral & Topical Formulations
Sterile Liquids
Lyophilization Services
Pharmaceutical Packaging
Product Development & Clinical Trial Services
Product Packaging

BRANDS/DIVISIONS/AFFILIATES:

DSM NV
Lyo-Advantage
Liquid-Advantage
DSM Pharmaceuticals Packaging Group
DSM Pharma Chemicals

CONTACTS: *Note: Officers with more than one job title may be intentionally listed here more than once.*

Hans Engels, CEO
Hans Engles, Pres.
Laura Parks, Sr. VP-Mktg. & Sales
Feike Sijbesma, Chmn.

Phone: 252-758-3436	Fax:
Toll-Free:	
Address: 5900 N.W. Greenville Blvd., Greenville, NC 27834 US	

GROWTH PLANS/SPECIAL FEATURES:

DSM Pharmaceuticals, Inc. is an outsourcing partner for both large and small pharmaceutical companies that specializes in custom chemical manufacturing of synthetic drugs and biopharmaceuticals. The company is part of DSM Pharmaceutical Products, a division of DSM NV. The firm also provides products and services from early and preclinical development to initial synthesis of promising new drugs and dosage formulation and packaging. Company products and services fall under five categories: manufacturing; orals and topicals; sterile liquids; lyophilization (freeze drying); and packaging. DSM's manufacturing services convert active pharmaceutical ingredients (APIs) and excipients processed by DSM Pharma Chemicals and other suppliers into finished dosage forms such as solid-dose creams, ointments, steriles and liquids. Its oral and topical products include creams and ointments, while services in this area include blending and granulation, coating, tablet compression, capsule filing, drying and all related packaging. Sterile liquid operations include the manufacturing, aseptic filling, terminal sterilization, cold storage and packaging of sterile liquids. The company's steriles processing facilities use state-of-the-art Liquid-Advantage and Lyo-Advantage systems, both developed in-house. These systems provide a precise level of manufacturing and environment control. The firm offers cutting edge lyophilization services, which involve freeze-drying of biological substances. The DSM Pharmaceuticals Packaging Group is capable of producing high-quality specified package engineering, graphics design and testing services. Its offerings include bottles, liners, stoppers, dosage cups and droppers and thermo films, among various other packaging items. Other services include clinical trial management and the development and manufacturing of finished dosage forms for pharmaceutical companies.

FINANCIALS: Sales and profits are in thousands of dollars—add 000 to get the full amount. 2009 Note: Financial information for 2009 was not available for all companies at press time.

2009 Sales: $	2009 Profits: $	**U.S. Stock Ticker: Subsidiary**	
2008 Sales: $	2008 Profits: $	**Int'l Ticker:** Int'l Exchange:	
2007 Sales: $	2007 Profits: $	Employees:	
2006 Sales: $	2006 Profits: $	Fiscal Year Ends: 12/31	
2005 Sales: $	2005 Profits: $	Parent Company: DSM NV	

SALARIES/BENEFITS:

Pension Plan:	ESOP Stock Plan:	Profit Sharing:	Top Exec. Salary: $	Bonus: $
Savings Plan:	Stock Purch. Plan:		Second Exec. Salary: $	Bonus: $

OTHER THOUGHTS:

Apparent Women Officers or Directors: 1
Hot Spot for Advancement for Women/Minorities:

LOCATIONS: ("Y" = Yes)

West:	Southwest:	Midwest:	Southeast:	Northeast:	International:
				Y	Y

DU PONT AGRICULTURE & NUTRITION

www2.dupont.com/Agriculture/en_US/

Industry Group Code: 11511 Ranks within this company's industry group: Sales: 3 Profits: 3

Drugs:	Other:		Clinical:	Computers:	Services:	
Discovery:	AgriBio:	Y	Trials/Services:	Hardware:	Specialty Services:	
Licensing:	Genetic Data:		Labs:	Software:	Consulting:	
Manufacturing:	Tissue Replacement:		Equipment/Supplies:	Arrays:	Blood Collection:	
Generics:			Research/Development Svcs.:	Database Management:	Drug Delivery:	
			Diagnostics:		Drug Distribution:	

TYPES OF BUSINESS:

Agricultural Biotechnology Products & Chemicals Manufacturing
Insecticides
Herbicides
Fungicides
Genetically Modified Plants
Soy Products
Forage & Grain Additives

BRANDS/DIVISIONS/AFFILIATES:

Pioneer Hi-Bred International
DuPont Crop Protection
DuPont Nutrition and Health
Solae Company (The)
Supro
Agroproducts Corey, S.A. de C.V.
Nurish
Qualicon

CONTACTS: Note: Officers with more than one job title may be intentionally listed here more than once.

James C. Borel, Exec. VP-DuPont Agriculture & Nutrition
James C. Collins, Pres., DuPont Crop Protection
Craig F. Binetti, Pres., DuPont Nutrition & Health
Paul Schickler, Pres., Pioneer Hi-Bred

Phone: 302-774-1000	Fax:
Toll-Free: 888-638-7668	
Address: 1007 Market St., DuPont Bldg., Wilmington, DE 19898 US	

GROWTH PLANS/SPECIAL FEATURES:

DuPont Agriculture & Nutrition (DPAN) is a business unit of global chemical giant DuPont. The company oversees a number of business units covering many aspects of crop protection, optimization and additives. Pioneer Hi-Bred International develops advanced plant genetics, including seeds and forage and grain additives. DuPont Crop Protection produces herbicide, fungicide and insecticide products and services. DuPont Nutrition and Health provides soy protein and soy fiber ingredients under brand names including The Solae Company, a joint venture with Bunge; Supro; SuproSoy; and Nurish. DPAN has joint ventures in the U.S. and around the world, with projects such as an Agricultural products venture with Agroproducts Corey, S.A. de C.V. in Mexico; a crop protection venture with AO Khimprom in Russia; a biofuel production partnership with British Petroleum (BP); and soy-based ventures with General Mills/PTI in Minnesota, So Good in the U.K and Syngenta in Illinois. DPAN also has a microbial testing branch, Qualicon. Recently a number of new products have been approved for registration by the Environmental Protection Agency (EPA), including the cleaning and personal care product Zemea propanediol and the herbicide Herculex.

The company offers employees a comprehensive benefits package with dependent care spending accounts, flexible work practices and adoption assistance.

FINANCIALS: Sales and profits are in thousands of dollars—add 000 to get the full amount. 2009 Note: Financial information for 2009 was not available for all companies at press time.

2009 Sales: $8,287,000	2009 Profits: $1,224,000	U.S. Stock Ticker: Subsidiary
2008 Sales: $7,952,000	2008 Profits: $1,087,000	Int'l Ticker: Int'l Exchange:
2007 Sales: $6,842,000	2007 Profits: $894,000	Employees:
2006 Sales: $	2006 Profits: $	Fiscal Year Ends: 12/31
2005 Sales: $	2005 Profits: $	Parent Company: E I DU PONT DE NEMOURS & CO (DUPONT)

SALARIES/BENEFITS:

Pension Plan:	ESOP Stock Plan:	Profit Sharing:	Top Exec. Salary: $	Bonus: $
Savings Plan:	Stock Purch. Plan:		Second Exec. Salary: $	Bonus: $

OTHER THOUGHTS:

Apparent Women Officers or Directors: 1
Hot Spot for Advancement for Women/Minorities:

LOCATIONS: ("Y" = Yes)

West:	Southwest:	Midwest:	Southeast:	Northeast:	International:
Y	Y	Y	Y	Y	Y

DURECT CORP

www.durect.com

Industry Group Code: 325412A **Ranks within this company's industry group:** Sales: 10   Profits: 16

Drugs:		Other:	Clinical:	Computers:	Services:	
Discovery:		AgriBio:	Trials/Services:	Hardware:	Specialty Services:	
Licensing:	Y	Genetic Data:	Labs:	Software:	Consulting:	
Manufacturing:	Y	Tissue Replacement:	Equipment/Supplies:	Arrays:	Blood Collection:	
Generics:			Research/Development Svcs.:	Database Management:	Drug Delivery:	Y
			Diagnostics:		Drug Distribution:	

TYPES OF BUSINESS:

Drug Delivery Systems
Biodegradable Polymer Manufacturing
Pharmaceutical Products

BRANDS/DIVISIONS/AFFILIATES:

SABER
ORADUR
TRANSDUR
DURIN
MICRODUR
REMOXY
LACTEL
ALZET

CONTACTS: Note: Officers with more than one job title may be intentionally listed here more than once.

James E. Brown, CEO
James E. Brown, Pres.
Matthew J. Hogan, CFO
Felix Theeuwes, Chief Scientific Officer
Su IL Yum, Principal Engineer
Paula Mendenhall, Exec. VP-Admin.
Jean I. Liu, General Counsel/Sr. VP
Paula Mendenhall, Exec. VP-Oper.
Michael H. Arenberg, VP-Bus. Dev.
Jian Li, VP-Finance/Corp. Controller
Joseph Stauffer, Chief Medical Officer
Su IL Yum, Exec. VP-Pharmaceutical Systems R&D
Nacer E. Dean Abrouk, VP-Biostatistics
Andrew R. Miksztal, VP-Pharmaceutical R&D
Felix Theeuwes, Chmn.

Phone: 408-777-1417	Fax: 408-777-3577
Toll-Free:	
Address: 2 Results Way, Cupertino, CA 95014 US	

GROWTH PLANS/SPECIAL FEATURES:

Durect Corp. is a specialty pharmaceutical company focused on developing products that utilize its proprietary drug delivery platforms to treat pain, central nervous system disorders, cardiovascular disease and other chronic diseases. The company's pharmaceutical systems optimize therapy for a given disease by controlling the rate and duration of drug administration, as well as enabling delivery at the intended site of action. Its delivery platforms include: Saber, an injectable used for proteins, peptides and small molecules; Oradur, sustained-release oral gel-cap technology; Transdur, a transdermal patch; Duros, a miniature drug-dispensing pump system, licensed from ALZA Corporation; and the Durin and MICRODurin biodegradable input systems. The company markets two products: Alzet, an implantable osmotic pump for laboratory animal testing; and Lactel biodegradable polymers. Durect's lead development product is Remoxy, a chronic pain solution based on Oradur technology and developed in partnership with Pain Therapeutics, Inc. and King Pharmaceuticals, Inc. Other products in development include Posidur, a post-operative pain treatment, which is being developed in conjunction with Nycomed; Transdur-Sufentanil, a chronic pain treatment; Eladur, a drug for post-herpetic neuralgia; and two yet-undisclosed opioids based on Oradur. In August 2009, the company signed a development and license agreement with Orient Pharma Co., Ltd., for Oradur-ADHD, an Attention Deficit Hyperactivity Disorder (ADHD) therapy based on Oradur technology. In January 2010, the firm began a Phase III clinical trial with Posidur for post-surgical pain treatment. In April 2010, King Pharmaceuticals began a Phase IIb trial with Eladur for the treatment of back pain. In June 2010, Durect formed an agreement with Hospira, Inc. for the development and marketing of Posidur in the U.S. and Canada.

Durect offers its employees medical, dental and vision insurance; life insurance; a 401(k) plan; a stock purchase plan; stock options; an employee assistance program; subsidized health club memberships; and flexible spending accounts.

FINANCIALS: Sales and profits are in thousands of dollars—add 000 to get the full amount. 2009 Note: Financial information for 2009 was not available for all companies at press time.

2009 Sales: $24,293	2009 Profits: $-30,288	U.S. Stock Ticker: DRRX
2008 Sales: $27,101	2008 Profits: $-43,907	Int'l Ticker: Int'l Exchange:
2007 Sales: $30,675	2007 Profits: $-24,339	Employees: 127
2006 Sales: $21,894	2006 Profits: $-33,327	Fiscal Year Ends: 12/31
2005 Sales: $28,571	2005 Profits: $-18,128	Parent Company:

SALARIES/BENEFITS:

Pension Plan:	ESOP Stock Plan:	Profit Sharing:	Top Exec. Salary: $473,660	Bonus: $25,151
Savings Plan: Y	Stock Purch. Plan: Y		Second Exec. Salary: $461,965	Bonus: $24,530

OTHER THOUGHTS:

Apparent Women Officers or Directors: 2
Hot Spot for Advancement for Women/Minorities: Y

LOCATIONS: ("Y" = Yes)

West:	Southwest:	Midwest:	Southeast:	Northeast:	International:
Y			Y		

DUSA PHARMACEUTICALS INC

www.dusapharma.com

Industry Group Code: 325412 Ranks within this company's industry group: Sales: 100 Profits: 81

Drugs:		Other:		Clinical:	Computers:	Services:
Discovery:	Y	AgriBio:		Trials/Services:	Hardware:	Specialty Services:
Licensing:	Y	Genetic Data:		Labs:	Software:	Consulting:
Manufacturing:	Y	Tissue Replacement:		Equipment/Supplies:	Arrays:	Blood Collection:
Generics:				Research/Development Svcs.:	Database Management:	Drug Delivery:
				Diagnostics:		Drug Distribution:

TYPES OF BUSINESS:

Photodynamic Therapy Products
Dermatology Products

BRANDS/DIVISIONS/AFFILIATES:

Levulan
BLU-U
Kerastick
Psoriatec
Nicomide-T
ClindaReach
Sirius Labratories, Inc.

CONTACTS:
Note: Officers with more than one job title may be intentionally listed here more than once.

Robert F. Doman, CEO
Robert F. Doman, Pres.
Richard Christopher, CFO/VP-Finance
William F. O'Dell, Exec. VP-Sales & Mktg.
Stuart Marcus, VP-Scientific Affairs
Nanette W. Mantell, Sec.
Mark Carota, VP-Oper.
Michael J. Todisco, VP/Controller
Stuart Marcus, Chief Medical Officer
Scott Lundahl, VP-Regulatory Affairs & Intellectual Property
Jay M. Haft, Chmn.

Phone: 978-657-7500	Fax: 978-657-9193
Toll-Free:	
Address: 25 Upton Dr., Wilmington, MA 01887 US	

GROWTH PLANS/SPECIAL FEATURES:

DUSA Pharmaceuticals, Inc. is a dermatology company that develops and markets Levulan photodynamic therapy (PDT) and other products for common skin conditions. The company's marketed products include Levulan Kerastick 20% topical solution with PDT; the BLU-U brand light source; and ClindaReach. The firm's Levulan brand of aminolevulinic acid HCl is designed to be utilized when followed by exposure to light, either to treat a medical condition (Levulan PDT) or to detect a medical condition (Levulan photodetection, or Levulan PD). The Levulan Kerastick is a single-use, disposable applicator, which follows for uniform application of Levulan topical solution in standardized doses. The Levulan Kerastick 20% topical solution with PDT and the BLU-U light source are used for the treatment of non-hyperkeratotic actinic keratoses, precancerous skin lesions caused by chronic sun exposure of the face or scalp. BLU-U is also used for the treatment of moderate inflammatory acne vulgaris and general dermatological conditions. ClindaReach provides medication for acne vulgaris and supplies a unique system for applying this medication to all hard-to-reach areas, including the back. Other products include Nicomide-T and METED Shampoo, which are distributed by Sirius Labratories, Inc., a wholly-owned subsidiary of DUSA Pharmaceuticals, Inc. DUSA has an agreement with Stiefel Laboratories to market the Levulan Kerastick in Mexico, Central and South America; with Coherent-AMT to market its products in Canada; and with Daewoong Pharmaceutical Co., Ltd. and DNC Daewoong Derma & Plastic Surgery Network Company to market Levulan Kerastick in certain Asian territories.

The company offers its employees a 401(k) plan; health, dental, vision and life insurance; flexible spending accounts; tuition reimbursement; short- and long-term disability; and business travel insurance.

FINANCIALS:
Sales and profits are in thousands of dollars—add 000 to get the full amount. 2009 Note: Financial information for 2009 was not available for all companies at press time.

2009 Sales: $29,808	2009 Profits: $-2,508	**U.S. Stock Ticker:** DUSA
2008 Sales: $29,545	2008 Profits: $-6,250	**Int'l Ticker:** Int'l Exchange:
2007 Sales: $27,663	2007 Profits: $-14,714	**Employees:** 86
2006 Sales: $25,583	2006 Profits: $-31,350	**Fiscal Year Ends:** 12/31
2005 Sales: $11,337	2005 Profits: $-14,999	**Parent Company:**

SALARIES/BENEFITS:

Pension Plan:	ESOP Stock Plan:	Profit Sharing:	Top Exec. Salary: $417,000	Bonus: $126,000
Savings Plan: Y	Stock Purch. Plan:		Second Exec. Salary: $285,500	Bonus: $29,000

OTHER THOUGHTS:

Apparent Women Officers or Directors: 1
Hot Spot for Advancement for Women/Minorities:

LOCATIONS: ("Y" = Yes)

West:	Southwest:	Midwest:	Southeast:	Northeast:	International:
				Y	

DYAX CORP

www.dyax.com

Industry Group Code: 325412 Ranks within this company's industry group: Sales: 106 Profits: 146

Drugs:		Other:		Clinical:		Computers:		Services:	
Discovery:	Y	AgriBio:		Trials/Services:		Hardware:		Specialty Services:	
Licensing:	Y	Genetic Data:		Labs:		Software:	Y	Consulting:	
Manufacturing:		Tissue Replacement:		Equipment/Supplies:		Arrays:		Blood Collection:	
Generics:				Research/Development Svcs.:	Y	Database Management:		Drug Delivery:	
				Diagnostics:				Drug Distribution:	

TYPES OF BUSINESS:

Pharmaceuticals Discovery & Development
Proteins, Peptides & Antibodies
Drugs-Cancer
Drugs-Anti-Inflammatory
Drugs-Blood Clotting Regulation

BRANDS/DIVISIONS/AFFILIATES:

DX-88
DX-2400
DX-2240
WebPhage

CONTACTS: *Note: Officers with more than one job title may be intentionally listed here more than once.*

Gustav A. Christensen, CEO
Gustav A. Christensen, Pres.
George Migausky, CFO/Exec. VP
Ivana Magovcevic-Liebisch, General Counsel/Exec. VP-Corp. Dev.
William E. Pullman, Chief Dev. Officer/Exec. VP
Nicole P. Jones, Dir.-Corp. Comm.
Nicole P. Jones, Dir.-Investor Rel.
Henry E. Blair, Chmn.

Phone: 617-225-2500	Fax: 617-225-2501
Toll-Free:	
Address: 300 Technology Sq., Cambridge, MA 02139 US	

GROWTH PLANS/SPECIAL FEATURES:

Dyax Corp. is a biopharmaceutical company principally focused on the discovery, development and commercialization of biopharmaceuticals, with an emphasis on cancer and inflammatory conditions. Its lead product is DX-88 (ecallantide), used in the treatment of acute attacks of hereditary angioedema (HAE), a genetic disease that causes swelling of the abdomen, airways and extremities. DX-88 is approved for sale in the U.S. under the brand name KALBITOR. DX-88 is also being tested for the treatment of ACE inhibitor-induced angioedema, retinal diseases and the reduction of blood loss during surgery. These applications are primarily in Phase I development. After DX-88, its most advanced candidates are DX-2240 and DX-2400, both fully human monoclonal antibodies designed to attack cancerous tumors. The company has signed a license agreement granting sanofi-aventis S.A. the responsibility for the future development of DX-2240. Dyax identifies its pipeline compounds through its proprietary phage display methodology, called WebPhage, which utilizes phage display libraries of protein variations to identify possible therapeutic compounds. The company allows other companies to use WebPhage display technology through its Licensing and Funded Research Program. Over 70 companies have licensed this technology; 18 new compounds currently undergoing clinical trials have been derived, as well as one product that has received market approval. In early 2009, Dyax agreed to license DX-88 to Fovea Pharmaceuticals, SA for the treatment of retinal diseases in Europe. In March 2009, the company announced it would be eliminating about a third of its workforce, or 60 employees. In April 2010, the firm sold its royalty rights for Xyntha, for the treatment of hemophilia A, Paul Capital Healthcare.

Employees of the firm receive health and dental benefits; life insurance; flexible spending accounts; a 401(k) plan and profit sharing plan; stock options; tuition reimbursement; an employee assistance program; health club reimbursement; and a transportation subsidy.

FINANCIALS: Sales and profits are in thousands of dollars—add 000 to get the full amount. 2009 Note: Financial information for 2009 was not available for all companies at press time.

2009 Sales: $21,643	2009 Profits: $-62,419	U.S. Stock Ticker: DYAX	
2008 Sales: $43,429	2008 Profits: $-66,468	Int'l Ticker:	Int'l Exchange:
2007 Sales: $26,096	2007 Profits: $-56,309	Employees: 164	
2006 Sales: $12,776	2006 Profits: $-50,323	Fiscal Year Ends: 12/31	
2005 Sales: $19,859	2005 Profits: $-30,944	Parent Company:	

SALARIES/BENEFITS:

Pension Plan:	ESOP Stock Plan:	Profit Sharing: Y	Top Exec. Salary: $495,192	Bonus: $350,000
Savings Plan: Y	Stock Purch. Plan: Y		Second Exec. Salary: $365,500	Bonus: $191,996

OTHER THOUGHTS:

Apparent Women Officers or Directors: 3
Hot Spot for Advancement for Women/Minorities: Y

LOCATIONS: ("Y" = Yes)

West:	Southwest:	Midwest:	Southeast:	Northeast:	International:
				Y	Y

E I DU PONT DE NEMOURS & CO (DUPONT) www2.dupont.com

Industry Group Code: 325 Ranks within this company's industry group: Sales: 3 Profits: 3

Drugs:	Other:		Clinical:	Computers:	Services:
Discovery:	AgriBio:	Y	Trials/Services:	Hardware:	Specialty Services:
Licensing:	Genetic Data:		Labs:	Software:	Consulting:
Manufacturing:	Tissue Replacement:		Equipment/Supplies:	Arrays:	Blood Collection:
Generics:			Research/Development Svcs.:	Database Management:	Drug Delivery:
			Diagnostics:		Drug Distribution:

TYPES OF BUSINESS:

Chemicals Manufacturing
Polymers
Performance Coatings
Nutrition & Health Products
Electronics Materials
Agricultural Seeds
Fuel-Cell, Biofuels & Solar Panel Technology
Contract Research & Development

BRANDS/DIVISIONS/AFFILIATES:

Pioneer Hi-Bred International Inc
Teflon
Corian
Kevlar
Tyvek
Solae
DuPont Nomex on Demand
Meyrin Photovoltaic Application Laboratory

CONTACTS: *Note: Officers with more than one job title may be intentionally listed here more than once.*

Ellen J. Kullman, CEO
Nicholas C. Farandakis, CFO/Sr. VP
Cynthia C. Green, Chief Mktg. & Sales Officer/VP-Corp. Mktg. & Sales
W. Donald Johnson, Sr. VP-Human Resources
Uma Chowdhry, Chief Science Officer
Phuong Tram, CIO/VP-IT
Uma Chowdhry, CTO/Sr. VP
Thomas M. Connelly, Jr., Chief Innovation Officer/Exec. VP
Jeffrey A. Coe, Sr. VP-Eng.
Thomas L. Sager, General Counsel/Sr. VP-Legal
Jeffrey A. Coe, Sr. VP-Integrated Oper.
David G. Bills, Sr. VP-Corp. Strategy
Karen A. Fletcher, VP-DuPont Investor Rel.
Susan M. Stalnecker, VP-Finance/Treas.
Criag F. Binetti, Sr. VP-DuPont Nutrition & Health
Diane H. Gulyas, Pres., DuPont Performance Polymers
Terry Caloghiris, Pres., DuPont Performance Coatings
Linda J. Fisher, Chief Sustainability Officer
Ellen J. Kullman, Chmn.
Donald D. Wirth, VP-Global Oper.
Keith J. Smith, Chief Procurement Officer/VP-Sourcing & Logistics

Phone: 302-774-1000	Fax: 302-773-2631
Toll-Free: 800-441-7515	
Address: 1007 Market St., Wilmington, DE 19898 US	

GROWTH PLANS/SPECIAL FEATURES:

E. I. du Pont de Nemours & Co. (DuPont), founded in 1802, develops and manufactures products in the biotechnology, electronics, materials science, synthetic fibers and safety and security sectors. DuPont operates in seven segments: Agriculture and Nutrition (A&N); Electronic and Communications (E&C); Performance Coatings, Performance Chemicals; Performance Materials (PM); Safety and Protection (S&P) and Pharmaceuticals. A&N delivers seed products, insecticides, fungicides, herbicides, soy-based food ingredients, food quality diagnostic testing equipment and liquid food packaging systems. Brands include Pioneer seeds and Solae soy proteins. E&C provides a range of advanced materials for the electronics industry, flexographic printing, color communication systems and a range of fluoropolymer and fluorochemical products. The Performance Coatings segment supplies automotive liquid and powder coatings, and general industrial applications such as coatings for heavy equipment, pipes and appliances and electrical insulation. Brands include DuPont, Standox, Spies Hecker and Nason. PM manufactures polymer-based materials, which include engineered polymers, specialized resins and films for use in food packaging, sealants, adhesives, sporting goods and laminated safety glass. The S&P segment provides protective materials and safety consulting services. Significant brands include Teflon fluoropolymers, films, fabric protectors, fibers and dispersions; Corian surfaces; Kevlar high strength material; and Tyvek protective material. The Pharmaceuticals segment involves the worldwide manufacturing and marketing activities of the antihypertensive drugs, Cozaar and Hyzaar. The company has operations in over 70 countries around the world and maintains 75 R&D and customer services labs in 12 countries. In March 2009, the firm launched the DuPont Nomex on Demand smart fiber technology, a material designed for thermal liners in firefighter gear, that expands for thermal insulation in over 250 degree heat. In February 2010, the company opened the Meyrin Photovoltaic Application Laboratory at its European Technical Center in Geneva.

Employees are offered medical, dental and life insurance; disability coverage; dependent care spending accounts; and adoption assistance.

FINANCIALS: Sales and profits are in thousands of dollars—add 000 to get the full amount. 2009 Note: Financial information for 2009 was not available for all companies at press time.

2009 Sales: $26,109,000	2009 Profits: $1,755,000	U.S. Stock Ticker: DD
2008 Sales: $30,529,000	2008 Profits: $2,007,000	Int'l Ticker: Int'l Exchange:
2007 Sales: $29,378,000	2007 Profits: $2,988,000	Employees: 60,000
2006 Sales: $27,421,000	2006 Profits: $3,148,000	Fiscal Year Ends: 12/31
2005 Sales: $26,639,000	2005 Profits: $2,053,000	Parent Company:

SALARIES/BENEFITS:

Pension Plan: Y	ESOP Stock Plan:	Profit Sharing:	Top Exec. Salary: $1,369,500	Bonus: $1,932,000
Savings Plan: Y	Stock Purch. Plan:		Second Exec. Salary: $865,992	Bonus: $763,000

OTHER THOUGHTS:

Apparent Women Officers or Directors: 17
Hot Spot for Advancement for Women/Minorities: Y

LOCATIONS: ("Y" = Yes)

West:	Southwest:	Midwest:	Southeast:	Northeast:	International:
Y	Y	Y	Y	Y	Y

Note: Financial information, benefits and other data can change quickly and may vary from those stated here.

EISAI CO LTD

www.eisai.co.jp

Industry Group Code: 325414 Ranks within this company's industry group: Sales: 1   Profits: 2

Drugs:		Other:	Clinical:		Computers:	Services:
Discovery:		AgriBio:	Trials/Services:		Hardware:	Specialty Services:
Licensing:		Genetic Data:	Labs:		Software:	Consulting:
Manufacturing:	Y	Tissue Replacement:	Equipment/Supplies:	Y	Arrays:	Blood Collection:
Generics:			Research/Development Svcs.:		Database Management:	Drug Delivery:
			Diagnostics:	Y		Drug Distribution:

TYPES OF BUSINESS:

Pharmaceuticals Manufacturing
Over-the-Counter Pharmaceuticals
Pharmaceutical Production Equipment
Diagnostic Products
Food Additives
Personal Health Care Products
Vitamins & Nutritional Supplements

BRANDS/DIVISIONS/AFFILIATES:

Aricept
Aciphex/Pariet
Coretec
Myonal
Chocola
Travelmin
Juvelux
MGI Pharma Inc

CONTACTS: *Note: Officers with more than one job title may be intentionally listed here more than once.*

Haruo Naito, CEO
Haruo Naito, Pres.
Hideaki Matsui, CFO
Soichi Matsuno, Deputy Pres., Global Human Resources
Kentaro Yoshimatsu, Sr. VP-R&D/Pres., Eisai R&D Mgmt. Co., Ltd.
Kazuo Hirai, VP-Info. System & Corp. Mgmt. Planning
Takafumi Asano, VP-Prod.
Kenta Takahashi, General Counsel/VP/Sr. Dir.-Legal Dept.
Makoto Shiina, Exec. VP-Strategy
Akira Fujiyoshi, VP-Corp. Comm. Dept.
Nobuo Deguchi, Exec. VP-Internal Control & Intellectual Property
Hajime Shimizu, Chmn./CEO-Eisai Corp. N. America
Yutaka Tsuchiya, Chmn./Pres., Eisai Europe Ltd.
Yukio Akada, Chmn./Pres., Eisai China, Inc.
Masanori Tsuno, Pres., Eisai Medical Research, Inc.
Norihiko Tanikawa, Chmn.
Soichi Matsuno, Deputy Pres., Global Pharmaceuticals Bus.
Takafumi Asano, VP-Logistics & Transformation

Phone: 81-3-3817-3700	Fax: 81-3-3811-3077
Toll-Free:	
Address: 4-6-10 Koishikawa, Bunkyo-ku, Tokyo, 112-8088 Japan	

GROWTH PLANS/SPECIAL FEATURES:

Eisai Co., Ltd. primarily develops, manufactures and distributes medical products, operating in two segments: prescription pharmaceuticals, which represented 96.9% of its net sales, and other. These other products include over-the-counter pharmaceuticals, consumer health care products, food additives, pharmaceutical production equipment and diagnostic products. Its two largest pharmaceuticals, Aricept, a treatment for Alzheimer's dementia, and proton pump inhibitor (PPI) AcipHex/Pariet, a treatment for gastroesophageal reflux disease and ulcers, account for roughly 39.6% and 24% of sales, respectively. The firm's research teams are currently investigating new indications for Aricept including dementia associated with Parkinson's disease. Other prescription pharmaceuticals include Coretec, an agent for acute heart failure; Glakay, an osteoporosis treatment; Azeptin, an antiallergic agent; and Myonal, a muscle relaxant. In the consumer health care field, the firm's products include Seabond denture adhesive (manufactured by U.S.-based Combe Laboratories, Inc.); motion sickness remedy Travelmin; Breathe Right nasal strips (manufactured by GlaxoSmithKline plc); and Juvelux natural vitamin E preparation. Eisai has a long-standing leading position in the Japanese vitamin and nutritional supplement market, focusing on synthetic and natural vitamin E products and derivatives; it also markets a full line of vitamin-enriched dietary supplements under the brand name Chocola. The firm also markets diagnostic products and manufactures pharmaceutical production systems including continuous sterilization devices, inspection systems and ampoule packing machines. Geographically, North America generates 46.2% of sales; Japan, 42.6%; Europe, 7.4%; and Asia and other regions, 3.8%. The company maintains approximately 32 overseas subsidiaries, in Asia, Europe and North America; and 12 domestic subsidiaries.

FINANCIALS: Sales and profits are in thousands of dollars—add 000 to get the full amount. 2009 Note: Financial information for 2009 was not available for all companies at press time.

2009 Sales: $8,244,890	2009 Profits: $502,850	**U.S. Stock Ticker: ESALY**
2008 Sales: $7,304,620	2008 Profits: $-169,110	**Int'l Ticker: 4523** Int'l Exchange: Tokyo-TSE
2007 Sales: $6,705,770	2007 Profits: $702,310	Employees:
2006 Sales: $5,113,100	2006 Profits: $539,200	Fiscal Year Ends: 3/31
2005 Sales: $4,530,600	2005 Profits: $471,760	Parent Company:

SALARIES/BENEFITS:

Pension Plan: Y	ESOP Stock Plan:	Profit Sharing:	Top Exec. Salary: $	Bonus: $
Savings Plan:	Stock Purch. Plan:		Second Exec. Salary: $	Bonus: $

OTHER THOUGHTS:

Apparent Women Officers or Directors:
Hot Spot for Advancement for Women/Minorities:

LOCATIONS: ("Y" = Yes)

West:	Southwest:	Midwest:	Southeast:	Northeast:	International:
		Y		Y	Y

ELAN CORP PLC

www.elan.com

Industry Group Code: 325412 **Ranks within this company's industry group: Sales: 42 Profits: 164**

Drugs:		Other:	Clinical:	Computers:	Services:
Discovery:	Y	AgriBio:	Trials/Services:	Hardware:	Specialty Services:
Licensing:		Genetic Data:	Labs:	Software:	Consulting:
Manufacturing:	Y	Tissue Replacement:	Equipment/Supplies:	Arrays:	Blood Collection:
Generics:			Research/Development Svcs.:	Database Management:	Drug Delivery:
			Diagnostics:		Drug Distribution:

TYPES OF BUSINESS:

Drugs-Neurology
Acute Care Drugs
Pain Management Drugs
Autoimmune Disease Drugs
Drug Delivery Technologies

BRANDS/DIVISIONS/AFFILIATES:

TYSABRI
PRIALT
Azactam
Maxipime
NanoCrystal

CONTACTS: Note: Officers with more than one job title may be intentionally listed here more than once.

G. Kelly Martin, CEO
Carlos Paya, Pres.
Shane Cooke, CFO/Exec. VP
Kathleen Martorano, Exec. VP-Strategic Human Resources
Dale Schenk, Chief Scientific Officer/Exec. VP
Shane Cooke, Head-Elan Drug Tech.
John B. Moriarty Jr., General Counsel/Sr. VP
Johannes Roebers, Sr. VP/Head-Biologic Oper., Strategy & Planning
Menghis Bairu, Exec. VP/Head-Global Dev.
Mary Stutts, Exec. VP/Head-Corp. Rel.
Chris Burns, Sr. VP-Investor Rel.
Nigel Clerkin, Sr. VP-Finance/Controller
Karen Kim, Exec. VP-Special Projects
William F. Daniel, Exec. VP/Corp. Sec.
Ted Yednock, Exec. VP/Head-Global Research
Kyran McLaughlin, Chmn.
Menghis Bairu, Sr. VP/Head-Int'l

Phone: 353-1-709-4000	Fax: 353-1-709-4700
Toll-Free:	
Address: Treasury Bldg., Lower Grand Canal St., Dublin, Ireland 2 UK	

GROWTH PLANS/SPECIAL FEATURES:

Elan Corp. plc is a leading global biotechnology company with a focus on neuroscience. Elan is divided into two segments: BioNeurology and Elan Drug Technologies (EDT). The Irish company conducts its worldwide business primarily through subsidiaries in Ireland and the U.S. EDT, which controls its marketed products, accounts for nearly 34% of the company's revenues with the BioNeurology business generating the remaining 66%. The biopharmaceuticals division focuses its research and development on Alzheimer's disease, Parkinson's disease, multiple sclerosis, pain management and Crohn's disease. The company uses proprietary technologies to develop, market and license drug delivery products to its pharmaceutical clients. It does not manufacture any of its products. The company's marketed products are TYSABRI, an alpha four integrin antagonist for the treatment of recurring multiple sclerosis and Maxipime, which treats a number of disorders including urinary tract infections and pneumonia. Elan's drug delivery technologies have been incorporated into 35 marketed products. Elan also owns proprietary rights to NanoCrystal technology, which it licenses to Roche and Johnson & Johnson. In the European Union, the firm's primary product is TYSABRI. The company's Alzheimer drug under development is an amyloid beta aggregation inhibitor called ELND005 (AZD-103), which is currently in Phase II of clinical study. In July 2009, Johnson & Johnson announced that it would invest approximately $1.5 billion for an 18.4% stake in Elan. In September 2009, a subsidiary of Johnson & Johnson, JANSSEN Alzheimer Immunotherapy, acquired most of the assets and rights to Elan's Alzheimer's Immunotherapy program (AIP) including the drug bapineuzumab. Elan owns a 49.9% equity interest in JANSSEN, which means it will be entitled to 49.9% of the profits and a portion of the royalties earned when these AIP products are made public. In April 2010, the firm began exploring the effects of separating its EDT business and creating two distinct companies.

FINANCIALS: Sales and profits are in thousands of dollars—add 000 to get the full amount. 2009 Note: Financial information for 2009 was not available for all companies at press time.

2009 Sales: $1,113,000	2009 Profits: $-176,200	**U.S. Stock Ticker: ELN**
2008 Sales: $1,000,200	2008 Profits: $-71,000	**Int'l Ticker: DRX** Int'l Exchange: Dublin-ISE
2007 Sales: $759,400	2007 Profits: $-405,000	Employees: 1,321
2006 Sales: $560,400	2006 Profits: $-267,300	Fiscal Year Ends: 12/31
2005 Sales: $490,300	2005 Profits: $-383,600	Parent Company:

SALARIES/BENEFITS:

Pension Plan:	ESOP Stock Plan:	Profit Sharing:	Top Exec. Salary: $	Bonus: $
Savings Plan: Y	Stock Purch. Plan: Y		Second Exec. Salary: $	Bonus: $

OTHER THOUGHTS:

Apparent Women Officers or Directors: 4
Hot Spot for Advancement for Women/Minorities: Y

LOCATIONS: ("Y" = Yes)

West:	Southwest:	Midwest:	Southeast:	Northeast:	International:
Y			Y	Y	Y

Note: Financial information, benefits and other data can change quickly and may vary from those stated here.

ELI LILLY & COMPANY

www.lilly.com

Industry Group Code: 325412 Ranks within this company's industry group: Sales: 10   Profits: 12

Drugs:		Other:	Clinical:	Computers:	Services:	
Discovery:	Y	AgriBio:	Trials/Services:	Hardware:	Specialty Services:	Y
Licensing:		Genetic Data:	Labs:	Software:	Consulting:	
Manufacturing:	Y	Tissue Replacement:	Equipment/Supplies:	Arrays:	Blood Collection:	
Generics:			Research/Development Svcs.:	Database Management:	Drug Delivery:	
			Diagnostics:		Drug Distribution:	

TYPES OF BUSINESS:
Pharmaceuticals Discovery & Development
Veterinary Products

BRANDS/DIVISIONS/AFFILIATES:
Zyprexa
Prozac
Humalog
Gemzar
Coban
Cialis
Alimta
Effient

CONTACTS: *Note: Officers with more than one job title may be intentionally listed here more than once.*
John C. Lechleiter, CEO
John C. Lechleiter, Pres.
Derica W. Rice, CFO/Exec. VP
Susan Mahony, Sr. VP-Human Resources
Jan M. Lundberg, Exec. VP-Science
Michael C. Heim, CIO/Sr. VP-IT
Jan M. Lundberg, Exec. VP-Tech.
William F. Heath, Jr., Sr. VP-Prod. Dev. & Research
W. Darin Moody, VP-Corp. Eng. & Continuous Improvement
Frank Deane, Pres., Mfg.
Robert A. Armitage, Co-General Counsel/Sr. VP
Gino Santini, Sr. VP-Bus. Dev. & Corp. Strategy
Bart Peterson, Sr. VP-Corp. Affairs & Comm.
Thomas W. Grein, Treas./Sr. VP
Peter J. Johnson, VP-Corp. Strategic Planning
Elizabeth G. O'Farrell, Sr. VP/Controller
Alecia A. DeCoudreaux, VP/Co-General Counsel
Timothy J. Garnett, Chief Medical Officer/Sr. VP
John C. Lechleiter, Chmn.
Karim Bitar, Pres., European Oper.
Andreas Witzel, Pres., Supply Chain & Global Packaging

Phone: 317-276-2000	Fax:
Toll-Free: 800-545-5979	
Address: Lilly Corporate Ctr., Indianapolis, IN 46285 US	

GROWTH PLANS/SPECIAL FEATURES:
Eli Lilly & Co. researches, develops, manufactures and sells pharmaceuticals designed to treat a variety of conditions. Most of Eli Lilly's products are developed by its in-house research staff, which primarily directs its research efforts towards the search for products to prevent and treat cancer and diseases of the central nervous, endocrine and cardiovascular systems. The firm's other research lies in anti-infectives and products to treat animal diseases. Major brands include neuroscience products Zyprexa, Strattera, Prozac, Cymbalta and Permax; endocrine products Humalog, Humulin, Byetta and Actos; oncology products Erbitux, Gemzar and Alimta; animal health products Tylan, Rumensin and Coban; cardiovascular products Cialis, Effient, ReoPro and Xigris; and anti-infectives Ceclor and Vancocin. In the U.S., the company distributes pharmaceuticals primarily through independent wholesale distributors. The company manufactures and distributes its products through facilities in the U.S., Puerto Rico and 17 other countries, which are then sold to markets in roughly 128 countries throughout the world. The firm owns 12 production and distribution facilities in the U.S. and Puerto Rico. Major research and development facilities abroad are located in the U.K., France, Ireland, Brazil, Italy, Mexico and Spain. In December 2009, Eli Lilly acquired the worldwide commercialization and development rights to Incyte Corporation's INCB28050 oral treatment for inflammation as part of an exclusive licensing and collaboration agreement. Also in December 2009, the firm agreed to co-market LIVALO, a treatment for primary hyperlipidemia and mixed dyslipidemia developed by Kowa Pharmaceuticals America, Inc. In March 2010, the company agreed to acquire the European rights to certain products from the portfolio of Pfizer Animal Health.

Eli Lilly offers employees life, health, prescription drug and dental insurance; domestic partner benefits; an employee assistance program; paid maternity leave; a 401(k); flexible spending accounts; adoption assistance; and tuition reimbursement.

FINANCIALS: Sales and profits are in thousands of dollars—add 000 to get the full amount. 2009 Note: Financial information for 2009 was not available for all companies at press time.
2009 Sales: $21,836,000	2009 Profits: $4,328,800	**U.S. Stock Ticker: LLY**
2008 Sales: $20,371,900	2008 Profits: $-2,071,900	**Int'l Ticker:** Int'l Exchange:
2007 Sales: $18,633,500	2007 Profits: $2,953,000	Employees: 40,360
2006 Sales: $15,691,000	2006 Profits: $2,662,700	Fiscal Year Ends: 12/31
2005 Sales: $14,645,300	2005 Profits: $1,979,600	Parent Company:

SALARIES/BENEFITS:
Pension Plan: Y	ESOP Stock Plan:	Profit Sharing:	Top Exec. Salary: $1,483,333	Bonus: $3,551,100
Savings Plan: Y	Stock Purch. Plan:		Second Exec. Salary: $1,023,450	Bonus: $1,575,090

OTHER THOUGHTS:
Apparent Women Officers or Directors: 13
Hot Spot for Advancement for Women/Minorities: Y

LOCATIONS: ("Y" = Yes)
West:	Southwest:	Midwest:	Southeast:	Northeast:	International:
Y	Y	Y	Y	Y	Y

Note: Financial information, benefits and other data can change quickly and may vary from those stated here.

EMISPHERE TECHNOLOGIES INC

www.emisphere.com

Industry Group Code: 325412A Ranks within this company's industry group: Sales: 19 Profits: 14

Drugs:		Other:		Clinical:		Computers:		Services:	
Discovery:	Y	AgriBio:		Trials/Services:		Hardware:		Specialty Services:	
Licensing:		Genetic Data:		Labs:		Software:		Consulting:	
Manufacturing:	Y	Tissue Replacement:		Equipment/Supplies:		Arrays:		Blood Collection:	
Generics:				Research/Development Svcs.:	Y	Database Management:		Drug Delivery:	Y
				Diagnostics:				Drug Distribution:	

TYPES OF BUSINESS:

Drug Delivery Systems
Cardiovascular Diseases Oral Drugs
Osteoarthritis & Osteoporosis Oral Drugs
Growth Disorders Oral Drugs
Diabetes Oral Drugs
Asthma & Allergies Oral Drugs
Obesity Oral Drugs
Infectious Diseases & Oncology Oral Drugs

BRANDS/DIVISIONS/AFFILIATES:

Eligen Technology
Novartis Pharma AG

CONTACTS: Note: Officers with more than one job title may be intentionally listed here more than once.

Michael V. Novinski, CEO
Michael V. Novinski, Pres.
Michael R. Garone, CFO/VP
Daria M. Palestina, Dir.-Human Resources
Ronald Zesch, VP-Tech.
Ronald Zesch, VP-Mfg.
Patrick J. Osinski, General Counsel
Nicholas J. Hart, VP-Dev. & Strategy
Daria M. Palestina, Dir.-Comm.
M. Gary I. Riley, VP-Non-Clinical Dev. & Applied Biology
Michael R. Garone, Corp. Sec.

Phone: 973-532-8000	Fax: 973-532-8115
Toll-Free:	
Address: 240 Cedar Knolls Rd., Ste. 200, Cedar Knolls, NJ 07927 US	

GROWTH PLANS/SPECIAL FEATURES:

Emisphere Technologies, Inc. is a biopharmaceutical company that focuses on the delivery of therapeutic molecules using its proprietary Eligen technology. The company's product pipeline includes product candidates for the treatment of cardiovascular diseases, osteoarthritis, osteoporosis, growth disorders, diabetes, asthma/allergies, obesity, infectious diseases and oncology. Emisphere's Eligen technology is based on the use of proprietary, synthetic chemical compounds known as Emisphere delivery agents. These delivery agents facilitate and enable the transport of therapeutic macromolecules (such as proteins, peptides and polysaccharides) and poorly absorbed small molecules across biological membranes such as the small intestine. The Eligen technology is rapidly absorbed, metabolized and eliminated from the body. It does not accumulate in the organs and tissues and is considered safe at anticipated dose and dosing regimens. The company's lead product candidate is Salmon Calcitonin for Osteoarthritis and Osteoporosis, developed in conjunction with Novartis Pharma AG and its partner, Nordic Bioscience. The drug uses Emisphere's Eligen delivery technology to create an effective oral treatment as an alternative to existing injectable and intranasal options. Another product in development combines GLP-1 and PYY with Eligen technology to treat Type II diabetes. Finally, Emisphere is developing an oral system for vitamin B12, a process that promises to be less costly and better absorbed than current injection methods. Other products in development include oral deliveries for insulin, heparin, recombinant human growth hormone, acyclovir, PYY, gallium and recombinant parathyroid hormone. In early 2009, the firm announced an agreement with AAIPharma to combine its drug development services with Eligen technology.

Emisphere offers its employees tuition reimbursement; flex time; an employee assistance program; college savings plans; financial advisors; credit union membership; childcare/eldercare referral programs; and medical, dental, life and disability insurance.

FINANCIALS: Sales and profits are in thousands of dollars—add 000 to get the full amount. 2009 Note: Financial information for 2009 was not available for all companies at press time.

2009 Sales: $ 92	2009 Profits: $-21,243	U.S. Stock Ticker: EMIS
2008 Sales: $ 251	2008 Profits: $-24,388	Int'l Ticker: Int'l Exchange:
2007 Sales: $4,077	2007 Profits: $-16,928	Employees: 17
2006 Sales: $7,259	2006 Profits: $-41,766	Fiscal Year Ends: 12/31
2005 Sales: $3,540	2005 Profits: $-18,051	Parent Company:

SALARIES/BENEFITS:

Pension Plan:	ESOP Stock Plan:	Profit Sharing:	Top Exec. Salary: $550,000	Bonus: $269,969
Savings Plan: Y	Stock Purch. Plan: Y		Second Exec. Salary: $267,039	Bonus: $40,000

OTHER THOUGHTS:

Apparent Women Officers or Directors: 1
Hot Spot for Advancement for Women/Minorities:

LOCATIONS: ("Y" = Yes)

West:	Southwest:	Midwest:	Southeast:	Northeast:	International:
				Y	

ENCORIUM GROUP INC

www.encorium.com

Industry Group Code: 541712 **Ranks within this company's industry group:** Sales: 14　Profits: 8

Drugs:	Other:	Clinical:		Computers:		Services:	
Discovery:	AgriBio:	Trials/Services:	Y	Hardware:		Specialty Services:	
Licensing:	Genetic Data:	Labs:		Software:	Y	Consulting:	
Manufacturing:	Tissue Replacement:	Equipment/Supplies:		Arrays:		Blood Collection:	Y
Generics:		Research/Development Svcs.:		Database Management:		Drug Delivery:	
		Diagnostics:				Drug Distribution:	

TYPES OF BUSINESS:

Contract Research
Clinical Trial & Data Management
Disease Assessment Software
Biostatistical Analysis
Regulatory Affairs Services

BRANDS/DIVISIONS/AFFILIATES:

CONTACTS: *Note: Officers with more than one job title may be intentionally listed here more than once.*

Kai E. Lindevall, CEO
Kenneth M. Borrow, Pres.
Philip L. Calamia, Interim CFO
Kenneth M. Borrow, Chief Medical Officer
Eeva-Kaarina Koskelo, VP-Clinical Oper., Europe & Asia
Kenneth M. Borrow, Strategic Dev. Officer
Kai E. Lindevall, Chmn.
Kai E. Lindevall, Pres., Int'l Oper.

Phone: 358-20-751-8200	Fax: 358-20-751-8250
Toll-Free:	
Address: Keilaranta 10, Espoo, FI-02150 Finland	

GROWTH PLANS/SPECIAL FEATURES:

Encorium Group, Inc. is a contract research organization (CRO) that designs and manages clinical trials for the pharmaceutical, biotechnology and medical device industries. Encorium offers therapeutic expertise, experienced team management and advanced technologies, with the capacity to conduct clinical trials on a global basis. The company specializes in Phase I through IV clinical trials, offering services including strategic trial planning; project management; monitoring; data management and biostatistics; pharmacovigilance; medical writing; quality assurance; outsoucing of clinical staff; and medical device certification in the European Union. Encorium has clinical trial experience across a wide variety of therapeutic areas, such as cardiovascular, nephrology, endocrinology/metabolism, hematology, diabetes, neurology, oncology, immunology, vaccines, infectious diseases, gastroenterology, dermatology, hepatology, rheumatology, urology, ophthalmology, women's health and respiratory medicine. Encorium's clients consist of many of the largest companies in the pharmaceutical, biotechnology and medical device industries. Subsidiary Encorium Oy, a CRO based in Finland, manages the firm's Northern and Eastern European operations. In July 2009, the company sold its U.S. business to Pierrel Research USA, Inc, a wholly-owned subsidiary of Pierrel SpA, an international contract research organization, for $2.7 million. As a result of this transaction, the company's U.S. operations have been discontinued, and Encorium Group now shares a headquarters with its Finnish subsidiary, Encorium Oy.

FINANCIALS: Sales and profits are in thousands of dollars—add 000 to get the full amount. 2009 Note: Financial information for 2009 was not available for all companies at press time.

2009 Sales: $21,167	2009 Profits: $-3,870	**U.S. Stock Ticker: ENCO**
2008 Sales: $26,357	2008 Profits: $-21,073	**Int'l Ticker:**　Int'l Exchange:
2007 Sales: $36,802	2007 Profits: $-2,751	Employees: 161
2006 Sales: $17,684	2006 Profits: $- 494	Fiscal Year Ends: 12/31
2005 Sales: $12,727	2005 Profits: $-1,484	Parent Company:

SALARIES/BENEFITS:

Pension Plan:	ESOP Stock Plan:	Profit Sharing:	Top Exec. Salary: $349,953	Bonus: $
Savings Plan: Y	Stock Purch. Plan:		Second Exec. Salary: $311,750	Bonus: $

OTHER THOUGHTS:

Apparent Women Officers or Directors: 1
Hot Spot for Advancement for Women/Minorities: Y

LOCATIONS: ("Y" = Yes)

West:	Southwest:	Midwest:	Southeast:	Northeast:	International:
				Y	Y

ENDO PHARMACEUTICALS HOLDINGS INC

www.endo.com

Industry Group Code: 325412 Ranks within this company's industry group: Sales: 38 Profits: 31

Drugs:		Other:	Clinical:		Computers:		Services:	
Discovery:	Y	AgriBio:	Trials/Services:	Y	Hardware:		Specialty Services:	Y
Licensing:	Y	Genetic Data:	Labs:		Software:		Consulting:	
Manufacturing:	Y	Tissue Replacement:	Equipment/Supplies:		Arrays:		Blood Collection:	
Generics:			Research/Development Svcs.:		Database Management:		Drug Delivery:	
			Diagnostics:				Drug Distribution:	

TYPES OF BUSINESS:

Drugs-Pain Management
Pharmaceutical Preparations

BRANDS/DIVISIONS/AFFILIATES:

Endo Pharmaceuticals, Inc.
Lidoderm
Opana
Percocet
Frova
Aveed
Axomadol

CONTACTS: Note: Officers with more than one job title may be intentionally listed here more than once.

David P. Holveck, CEO
Julie H. McHugh, COO
David P. Holveck, Pres.
Alan G. Levin, CFO/Exec. VP
Larry Cunningham, Sr. VP-Human Resources
Ivan Gergel, Exec. VP-R&D
Caroline B. Manogue, Chief Legal Officer/Corp. Sec./Exec. VP
Robert J. Cobuzzi, Corp. VP-Corp. Dev.
Edward J. Sweeney, Principal Acct. Officer/VP/Controller
Colleen M. Craven, VP-Corp. Compliance & Bus. Practices
Roger H. Kimmel, Chmn.

Phone: 610-558-9800	Fax: 610-558-8979
Toll-Free:	
Address: 100 Endo Blvd., Chadds Ford, PA 19317 US	

GROWTH PLANS/SPECIAL FEATURES:

Endo Pharmaceuticals Holdings, Inc., through subsidiary Endo Pharmaceuticals, Inc., is a specialty pharmaceutical company engaged in the research, development, marketing and sale of branded and generic prescription pharmaceuticals used primarily to treat and manage pain, overactive bladder, central precocious puberty and prostate and bladder cancer. The company has a portfolio of branded products that includes such names as Lidoderm, a topical patch containing lidocaine; Opana and Opana ER, indicated for the relief of moderate to severe pain in patients requiring continuous opioid treatment; Percocet, indicated for the treatment of moderate to moderately severe pain; and Frova, a migraine treatment product. Branded products generate approximately 91% of Endo's net sales in 2009, with 52% of its net sales generated from Lidoderm. Its non-branded generic portfolio currently consists of products primarily focused in pain management. The firm focuses selectively on generics that have one or more barriers to market entry, such as complex formulation, regulatory or legal challenges or difficulty in raw material sourcing. Endo's primary development products include Aveed for the treatment of male hypogonadism; Axomadol, which is currently in Phase II development for the treatment of moderate to severe chronic pain and diabetic peripheral neuropathic pain; and octreotide implant, which is currently in Phase III clinical trials for the treatment of acromegaly and carcinoid syndrome. Endo sells its products to a limited number of large pharmacy chains and through wholesale drug distributors, which supply them to pharmacies, hospitals, governmental agencies and physicians. In May 2010, the firm agreed to acquire HealthTronics for $223 million.

Employees are offered medical, dental and health insurance; flexible spending accounts; a 401(k) plan; life insurance; short- and long-term disability; financial planning and a legal assistance plan; educational assistance; parenting benefits; an employee assistance program; and a matching gift program.

FINANCIALS: Sales and profits are in thousands of dollars—add 000 to get the full amount. 2009 Note: Financial information for 2009 was not available for all companies at press time.

2009 Sales: $1,460,841	2009 Profits: $266,336	U.S. Stock Ticker: ENDP
2008 Sales: $1,260,536	2008 Profits: $255,336	Int'l Ticker: Int'l Exchange:
2007 Sales: $1,085,608	2007 Profits: $227,440	Employees: 1,487
2006 Sales: $909,659	2006 Profits: $137,839	Fiscal Year Ends: 12/31
2005 Sales: $820,164	2005 Profits: $202,295	Parent Company:

SALARIES/BENEFITS:

Pension Plan:	ESOP Stock Plan:	Profit Sharing:	Top Exec. Salary: $833,333	Bonus: $739,200
Savings Plan: Y	Stock Purch. Plan:		Second Exec. Salary: $592,729	Bonus: $360,746

OTHER THOUGHTS:

Apparent Women Officers or Directors: 4
Hot Spot for Advancement for Women/Minorities: Y

LOCATIONS: ("Y" = Yes)

West:	Southwest:	Midwest:	Southeast:	Northeast:	International:
				Y	

Note: Financial information, benefits and other data can change quickly and may vary from those stated here.

ENTREMED INC

www.entremed.com

Industry Group Code: 325412 Ranks within this company's industry group: Sales: 129 Profits: 95

Drugs:		Other:	Clinical:	Computers:	Services:
Discovery:	Y	AgriBio:	Trials/Services:	Hardware:	Specialty Services:
Licensing:	Y	Genetic Data:	Labs:	Software:	Consulting:
Manufacturing:		Tissue Replacement:	Equipment/Supplies:	Arrays:	Blood Collection:
Generics:			Research/Development Svcs.:	Database Management:	Drug Delivery:
			Diagnostics:		Drug Distribution:

TYPES OF BUSINESS:

Drugs-Angiogenesis
Drugs-Cancer & Rheumatoid Arthritis

BRANDS/DIVISIONS/AFFILIATES:

Panzem 2ME2
Panzem NCD
ENMD-2076
MKC-1
ENMD-1198

CONTACTS: Note: Officers with more than one job title may be intentionally listed here more than once.

Cynthia Wong Hu, COO
Mark R. Bray, VP-Research
Cynthia Wong Hu, General Counsel/VP/Corp. Sec./Chief Legal Officer
Ginny Dunn, Associate Dir.-Corp. Comm.
Ginny Dunn, Associate Dir.-Investor Rel.
Kathy R. Wehmeir-Davis, Principal Acct. Officer/Controller
Carolyn F. Sidor, Chief Medical Officer/VP
Michael Tarnow, Chmn.

Phone: 240-864-2600	Fax: 240-864-2601
Toll-Free:	
Address: 9640 Medical Center Dr., Rockville, MD 20850 US	

GROWTH PLANS/SPECIAL FEATURES:

EntreMed, Inc. is a clinical-stage biopharmaceutical company that develops multi-mechanism oncology drugs to treat cancer and inflammatory disease by targeting diseases cells and their nourishing blood vessels. The company's four clinical product candidates, Panzem 2ME2, MKC-1, ENMD-1198 and ENMD-2076, are aimed at controlling angiogenesis, cell cycle regulation and inflammation to prevent the progression of cancer. ENMD 2076, the only candidate currently under active clinical evaluation, is an oral drug, that acts as an Aurora A/angiogenic kinase inhibitor; it is potent against Aurora A and multiple tyrosine kinases linked to cancer and inflammatory diseases. Panzem 2ME2 works by inhibiting angiogenesis, disrupting microtubule formation, down regulating hypoxia inducible factor-one alpha and inducing apoptosis. In recent years, the FDA has approved the use of predecessor Panzem NMD, which utilizes NanoCrystal technology through its licensing agreement with Elan Corporation, for the treatment of glioblastoma multiforme, multiple myeloma and ovarian cancer. MKC-1 arrests cellular mitosis by inhibiting a novel intracellular target involved in cell division during cellular trafficking in metastatic breast cancer patients, hematological cancers and non-small cell lung cancer. ENMD-1198 inhibits tumor growth through the modification of the chemical structure of 2-methoxyestradiol, which increases the molecule's anti-tumor properties, anti-angiogenic properties and metabolic rate. ENMD-2076, MKC-1 and Panzem NMD are in Phase II of clinical trial, and ENMD-1198 remains in Phase I. Both Panzem and ENMD-2076 have been accepted as Investigational New Drugs (IND) by the FDA. In February 2010, ENMD-2076 received orphan drug designation from the FDA.

EntreMed offers employees tuition reimbursement; an employee assistance plan; disability, health, dental, and vision insurance, a 401(k) plan; company-paid life insurance and supplemental coverage of home and auto insurance; and a 529 savings plan.

FINANCIALS: Sales and profits are in thousands of dollars—add 000 to get the full amount. 2009 Note: Financial information for 2009 was not available for all companies at press time.

2009 Sales: $5,284	2009 Profits: $-8,216	**U.S. Stock Ticker: ENMD**
2008 Sales: $7,477	2008 Profits: $-23,862	Int'l Ticker: Int'l Exchange:
2007 Sales: $7,396	2007 Profits: $-22,411	Employees: 14
2006 Sales: $6,894	2006 Profits: $-49,889	Fiscal Year Ends: 12/31
2005 Sales: $5,918	2005 Profits: $-16,313	Parent Company:

SALARIES/BENEFITS:

Pension Plan:	ESOP Stock Plan:	Profit Sharing:	Top Exec. Salary: $300,000	Bonus: $
Savings Plan: Y	Stock Purch. Plan:		Second Exec. Salary: $300,000	Bonus: $

OTHER THOUGHTS:

Apparent Women Officers or Directors: 5
Hot Spot for Advancement for Women/Minorities: Y

LOCATIONS: ("Y" = Yes)

West:	Southwest:	Midwest:	Southeast:	Northeast:	International:
				Y	

ENZO BIOCHEM INC

www.enzo.com

Industry Group Code: 325412 Ranks within this company's industry group: Sales: 76   Profits: 123

Drugs:	Other:		Clinical:		Computers:	Services:	
Discovery:	AgriBio:		Trials/Services:		Hardware:	Specialty Services:	Y
Licensing:	Genetic Data:	Y	Labs:	Y	Software:	Consulting:	
Manufacturing: Y	Tissue Replacement:		Equipment/Supplies:	Y	Arrays:	Blood Collection:	
Generics:			Research/Development Svcs.:		Database Management:	Drug Delivery:	
			Diagnostics:	Y		Drug Distribution:	

TYPES OF BUSINESS:

Pharmaceutical Discovery & Development
Genetic Analysis Products
Therapeutic Products
Diagnostic Products
Drug Development
Genetics Research
Clinical Laboratories
Biomedical Research Products & Tools

BRANDS/DIVISIONS/AFFILIATES:

Enzo Life Sciences, Inc.
Enzo Therapeutics, Inc.
Enzo Clinical Labs, Inc.

CONTACTS: Note: Officers with more than one job title may be intentionally listed here more than once.

Elazar Rabbani, CEO
Barry W. Weiner, Pres.
Natalie Bogdanos, General & Patent Counsel
David C. Goldberg, VP-Corp. Dev.
Andrew R. Crescenzo, Sr. VP-Finance
Kevin Krenitsky, Pres., Enzo Clinical Labs
Andrew Whiteley, COO-Enzo Life Sciences
Carl W. Balezentis, Pres., Enzo Life Sciences
Christine T. Fischette, Pres., Enzo Therapeutics
Elazar Rabbani, Chmn.

Phone: 212-583-0100	Fax:
Toll-Free:	
Address: 527 Madison Ave., New York, NY 10022 US	

GROWTH PLANS/SPECIAL FEATURES:

Enzo Biochem, Inc. is a life sciences and biotechnology company that employs a variety of genetic processes to develop research tools, diagnostics and therapeutics, and serves as a diagnostic services provider to the medical community. The firm has approximately 249 patents and roughly 200 pending patent applications. Enzo is primarily focused on developing technologies for gene regulation and modification for use in the treatment of viral and immunological disorders. Operations are conducted under three subsidiary companies: Enzo Life Sciences, Inc.; Enzo Therapeutics, Inc.; and Enzo Clinical Labs, Inc. Enzo Life Sciences manufactures, develops and markets functional biology and cellular biochemistry products and tools to research and pharmaceutical customers. Enzo Therapeutics works in collaboration with Enzo Life Sciences to develop multiple approaches in the areas of gastrointestinal, infectious, ophthalmic and metabolic diseases. This sector has also generated several clinical and preclinical pipelines in its effort to develop treatments for diseases and conditions where current treatment options are either ineffective, costly or cause undesired side effects, including HIV, Autoimmune Uveitis and Crohn's Disease. Enzo's therapeutic technologies deal with gene regulation, in which genes must be delivered into cells efficiently and without eliciting an unfavorable immune response. Enzo Clinical Labs, located in New York and New Jersey, offer a variety of routine clinical laboratory tests and procedures for general patient care. Located in Farmingdale, New York, Enzo Clinical Labs offers a full-service clinical laboratory with 30 patient service centers, a rapid response laboratory and a phlebotomy department.

Enzo offers its employees medical, prescription and dental coverage; a 401(k) plan; short- and long-term disability; supplemental cancer care insurance; life insurance; and paid entitlements.

FINANCIALS: Sales and profits are in thousands of dollars—add 000 to get the full amount. 2009 Note: Financial information for 2009 was not available for all companies at press time.

2009 Sales: $89,572	2009 Profits: $-23,564	U.S. Stock Ticker: ENZ
2008 Sales: $77,795	2008 Profits: $-10,653	Int'l Ticker: Int'l Exchange:
2007 Sales: $52,908	2007 Profits: $-13,260	Employees: 614
2006 Sales: $39,826	2006 Profits: $-15,667	Fiscal Year Ends: 7/31
2005 Sales: $43,403	2005 Profits: $3,004	Parent Company:

SALARIES/BENEFITS:

Pension Plan:	ESOP Stock Plan:	Profit Sharing:	Top Exec. Salary: $534,582	Bonus: $465,000
Savings Plan: Y	Stock Purch. Plan:		Second Exec. Salary: $474,174	Bonus: $340,000

OTHER THOUGHTS:

Apparent Women Officers or Directors: 2
Hot Spot for Advancement for Women/Minorities: Y

LOCATIONS: ("Y" = Yes)

West:	Southwest:	Midwest:	Southeast:	Northeast: Y	International:

Note: Financial information, benefits and other data can change quickly and may vary from those stated here.

ENZON PHARMACEUTICALS INC

www.enzon.com

Industry Group Code: 325412 Ranks within this company's industry group: Sales: 66   Profits: 71

Drugs:		Other:	Clinical:	Computers:	Services:
Discovery:	Y	AgriBio:	Trials/Services:	Hardware:	Specialty Services:
Licensing:	Y	Genetic Data:	Labs:	Software:	Consulting:
Manufacturing:		Tissue Replacement:	Equipment/Supplies:	Arrays:	Blood Collection:
Generics:			Research/Development Svcs.:	Database Management:	Drug Delivery:
			Diagnostics:		Drug Distribution:

TYPES OF BUSINESS:

Pharmaceutical Development
Drugs-Oncology & Hematology
Drugs-Transplantation
Drugs-Infectious Diseases
Single-Chain Antibody Technology
Polyethylene Glycol Technology

BRANDS/DIVISIONS/AFFILIATES:

Adagen
Oncaspar
PEGINTRON
Macugen
Cimzia
Pegasys
PEG-SN38

CONTACTS: Note: Officers with more than one job title may be intentionally listed here more than once.

Ralph del Campo, COO
Craig Tooman, CFO
Paul Davit, Exec. VP-Human Resources
Ivan Horak, Chief Scientific Officer/Exec. VP-R&D
Ralph del Campo, Exec. VP-Tech. Oper.
Paul S. Davit, Corp. Sec.
Craig Tooman, Exec. VP-Finance
Alexander J. Denner, Chmn.

Phone: 908-541-8600	Fax:
Toll-Free:	
Address: 685 Route 202/206, Bridgewater, NJ 08807 US	

GROWTH PLANS/SPECIAL FEATURES:

Enzon, Inc. is a biopharmaceutical company that develops and commercializes pharmaceutical products for patients with cancer. The firm advances its products through continued research and development of its proprietary PEGylation technology platform, its Customized Linker Technology and the Locked Nucleic Acid (LNA) technology. Enzon's polyethylene glycol (PEG) technology is used to improve the delivery, safety and efficacy of proteins and small molecules with known therapeutic value. There are six marketed biologic products that utilize the firm's PEG platform: Adagen; Oncaspar; PEGINTRON; Macugen; Cimzia; and Pegasys. The firm uses its Customized Linker Technology to generate compounds with enhanced therapeutic value over their unmodified forms. Enzon utilizes linkers to release the native molecule at a controlled rate. The firm is currently conducting research on PEG-SN38, a PEGylated conjugate of SN38, an active metabolite of the cancer drug irinotecan. PEG-SN38 allows for parenteral delivery, increased solubility, higher exposure and longer half-life. The firm's LNA technology-based programs include eight mRNA antagonists that the firm intends to develop through a license and collaboration agreement with Santaris. Enzon holds rights worldwide, other than Europe, to develop and commercialize mRNA antagonists directed against the HIF-1 alpha and Survivin mRNA targets. The firm recently presented data on five additional LNA targets: Beta-Catenin; HER3; PI3K; Androgen Receptor; and GLI2. In October 2009, the firm announced that the European Commission approved Cimzia for the treatment of adult patients with moderate to severe rheumatoid arthritis. In January 2010, Enzon sold its pharmaceutical business, which consisted of its products segment and contract manufacturing segment, to the sigma-tau Group for $300 million.

Enzon offers its employees medical, dental and vision coverage, along with a prescription plan; dependent and healthcare flexible spending accounts; life and disability insurance; a 401(k) plan; tuition reimbursement; and an employee assistance plan.

FINANCIALS: Sales and profits are in thousands of dollars—add 000 to get the full amount. 2009 Note: Financial information for 2009 was not available for all companies at press time.

2009 Sales: $184,622	2009 Profits: $ 683	U.S. Stock Ticker: ENZN
2008 Sales: $196,938	2008 Profits: $-2,715	Int'l Ticker: Int'l Exchange:
2007 Sales: $185,601	2007 Profits: $83,053	Employees: 317
2006 Sales: $185,653	2006 Profits: $21,309	Fiscal Year Ends: 12/31
2005 Sales: $166,250	2005 Profits: $-89,606	Parent Company:

SALARIES/BENEFITS:

Pension Plan:	ESOP Stock Plan:	Profit Sharing:	Top Exec. Salary: $855,000	Bonus: $375,000
Savings Plan: Y	Stock Purch. Plan:		Second Exec. Salary: $533,663	Bonus: $162,500

OTHER THOUGHTS:

Apparent Women Officers or Directors:
Hot Spot for Advancement for Women/Minorities:

LOCATIONS: ("Y" = Yes)

West:	Southwest:	Midwest:	Southeast:	Northeast:	International:
				Y	

ERESEARCH TECHNOLOGY INC

www.ert.com

Industry Group Code: 511210D Ranks within this company's industry group: Sales: 3   Profits: 2

Drugs:	Other:	Clinical:		Computers:		Services:	
Discovery:	AgriBio:	Trials/Services:		Hardware:		Specialty Services:	Y
Licensing:	Genetic Data:	Labs:	Y	Software:	Y	Consulting:	Y
Manufacturing:	Tissue Replacement:	Equipment/Supplies:		Arrays:		Blood Collection:	
Generics:		Research/Development Svcs.:		Database Management:		Drug Delivery:	
		Diagnostics:				Drug Distribution:	

TYPES OF BUSINESS:

Software-Clinical Research Technology
Technology Consulting Services
Cardiac Safety Services
Service-Business Services, NEC
Service-Testing Laboratories

BRANDS/DIVISIONS/AFFILIATES:

EXPeRT
ePRO Solutions
My Study Portal

CONTACTS: Note: Officers with more than one job title may be intentionally listed here more than once.

Michael J. McKelvey, CEO
Michael J. McKelvey, Pres.
Keith D. Schneck, CFO/Exec. VP
John Blakeley, Exec. VP-Mktg. & Sales
Valerie Mattern, VP-Human Resources
Joel Morganroth, Chief Scientific Officer
Thomas P. Devine, Chief Dev. Officer/Exec. VP
Amy Furlong, Exec. VP-Cardiac Safety Oper.
Tom Devine, Chief Dev. Officer/Exec. VP
John Blakeley, Contact-Investor Rel.
Robert Brown, Sr. VP-Strategic Mktg., Planning, & Partnerships
Jeffrey S. Litwin, Chief Medical Officer/Exec. VP
George Tiger, Sr. VP-Global Sales
Joel Morganroth, Chmn.
Miki Stevens, Dir.-Bus. Dev., Japan

Phone: 215-972-0420	Fax: 215-972-0414
Toll-Free:	
Address: 1818 Market St., Ste. 1000, Philadelphia, PA 19103 US	

GROWTH PLANS/SPECIAL FEATURES:

eResearch Technology, Inc. (ERT) is a provider of technology and service solutions that enable the pharmaceutical, biotechnology and medical device companies to collect, interpret and distribute cardiac safety and clinical data more efficiently. The company also provides centralized electrocardiograph (ECG) services (through its EXPeRT Cardiac Safety Intelligent Data Management system) and clinical research technology and services, which include the development, marketing and support of clinical research technology. Clinical trial sponsors and clinical research organizations use the cardiac safety services during clinical trials. ERT's clinical research technology and services include the licensing of its proprietary software products and the provision of maintenance and consulting services supporting its proprietary software products. The service's My Study Portal supports access to data without a software footprint, as it utilizes the web rather than installed programs. The company's electronic patient reported outcomes (ePRO) business allows users to collect, monitor and store patient information in a centralized database and utilize expert consultants to extrapolate strategies. In June 2009, OmniComm Systems, Inc. acquired the electronic data capture (EDC) operations of ERT for $1.15 million. In June 2010, ERT acquired the Research Services division of CareFusion Corporation, a respiratory diagnostics services and manufacturing firm, for $80.8 million.

eResearch Technology's benefits include medical, vision and dental coverage; spending accounts; short and long-term disability coverage; tuition reimbursement; and transit credits.

FINANCIALS: Sales and profits are in thousands of dollars—add 000 to get the full amount. 2009 Note: Financial information for 2009 was not available for all companies at press time.

2009 Sales: $93,823	2009 Profits: $10,687	U.S. Stock Ticker: ERES
2008 Sales: $133,140	2008 Profits: $25,002	Int'l Ticker: Int'l Exchange:
2007 Sales: $98,698	2007 Profits: $15,252	Employees: 353
2006 Sales: $86,368	2006 Profits: $8,310	Fiscal Year Ends: 12/31
2005 Sales: $86,847	2005 Profits: $15,365	Parent Company:

SALARIES/BENEFITS:

Pension Plan:	ESOP Stock Plan:	Profit Sharing:	Top Exec. Salary: $515,000	Bonus: $95,000
Savings Plan: Y	Stock Purch. Plan: Y		Second Exec. Salary: $298,700	Bonus: $47,500

OTHER THOUGHTS:

Apparent Women Officers or Directors: 2
Hot Spot for Advancement for Women/Minorities: Y

LOCATIONS: ("Y" = Yes)

West:	Southwest:	Midwest:	Southeast:	Northeast:	International:
				Y	Y

EUSA PHARMA (USA) INC

www.eusapharma.com

Industry Group Code: 325412 Ranks within this company's industry group: Sales: Profits:

Drugs:		Other:		Clinical:		Computers:		Services:	
Discovery:	Y	AgriBio:		Trials/Services:		Hardware:		Specialty Services:	
Licensing:	Y	Genetic Data:		Labs:		Software:		Consulting:	
Manufacturing:		Tissue Replacement:		Equipment/Supplies:		Arrays:		Blood Collection:	
Generics:				Research/Development Svcs.:		Database Management:		Drug Delivery:	
				Diagnostics:	Y			Drug Distribution:	

TYPES OF BUSINESS:
Drugs-Oncology
Diagnostic Imaging Agents

BRANDS/DIVISIONS/AFFILIATES:
ProstaScint
Caphosol
Quadramet
EUSA Pharma Inc

CONTACTS: *Note: Officers with more than one job title may be intentionally listed here more than once.*
Jim Mitchum, Pres.
Rolf Stahel, Chmn.

Phone: 215-867-4900	**Fax:** 215-579-0384
Toll-Free: 800-833-3533	
Address: 1717 Langhorne-Newtown Rd., Ste. 201, Langhorne, PA 19047 US	

GROWTH PLANS/SPECIAL FEATURES:
EUSA Pharma (USA) Inc., formerly Cytogen Corp., develops and commercializes oncology products. The firm recently became a wholly-owned subsidiary of EUSA Pharma, an international specialty pharmaceutical company focused on licensing, developing and marketing late-stage oncology, pain control and critical care products. EUSA Pharma (USA) Inc. is responsible for the group's North American operations. Its FDA-approved and marketed products include ProstaScint, a monoclonal antibody-based agent used to image the extent and spread of prostate cancer; QuadraMet, a therapeutic agent marketed for the relief of pain in prostate, breast and other cancers that have spread to the bone; and Caphosol, an advanced electrolyte solution for the treatment of oral mucositis and dry mouth. While ProstaScint and QuadraMet are marketed and sold exclusively in the U.S., Caphosol is available in the U.S. and in Europe through EUSA Pharma.

FINANCIALS: Sales and profits are in thousands of dollars—add 000 to get the full amount. 2009 Note: Financial information for 2009 was not available for all companies at press time.

			U.S. Stock Ticker: Subsidiary
2009 Sales: $	2009 Profits: $		
2008 Sales: $	2008 Profits: $		Int'l Ticker: Int'l Exchange:
2007 Sales: $	2007 Profits: $		Employees:
2006 Sales: $17,307	2006 Profits: $-15,103		Fiscal Year Ends: 12/31
2005 Sales: $15,946	2005 Profits: $-26,289		Parent Company: EUSA PHARMA INC

SALARIES/BENEFITS:

Pension Plan:	ESOP Stock Plan:	Profit Sharing:	Top Exec. Salary: $	Bonus: $
Savings Plan:	Stock Purch. Plan:		Second Exec. Salary: $	Bonus: $

OTHER THOUGHTS:
Apparent Women Officers or Directors: 1
Hot Spot for Advancement for Women/Minorities:

LOCATIONS: ("Y" = Yes)

West:	Southwest:	Midwest:	Southeast:	Northeast:	International:
				Y	

EVOTEC AG

www.evotec.com

Industry Group Code: 541712 **Ranks within this company's industry group:** Sales: 9   Profits: 17

Drugs:		Other:		Clinical:		Computers:		Services:	
Discovery:	Y	AgriBio:		Trials/Services:		Hardware:		Specialty Services:	Y
Licensing:	Y	Genetic Data:		Labs:	Y	Software:		Consulting:	
Manufacturing:	Y	Tissue Replacement:		Equipment/Supplies:		Arrays:		Blood Collection:	
Generics:				Research/Development Svcs.:	Y	Database Management:	Y	Drug Delivery:	
				Diagnostics:				Drug Distribution:	

TYPES OF BUSINESS:

Drug Discovery & Development Services
Assay Development
High-Throughput Screening
Chemical Compound Libraries
Medicinal Chemistry
Manufacturing Services
Drug Discovery Hardware & Software

BRANDS/DIVISIONS/AFFILIATES:

Zebrafish
EVOlution
Research Support International Private Limited

CONTACTS: *Note: Officers with more than one job title may be intentionally listed here more than once.*

Werner Lanthaler, CEO
Mario Polywka, COO
Klaus Maleck, CFO
Charmion Gillmore, Sr. VP-Human Resources & Internal Comm.
John Kemp, Chief R&D Officer
Mark Ashton, Exec. VP-Bus. Dev.
Doris Kynast, Corp. Comm.
Doris Kynast, Investor Rel.
David Brister, Chief Bus. Officer
Flemming Ornskov, Chmn.

Phone: 49-40-5-60-81-0	Fax: 49-40-5-60-81-222
Toll-Free:	
Address: Schnackenburgallee 114, Hamburg, 22525 Germany	

GROWTH PLANS/SPECIAL FEATURES:

Evotec AG provides investigational new drug programs for its biotechnology and pharmaceutical company clients. Evotec's offers collaborative, contract research to companies involved in drug discovery; with specific focus on the central nervous system (CNS) and related diseases. The firm's capabilities include assay development; high-throughput screening; compound libraries; medicinal chemistry; chemical and pharmaceutical development; formulation and manufacture; and fragment-based drug discovery (FBDD). Assay development identifies chemical compounds that interact with a selected biological target. High-throughput screening takes the assay format and uses it to test up to 100,000 chemical compounds per day in Evotec's, and its clients' chemical libraries. Primary screens determine whether a compound shows biological activity worth exploring. The company fine-tunes its chemical library findings into pre-clinical candidates, and then develops processes for manufacturing enough of the test drug for Phase I-III clinical trials. Evotec's FBDD platform, EVOlution, utilizes fragments of molecules to find starting points for drug discovery. In addition to contract services Evotec conducts internal research to discover compounds for the treatment of major CNS related disorders including sleep disorders and Alzheimer's disease, as well as inflammation and pain. The company, in partnership with Roche, is developing the EVT 100 compound family for treatment resistant depression, pain, Alzheimer's disease and other indications. For the treatment of insomnia, Evotec has developed EVT 201, which has completed two Phase II trials with positive results. The company also has several drug candidates in pre-clinical development. Furthest along are its H3 receptor antagonists, targeted at treating narcolepsy and improving cognition in patients with Alzheimer's disease. In May 2009, the company acquired the Zebrafish drug screening technology from Summit Corporation plc. In August 2009, Evotec acquired a controlling stake in Research Support International Private Limited. In July 2010, EVT-101 began a Phase II study for the treatment of depression.

FINANCIALS: Sales and profits are in thousands of dollars—add 000 to get the full amount. 2009 Note: Financial information for 2009 was not available for all companies at press time.

2009 Sales: $56,570	2009 Profits: $-60,340	U.S. Stock Ticker: EVT
2008 Sales: $56,300	2008 Profits: $-110,400	Int'l Ticker: EVT Int'l Exchange: Frankfurt-Euronext
2007 Sales: $46,700	2007 Profits: $-15,700	Employees: 485
2006 Sales: $57,600	2006 Profits: $-39,100	Fiscal Year Ends: 12/31
2005 Sales: $101,630	2005 Profits: $-42,780	Parent Company:

SALARIES/BENEFITS:

Pension Plan:	ESOP Stock Plan:	Profit Sharing:	Top Exec. Salary: $	Bonus: $
Savings Plan:	Stock Purch. Plan:		Second Exec. Salary: $	Bonus: $

OTHER THOUGHTS:

Apparent Women Officers or Directors: 1
Hot Spot for Advancement for Women/Minorities:

LOCATIONS: ("Y" = Yes)

West:	Southwest:	Midwest:	Southeast:	Northeast:	International:
					Y

Note: Financial information, benefits and other data can change quickly and may vary from those stated here.

EXELIXIS INC

www.exelixis.com

Industry Group Code: 325412 **Ranks within this company's industry group:** Sales: 67   Profits: 162

Drugs:		Other:		Clinical:	Computers:	Services:
Discovery:	Y	AgriBio:	Y	Trials/Services:	Hardware:	Specialty Services:
Licensing:	Y	Genetic Data:	Y	Labs:	Software:	Consulting:
Manufacturing:		Tissue Replacement:	Y	Equipment/Supplies:	Arrays:	Blood Collection:
Generics:				Research/Development Svcs.:	Database Management:	Drug Delivery:
				Diagnostics:		Drug Distribution:

TYPES OF BUSINESS:
Genetic Research & Drug Development
Crop Protection Products
Genomics
Anti-Cancer Compounds

BRANDS/DIVISIONS/AFFILIATES:
Exelixis Plant Sciences Inc

CONTACTS: Note: Officers with more than one job title may be intentionally listed here more than once.
George A. Scangos, CEO
George A. Scangos, Pres.
Frank Karbe, CFO/Exec. VP
Michael Morrissey, Pres., R&D
Pamela A. Simonton, General Counsel/Exec. VP
Lupe M. Rivera, Sr. VP-Oper.
Frances K. Heller, Exec. VP-Bus. Dev.
D. Ry Wagner, VP-Research, Exelixis Plant Sciences
Gisela M. Schwab, Chief Medical Officer/Exec. VP
Peter Lamb, Chief Scientific Officer/Sr. VP-Discovery Research
Stelios Papadopoulos, Chmn.

Phone: 650-837-7000	Fax: 650-837-8300
Toll-Free:	
Address: 210 E. Grand Ave., South San Francisco, CA 94083 US	

GROWTH PLANS/SPECIAL FEATURES:

Exelixis, Inc. is a biotechnology company that focuses on the discovery and development of potential new drug therapies for cancer and other life-threatening diseases. The company has developed an integrated research and discovery platform utilizing proprietary technologies such as medicinal chemistry, bioinformatics, structural biology and early in vivo testing to provide an efficient and cost-effective process in gene analysis and drug development. Exelixis' translational research group employs knowledge generated in the discovery process to identify targeted patient populations for possible gene mutations or gene variants that impact response to therapy. The company's clinical program then conducts clinical trials to move candidate compounds through clinical registration phases in order to market newly discovered treatments. Exelixis Plant Sciences, a subsidiary of Exelixis, is working in collaboration with agricultural companies in plant biotechnology and crop protection. Research areas in crop protection focus on chemical products such as herbicides, insecticides and nematides designed specifically to target implicated crops. The plant biotechnology sector aims to develop crops with higher yields and improved nutritional profiles in oil content and protein composition. Exelixis has entered into research collaborations with Bristol-Myers Squibb to identify novel targets for new drugs in the fields of oncology and cardiovascular disease. The firm collaborates its research with GlaxoSmithKline in therapeutic areas such as vascular biology, inflammatory disease and oncology.

The company offers its employees medical, dental and vision insurance; domestic partner benefits; a flexible spending account; a 401(k) plan; life insurance; AD&D insurance; an employee assistance program; tuition assistance; a college saving plan; business travel accident insurance; a legal assistance plan; pet insurance; a stock purchase program; an employee discount program; subsidized cafeteria meals; and subsidies for fitness center memberships.

FINANCIALS: Sales and profits are in thousands of dollars—add 000 to get the full amount. 2009 Note: Financial information for 2009 was not available for all companies at press time.

2009 Sales: $151,759	2009 Profits: $-135,220	**U.S. Stock Ticker:** EXEL	
2008 Sales: $117,859	2008 Profits: $-162,854	**Int'l Ticker:**	**Int'l Exchange:**
2007 Sales: $113,470	2007 Profits: $-86,381	**Employees:** 676	
2006 Sales: $98,670	2006 Profits: $-101,492	**Fiscal Year Ends:** 12/31	
2005 Sales: $75,961	2005 Profits: $-84,404	**Parent Company:**	

SALARIES/BENEFITS:

Pension Plan:	ESOP Stock Plan:	Profit Sharing:	Top Exec. Salary: $848,731	Bonus: $255,000
Savings Plan: Y	Stock Purch. Plan: Y		Second Exec. Salary: $483,612	Bonus: $121,157

OTHER THOUGHTS:
Apparent Women Officers or Directors: 4
Hot Spot for Advancement for Women/Minorities: Y

LOCATIONS: ("Y" = Yes)

West:	Southwest:	Midwest:	Southeast:	Northeast:	International:
Y					Y

FLAMEL TECHNOLOGIES SA

www.flamel-technologies.fr

Industry Group Code: 325412A Ranks within this company's industry group: Sales: 8 Profits: 11

Drugs:		Other:		Clinical:		Computers:		Services:	
Discovery:		AgriBio:		Trials/Services:		Hardware:		Specialty Services:	
Licensing:	Y	Genetic Data:		Labs:		Software:		Consulting:	
Manufacturing:	Y	Tissue Replacement:		Equipment/Supplies:	Y	Arrays:		Blood Collection:	
Generics:				Research/Development Svcs.:		Database Management:		Drug Delivery:	Y
				Diagnostics:				Drug Distribution:	

TYPES OF BUSINESS:

Drug Delivery Systems
Extended-Release Formulations
Collagen-Based Biomaterials
Long-Acting Insulin

BRANDS/DIVISIONS/AFFILIATES:

Medusa
Micropump
Asacard
Genvir
Colcys
Basulin
Trigger-Lock
Metformin XL

CONTACTS: *Note: Officers with more than one job title may be intentionally listed here more than once.*

Stephen H. Willard, CEO
Raphael Jorda, COO/Exec. VP
Sian Crouzet, Principal Financial Officer
Remi Meyrueix, Dir.-Physico-Chemistry & Nanotech Dept.
Rafael Jorda, Dir.-Mfg. & Dev.
Charles Marlio, Dir.-Strategic Planning
Charles Marlio, Dir.-Investor Rel.
Christian Kalita, Chief Pharmacist
Catherine Castan, Dir.-R&D Head-Micropump Team
Roger Kravtzoff, Dir.-Preclinical & Early Clinical Dev.
Jeffery S. Vick, Chief Bus. Officer
Elie Vannier, Chmn.
Nigel McWilliam, Dir.-Bus. Dev., U.S.
David Weber, Dir.-Purchasing Oper.

Phone: 33-472-783-434	Fax: 33-472-783-435
Toll-Free:	
Address: 33 Ave. du Docteur Georges Levy, Venissieux Cedex, 69693 France	

GROWTH PLANS/SPECIAL FEATURES:

Flamel Technologies S.A. is a drug delivery company that is engaged in the development of small molecule and protein therapeutic products for the biotechnology and pharmaceutical industries. Named after a 14th Century French alchemist, the company operates in a 103,900-square-foot pharmaceutical production facility in Pessac, France. Flamel currently holds two patented drug delivery platforms: Medusa and Micropump. Medusa consists of an injectable 20-50 nanometer diameter self-assembled poly-amino-acid nanogel system with a long duration, which enables the controlled delivery of fully human and non-denatured proteins. Medusa technology is currently being tested for FT-105, a long-acting basal insulin for the treatment of type II diabetes; and in IFN alpha-2b XL for the treatment of hepatitis B, hepatitis C and some forms of cancers. Micropump is a drug delivery system permitting either the extended or delayed and extended controlled delivery of small molecule drugs. Since the active ingredients are enclosed within each microparticle, drugs can easily be delivered in suspensions or syrups that completely mask any taste of the active pharmaceutical compound. Micropump technology-based products include Coreg CR, a beta-blocker that can be taken once daily to treat hypertension, heart failure and cardiac arrest, and Trigger Lock, a technology that may prevent drug abusers from misusing opioid pain medications, by preventing tablet crushing and other methods used to extract the active ingredients. Flamel operates primarily through licensing agreements and collaborations with other pharmaceutical companies such as GlaxoSmithKline (GSK), Merck and Servier Monde.

FINANCIALS: Sales and profits are in thousands of dollars—add 000 to get the full amount. 2009 Note: Financial information for 2009 was not available for all companies at press time.

2009 Sales: $42,100	2009 Profits: $-11,400	**U.S. Stock Ticker:** FLML
2008 Sales: $38,619	2008 Profits: $-12,084	**Int'l Ticker:** FL3 Int'l Exchange: Frankfurt-Euronext
2007 Sales: $36,654	2007 Profits: $-37,737	Employees:
2006 Sales: $23,020	2006 Profits: $-35,201	Fiscal Year Ends: 12/31
2005 Sales: $23,598	2005 Profits: $-27,377	Parent Company:

SALARIES/BENEFITS:

Pension Plan:	ESOP Stock Plan:	Profit Sharing:	Top Exec. Salary: $	Bonus: $
Savings Plan:	Stock Purch. Plan:		Second Exec. Salary: $	Bonus: $

OTHER THOUGHTS:

Apparent Women Officers or Directors: 2
Hot Spot for Advancement for Women/Minorities: Y

LOCATIONS: ("Y" = Yes)

West:	Southwest:	Midwest:	Southeast:	Northeast:	International:
					Y

FOREST LABORATORIES INC

www.frx.com

Industry Group Code: 325412 Ranks within this company's industry group: Sales: 27 Profits: 20

Drugs:		Other:		Clinical:		Computers:		Services:	
Discovery:	Y	AgriBio:		Trials/Services:		Hardware:		Specialty Services:	
Licensing:		Genetic Data:		Labs:		Software:		Consulting:	
Manufacturing:	Y	Tissue Replacement:		Equipment/Supplies:		Arrays:		Blood Collection:	
Generics:				Research/Development Svcs.:		Database Management:		Drug Delivery:	
				Diagnostics:				Drug Distribution:	

TYPES OF BUSINESS:
Drugs, Manufacturing
Over-the-Counter Pharmaceuticals
Generic Pharmaceuticals
Antidepressants
Asthma Medications
Cardiovascular Products
OB/Gyn Products
Endocrinology Products

BRANDS/DIVISIONS/AFFILIATES:
Lexapro
Namenda
Bystolic
Forest Research Institute
Forest Pharmaceuticals, Inc.
Forest Laboratories Europe
Inwood Laboratories
Cerexa, Inc.

CONTACTS: *Note: Officers with more than one job title may be intentionally listed here more than once.*
Howard Solomon, CEO
Lawrence S. Olanoff, COO
Lawrence S. Olanoff, Pres.
Francis I. Perier, Jr., CFO/Sr. VP-Finance
Elaine Hochberg, Sr. VP-Mktg./Chief Commercial Officer
Bernard J. McGovern, VP-Human Resources
Marco Taglietti, VP-R&D/Pres., Forest Research Institute
Kevin Walsh, VP-Info. Systems
Kevin Walsh, VP-Mfg. Oper.
Herschel S. Weinstein, General Counsel/VP
Richard S. Overton, VP-Oper. & Facilities
David F. Solomon, VP-Bus. Dev. & Strategic Planning
Frank Murdolo, VP-Investor Rel.
Ralph Kleinman, VP-Tax/Treas.
Raymond Stafford, Exec. VP-Global Mktg.
Rita Weinberger, Controller/VP
William J. Meury, VP-Mktg.
William J. Candee III, Sec.
Howard Solomon, Chmn.
Raymond Stafford, CEO-Forest Laboratories Europe

Phone: 212-421-7850	Fax:
Toll-Free: 800-947-5227	
Address: 909 3rd Ave., New York, NY 10022 US	

GROWTH PLANS/SPECIAL FEATURES:
Forest Laboratories, Inc. develops, manufactures and sells pharmaceutical products. It currently covers six therapeutic areas, developing treatments for respiratory, pain management, obstetrics and gynecology (ob/gyn) and pediatric, endocrinology, central nervous system and cardiovascular conditions. Forest's four principal brands are Lexapro, an antidepressant; Bystolic, a hypertension treatment; Namenda, a therapy for moderate or severe Alzheimer's disease; and Savella, a dual reuptake inhibitor for the treatment of fibromyalgia. Other products include Tiazac, for hypertension; Celexa, an antidepressant; Campral, used to treat alcohol dependence; Armour Thyroid, Levothroid and Thyrolar, for treating hypothyroidism; Cervidil, used to prepare the cervix before inducing labor; Aerobid, an asthma medication; AeroChamber Plus, an inhalant delivery system for asthma medications; and Combunox, a pain medication. Forest markets directly to physicians, as well as to pharmacies, hospitals, managed care and other healthcare organizations. The firm's principal customers include McKesson Drug Company; Cardinal Health, Inc.; and AmeriSource Bergen Corporation. Forest Research Institute, Forest's scientific division, maintains labs on Long Island and in New Jersey. Subsidiary Forest Pharmaceuticals, Inc. manufactures and distributes Forest's branded prescription products in the U.S. Subsidiary Forest Laboratories Europe distributes prescription and over-the-counter drugs in Europe, the Middle East, Australia and Asia. Subsidiary Inwood Laboratories manufactures and supplies generic versions of Forest's medications. Cerexa, Inc. develops and commercializes treatments for life-threatening infections. In January 2009, Savella, a selective serotonin and norepinephrine inhibitor, was approved by the FDA for the management of fibromyalgia, a chronic pain condition. In December 2009, Forest agreed to acquire additional rights to NXL 104 a beta-lactamase inhibitor.

Employees at Forest receive medical, dental, vision and prescription coverage; financial assistance for adoption and fertility treatment; flexible spending accounts; life insurance; short and long-term disability insurance; a commuter benefit program; a profit sharing plan; and a 401(k) plan with company match.

FINANCIALS: Sales and profits are in thousands of dollars—add 000 to get the full amount. 2009 Note: Financial information for 2009 was not available for all companies at press time.

2009 Sales: $3,636,055	2009 Profits: $767,743	**U.S. Stock Ticker:** FRX	
2008 Sales: $3,501,802	2008 Profits: $967,933	**Int'l Ticker:** Int'l Exchange:	
2007 Sales: $3,183,324	2007 Profits: $454,103	Employees: 5,225	
2006 Sales: $2,793,934	2006 Profits: $708,514	Fiscal Year Ends: 11/30	
2005 Sales: $3,052,408	2005 Profits: $838,805	Parent Company:	

SALARIES/BENEFITS:
Pension Plan:	ESOP Stock Plan:	Profit Sharing: Y	Top Exec. Salary: $1,217,500	Bonus: $700,000
Savings Plan: Y	Stock Purch. Plan:		Second Exec. Salary: $798,750	Bonus: $440,000

OTHER THOUGHTS:
Apparent Women Officers or Directors: 7
Hot Spot for Advancement for Women/Minorities: Y

LOCATIONS: ("Y" = Yes)
West:	Southwest:	Midwest:	Southeast:	Northeast:	International:
Y		Y		Y	Y

Note: Financial information, benefits and other data can change quickly and may vary from those stated here.

GALDERMA PHARMA SA

www.galderma.com

Industry Group Code: 325412 **Ranks within this company's industry group:** Sales: Profits:

Drugs:		Other:	Clinical:		Computers:		Services:	
Discovery:	Y	AgriBio:	Trials/Services:		Hardware:		Specialty Services:	
Licensing:		Genetic Data:	Labs:	Y	Software:		Consulting:	
Manufacturing:	Y	Tissue Replacement:	Equipment/Supplies:		Arrays:		Blood Collection:	
Generics:			Research/Development Svcs.:		Database Management:		Drug Delivery:	
			Diagnostics:				Drug Distribution:	

TYPES OF BUSINESS:

Dermatological Pharmaceuticals
Dermatological Product Research and Development

BRANDS/DIVISIONS/AFFILIATES:

Nestle SA
Loreal SA
Cetaphil
Clobex
Differin
Epiduo
Loceryl
Rozex

CONTACTS: *Note: Officers with more than one job title may be intentionally listed here more than once.*

Humberto C. Antunes, CEO
Humberto C. Antunes, Pres.
Annika Ohlin, Head-Comm.
Francois Fournier, Pres., Galderma Laboratories, North America
Dale Weiss, Contact-Media
Albert Draaijer, Pres., Galderma Laboratories, Americas

Phone: 41-21-642-78-00	Fax: 41-21-642-78-01
Toll-Free:	
Address: 2 Ave. de Gratta-Paille, World Trade Center, Lausanne, 1018 Switzerland	

GROWTH PLANS/SPECIAL FEATURES:

Galderma Pharma S.A. is engaged in the research, development, manufacture and distribution of dermatological products worldwide. The firm, a joint venture between Nestle and L'Oreal, has a portfolio that includes treatment for skin conditions such as acne, dry skin, fungal nail infections, pigmentary disorders, skin cancers, rosacea, psoriasis/SRD (Steroid-Responsive Dermatoses) and skin senescence. Galderma's products vary from topical and oral treatments to photodynamic therapy and include key brands such as Cetaphil, Clobex, Differin, Epiduo, Loceryl, Rozex, Metvix, Oracea, Silkis, Pandel, Alcortin, Novacort, Vectical and Tri-Luma. The company's primary research and development department, located in Sophia Antipolis, France, uses technologies such as human genome analysis, robotic screening, pharmaco-dynamic modeling, computer-assisted imaging and miniaturized human biological assaying. Galderma also houses research and development teams in Princeton, New Jersey and Tokyo, Japan, as well as manufacturing facilities in Alby-sur-Cheran, France; Montreal, Quebec; and Hortolandia, Brazil. The company holds operating subsidiaries around the globe, including Galderma Australia (PTY) Ltd. in Australia; Galderma Laboratories L.P. in the U.S.; Galderma Canada, Inc.; Galderma Argentina S.A.; Galderma Brasil LTDA; and Galderma (Japan), among many others. The company's products are distributed in over 65 countries. The most recent products from Galderma's pipeline are Vectical, a vitamin D3 product for plaque psoriasis; Differin acne lotion in the U.S.; Azzalure and Oracea in Europe; and Epiduo Gel, a topical acne treatment. Recently, Galderma acquired Collagenex Pharmaceuticals, Inc., a specialty pharmaceutical company focused on dermatological therapies. CollaGenex's chief products include Oracea for the treatment of inflammatory lesions of rosacea in adults; Pandel, used for the treatment of dermatitis and psoriasis; Alcortin, a mild dermatosis gel with combined anti-fungal, anti-bacterial and anti-inflammatory effects; and Novacort, a topical corticosteroid with anti-inflammatory and anesthetic treatment for dermatoses.

FINANCIALS: Sales and profits are in thousands of dollars—add 000 to get the full amount. 2009 Note: Financial information for 2009 was not available for all companies at press time.

2009 Sales: $	2009 Profits: $	U.S. Stock Ticker: Joint Venture
2008 Sales: $	2008 Profits: $	Int'l Ticker: Int'l Exchange:
2007 Sales: $	2007 Profits: $	Employees:
2006 Sales: $	2006 Profits: $	Fiscal Year Ends:
2005 Sales: $	2005 Profits: $	Parent Company:

SALARIES/BENEFITS:

Pension Plan:	ESOP Stock Plan:	Profit Sharing:	Top Exec. Salary: $	Bonus: $
Savings Plan:	Stock Purch. Plan:		Second Exec. Salary: $	Bonus: $

OTHER THOUGHTS:

Apparent Women Officers or Directors:
Hot Spot for Advancement for Women/Minorities:

LOCATIONS: ("Y" = Yes)

West:	Southwest:	Midwest:	Southeast:	Northeast:	International:
	Y			Y	Y

GE HEALTHCARE

www.gehealthcare.com

Industry Group Code: 33911 Ranks within this company's industry group: Sales: 1 Profits: 1

Drugs:		Other:	Clinical:		Computers:		Services:	
Discovery:	Y	AgriBio:	Trials/Services:		Hardware:	Y	Specialty Services:	Y
Licensing:		Genetic Data:	Labs:		Software:	Y	Consulting:	
Manufacturing:		Tissue Replacement:	Equipment/Supplies:	Y	Arrays:	Y	Blood Collection:	
Generics:			Research/Development Svcs.:	Y	Database Management:		Drug Delivery:	
			Diagnostics:	Y			Drug Distribution:	

TYPES OF BUSINESS:

Medical Imaging & Information Technology
Magnetic Resonance Imaging Systems
Patient Monitoring Systems
Clinical Information Systems
Nuclear Medicine
Surgery & Vascular Imaging
X-Ray & Ultrasound Bone Densitometers
Clinical & Business Services

BRANDS/DIVISIONS/AFFILIATES:

General Electric Co (GE)
Innova EPVision

CONTACTS: *Note: Officers with more than one job title may be intentionally listed here more than once.*

John Dineen, CEO
John Dineen, Pres.
Frank Schulkes, CFO/Exec. VP
Jean-Michel Cossery, Chief Mktg. Officer
Mike Hanley, VP-Human Resources
Russel P. Meyer, CIO
Michael Harsh, CTO
Michael A. Jones, Exec. VP-Bus. Dev.
Lynne Gailey, Exec. VP-Global Comm.
Pete McCabe, CEO/Pres., Surgery
Peter Ehrenheim, CEO/Pres., Life Sciences
Vishal Wanchoo, CEO/Pres., IT
Pascale Witz, CEO/Pres., Medical Diagnostics
Reinaldo Garcia, CEO-EMEA
Brian Masterson, VP-Supply Chain

Phone: 44-1494-544-000	Fax:
Toll-Free:	
Address: Pollards Wood, Nightingales Ln., Chalfont St. Giles, HP8 4SP UK	

GROWTH PLANS/SPECIAL FEATURES:

GE Healthcare, a subsidiary of General Electric Co. (GE) and a part of GE's Technology Infrastructure division, is a global leader in medical imaging and information technologies, patient monitoring systems and health care services. The company operates through five divisions: healthcare systems; life systems; medical diagnostics; healthcare IT; and surgery. GE Healthcare's healthcare systems business provides technologies and services for clinicians and healthcare administrators. GE Healthcare's life sciences segment offers drug discovery, biopharmaceutical manufacturing and cellular technologies, enabling scientists and specialists around the world to discover new ways to predict, diagnose and treat disease earlier. The segment also makes systems and equipment for the purification of biopharmaceuticals. The firm's medical diagnostics business researches, manufactures and markets agents used during medical scanning procedures to highlight organs, tissue and functions inside the human body. The healthcare IT business provides clinical and financial information technology solutions including enterprise and departmental IT products, revenue cycle management and practice applications, to help customers streamline healthcare costs and improve the quality of care. GE Healthcare's surgery business provides tools and technologies for general surgery, urology, cardiology, orthopedics, neurosurgery pain management and more.

FINANCIALS: Sales and profits are in thousands of dollars—add 000 to get the full amount. 2009 Note: Financial information for 2009 was not available for all companies at press time.

2009 Sales: $16,015,000	2009 Profits: $2,420,000	U.S. Stock Ticker: Subsidiary
2008 Sales: $17,392,000	2008 Profits: $2,851,000	Int'l Ticker: Int'l Exchange:
2007 Sales: $16,997,000	2007 Profits: $3,056,000	Employees: 46,000
2006 Sales: $16,560,000	2006 Profits: $3,142,000	Fiscal Year Ends: 12/31
2005 Sales: $15,016,000	2005 Profits: $2,601,000	Parent Company: GENERAL ELECTRIC CO (GE)

SALARIES/BENEFITS:

Pension Plan: Y	ESOP Stock Plan:	Profit Sharing:	Top Exec. Salary: $	Bonus: $
Savings Plan: Y	Stock Purch. Plan:		Second Exec. Salary: $	Bonus: $

OTHER THOUGHTS:

Apparent Women Officers or Directors: 2
Hot Spot for Advancement for Women/Minorities: Y

LOCATIONS: ("Y" = Yes)

West:	Southwest:	Midwest:	Southeast:	Northeast:	International:
Y		Y		Y	Y

GENENCOR INTERNATIONAL INC

www.genencor.com

Industry Group Code: 325414 Ranks within this company's industry group: Sales: Profits:

Drugs:		Other:		Clinical:	Computers:	Services:	
Discovery:	Y	AgriBio:		Trials/Services:	Hardware:	Specialty Services:	Y
Licensing:		Genetic Data:	Y	Labs:	Software:	Consulting:	
Manufacturing:	Y	Tissue Replacement:	Y	Equipment/Supplies:	Arrays:	Blood Collection:	
Generics:				Research/Development Svcs.:	Database Management:	Drug Delivery:	
				Diagnostics:		Drug Distribution:	

TYPES OF BUSINESS:

Biological Manufacturing
Biotech Research & Discovery
Enzyme-Based Products
Protein-Based Products
Bioethanol Technology Research & Development
Enzymes for Ethanol Production

BRANDS/DIVISIONS/AFFILIATES:

Danisco A/S
Multifect Protex
Accellerase
Optimax
Gensweet

CONTACTS: Note: Officers with more than one job title may be intentionally listed here more than once.

Tjerk De Ruiter, CEO
Jim Sjoerdsma, VP-Human Resources
Michael V. Arbige, Exec. VP-Tech.
Soonhee Jang, VP-Intellectual Property/Chief IP Counsel
Philippe Lavielle, Exec. VP-Bus. Dev.
Andrew Ashworth, VP-Finance
James Laughton, Exec. VP-Animal Nutrition/Food & Beverage Enzymes
Glenn Nedwin, Exec. VP-Technical Enzymes
Ken Herfert, Sr. VP-Supply

Phone: 585-256-5200	Fax: 585-256-5265
Toll-Free:	
Address: 3490 Wilston Pl., Rochester, NY 14623 US	

GROWTH PLANS/SPECIAL FEATURES:

Genencor International, Inc., the biotechnology division of Danish firm Danisco A/S, discovers, develops and sells biocatalysts and other biochemicals. It delivers 250 products to customers in 80 countries throughout the world. The firm operates three research and development centers located in the U.S and Europe and seven manufacturing centers located worldwide. Its products serve the following sectors: Agri Processing, Industrial Processing and Consumer Products. Agri Processing products serve customers who process agricultural raw materials to produce animal feeds, food, food ingredients and renewable fuels. Specific enzymes include Multifect Protex, used in rice and soybean processing; Accellerase, used to enhance the production of ethanol from biomass; Optimax, which converts starches into glucose; and Gensweet, which coverts glucose into fructose. Industrial Processing products include enzyme-based solutions for use in the textiles industry; wastewater treatment; oil and gas production; medical instrument cleaning; and as decontamination agents in chemical and biological warfare attack situations. Consumer Products applications include enzymes-based solutions for use in fabric and household care products and personal care items such as contact lens cleaner, whitening toothpaste and diabetic test kits. The firm maintains strategic alliances with several companies including: Procter and Gamble; DuPont; and Eastman Chemical. In 2009, the company introduced many new products including: Gentle Power Bleach, a low temperature peroxide bleach system for textiles developed in collaboration with Huntsman, a worldwide marketer and manufacturer of chemicals for a variety of industries; Accellerase XY, XC and BG accessory enzyme products which aid in the ethanol production from biomass feedstocks; PuraFast, a performance ingredient for laundry detergents which improves cleaning power in shorter cycles and lower temperatures; and PrimaGreen EcoLight 1 for more sustainable denim bleaching.

FINANCIALS: Sales and profits are in thousands of dollars—add 000 to get the full amount. 2009 Note: Financial information for 2009 was not available for all companies at press time.

2009 Sales: $	2009 Profits: $	U.S. Stock Ticker: Subsidiary
2008 Sales: $	2008 Profits: $	Int'l Ticker: Int'l Exchange:
2007 Sales: $90,700	2007 Profits: $	Employees:
2006 Sales: $	2006 Profits: $	Fiscal Year Ends: 12/31
2005 Sales: $	2005 Profits: $	Parent Company: DANISCO A/S

SALARIES/BENEFITS:

Pension Plan:	ESOP Stock Plan:	Profit Sharing:	Top Exec. Salary: $	Bonus: $
Savings Plan:	Stock Purch. Plan:		Second Exec. Salary: $	Bonus: $

OTHER THOUGHTS:

Apparent Women Officers or Directors: 1
Hot Spot for Advancement for Women/Minorities:

LOCATIONS: ("Y" = Yes)

West:	Southwest:	Midwest:	Southeast:	Northeast:	International:
Y		Y		Y	Y

Note: Financial information, benefits and other data can change quickly and may vary from those stated here.

GENENTECH INC

www.gene.com

Industry Group Code: 325412 Ranks within this company's industry group: Sales: Profits:

Drugs:		Other:	Clinical:	Computers:	Services:
Discovery:	Y	AgriBio:	Trials/Services:	Hardware:	Specialty Services:
Licensing:		Genetic Data:	Labs:	Software:	Consulting:
Manufacturing:	Y	Tissue Replacement:	Equipment/Supplies:	Arrays:	Blood Collection:
Generics:			Research/Development Svcs.:	Database Management:	Drug Delivery:
			Diagnostics:		Drug Distribution:

TYPES OF BUSINESS:

Drug Development & Manufacturing
Genetically Engineered Drugs

BRANDS/DIVISIONS/AFFILIATES:

Avastin
TNKase
Herceptin
Rituxan
Activase
Pulmozyme
Nutropin

CONTACTS: *Note: Officers with more than one job title may be intentionally listed here more than once.*

Ian T. Clark, CEO
Steve Krognes, CFO/Sr. VP
Denise Smith-Hams, Sr. VP-Human Resources
Richard H. Scheller, Exec. VP-Research & Early Dev.
Frederick C. Kentz, Sec./Sr. VP
Hal Barron, Head-Global Dev.
Frederick C. Kentz, Chief Compliance Officer
Hal Barron, Chief Medical Officer
Marc Tessier-Lavigne, Chief Scientific Officer/Exec. VP
Troy Cox, Sr. VP-Metabolism & Primary Care
Arthur D. Levinson, Chmn.

Phone: 650-225-1000	**Fax:** 650-225-6000
Toll-Free:	
Address: 1 DNA Way, South San Francisco, CA 94080-4990 US	

GROWTH PLANS/SPECIAL FEATURES:

Genentech, Inc., a wholly-owned subsidiary of the Roche Group, makes medicines by splicing genes into fast-growing bacteria that then produce therapeutic proteins and combat diseases on a molecular level. Genentech uses cutting-edge technologies such as computer visualization of molecules, micro arrays and sensitive assaying techniques to develop, manufacture and market pharmaceuticals for unmet medical needs. Genentech's research is directed toward the oncology, immunology and vascular biology fields. The company's products consist of a variety of cardio-centric medications, as well as cancer, growth hormone deficiency (GHD) and cystic fibrosis treatments. Biotechnology products offered by Genentech include Herceptin, used to treat metastatic breast cancers; Avastin, used to inhibit angiogenesis of solid-tumor cancers; Nutropin, a growth hormone for the treatment of GHD in children and adults; TNKase, for the treatment of acute myocardial infarction; and Pulmozyme, for the treatment of cystic fibrosis. The company also produces the Rituxan antibody, used for the treatment of patients with non-Hodgkin's lymphoma. Through its long-standing Genentech Access to Care Foundation, Genentech assists those without sufficient health insurance to receive its medicines. In March 2009, Roche Group completed its acquisition of the company, and as part of the merger agreement the two companies combined their U.S. pharmaceutical operations.

For the last ten years, the company has been named to Fortune Magazine's 100 Best Companies to Work For. Every Friday evening, Genentech hosts socials called Ho-Hos, providing free food, beverages and a chance to socialize with co-workers.

FINANCIALS: Sales and profits are in thousands of dollars—add 000 to get the full amount. 2009 Note: Financial information for 2009 was not available for all companies at press time.

2009 Sales: $	2009 Profits: $	**U.S. Stock Ticker: Subsidiary**
2008 Sales: $13,418,000	2008 Profits: $3,427,000	**Int'l Ticker:** Int'l Exchange:
2007 Sales: $11,724,000	2007 Profits: $2,769,000	Employees: 11,186
2006 Sales: $9,284,000	2006 Profits: $2,113,000	Fiscal Year Ends: 12/31
2005 Sales: $6,633,372	2005 Profits: $1,278,991	Parent Company: ROCHE GROUP

SALARIES/BENEFITS:

Pension Plan:	ESOP Stock Plan:	Profit Sharing:	Top Exec. Salary: $	Bonus: $
Savings Plan: Y	Stock Purch. Plan:		Second Exec. Salary: $	Bonus: $

OTHER THOUGHTS:

Apparent Women Officers or Directors: 12
Hot Spot for Advancement for Women/Minorities: Y

LOCATIONS: ("Y" = Yes)

West:	Southwest:	Midwest:	Southeast:	Northeast:	International:
Y					Y

Note: Financial information, benefits and other data can change quickly and may vary from those stated here.

GENEREX BIOTECHNOLOGY

www.generex.com

Industry Group Code: 325412A Ranks within this company's industry group: Sales: 17   Profits: 19

Drugs:		Other:		Clinical:	Computers:	Services:	
Discovery:	Y	AgriBio:		Trials/Services:	Hardware:	Specialty Services:	
Licensing:	Y	Genetic Data:		Labs:	Software:	Consulting:	
Manufacturing:	Y	Tissue Replacement:		Equipment/Supplies:	Arrays:	Blood Collection:	
Generics:				Research/Development Svcs.:	Database Management:	Drug Delivery:	Y
				Diagnostics:		Drug Distribution:	

TYPES OF BUSINESS:

Drug Delivery Systems
Buccal Drug Delivery Systems
Diabetes Treatment-Insulin
Infectious & Autoimmune Disease Treatments
Large-Molecule Drug Delivery Systems

BRANDS/DIVISIONS/AFFILIATES:

RapidMist
Oral-Lyn
Glucose RapidSpray
Antigen Express, Inc.
BaBOOM! Energy Spray
Crave-NX 7-Day Diet Aid Spray
SIA Generex Biotechnology Baltic

CONTACTS: Note: Officers with more than one job title may be intentionally listed here more than once.

Anna E. Gluskin, CEO
Rose C. Perri, COO
Anna E. Gluskin, Pres.
Rose C. Perri, CFO/Corp. Sec.
Mark Fletcher, General Counsel/Exec. VP
William D. Abajian, Sr. Exec. Advisor-Global Bus. Dev. & Alliances
Rose C. Perri, Treas.
Eric von Hofe, VP/Pres., Antigen Express
Gerald Bernstein, VP-Medical Affairs
Anna E. Gluskin, Chmn.

Phone: 416-364-2551	Fax: 416-364-9363
Toll-Free:	
Address: 33 Harbour Square, Ste. 202, Toronto, ON M5J 2G2 Canada	

GROWTH PLANS/SPECIAL FEATURES:

Generex Biotechnology Corporation primarily researches and develops drug delivery systems and technologies. It currently focuses on formulations that use its RapidMist hand-held aerosol applicator to administer large-molecule drugs through the buccal mucosa, primarily the inner cheek walls. For instance, Oral-Lyn is a formulation of human insulin designed for buccal delivery; it is commercially approved in Ecuador, India, Algeria and Lebanon, and is undergoing Phase III studies in the U.S. and Canada. The firm's Glucose RapidSpray product offers a fat-free, low-calorie glucose supplement for those who require extra glucose in their diet. The company has also begun development on buccal formulations of morphine and of the synthetic opioid analgesic fentanyl, which is around 80 times stronger than morphine. However, these projects are on hold until the late stage trials for its oral insulin products are completed for the U.S., Canada and Europe. In general, these oral formulations are designed to ease patients onto medication without invasive procedures, such as injections, and with the added benefit of being self-administered. Generex also owns Antigen Express, Inc., a Massachusetts-based company that develops treatments, primarily vaccine formulations, for malignant, infectious, autoimmune and allergic diseases. Diseases currently being studied and tested include melanoma, breast cancer, prostate cancer, HIV, influenza, avian influenza (or bird flu), H1N1 (swine flu) and Type-1 diabetes. Additional Generex products include BaBOOM! Energy Spray, designed to enhance energy levels for sports, work, travel and overall fatigue, with ingredients including glucose, caffeine, ginseng and Vitamins B and C. In February 2009, the company launched Crave-NX 7-Day Diet Aid Spray, designed to help control cravings and support customers' weight loss efforts. In September 2009, the firm established SIA Generex Biotechnology Baltic, a new European subsidiary. In October 2009, Generex opened a representative office in Latvia, its first European location.

FINANCIALS: Sales and profits are in thousands of dollars—add 000 to get the full amount. 2009 Note: Financial information for 2009 was not available for all companies at press time.

2009 Sales: $1,119	2009 Profits: $-45,812	**U.S. Stock Ticker: GNBT**
2008 Sales: $ 125	2008 Profits: $-36,229	**Int'l Ticker:** Int'l Exchange:
2007 Sales: $ 180	2007 Profits: $-23,505	Employees: 43
2006 Sales: $ 175	2006 Profits: $-67,967	Fiscal Year Ends: 7/31
2005 Sales: $ 392	2005 Profits: $-24,002	Parent Company:

SALARIES/BENEFITS:

Pension Plan:	ESOP Stock Plan:	Profit Sharing:	Top Exec. Salary: $525,000	Bonus: $
Savings Plan:	Stock Purch. Plan:		Second Exec. Salary: $420,000	Bonus: $

OTHER THOUGHTS:

Apparent Women Officers or Directors: 3
Hot Spot for Advancement for Women/Minorities: Y

LOCATIONS: ("Y" = Yes)

West:	Southwest:	Midwest:	Southeast:	Northeast:	International:
				Y	Y

GENOMIC HEALTH INC

www.genomichealth.com

Industry Group Code: 325413 Ranks within this company's industry group: Sales: 7 Profits: 16

Drugs:	Other:		Clinical:		Computers:	Services:	
Discovery:	AgriBio:		Trials/Services:		Hardware:	Specialty Services:	Y
Licensing:	Genetic Data:	Y	Labs:	Y	Software:	Consulting:	
Manufacturing:	Tissue Replacement:		Equipment/Supplies:		Arrays:	Blood Collection:	
Generics:			Research/Development Svcs.:		Database Management:	Drug Delivery:	
			Diagnostics:	Y		Drug Distribution:	

TYPES OF BUSINESS:

Genomic-Based Cancer Diagnostic Test Development

BRANDS/DIVISIONS/AFFILIATES:

Oncotype DX

CONTACTS: Note: Officers with more than one job title may be intentionally listed here more than once.

Kim Popovits, CEO
Brad Cole, COO
Kim Popovits, Pres.
Brad Cole, CFO
Tricia Tomlinson, Sr. VP-Human Resources
Joffre B. Baker, Chief Scientific Officer
Paul Aldridge, CIO
Kathy Hibbs, General Counsel/Sr. VP
Laura Leber, VP-Corp. Comm.
Dean Schorno, Sr. VP-Finance
Steven Shak, Chief Medical Officer
David Logan, Sr. VP-Worldwide Commercialization
Randal W. Scott, Chmn.

Phone: 650-556-9300	Fax: 650-556-1132
Toll-Free: 866-662-6897	
Address: 101 Galveston Dr., Redwood City, CA 94063 US	

GROWTH PLANS/SPECIAL FEATURES:

Genomic Health, Inc. focuses on the development and commercialization of genomic-based clinical diagnostic tests for cancer. Oncotype DX, its first and only marketed test product, is designed to help physicians determine the most effective treatment type for patients with node-negative, estrogen-receptor positive early stage breast cancer. Unlike comparable test assay systems that use fresh or frozen samples, Oncotype DX uses tissue samples that are chemically preserved and sealed in paraffin wax, thus allowing for easier handling and transportation. Hence, unlike other tests that may require a frozen sample shipped on dry ice, the Oncotype DX sample can be sent via regular overnight mail. All samples are processed at the firm's laboratory in Redwood City, California, and most physicians receive the results of their test within 10-14 days. Oncotype DX is offered as a laboratory service that tests 21 specific genes in a tumor sample to provide tumor-specific information (or the oncotype of the tumor), mainly to calculate the probability of recurrence and the potential efficacy of chemotherapy. This technology compares the genetic makeup of the tissue sample with archived information from other cancer patients, in order to facilitate a correlation between the patient's condition and known clinical outcomes. It expresses the resulting profile as a single quantitative score, called the Recurrence Score, which ranges from 0-100. Higher Recurrence Scores indicate a more aggressive tumor, lower scores indicate less aggressive tumors. The company is also developing tests to detect prostate, renal and other types of breast and colon cancers.

Genomic employees receiver medical, dental, vision, and disability coverage; a 401(k) plan; reimbursement accounts; employee assistance plans; and discounted gym memberships.

FINANCIALS: Sales and profits are in thousands of dollars—add 000 to get the full amount. 2009 Note: Financial information for 2009 was not available for all companies at press time.

2009 Sales: $194,548	2009 Profits: $-9,411	U.S. Stock Ticker: GHDX
2008 Sales: $110,579	2008 Profits: $-16,089	Int'l Ticker: Int'l Exchange:
2007 Sales: $64,027	2007 Profits: $-27,292	Employees:
2006 Sales: $29,174	2006 Profits: $-28,920	Fiscal Year Ends: 12/31
2005 Sales: $5,202	2005 Profits: $-31,361	Parent Company:

SALARIES/BENEFITS:

Pension Plan:	ESOP Stock Plan:	Profit Sharing:	Top Exec. Salary: $420,000	Bonus: $37,000
Savings Plan: Y	Stock Purch. Plan: Y		Second Exec. Salary: $380,000	Bonus: $33,400

OTHER THOUGHTS:

Apparent Women Officers or Directors: 4
Hot Spot for Advancement for Women/Minorities: Y

LOCATIONS: ("Y" = Yes)

West:	Southwest:	Midwest:	Southeast:	Northeast:	International:
Y					

GEN-PROBE INC

www.gen-probe.com

Industry Group Code: 325413 Ranks within this company's industry group: Sales: 4 Profits: 5

Drugs:	Other:	Clinical:	Computers:	Services:
Discovery:	AgriBio:	Trials/Services:	Hardware:	Specialty Services:
Licensing:	Genetic Data:	Labs:	Software:	Consulting:
Manufacturing:	Tissue Replacement:	Equipment/Supplies:	Arrays:	Blood Collection:
Generics:		Research/Development Svcs.:	Database Management:	Drug Delivery:
		Diagnostics: Y		Drug Distribution:

TYPES OF BUSINESS:

Medical Diagnostics Products
Diagnostic Tests
Blood Screening Assays
Services-Commercial Physical Research
Services-Commercial Biological Research

BRANDS/DIVISIONS/AFFILIATES:

TIGRIS
Tepnel Life Sciences, plc
Procleix
Roka Bioscience, Inc.
Prodesse, Inc.
Pacific Biosciences

CONTACTS: Note: Officers with more than one job title may be intentionally listed here more than once.

Carl W. Hull, CEO
Carl W. Hull, Pres.
Herm Rosenman, CFO/Sr. VP-Finance
Stephen J. Kondor, Sr. VP-Mktg. & Sales
Diana De Walt, Sr. VP-Human Resources
Daniel L. Kacian, Chief Scientist/Exec. VP
Brad Phillips, CIO/VP
R. William Bowen, General Counsel/Sr. VP/Sec.
Jorgine Ellerbrock, Sr. VP-Oper.
Paul Gargan, Sr. VP-Bus. Dev.
Michael Watts, VP-Corp. Comm.
Michael Watts, VP-Investor Rel.
Kevin Herde, VP-Finance/Corp. Controller
Eric Lai, Sr. VP-R&D
Eric Tardif, Sr. VP-Corp. Strategy
Christina Yang, Sr. VP-Clinical, Regulatory & Quality
Tammy J. Brach, VP-Program Mgmt.
Henry L. Nordhoff, Chmn.
Brian B. Hansen, VP-North American Sales

Phone: 858-410-8000	Fax: 800-288-3141
Toll-Free: 800-523-5001	
Address: 10210 Genetic Center Dr., San Diego, CA 92121 US	

GROWTH PLANS/SPECIAL FEATURES:

Gen-Probe, Inc. develops, manufactures and markets nucleic acid testing (NAT) products for clinical diagnosis of human diseases and screening of human blood donations. Gen-Probe has received U.S. Food and Drug Administration (FDA) approval for over 60 products to detect infectious microorganisms causing sexually transmitted diseases, tuberculosis, strep throat, pneumonia and fungal infections. It has developed and commercialized one of the first fully automated, integrated, high-throughput NAT instrument systems, the TIGRIS instrument. The company also developed, and now manufactures, the Procleix assay, the only FDA-approved blood screening assay for the simultaneous detection of HIV-1 and HCV (hepatitis C virus); and the Procleix Ultrio assay, which is used to detect the hepatitis B virus in addition to HIV-1 and HCV. The firm also designs and develops, often with outside vendors, a range of clinical diagnostics instruments for use with its assays. The company is currently developing NAT assays and instruments for the detection of human papillomavirus (HPV), gene-based markers for prostate cancer, Trichomonas, different types of influenza virus and human leukocyte antigens. In April 2009, Gen-Probe acquired Tepnel Life Sciences plc, a U.K.-based molecular diagnostics and pharmaceutical services company. In September 2009, the firm spun off its industrial testing assets into an independent company, Roka Bioscience, Inc., of which it owns 19.9%. In October 2009, Gen-Probe acquired Prodesse, Inc., which provides molecular testing services for influenza and other infectious diseases. In December 2009, the firm sold its BioKits food safety testing business to Neogen Corporation. In June 2010, the company announced a $50 million strategic investment in Pacific Biosciences, a private sequencing company.

Gen-Probe offers its employees medical, dental and vision insurance, flexible spending accounts, an employee assistance program, tuition reimbursement, wellness programs and on-site car wash, gym and cafeteria facilities.

FINANCIALS: Sales and profits are in thousands of dollars—add 000 to get the full amount. 2009 Note: Financial information for 2009 was not available for all companies at press time.

2009 Sales: $489,302	2009 Profits: $91,783	U.S. Stock Ticker: GPRO
2008 Sales: $472,695	2008 Profits: $106,954	Int'l Ticker: Int'l Exchange:
2007 Sales: $403,014	2007 Profits: $86,140	Employees: 991
2006 Sales: $354,764	2006 Profits: $59,498	Fiscal Year Ends: 12/31
2005 Sales: $305,965	2005 Profits: $60,089	Parent Company:

SALARIES/BENEFITS:

Pension Plan:	ESOP Stock Plan:	Profit Sharing:	Top Exec. Salary: $611,507	Bonus: $150,000
Savings Plan: Y	Stock Purch. Plan: Y		Second Exec. Salary: $586,700	Bonus: $465,773

OTHER THOUGHTS:

Apparent Women Officers or Directors: 6
Hot Spot for Advancement for Women/Minorities: Y

LOCATIONS: ("Y" = Yes)

West:	Southwest:	Midwest:	Southeast:	Northeast:	International:
Y				Y	Y

Note: Financial information, benefits and other data can change quickly and may vary from those stated here.

GENTA INC

www.genta.com

Industry Group Code: 325412 Ranks within this company's industry group: Sales: 149 Profits: 156

Drugs:		Other:	Clinical:	Computers:	Services:
Discovery:	Y	AgriBio:	Trials/Services:	Hardware:	Specialty Services:
Licensing:		Genetic Data:	Labs:	Software:	Consulting:
Manufacturing:		Tissue Replacement:	Equipment/Supplies:	Arrays:	Blood Collection:
Generics:			Research/Development Svcs.:	Database Management:	Drug Delivery:
			Diagnostics:		Drug Distribution:

TYPES OF BUSINESS:

Anticancer & Related Diseases Drugs
Antisense Drugs

BRANDS/DIVISIONS/AFFILIATES:

Genasense
Ganite
Tesetaxel

CONTACTS: *Note: Officers with more than one job title may be intentionally listed here more than once.*

Raymond P. Warrell, Jr., CEO
Loretta M. Itri, Chief Medical Officer/Pres., Pharmaceutical Dev.
Edward C. Spindler, Jr., Associate VP-Tech. Dev.
Gary Siegel, VP-Finance
Jane Z. Wu, Associate VP-Data Mgmt. & Statistics
Edward C. Spindler, Jr., Associate VP-Project Mgmt.
Raymond P. Warrell, Jr., Chmn.

Phone: 908-286-9800	Fax: 908-464-1701
Toll-Free:	
Address: 200 Connell Dr., Berkeley Heights, NJ 07922 US	

GROWTH PLANS/SPECIAL FEATURES:

Genta, Inc. is a biopharmaceutical company engaged in pharmaceutical research and development of drugs for the treatment of cancer and related diseases. The company's research portfolio consists of two major programs: DNA/RNA medicines and small molecules. The DNA/RNA medicines program includes drugs that are based on using modifications of either DNA or RNA as drugs that can be used to treat disease. The program includes technologies such as antisense, decoys and small interfering or micro RNAs. The lead drug from this program is an investigational antisense compound known as Genasense. Genasense is designed to block the production of Bcl-2, a protein that fortifies cancer cells against treatment, and is being developed primarily as a means of amplifying the cytotoxic effects of other anticancer treatments. The small molecules program includes drugs that are based on gallium-containing compounds. The lead drug from this program is Ganite, which is FDA approved for the treatment of patients with symptomatic cancer-related hypercalcemia that is resistant to hydration. The firm is engaged in developing new formulations of gallium-containing compounds that may be orally absorbed. The company is also developing Tesetaxel, which is an oral version of the prominent anti-cancer taxane, currently available only in intravenous forms. Genta owns 65 patents, and has 66 pending patents applications in the U.S. and foreign countries.

Genta offers its employees medical, prescription, vision and dental coverage; disability plans; life insurance; a 401(k) program; flexible spending accounts; an employee assistance program; and educational assistance.

FINANCIALS: Sales and profits are in thousands of dollars—add 000 to get the full amount. 2009 Note: Financial information for 2009 was not available for all companies at press time.

2009 Sales: $ 218	2009 Profits: $-86,301	**U.S. Stock Ticker: GETA**
2008 Sales: $ 363	2008 Profits: $-505,838	**Int'l Ticker:** Int'l Exchange:
2007 Sales: $ 580	2007 Profits: $-23,320	Employees: 16
2006 Sales: $ 708	2006 Profits: $-56,781	Fiscal Year Ends: 12/31
2005 Sales: $26,585	2005 Profits: $-2,203	Parent Company:

SALARIES/BENEFITS:

Pension Plan:	ESOP Stock Plan:	Profit Sharing:	Top Exec. Salary: $483,683	Bonus: $
Savings Plan: Y	Stock Purch. Plan:		Second Exec. Salary: $422,123	Bonus: $

OTHER THOUGHTS:

Apparent Women Officers or Directors: 2
Hot Spot for Advancement for Women/Minorities: Y

LOCATIONS: ("Y" = Yes)

West:	Southwest:	Midwest:	Southeast:	Northeast: Y	International:

GENVEC INC

www.genvec.com

Industry Group Code: 325412 Ranks within this company's industry group: Sales: 114 Profits: 113

Drugs:		Other:		Clinical:	Computers:	Services:
Discovery:	Y	AgriBio:		Trials/Services:	Hardware:	Specialty Services:
Licensing:		Genetic Data:	Y	Labs:	Software:	Consulting:
Manufacturing:	Y	Tissue Replacement:		Equipment/Supplies:	Arrays:	Blood Collection:
Generics:				Research/Development Svcs.:	Database Management:	Drug Delivery:
				Diagnostics:		Drug Distribution:

TYPES OF BUSINESS:

Gene-Based Therapeutic Drugs & Vaccines
Cancer Drugs

BRANDS/DIVISIONS/AFFILIATES:

TNFerade
TherAtoh

CONTACTS: *Note: Officers with more than one job title may be intentionally listed here more than once.*

Paul H. Fischer, CEO
Paul H. Fischer, Pres.
Douglas J. Swirsky, CFO/Sr. VP
Douglas J. Swirsky, Corp. Sec.
Douglas J. Swirsky, Treas.
Milan Kovacevic, VP-Clinical Oper.
Bryan T. Butman, Sr. VP-Vector Oper.
Zola P. Horovitz, Chmn.

Phone: 240-632-0740	Fax: 240-632-0735
Toll-Free: 877-943-6832	
Address: 65 W. Watkins Mill Rd., Gaithersburg, MD 20878 US	

GROWTH PLANS/SPECIAL FEATURES:

GenVec, Inc. is a biopharmaceutical company that develops gene-based therapeutic drugs and vaccines. The company's lead product candidate, TNFerade biologic, is being developed for use in the treatment of cancer. The drug is an adenovector, or DNA carrier, and is administered directly into tumors. TNFerade is currently the subject of clinical trials to assess the effectiveness of using TNFerade in combination with standard care treatment for patients in cases of prostate; head and neck; esophageal; and rectal cancer. GenVec is also developing TherAtoh, a therapy for delivering the human atonal gene to trigger the production of therapeutic proteins by cells in the inner ear. GenVec and its collaborators also have multiple vaccines in development, which use the firm's adenovector technology. The company has collaborations with the National Institute of Allergy and Infectious Diseases (NIAID) to develop vaccines for the HIV virus; the U.S. Naval Medical Research Center to develop vaccines for malaria; and with the U.S. Department of Homeland Security and the U.S. Department of Agriculture to develop a vaccine for food-and-mouth disease. The firm has access to over 268 issued, allowed or pending patents worldwide. In March 2010, the company discontinued its Phase III clinical trials for TNFerade in pancreatic cancer due to disappointing results.

The company offers its employees medical, dental, prescription and vision insurance; life, AD&D and short- and long-term disability insurance; a 401(k); an employee assistance program; tuition reimbursement; and an employee stock purchase program. Additional benefits include car washes, dry cleaning services and credit union membership.

FINANCIALS: Sales and profits are in thousands of dollars—add 000 to get the full amount. 2009 Note: Financial information for 2009 was not available for all companies at press time.

2009 Sales: $13,857	2009 Profits: $-18,362	**U.S. Stock Ticker: GNVC**
2008 Sales: $15,121	2008 Profits: $-26,063	**Int'l Ticker:** Int'l Exchange:
2007 Sales: $14,047	2007 Profits: $-18,708	Employees: 90
2006 Sales: $18,923	2006 Profits: $-19,272	Fiscal Year Ends: 12/31
2005 Sales: $26,554	2005 Profits: $-13,992	Parent Company:

SALARIES/BENEFITS:

Pension Plan:	ESOP Stock Plan:	Profit Sharing:	Top Exec. Salary: $410,000	Bonus: $205,000
Savings Plan: Y	Stock Purch. Plan: Y		Second Exec. Salary: $304,950	Bonus: $85,000

OTHER THOUGHTS:

Apparent Women Officers or Directors:
Hot Spot for Advancement for Women/Minorities:

LOCATIONS: ("Y" = Yes)

West:	Southwest:	Midwest:	Southeast:	Northeast:	International:
				Y	

GENZYME BIOSURGERY

www.genzyme.com/business/biosurgery/biosurg_home.asp

Industry Group Code: 33911 Ranks within this company's industry group: Sales: Profits:

Drugs:	Other:		Clinical:		Computers:	Services:
Discovery:	AgriBio:		Trials/Services:		Hardware:	Specialty Services:
Licensing:	Genetic Data:		Labs:		Software:	Consulting:
Manufacturing:	Tissue Replacement:	Y	Equipment/Supplies:	Y	Arrays:	Blood Collection:
Generics:			Research/Development Svcs.:		Database Management:	Drug Delivery:
			Diagnostics:			Drug Distribution:

TYPES OF BUSINESS:

Equipment-Surgery & Orthopedic Products
Burn Treatment Products
Biomaterials & Biotherapeutics

BRANDS/DIVISIONS/AFFILIATES:

Genzyme Corp.
Synvisc
Seprafilm
Seprapack
Sepragel ENT
Carticel
Epicel
HyluMed

CONTACTS: Note: Officers with more than one job title may be intentionally listed here more than once.

Alison Lawton, Gen. Mgr./Sr. VP
Henri A. Termeer, Chmn./CEO/Pres., Genzyme Corp.

Phone: 617-768-1000	Fax: 617-591-5986
Toll-Free:	
Address: 55 Cambridge Pkwy., Cambridge, MA 02142 US	

GROWTH PLANS/SPECIAL FEATURES:

Genzyme Biosurgery (GB), a division of Genzyme Corp. develops and markets a portfolio of devices, biomaterials and biotherapeutics primarily for the general surgery market. GB's products are used for osteoarthritis relief, adhesion prevention, hernia repair, cartilage repair and burn treatment. It derives substantially all of its revenue from the following products: Synvisc and Synvisc-One (hylan G-F 20); Jonexa (hylastan SGL-80); and Sepra. Synvisc and Synvisc-One are biomaterial-based products derived from hyaluronan that are used to treat the pain associated with osteoarthritis (OA) of the knee. Synvisc-One, a single-injection product, has been approved by the U.S., the European Union, and various Asian and Latin American countries. Synvisc is sold commercially in approximately 40 countries, and Synvisc-One is sold in 20 countries. Jonexa is a bacterially fermented product derived from hyaluronan that is indicated for the treatment of pain associated with OA of the knee and is administered in one or two injections. The Sepra family of products consists of the Sepramesh IP, which is manufactured and marketed by Davol, Inc. under a license agreement; Seprafilm, one of the only FDA-approved products clinically proven to reduce the incidence, extent and severity of postsurgical adhesions following abdominal and pelvic surgery; Seprapack, which is indicated for use in patients undergoing nasal/sinus surgery; and Sepragel ENT. The firm also produces Carticel, which helps repair damaged knee cartilage; Epicel, a permanent skin replacement product for patients with life-threatening burns; and the HyluMed product line, which consists of medical-grade and sterile HA powder. The company markets its products directly to physicians and hospital administrators throughout the U.S. and Europe. It also uses Genzyme Corp.'s network of distributors to sell certain products in the U.S., Europe, Asia and Latin America.

FINANCIALS: Sales and profits are in thousands of dollars—add 000 to get the full amount. 2009 Note: Financial information for 2009 was not available for all companies at press time.

2009 Sales: $	2009 Profits: $	U.S. Stock Ticker: Subsidiary
2008 Sales: $	2008 Profits: $	Int'l Ticker: Int'l Exchange:
2007 Sales: $	2007 Profits: $	Employees:
2006 Sales: $	2006 Profits: $	Fiscal Year Ends: 12/31
2005 Sales: $	2005 Profits: $	Parent Company: GENZYME CORP

SALARIES/BENEFITS:

Pension Plan:	ESOP Stock Plan:	Profit Sharing:	Top Exec. Salary: $	Bonus: $
Savings Plan: Y	Stock Purch. Plan: Y		Second Exec. Salary: $	Bonus: $

OTHER THOUGHTS:

Apparent Women Officers or Directors: 1
Hot Spot for Advancement for Women/Minorities: Y

LOCATIONS: ("Y" = Yes)

West:	Southwest:	Midwest:	Southeast:	Northeast:	International:
				Y	

GENZYME CORP

www.genzyme.com

Industry Group Code: 325412 Ranks within this company's industry group: Sales: 22   Profits: 27

Drugs:		Other:		Clinical:		Computers:		Services:	
Discovery:	Y	AgriBio:		Trials/Services:		Hardware:		Specialty Services:	
Licensing:		Genetic Data:	Y	Labs:		Software:		Consulting:	
Manufacturing:	Y	Tissue Replacement:		Equipment/Supplies:	Y	Arrays:		Blood Collection:	
Generics:				Research/Development Svcs.:		Database Management:		Drug Delivery:	
				Diagnostics:	Y			Drug Distribution:	

TYPES OF BUSINESS:

Pharmaceuticals Discovery & Development
Genetic Disease Treatments
Surgical Products
Diagnostic Products
Genetic Testing Services
Oncology Products
Biomaterials
Medical Devices

BRANDS/DIVISIONS/AFFILIATES:

Renagel
Cerezyme
Fabrazyme
Mozobil
Thyrogen
Campath/MabCampath
Fludara
Leukine

CONTACTS: Note: Officers with more than one job title may be intentionally listed here more than once.

Henri A. Termeer, CEO
David Meeker, COO/Exec. VP
Henri A. Termeer, Pres.
Michael S. Wyzga, CFO/Exec. VP-Finance
Zoltan Csimma, Chief Human Resources Officer/Sr. VP
Alan E. Smith, Chief Scientific Officer/Sr. VP-Research
Scott Canute, Pres., Global Mfg.
Thomas J. DesRosier, Chief Legal Officer/General Counsel/Sr. VP
Scott Canute, Pres., Corp. Oper.
Peter Worth, Exec. VP-Corp. Dev & Legal/Sec.
Mary McGrane, Sr. VP-Gov't Rel.
Jason A. Amello, Chief Acct. Officer/Corp. Controller
Mark J. Enyedy, Sr. VP/Pres., Oncology & Multiple Sclerosis
John Butler, Sr. VP/Pres., Cardiometabolic & Renal
Donald E. Pogorzelski, VP/Pres., Diagnostic Prod.
Richard H. Douglas, Sr. VP-Corp. Dev.
Henri A. Termeer, Chmn.
Sandford D. Smith, Pres., Int'l Group

Phone: 617-252-7500	Fax: 617-252-7600
Toll-Free:	
Address: 500 Kendall St., Cambridge, MA 02142 US	

GROWTH PLANS/SPECIAL FEATURES:

Genzyme Corporation is a major biotech drug manufacturer operating through four major units: Genetic Diseases; Cardiometabolic and Renal; Biosurgery; and Hematologic Oncology. The Genetic Diseases segment develops therapeutic products to treat patients suffering from genetic and other chronic debilitating diseases including lysosomal disorders (LSDs). Products include Cerezyme, an enzyme replacement treatment for Type 1 Gaucher disease and Fabrazyme for Fabry disease. The Cardiometabolic and Renal unit produces products for patients suffering from renal disease, including chronic renal failure, and endocrine and cardiovascular diseases. Products include Renagel, a calcium-free, metal-free phosphate binder for patients with Chronic Kidney disease on hemodialysis; and Thyrogen, an injection used as a diagnostic in follow-up screenings of cancer patients with thyroid cancer. The Biosurgery division develops biotherapeutics and biomaterial-based products for the orthopedics sector and broader surgical areas. Its main products are Synvisc, a lubricant and pain reducer for the knee joint in patients with osteoarthritic knees; and the MACI implant, which uses the patient's own culture cartilage cells to repair a damaged knee joint. The Hematologic Oncology segment focuses on the treatment of cancer. Its main products consists of Mozobil, an injection used to mobilize stem cells in the blood stream for collection and for transplantation in patients with non-Hodgkin's lymphoma and multiple myeloma; Campath, for the treatment of B-cell chronic lymphocytic leukemia; Fludara for chronic lymphocytic leukemia, Leukine, for acute myelogenous leukemia (AML) and for autologous bone marrow transplants; and Clolar, a treatment for children with acute lymphoblastic leukemia. In June 2009, the company acquired the rights to Campath/MabCampath multiple sclerosis drug candidate; as well as the marketing and distribution rights to Fludara and Leukine hematologic oncology drugs from Bayer HealthCare.

Employees are offered medical and dental insurance; life insurance; flexible spending accounts; disability coverage; financial education programs; adoption leave; tuition reimbursement; and an employee assistance program.

FINANCIALS: Sales and profits are in thousands of dollars—add 000 to get the full amount. 2009 Note: Financial information for 2009 was not available for all companies at press time.

2009 Sales: $4,515,525	2009 Profits: $422,300	**U.S. Stock Ticker: GENZ**
2008 Sales: $4,605,039	2008 Profits: $421,081	**Int'l Ticker:** Int'l Exchange:
2007 Sales: $3,813,519	2007 Profits: $480,193	Employees: 12,000
2006 Sales: $3,187,013	2006 Profits: $-16,797	Fiscal Year Ends: 12/31
2005 Sales: $2,734,842	2005 Profits: $441,489	Parent Company:

SALARIES/BENEFITS:

Pension Plan:	ESOP Stock Plan:	Profit Sharing:	Top Exec. Salary: $1,578,514	Bonus: $1,962,725
Savings Plan: Y	Stock Purch. Plan: Y		Second Exec. Salary: $734,331	Bonus: $489,500

OTHER THOUGHTS:

Apparent Women Officers or Directors: 5
Hot Spot for Advancement for Women/Minorities: Y

LOCATIONS: ("Y" = Yes)

West:	Southwest:	Midwest:	Southeast:	Northeast:	International:
Y	Y	Y	Y	Y	Y

Note: Financial information, benefits and other data can change quickly and may vary from those stated here.

GENZYME ONCOLOGY

www.genzymeoncology.com

Industry Group Code: 325412　Ranks within this company's industry group: Sales:　　Profits:

Drugs:		Other:		Clinical:	Computers:	Services:
Discovery:	Y	AgriBio:		Trials/Services:	Hardware:	Specialty Services:
Licensing:		Genetic Data:	Y	Labs:	Software:	Consulting:
Manufacturing:	Y	Tissue Replacement:		Equipment/Supplies:	Arrays:	Blood Collection:
Generics:				Research/Development Svcs.:	Database Management:	Drug Delivery:
				Diagnostics:		Drug Distribution:

TYPES OF BUSINESS:

Drugs-Gene-Based
Cancer Vaccines
Angiogenesis Inhibitors
Drug Discovery Platforms

BRANDS/DIVISIONS/AFFILIATES:

Genzyme Corp
Tasidotin Hydrochloride
Campath
Clolar
Mozobil
Fludara
Leukine

CONTACTS: *Note: Officers with more than one job title may be intentionally listed here more than once.*

Mark J. Enyedy, Pres.
Frederic J. Vinick, Sr. VP-Drug Discovery
Terry L. Murdock, Sr. VP/Gen. Mgr.-Prod.

Phone: 617-761-8777	Fax: 617-761-8918
Toll-Free:	
Address: 55 Cambridge Pkwy., Cambridge, MA 02142 US	

GROWTH PLANS/SPECIAL FEATURES:

Genzyme Oncology, a division of Genzyme Corp., creates cancer vaccines and angiogenesis inhibitors through the integration of its genomics, gene and cell therapy; small-molecule drug discovery; and protein therapeutic capabilities. The company's research candidates consist of Tasidotin Hydrochloride, GC1008, Prochymal and Topoisomeric I Inhibitor. Tasidotin Hydrochloride is a synthetic dolastatin analog currently being studied for treatment of advanced, refractory neoplasms. GC1008 is a human antibody which combats all forms of Transforming Growth Factor, which is itself responsible for a patient's inability to produce an immune response to cancer tumors. Prochymal is a late-stage adult stem cell treatment designed to control inflammation while promoting tissue regeneration. It is in clinical trials for treatment of graft vs. host disease and Crohn's disease. The Topoisomerase I Inhibitor, currently in Phase I development, prevents Topoisomeric I enzymes from replicating DNA. The company's acquisition of ILEX Oncology, Inc. gave the firm its marketed cancer drugs Campath and Clolar. Campath is indicated for the treatment of B-cell chronic lymphocytic leukemia, while Clolar is used for the treatment of children with refractory or relapsed acute lymphoblastic leukemia. Additional products include, Mozobil, Fludara and Leukine, which are designed for various hematological oncology treatments.

FINANCIALS: Sales and profits are in thousands of dollars—add 000 to get the full amount. 2009 Note: Financial information for 2009 was not available for all companies at press time.

2009 Sales: $	2009 Profits: $	**U.S. Stock Ticker: Subsidiary**
2008 Sales: $	2008 Profits: $	**Int'l Ticker:**　Int'l Exchange:
2007 Sales: $	2007 Profits: $	Employees:
2006 Sales: $	2006 Profits: $	Fiscal Year Ends: 12/31
2005 Sales: $	2005 Profits: $	Parent Company: GENZYME CORP

SALARIES/BENEFITS:

Pension Plan:	ESOP Stock Plan:	Profit Sharing:	Top Exec. Salary: $	Bonus: $
Savings Plan: Y	Stock Purch. Plan: Y		Second Exec. Salary: $	Bonus: $

OTHER THOUGHTS:

Apparent Women Officers or Directors:
Hot Spot for Advancement for Women/Minorities:

LOCATIONS: ("Y" = Yes)

West:	Southwest:	Midwest:	Southeast:	Northeast:	International:
				Y	

GERON CORPORATION

www.geron.com

Industry Group Code: 325412 Ranks within this company's industry group: Sales: 140 Profits: 149

Drugs:		Other:		Clinical:		Computers:		Services:	
Discovery:	Y	AgriBio:		Trials/Services:		Hardware:		Specialty Services:	
Licensing:	Y	Genetic Data:		Labs:		Software:		Consulting:	
Manufacturing:		Tissue Replacement:	Y	Equipment/Supplies:		Arrays:		Blood Collection:	
Generics:				Research/Development Svcs.:		Database Management:		Drug Delivery:	
				Diagnostics:				Drug Distribution:	

TYPES OF BUSINESS:

Drug Discovery & Development
Telomerase Technologies
Human Stem Cell Technologies

BRANDS/DIVISIONS/AFFILIATES:

GRNCM1
GRN163L
GRNVAC1
GRNOPC1

CONTACTS: *Note: Officers with more than one job title may be intentionally listed here more than once.*

Thomas B. Okarma, CEO
Thomas B. Okarma, Pres.
David L. Greenwood, CFO/Exec. VP/Sec.
Jane S. Lebkowski, Chief Scientific Officer
David J. Earp, Chief Patent Counsel
Katharine E. Spink, VP-Oper.
David J. Earp, Sr. VP-Bus. Dev.
David L. Greenwood, Treas.
Stephen M. Kelsey, Chief Medical Officer/Exec. VP-Oncology
Katharine E. Spink, VP-Regenerative Medicine
Alexander E. Barkas, Chmn.

Phone: 650-473-7700	Fax: 650-473-7750
Toll-Free:	
Address: 230 Constitution Dr., Menlo Park, CA 94025 US	

GROWTH PLANS/SPECIAL FEATURES:

Geron Corp. develops biopharmaceuticals for the treatment of cancer and chronic degenerative diseases such as spinal cord injuries, heart failure and diabetes. The company's therapies are based on telomerase and human embryonic stem cell technologies. The company is advancing telomerase targeted therapies, which enable cell division, protect chromosomes from degradation and act as molecular clocks for cellular aging. The enzyme telomerase restores telomere length, which shortens as cells multiply, extending a cell's ability to replicate. The company seeks to use human embryonic stem cells (hESC) as a potential source for manufacturing replacement cells and tissues for organ repair. The firm is now testing hESC-derived therapeutic cell types for oncology and regenerative applications. The company has two oncology candidates in clinical trials: GRN163L, a telomerase inhibitor drug, in patients with advanced non-small cell lung cancer and other ailments; and GRNVAC1, a developmental vaccine for leukemia. In regenerative medicine, the Geron is testing GRNOPC1, a targeted treatment for spinal cord injuries, and has multiple others in developmental stages, including GRNCM1, a treatment for patients with myocardial disease; GRNIC1 for diabetes, Osteoblasts and Chondrocytes for osteoporosis; and research towards a better understanding of hESC treatments in humans. The firm owns or licenses over 184 U.S. and 367 foreign patents, with more than 341 pending applications worldwide.

Geron Corp. offers its employees medical, dental, vision, life and AD&D insurance; long-term disability; 401(k) plan; and employee stock purchase plan; and flexible spending accounts.

FINANCIALS: Sales and profits are in thousands of dollars—add 000 to get the full amount. 2009 Note: Financial information for 2009 was not available for all companies at press time.

2009 Sales: $1,726	2009 Profits: $-70,184	**U.S. Stock Ticker: GERN**
2008 Sales: $2,803	2008 Profits: $-62,021	**Int'l Ticker:** Int'l Exchange:
2007 Sales: $7,622	2007 Profits: $-36,697	Employees: 172
2006 Sales: $3,277	2006 Profits: $-31,365	Fiscal Year Ends: 12/31
2005 Sales: $6,158	2005 Profits: $-33,689	Parent Company:

SALARIES/BENEFITS:

Pension Plan:	ESOP Stock Plan:	Profit Sharing:	Top Exec. Salary: $535,000	Bonus: $282,500
Savings Plan: Y	Stock Purch. Plan: Y		Second Exec. Salary: $415,000	Bonus: $164,300

OTHER THOUGHTS:

Apparent Women Officers or Directors: 3
Hot Spot for Advancement for Women/Minorities: Y

LOCATIONS: ("Y" = Yes)

West:	Southwest:	Midwest:	Southeast:	Northeast:	International:
Y					

GILEAD SCIENCES INC

www.gilead.com

Industry Group Code: 325412 Ranks within this company's industry group: Sales: 19   Profits: 13

Drugs:		Other:		Clinical:		Computers:		Services:	
Discovery:	Y	AgriBio:		Trials/Services:		Hardware:		Specialty Services:	
Licensing:	Y	Genetic Data:		Labs:		Software:		Consulting:	
Manufacturing:	Y	Tissue Replacement:		Equipment/Supplies:		Arrays:		Blood Collection:	
Generics:				Research/Development Svcs.:	Y	Database Management:		Drug Delivery:	
				Diagnostics:				Drug Distribution:	

TYPES OF BUSINESS:

Viral & Bacterial Infections Drugs
Respiratory & Cardiopulmonary Diseases Drugs

BRANDS/DIVISIONS/AFFILIATES:

Vistide
Truvada
Emtriva
Atripla
Hepsera
Viread
Lexiscan
Ranexa

CONTACTS: *Note: Officers with more than one job title may be intentionally listed here more than once.*

John C. Martin, CEO
John F. Milligan, COO
John F. Milligan, Pres.
Robin L. Washington, CFO/Sr. VP
Kristen M. Metza, Sr. VP-Human Resources
Norbert W. Bischofberger, Exec. VP-R&D/Chief Scientific Officer
Anthony D. Caracciolo, Sr. VP-Mfg.
Anthony D. Caracciolo, Sr. VP-Oper.
Gregg H. Alton, Exec. VP-Corp. & Medical Affairs
Kevin Young, Exec. VP-Commercial Oper.
A. Bruce Montgomery, Sr. VP/Head-Respiratory Therapeutics
Seigo Izump, Sr. VP-Cardiovascular Therapeutics
John C. Martin, Chmn.
Paul Carter, Sr. VP-Int'l Commercial Oper.

Phone: 650-574-3000	Fax: 650-578-9264
Toll-Free: 800-445-3235	
Address: 333 Lakeside Dr., Foster City, CA 94404 US	

GROWTH PLANS/SPECIAL FEATURES:

Gilead Sciences, Inc. is a biopharmaceutical company that discovers, develops and commercializes therapeutics for the treatment of life-threatening diseases such as viral and bacterial infections. The company expanded its efforts to include respiratory and cardiopulmonary diseases. The firm maintains research, development, manufacturing, sales and marketing facilities in the U.S., Europe and Australia and operates marketing subsidiaries in another 12 countries. Gilead currently has 10 products on the market: Viread, Truvada and Emtriva, which are oral medicines used as part of a combination therapy to treat HIV; Atripla, an oral formulation for treatment of HIV; Hespera, an oral medication used for treatment of Hepatitis B; AmBisome, an antifungal agent to treat serious invasive fungal infections; Vistide, an antiviral medication for the treatment of cytomegalovirus retinitis in patients with AIDS; Ranexa, a treatment for chronic angina; Letairis, for the treatment of pulmonary arterial hypertension; and Cayston, an inhaled antibiotic as a treatment to improve respiratory systems in cystic fibrosis patients with Pseudomonas aeruginosa. The firm has a number of products in development, including treatments for hypertension, heart failure, hepatitis C, HIV/AIDS, and others. The company also derives revenues from licensing agreements for Macugen, a macular degeneration treatment developed by OSI Pharmaceuticals, Inc.; Lexiscan, an used as a pharmacologic stress agent in radionuclide myocardial perfusion imaging ;and Tamiflu, an influenza medication sold by F. Hoffman-LaRoche. In April 2009, the firm acquired CV Therapeutics.

The company offers its employees medical, vision, dental, disability, life and AD&D insurance; a 401(k); a stock purchase plan; an entertainment discount program; an employee assistance plan; and tuition reimbursement.

FINANCIALS: Sales and profits are in thousands of dollars—add 000 to get the full amount. 2009 Note: Financial information for 2009 was not available for all companies at press time.

2009 Sales: $7,011,383	2009 Profits: $2,635,755	**U.S. Stock Ticker: GILD**
2008 Sales: $5,335,750	2008 Profits: $1,978,899	**Int'l Ticker:** Int'l Exchange:
2007 Sales: $4,230,045	2007 Profits: $1,584,902	Employees: 3,852
2006 Sales: $3,026,139	2006 Profits: $-1,189,957	Fiscal Year Ends: 12/31
2005 Sales: $2,028,400	2005 Profits: $813,914	Parent Company:

SALARIES/BENEFITS:

Pension Plan:	ESOP Stock Plan:	Profit Sharing:	Top Exec. Salary: $1,242,095	Bonus: $2,187,500
Savings Plan: Y	Stock Purch. Plan: Y		Second Exec. Salary: $805,008	Bonus: $850,500

OTHER THOUGHTS:

Apparent Women Officers or Directors: 4
Hot Spot for Advancement for Women/Minorities: Y

LOCATIONS: ("Y" = Yes)

West:	Southwest:	Midwest:	Southeast:	Northeast:	International:
Y				Y	Y

GLAXOSMITHKLINE PLC

www.gsk.com

Industry Group Code: 325412 Ranks within this company's industry group: Sales: 5 Profits: 5

Drugs:		Other:	Clinical:	Computers:	Services:
Discovery:	Y	AgriBio:	Trials/Services:	Hardware:	Specialty Services:
Licensing:		Genetic Data:	Labs:	Software:	Consulting:
Manufacturing:	Y	Tissue Replacement:	Equipment/Supplies:	Arrays:	Blood Collection:
Generics:			Research/Development Svcs.:	Database Management:	Drug Delivery:
			Diagnostics:		Drug Distribution:

TYPES OF BUSINESS:

Prescription Medications
Asthma Drugs
Respiratory Drugs
Antibiotics
Antivirals
Dermatological Drugs
Over-the-Counter & Nutritional Products

BRANDS/DIVISIONS/AFFILIATES:

Lanoxin
Lamictal
Adoair
Alvedon
Zantac
Nicorette
Ceravix
Genelabs Technologies

CONTACTS: Note: Officers with more than one job title may be intentionally listed here more than once.

Andrew Witty, CEO
Julian Heslop, CFO
Moncef Slaoui, Chmn.-R&D
Bill Louv, CIO
David Pulman, Pres., Global Mfg. & Supply
Daniel Phelan, Chief of Staff
Simon Bicknell, Company Sec./Compliance Officer/Sr. VP
David Redfern, Chief Strategy Officer
Duncan Learmouth, Sr. VP-Global Comm.
John Clarke, Pres., Consumer Healthcare
Deirdre Connelly, Pres., North American Pharmaceuticals
Eddie Gray, Pres., Pharmaceuticals Europe
Christopher Gent, Chmn.
Marc Dunoyer, Pres., Pharmaceuticals Asia Pacific & Japan

Phone: 44-20-8047-5000	Fax: 44-20-8047-7807
Toll-Free: 888-825-5249	
Address: 980 Great West Rd., Brentford, Middlesex, TW8 9GS UK	

GROWTH PLANS/SPECIAL FEATURES:

GlaxoSmithKline (GSK) is a leading research-based pharmaceutical company formed from the merger of Glaxo Wellcome and SmithKline Beecham. Its subsidiaries consist of global drug and health companies engaged in the creation, discovery, development, manufacturing and marketing of pharmaceuticals and other consumer health products. GSK operates in two segments: pharmaceuticals and consumer health care. The pharmaceuticals segment includes prescription medications and vaccines. GSK designs prescription medications for the treatment many conditions including heart and circulatory conditions, cancer, and malaria. GSK's vaccines are designed to treat life-threatening illnesses such as hepatitis A, diphtheria, influenza and bacterial meningitis. The consumer health care division is divided into three segments: over-the-counter, oral healthcare and nutritional healthcare. Products from the consumer health care segment include over-the-counter medications such as Citrucel and Nicorette; oral care products such as Aquafresh; and nutritional products such as Boost. Research and development operations take place at 17 sites in four countries. Its research areas include neurosciences, oncology, infectious diseases, biopharmaceuticals, respiratory, metabolic pathways and immunoinflammation. In December 2009, GSK and NanoBio Corporation signed an exclusive agreement in the U.S. and Canada for the over-the-counter use of NB-001, a new cold sore treatment. Also in December 2009, GSK announced plans to form a strategic alliance with Intercell. Under this alliance, the two companies will focus on developing and commercializing patch-based, needle-free vaccines. In March 2010, GSK announced a new strategic alliance with Isis Pharmaceuticals Inc. Under this alliance, GSK will apply Isis' antisense drug discovery platform to develop new therapeutics for rare, serious and infectious diseases. Also in March 2010, GSK signed an agreement with the GAVI Alliance. Under this alliance, over a ten-year period GSK will supply up to 300 million doses of Synflorix to GAVI to prevent children in developing countries from being diagnosed with pneumococcal disease.

FINANCIALS: Sales and profits are in thousands of dollars—add 000 to get the full amount. 2009 Note: Financial information for 2009 was not available for all companies at press time.

2009 Sales: $43,608,100	2009 Profits: $8,502,420	**U.S. Stock Ticker:** GSK
2008 Sales: $36,127,200	2008 Profits: $10,594,000	**Int'l Ticker:** GSK Int'l Exchange: London-LSE
2007 Sales: $33,700,100	2007 Profits: $11,264,500	Employees:
2006 Sales: $45,595,800	2006 Profits: $10,793,000	Fiscal Year Ends: 12/31
2005 Sales: $37,783,631	2005 Profits: $8,400,952	Parent Company:

SALARIES/BENEFITS:

Pension Plan: Y	ESOP Stock Plan: Y	Profit Sharing:	Top Exec. Salary: $	Bonus: $
Savings Plan: Y	Stock Purch. Plan:		Second Exec. Salary: $	Bonus: $

OTHER THOUGHTS:

Apparent Women Officers or Directors: 3
Hot Spot for Advancement for Women/Minorities: Y

LOCATIONS: ("Y" = Yes)

West:	Southwest:	Midwest:	Southeast:	Northeast:	International:
Y	Y	Y	Y	Y	Y

GTC BIOTHERAPEUTICS INC

www.gtc-bio.com

Industry Group Code: 325414 Ranks within this company's industry group: Sales: 7   Profits: 8

Drugs:		Other:		Clinical:	Computers:	Services:	
Discovery:	Y	AgriBio:	Y	Trials/Services:	Hardware:	Specialty Services:	
Licensing:	Y	Genetic Data:	Y	Labs:	Software:	Consulting:	
Manufacturing:	Y	Tissue Replacement:		Equipment/Supplies:	Arrays:	Blood Collection:	
Generics:				Research/Development Svcs.:	Database Management:	Drug Delivery:	
				Diagnostics:		Drug Distribution:	

TYPES OF BUSINESS:
Recombinant Proteins
Drugs-Anticoagulants
Transgenic Animals

BRANDS/DIVISIONS/AFFILIATES:
Atryn

CONTACTS: *Note: Officers with more than one job title may be intentionally listed here more than once.*
William Heiden, CEO
William Heiden, Pres.
John B. Green, CFO/Sr. VP
Harry M. Meade, Sr. VP-R&D
Carol A. Ziomek, VP-Prod. Dev.
Daniel S. Woloshen, General Counsel/Sr. VP
John B. Green, Treas.
Richard A. Scotland, Sr. VP-Regulatory Affairs
William Heiden, Chmn.

Phone: 508-620-9700	Fax: 508-370-3797
Toll-Free:	
Address: 175 Crossing Blvd., Framingham, MA 01702 US	

GROWTH PLANS/SPECIAL FEATURES:

GTC Biotherapeutics, Inc. (GTC) applies transgenic technology to develop recombinant proteins and monoclonal antibodies for human therapeutic uses. The company uses transgenic animals that express specific recombinant proteins in their milk. The firm generates transgenic animals through microinjection and nuclear transfer. GTC uses goats in most of its commercial development programs due to the relatively short gestation times and relatively high milk production volume of the animals. The company's leading product is ATryn for patients with hereditary antithrombin deficiency undergoing surgical procedures. Antithrombin is an important protein found in the bloodstream with anticoagulant and anti-inflammatory properties. The drug is in clinical trials for disseminated intravascular coagulation, which is an acquired deficiency of antithrombin that occurs in sepsis. The company is also using transgenic methods to produce monoclonal antibodies (MAbs), including potential therapeutic and follow-on biologics. This line of research could lead to treatments for cancer and autoimmune diseases. Other transgenic projects currently under development include a recombinant human coagulation factor for the treatment of hemophilia; a second recombinant human coagulation factor for type B hemophilia; another to treat autoimmune disorders; and an elastase inhibitor to treat emphysema and several other respiratory disorders. In May 2009, GTC launched ATryn in the U.S. with its commercial partner, Lundbeck.

FINANCIALS: Sales and profits are in thousands of dollars—add 000 to get the full amount. 2009 Note: Financial information for 2009 was not available for all companies at press time.

2009 Sales: $2,826	2009 Profits: $-26,953	U.S. Stock Ticker: GTCB
2008 Sales: $16,656	2008 Profits: $-22,665	Int'l Ticker: Int'l Exchange:
2007 Sales: $13,896	2007 Profits: $-36,321	Employees: 109
2006 Sales: $6,128	2006 Profits: $-33,345	Fiscal Year Ends: 12/31
2005 Sales: $4,152	2005 Profits: $-30,112	Parent Company:

SALARIES/BENEFITS:

Pension Plan:	ESOP Stock Plan:	Profit Sharing:	Top Exec. Salary: $452,308	Bonus: $137,904
Savings Plan: Y	Stock Purch. Plan: Y		Second Exec. Salary: $288,668	Bonus: $65,435

OTHER THOUGHTS:
Apparent Women Officers or Directors: 2
Hot Spot for Advancement for Women/Minorities: Y

LOCATIONS: ("Y" = Yes)

West:	Southwest:	Midwest:	Southeast:	Northeast:	International:
				Y	

HARBOR BIOSCIENCES INC

www.holliseden.com

Industry Group Code: 325412 Ranks within this company's industry group: Sales: Profits: 106

Drugs:	Other:	Clinical:	Computers:	Services:
Discovery: Y	AgriBio:	Trials/Services:	Hardware:	Specialty Services:
Licensing:	Genetic Data:	Labs:	Software:	Consulting:
Manufacturing:	Tissue Replacement:	Equipment/Supplies:	Arrays:	Blood Collection:
Generics:		Research/Development Svcs.:	Database Management:	Drug Delivery:
		Diagnostics:		Drug Distribution:

TYPES OF BUSINESS:
Drugs-Immune System Regulation
Drugs-Infectious Diseases
Drugs-Hormonal Imbalances

BRANDS/DIVISIONS/AFFILIATES:
Triolex
Apoptone
Hormonal Signaling Technology Platform

CONTACTS: *Note: Officers with more than one job title may be intentionally listed here more than once.*
James M. Frincke, CEO
James M. Frincke, Pres.
Robert W. Weber, CFO
Robert L. Marsella, Sr. VP-Mktg.
Christopher L. Reading, Chief Scientific Officer
Dwight R. Stickney, Chief Medical Officer
Salvatore J. Zizza, Chmn.

Phone: 858-587-9333	Fax: 858-558-6470

Toll-Free:

Address: 4435 Eastgate Mall, Ste. 400, San Diego, CA 92121 US

GROWTH PLANS/SPECIAL FEATURES:

Harbor BioSciences, Inc., formerly Hollis-Eden Pharmaceuticals, Inc., is a development-stage pharmaceutical company engaged in the discovery, development and commercialization of products for the treatment of diseases and disorders against which the body is unable to mount an appropriate immune response. It has focused its initial technology efforts on a series of adrenal steroid hormones and hormone analogs, derived from its Hormonal Signaling Technology Platform. The compounds are safe, cost-effective to manufacture and unlikely to produce resistance. The firm is currently focused on the development of two clinical drug development candidates: Triolex (HE3286) and Apoptone (HE3235). Triolex is a next-generation compound currently in a multi-dose Phase I/II clinical trial for the treatment of metabolic and autoimmune disorders and is cleared for clinical trials for type II diabetes, rheumatoid arthritis and ulcerative colitis, with developments for cystic fibrosis in preclinical stages. Apoptone is a next-generation compound selected for clinical development for prostate cancer. In preclinical models of prostate and breast cancer, Apoptone has been shown to reduce the incidence, growth and progression of tumors. In February 2010, the company changed its name from Hollis-Eden Pharmaceuticals, Inc. to Harbor BioSciences, Inc. Also in February 2010, the firm received notice that it has been granted allowance to patent Apoptone in prostate cancer, breast cancer and benign prostatic hypertrophy.

FINANCIALS: Sales and profits are in thousands of dollars—add 000 to get the full amount. 2009 Note: Financial information for 2009 was not available for all companies at press time.

2009 Sales: $	2009 Profits: $-15,626	**U.S. Stock Ticker: HRBR**
2008 Sales: $	2008 Profits: $-21,565	**Int'l Ticker:** Int'l Exchange:
2007 Sales: $ 645	2007 Profits: $-23,121	Employees: 19
2006 Sales: $ 444	2006 Profits: $-30,231	Fiscal Year Ends: 12/31
2005 Sales: $ 56	2005 Profits: $-29,441	Parent Company:

SALARIES/BENEFITS:

Pension Plan:	ESOP Stock Plan:	Profit Sharing:	Top Exec. Salary: $371,394	Bonus: $
Savings Plan: Y	Stock Purch. Plan:		Second Exec. Salary: $325,000	Bonus: $

OTHER THOUGHTS:
Apparent Women Officers or Directors:
Hot Spot for Advancement for Women/Minorities:

LOCATIONS: ("Y" = Yes)

West:	Southwest:	Midwest:	Southeast:	Northeast:	International:
Y					

HARVARD BIOSCIENCE INC

www.harvardbioscience.com

Industry Group Code: 3345 Ranks within this company's industry group: Sales: 4   Profits: 3

Drugs:	Other:	Clinical:		Computers:		Services:
Discovery:	AgriBio:	Trials/Services:		Hardware:		Specialty Services:
Licensing:	Genetic Data:	Labs:		Software:		Consulting:
Manufacturing:	Tissue Replacement:	Equipment/Supplies:	Y	Arrays:		Blood Collection:
Generics:		Research/Development Svcs.:	Y	Database Management:		Drug Delivery:
		Diagnostics:				Drug Distribution:

TYPES OF BUSINESS:

Apparatus & Scientific Instruments

BRANDS/DIVISIONS/AFFILIATES:

Hoefer, Inc.
BTX Molecular Delivery Systems
GE Healthcare
Biochrom Ltd.
Asys HiTech GmbH
Warner Instruments
Harvard Apparatus
Denville Scientific Inc.

CONTACTS: *Note: Officers with more than one job title may be intentionally listed here more than once.*

Chane Graziano, CEO
Susan Luscinski, COO
David Green, Pres.
Thomas McNaughton, CFO
Chane Graziano, Chmn.

Phone: 508-893-8999	Fax: 508-429-5732
Toll-Free: 800-272-2775	
Address: 84 October Hill Rd., Holliston, MA 01746 US	

GROWTH PLANS/SPECIAL FEATURES:

Harvard Bioscience, Inc. is a global developer, manufacturer and marketer of a broad range of specialized products, primarily apparatus and scientific instruments used to advance life science research at pharmaceutical and biotechnology companies, universities and government laboratories. Products are targeted toward two major areas of application: ADMET testing and molecular biology and liquid handling. ADMET testing is used to identify compounds that have toxic side effects or undesirable physiological or pharmacological properties. These pharmacological properties consist of absorption, distribution, metabolism and elimination. The company's products in this area include absorption diffusion chambers, 96 well equilibrium dialysis plates, organ testing systems, precision infusion pumps, cell injection systems, ventilators and electroporation products. The molecular biology products are mainly scientific instruments such as spectrophotometers and plate readers that analyze light to detect and quantify a wide range of molecular and cellular processes or apparatus such as gel electrophoresis units. These products can quantify the amount of DNA, RNA or protein in a sample; can use chromatography to separate the amino acids in a sample; and can use the method of gel electrophoresis to separate and purify DNA, RNA and proteins. Most of these molecular biology products are sold through GE Healthcare. Harvard Bioscience operates though its 20 wholly-owned subsidiaries; which have the rights to numerous trade names and trademarks, notably Hoefer, Warner, BTX and Biochrom, Harvard Bioscience has manufacturing operations in the U.S., the U.K., Germany, and Spain with sales facilities in France, Germany and Canada. The company sells its products to thousands of researchers in over 100 countries primarily through its 8,500 page catalog, and secondarily through its web site, a field sales organization and through other distributors. Customers primarily include research scientists at pharmaceutical and biotechnology companies, universities and government laboratories, including the U.S. National Institutes of Health.

FINANCIALS: Sales and profits are in thousands of dollars—add 000 to get the full amount. 2009 Note: Financial information for 2009 was not available for all companies at press time.

2009 Sales: $85,772	2009 Profits: $7,233	U.S. Stock Ticker: HBIO
2008 Sales: $88,049	2008 Profits: $1,673	Int'l Ticker: Int'l Exchange:
2007 Sales: $83,407	2007 Profits: $-1,354	Employees: 348
2006 Sales: $76,181	2006 Profits: $-2,341	Fiscal Year Ends: 12/31
2005 Sales: $67,431	2005 Profits: $-31,877	Parent Company:

SALARIES/BENEFITS:

Pension Plan:	ESOP Stock Plan:	Profit Sharing:	Top Exec. Salary: $535,500	Bonus: $56,000
Savings Plan: Y	Stock Purch. Plan: Y		Second Exec. Salary: $441,000	Bonus: $56,000

OTHER THOUGHTS:

Apparent Women Officers or Directors: 1
Hot Spot for Advancement for Women/Minorities:

LOCATIONS: ("Y" = Yes)

West:	Southwest:	Midwest:	Southeast:	Northeast:	International:
Y				Y	Y

HEMAGEN DIAGNOSTICS INC

www.hemagen.com

Industry Group Code: 325413 Ranks within this company's industry group: Sales: 19 Profits: 12

Drugs:		Other:		Clinical:		Computers:		Services:	
Discovery:		AgriBio:		Trials/Services:		Hardware:		Specialty Services:	
Licensing:	Y	Genetic Data:		Labs:		Software:		Consulting:	
Manufacturing:	Y	Tissue Replacement:		Equipment/Supplies:	Y	Arrays:		Blood Collection:	
Generics:				Research/Development Svcs.:		Database Management:		Drug Delivery:	
				Diagnostics:	Y			Drug Distribution:	

TYPES OF BUSINESS:
Medical Diagnostics Products
Clinical Chemical Analysis Products
Clinical Chemistry Reagents

BRANDS/DIVISIONS/AFFILIATES:
Hemagen Diagnosticos Comerico
Analyst
Virgo
VET-16
VetFlex7
VetFlex

CONTACTS: Note: Officers with more than one job title may be intentionally listed here more than once.
William P. Hales, CEO
William P. Hales, Pres.
Catherine M. Davidson, CFO
Catherine M. Davidson, Controller
William P. Hales, Chmn.

Phone: 443-367-5500	Fax: 443-367-5527
Toll-Free: 800-436-2436	
Address: 9033 Red Branch Rd., Columbia, MD 21045 US	

GROWTH PLANS/SPECIAL FEATURES:
Hemagen Diagnostics, Inc. develops, manufactures and markets proprietary medical diagnostic test kits for both human and veterinary subjects. Hemagen has two different product lines: the Virgo line and the Analyst line. The Virgo product line of diagnostic test kits is used to aid in the diagnosis of certain autoimmune and infectious diseases, using enzyme-linked immunosorbent assay (ELISA), immunoflourescence and hemagglutination technology. The company markets its Virgo product line in South America through Hemagen Diagnosticos Comerico, Importacao e Exportacao, Ltd. (HDC), its wholly-owned subsidiary. The Analyst product line is an FDA-cleared Benchtop Clinical Chemistry Analyzer System, including consumables that are used to measure important constituents in human and animal blood. Hemagen markets four FDA-cleared rotor types for use on the Analyst clinical chemistry analyzer, two general chemistry rotors, a glucose test and a lipid screen test. It also sells four rotors for the veterinary marketplace: VET-16, VetFlex7, VetFlex and T4 rotors. Hemagen currently offers approximately 68 test kits that have been cleared by the FDA for sale in the U.S. Hemagen sells its products directly and through distributors to clinical laboratories, hospitals, veterinary offices and research organizations. The company has relationships with over 30 distributors in various countries worldwide.

Hemagen Diagnostics, Inc. offers its employees a 401(k) plan; medical, dental and disability coverage; paid holidays and comprehensive leave; and an employee stock ownership plan.

FINANCIALS: Sales and profits are in thousands of dollars—add 000 to get the full amount. 2009 Note: Financial information for 2009 was not available for all companies at press time.

2009 Sales: $5,419	2009 Profits: $- 808	U.S. Stock Ticker: HMGN
2008 Sales: $6,375	2008 Profits: $ 427	Int'l Ticker: Int'l Exchange:
2007 Sales: $4,487	2007 Profits: $- 850	Employees: 25
2006 Sales: $7,250	2006 Profits: $ 313	Fiscal Year Ends: 9/30
2005 Sales: $7,586	2005 Profits: $-1,337	Parent Company:

SALARIES/BENEFITS:
Pension Plan:	ESOP Stock Plan: Y	Profit Sharing:	Top Exec. Salary: $172,500	Bonus: $
Savings Plan: Y	Stock Purch. Plan:		Second Exec. Salary: $115,000	Bonus: $

OTHER THOUGHTS:
Apparent Women Officers or Directors: 1
Hot Spot for Advancement for Women/Minorities:

LOCATIONS: ("Y" = Yes)
West:	Southwest:	Midwest:	Southeast:	Northeast:	International:
				Y	Y

HEMISPHERX BIOPHARMA INC
www.hemispherx.net

Industry Group Code: 325412 Ranks within this company's industry group: Sales: 141   Profits: 103

Drugs:		Other:		Clinical:	Computers:	Services:
Discovery:	Y	AgriBio:		Trials/Services:	Hardware:	Specialty Services:
Licensing:		Genetic Data:	Y	Labs:	Software:	Consulting:
Manufacturing:	Y	Tissue Replacement:		Equipment/Supplies:	Arrays:	Blood Collection:
Generics:				Research/Development Svcs.:	Database Management:	Drug Delivery:
				Diagnostics:		Drug Distribution:

TYPES OF BUSINESS:
RNA-Related Drugs
HIV Treatments
Antivirals

BRANDS/DIVISIONS/AFFILIATES:
Ampligen
Alferon LDO
Alferon N

CONTACTS: *Note: Officers with more than one job title may be intentionally listed here more than once.*
William A. Carter, CEO
William A. Carter, Pres.
Charles T. Bernhardt, CFO
Carol A. Smith, VP-Mfg. Quality Dev.
Ransom W. Etheridge, General Counsel
Wayne S. Springate, VP-Oper.
Charles T. Bernhardt, Chief Acct. Officer
Carol A. Smith, VP-Process Dev.
David R. Strayer, Medical Dir.-Regulatory Affairs
Katalin Ferencz-Biro, Sr. VP-Regulatory Affairs
Russel Lander, VP-Quality Assurance
William A. Carter, Chmn.

Phone: 215-988-0080 **Fax:** 215-988-1739
Toll-Free:
Address: 1617 JFK Blvd., 1 Penn Ctr., 6th Fl., Philadelphia, PA 19103 US

GROWTH PLANS/SPECIAL FEATURES:
Hemispherx Biopharma, Inc. is a biopharmaceutical company that develops and manufactures nucleic acid drugs for the treatment of viral and immune-based disorders. The company's proprietary drug technology uses specifically configured RNA. Hemispherx's flagship products include the antiviral/immunotherapeutic drugs Ampligen, Alferon N Injection and Alferon Low Dose Oral. Ampligen is a synthetic double-stranded RNA configured to address a variety of chronic diseases and viral disorders such as HIV and chronic fatigue syndrome. Alferon, in either its injection form (N) or its low dose oral form (LDO), is a purified, natural source alpha interferon (interferon alfa-n3, human leukocyte derived). Interferon assists the body's defenses against potentially dangerous foreign substances. Alferon N is also in development for treating avian flu, West Nile Virus and other diseases. Alferon LDO is currently in early stage development targeting influenza and viral diseases. In addition to these flagship drugs and projects, Hemispherx has a patent portfolio of 38 patents issued worldwide, with 46 more pending.

FINANCIALS: Sales and profits are in thousands of dollars—add 000 to get the full amount. 2009 Note: Financial information for 2009 was not available for all companies at press time.
2009 Sales: $1,083	2009 Profits: $-12,446	**U.S. Stock Ticker:** HEB
2008 Sales: $ 933	2008 Profits: $-19,399	**Int'l Ticker:** Int'l Exchange:
2007 Sales: $1,059	2007 Profits: $-18,139	Employees: 44
2006 Sales: $ 933	2006 Profits: $-19,399	Fiscal Year Ends: 12/31
2005 Sales: $1,083	2005 Profits: $-13,213	Parent Company:

SALARIES/BENEFITS:
Pension Plan:	ESOP Stock Plan:	Profit Sharing:	Top Exec. Salary: $664,624	Bonus: $
Savings Plan: Y	Stock Purch. Plan:		Second Exec. Salary: $259,164	Bonus: $

OTHER THOUGHTS:
Apparent Women Officers or Directors: 1
Hot Spot for Advancement for Women/Minorities: Y

LOCATIONS: ("Y" = Yes)
West:	Southwest:	Midwest:	Southeast:	Northeast:	International:
				Y	Y

HESKA CORP

www.heska.com

Industry Group Code: 325412B Ranks within this company's industry group: Sales: 1 Profits: 1

Drugs:		Other:		Clinical:		Computers:		Services:	
Discovery:		AgriBio:		Trials/Services:		Hardware:		Specialty Services:	
Licensing:		Genetic Data:		Labs:		Software:		Consulting:	
Manufacturing:	Y	Tissue Replacement:		Equipment/Supplies:	Y	Arrays:		Blood Collection:	
Generics:				Research/Development Svcs.:		Database Management:		Drug Delivery:	
				Diagnostics:	Y			Drug Distribution:	

TYPES OF BUSINESS:

Drugs-Animal Health & Pet Care
Veterinary Diagnostics Products & Services
Animal Dietary Supplements
Veterinary Vaccines

BRANDS/DIVISIONS/AFFILIATES:

Core Companion Animal Health
HESKA Feline UltraNasal FVRCP Vaccine
VitalPath Blood Gas and Electrolyte Analyzer
DRI-CHEM Veterinary Chemistry Analyzer
HEMATRUE Veterinary Hematology Analyzer
VET/IV 2.2 Infusion Pump
ALLERCEPT
Roche Diagnostics Corp.

CONTACTS: Note: Officers with more than one job title may be intentionally listed here more than once.

Robert B. Grieve, CEO
Michael J. McGinley, COO
Michael J. McGinley, Pres.
Jason A. Napolitano, CFO/Exec. VP
G. Lynn Snodgrass, VP-Sales
Nancy Wisnewski, VP-Prod. Dev.
Jason A. Napolitano, Corp. Sec.
Claudine Zachara, VP-Comm.
Michael A. Bent, Controller/Principle Acct. Officer/VP
Laurie Peterson, Gen. Mgr.-Heska Des Moines
Claudine Zachara, VP-Mktg.
Nancy Wisnewski, VP-Tech. Customer Service
Robert B. Grieve, Chmn.
Donald L. Wassom, Managing Dir.-Heska Fribourg/Dir.-Global Allergy

Phone: 970-493-7272	Fax: 970-619-3003
Toll-Free: 800-464-3752	
Address: 3760 Rocky Mountain Ave., Loveland, CO 80538 US	

GROWTH PLANS/SPECIAL FEATURES:

Heska Corp. focuses on the discovery, development, manufacture, marketing and support of animal health care products. The company uses biotechnology to create a broad range of diagnostic, therapeutic and vaccine products for dogs, cats and other animals. The business is divided into two segments: Core Companion Animal Health (CCA); and Other Vaccines, Pharmaceuticals and Products (OVP). The CCA segment includes diagnostic instruments and supplies, single-use diagnostic tests, vaccines and pharmaceuticals, primarily for canine and feline use. Its line of veterinary diagnostic instruments includes the DRI-CHEM Veterinary Chemistry Analyzer, which analyzes blood chemistry and electrolytes; the HEMATRUE Veterinary Hematology Analyzer, a blood analyzer focused on white blood cell count, red blood cell count, platelet count and hemoglobin levels; the VitalPath Blood Gas and Electrolyte Analyzer, which analyzes blood gases, electrolytes and hemocrit; and the VET/IV 2.2 infusion pump for regulated infusion of fluids, drugs or nutritional products. Heska's point-of-care diagnostic tests include tests for heartworms and early renal damage detection products. The E.R.D.-Healthscreen Urine Tests can identify dogs and cats at risk for kidney disease before the majority of kidney function is lost. The company also sells a variety of laboratory diagnostics products, including the ALLERCEPT Definitive Allergy Panels for the determination of specific allergies in dogs, cats and horses. This business segment also markets heartworm prevention pharmaceuticals, hypothyroid treatment supplements and nutritional supplements, immunotherapy treatments for allergies and the Feline Ultranasal FVRCP Vaccine for the prevention of disease caused by respiratory viruses in cats. The OVP business manufactures private-label vaccines and pharmaceutical products that are marketed and distributed by third parties primarily for cattle, although some products are also manufactured for small mammals and fish. In November 2009, Heska entered an agreement with Roche Diagnostics Corporation for the supply of VitalPath blood gas products.

FINANCIALS: Sales and profits are in thousands of dollars—add 000 to get the full amount. 2009 Note: Financial information for 2009 was not available for all companies at press time.

2009 Sales: $75,678	2009 Profits: $2,242	U.S. Stock Ticker: HSKA
2008 Sales: $81,653	2008 Profits: $- 850	Int'l Ticker: Int'l Exchange:
2007 Sales: $82,335	2007 Profits: $34,808	Employees: 276
2006 Sales: $75,060	2006 Profits: $1,828	Fiscal Year Ends: 12/31
2005 Sales: $69,437	2005 Profits: $ 282	Parent Company:

SALARIES/BENEFITS:

Pension Plan:	ESOP Stock Plan:	Profit Sharing:	Top Exec. Salary: $420,000	Bonus: $180,182
Savings Plan:	Stock Purch. Plan:		Second Exec. Salary: $243,000	Bonus: $72,974

OTHER THOUGHTS:

Apparent Women Officers or Directors: 4
Hot Spot for Advancement for Women/Minorities: Y

LOCATIONS: ("Y" = Yes)

West:	Southwest:	Midwest:	Southeast:	Northeast:	International:
Y		Y			Y

Note: Financial information, benefits and other data can change quickly and may vary from those stated here.

HI-TECH PHARMACAL CO INC

www.hitechpharm.com

Industry Group Code: 325412 Ranks within this company's industry group: Sales: 72   Profits: 60

Drugs:	Other:	Clinical:	Computers:	Services:	
Discovery:	AgriBio:	Trials/Services:	Hardware:	Specialty Services:	Y
Licensing:	Genetic Data:	Labs:	Software:	Consulting:	
Manufacturing: Y	Tissue Replacement:	Equipment/Supplies:	Arrays:	Blood Collection:	
Generics:		Research/Development Svcs.:	Database Management:	Drug Delivery:	
		Diagnostics:		Drug Distribution:	

TYPES OF BUSINESS:

Drugs-Generic
Nutritional Products
Over-the-Counter Products
Ophthalmic Products
Manufacturing Contract Services
Inhalation Products
Diabetes Products
Generic Drugs

BRANDS/DIVISIONS/AFFILIATES:

Diabetic Tussin
DiabetiSweet
DiabetDerm
Multi-Betic
Zostrix
ECR Pharmaceuticals
Midlothian Laboratories
Health Care Products Division

CONTACTS: Note: Officers with more than one job title may be intentionally listed here more than once.

David S. Seltzer, CEO
David S. Seltzer, Pres.
William Peters, CFO/VP
Kamel Egbaria, Chief Scientific Officer
David S. Seltzer, Sec.
David S. Seltzer, Treas.
Gary M. April, Pres., Health Care Prod. Div.
Bryce M. Harvey, Pres., Midlothian Laboratories Div.
David S. Caskey, VP-Pharmaceutical Oper., ECR Pharmaceuticals, Inc.
David S. Seltzer, Chmn.

Phone: 631-789-8228	Fax: 631-789-8429
Toll-Free:	
Address: 369 Bayview Ave., Amityville, NY 11701-2802 US	

GROWTH PLANS/SPECIAL FEATURES:

Hi-Tech Pharmacal Co., Inc. is a manufacturer and marketer of prescription, over-the-counter (OTC) and nutritional products that are sold in liquid, inhalation and cream forms. A wide range of products are produced for various disease states, including asthma, bronchial disorders, dermatological disorders, allergies, pain, stomach, oral care, neurological disorders and other conditions. The firm divides its products into three main lines: generic, branded prescription and OTC brands. The generic products division, operating as Midlothian Laboratories, primarily includes prescription items such as oral solutions and suspensions, topical creams and ointments as well as nasal sprays. This division also manufactures ophthalmic, optic and inhalation products and provides sterile manufacturing contract services. Hi-Tech's top five selling generic products are Dorzolamide, Fluticasone propionate, sulfamethoxazole, pediatric multivitamins and Lactulose. The company's branded prescription products, gained through the 2009 acquisition of the E. Claiborne Robins Company, Inc. (ECR), include antihistamines, corticosteroids and analgesic tablets. The firm's OTC brands division, also named the Health Care Products Division, develops and markets a line of branded products primarily for people with diabetes, including Diabetic Tussin, a line of cough medications; DiabetiSweet, sugar substitutes that can be used for baking and cooking; Multi-betic, a daily multi-vitamin; and DiabetiDerm, a diabetic skin care line. This division also sells Zostrix, a brand of capsaisin products for pain and arthritis. In 2009, sales of generic pharmaceuticals represented 88% of total sales, prescriptions accounted for 3% and the health care products line of over-the-counter products accounted for 9%. The company's customers consist of generic distributors, drug wholesalers, chain drug stores, mass merchandise chains, certain government agencies and mail-order pharmacies. In March 2010, the firm acquired the magnesium supplement Mag-Ox product line from Blaine Company, Inc.

FINANCIALS: Sales and profits are in thousands of dollars—add 000 to get the full amount. 2009 Note: Financial information for 2009 was not available for all companies at press time.

2009 Sales: $108,651	2009 Profits: $9,817	**U.S. Stock Ticker: HITK**
2008 Sales: $62,017	2008 Profits: $-5,098	**Int'l Ticker:** Int'l Exchange:
2007 Sales: $58,898	2007 Profits: $-2,036	Employees: 366
2006 Sales: $78,020	2006 Profits: $11,453	Fiscal Year Ends: 4/30
2005 Sales: $67,683	2005 Profits: $8,288	Parent Company:

SALARIES/BENEFITS:

Pension Plan:	ESOP Stock Plan:	Profit Sharing:	Top Exec. Salary: $442,000	Bonus: $
Savings Plan: Y	Stock Purch. Plan:		Second Exec. Salary: $251,000	Bonus: $45,000

OTHER THOUGHTS:

Apparent Women Officers or Directors: 3
Hot Spot for Advancement for Women/Minorities: Y

LOCATIONS: ("Y" = Yes)

West:	Southwest:	Midwest:	Southeast:	Northeast:	International:
				Y	

HOSPIRA INC

www.hospira.com

Industry Group Code: 325412A Ranks within this company's industry group: Sales: 1 Profits: 1

Drugs:		Other:	Clinical:		Computers:		Services:	
Discovery:	Y	AgriBio:	Trials/Services:		Hardware:	Y	Specialty Services:	
Licensing:	Y	Genetic Data:	Labs:		Software:	Y	Consulting:	
Manufacturing:	Y	Tissue Replacement:	Equipment/Supplies:	Y	Arrays:		Blood Collection:	
Generics:			Research/Development Svcs.:		Database Management:		Drug Delivery:	
			Diagnostics:				Drug Distribution:	

TYPES OF BUSINESS:

Pharmaceutical Development
Generic Pharmaceuticals
Medication Delivery Systems
Anesthetics
Injectable Medications
Diagnostic Imaging Agents
Contract Manufacturing

BRANDS/DIVISIONS/AFFILIATES:

One2One
TheraDoc
LifeCare PCA
Symbiq
Retacrit
GemStar
LifeShield
CLAVE

CONTACTS: *Note: Officers with more than one job title may be intentionally listed here more than once.*

Christopher B. Begley, CEO
Terrence C. Kearney, COO
Thomas E. Werner, CFO
Anil G. D'Souza, VP-Global Mktg.
Gail Denham, VP-Int'l Human Resources Oper. & Rewards
Sumant Ramachandra, Chief Science Officer/Sr. VP-R&D & Medical Affairs
Daphne E. Jones, CIO/Sr. VP
Brian J. Smith, General Counsel/Sr. VP/Sec.
Anil G. D'Souza, VP-Corp. Dev.
Thomas E. Werner, Sr. VP-Finance
Ken Meyers, Sr. VP-Organizational Transformation & People Dev.
Ron Squarer, Chief Commercial Officer
Christopher B. Begley, Chmn.

Phone: 224-212-2000	Fax: 224-212-3350
Toll-Free: 877-946-7747	
Address: 275 N. Field Dr., Lake Forest, IL 60045 US	

GROWTH PLANS/SPECIAL FEATURES:

Hospira, Inc. is a global specialty pharmaceutical and medication delivery company. The company's primary operations involve the research and development of generic pharmaceuticals, pharmaceuticals based on proprietary pharmaceuticals whose patents have expired. The company's activities include the development, manufacture and marketing of generic acute-care and oncology injectables, as well as integrated infusion therapy and medication management systems. Hospira divides its products into four categories: specialty injectable pharmaceuticals, other pharmaceuticals, medication management systems and other devices. The specialty injectable pharmaceuticals segment consists of approximately 200 injectable generic drugs; the proprietary sedation drug Precedex; and Retacrit, a biogeneric version of erythropoietin, used primarily in the treatment of anemia in dialysis and in certain oncology applications. The other pharmaceuticals segment consists primarily of large volume I.V. solutions and nutritional products. Through its One2One manufacturing services group, this segment also offers contract manufacturing services to proprietary pharmaceutical and biotechnology companies for formulation development, filling and finishing of injectable and oral drugs worldwide. The medication management systems segment's products include electronic drug delivery pumps under the Symbiq, LifeCare PCA, GemStar and Plum brands; safety software; and administration sets that are used to deliver I.V. fluids and medications. The other devices segment includes gravity administration sets and other device products, which include needlestick safety products. Products in this segment are sold under the LifeShield, CLAVE and MicroClave brands. In August 2009, the company sold the commercial rights and physical assets of its critical care product line to ICU Medical, Inc. In December 2009, the firm acquired TheraDoc, a developer of hospital surveillance systems. In March 2010, the company acquired the generic injectable pharmaceuticals business of Indian firm Orchid Chemicals & Pharmaceuticals Ltd. for $400 million.

Employees are offered medical and dental insurance; flexible spending accounts; an employee assistance plan; disability coverage; life insurance; adoption assistance; and tuition assistance.

FINANCIALS: Sales and profits are in thousands of dollars—add 000 to get the full amount. 2009 Note: Financial information for 2009 was not available for all companies at press time.

2009 Sales: $3,879,300	2009 Profits: $403,900	U.S. Stock Ticker: HSP
2008 Sales: $3,629,500	2008 Profits: $320,900	Int'l Ticker: Int'l Exchange:
2007 Sales: $3,436,238	2007 Profits: $136,758	Employees: 13,500
2006 Sales: $2,688,505	2006 Profits: $237,679	Fiscal Year Ends: 12/31
2005 Sales: $2,626,696	2005 Profits: $235,638	Parent Company:

SALARIES/BENEFITS:

Pension Plan: Y	ESOP Stock Plan:	Profit Sharing:	Top Exec. Salary: $1,050,000	Bonus: $2,100,000
Savings Plan: Y	Stock Purch. Plan:		Second Exec. Salary: $625,000	Bonus: $1,000,000

OTHER THOUGHTS:

Apparent Women Officers or Directors: 3
Hot Spot for Advancement for Women/Minorities: Y

LOCATIONS: ("Y" = Yes)

West:	Southwest:	Midwest:	Southeast:	Northeast:	International:
Y	Y	Y		Y	Y

HUMAN GENOME SCIENCES INC

www.hgsi.com

Industry Group Code: 325412 Ranks within this company's industry group: Sales: 60   Profits: 66

Drugs:		Other:		Clinical:	Computers:	Services:
Discovery:	Y	AgriBio:		Trials/Services:	Hardware:	Specialty Services:
Licensing:		Genetic Data:	Y	Labs:	Software:	Consulting:
Manufacturing:	Y	Tissue Replacement:		Equipment/Supplies:	Arrays:	Blood Collection:
Generics:				Research/Development Svcs.:	Database Management:	Drug Delivery:
				Diagnostics:		Drug Distribution:

TYPES OF BUSINESS:

Oncology, Immunology & Infectious Diseases Drugs

BRANDS/DIVISIONS/AFFILIATES:

Raxibacumab
BENLYSTA
Abthrax
Darapladib
Syncria
TRAIL Receptor Antibodies
ZALBIN
Mapatumumab

CONTACTS: *Note: Officers with more than one job title may be intentionally listed here more than once.*

H. Thomas Watkins, CEO
H. Thomas Watkins, Pres.
David P. Southwell, CFO/Exec. VP
Barry A. Labinger, Chief Commercial Officer/Exec. VP
Susan Bateson, Sr. VP-Human Resources
David C. Stump, Exec. VP-R&D
Joseph A. Morin, VP-Eng.
Randy Maddox, VP-Mfg. Oper.
James H. Davis, General Counsel/Exec. VP/Sec.
Curran M. Simpson, Sr. VP-Oper.
Sally D. Bolmer, Sr. VP-Dev. & Regulatory Affairs
Jerry Parrott, VP-Corp. Comm. & Public Policy
William W. Freimuth, VP-Clinical Research, Immunology
Ann L. Wang, VP-Clinical Oper.
Gilles Gallant, VP-Clinical Research, Oncology
Daniel J. Odenheimer, VP-Clinical Research, Gen. Medicine
Argeris N. Karabelas, Chmn.

Phone: 301-309-8504	Fax: 301-309-8512
Toll-Free:	
Address: 14200 Shady Grove Rd., Rockville, MD 20850 US	

GROWTH PLANS/SPECIAL FEATURES:

Human Genome Sciences, Inc. (HGS) is a commercially focused drug development company with several products in late-stage development or clinical trials, including BENLYSTA for lupus and other autoimmune diseases; ZALBIN for Hepatitis C; Darapladib for heart disease; Raxibacumab for inhaled anthrax; Syncria for diabetes; and Mapatumumab and HGS1029 for cancer. The company's research focuses on treatments for oncology and immunology. The firm's partners conduct clinical trials of additional drugs to treat cardiovascular, metabolic and central nervous system diseases and advanced a number of products derived from the company's technology to clinical development. HGS works with GlaxoSmithKline to develop and market BENLYSTA, Darapladib and Syncria. Raxibacumab, or ABthrax, is being developed under a contract with the U.S. government based on highly successfully efficacy studies. HGS began manufacturing on schedule and has worked to deliver 20,000 doses to the Strategic National Stockpile. HGS's leading products still in early to mid development are based on its TRAIL receptor antibodies for cancer. HGS has over 500 U.S. patents covering genes and proteins. In March 2010, the company and BioInvent International AB announced a collaborative agreement to develop monoclonal antibodies that target specific antigens.

The company offers its employees medical, dental and vision insurance; flexible spending accounts; life and AD&D insurance; short- and long-term disability; a 401(k) plan; an employee stock purchase plan; education assistance; and ongoing training through employee development programs.

FINANCIALS: Sales and profits are in thousands of dollars—add 000 to get the full amount. 2009 Note: Financial information for 2009 was not available for all companies at press time.

2009 Sales: $275,749	2009 Profits: $5,659	**U.S. Stock Ticker: HGSI**
2008 Sales: $48,422	2008 Profits: $-268,891	**Int'l Ticker:** Int'l Exchange:
2007 Sales: $41,851	2007 Profits: $-284,371	Employees: 850
2006 Sales: $25,755	2006 Profits: $-251,173	Fiscal Year Ends: 12/31
2005 Sales: $19,113	2005 Profits: $-239,439	Parent Company:

SALARIES/BENEFITS:

Pension Plan:	ESOP Stock Plan: Y	Profit Sharing:	Top Exec. Salary: $700,000	Bonus: $1,050,000
Savings Plan: Y	Stock Purch. Plan: Y		Second Exec. Salary: $479,231	Bonus: $650,000

OTHER THOUGHTS:

Apparent Women Officers or Directors: 5
Hot Spot for Advancement for Women/Minorities: Y

LOCATIONS: ("Y" = Yes)

West:	Southwest:	Midwest:	Southeast:	Northeast:	International:
				Y	

HYCOR BIOMEDICAL INC

www.hycorbiomedical.com

Industry Group Code: 325413 Ranks within this company's industry group: Sales: Profits:

Drugs:	Other:	Clinical:		Computers:		Services:
Discovery:	AgriBio:	Trials/Services:		Hardware:	Y	Specialty Services:
Licensing:	Genetic Data:	Labs:		Software:		Consulting:
Manufacturing:	Tissue Replacement:	Equipment/Supplies:	Y	Arrays:		Blood Collection:
Generics:		Research/Development Svcs.:		Database Management:		Drug Delivery:
		Diagnostics:	Y			Drug Distribution:

TYPES OF BUSINESS:

Medical Diagnostics Products
Allergy & Autoimmune Diagnostics
Urinalysis Products

BRANDS/DIVISIONS/AFFILIATES:

HY-TEC
Autostat II Elisa
Hycor Ultra-Sensitive EIA System
HY-TEC 288
KOVA Urinalysis
KOVA-Trol
KOVA Refractrol SP
Linden LLC

CONTACTS: Note: Officers with more than one job title may be intentionally listed here more than once.

Dick Aderman, CEO
Dick Aderman, Pres.
Richard Hockins, VP-Mktg. & Sales
Richard L. Novak, Chmn.

Phone:	Fax: 714-933-3222
Toll-Free: 800-382-2527	
Address: 7272 Chapman Ave., Garden Grove, CA 92841 US	

GROWTH PLANS/SPECIAL FEATURES:

Hycor Biomedical, Inc., a subsidiary of Linden LLC, researches, develops and manufactures medical diagnostic products for clinical laboratories and specialty physicians. The firm has a particular focus on urinalysis, allergy products and autoimmune testing. In the urinalysis segment, the company's KOVA products include: KOVA Urinalysis, the market leader in Standardized Microscopic Urinalysis; KOVA-Trol tri-level, lyophilized urine dipstick chemistry control, which is stable for 5-7 days at room temperature and for four months when frozen; KOVA Liqua-Trol, a ready-to-use bi-level liquid control, useable with most brands of urine chemistry dipsticks; and KOVA Refractrol SP, a tri-level control product with a 24-month shelf life designed for use with temperature compensated and non-compensated refractometers. Hycor's allergy diagnostic product line, which includes HY-TEC specific IgE tests; and Hycor Ultra-Sensitive EIA system for allergen specific IgE testing, features radio-immunoassays and enzymatic immunoassays that test for reactions to more than 1,000 allergens. The company's autoimmune products division manufactures devices (branded as Autostat II Elisa) that test for disorders such as systemic lupus erythematosus and rheumatoid arthritis. Hycor also produces HY-TEC 288, an automated consolidated workstation capable of storing up to 50 allergy and 100 autoimmune samples, conducting 288 tests per run, providing bar-coded sample identification, high precision robotic liquid handling and real-time incubation control. In addition, the firm tests for infectious diseases, such as, respiratory pathogens; ticks; herpes; gastrointestinal pathogens; mycosis; and basic reagents, that include varicella, borrelia, diphtheria, tetanus, and echinococcus. Lastly, the firm makes vaccines for diphtheria, poliomyelitis and tetanus. The firm has offices in the Netherlands, Scotland and the U.S. The firm recently announced the release of the Extended Range Calibrator Set for the HY-TEC 288 Allergy System. In May 2010, Linden LLC acquired the company from Agilent Technologies.

FINANCIALS: Sales and profits are in thousands of dollars—add 000 to get the full amount. 2009 Note: Financial information for 2009 was not available for all companies at press time.

2009 Sales: $	2009 Profits: $	U.S. Stock Ticker: Subsidiary
2008 Sales: $	2008 Profits: $	Int'l Ticker: Int'l Exchange:
2007 Sales: $19,800	2007 Profits: $	Employees:
2006 Sales: $	2006 Profits: $	Fiscal Year Ends: 12/31
2005 Sales: $	2005 Profits: $	Parent Company: LINDEN LLC

SALARIES/BENEFITS:

Pension Plan:	ESOP Stock Plan:	Profit Sharing:	Top Exec. Salary: $	Bonus: $
Savings Plan:	Stock Purch. Plan:		Second Exec. Salary: $	Bonus: $

OTHER THOUGHTS:

Apparent Women Officers or Directors:
Hot Spot for Advancement for Women/Minorities:

LOCATIONS: ("Y" = Yes)

West:	Southwest:	Midwest:	Southeast:	Northeast:	International:
Y					Y

I3 CANREG

www.canreginc.com

Industry Group Code: 541690 Ranks within this company's industry group: Sales: Profits:

Drugs:	Other:	Clinical:	Computers:	Services:	
Discovery:	AgriBio:	Trials/Services:	Hardware:	Specialty Services:	Y
Licensing:	Genetic Data:	Labs:	Software:	Consulting:	Y
Manufacturing:	Tissue Replacement:	Equipment/Supplies:	Arrays:	Blood Collection:	
Generics:		Research/Development Svcs.:	Database Management:	Drug Delivery:	
		Diagnostics:		Drug Distribution:	

TYPES OF BUSINESS:
Regulatory Affairs Consulting

BRANDS/DIVISIONS/AFFILIATES:
i3

CONTACTS: *Note: Officers with more than one job title may be intentionally listed here more than once.*
Anne Tomalin, Pres.
Lynda Rattenbury, Exec. Dir.-Human Resources
Harold DeVenne, VP-Oper.
Darrell Ethell, Exec Dir.-Bus. Dev.
Lynda Rattenbury, Exec. Dir.-Comm. Affairs
Harold DeVenne, VP-Finance
Patricia Anderson, VP-Regulatory Svcs.
Mary Speagle, Exec. Dir.-Canadian Regulatory Affairs
Stuart Wright, Exec. Dir.-CMC & Quality
Bernard Chiasson, Exec. Dir.-US Bus. & Medical Svcs.

Phone: 905-689-3980	**Fax:** 905-689-1465
Toll-Free: 866-722-6734	
Address: 4 Innovation Drive, Dundas, ON L9H 7P3 Canada	

GROWTH PLANS/SPECIAL FEATURES:

i3 CanReg, Inc., formerly CanReg, Inc., is a regulatory affairs consulting company, focusing on the pharmaceutical, biotechnology and medical device industries. Other clients include venture capitalists, contract research organizations, natural health product and food companies, cosmetic manufacturers and governments. The firm's more than 100 consultants and staff serve companies at all levels in Canada, the U.S. and Europe, serving either as the global regulatory affairs department or simply augmenting internal staff. The firm's Regulatory Operations Team works with the FDA, Health Canada and European regulatory agencies in both electronic and paper form. In Canada, i3 CanReg provides New Drug Submissions, Abbreviated New Drug Submissions, Clinical Trial Applications, DIN applications, Medical Device License Applications, Natural Health Product (NHP) Submissions, Notifiable Changes, Common Drug Review strategy and submissions, Provincial and Private Payer Formulary submissions and Patented Medicine Prices Review Board reporting, as well as a Health Canada Liaison and Canadian Agent services. For the U.S., i3 CanReg provides Investigational New Drug Applications, New Drug Applications, Abbreviated New Drug Applications, Biologic License Applications, Orphan Drug Applications and Medical Device Applications, as well as a liaison with the FDA, U.S. Agent service and structured product labeling. In Europe, the firm offers Marketing Authorization applications (centralized, de-centralized or national), Clinical Trial Submissions and Orphan Drug applications, as well as agency liaisons and European Agent services. Globally, the company provides regulatory assessments and strategy development, quality and compliance services, clinical compliance services, pharmacovigilance, medical writing, due diligence, regulatory operations (electronic and paper) and training. In December 2009, the firm was acquired by i3, a global pharmaceutical services company. Its core business has remained unaffected by the acquisition.

FINANCIALS: Sales and profits are in thousands of dollars—add 000 to get the full amount. 2009 Note: Financial information for 2009 was not available for all companies at press time.

2009 Sales: $	2009 Profits: $	**U.S. Stock Ticker: Subsidiary**
2008 Sales: $	2008 Profits: $	**Int'l Ticker:** Int'l Exchange:
2007 Sales: $	2007 Profits: $	Employees:
2006 Sales: $	2006 Profits: $	Fiscal Year Ends:
2005 Sales: $	2005 Profits: $	Parent Company: I3

SALARIES/BENEFITS:

Pension Plan:	ESOP Stock Plan:	Profit Sharing:	Top Exec. Salary: $	Bonus: $
Savings Plan:	Stock Purch. Plan:		Second Exec. Salary: $	Bonus: $

OTHER THOUGHTS:
Apparent Women Officers or Directors: 4
Hot Spot for Advancement for Women/Minorities: Y

LOCATIONS: ("Y" = Yes)

West:	Southwest:	Midwest:	Southeast:	Northeast:	International:
				Y	Y

ICON PLC

www.iconplc.com

Industry Group Code: 541712 Ranks within this company's industry group: Sales: 5 Profits: 4

Drugs:	Other:	Clinical:		Computers:		Services:	
Discovery:	AgriBio:	Trials/Services:	Y	Hardware:	Y	Specialty Services:	
Licensing:	Genetic Data:	Labs:	Y	Software:		Consulting:	Y
Manufacturing:	Tissue Replacement:	Equipment/Supplies:		Arrays:		Blood Collection:	
Generics:		Research/Development Svcs.:	Y	Database Management:		Drug Delivery:	
		Diagnostics:				Drug Distribution:	

TYPES OF BUSINESS:

Clinical Trial Research Services
Data Management & Analysis
Laboratory Services
Regulatory Affairs Consulting
Interactive Voice Response Systems
Strategic Consulting & Marketing

BRANDS/DIVISIONS/AFFILIATES:

ICON Central Laboratories
ICON Clinical Research
DOCS
ICON Development Solutions
ICON Medical Imaging
Prevalere Life Sciences Inc
Beacon Bioscience Inc

CONTACTS: Note: Officers with more than one job title may be intentionally listed here more than once.

Peter Gray, CEO
Ciaran Murray, CFO
Simon Holmes, VP-Mktg. & Market Dev.
Eimear Kenny, VP-Strategic Human Resources
Alan Morgan, Pres., Early Clinical Research & Labs
Michael McGrath, VP-IT
Ciaran Murray, Sec.
Niamh Murphy, Mgr.-Corp. Comm.
Mark Quigley, Sr. VP-Corp. Quality & Compliance
Tom O'Leary, Pres., Central Labs
John Hubbard, Pres., Clinical Research
Ted Gastineau, CEO-Medical Imaging
Bruce Given, Chmn.

Phone: 353-291-2000	Fax: 353-291-2700
Toll-Free:	
Address: S. County Business Park, Leopardstown, Dublin, Ireland 18 UK	

GROWTH PLANS/SPECIAL FEATURES:

ICON plc is a contract research organization serving the pharmaceutical, medical device and biotechnology industries. The company has 71 locations in 38 countries and operates in five divisions: ICON Central Laboratories; ICON Clinical Research; ICON Development Solutions; ICON Medical Imaging; and DOCS. ICON Central Laboratories offers services ranging from complex microbiology and chemistry to simple safety tests. ICON Clinical Research, ICON's core service, specializes in planning management, execution and analysis of Phase IIb-IV clinical trials. It has conducted over 1,000 clinical studies in over 31,000 centers worldwide in areas such as the central nervous system (CNS), oncology, cardiology, transplant dermatology, pediatrics, respiratory and urology. Research focuses include clinical chemistry, hematology, toxicology, infectious disease, immunology/serology, flow cytometry, pharmacogenomics and molecular diagnostics. ICON Development Solutions specializes in the strategy and delivery of early drug development. This division has expertise in early-phase clinical research; non-clinical development; and bioanalytical modeling and simulation. ICON Medical Imaging provides image based product development services to the biotechnology, medical device and pharmaceutical industries. Lastly, ICON Contracting Solutions and DOCS International combined to form DOCS, the company's staffing and consulting division. DOCS assists clients with staffing services, direct-hire recruitment support, and training programs. ICON also offers technological products which include ICOPhone, an Interactive Voice Recognition (IVR) system used to increase accuracy, efficiency and cost effectiveness of clinical trials; ICONet, an online tool offering access to case report form (CRF) pages, status reports and project management documentation; and Electronic Data Capture (EDC) Technology, which provides electronic data collection and data management for clinical trials. In March 2010, ICON opened an office in Manila, Philippines.

Employees are offered medical, dental and vision insurance; flexible spending accounts; short- and long-term disability coverage; life and AD&D insurance; a 401(k) plan with company match; and paid vacation time.

FINANCIALS: Sales and profits are in thousands of dollars—add 000 to get the full amount. 2009 Note: Financial information for 2009 was not available for all companies at press time.

2009 Sales: $887,612	2009 Profits: $91,558	U.S. Stock Ticker: ICLR
2008 Sales: $865,248	2008 Profits: $78,120	Int'l Ticker: IJF Int'l Exchange: Dublin-ISE
2007 Sales: $630,722	2007 Profits: $55,142	Employees: 7,170
2006 Sales: $649,826	2006 Profits: $38,304	Fiscal Year Ends: 12/31
2005 Sales: $326,658	2005 Profits: $13,545	Parent Company:

SALARIES/BENEFITS:

Pension Plan:	ESOP Stock Plan:	Profit Sharing:	Top Exec. Salary: $778,753	Bonus: $454,273
Savings Plan: Y	Stock Purch. Plan:		Second Exec. Salary: $648,961	Bonus: $502,961

OTHER THOUGHTS:

Apparent Women Officers or Directors: 1
Hot Spot for Advancement for Women/Minorities: Y

LOCATIONS: ("Y" = Yes)

West:	Southwest:	Midwest:	Southeast:	Northeast:	International:
Y	Y	Y	Y	Y	Y

IDERA PHARMACEUTICALS INC

www.iderapharma.com

Industry Group Code: 325412 Ranks within this company's industry group: Sales: 96   Profits: 62

Drugs:		Other:		Clinical:	Computers:	Services:
Discovery:	Y	AgriBio:		Trials/Services:	Hardware:	Specialty Services:
Licensing:	Y	Genetic Data:	Y	Labs:	Software:	Consulting:
Manufacturing:		Tissue Replacement:		Equipment/Supplies:	Arrays:	Blood Collection:
Generics:				Research/Development Svcs.:	Database Management:	Drug Delivery:
				Diagnostics:		Drug Distribution:

TYPES OF BUSINESS:

Drugs-Targeted Immune Therapies
Drugs-Cancer
Drugs-Infectious Diseases
Drugs-Respiratory Diseases
Drugs-Autoimmune Diseases

BRANDS/DIVISIONS/AFFILIATES:

IMO-2125
IMO-2055
IMO-3100
IMO-2134

CONTACTS: Note: Officers with more than one job title may be intentionally listed here more than once.

Sudhir Agrawal, CEO
Sudhir Agrawal, Pres.
Louis J. Arcudi III, CFO
Sudhir Agrawal, Chief Scientific Officer
Timothy M. Sullivan, VP-Dev. Programs & Alliance Mgmt.
Louis J. Arcudi III, Sec.
Rahul Jasuja, Dir.-Corp. Dev.
Louis J. Arcudi III, Treas.
Nicola La Monica, VP-Biology
Robert D. Arbeit, VP-Clinical Dev.
Ekambar R. Kandimalla, VP-Discovery
Steven J. Ritter, VP-Intellectual Property & Contracts
James B. Wyngaarden, Chmn.

Phone: 617-679-5500	Fax: 617-679-5592
Toll-Free:	
Address: 167 Sidney St., Cambridge, MA 02139 US	

GROWTH PLANS/SPECIAL FEATURES:

Idera Pharmaceuticals, Inc. is a biotechnology company engaged in the discovery and development of therapeutics that modulate immune responses through Toll-like Receptors (TLR) for the treatment of multiple diseases, including infectious diseases, autoimmune and inflammatory diseases, cancer, asthma and allergies and as vaccine adjuvants. The firm focuses on DNA- and RNA-based drug candidates, which are used to stimulate or block an immune response through the targeted TLR. Idera's research aims to modulate the immune system by using DNA and RNA compounds to interact with the TLR 7, 8 and 9 receptors. The firm maintains collaboration agreements with other pharmaceutical companies, allowing it to conduct a variety of different research programs simultaneously. The company's lead drug candidate, IMO-2055, is a cancer treatment being developed in conjunction with Merck Pharmaceuticals. IMO-2055 is in ongoing clinical trials for the treatment of squamous cell carcinoma of the head and neck, non-small cell lung cancer and colorectal cancer. Other drugs in development include IMO-2134, a treatment for asthma and allergies; IMO-2125, a treatment for hepatitis C; IMO-3100, a treatment for autoimmune diseases such as lupus, rheumatoid arthritis, multiple sclerosis, psoriasis, colitis, pulmonary inflammation and hyperlipidemia; and vaccines for cancer, infectious diseases and Alzheimer's Disease (also being developed with Merck). The company claims 66 U.S. patents or patent applications and 205 corresponding worldwide patents. In addition to other pharmaceutical companies, Idera also works with academic institutions. In November 2009, the firm announced the termination of its research collaboration with Novartis concerning IMO-2055.

FINANCIALS: Sales and profits are in thousands of dollars—add 000 to get the full amount. 2009 Note: Financial information for 2009 was not available for all companies at press time.

2009 Sales: $34,518	2009 Profits: $7,546	**U.S. Stock Ticker: IDRA**
2008 Sales: $26,450	2008 Profits: $1,509	**Int'l Ticker:** Int'l Exchange:
2007 Sales: $8,124	2007 Profits: $-13,208	Employees: 37
2006 Sales: $2,421	2006 Profits: $-16,525	Fiscal Year Ends: 12/31
2005 Sales: $2,467	2005 Profits: $-13,706	Parent Company:

SALARIES/BENEFITS:

Pension Plan:	ESOP Stock Plan:	Profit Sharing:	Top Exec. Salary: $510,000	Bonus: $357,000
Savings Plan:	Stock Purch. Plan:		Second Exec. Salary: $278,000	Bonus: $65,000

OTHER THOUGHTS:

Apparent Women Officers or Directors: 2
Hot Spot for Advancement for Women/Minorities:

LOCATIONS: ("Y" = Yes)

West:	Southwest:	Midwest:	Southeast:	Northeast:	International:
				Y	

IDEXX LABORATORIES INC

www.idexx.com

Industry Group Code: 325413 Ranks within this company's industry group: Sales: 2 Profits: 3

Drugs:	Other:	Clinical:		Computers:		Services:	
Discovery:	AgriBio:	Trials/Services:		Hardware:		Specialty Services:	
Licensing:	Genetic Data:	Labs:	Y	Software:	Y	Consulting:	Y
Manufacturing:	Tissue Replacement:	Equipment/Supplies:	Y	Arrays:		Blood Collection:	
Generics:		Research/Development Svcs.:		Database Management:		Drug Delivery:	
		Diagnostics:	Y			Drug Distribution:	

TYPES OF BUSINESS:

Veterinary Laboratory Testing & Consulting
Point-of-Care Diagnostic Products
Veterinary Pharmaceuticals
Information Management Software
Food & Water Testing Products

BRANDS/DIVISIONS/AFFILIATES:

VetTest
VetLyte
VetStat
LaserCyte
SNAPshot DX
Coag Dx
Colisure
ProCyte Dx Hematology Analyzer

CONTACTS: Note: Officers with more than one job title may be intentionally listed here more than once.

Jonathan W. Ayers, CEO
Jonathan W. Ayers, Pres.
Merilee Raines, CFO/VP
William E. Brown, III, Chief Scientific Officer/Sr. VP
Conan R. Deady, General Counsel/VP/Sec.
Dan Meyaard, VP-Worldwide Oper.
Merilee Raines, Treas.
James Polewaczyck, VP-Rapid Assay & Digital Radiography
Michael Williams, VP-Instrument Diagnostics
Thomas J. Dupree, VP-Companion Animal Group
Johnny D. Powers, VP-IDEXX Reference Laboratories
Jonathan W. Ayers, Chmn.
Ali Naqui, VP-Int'l

Phone: 207-556-0300	Fax: 207-556-4286
Toll-Free: 800-548-6733	
Address: 1 Idexx Dr., Westbrook, ME 04092-2041 US	

GROWTH PLANS/SPECIAL FEATURES:

IDEXX Laboratories, Inc. develops, manufactures and distributes products and provides services for the veterinary and the food and water testing markets. The company operates in two business segments: The Companion Animal Group, which provides products and services for the veterinary market; Water Quality; and the Production Animal Segment, which provides products for production animal health. The company also operated two smaller segments: Dairy, comprising products for dairy quality, and OPTI Medical, comprising products for the human medical diagnostic market. Its primary business focus is on animal health. IDEXX currently markets an integrated and flexible suite of in-house laboratory analyzers for use in veterinary practices, which is referred to as the VetLab suite of analyzers. The suite includes several instrument systems, as well as associated proprietary consumable products such as VetTest, VetLyte, VetStat and LaserCyte analyzers; the IDEXX SNAPshot Dx; and the Coag Dx Analyzer, among other offerings. In addition, it also provides assay kits, software and instrumentation for accurate assessment of infectious disease in production animals, such as cattle, swine and poultry. The company currently offers commercial veterinary laboratory and consulting services throughout the U.S. The Water Quality segment's products include Colilert-18 and Colisure tests, which simultaneously detect total coliforms and E. coli in water. IDEXX's principal product for use in testing for antibiotic residue in milk is the SNAP Beta-Lactam test. In April 2010, the firm launched ProCyte Dx Hematology Analyzer, which offers an advanced five-part white blood cell differential and a complete red blood cell analysis within two minutes.

IDEXX offers its employee health, dental and life insurance; a 401(k) plan; short- and long-term disability programs; flexible spending accounts; sick days; vacation and paid holidays; an employee stock purchase plan; and employee assistance programs.

FINANCIALS: Sales and profits are in thousands of dollars—add 000 to get the full amount. 2009 Note: Financial information for 2009 was not available for all companies at press time.

2009 Sales: $1,031,633	2009 Profits: $122,225	U.S. Stock Ticker: IDXX
2008 Sales: $1,024,030	2008 Profits: $116,169	Int'l Ticker: Int'l Exchange:
2007 Sales: $922,555	2007 Profits: $94,014	Employees: 4,800
2006 Sales: $739,117	2006 Profits: $93,678	Fiscal Year Ends: 12/31
2005 Sales: $638,095	2005 Profits: $78,254	Parent Company:

SALARIES/BENEFITS:

Pension Plan:	ESOP Stock Plan:	Profit Sharing:	Top Exec. Salary: $726,923	Bonus: $825,000
Savings Plan: Y	Stock Purch. Plan: Y		Second Exec. Salary: $321,923	Bonus: $240,000

OTHER THOUGHTS:

Apparent Women Officers or Directors: 2
Hot Spot for Advancement for Women/Minorities: Y

LOCATIONS: ("Y" = Yes)

West:	Southwest:	Midwest:	Southeast:	Northeast:	International:
Y	Y	Y	Y	Y	Y

Note: Financial information, benefits and other data can change quickly and may vary from those stated here.

ILLUMINA INC

www.illumina.com

Industry Group Code: 3345 Ranks within this company's industry group: Sales: 3   Profits: 2

Drugs:	Other:		Clinical:		Computers:		Services:	
Discovery:	AgriBio:		Trials/Services:		Hardware:	Y	Specialty Services:	Y
Licensing:	Genetic Data:	Y	Labs:		Software:	Y	Consulting:	
Manufacturing:	Tissue Replacement:		Equipment/Supplies:	Y	Arrays:	Y	Blood Collection:	
Generics:			Research/Development Svcs.:		Database Management:		Drug Delivery:	
			Diagnostics:				Drug Distribution:	

TYPES OF BUSINESS:

Instruments-Genetic Variation Measurement
Array Technology
Digital Microbead Technology
Software
Genotyping Services

BRANDS/DIVISIONS/AFFILIATES:

BeadArray Technology
Genome Analyzer
iScan System
Oligator
VeraCode
Illumina Sequencing
BeadXpress Reader
HiSeq 2000

CONTACTS: *Note: Officers with more than one job title may be intentionally listed here more than once.*

Jay T. Flatley, CEO
Jay T. Flatley, Pres.
Christian Henry, CFO/Sr. VP
Mark Lewis, Sr. VP-Product Dev.
Mostafa Ronaghi, CTO/Sr. VP
Christian G. Cabou, General Counsel/Sr. VP
Bill Bonnar, Sr. VP-Oper.
Jorge Velarde, VP-Bus. Dev.
Peter J. Fromen, Sr. Dir.-Investor Rel.
Christian Henry, Gen. Mgr.-Life Sciences
Gregory F. Heath, Sr. VP/Gen. Mgr.-Diagnostics Bus. Unit
Tristan Orpin, Sr. VP/Chief Commercial Officer
Emily Winn-Deen, VP-Diagnostics Dev.
William H. Rastetter, Chmn.
Elizabeth Brady, VP-Global Supply Chain

Phone: 858-202-4500	Fax: 858-202-4766
Toll-Free: 800-809-4566	
Address: 9885 Towne Centre Dr., San Diego, CA 92121 US	

GROWTH PLANS/SPECIAL FEATURES:

Illumina, Inc. develops tools for the large-scale analysis of genetic variation and function. The firm's tools provide genomic information that can be used to improve drugs and therapies, customize diagnoses and treatments and cure disease. The company deploys three primary systems for analyzing genetic variation and function: the Genome Analyzer, based on Illumina Sequencing technology, provides DNA sequencing services; the iScan System, based on BeadArray technology, performs multiple assays simultaneously; and the BeadXpress Reader, based on VeraCode technology acquired from the CyVera Corporation, is similar to the BeadArray but is used in lower multiplex projects. These use fiber optics to achieve a level of array miniaturization that allows experimentation to be easily scaled up. The firm also markets a series of sequencing kits and BeadChips for use in their instrumentation. Illumina arranges its arrays in patterns that match the wells of industry-standard microtiter plates, allowing higher throughput than other technologies. The company's other, complementary technology, Oligator, permits parallel synthesis of the millions of pieces of DNA necessary to perform large-scale genetic analysis. Illumina additionally provides genotyping services for other companies, as well as software, benchtop and production systems, installation and certain warranty services for its products. Illumina is one of six U.S. research groups participating in the International HapMap Project, a global consortium aimed at creating a detailed map of genetic variation. The firm is also one of three participating in the 1000 Genomes Project, which will build on HapMap. In October 2009, Illumina introduced cBot, a workflow automation solution for clonal amplification of sequencing libraries. In January 2010, the company released HiSeq 2000, a sequencer that allows researchers to obtain 30-fold coverage of two human genomes in one run.

Illumina offers its employees medical, disability, dental, vision and prescription drug plans; flexible spending accounts; and a stock purchase plan.

FINANCIALS: Sales and profits are in thousands of dollars—add 000 to get the full amount. 2009 Note: Financial information for 2009 was not available for all companies at press time.

2009 Sales: $666,324	2009 Profits: $72,281	**U.S. Stock Ticker: ILMN**
2008 Sales: $573,225	2008 Profits: $50,477	**Int'l Ticker:** Int'l Exchange:
2007 Sales: $366,799	2007 Profits: $-278,359	Employees:
2006 Sales: $184,586	2006 Profits: $39,968	Fiscal Year Ends: 12/31
2005 Sales: $73,501	2005 Profits: $-20,874	Parent Company:

SALARIES/BENEFITS:

Pension Plan:	ESOP Stock Plan:	Profit Sharing:	Top Exec. Salary: $749,162	Bonus: $498,731
Savings Plan: Y	Stock Purch. Plan: Y		Second Exec. Salary: $390,203	Bonus: $159,600

OTHER THOUGHTS:

Apparent Women Officers or Directors: 3
Hot Spot for Advancement for Women/Minorities: Y

LOCATIONS: ("Y" = Yes)

West:	Southwest:	Midwest:	Southeast:	Northeast:	International:
Y					Y

IMCLONE SYSTEMS INC

www.imclone.com

Industry Group Code: 325412 Ranks within this company's industry group: Sales: Profits:

Drugs:		Other:	Clinical:		Computers:		Services:
Discovery:	Y	AgriBio:	Trials/Services:		Hardware:		Specialty Services:
Licensing:		Genetic Data:	Labs:	Y	Software:		Consulting:
Manufacturing:	Y	Tissue Replacement:	Equipment/Supplies:		Arrays:		Blood Collection:
Generics:			Research/Development Svcs.:		Database Management:		Drug Delivery:
			Diagnostics:				Drug Distribution:

TYPES OF BUSINESS:

Drugs-Cancer

BRANDS/DIVISIONS/AFFILIATES:

Erbitux
Eli Lilly & Co
Bristol-Myers Squibb Company
Alexandria Center for Science and Technology

CONTACTS: Note: Officers with more than one job title may be intentionally listed here more than once.

Bernhard Ehmer, Pres.
Greg Reynolds, VP-Human Resources
Larry Witte, Sr. VP-Research
Greg Mayes, General Counsel/VP
Richard P. Crowley, Sr. VP-Biopharmaceutical Oper.
Andreas Harstrick, Sr. VP-Global Strategy
Tracy Henrikson, Sr. Dir.-Corp. Comm.
Hagop Youssoufian, Chief Medical Officer
Hagop Youssoufian, Sr. VP-Global Clinical Sciences

Phone: 908-218-9588	Fax: 908-704-8325
Toll-Free:	
Address: 33 ImClone Dr., Branchburg, NJ 08876 US	

GROWTH PLANS/SPECIAL FEATURES:

ImClone Systems, Inc. is a biopharmaceutical company that is engaged in the research and development of cancer treatments through growth factor blockers and angiogenesis inhibitors. The firm is a wholly-owned subsidiary of global pharmaceutical company Eli Lilly & Co. ImClone's commercially available product is Erbitux (Cetuximab), a growth factor blocker that has been approved by the U.S. Food and Drug Administration (FDA) for the treatment of metastatic colorectal cancer and of squamous cell carcinoma of the head and neck, or head and neck cancer. Erbitux works by binding specifically to EGFR on both normal and tumor cells, and competitively inhibits the binding of the epidermal growth factor and other growth stimulatory ligands, such as transforming growth factor-alpha. ImClone is currently working in collaboration with its partner, Bristol-Myers Squibb Company, in development trials of Erbitux for broader use in colorectal, head and neck, lung, pancreatic, esophageal, stomach, prostate, bladder and other cancers. In addition to the development and commercialization of Erbitux, the company has developed several investigational agents that address nearly all major solid tumors and are in mid and late stages of clinical development. Erbitux recently received marketing authorization in Japan for use in treating patients with advanced or metastatic colorectal cancer. In July 2009, ImClone agreed to move its research and development activities to the new Alexandria Center for Science and Technology at East River Science Park in New York City.

FINANCIALS: Sales and profits are in thousands of dollars—add 000 to get the full amount. 2009 Note: Financial information for 2009 was not available for all companies at press time.

2009 Sales: $	2009 Profits: $	U.S. Stock Ticker: Subsidiary
2008 Sales: $	2008 Profits: $	Int'l Ticker: Int'l Exchange:
2007 Sales: $590,833	2007 Profits: $39,799	Employees:
2006 Sales: $677,847	2006 Profits: $370,674	Fiscal Year Ends: 12/31
2005 Sales: $383,673	2005 Profits: $86,496	Parent Company: ELI LILLY & COMPANY

SALARIES/BENEFITS:

Pension Plan:	ESOP Stock Plan:	Profit Sharing:	Top Exec. Salary: $	Bonus: $
Savings Plan:	Stock Purch. Plan:		Second Exec. Salary: $	Bonus: $

OTHER THOUGHTS:

Apparent Women Officers or Directors: 1
Hot Spot for Advancement for Women/Minorities: Y

LOCATIONS: ("Y" = Yes)

West:	Southwest:	Midwest:	Southeast:	Northeast:	International:
				Y	

IMMTECH PHARMACEUTICALS INC
www.immtechpharma.com

Industry Group Code: 325412 Ranks within this company's industry group: Sales: 136   Profits: 92

Drugs:		Other:	Clinical:	Computers:	Services:	
Discovery:	Y	AgriBio:	Trials/Services:	Hardware:	Specialty Services:	Y
Licensing:		Genetic Data:	Labs:	Software:	Consulting:	
Manufacturing:		Tissue Replacement:	Equipment/Supplies:	Arrays:	Blood Collection:	
Generics:			Research/Development Svcs.:	Database Management:	Drug Delivery:	
			Diagnostics:		Drug Distribution:	

TYPES OF BUSINESS:
Pharmaceuticals Discovery & Development
Drugs-Antifungal
Drugs-Infectious Diseases
Drugs-Antibacterial

BRANDS/DIVISIONS/AFFILIATES:
Gold Avenue Ltd.

CONTACTS: Note: Officers with more than one job title may be intentionally listed here more than once.
Eric L. Sorkin, CEO
Gary C. Parks, CFO
Gary C. Parks, Sec.
Gary C. Parks, Treas.
Cecilia Chan, Vice Chmn.
Norman A. Abood, Sr. VP-Discovery Programs
Eric L. Sorkin, Chmn.

Phone: 212-791-2911	Fax: 212-791-2917
Toll-Free: 877-898-8038	
Address: 1 North End Ave., New York, NY 10282 US	

GROWTH PLANS/SPECIAL FEATURES:
Immtech Pharmaceuticals, Inc. focuses on the discovery and commercialization of therapeutics for the treatment of infectious diseases. In order to capitalize on the opportunities arising in the global healthcare market, the firm emphasizes the development of operations in China and other emerging markets. The company's drug candidates are produced using its proprietary library of aromatic cation compounds. These compounds have at least one positively charged end and at least one benzene ring in their structure, and many of them demonstrate the potential to prevent growth in pathogens. Immtech focuses on the composition of cations with links of different length, shape, flexibility and curvature, in order to target DNA and other receptors. The firm has also developed a pro-drug technology that improves the efficacy of aromatic cationic compounds by enhancing their ability to be absorbed and transported by the body. The company's drug development section currently has programs researching treatments in three primary areas: Hepatitis C (HCV), fungal diseases and multi-drug-resistant strains of bacterial infections. Within the category of bacterial infections, the company is developing compounds that target Staphylococcus aureus, methicillin-resistant Staphylococcus aureus (MRSA) and vancomycin-resistant enterococcus (VRE). Immtech has also generated data indicating that a subset of compounds cures bovine viral diarrhea virus (BVDV). In January 2009, the firm agreed to sell its Chinese subsidiary, Immtech Life Science Limited, for approximately $2 million. In April 2009, the company invested in Gold Avenue, Ltd., a firm involved in Chinese tin production.

FINANCIALS: Sales and profits are in thousands of dollars—add 000 to get the full amount. 2009 Note: Financial information for 2009 was not available for all companies at press time.
2009 Sales: $2,383	2009 Profits: $-6,502	U.S. Stock Ticker: IMMP
2008 Sales: $9,717	2008 Profits: $-10,513	Int'l Ticker: Int'l Exchange:
2007 Sales: $4,318	2007 Profits: $-11,133	Employees:
2006 Sales: $3,575	2006 Profits: $-15,526	Fiscal Year Ends: 3/31
2005 Sales: $5,931	2005 Profits: $-13,433	Parent Company:

SALARIES/BENEFITS:
Pension Plan:	ESOP Stock Plan:	Profit Sharing:	Top Exec. Salary: $375,000	Bonus: $
Savings Plan:	Stock Purch. Plan:		Second Exec. Salary: $235,000	Bonus: $

OTHER THOUGHTS:
Apparent Women Officers or Directors: 2
Hot Spot for Advancement for Women/Minorities: Y

LOCATIONS: ("Y" = Yes)
West:	Southwest:	Midwest:	Southeast:	Northeast:	International:
		Y		Y	

Note: Financial information, benefits and other data can change quickly and may vary from those stated here.

IMMUCELL CORPORATION

www.immucell.com

Industry Group Code: 325412B Ranks within this company's industry group: Sales: 2 Profits: 2

Drugs:		Other:		Clinical:		Computers:		Services:	
Discovery:	Y	AgriBio:	Y	Trials/Services:		Hardware:		Specialty Services:	
Licensing:	Y	Genetic Data:		Labs:		Software:		Consulting:	
Manufacturing:	Y	Tissue Replacement:		Equipment/Supplies:	Y	Arrays:		Blood Collection:	
Generics:				Research/Development Svcs.:		Database Management:		Drug Delivery:	
				Diagnostics:	Y			Drug Distribution:	

TYPES OF BUSINESS:

Diagnostic Tests & Products
Animal Health Drugs

BRANDS/DIVISIONS/AFFILIATES:

Wipe Out Dairy Wipes
California Mastitis Test
Mastik
MastOut
First Defense
RJT

CONTACTS: *Note: Officers with more than one job title may be intentionally listed here more than once.*

Michael F. Brigham, CEO
Michael F. Brigham, Pres.
Bobbi Jo Kunde, Dir.-Sales & Mktg.
Joseph H. Crabb, Chief Scientific Officer/VP
Michael F. Brigham, Sec.
Michael F. Brigham, Treas.
Joseph H. Crabb, Chmn.

Phone: 207-878-2770	Fax: 207-878-2117
Toll-Free: 800-466-8235	
Address: 56 Evergreen Dr., Portland, ME 04103 US	

GROWTH PLANS/SPECIAL FEATURES:

ImmuCell Corp. is a biotechnology company serving veterinarians and producers in the dairy and beef industries with products that improve animal health and productivity. The company sells diagnostic tests and products for therapeutic and preventive use against certain infectious diseases in animals and humans. The firm's leading product, First Defense, is an oral medicine manufactured from cows' colostrum using its proprietary vaccine and milk protein purification technologies. The drug targets bovine enteritis (calf scours), which causes diarrhea and dehydration in newborn calves and often leads to serious sickness or death. First Defense is sold primarily through major veterinarian distributors. ImmuCell also sells products designed to aid in the management of mastitis (inflammation of the mammary gland) caused by bacterial infections. The company's second leading source of product sales is Wipe Out Dairy Wipes, which consist of pre-moistened, biodegradable towelettes that are impregnated with Nisin, a natural antibacterial peptide, to prepare the teat area of a cow in advance of milking. Wipe Out Dairy Wipes are sold directly to dairy producers. The firm also offers MASTiK (Mastitis Antibiotic Susceptibility Test Kit), which helps veterinarians and producers select the antibiotic most likely to be effective in the treatment of individual cases of mastitis; California Mastitis Test (CMT), which can be performed at cow-side for early detection of mastitis; and RJT (Rapid Johne's Test) that can identify cattle with symptomatic Johne's disease. The company is currently developing a treatment called Mast Out, which is designed to treat Mastitis, and investigating therapies that could prevent scours in calves caused by enteric pathogens other than E. coli K99 and bovine coronavirus.

FINANCIALS: Sales and profits are in thousands of dollars—add 000 to get the full amount. 2009 Note: Financial information for 2009 was not available for all companies at press time.

2009 Sales: $4,509	2009 Profits: $- 216	U.S. Stock Ticker: ICCC
2008 Sales: $4,634	2008 Profits: $- 469	Int'l Ticker: Int'l Exchange:
2007 Sales: $6,069	2007 Profits: $ 662	Employees: 31
2006 Sales: $4,801	2006 Profits: $ 647	Fiscal Year Ends: 12/31
2005 Sales: $4,983	2005 Profits: $ 708	Parent Company:

SALARIES/BENEFITS:

Pension Plan:	ESOP Stock Plan:	Profit Sharing:	Top Exec. Salary: $187,380	Bonus: $1,000
Savings Plan: Y	Stock Purch. Plan:		Second Exec. Salary: $187,380	Bonus: $1,000

OTHER THOUGHTS:

Apparent Women Officers or Directors: 2
Hot Spot for Advancement for Women/Minorities:

LOCATIONS: ("Y" = Yes)

West:	Southwest:	Midwest:	Southeast:	Northeast:	International:
				Y	

IMMUNOGEN INC

www.immunogen.com

Industry Group Code: 325412 Ranks within this company's industry group: Sales: 102   Profits: 136

Drugs:		Other:	Clinical:	Computers:	Services:
Discovery:	Y	AgriBio:	Trials/Services:	Hardware:	Specialty Services:
Licensing:	Y	Genetic Data:	Labs:	Software:	Consulting:
Manufacturing:		Tissue Replacement:	Equipment/Supplies:	Arrays:	Blood Collection:
Generics:			Research/Development Svcs.:	Database Management:	Drug Delivery:
			Diagnostics:		Drug Distribution:

TYPES OF BUSINESS:

Anticancer Biopharmaceuticals
Drugs-Oncology

BRANDS/DIVISIONS/AFFILIATES:

Biogen Idec, Inc.
Genetech, Inc.
sanofi-aventis Group
Biotest AG

CONTACTS: Note: Officers with more than one job title may be intentionally listed here more than once.

Daniel M. Junius, CEO
Daniel M. Junius, Pres.
Gregory Perry, CFO/Sr. VP
John M. Lambert, Chief Science Officer/Exec. VP-R&D
Craig Barrows, General Counsel/VP/Sec.
Peter Williams, VP-Bus. Dev.
Gregory Perry, Treas.
James O'Leary, Chief Medical Officer/VP
Suzanne Cadden, VP-Regulatory Affairs & Quality
Godfrey Amphlett, VP-Process & Analytical Dev.
Robert J. Lutz, VP-Translational R&D
Stephen C. McCluski, Chmn.

Phone: 781-895-0600	Fax: 781-895-0611
Toll-Free:	
Address: 830 Winter St., Waltham, MA 02451-1477 US	

GROWTH PLANS/SPECIAL FEATURES:

ImmunoGen develops targeted therapeutics for the treatment of cancer. The company's Tumor-Activated Prodrug (TAP) technology uses antibodies to deliver a cytotoxic agent specifically to cancer cells. The TAP technology is designed to enable the creation of potent, well-tolerated anticancer products. TAP compounds are designed to bind to and destroy certain targets using monoclonal antibodies, for instance antigens on cancer cells. These cell-killing agents (CKAs) are believed to be far more powerful and potent than traditional chemotherapy techniques. The company then licenses this technology to a variety of pharmaceutical companies. ImmunoGen's collaborative partners include Biogen Idec, Inc.; Genentech, Inc.; the sanofi-aventis Group; and Biotest AG. The company's lead collaborative product in development is trastuzumab-DM1 (T-DM1), a CKA that targets HER2. The drug is currently in testing for indications of metastatic breast cancer. The firm has a licensing agreement for T-DM1 with Roche Group outside the U.S. and with its subsidiary Genentech for American sales. Other products in development include SAR3419, SAR566658 and SAR650984 in development for various solid and liquid cancers by sanofi-aventis; IMGN901 and IMGN388 for myeloma and other cancers, which are wholly-owned by ImmunoGen; BIIB015, in development for solid tumors by Biogen Idec; and BT-062, in development for myeloma by Biotest. All are TAP compounds. In July 2010, the firm, in collaboration with Genentech, submitted a Biologics License Application (BLA) to the FDA for marketing approval of T-DM1 for breast cancer treatment.

The company offers its employees health and dental coverage; life and AD&D insurance; a 401(k) plan; tuition reimbursement; and stock options.

FINANCIALS: Sales and profits are in thousands of dollars—add 000 to get the full amount. 2009 Note: Financial information for 2009 was not available for all companies at press time.

2009 Sales: $27,988	2009 Profits: $-31,937	**U.S. Stock Ticker:** IMGN
2008 Sales: $40,249	2008 Profits: $-32,020	**Int'l Ticker:** Int'l Exchange:
2007 Sales: $38,212	2007 Profits: $-18,987	Employees: 210
2006 Sales: $32,088	2006 Profits: $-17,834	Fiscal Year Ends: 6/30
2005 Sales: $35,718	2005 Profits: $-10,951	Parent Company:

SALARIES/BENEFITS:

Pension Plan:	ESOP Stock Plan:	Profit Sharing:	Top Exec. Salary: $482,227	Bonus: $196,508
Savings Plan: Y	Stock Purch. Plan:		Second Exec. Salary: $409,500	Bonus: $141,656

OTHER THOUGHTS:

Apparent Women Officers or Directors: 1
Hot Spot for Advancement for Women/Minorities:

LOCATIONS: ("Y" = Yes)

West:	Southwest:	Midwest:	Southeast:	Northeast:	International:
				Y	

IMMUNOMEDICS INC

www.immunomedics.com

Industry Group Code: 325413 Ranks within this company's industry group: Sales: 17 Profits: 11

Drugs:		Other:		Clinical:		Computers:		Services:	
Discovery:	Y	AgriBio:		Trials/Services:		Hardware:		Specialty Services:	
Licensing:	Y	Genetic Data:		Labs:		Software:		Consulting:	
Manufacturing:	Y	Tissue Replacement:		Equipment/Supplies:		Arrays:		Blood Collection:	
Generics:				Research/Development Svcs.:		Database Management:		Drug Delivery:	
				Diagnostics:	Y			Drug Distribution:	

TYPES OF BUSINESS:

Monoclonal Antibody-Based Products
Cancer & Autoimmune Products

BRANDS/DIVISIONS/AFFILIATES:

Epratuzumab
Veltuzumab
Yttrium Y 90
Milatuzumab
Yttrium Y 90 epratuzumab tetraxetan
Labetuzumab
LeukoScan
Dock-and-Lock

CONTACTS: Note: Officers with more than one job title may be intentionally listed here more than once.

Cynthia L. Sullivan, CEO
Cynthia L. Sullivan, Pres.
Gerald G. Gorman, CFO
David M. Goldenberg, Chief Scientific Officer/Chief Medical Officer
Gerard G. Gorman, Sr. VP-Bus. Dev.
Gerard G. Gorman, Sr. VP-Finance
David M. Goldenberg, Chmn.

Phone: 973-605-8200	Fax: 973-605-8282
Toll-Free:	
Address: 300 American Rd., Morris Plains, NJ 07950 US	

GROWTH PLANS/SPECIAL FEATURES:

Immunomedics, Inc. is a biopharmaceutical company focused on the development of monoclonal antibody-based products for the targeted treatment of cancer, autoimmune diseases and other diseases. The company creates humanized antibodies that can be used either alone in unlabeled form or conjugated with radioactive isotopes, chemotherapeutics or toxins to create targeted agents. Immunomedics has a number of therapeutic product candidates in clinical trials including epratuzumab, the firm's most advanced therapeutic product candidate that targets CD22; veltuzumab, an antigen that is expressed on B-lymphocytes; yttrium Y 90 for patients with pancreatic cancer; milatuzumab, which is currently in Phase I/II multicenter clinical trials to evaluate its safety in patients with multiple myeloma; yttrium Y 90 epratuzumab tetraxetan, a radiolabeled antibody product; and labetuzumab, a solid tumor therapeutic product candidate. Diagnostic imaging products manufactured by Immunomedics include LeukoScan for the detection of bone infections. LeukoScan has been approved in Canada, Europe and Australia. The company also offers ImmuSTRIP, a laboratory test kit for detection of human anti-murine antibodies (HAMA). The firm's majority-owned subsidiary, IBC Pharmaceuticals, Inc., develops radioimmunotherapeutics with target-specific antibodies. Together with IBC, Immunomedics developed the Dock-and-Lock method (DNL), which has the potential to make a considerable number of bioactive molecules of increasing complexity. The company has 137 U.S. patents and more than 200 issued foreign patents. In July 2009, the company entered into a partnership and cross-licensing agreement with Alexis Biotech Ltd. to jointly develop targeted vaccines against cancers that include melanoma and chronic lymphocytic leukemia. The development will combine Immunomedics' DNL technology with Alexis Biotech's HLA-antibody targeting technology.

The company offers its employees medical, dental and vision insurance; flexible spending accounts; life insurance; a 401(k) plan; and tuition reimbursement.

FINANCIALS: Sales and profits are in thousands of dollars—add 000 to get the full amount. 2009 Note: Financial information for 2009 was not available for all companies at press time.

2009 Sales: $30,021	2009 Profits: $2,274	**U.S. Stock Ticker:** IMMU
2008 Sales: $3,651	2008 Profits: $-22,909	**Int'l Ticker:** Int'l Exchange:
2007 Sales: $8,506	2007 Profits: $-16,655	Employees: 120
2006 Sales: $4,353	2006 Profits: $-28,764	Fiscal Year Ends: 6/30
2005 Sales: $3,813	2005 Profits: $-26,758	Parent Company:

SALARIES/BENEFITS:

Pension Plan:	ESOP Stock Plan:	Profit Sharing:	Top Exec. Salary: $532,000	Bonus: $160,000
Savings Plan: Y	Stock Purch. Plan:		Second Exec. Salary: $500,000	Bonus: $200,000

OTHER THOUGHTS:

Apparent Women Officers or Directors: 2
Hot Spot for Advancement for Women/Minorities: Y

LOCATIONS: ("Y" = Yes)

West:	Southwest:	Midwest:	Southeast:	Northeast:	International:
				Y	Y

Note: Financial information, benefits and other data can change quickly and may vary from those stated here.

IMPAX LABORATORIES INC

www.impaxlabs.com

Industry Group Code: 325412A Ranks within this company's industry group: Sales: 3 Profits: 4

Drugs:	Other:	Clinical:	Computers:	Services:
Discovery:	AgriBio:	Trials/Services:	Hardware:	Specialty Services:
Licensing:	Genetic Data:	Labs:	Software:	Consulting:
Manufacturing: Y	Tissue Replacement:	Equipment/Supplies:	Arrays:	Blood Collection:
Generics:		Research/Development Svcs.:	Database Management:	Drug Delivery: Y
		Diagnostics:		Drug Distribution:

TYPES OF BUSINESS:

Drug Delivery Systems
Drugs-Generic
Pharmaceutical Discovery & Development
Drugs-Central Nervous System

BRANDS/DIVISIONS/AFFILIATES:

Global Pharmaceuticals
IMPAX Pharmaceuticals
Cabergoline
Dronabinol

CONTACTS: *Note: Officers with more than one job title may be intentionally listed here more than once.*

Larry Hsu, CEO
Larry Hsu, Pres.
Arthur A. Koch, Jr., CFO
Charles V. Hildenbrand, Sr. VP-Oper.
Arthur A. Koch, Jr., Sr. VP-Finance
Michael J. Nestor, Pres., IMPAX Pharmaceuticals
Robert L. Burr, Chmn.
Christopher Mengler, Pres., Global Pharmaceuticals

Phone: 510-476-2000	Fax: 510-471-3200
Toll-Free:	
Address: 30831 Huntwood Ave., Hayward, CA 94544 US	

GROWTH PLANS/SPECIAL FEATURES:

IMPAX Laboratories, Inc. is a technology-based, specialty pharmaceutical company focused on the development and commercialization of generic and brand-name pharmaceuticals. In the generic pharmaceuticals market, the firm focuses on controlled-release generic versions of brand-name pharmaceuticals covering a range of therapeutic areas with technically challenging drug-delivery mechanisms or limited competition. The company also develops specialty generic pharmaceuticals that present barriers to entry by competitors, such as special handling requirements, difficulty in raw materials sourcing or complex formulation or development characteristics. In the brand-name pharmaceuticals market, IMPAX develops products for the treatment of central nervous system disorders. The company markets generic products through its Global Pharmaceuticals division and intends to market branded products through its IMPAX Pharmaceuticals division. The firm markets 83 generic pharmaceuticals representing dosage variations of 26 different pharmaceutical compounds through its Global Pharmaceuticals division and another 16 products representing dosage variations of four different pharmaceutical compounds through its alliance agreements' partners. It has 57 abbreviated new drug applications (ANDAs) approved by the U.S. Food and Drug Administration (FDA), including generic versions of Brethine, Florinef, Minocine, Claritin-D 12-hour, Claritin-D 24-hour, Wellbutrin SR and Prilosec; 32 applications pending at the FDA, including five tentatively approved; and 52 products in various stages of development. IMPAX also has one branded pharmaceutical product for which it has completed a Phase III clinical study, a second product in two Phase III clinical trials and other programs in the early exploratory phase. In December 2009, IMPAX acquired Watson Pharmaceuticals' ANDA for Cabergoline, the generic equivalent to Dostinex, and Cobalt Pharmaceuticals' pending ANDA for Dronabinol, the generic equivalent to Marinol.

IMPAX offers its employees flexible spending accounts; an employee assistance program; medical, dental, life and long-term disability insurance; supplemental insurance options; a 401(k) plan; and an employee stock purchase plan.

FINANCIALS: Sales and profits are in thousands of dollars—add 000 to get the full amount. 2009 Note: Financial information for 2009 was not available for all companies at press time.

2009 Sales: $358,409	2009 Profits: $50,061	**U.S. Stock Ticker:** IPXL
2008 Sales: $210,071	2008 Profits: $15,987	**Int'l Ticker:** Int'l Exchange:
2007 Sales: $273,753	2007 Profits: $125,410	Employees: 801
2006 Sales: $135,246	2006 Profits: $-12,044	Fiscal Year Ends: 12/31
2005 Sales: $	2005 Profits: $	Parent Company:

SALARIES/BENEFITS:

Pension Plan:	ESOP Stock Plan:	Profit Sharing:	Top Exec. Salary: $585,144	Bonus: $715,500
Savings Plan: Y	Stock Purch. Plan: Y		Second Exec. Salary: $459,038	Bonus: $389,375

OTHER THOUGHTS:

Apparent Women Officers or Directors:
Hot Spot for Advancement for Women/Minorities:

LOCATIONS: ("Y" = Yes)

West:	Southwest:	Midwest:	Southeast:	Northeast:	International:
Y				Y	

Note: Financial information, benefits and other data can change quickly and may vary from those stated here.

IMS HEALTH INC

www.imshealth.com

Industry Group Code: 541910 Ranks within this company's industry group: Sales: 1 Profits: 1

Drugs:	Other:	Clinical:	Computers:		Services:	
Discovery:	AgriBio:	Trials/Services:	Hardware:		Specialty Services:	Y
Licensing:	Genetic Data:	Labs:	Software:	Y	Consulting:	Y
Manufacturing:	Tissue Replacement:	Equipment/Supplies:	Arrays:		Blood Collection:	
Generics:		Research/Development Svcs.:	Database Management:	Y	Drug Delivery:	
		Diagnostics:			Drug Distribution:	

TYPES OF BUSINESS:

Market Research - Pharmaceuticals
Pharmaceutical Sales Tracking
Health Care Databases
Software-Sales Management & Market Research
Physician Profiling
Industry Audits
Prescription Tracking Reporting Services

BRANDS/DIVISIONS/AFFILIATES:

MIDAS
TPG Capital
Canadian Pension Plan Investment Board
TPG Inc

CONTACTS: *Note: Officers with more than one job title may be intentionally listed here more than once.*

David R. Carlucci, CEO
Giles V. J. Pajot, COO
Leslye G. Katz, CFO/Sr. VP
Karla L. Packer, Sr. VP-Human Resources
Harvey A. Ashman, General Counsel/Sr. VP-External Affairs
Kevin S. McKay, Sr. VP-Dev. & Customer Delivery
Murray L. Aitken, Sr. VP-Healthcare Insight
Tatsuyuki Saeki, Pres., Japan
Kevin Knightly, Sr. VP-Pharma Bus. Mgmt.
Deborah Kobewka, Gen. Mgr.-Asia/Pacific
David R. Carlucci, Chmn.
Adel Al-Saleh, Pres., EMEA

Phone: 203-845-5200	Fax:
Toll-Free:	
Address: 901 Main Ave., Ste. 612, Norwalk, CT 06851-1187 US	

GROWTH PLANS/SPECIAL FEATURES:

IMS Health, Inc., formerly TPG, Inc., is a business intelligence firm serving the healthcare and pharmaceutical industries. IMS Health serves clients in more than 100 countries from its 75 offices, located on six continents. It offers services in three areas: commercial effectiveness; product and portfolio offerings; and new business areas. Its commercial effectiveness business represented approximately 51% of its worldwide revenue in 2009. Its offerings in this area consist of intelligence designed to support the planning, development and execution of critical business processes such as sales territory reporting; prescription tracking reporting; sales and account management consulting services; promotional audits and promotion management consulting; and launch and brand management offerings. The company's product and portfolio management business represented 31% of its 2009 revenues. It offers in-depth business intelligence, analysis and forecasting to help customers identify and optimize their pharmaceutical product portfolios. Services in this area include pharmaceutical, medical, hospital and prescription audits; MIDAS, an online data analysis tool that provides access to the firm's databases of market trend information; and consulting, among others. New business services offered by IMS Health, which accounted for approximately 18% of 2009 revenues, include pricing and market access consulting; health economics and outcomes research offerings; pricing and reimbursement offerings; managed markets information and consulting; consumer health information; payer intelligence; and intelligence, analytics, tools and services aimed at government healthcare operations and decision making. The firm has a strong international presence, with approximately 63% of its total revenue in 2009 coming from outside the U.S. A vast majority of its revenue, approximately 85%, comes from sales to traditional pharmaceutical companies. In February 2010, IMS Health, Inc. agreed to be acquired by TPG Capital, a private equity firm, and the Canadian Pension Plan Investment Board for $5.8 billion.

FINANCIALS: Sales and profits are in thousands of dollars—add 000 to get the full amount. 2009 Note: Financial information for 2009 was not available for all companies at press time.

2009 Sales: $2,189,745	2009 Profits: $258,455	U.S. Stock Ticker: Private
2008 Sales: $2,329,528	2008 Profits: $311,250	Int'l Ticker: Int'l Exchange:
2007 Sales: $2,192,571	2007 Profits: $234,040	Employees: 7,250
2006 Sales: $1,958,588	2006 Profits: $315,511	Fiscal Year Ends: 12/31
2005 Sales: $1,754,791	2005 Profits: $284,091	Parent Company: TPG CAPITAL

SALARIES/BENEFITS:

Pension Plan:	ESOP Stock Plan:	Profit Sharing:	Top Exec. Salary: $850,000	Bonus: $567,500
Savings Plan:	Stock Purch. Plan:		Second Exec. Salary: $725,000	Bonus: $364,400

OTHER THOUGHTS:

Apparent Women Officers or Directors: 3
Hot Spot for Advancement for Women/Minorities: Y

LOCATIONS: ("Y" = Yes)

West:	Southwest:	Midwest:	Southeast:	Northeast:	International:
Y	Y			Y	Y

INCYTE CORP

www.incyte.com

Industry Group Code: 325412 Ranks within this company's industry group: Sales: 126   Profits: 166

Drugs:		Other:	Clinical:	Computers:	Services:
Discovery:	Y	AgriBio:	Trials/Services:	Hardware:	Specialty Services:
Licensing:	Y	Genetic Data:	Labs:	Software:	Consulting:
Manufacturing:		Tissue Replacement:	Equipment/Supplies:	Arrays:	Blood Collection:
Generics:			Research/Development Svcs.:	Database Management:	Drug Delivery:
			Diagnostics:		Drug Distribution:

TYPES OF BUSINESS:
Drug Discovery & Development
Drug Development
Drug Research

BRANDS/DIVISIONS/AFFILIATES:
INCB18424
INCB28050
INCB13739
INCB7839

CONTACTS: *Note: Officers with more than one job title may be intentionally listed here more than once.*
Paul A. Friedman, CEO
Paul A. Friedman, Pres.
David C. Hastings, CFO/Exec. VP
Paula J. Swain, Exec. VP-Human Resources
Brian W. Metcalf, Chief Drug Discovery Scientist/Exec. VP
Patricia A. Schreck, General Counsel/Exec. VP
Dan Maravei, Dir.-Bus. Dev.
Pamela M. Murphy, VP-Corp. Comm.
Pamela M. Murphy, VP-Investor Rel.
Patricia S. Andrews, Chief Commercial Officer/Exec. VP
Robert Newton, VP-Drug Discovery Biology
Richard U. DeSchutter, Chmn.

Phone: 302-498-6700	Fax: 302-425-2750
Toll-Free:	
Address: Rte. 141 & Henry Clay Rd., Bldg. E336, Wilmington, DE 19880 US	

GROWTH PLANS/SPECIAL FEATURES:

Incyte Corp., formerly Incyte Genomics, discovers and develops small molecule drugs focusing on oncology, diabetes, autoimmune diseases and inflammatory diseases. The company currently has multiple products in its pipeline, all within various stages of preclinical and clinical studies. The firm currently has three target inhibitors in its pipeline: Janus Kinase (JAK), HSD1 and Sheddase. The JAK program has been given the highest priority, and products in development are; INCB18424 (oral) for myelofibrosis, polycythemia vera/essential thrombocythemia and rheumatoid arthritis; INCB18424 (topical) for psoriasis; and INCB28050 for rheumatoid arthritis and autoimmune diseases. The drugs in development are designed to influence inflammatory cytokines that lead to the aforementioned conditions. HSD1 product INCB13739 is being developed for type II diabetes and is designed to reduce insulin resistance caused by cortisol. The Sheddase program, INCB7839, aims to treat breast cancer by inhibiting HER pathway signaling. Other programs in very early developmental stages include, c-MET and IDO, which are geared towards solid cancers and general oncology. In May 2009, the company initiated plans for Phase III trials of INC18424, the developmental treatment for myelofibrosis. In December 2009, the firm entered into an exclusive worldwide license agreement with Eli Lilly and Company for development and commercialization of JAK inhibitor INCB28050. Incyte also signed license agreements with Novartis for INCB18424 and INCB28050.

Incyte offers its employees medical, vision and dental plans; life insurance; disability coverage; a 401(k) plan; tuition reimbursement; employee discounts; referral and spot bonuses; and variable compensation plans.

FINANCIALS: Sales and profits are in thousands of dollars—add 000 to get the full amount. 2009 Note: Financial information for 2009 was not available for all companies at press time.

2009 Sales: $9,265	2009 Profits: $-211,870	**U.S. Stock Ticker: INCY**
2008 Sales: $3,919	2008 Profits: $-178,920	**Int'l Ticker:** Int'l Exchange:
2007 Sales: $34,440	2007 Profits: $-86,881	Employees: 221
2006 Sales: $27,643	2006 Profits: $-74,166	Fiscal Year Ends: 12/31
2005 Sales: $7,846	2005 Profits: $-103,043	Parent Company:

SALARIES/BENEFITS:

Pension Plan:	ESOP Stock Plan:	Profit Sharing:	Top Exec. Salary: $590,554	Bonus: $611,223
Savings Plan: Y	Stock Purch. Plan: Y		Second Exec. Salary: $399,562	Bonus: $275,698

OTHER THOUGHTS:
Apparent Women Officers or Directors: 5
Hot Spot for Advancement for Women/Minorities: Y

LOCATIONS: ("Y" = Yes)

West:	Southwest:	Midwest:	Southeast:	Northeast:	International:
				Y	

INFINITY PHARMACEUTICALS INC   www.infi.com

Industry Group Code: 541712 Ranks within this company's industry group: Sales: 10   Profits: 13

Drugs:	Other:	Clinical:	Computers:	Services:
Discovery: Y	AgriBio:	Trials/Services:	Hardware:	Specialty Services:
Licensing:	Genetic Data:	Labs:	Software:	Consulting:
Manufacturing:	Tissue Replacement:	Equipment/Supplies:	Arrays:	Blood Collection:
Generics:		Research/Development Svcs.:	Database Management:	Drug Delivery: Y
		Diagnostics:		Drug Distribution:

TYPES OF BUSINESS:
Research & Development Services
Drug Discovery Services
Drugs-Cancer

BRANDS/DIVISIONS/AFFILIATES:
IPI-504
IPI-493
IPI-926

CONTACTS: Note: Officers with more than one job title may be intentionally listed here more than once.
Adelene Q. Perkins, CEO
Adelene Q. Perkins, Pres.
Winselow S. Tucker, Jr., VP-Mktg.
Jeanette W. Kohlbrenner, VP-Human Resources
Vito J. Palombella,, Chief Scientific Officer
John J. Keilty, VP-IT & Informatics
Jeffrey K. Tong, VP-Prod. Dev.
Gerald E. Quirk, General Counsel
Jeffrey K. Tong, VP-Bus. Dev.
Gerald E. Quirk, VP-Corp. Affairs
Christopher M. Lindblom, Sr. Dir.-Finance/Controller
Tamara Toole, VP-Regulatory Affairs & Quality Assurance
Michael S. Curtis, VP-Pharmaceutical Dev.
Julian Adams, Pres., R&D
Steven H. Holtzman, Chmn.

Phone: 617-453-1000	Fax: 617-453-1001
Toll-Free:	
Address: 780 Memorial Dr., Cambridge, MA 02139 US	

GROWTH PLANS/SPECIAL FEATURES:
Infinity Pharmaceuticals, Inc. (IPI) discovers, develops and delivers medicines for the treatment of cancer and related conditions. IPI's discovery program has generated four clinical stage drug candidates spanning programs in the inhibition of heat shock protein 90 (Hsp90) system, the Hedgehog signaling pathway and fatty acid amide hydrolase (FAAH). The company's lead product candidate, IPI-504, was developed in collaboration with AstraZeneca/MedImmune (AZ/MI) and is an intravenously-administered small molecule inhibitor of Hsdp90. IPI-504 inhibits heat shock protein 90 (Hsp90), which is believed to have broad therapeutic potential for patients with solid and hematological tumors. It is currently undergoing a Phase II clinical trial in combination with Herceptin in patients with HER2-positive metastatic breast cancer; a Phase II clinical trial in patients with advanced non-small cell lung cancer; and a Phase I clinical trial in combination with Taxotere in patients with advanced solid tumors. The company's next most advanced program, IPI-926, was also developed in collaboration with AZ/MI and is directed against the Hedgehog cell pathway, which normally regulates tissue and organ formation during embryonic development. When abnormally activated during adulthood, the Hedgehog pathway is believed to play a central role in the proliferation and survival of certain cancer-causing cells, and is implicated in many of the most deadly cancers. The firm is also enrolling patients in a Phase I clinical trial of IPI-493, an orally-delivered inhibitor of Hsp90, in patients with advanced solid tumors. IPI has also developed compounds that target the Bcl-2 and Bcl-xL anti-apoptotic proteins, which play an important role in resistance to chemotherapy.

Infinity offers its employees health, dental, vision, AD&D and life insurance; a 401(k) plan; healthcare and dependent care flexible spending accounts; stock options; discounted auto and home insurance; a life-balance support program; fitness room access; a tuition reimbursement program; transportation and parking assistance; parental leave; and backup childcare.

FINANCIALS: Sales and profits are in thousands of dollars—add 000 to get the full amount. 2009 Note: Financial information for 2009 was not available for all companies at press time.

2009 Sales: $49,539	2009 Profits: $-32,505	U.S. Stock Ticker: INFI
2008 Sales: $83,441	2008 Profits: $23,654	Int'l Ticker: Int'l Exchange:
2007 Sales: $24,536	2007 Profits: $-16,898	Employees: 179
2006 Sales: $18,495	2006 Profits: $-28,448	Fiscal Year Ends: 12/31
2005 Sales: $ 522	2005 Profits: $-36,369	Parent Company:

SALARIES/BENEFITS:

Pension Plan:	ESOP Stock Plan:	Profit Sharing:	Top Exec. Salary: $520,000	Bonus: $208,000
Savings Plan: Y	Stock Purch. Plan: Y		Second Exec. Salary: $410,000	Bonus: $187,000

OTHER THOUGHTS:
Apparent Women Officers or Directors: 3
Hot Spot for Advancement for Women/Minorities: Y

LOCATIONS: ("Y" = Yes)

West:	Southwest:	Midwest:	Southeast:	Northeast:	International:
				Y	

INSITE VISION INC

Industry Group Code: 325412A Ranks within this company's industry group: Sales: 13   Profits: 13

Drugs:		Other:		Clinical:	Computers:		Services:	
Discovery:	Y	AgriBio:		Trials/Services:	Hardware:		Specialty Services:	
Licensing:	Y	Genetic Data:		Labs:	Software:		Consulting:	
Manufacturing:	Y	Tissue Replacement:		Equipment/Supplies:	Arrays:		Blood Collection:	
Generics:				Research/Development Svcs.:	Database Management:		Drug Delivery:	
				Diagnostics:			Drug Distribution:	

TYPES OF BUSINESS:

Drug Delivery Systems
Drugs-Ophthalmic
Eye Disease Diagnostics
Ear Infection Diagnostics

BRANDS/DIVISIONS/AFFILIATES:

DuraSite
AzaSite
AzaSite Xtra
AzaSite Plus
AquaSite
Besivance
ISV-305
ISV-502

CONTACTS: *Note: Officers with more than one job title may be intentionally listed here more than once.*

Louis Drapeau, CEO
Louis Drapeau, CFO/VP
Kamran Hosseini, Chief Medical Officer
Surendra Patel, VP-Oper.
Lyle M. Bowman, VP-Dev.
Surendra Patel, VP-Quality
Kamran Hosseini, VP-Clinical Affairs
Evan Melrose, Chmn.

Phone: 510-865-8800	Fax: 510-865-5700
Toll-Free:	
Address: 965 Atlantic Ave., Alameda, CA 94501 US	

GROWTH PLANS/SPECIAL FEATURES:

InSite Vision, Inc. is an ophthalmic product development company focusing on therapies that treat ocular infections and related diseases. Products are based on proprietary DuraSite eye drop drug delivery technology, a patented system that allows medication to stay in the eye for several hours rather than a few minutes, reducing the number of doses needed and lowering the potential for adverse side effects, as well as reducing the probability of bacterial resistance. InSite's lead product, AzaSite (a DuraSite formulation of azithromycin, a broad spectrum antibiotic) has been approved by the FDA for the treatment of bacterial conjunctivitis (pink eye). Inspire Pharmaceuticals commercialized AzaSite in the U.S. and Canada. InSite has also entered a licensing and distribution agreement with Nitten Pharmaceutical Co., Ltd., to market AzaSite in Japan and Taiwan. Two other products are also in development for ocular infections in the AzaSite family: AzaSite Plus, currently in Phase III trials; and AzaSite Xtra, currently in preclinical trials. Other products under development include ISV-502, for the treatment of blepharitis, an infection of the eyelid; ISV-305, for ocular inflammation; ISV-303, for post-operative inflammation and eye pain; and ISV-405, for eye infections. InSite's first product utilizing the DuraSite technology, AquaSite dry eye treatment, was launched as an over-the-counter medication by CIBA Vision. Besivance, a treatment for bacterial infection, was sold to and commercialized by Bausch and Lomb, Inc. In March 2009, InSite received regulatory approval to market Azasite in Canada. Also in March 2009, the firm announced a broad restructuring program in which it eliminated about 52% of its workforce.

InSite offers its employees medical, dental, vision, life and health insurance; short- and long-term disability coverage; an employee assistance program; travel assistance; a 401(k) plan; stock options; an employee stock purchase plan; and a health club reimbursement program.

FINANCIALS: Sales and profits are in thousands of dollars—add 000 to get the full amount. 2009 Note: Financial information for 2009 was not available for all companies at press time.

2009 Sales: $9,798	2009 Profits: $-14,155	**U.S. Stock Ticker: INSV**
2008 Sales: $13,706	2008 Profits: $-21,310	**Int'l Ticker:** Int'l Exchange:
2007 Sales: $23,761	2007 Profits: $5,535	Employees: 10
2006 Sales: $ 2	2006 Profits: $-16,611	Fiscal Year Ends: 12/31
2005 Sales: $ 4	2005 Profits: $-15,215	Parent Company:

SALARIES/BENEFITS:

Pension Plan:	ESOP Stock Plan:	Profit Sharing:	Top Exec. Salary: $263,750	Bonus: $55,388
Savings Plan: Y	Stock Purch. Plan: Y		Second Exec. Salary: $255,000	Bonus: $121,400

OTHER THOUGHTS:

Apparent Women Officers or Directors:
Hot Spot for Advancement for Women/Minorities:

LOCATIONS: ("Y" = Yes)

West:	Southwest:	Midwest:	Southeast:	Northeast:	International:
Y					

INSMED INCORPORATED

www.insmed.com

Industry Group Code: 325412 Ranks within this company's industry group: Sales: 124 Profits: 40

Drugs:		Other:	Clinical:	Computers:	Services:
Discovery:	Y	AgriBio:	Trials/Services:	Hardware:	Specialty Services:
Licensing:	Y	Genetic Data:	Labs:	Software:	Consulting:
Manufacturing:		Tissue Replacement:	Equipment/Supplies:	Arrays:	Blood Collection:
Generics:			Research/Development Svcs.:	Database Management:	Drug Delivery:
			Diagnostics:		Drug Distribution:

TYPES OF BUSINESS:

Drugs-Metabolic & Endocrine Diseases
Drugs-Cancer
Drugs-Amyotrophic Lateral Sclerosis

BRANDS/DIVISIONS/AFFILIATES:

IPLEX

CONTACTS: Note: Officers with more than one job title may be intentionally listed here more than once.

Kevin P. Tully, CFO/Exec. VP
Nicholas A. LaBella, Jr., Chief Scientific Officer
W. McIlwaine Thompson, Jr., Sec.
Steve Glover, Chief Business Officer
Glen Kelley, VP-Regulatory Affairs
Melvin Sharoky, Chmn.

Phone: 804-565-3000   Fax: 804-565-3500
Toll-Free:
Address: 8720 Stony Point Pkwy., Ste. 200, Richmond, VA 23235 US

GROWTH PLANS/SPECIAL FEATURES:

Insmed, Inc. is a biologics manufacturing and development stage company with expertise in recombinant protein drug development. In the proprietary protein field, the firm's lead product, IPLEX is FDA-approved. It is a composition of recombinant human IGF-1 and IGF binding protein 3. IPLEX is now being studied as a treatment for several medical conditions including ALS (Lou Gehrig's disease). The company has acquired non-exclusive patent rights to the use of an insulin-like growth factor (IGF-I) therapy for the treatment of extreme or severe insulin-resistant diabetes from Fujisawa Pharmaceutical Co., Ltd. Under the terms of the agreement, Insmed will obtain worldwide rights in territories, excluding Japan and including the U.S. and Europe. Because its Boulder, Colorado facility was sold as part of a transaction with Merck, the firm has lost the ability to manufacture IPLEX. The company has two oncology compounds, rhIGFBP-3 and INSM-18, in development. Preclinical models show that one or both treatments interact with the IGF-1 system to reduce tumor growth in models of breast, prostate, lung, colorectal and head and neck cancers. Taken orally, these compounds have shown capability in blocking tumor-associated IGF signaling and preventing tumor growth. An initial, Phase I trial has been completed with INSM-18, but without a partner the firm has been unable to initiate a clinical trial for rhIGFBP-3.

FINANCIALS: Sales and profits are in thousands of dollars—add 000 to get the full amount. 2009 Note: Financial information for 2009 was not available for all companies at press time.

2009 Sales: $10,373	2009 Profits: $118,350	U.S. Stock Ticker: INSM
2008 Sales: $11,699	2008 Profits: $-15,667	Int'l Ticker: Int'l Exchange:
2007 Sales: $7,581	2007 Profits: $-19,962	Employees: 15
2006 Sales: $1,025	2006 Profits: $-56,139	Fiscal Year Ends: 12/31
2005 Sales: $ 131	2005 Profits: $-40,929	Parent Company:

SALARIES/BENEFITS:

Pension Plan:	ESOP Stock Plan:	Profit Sharing:	Top Exec. Salary: $315,000	Bonus: $110,250
Savings Plan:	Stock Purch. Plan:		Second Exec. Salary: $275,000	Bonus: $110,250

OTHER THOUGHTS:

Apparent Women Officers or Directors:
Hot Spot for Advancement for Women/Minorities:

LOCATIONS: ("Y" = Yes)

West:	Southwest:	Midwest:	Southeast:	Northeast: Y	International:

Note: Financial information, benefits and other data can change quickly and may vary from those stated here.

INSPIRE PHARMACEUTICALS INC
www.inspirepharm.com

Industry Group Code: 325412 Ranks within this company's industry group: Sales: 75   Profits: 140

Drugs:		Other:		Clinical:	Computers:		Services:	
Discovery:	Y	AgriBio:		Trials/Services:	Hardware:		Specialty Services:	
Licensing:	Y	Genetic Data:		Labs:	Software:		Consulting:	
Manufacturing:	Y	Tissue Replacement:		Equipment/Supplies:	Arrays:		Blood Collection:	
Generics:				Research/Development Svcs.:	Database Management:		Drug Delivery:	
				Diagnostics:			Drug Distribution:	

TYPES OF BUSINESS:
Drugs-Respiratory & Ocular
Cystic Fibrosis Treatment
Retinal Disease Treatment
Dry Eye Treatment
Allergic Conjunctivitis
Cardiovascular Disease Treatment

BRANDS/DIVISIONS/AFFILIATES:
Restasis
Elestat
AzaSite
Prolacria
Diquafosol tetrasodium
Denufosol Tetrasodium

CONTACTS:
Note: Officers with more than one job title may be intentionally listed here more than once.
Adrian Adams, CEO
Adrian Adams, Pres.
Thomas R. Staab, II, CFO/Exec. VP
Francisca Yanez, VP-Human Resources
Andrew I. Koven, Chief Admin. Officer/Exec. VP
Joseph M. Spagnardi, General Counsel/Sr. VP/Sec.
Cara Amoroso, Mgr.-Corp. Comm.
R. Kim Brazzell, Exec. VP-Medical & Scientific Affairs
Andrew I. Koven, Chief Legal Officer
Joseph K. Schachle, Exec. VP-Pulmonary Bus.
Gerald St. Peter, Sr. VP-Ophthalmology Bus.
Kenneth B. Lee, Jr., Chmn.

Phone: 919-941-9777	Fax: 919-941-9797
Toll-Free:	
Address: 4222 Emperor Blvd., Ste. 200, Durham, NC 27703-8466 US	

GROWTH PLANS/SPECIAL FEATURES:

Inspire Pharmaceuticals, Inc. is a biopharmaceutical company dedicated to discovering, developing and commercializing prescription pharmaceutical products for ophthalmic and pulmonary diseases. Inspire's ophthalmic products and product candidates are concentrated in the allergic conjunctivitis, dry eye disease, cystic fibrosis, acute cardiac care, seasonal allergic rhinitis and glaucoma indications. The firm's product portfolio includes AzaSite, Elestat and Restasis. AzaSite is a topical anti-infective that is used to treat blephartis and is in Phase II development. Elestat is a topical antihistamine developed by Allergan for the prevention of ocular itching associated with allergic conjunctivitis. Restasis is used to treat dry eye disease. Products in clinical development include denufosol tetrasodium for cystic fibrosis, currently in Phase III trials; and INS115644 and INS117548 for glaucoma, both currently in Phase I; and Prolacria (diquafosol tetrasodium) for dry eye, which has completed Phase III testing and has received two approvable letters from the FDA. The firm recently entered into an agreement with Santen Pharmaceutical Co., Ltd., which entitled Santen to develop DE-089, a different formulation of diquafosol tetrasodium, in Japan. In April 2010, Inspire announced that the Japanese Ministry of Health, Labour and Welfare granted approval for DIQUAS Opthalmic Solution 3% (DE-809).

Inspire offers its employees medical, dental, vision, life and AD&D insurance; short- and long-term disability coverage; a 401(k) plan; flexible spending accounts; an employee assistance program; stock options; wellness benefits program; and activities including luncheons, winter holiday celebrations, monthly on-site massage therapists, weekly on-site yoga and pilates classes and corporate cycling and softball teams.

FINANCIALS:
Sales and profits are in thousands of dollars—add 000 to get the full amount. 2009 Note: Financial information for 2009 was not available for all companies at press time.

2009 Sales: $92,159	2009 Profits: $-39,976	**U.S. Stock Ticker:** ISPH
2008 Sales: $70,498	2008 Profits: $-51,603	**Int'l Ticker:** Int'l Exchange:
2007 Sales: $48,665	2007 Profits: $-63,740	Employees: 240
2006 Sales: $37,059	2006 Profits: $-42,115	Fiscal Year Ends: 12/31
2005 Sales: $23,266	2005 Profits: $-31,847	Parent Company:

SALARIES/BENEFITS:

Pension Plan:	ESOP Stock Plan:	Profit Sharing:	Top Exec. Salary: $467,943	Bonus: $421,149
Savings Plan: Y	Stock Purch. Plan:		Second Exec. Salary: $330,000	Bonus: $135,300

OTHER THOUGHTS:
Apparent Women Officers or Directors: 3
Hot Spot for Advancement for Women/Minorities: Y

LOCATIONS: ("Y" = Yes)

West:	Southwest:	Midwest:	Southeast:	Northeast:	International:
				Y	

Note: Financial information, benefits and other data can change quickly and may vary from those stated here.

INSTITUT STRAUMANN AG

www.straumann.com

Industry Group Code: 33911 Ranks within this company's industry group: Sales: 6 Profits: 5

Drugs:	Other:	Clinical:	Computers:	Services:
Discovery:	AgriBio:	Trials/Services:	Hardware:	Specialty Services:
Licensing:	Genetic Data:	Labs: Y	Software: Y	Consulting:
Manufacturing:	Tissue Replacement: Y	Equipment/Supplies:	Arrays:	Blood Collection:
Generics:		Research/Development Svcs.:	Database Management:	Drug Delivery:
		Diagnostics:		Drug Distribution:

TYPES OF BUSINESS:

Dental Implants
Dental Tissue Regeneration
CAD/CAM Elements & Equipment

BRANDS/DIVISIONS/AFFILIATES:

SLActive
BoneCeramic
etkon AG
Straumann CADCAM GmbH
Ormedent spol. s.r.o.
IVS Solutions AG
Allograft
IPS e.max

CONTACTS: Note: Officers with more than one job title may be intentionally listed here more than once.

Beat Spalinger, CEO
Beat Spalinger, Pres.
Wolf-Ruediger Daetz, CFO
Franz Maier, Exec. VP-Sales
Sandro Matter, Exec. VP-Prod.
Wolf-Ruediger Daetz, VP-Oper.
Mark Hill, VP-Corp. Comm.
Fabian Hildbrand, VP-Investor Rel.
Wolf-Ruediger Daetz, VP-Finance
Gilbert Achermann, Chmn.

Phone: 41-61-965-11-11	Fax: 41-61-965-11-01
Toll-Free:	
Address: Peter Merian-Weg 12, Basel, CH 4052 Switzerland	

GROWTH PLANS/SPECIAL FEATURES:

Institut Straumann AG (Straumann) is a world leader in implant and restorative dentistry products and a major provider of dental tissue regeneration products. It is a subsidiary of Straumann Holding AG, parent company of the Straumann Group. Straumann operates four geographical segments: Europe, (63% of net revenue in 2009); North America, (21%); Asia/Pacific, (13%) and the rest of the world, (3%). It maintains subsidiaries and distributors in more than 60 countries, with its most important markets being Germany and the U.S. The company offers three basic products: implants, regenerative systems and computer aided design and manufacturing (CAD/CAM) technology. Implants are designed to mimic natural teeth as closely as possible, being more durable and supporting themselves better than conventional bridges. They include devices inserted at the bone or soft tissue level, surgical tools and 3D modeling software used to plan surgery. Straumann's most recent product, Roxolid, is a high-performance implant material combining high tensile and fatigue strengths with osseointegration. The firm also offers SLActive, an implant surface technology that decreases the healing time after surgery and increases general stability of the dental implants. Regenerative systems include products that support or repair oral structures, such as Straumann Allograft. This product, along with BoneCeramic, a synthetic bone graft substitute that augments the patient's jaw bone, provides stability for synthetic implants. Lastly, CAD/CAM products, implemented following the acquisition of etkon AG, are divided into prosthetic elements (crowns, inlays, overlays or bridges) and manufacturing equipment (software, scanners and milling units). The newest addition to Straumann's CAD/CAM line, IPS e.max, is a ceramic-based prosthetic, which provides increased durability and aesthetic design. In 2009, the company entered the field of computer-guided implant surgery when it acquired the German firm IVS Solutions AG.

FINANCIALS: Sales and profits are in thousands of dollars—add 000 to get the full amount. 2009 Note: Financial information for 2009 was not available for all companies at press time.

2009 Sales: $705,990	2009 Profits: $140,350	**U.S. Stock Ticker:**
2008 Sales: $726,850	2008 Profits: $7,650	**Int'l Ticker: STMN** Int'l Exchange: Zurich-SWX
2007 Sales: $678,000	2007 Profits: $167,100	Employees: 2,040
2006 Sales: $569,200	2006 Profits: $134,600	Fiscal Year Ends: 12/31
2005 Sales: $395,800	2005 Profits: $121,500	Parent Company: STRAUMANN HOLDING AG

SALARIES/BENEFITS:

Pension Plan:	ESOP Stock Plan:	Profit Sharing:	Top Exec. Salary: $478,735	Bonus: $109,844
Savings Plan:	Stock Purch. Plan:		Second Exec. Salary: $	Bonus: $

OTHER THOUGHTS:

Apparent Women Officers or Directors:
Hot Spot for Advancement for Women/Minorities:

LOCATIONS: ("Y" = Yes)

West:	Southwest:	Midwest:	Southeast:	Northeast:	International:
				Y	Y

INTEGRA LIFESCIENCES HOLDINGS CORP   www.integra-ls.com

Industry Group Code: 33911 Ranks within this company's industry group: Sales: 7   Profits: 6

Drugs:	Other:	Clinical:	Computers:	Services:
Discovery:	AgriBio:	Trials/Services:	Hardware:	Specialty Services:
Licensing:	Genetic Data:	Labs:	Software:	Consulting:
Manufacturing: Y	Tissue Replacement:	Equipment/Supplies: Y	Arrays:	Blood Collection:
Generics:		Research/Development Svcs.:	Database Management:	Drug Delivery:
		Diagnostics:		Drug Distribution:

TYPES OF BUSINESS:
Medical Equipment Manufacturing
Implants & Biomaterials
Absorbable Medical Products
Tissue Regeneration Technology
Neurosurgery Products
Skin Replacement Products

BRANDS/DIVISIONS/AFFILIATES:
DuraGen Plus Dural Regeneration Matrix
Suturable DuraGen Dural Regeneration Matrix
DuraGen XS Dural Regeneration Matrix
AccuDrain External Ventricular Drainage Systems
Jarit
Luxtec
Miltex
Total Wrist Fusion System

CONTACTS: Note: Officers with more than one job title may be intentionally listed here more than once.
Stuart M. Essig, CEO
Gerard S. Carlozzi, COO/Exec. VP
Stuart M. Essig, Pres.
John B. Henneman III, CFO/Exec. VP-Finance
Deborah A. Leonetti, Chief Mktg. Officer
Richard D. Gorelick, Sr. VP-Human Resources
Simon J. Archibald, Chief Scientific Officer
Gabrielle Wolfson, CIO
John B. Henneman, III, Exec. VP-Admin.
Richard D. Gorelick, General Counsel/Sr. VP/Sec.
James A. Oti, Sr. VP-Global Oper.
Maria Platsis, VP-Corp. Dev.
Gianna Sabella, Dir.-Corp. Comm.
Angela Steinway, Mgr.-Investor Rel.
Jerry Corbin, Corp. Controller/VP
Nora Brennan, Treas./VP
Judith E. O'Grady, Sr. VP-Quality, Regulatory & Clinical Affairs
Brian Baker, Pres., Integra Pain Management
Richard E. Caruso, Chmn.
Eric Fourcault, Pres., EMEA Div.

Phone: 609-275-0500	Fax: 609-275-5363
Toll-Free: 800-654-2873	
Address: 311 Enterprise Dr., Plainsboro, NJ 08536 US	

GROWTH PLANS/SPECIAL FEATURES:

Integra Lifesciences Holdings Corporation develops, manufactures and markets surgical implants and medical instruments primarily for use in neurosurgery, orthopedics and general surgery. The company divides its products into three groups: neurosciences, orthopedics and medical instruments. The neurosciences group's products include duraplasty products such as the DuraGen Plus, Suturable DuraGen and DuraGen XS Dural Regeneration Matrices; tissue ablation equipment, which features the company's CUSA tissue ablation system; cerebral spinal fluid (CSF) management devices; cranial stabilization equipment, which features the MAYFIELD system; and intracranial monitoring equipment including the Camino ICP monitor, the LICOX brain tissue monitoring system and the AccuDrain External Ventricular Drainage Systems. The orthopedics group includes specialty metal implants for surgery of the extremities and spine; orthobiologics products for repair and grafting of bone; dermal regeneration products; tissue engineered wound dressings; and nerve and tendon repair products. The medical instruments group includes operating room instrumentation and surgical lighting. The Jarit instrument line offers reusable surgical instruments for laparoscopic, general, cardiovascular, neuro, gynecological, and orthopedic surgical specialties. Luxtec provides products for the surgical illumination market such as Xenon illumination systems, fiber optic cables and surgical loupes. Miltex provides dental instruments for oral surgery and periodontal and endodontic instrumentation. In March 2010, Integra launched the Integra Spine's Paramount Pedical Screw Fixation System for minimally invasive spine surgery; the Integra Mozaik Moldable Morsels bone void filler; and the Total Wrist Fusion System, a solution for wrist arthrodesis that provides wrist fixation while decreasing soft tissue irritation.

Integra offers its employees medical, dental and vision plans; flexible spending accounts; life and AD&D insurance; dependent life insurance; short- and long-term disability; an employee assistance program; an employee stock purchase plan; a 401(k) plan; and educational assistance.

FINANCIALS: Sales and profits are in thousands of dollars—add 000 to get the full amount. 2009 Note: Financial information for 2009 was not available for all companies at press time.

2009 Sales: $682,487	2009 Profits: $50,955	U.S. Stock Ticker: IART
2008 Sales: $654,604	2008 Profits: $27,727	Int'l Ticker: Int'l Exchange:
2007 Sales: $550,459	2007 Profits: $25,749	Employees: 3,000
2006 Sales: $419,297	2006 Profits: $29,407	Fiscal Year Ends: 12/31
2005 Sales: $277,935	2005 Profits: $37,194	Parent Company:

SALARIES/BENEFITS:

Pension Plan:	ESOP Stock Plan:	Profit Sharing:	Top Exec. Salary: $614,808	Bonus: $
Savings Plan: Y	Stock Purch. Plan: Y		Second Exec. Salary: $457,404	Bonus: $

OTHER THOUGHTS:
Apparent Women Officers or Directors: 8
Hot Spot for Advancement for Women/Minorities: Y

LOCATIONS: ("Y" = Yes)

West:	Southwest:	Midwest:	Southeast:	Northeast:	International:
Y		Y		Y	Y

INTELLIPHARMACEUTICS INTERNATIONAL INC

www.intellipharmaceutics.com
Industry Group Code: 325412 Ranks within this company's industry group: Sales: 143   Profits: 78

Drugs:		Other:	Clinical:	Computers:	Services:
Discovery:	Y	AgriBio:	Trials/Services:	Hardware:	Specialty Services:
Licensing:		Genetic Data:	Labs:	Software:	Consulting:
Manufacturing:		Tissue Replacement:	Equipment/Supplies:	Arrays:	Blood Collection:
Generics:			Research/Development Svcs.:	Database Management:	Drug Delivery:
			Diagnostics:		Drug Distribution:

TYPES OF BUSINESS:
Biopharmaceuticals Development
Controlled-Release Technologies
Generic Pharmaceuticals
Novel Oral Solid Drugs

BRANDS/DIVISIONS/AFFILIATES:
Dexmethylphenidate Hydrochloride (Generic Focalin)
Venlafaxine Hydrochloride (Generic Effexor)
Cardvedilol Phosphate (Generic Coreg CR)
Pantoprazole Sodium (Generic Protonix DR)
Hypermatrix
IntelliGIT
IntelliMatrix
Rexista

CONTACTS: *Note: Officers with more than one job title may be intentionally listed here more than once.*
Isa Odidi, CEO
Amina Odidi, COO
Amina Odidi, Pres.
Graham D. Neil, CFO
Patrick N. Yat, VP-Chemistry & Analytical Svcs.
John N. Allport, VP-Legal Affairs & Licensing
Graham D. Neil, VP-Finance
Isa Odidi, Chmn.

Phone: 41-798-3001	Fax: 416-798-3007
Toll-Free:	
Address: 30 Worcester Rd., Toronto, ON M9W 5X2 Canada	

GROWTH PLANS/SPECIAL FEATURES:
IntelliPharmaceutics International, Inc. researches, develops and manufactures controlled and targeted novel oral solid drugs for use in the treatment of neurological disorders, cardiovascular diseases, gastrointestinal issues, as well as pain and infection. The company's patented Hypermatrix technology is a unique multidimensional controlled-release drug delivery platform that can be applied to the development of a wide range of existing and new pharmaceuticals. Included within the group of Drug Delivery Engine technologies based on the Hypermatrix platform are IntelliGIT, IntelliMatrix, IntelliOsmotics, Intellipellets, IntelliShuttle; and Intellifoam, all of which provide different forms of controlled-release. The company's products in development fall into two categories: novel drugs that follow the New Drug Application (NDA) FDA regulatory pathway and controlled-released generics, which follow the Abbreviated New Drug Application (ANDA) pathway. On the NDA pathway is the company's Rexista product, an abuse- and alcohol-resistant controlled-release oral oxycodone formulation for the relief of pain. Rexista is currently in pre-clinical development. On the ANDA pathway IntelliPharmaceutics has four drugs, including Dexmethylphenidate Hydrochloride, a generic form of Focalin XR, used in the treatment of Attention Deficit Hyperactivity Disorder (ADHD); Venlafaxine Hydrochloride, a generic form of Effexor XR, for the treatment of clinical depression; Cardvedilol Phosphate, a generic form of Coreg CR, used for the treatment of hypertension and heart conditions; and Pantoprazole Sodium, a generic form of Protonix DR, for the treatment of Gastroesophageal Reflux Disease. Generic Focalin XR, Generic Protonix DR and Generic Effexor XR are awaiting FDA approval and IntelliPharmaceutics is seeking licensing agreements for the therapeutics. Generic Coreg CR is in late-stage clinical development.

FINANCIALS: Sales and profits are in thousands of dollars—add 000 to get the full amount. 2009 Note: Financial information for 2009 was not available for all companies at press time.

2009 Sales: $ 700	2009 Profits: $-2,000	**U.S. Stock Ticker: IPCI**
2008 Sales: $1,300	2008 Profits: $-3,800	**Int'l Ticker: VAS** Int'l Exchange: Toronto-TSX
2007 Sales: $2,300	2007 Profits: $-1,300	Employees: 23
2006 Sales: $	2006 Profits: $-61,100	Fiscal Year Ends: 11/30
2005 Sales: $90,700	2005 Profits: $-90,300	Parent Company:

SALARIES/BENEFITS:

Pension Plan:	ESOP Stock Plan:	Profit Sharing:	Top Exec. Salary: $383,481	Bonus: $
Savings Plan:	Stock Purch. Plan: Y		Second Exec. Salary: $	Bonus: $

OTHER THOUGHTS:
Apparent Women Officers or Directors:
Hot Spot for Advancement for Women/Minorities:

LOCATIONS: ("Y" = Yes)

West:	Southwest:	Midwest:	Southeast:	Northeast:	International:
					Y

INTERMUNE INC

www.intermune.com

Industry Group Code: 325412 Ranks within this company's industry group: Sales: 86   Profits: 161

Drugs:		Other:		Clinical:	Computers:	Services:
Discovery:	Y	AgriBio:		Trials/Services:	Hardware:	Specialty Services:
Licensing:	Y	Genetic Data:		Labs:	Software:	Consulting:
Manufacturing:		Tissue Replacement:		Equipment/Supplies:	Arrays:	Blood Collection:
Generics:				Research/Development Svcs.:	Database Management:	Drug Delivery:
				Diagnostics:		Drug Distribution:

TYPES OF BUSINESS:

Drugs-Pulmonology & Hepatology
Infectious Disease & Cancer Treatments
Drugs-Osteoporosis
Drugs-Chronic Granulomatous Disease

BRANDS/DIVISIONS/AFFILIATES:

Actimmune
Pirfenidone
ITMN-191
Esbriet
KDL GmbH
Marnac Inc

CONTACTS: *Note: Officers with more than one job title may be intentionally listed here more than once.*

Daniel G. Welch, CEO
Daniel G. Welch, Pres.
John Hodgman, CFO/Sr. VP
Howard Simon, Sr. VP-Human Resources & Corp. Svcs.
Williamson Bradford, Sr. VP-Clinical Science & Biometrics
Scott Seiwert, Sr. VP-Research & Tech Dev.
Robin Steele, General Counsel/Sr. VP/Corp. Sec.
Steven Porter, Chief Medical Officer/Sr. VP-Clinical Affairs
Marianne Armstrong, Chief Regulatory & Drug Safety Officer
Alan H. Cohen, Sr. VP-Medical Affairs
Daniel G. Welch, Chmn.
Giacomo Di Nepi, Managing Dir./Sr. VP-Europe

Phone: 415-466-2200	Fax: 415-466-2300
Toll-Free:	
Address: 3280 Bayshore Blvd., Brisbane, CA 94005 US	

GROWTH PLANS/SPECIAL FEATURES:

InterMune, Inc. is a biotechnology company that develops and commercializes products and therapies for pulmonary and hepatic diseases.   The company's core and sole approved product is Actimmune, which is used for the treatment of severe, malignant osteopetrosis and chronic granulomatous disease. The active ingredient in Actimmune, interferon gamma-1b, consists of a bioengineered version of a human protein that acts as a potent stimulator of the immune system. The company's developing drug pipeline falls into two categories, pulmonology and hepatology. InterMune is currently developing one pulmonary therapy for the treatment of idiopathic pulmonary fibrosis (IPF), pirfenidone. Pirfenidone, currently in Phase III trials and granted fast track treatment by the FDA, is an orally available small molecule that shows potential in inhibiting collagen synthesis and decreasing fibroblast proliferation. The company, along with co-licensor KDL GmbH, bought back a former license agreement for the drug from Marnac, Inc. The company plans to market Pirfenidone under the Esbriet trademark name. In hepatology, the company is developing protease inhibitor compound RG7227 (ITMN-191) ITMN-191 as a treatment for hepatitis C virus (HCV) infections and a protease inhibitor program. In February 2009, the firm was issued a U.S. patent for ITMN-191. In August 2009, ITMN-191 entered into a Phase II research program.

Employees of the firm are offered medical, dental and vision plans; short- and long-term disability coverage; life and AD&D coverage; a 401(k) plan; tuition reimbursement; stock options; flexible spending accounts; a 529 college savings plan; and employee referral benefits.

FINANCIALS: Sales and profits are in thousands of dollars—add 000 to get the full amount. 2009 Note: Financial information for 2009 was not available for all companies at press time.

2009 Sales: $48,700	2009 Profits: $-116,020	U.S. Stock Ticker: ITMN
2008 Sales: $48,152	2008 Profits: $-97,739	Int'l Ticker: Int'l Exchange:
2007 Sales: $66,692	2007 Profits: $-89,602	Employees: 121
2006 Sales: $90,784	2006 Profits: $-107,206	Fiscal Year Ends: 12/31
2005 Sales: $110,496	2005 Profits: $-5,235	Parent Company:

SALARIES/BENEFITS:

Pension Plan:	ESOP Stock Plan:	Profit Sharing:	Top Exec. Salary: $643,732	Bonus: $608,327
Savings Plan: Y	Stock Purch. Plan: Y		Second Exec. Salary: $342,002	Bonus: $143,641

OTHER THOUGHTS:

Apparent Women Officers or Directors: 1
Hot Spot for Advancement for Women/Minorities:

LOCATIONS: ("Y" = Yes)

West:	Southwest:	Midwest:	Southeast:	Northeast:	International:
Y					

INTERNATIONAL ISOTOPES INC

www.intisoid.com

Industry Group Code: 325 Ranks within this company's industry group: Sales: 7 Profits: 7

Drugs:	Other:		Clinical:		Computers:		Services:	
Discovery:	AgriBio:		Trials/Services:		Hardware:		Specialty Services:	Y
Licensing:	Genetic Data:		Labs:		Software:		Consulting:	
Manufacturing: Y	Tissue Replacement:		Equipment/Supplies:	Y	Arrays:		Blood Collection:	
Generics:			Research/Development Svcs.:		Database Management:		Drug Delivery:	
			Diagnostics:				Drug Distribution:	

TYPES OF BUSINESS:

Radioactive Isotopes Manufacturing
Nuclear Medicine Standards Publishing
Assay Analysis Services
Fluorine Gas
Depleted Uranium Oxide Products
Gemstone Processing

BRANDS/DIVISIONS/AFFILIATES:

International Isotopes Idaho, Inc.
International Isotopes Flourine Products, Inc.
International Isotopes Transportation Svcs., Inc.
RadQual LLC

CONTACTS: Note: Officers with more than one job title may be intentionally listed here more than once.

Steve T. Laflin, CEO
Steve T. Laflin, Pres.
Laurie A. McKenzie-Carter, CFO
Darin Lords, Mgr.-Cobalt Prod.
Chuckie Neitzel, Dir.-Admin. & Customer Service
John Miller, Radiation & Industrial Safety Officer
Jody Henley, Mgr.-Quality Assurance
Ralph M. Richart, Chmn.

Phone: 208-524-5300	Fax: 208-524-1411
Toll-Free: 800-699-3108	
Address: 4137 Commerce Cir., Idaho Falls, ID 83401 US	

GROWTH PLANS/SPECIAL FEATURES:

International Isotopes, Inc. (INIS) manufactures high purity fluoride gas and depleted uranium oxide products using the Fluorine Extraction Process (FEP). Subsidiaries of the firm include International Isotopes Idaho, Inc.; International Isotopes Flourine Products, Inc.; and International Isotopes Transportation Services, Inc. (IITS). The company operates in six segments: nuclear medicine standards (NMS); cobalt products; radiochemical products; fluorine products; radiological services; and transportation. The NMS segment consists of the manufacture of sources and standards associated with SPECT (single photon emission computed tomography), patient positioning and calibration or operational testing of dose measuring equipment for the nuclear pharmacy business. These items include flood sources; dose calibrators; rod sources; flexible and rigid rulers; spot markers; pen point markers; and specialty items. INIS is an exclusive manufacturer of these products for RadQual LLC. In the cobalt products segment, the firm manufactures bulk cobalt; fabricates cobalt capsules for teletherapy or irradiation devices; and recycles expended cobalt sources. In the radiochemical products segment, the company produces and distributes various isotopically pure radiochemicals for medical, industrial and research applications. The fluorine products segment uses seven patents for the FEP to produce high purity fluoride products such as germanium tetrafluoride, silicon tetrafluoride and boron trifluoride. The radiological services segment includes a variety of services, the largest of which is processing gemstones that have undergone irradiation for color enhancement. INIS has an exclusive contract with one customer for gemstone processing that accounts for most of the segment's sales. The transportation segment, through IITS, provides transportation for INIS's products and offers for-hire transportation services of hazardous and non-hazardous cargo materials. In March 2009, the company selected a site in Lea County, New Mexico for the construction of its depleted uranium de-conversion and fluorine extraction processing facility, which will be one of the first commercial uranium de-conversion facilities in the U.S.

FINANCIALS: Sales and profits are in thousands of dollars—add 000 to get the full amount. 2009 Note: Financial information for 2009 was not available for all companies at press time.

2009 Sales: $6,123	2009 Profits: $-4,633	**U.S. Stock Ticker:** INIS
2008 Sales: $5,602	2008 Profits: $-2,167	**Int'l Ticker:** Int'l Exchange:
2007 Sales: $4,691	2007 Profits: $-1,719	Employees: 25
2006 Sales: $4,470	2006 Profits: $-1,037	Fiscal Year Ends: 12/31
2005 Sales: $2,985	2005 Profits: $- 983	Parent Company:

SALARIES/BENEFITS:

Pension Plan:	ESOP Stock Plan:	Profit Sharing:	Top Exec. Salary: $170,549	Bonus: $35,000
Savings Plan:	Stock Purch. Plan:		Second Exec. Salary: $85,838	Bonus: $7,500

OTHER THOUGHTS:

Apparent Women Officers or Directors: 2
Hot Spot for Advancement for Women/Minorities:

LOCATIONS: ("Y" = Yes)

West:	Southwest:	Midwest:	Southeast:	Northeast:	International:
Y	Y				

IRONWOOD PHARMACEUTICALS INC   www.ironwoodpharma.com

Industry Group Code: 325412 Ranks within this company's industry group: Sales: 95 Profits: 150

Drugs:		Other:	Clinical:	Computers:	Services:
Discovery:	Y	AgriBio:	Trials/Services:	Hardware:	Specialty Services:
Licensing:	Y	Genetic Data:	Labs:	Software:	Consulting:
Manufacturing:		Tissue Replacement:	Equipment/Supplies:	Arrays:	Blood Collection:
Generics:			Research/Development Svcs.:	Database Management:	Drug Delivery:
			Diagnostics:		Drug Distribution:

TYPES OF BUSINESS:
Pharmaceuticals Discovery & Development

GROWTH PLANS/SPECIAL FEATURES:

Ironwood Pharmaceutical, Inc. is an entrepreneurial pharmaceutical firm that discovers, develops and plans to market innovative medicines. The company's leading product candidate, linaclotide, is a potential treatment for patients with irritable bowel syndrome with constipation (IBS-C) or chronic constipation (CC). Linaclotide is currently in Phase III clinical trials. Ironwood Pharmaceutical has also developed IW-6118, a novel mechanism agent that treats pain by inhibiting the enzyme Fatty Acid Amide Hydrolase (FAAH). IW-6118 is currently in Phase I clinical trials. In addition, the firm is conducting preclinical research on roughly eight potential therapies for inflammation, gastrointestinal pain and cardiovascular indications. The company's majority-owned subsidiary, Microbia, Inc., is a biomanufacturing firm based in Lexington, Massachusetts, which manufactures specialty ingredients and industrial biomaterials from renewable resources. In early 2009, Ironwood Pharmaceutical formed a licensing partnership with Almirall to develop and market linaclotide in Europe (including Turkey and the Commonwealth of Independent States) for the treatment of IBS-C and other gastrointestinal conditions. In November 2009, Ironwood Pharmaceutical agreed to licensing linaclotide to Astellas, allowing Astellas to develop and market linaclotide for the treatment of IBS-C and other gastrointestinal conditions in the Philippines, Taiwan, Japan, South Korea, Thailand and Indonesia.

BRANDS/DIVISIONS/AFFILIATES:
linaclotide
Microbia Inc

CONTACTS: *Note: Officers with more than one job title may be intentionally listed here more than once.*
Peter M. Hecht, CEO
Michael J. Higgins, COO
Michael J. Higgins, CFO
Thomas McCourt, Sr. VP-Mktg. & Sales
Mark G. Currie, Sr. VP-R&D/Chief Scientific Officer
Halley Gilbert, General Counsel/VP-Legal Affairs
James J. O'Mara, VP-Bus. Dev.
James M. DeTore, VP-Finance
Brian M. Cali, VP-Program Mgmt.
Jeffrey M. Johnston, VP-Clinical Dev./Chief Medical Officer
Thomas McCourt, Chief Commercial Officer
G. Todd Milne, VP-Biology
Joseph C. Cook, Jr., Chmn.

Phone: 617-621-7722	Fax: 617-494-0480
Toll-Free:	
Address: 301 Binney St., 2nd Fl., Cambridge, MA 02142 US	

FINANCIALS: Sales and profits are in thousands of dollars—add 000 to get the full amount. 2009 Note: Financial information for 2009 was not available for all companies at press time.

2009 Sales: $36,102	2009 Profits: $-71,185	U.S. Stock Ticker: IRWD
2008 Sales: $22,216	2008 Profits: $-53,874	Int'l Ticker: Int'l Exchange:
2007 Sales: $10,464	2007 Profits: $-52,752	Employees:
2006 Sales: $	2006 Profits: $	Fiscal Year Ends: 12/31
2005 Sales: $	2005 Profits: $	Parent Company:

SALARIES/BENEFITS:

Pension Plan:	ESOP Stock Plan:	Profit Sharing:	Top Exec. Salary: $315,000	Bonus: $8,000
Savings Plan:	Stock Purch. Plan:		Second Exec. Salary: $265,000	Bonus: $5,000

OTHER THOUGHTS:
Apparent Women Officers or Directors: 3
Hot Spot for Advancement for Women/Minorities: Y

LOCATIONS: ("Y" = Yes)

West:	Southwest:	Midwest:	Southeast:	Northeast:	International:
				Y	

ISIS PHARMACEUTICALS INC

www.isispharm.com

Industry Group Code: 325412 Ranks within this company's industry group: Sales: 70 Profits: 38

Drugs:		Other:		Clinical:	Computers:	Services:
Discovery:	Y	AgriBio:		Trials/Services:	Hardware:	Specialty Services:
Licensing:	Y	Genetic Data:	Y	Labs:	Software:	Consulting:
Manufacturing:	Y	Tissue Replacement:		Equipment/Supplies:	Arrays:	Blood Collection:
Generics:				Research/Development Svcs.:	Database Management:	Drug Delivery:
				Diagnostics:		Drug Distribution:

TYPES OF BUSINESS:

Drugs-Antisense Technology
Antisense Technology
Biosensors
Small-Molecule Drugs

BRANDS/DIVISIONS/AFFILIATES:

Vitravene
Mipomersen
BMS-PCSK9
ISIS-APOCIII
ISIS 113715
ISIS-EIF4E
ACHN-490
ISIS-FXIR

CONTACTS: Note: Officers with more than one job title may be intentionally listed here more than once.

Stanley T. Crooke, CEO
B. Lynne Parshall, COO
B. Lynne Parshall, CFO/Exec. VP
C. Frank Bennett, Sr. VP-Research
B. Lynne Parshall, Sec.
Richard S. Geary, Sr. VP-Dev.
Stanley T. Crooke, Chmn.

Phone: 760-931-9200	Fax: 760-603-2700
Toll-Free:	
Address: 1896 Rutherford Rd., Carlsbad, CA 92008-7208 US	

GROWTH PLANS/SPECIAL FEATURES:

Isis Pharmaceuticals, Inc. (Isis) is a pioneer in the area of RNA-based antisense technology, which involves direct application and interaction of gene sequence information to combat various diseases. The company successfully developed Vitravene, the first antisense drug to achieve marketing clearance, as a treatment for cytomegalovirus retinitis for patients living with AIDS. Isis subsequently licensed Vitravene to Novartis AG. The firm divides its drugs into four franchises: cardiovascular, metabolic, cancer and neurodegenerative. The cardiovascular franchise includes mipomersen, which is a potential treatment to reduce LDL-C in patients with high cholesterol and is currently in Phase 3 development; ISIS-CRP, which targets CRP, a protein produced by the liver; BMS-PCSK9, a drug that targets proprotein convertase subtilisin/kexin type 9, a protein involved in the metabolism of cholesterol and LDL; ISI-FXI to treat clotting disorders; and ISIS-APOCIII to lower triglycerides. The metabolic franchise includes ISIS 113715 to treat type 2 diabetes; ISIS-GCGR; ISIS-GCCR, which targets the glucocorticoid receptor; and ISIS-SGLT2, which targets sodium glucose co-transporter type 2. The cancer franchise includes OGX-011, which targets clusterin; LY2181308, which targets survivin; ISIS-EIF4E, which targets eukaryotic initiation factor 4-E; and OGX-427, which targets heat shock protein 27 (Hsp27). The firm's other drugs in development include ACHN-490 to treat multi-drug resistant gram-negative bacterial infections; AIR645; Alicaforsen, which targets intercellular adhesion molecule 1; ATL/TV1102; ATL1103 to reduce insulin-like IGF-1 in the liver; EXC 001 to treat fibrotic diseases; and iCo-007, which targets c-Raf kinase. Isis is the owner or exclusive licensee of approximately 1,600 RNA-based drug discovery and development patents. In December 2009, Isis added two new drugs to its development pipeline: ISIS-FXIR to treat clotting disorders, and ISIS-SMNR to treat spinal muscular atrophy.

Isis offers its employees medical, vision and dental coverage; flexible spending accounts; life, long-term disability and AD&D coverage; a 401(k) plan; an employee stock purchase plan; and a 529 college savings plan.

FINANCIALS: Sales and profits are in thousands of dollars—add 000 to get the full amount. 2009 Note: Financial information for 2009 was not available for all companies at press time.

2009 Sales: $121,600	2009 Profits: $150,672	U.S. Stock Ticker: ISIS
2008 Sales: $107,190	2008 Profits: $-22,906	Int'l Ticker: Int'l Exchange:
2007 Sales: $58,344	2007 Profits: $-40,079	Employees: 250
2006 Sales: $14,859	2006 Profits: $-45,903	Fiscal Year Ends: 12/31
2005 Sales: $40,133	2005 Profits: $-72,401	Parent Company:

SALARIES/BENEFITS:

Pension Plan:	ESOP Stock Plan:	Profit Sharing:	Top Exec. Salary: $683,649	Bonus: $361,650
Savings Plan: Y	Stock Purch. Plan: Y		Second Exec. Salary: $596,613	Bonus: $276,157

OTHER THOUGHTS:

Apparent Women Officers or Directors: 1
Hot Spot for Advancement for Women/Minorities:

LOCATIONS: ("Y" = Yes)

West:	Southwest:	Midwest:	Southeast:	Northeast:	International:
Y					

JAZZ PHARMACEUTICALS

www.jazzpharmaceuticals.com

Industry Group Code: 325412 Ranks within this company's industry group: Sales: 69   Profits: 93

Drugs:		Other:	Clinical:	Computers:	Services:
Discovery:	Y	AgriBio:	Trials/Services:	Hardware:	Specialty Services:
Licensing:	Y	Genetic Data:	Labs:	Software:	Consulting:
Manufacturing:		Tissue Replacement:	Equipment/Supplies:	Arrays:	Blood Collection:
Generics:			Research/Development Svcs.:	Database Management:	Drug Delivery:
			Diagnostics:		Drug Distribution:

TYPES OF BUSINESS:

Pharmaceuticals Discovery & Development
Neurological & Psychiatric Therapeutics

BRANDS/DIVISIONS/AFFILIATES:

Xyrem
Luvox CR
JZP-6
JZP-8
JZP-4
JZP-7

CONTACTS: *Note: Officers with more than one job title may be intentionally listed here more than once.*

Bruce C. Cozadd, CEO
Robert M. Myers, Pres.
Kathryn E. Falberg, CFO/Sr. VP
Edwin W. Luker, VP-Sales
Heather McGaughey, VP-Human Resources
Mark G. Eller, Sr. VP-Research & Clinical Dev.
Michael DesJardin, Sr. VP-Prod. Dev.
Carol A. Gamble, General Counsel/Sr. VP/Corp. Sec.
Joel M. Rothman, VP-Dev. Oper.
Diane R. Guinta, VP-Clinical Dev. & Scientific Affairs
Janne L. T. Wissel, Chief Regulatory & Compliance Officer/Sr. VP
P. J. Honerkamp, VP/Deputy General Counsel
Annette L. Madrid, Chief Medical Officer/VP-Clinical Medicine
Bruce C. Cozadd, Chmn.

Phone: 650-496-3777	Fax: 650-496-3781
Toll-Free:	
Address: 3180 Porter Dr., Palo Alto, CA 94304 US	

GROWTH PLANS/SPECIAL FEATURES:

Jazz Pharmaceuticals, Inc. develops and commercializes products that address unmet medical needs in neurology and psychiatry. The firm's portfolio includes two marketed products and several product candidates in various stages of clinical development. Its marketed products are Xyrem and Luvox CR. Xyrem is one of the only products approved by the U.S. Food and Drug Administration (FDA) for the treatment of both excessive daytime sleepiness and cataplexy in patients with narcolepsy. Luvox CR, a once-daily extended-release formulation of selective serotonin reuptake inhibitor fluvoxamine, was recently approved by the FDA for the treatment of both obsessive-compulsive disorder and social anxiety disorder. Jazz is also developing sodium oxybate (JZP-6), the active pharmaceutical ingredient in Xyrem, for the treatment of fibromyalgia. The firm's other product candidates include oral tablet forms of sodium oxybate; JZP-8 (intranasal clonazepam) for the treatment of recurrent acute repetitive seizures in epilepsy patients who continue to have seizures while on stable anti-epileptic regimens; JZP-4 (elpetrigine) for the treatment of epilepsy and bipolar disorder; and JZP-7 (ropinirole gel) for the treatment of restless legs syndrome. In June 2009, the firm presented Phase III data demonstrating that JZP-6 decreased fatigue and pain and improved daily function and patients' impression of change.

Jazz offers its employees medical and dependent care flexible spending accounts; a choice of Preferred Provider Organization (PPO) and Health Maintenance Organization (HMO) plans; a 401(k) plan; dental and vision insurance; and life, AD&D and disability insurance.

FINANCIALS: Sales and profits are in thousands of dollars—add 000 to get the full amount. 2009 Note: Financial information for 2009 was not available for all companies at press time.

2009 Sales: $128,449	2009 Profits: $-6,836	**U.S. Stock Ticker: JAZZ**
2008 Sales: $67,514	2008 Profits: $-184,339	**Int'l Ticker:** Int'l Exchange:
2007 Sales: $65,303	2007 Profits: $-138,778	Employees: 228
2006 Sales: $44,856	2006 Profits: $-59,391	Fiscal Year Ends: 12/31
2005 Sales: $21,442	2005 Profits: $-85,156	Parent Company:

SALARIES/BENEFITS:

Pension Plan:	ESOP Stock Plan:	Profit Sharing:	Top Exec. Salary: $442,729	Bonus: $205,300
Savings Plan: Y	Stock Purch. Plan:		Second Exec. Salary: $420,024	Bonus: $193,900

OTHER THOUGHTS:

Apparent Women Officers or Directors: 6
Hot Spot for Advancement for Women/Minorities: Y

LOCATIONS: ("Y" = Yes)

West:	Southwest:	Midwest:	Southeast:	Northeast:	International:
Y					

JOHNSON & JOHNSON

www.jnj.com

Industry Group Code: 325412 **Ranks within this company's industry group:** Sales: 1 Profits: 2

Drugs:		Other:	Clinical:		Computers:		Services:	
Discovery:	Y	AgriBio:	Trials/Services:		Hardware:		Specialty Services:	
Licensing:		Genetic Data:	Labs:		Software:		Consulting:	
Manufacturing:	Y	Tissue Replacement:	Equipment/Supplies:	Y	Arrays:		Blood Collection:	
Generics:			Research/Development Svcs.:		Database Management:		Drug Delivery:	
			Diagnostics:	Y			Drug Distribution:	

TYPES OF BUSINESS:

Personal Health Care & Hygiene Products
Sterilization Products
Surgical Products
Pharmaceuticals
Skin Care Products
Baby Care Products
Contact Lenses
Medical Equipment

BRANDS/DIVISIONS/AFFILIATES:

Alza Corp
Cordis Corp
DePuy Inc
Ethicon Inc
LifeScan Inc
Mentor Corp
Elan Corp PLC
Acclarent, Inc.

CONTACTS: Note: Officers with more than one job title may be intentionally listed here more than once.

William C. Weldon, CEO
Dominic J. Caruso, CFO
Russell C. Deyo, VP-Human Resources
Russell C. Deyo, General Counsel
Dominic J. Caruso, VP-Finance
Colleen Goggins, Chmn.-Consumer Group
Sherilyn S. McCoy, Chmn.-Pharmaceutical Group
Alex Gorsky, Chmn.-Medical Devices & Diagnostics Group
William C. Weldon, Chmn.

Phone: 732-524-0400	Fax: 732-214-0332
Toll-Free:	
Address: 1 Johnson & Johnson Plz., New Brunswick, NJ 08933 US	

GROWTH PLANS/SPECIAL FEATURES:

Johnson & Johnson, founded in 1886, is one of the world's most comprehensive and well-known manufacturers of health care products. The firm owns more than 250 companies in over 57 countries and markets its products in almost every country in the world. Johnson & Johnson's worldwide operations are divided into three segments: consumer health care; pharmaceuticals; and medical devices and diagnostics. The company's principal consumer goods are personal care and hygiene products, including nonprescription drugs, adult skin and hair care, baby care, oral care, first aid and sanitary protection products. Major consumer brands include Mylanta, Band-Aid, Listerine, Tylenol, Aveeno and Monistat. The pharmaceutical segment covers a wide spectrum of health fields, including antifungal, anti-infective, cardiovascular, dermatology, immunology, pain management, psychotropic and women's health. Among its pharmaceutical products are Risperdal, an antipsychotic used to treat schizophrenia, and Remicade for the treatment of Crohn's disease and rheumatoid arthritis. In the medical devices and diagnostics segment, Johnson & Johnson makes a number of products including suture and mechanical wound closure products, surgical instruments, disposable contact lenses, joint reconstruction products and intravenous catheters. Subsidiaries of the company include Cordis LLC; DePuy, Inc.; Diabetes Diagnostics, Inc.; Ethicon Endo-Surgery, Inc.; LifeScan, Inc.; Neutrogena Corporation; SurgRx, Inc.; The Tylenol Company; and McNeil Nutritionals, LLC, which oversees the Splenda sweetener brand. Also in July 2009, Johnson & Johnson invested approximately $1 billion for an 18.4% stake in Elan Corporation, an Irish biotechnology firm. In September 2009, the company acquired Elan's Alzheimer's drug development program. In January 2010, the firm's subsidiary Ethicon, Inc. acquired Acclarent, Inc., a California-based medical device company focused on sinus surgery technologies, for approximately $785 million.

Johnson & Johnson offers its employees benefits that include medical coverage; an employee assistance program; health assessments and health counseling; and on-site fitness centers and fitness classes at certain locations.

FINANCIALS: Sales and profits are in thousands of dollars—add 000 to get the full amount. 2009 Note: Financial information for 2009 was not available for all companies at press time.

2009 Sales: $61,897,000	2009 Profits: $12,266,000	**U.S. Stock Ticker:** JNJ
2008 Sales: $63,747,000	2008 Profits: $12,949,000	**Int'l Ticker:** Int'l Exchange:
2007 Sales: $61,095,000	2007 Profits: $10,576,000	**Employees:** 115,500
2006 Sales: $53,324,000	2006 Profits: $11,053,000	**Fiscal Year Ends:** 12/31
2005 Sales: $50,514,000	2005 Profits: $10,060,000	**Parent Company:**

SALARIES/BENEFITS:

Pension Plan:	ESOP Stock Plan:	Profit Sharing:	Top Exec. Salary: $1,802,500	Bonus: $12,831,146
Savings Plan: Y	Stock Purch. Plan:		Second Exec. Salary: $831,838	Bonus: $3,608,760

OTHER THOUGHTS:

Apparent Women Officers or Directors: 4
Hot Spot for Advancement for Women/Minorities: Y

LOCATIONS: ("Y" = Yes)

West:	Southwest:	Midwest:	Southeast:	Northeast:	International:
Y	Y	Y	Y	Y	Y

Note: Financial information, benefits and other data can change quickly and may vary from those stated here.

JUBILANT BIOSYS LTD

www.jubilantbiosys.com

Industry Group Code: 325414 Ranks within this company's industry group: Sales: Profits:

Drugs:		Other:	Clinical:	Computers:		Services:	
Discovery:	Y	AgriBio:	Trials/Services:	Hardware:		Specialty Services:	Y
Licensing:		Genetic Data:	Labs:	Software:	Y	Consulting:	
Manufacturing:		Tissue Replacement:	Equipment/Supplies:	Arrays:		Blood Collection:	
Generics:			Research/Development Svcs.:	Database Management:		Drug Delivery:	
			Diagnostics:			Drug Distribution:	

TYPES OF BUSINESS:

Biotech Drug Discovery & Development
Medicinal Chemistry
Structural Biology
Discovery Biology
Molecular Modeling
Pharmaceutical IT

BRANDS/DIVISIONS/AFFILIATES:

Jubilant Organosys Ltd
Jubilant Group
Jubilant Discovery Research Center
iAnnotate
Legend

CONTACTS: *Note: Officers with more than one job title may be intentionally listed here more than once.*

Sridhar (Sri) Mosur, CEO
Kailash Swarna, COO/Sr. VP
Sridhar (Sri) Mosur, Pres.
Kankana Barua, VP/Head-Global Human Resources
Raman Govindarajan, Sr. VP-Discovery Biology & Translational Medicine
Amit Kumar Rustagi, Head-IT
Warren Stern, Sr. VP-Drug Dev.
Melwyn A. Abreo, Dir.-Bus. Dev.
Sridhar (Sri) Mosur, CEO/Pres., Jubilant Discovery Services, Inc.
Pankaj Garg, CFO-Global Drug Discovery & Dev.
Jonathan P. Northrup, COO-Drug Dev.
Sriram Rajagopal, VP-Biology
Hari S. Bhartia, Chmn.

Phone: 91-120-251-6601	Fax: 91-120-251-6628
Toll-Free:	
Address: 1A, Sec 16A, Noida, 201301 India	

GROWTH PLANS/SPECIAL FEATURES:

Jubilant Biosys Ltd. is an Indian drug discovery firm that primarily operates in collaboration with major pharmaceutical and biotech companies. It is a subsidiary of Jubilant Organosys Ltd., the primary company in the pharmaceuticals branch of Indian conglomerate the Jubilant Group. The company's primary facility is the Jubilant Discovery Research Center, a 125,000-square-foot lab in Bangalore, India, specializing in medicinal chemistry, structural biology, discovery biology, molecular modeling, pharmacology, toxicology, ADME (adsorption, distribution, metabolism and excretion) and pharmaceutical information technology (IT). This facility is capable of assisting clients from the initial target identification of a new drug through to the investigational new drug (IND) application. Specifically, the facility's integrated drug discovery services include the following divisions. Computational chemistry services comprising computer modeling and analysis tools, including databases of drugs and other chemicals. Structural biology services primarily relate to the crystallography of protein structures, including imaging proteins, purifying samples and generating expressions of protein variations. Medicinal chemistry services include tailoring chemicals to hit specific targets, designing novel chemicals and providing biological profiles of new chemicals. Drug metabolism and pharmacokinetics (DMPK) services include analyzing the ADME properties of a drug, such as its toxicology potential and pathways it follows inside the body. In-vivo biology services comprise preclinical efficacy and safety testing for inflammation, cancer, pain and other therapeutic areas. Lastly, pharmaceutical IT products services include application development and maintenance (ADM), remote IT infrastructure support and proprietary products designed for the pharmaceuticals market, including protein imaging solution iAnnotate and screening and data management program Legend. In May 2009, the company agreed to assist AstraZeneca research new preclinical candidates. In June 2009, the firm partnered with U.S.-based Endo Pharmaceuticals to develop new oncology treatments. In January 2010, Jubilant entered a co-marketing agreement with DiscoveRx Corp., gaining access to its PathHunter and cAMPHunter cell lines.

FINANCIALS: Sales and profits are in thousands of dollars—add 000 to get the full amount. 2009 Note: Financial information for 2009 was not available for all companies at press time.

2009 Sales: $	2009 Profits: $	**U.S. Stock Ticker: Subsidiary**	
2008 Sales: $	2008 Profits: $	**Int'l Ticker:** Int'l Exchange:	
2007 Sales: $	2007 Profits: $	Employees:	
2006 Sales: $	2006 Profits: $	Fiscal Year Ends:	
2005 Sales: $	2005 Profits: $	Parent Company: JUBILANT ORGANOSYS LTD	

SALARIES/BENEFITS:

Pension Plan:	ESOP Stock Plan:	Profit Sharing:	Top Exec. Salary: $	Bonus: $
Savings Plan:	Stock Purch. Plan:		Second Exec. Salary: $	Bonus: $

OTHER THOUGHTS:

Apparent Women Officers or Directors: 1
Hot Spot for Advancement for Women/Minorities:

LOCATIONS: ("Y" = Yes)

West:	Southwest:	Midwest:	Southeast:	Northeast:	International:
				Y	Y

Note: Financial information, benefits and other data can change quickly and may vary from those stated here.

KENDLE INTERNATIONAL INC

www.kendle.com

Industry Group Code: 325412 Ranks within this company's industry group: Sales: 47 Profits: 53

Drugs:	Other:	Clinical:		Computers:		Services:	
Discovery:	AgriBio:	Trials/Services:	Y	Hardware:		Specialty Services:	Y
Licensing:	Genetic Data:	Labs:		Software:	Y	Consulting:	Y
Manufacturing:	Tissue Replacement:	Equipment/Supplies:		Arrays:		Blood Collection:	
Generics:		Research/Development Svcs.:	Y	Database Management:		Drug Delivery:	
		Diagnostics:				Drug Distribution:	

TYPES OF BUSINESS:

Pharmaceutical Development-Clinical Trials
Statistical Analysis
Medical Education Services
Regulatory Assistance
Consulting Services
Clinical Trial Software
Clinical Data Management

BRANDS/DIVISIONS/AFFILIATES:

CONTACTS: *Note: Officers with more than one job title may be intentionally listed here more than once.*

Candace Kendle, CEO
Stephen A. Cutler, COO/Sr. VP
Keith A. Cheesman, CFO/Sr. VP
Simon S. Higginbotham, Chief Mktg. Officer/Sr. VP
Christopher C. Bergen, Chief Admin. Officer/Exec. VP
Jarrod B. Pontius, Chief Legal Officer/Sec./VP
Michael Lawson, Dir.-Investor Rel.
J. Michael Sprafka, VP-Kendle Consulting
John Needham, Global Head-Patient Enrollment Strategy
Candace Kendle, Chmn.

Phone: 513-381-5550	Fax: 513-381-5870
Toll-Free: 800-733-1572	
Address: 441 Vine St., Carew Tower, Ste. 1200, Cincinnati, OH 45202 US	

GROWTH PLANS/SPECIAL FEATURES:

Kendle International, Inc. is a global clinical research organization that provides a broad range of Phase I-IV global clinical development services to the biopharmaceutical industry. The company supplements the research and development activities of biopharmaceutical companies by offering clinical research services and information technology designed to reduce the time and expense of drug development. The firm operates in two segments: early stage, which handles all Phase I testing services; and late stage, which handles all Phase II-IV services. Early stage operations include exploratory medicine and proof-of-concept studies that are conducted through a state-of-the-art clinical pharmacology unit in the Netherlands, where it offers Phase I clinical trials with drugs under development. Kendle's late stage services include clinical and data monitoring; project management and late phase services; regulatory, site and medical affairs; and biostatistics. The clinical and data monitoring unit provides services such as clinical monitoring, investigator recruitment, patient recruitment, data management and study reports for clinical trials around the world. The project management unit monitors all phases of trials and reports on health economics, outcomes research, observational studies and scientific events, and operates as a supplier of medical education services. The regulatory, site and medical affairs group designs trial protocols and reviews programs for regulatory compliance and safety issues. The biostatistics unit supplies statistical analysis and interpretation. Kendle operates over 50 offices in 30 countries around the world.

FINANCIALS: Sales and profits are in thousands of dollars—add 000 to get the full amount. 2009 Note: Financial information for 2009 was not available for all companies at press time.

2009 Sales: $551,912	2009 Profits: $15,237	**U.S. Stock Ticker: KNDL**	
2008 Sales: $678,581	2008 Profits: $29,397	**Int'l Ticker:** Int'l Exchange:	
2007 Sales: $568,818	2007 Profits: $18,687	Employees: 3,640	
2006 Sales: $373,936	2006 Profits: $8,530	Fiscal Year Ends: 12/31	
2005 Sales: $250,639	2005 Profits: $10,674	Parent Company:	

SALARIES/BENEFITS:

Pension Plan:	ESOP Stock Plan:	Profit Sharing: Y	Top Exec. Salary: $572,140	Bonus: $54,600
Savings Plan: Y	Stock Purch. Plan:		Second Exec. Salary: $410,034	Bonus: $30,100

OTHER THOUGHTS:

Apparent Women Officers or Directors: 1
Hot Spot for Advancement for Women/Minorities:

LOCATIONS: ("Y" = Yes)

West:	Southwest:	Midwest:	Southeast:	Northeast:	International:
Y		Y		Y	Y

KERYX BIOPHARMACEUTICALS INC

www.keryx.com

Industry Group Code: 325412 Ranks within this company's industry group: Sales: 105   Profits: 58

Drugs:		Other:	Clinical:	Computers:	Services:
Discovery:	Y	AgriBio:	Trials/Services:	Hardware:	Specialty Services:
Licensing:	Y	Genetic Data:	Labs:	Software:	Consulting:
Manufacturing:		Tissue Replacement:	Equipment/Supplies:	Arrays:	Blood Collection:
Generics:			Research/Development Svcs.:	Database Management:	Drug Delivery:
			Diagnostics:		Drug Distribution:

TYPES OF BUSINESS:

Pharmaceuticals Development
Drugs-Cancer
Drugs-Renal Disease

BRANDS/DIVISIONS/AFFILIATES:

ACCESS Oncology, Inc.
Accumin Diagnostics, Inc.
Neryx Biopharmaceuticals, Inc.
KRX-0401
Zerenex

CONTACTS: Note: Officers with more than one job title may be intentionally listed here more than once.

Ron Bentsur, CEO
James F. Oliviero, CFO
James F. Oliviero, Corp. Sec.
James F. Oliviero, Treas.
Michael P. Tarnok, Chmn.

Phone: 212-531-5965	Fax: 212-531-5961
Toll-Free:	
Address: 750 Lexington Ave., 20th Fl., New York, NY 10022 US	

GROWTH PLANS/SPECIAL FEATURES:

Keryx Biopharmaceuticals, Inc. is a biopharmaceutical company focused on the development and commercialization of pharmaceutical products for the treatment of life-threatening diseases, including renal disease and cancer. Keryx's product portfolio includes KRX-0401 and Zerenex. KRX-0401 (perifosine) is an oral anti-cancer agent that inhibits Akt activation in the phosphoinositide 3-kinase (PI3K) pathway. It also affects a number of other key signal transduction pathways, including the JNK pathway, which is associated with programmed cell death, cell growth cell differentiation and cell survival. Preliminary data suggests that KRX-0401 would be most effective as a complementary treatment along with other anti-cancer agents. The drug appears to have less toxic effects than other cancer treatments, which means fewer side effects. It is currently in a Phase III trial for multiple myeloma; a Phase III trial for colorectal cancer; and Phase two trials for other forms of cancer. Zerenex (ferric citrate) is an oral, iron-based compound that has the capacity to bind to phosphate and form non-absorbable complexes. It has completed five Phase II clinical studies as a treatment for hyperphosphatemia (elevated phosphate levels) in patients with ESRD, and a Phase III program under SPA is pending commencement. It is also in Phase II development in Japan by the firm's Japanese partner, JT and Torii. Keryx owns three subsidiaries located in the U.S.: ACCESS Oncology, Inc.; Neryx Biopharmaceuticals, Inc.; and Accumin Diagnostics, Inc.

FINANCIALS: Sales and profits are in thousands of dollars—add 000 to get the full amount. 2009 Note: Financial information for 2009 was not available for all companies at press time.

2009 Sales: $25,194	2009 Profits: $10,485	U.S. Stock Ticker: KERX
2008 Sales: $1,283	2008 Profits: $-52,881	Int'l Ticker: Int'l Exchange:
2007 Sales: $ 983	2007 Profits: $-90,062	Employees: 15
2006 Sales: $ 431	2006 Profits: $-73,764	Fiscal Year Ends: 12/31
2005 Sales: $ 574	2005 Profits: $-26,895	Parent Company:

SALARIES/BENEFITS:

Pension Plan:	ESOP Stock Plan:	Profit Sharing:	Top Exec. Salary: $273,085	Bonus: $
Savings Plan:	Stock Purch. Plan:		Second Exec. Salary: $240,000	Bonus: $100,000

OTHER THOUGHTS:

Apparent Women Officers or Directors:
Hot Spot for Advancement for Women/Minorities:

LOCATIONS: ("Y" = Yes)

West:	Southwest:	Midwest:	Southeast:	Northeast:	International:
				Y	

KING PHARMACEUTICALS INC

www.kingpharm.com

Industry Group Code: 325412 Ranks within this company's industry group: Sales: 34   Profits: 43

Drugs:		Other:		Clinical:		Computers:		Services:	
Discovery:		AgriBio:		Trials/Services:		Hardware:		Specialty Services:	
Licensing:	Y	Genetic Data:		Labs:		Software:		Consulting:	
Manufacturing:	Y	Tissue Replacement:		Equipment/Supplies:		Arrays:		Blood Collection:	
Generics:				Research/Development Svcs.:	Y	Database Management:		Drug Delivery:	
				Diagnostics:				Drug Distribution:	

TYPES OF BUSINESS:

Pharmaceuticals Acquisition & Manufacturing
Prescription Pharmaceuticals-Diversified
Animal Feed Additives

BRANDS/DIVISIONS/AFFILIATES:

Monarch Pharmaceuticals, Inc.
King Pharmaceuticals Research and Development, Inc
Meridian Medical Technologies, Inc.
Alpharma, Inc.
Alpharma Pharmaceuticals, Inc.
Alpharma Animal Health

CONTACTS: Note: Officers with more than one job title may be intentionally listed here more than once.

Brian A. Markison, CEO
Brian A. Markison, Pres.
Joseph Squicciarino, CFO
David J. Whitehead, Exec. VP-Sales
Janet Tuffy, Exec. VP-Human Resources
Eric G. Carter, Chief Science Officer
Bradley Knoll, Exec. VP-Tech. Oper.
Lynn F. Palmer, Exec. VP-Eng.
James W. Elrod, Chief Legal Officer/Corp. Sec.
Jack Howarth, VP-Investor Rel.
Dennis P. O'Brian, Exec. VP/Pres., Meridian
Frederick Brouillette, Jr., Chief Compliance Officer
Eric J. Bruce, Pres., Alpharma Animal Health
Linda C. Wase, Exec. VP-Medical Affairs
Brian A. Markison, Chmn.

Phone: 423-989-8000	Fax: 423-274-8677
Toll-Free: 800-776-3637	
Address: 501 5th St., Bristol, TN 37620 US	

GROWTH PLANS/SPECIAL FEATURES:

King Pharmaceuticals, Inc. (KPI) develops, manufactures and markets branded prescription pharmaceutical products and animal health products. KPI operates in four segments: branded prescription pharmaceuticals; animal health products; the Meridian Auto-Injector business; and royalties and other. Branded pharmaceutical products account for approximately 63% of total net revenues. Key products in the company's branded pharmaceutical segment include neuroscience products, including Skelaxin, Avinza, Embeda and the Flector patch; hospital-use products, including Thrombin-JMI; and legacy products, including Levoxyl, Cytomel, Bicillan and Altace, which are used to treat hypothyroidism, infection and hypertension, respectively. Notable products include Embeda, a long-acting Schedule II opioid analgesic for pain management; Avinza, an extended release formulation of morphine for severe pain; and Thrombin-JMI, which controls minor bleeding during surgery. The animal health segment, operating through Alpharma, Inc., manufactures and develops medicated feed additives (MFAs) and water-soluble therapeutics for poultry, cattle and swine, including antibiotic, anticoccidial and antibacterial products. The Meridian Auto-Injector segment manufactures and markets pharmaceuticals delivered via an auto-injector. Its products consist of EpiPen, an epinephrine administration tool for emergency treatment of anaphylaxis; and nerve gas antidotes, sold to the Department of Defense. KPI's royalties and other segment receives royalties from the sale of Adenoscan, which function as an imaging agent in cardiac stress testing. KPI's wholly owned subsidiaries include Monarch Pharmaceuticals, Inc.; Alpharma, Inc.; King Pharmaceuticals Research and Development, Inc.; Meridian Medical Technologies, Inc.; and Alpharma Pharmaceuticals, Inc. In February 2009, the company announced plans to reduce its workforce by about 22%. In October 2009, the firm licensed the right to market and sell CYANOKIT, for the treatment of cyanide poisoning, in the U.S.

KPI offers its employees benefits including medical insurance; vision, dental and prescription coverage; disability plans; flexible spending accounts; a travel assistance program; tuition reimbursement; adoption benefits; and a 401(k) plan.

FINANCIALS: Sales and profits are in thousands of dollars—add 000 to get the full amount. 2009 Note: Financial information for 2009 was not available for all companies at press time.

2009 Sales: $1,776,500	2009 Profits: $91,953	U.S. Stock Ticker: KG
2008 Sales: $1,565,061	2008 Profits: $-333,063	Int'l Ticker: Int'l Exchange:
2007 Sales: $2,136,882	2007 Profits: $182,981	Employees: 2,640
2006 Sales: $1,988,500	2006 Profits: $288,949	Fiscal Year Ends: 12/31
2005 Sales: $1,772,881	2005 Profits: $117,833	Parent Company:

SALARIES/BENEFITS:

Pension Plan:	ESOP Stock Plan:	Profit Sharing:	Top Exec. Salary: $990,000	Bonus: $1,198,761
Savings Plan: Y	Stock Purch. Plan:		Second Exec. Salary: $600,000	Bonus: $508,565

OTHER THOUGHTS:

Apparent Women Officers or Directors: 5
Hot Spot for Advancement for Women/Minorities: Y

LOCATIONS: ("Y" = Yes)

West:	Southwest:	Midwest:	Southeast:	Northeast:	International:
Y		Y	Y	Y	Y

KV PHARMACEUTICAL CO

www.kvpharmaceutical.com

Industry Group Code: 325412 Ranks within this company's industry group: Sales: 57   Profits: 168

Drugs:		Other:	Clinical:		Computers:		Services:	
Discovery:	Y	AgriBio:	Trials/Services:		Hardware:		Specialty Services:	
Licensing:		Genetic Data:	Labs:		Software:		Consulting:	
Manufacturing:	Y	Tissue Replacement:	Equipment/Supplies:		Arrays:		Blood Collection:	
Generics:			Research/Development Svcs.:		Database Management:		Drug Delivery:	Y
			Diagnostics:				Drug Distribution:	

TYPES OF BUSINESS:

Pharmaceutical Products
Drug Delivery & Formulation Technologies
Specialty Ingredients
Taste Masking Systems
Branded Prescription Pharmaceuticals

BRANDS/DIVISIONS/AFFILIATES:

ETHEX Corporation
Ther-Rx Corporation
Clindesse
Evamist
SITE RELEASE
Gynazole-1

CONTACTS: *Note: Officers with more than one job title may be intentionally listed here more than once.*

Gregory J. Divis, Jr., Interim CEO
Gregory J. Divis, Jr., Interim Pres.
Thomas S. McHugh, CFO
Michael S. Anderson, VP-Dev. & Industry Presence
Thomas S. McHugh, Chief Acct. Officer/Interim Treas.
Gregory J. Divis, Jr., Pres., Ther-Rx Corporation
Terry B. Hatfield, Chmn.

Phone: 314-645-6600	Fax: 314-646-3751
Toll-Free:	
Address: 2503 S. Hanley Rd., St. Louis, MO 63144 US	

GROWTH PLANS/SPECIAL FEATURES:

KV Pharmaceutical Co. is a pharmaceutical company that develops, manufactures, acquires and markets branded and generic/non-branded prescription pharmaceutical products. The company develops a wide variety of drug delivery and formulation technologies, which are primarily focused in four areas: SITE RELEASE bioadhesives; tastemasking; oral controlled release; and oral quick dissolving tablets. The firm incorporates these technologies in the products it markets to control and improve the absorption and utilization of active pharmaceutical compounds.   KV has a broad range of dosage form capabilities including tablets, capsules, creams, liquids and ointments.   The company manufactures and markets these specialty pharmaceutical products through two wholly-owned subsidiaries: ETHEX Corporation, which targets generic and non-branded market segments; and Ther-Rx Corporation, which conducts the company's branded pharmaceutical operations. ETHEX, the company's major subsidiary, markets more than 160 products in its cardiovascular, pain management, women's health, respiratory, gastrointestinal, dermatological, anti-anxiety, digestive enzyme and dental categories. Its products include four strengths of metoprolol, the generic version of AstraZeneca's Toprol-XL. Ther-Rx, the firm's brand name pharmaceutical products subsidiary, generally focuses on the women's health therapeutic category. It offers prescription pharmaceuticals in two areas: estrogen therapy and anti-infectives. In the estrogen therapy category, Ther-Rx offers Evamist, a dose spray applicator for women suffering from moderate to severe hot flashes. The anti-infectives category includes Clindesse and Gynazole-1, products for treating vaginal yeast infections and bacterial vaginosis. In March 2010, the firm reduced its work force by 289 employees, or 42%, to lower its operating costs. In June 2010, the company sold Particle Dynamics, Inc., one of its wholly-owned subsidiaries, to an investor group led by Edgewater Capital Partners for $24.6 million.

The firm offers employees health, dental and life insurance; an employee stock option plan; a retirement plan; a 401(k) plan; short- and long-term disability; a profit sharing plan; educational assistance; and an employee assistance program.

FINANCIALS: Sales and profits are in thousands of dollars—add 000 to get the full amount. 2009 Note: Financial information for 2009 was not available for all companies at press time.

2009 Sales: $312,327	2009 Profits: $-313,627	**U.S. Stock Ticker: KV**
2008 Sales: $577,623	2008 Profits: $86,438	**Int'l Ticker:** Int'l Exchange:
2007 Sales: $424,307	2007 Profits: $58,559	Employees: 1,590
2006 Sales: $367,618	2006 Profits: $11,416	Fiscal Year Ends: 3/31
2005 Sales: $303,493	2005 Profits: $33,269	Parent Company:

SALARIES/BENEFITS:

Pension Plan:	ESOP Stock Plan:	Profit Sharing: Y	Top Exec. Salary: $1,087,917	Bonus: $
Savings Plan: Y	Stock Purch. Plan:		Second Exec. Salary: $471,700	Bonus: $

OTHER THOUGHTS:

Apparent Women Officers or Directors:
Hot Spot for Advancement for Women/Minorities:

LOCATIONS: ("Y" = Yes)

West:	Southwest:	Midwest:	Southeast:	Northeast:	International:
		Y			

Note: Financial information, benefits and other data can change quickly and may vary from those stated here.

LA JOLLA PHARMACEUTICAL

www.ljpc.com

Industry Group Code: 325412 Ranks within this company's industry group: Sales: Profits: 97

Drugs:		Other:	Clinical:	Computers:	Services:
Discovery:	Y	AgriBio:	Trials/Services:	Hardware:	Specialty Services:
Licensing:		Genetic Data:	Labs:	Software:	Consulting:
Manufacturing:		Tissue Replacement:	Equipment/Supplies:	Arrays:	Blood Collection:
Generics:			Research/Development Svcs.:	Database Management:	Drug Delivery:
			Diagnostics:		Drug Distribution:

TYPES OF BUSINESS:

Drugs-Autoimmune Diseases
Lupus Treatments
Small-Molecule Therapeutics

BRANDS/DIVISIONS/AFFILIATES:

Riquent
SSAO Inhibitors
BioMarin Pharma

CONTACTS: *Note: Officers with more than one job title may be intentionally listed here more than once.*

Deirdre Gillespie, CEO
Deirdre Gillespie, Pres.
Gail A. Sloan, Corp. Sec.
Gail A. Sloan, VP-Finance
Craig R. Smith, Chmn.

Phone: 858-452-6600	Fax: 858-626-2851
Toll-Free:	
Address: 4365 Executive Dr., Ste. 300, San Diego, CA 92121 US	

GROWTH PLANS/SPECIAL FEATURES:

La Jolla Pharmaceutical Company (LJPC) is a biopharmaceutical company focused on the research and development of therapeutic products for the treatment of certain life-threatening antibody-mediated diseases, especially autoimmune conditions such as lupus. The company was previously involved in the development of Riquent, a clinical drug candidate for the treatment of lupus kidney disease. The firm signed a commercialization agreement with BioMarin CF Limited, a subsidiary of BioMarin Pharmaceuticals, Inc., for global sales rights. However, Phase III trials indicated that the drug produced unsafe side effects, and in March 2009 the company decided to cease development of Riquent. The firm's other asset is a preclinical stage study on the use of SSAO (semicarbazide-sensitive amine oxidase) inhibitors for use in the treatment of stroke, ulcerative colitis, and other autoimmune disorders. The company has stated that this asset would likely be out-licensed as it does not currently have the resources to devote to the product's development. In February 2009, La Jolla laid off 75 employees, most of its staff. In October 2009, wholly-owned subsidiary La Jolla Ltd. was dissolved. In December 2009, wholly-owned subsidiary, Jewel Merger Sub, Inc., was consolidated into LJPC.

FINANCIALS: Sales and profits are in thousands of dollars—add 000 to get the full amount. 2009 Note: Financial information for 2009 was not available for all companies at press time.

2009 Sales: $	2009 Profits: $-8,634	U.S. Stock Ticker: LJPC.PK	
2008 Sales: $	2008 Profits: $-62,854	Int'l Ticker: Int'l Exchange:	
2007 Sales: $	2007 Profits: $-53,076	Employees: 94	
2006 Sales: $	2006 Profits: $-39,445	Fiscal Year Ends: 12/31	
2005 Sales: $	2005 Profits: $-27,363	Parent Company:	

SALARIES/BENEFITS:

Pension Plan:	ESOP Stock Plan: Y	Profit Sharing:	Top Exec. Salary: $	Bonus: $
Savings Plan:	Stock Purch. Plan:		Second Exec. Salary: $	Bonus: $

OTHER THOUGHTS:

Apparent Women Officers or Directors: 2
Hot Spot for Advancement for Women/Minorities: Y

LOCATIONS: ("Y" = Yes)

West:	Southwest:	Midwest:	Southeast:	Northeast:	International:
Y					

LANNETT COMPANY INC

www.lannett.com

Industry Group Code: 325412   **Ranks within this company's industry group:**   Sales: 71   Profits: 63

Drugs:		Other:	Clinical:	Computers:	Services:	
Discovery:		AgriBio:	Trials/Services:	Hardware:	Specialty Services:	
Licensing:		Genetic Data:	Labs:	Software:	Consulting:	
Manufacturing:	Y	Tissue Replacement:	Equipment/Supplies:	Arrays:	Blood Collection:	
Generics:			Research/Development Svcs.:	Database Management:	Drug Delivery:	
			Diagnostics:		Drug Distribution:	Y

TYPES OF BUSINESS:

Drugs-Generic
Drug Delivery System Development

BRANDS/DIVISIONS/AFFILIATES:

Acetazolamide
Butalbital
Digoxin
Unithroid
Levothyroxine Sodium
Primidone
Hydromorphone
Phentermine

CONTACTS: *Note: Officers with more than one job title may be intentionally listed here more than once.*

Arthur P. Bedrosian, CEO
Arthur P. Bedrosian, Pres.
Keith Ruck, CFO
Kevin Smith, VP-Mktg. & Sales
Stephen Kovary, VP-Oper.
Keith Ruck, Controller
Ernest J. Sabo, Chief Compliance Officer/VP-Regulatory Affairs
Ronald A. West, Vice Chmn.
William Schreck, Sr. VP/Gen. Mgr.
William Farber, Chmn.

Phone: 215-333-9000	Fax: 215-333-9004
Toll-Free: 800-325-9994	
Address: 9000 State Rd., Philadelphia, PA 19136 US	

GROWTH PLANS/SPECIAL FEATURES:

Lannett Company, Inc. develops, manufactures, markets and distributes pharmaceutical products sold under generic names, marketing them primarily to drug wholesalers, retail drug chains, repackagers, distributors and government agencies. It manufactures and/or distributes 31 products, including Acetazolamide for glaucoma; Butalbital with aspirin, caffeine and codeine for migraines; Clindamycin, an antibiotic; Digoxin for congestive heart failure; Dicyclomine, a treatment for irritable bowels; Levothyroxine Sodium for thyroid deficiency; Phentermine, a pill for weight loss; Primidone for epilepsy; Terbutaline, a medication for bronchospasms; and Unithroid for thyroid deficiency. All of the products currently manufactured and sold by the company are prescription products. The company's key products are Butalbital, Digoxin, Primidone and Levothyroxine, contributing more than 72% of 2009 revenue. Lannett has a wholly-owned subsidiary, Cody Laboratories, Inc., in Cody, Wyoming. Lannett's services also include granulation, blending, encapsulation, coating and packaging. In December 2009, the company received FDA approval for its Hydromorphone Hydrochloride Tablets, the generic equivalent of Purdue Pharmaceuticals' Dilaudid Tablets. In April 2010, Lannett received FDA approval for its Ondansetron Injection USP Multi-Dose Vials, the generic equivalent of GlaxoSmithKline's Zofran Injection.

Lannett offers its employees medical, dental, vision and prescription drug coverage; a 401(k) plan; an employee stock purchase plan; and tuition reimbursement.

FINANCIALS: Sales and profits are in thousands of dollars—add 000 to get the full amount. 2009 Note: Financial information for 2009 was not available for all companies at press time.

2009 Sales: $119,002	2009 Profits: $6,534	U.S. Stock Ticker: LCI
2008 Sales: $72,403	2008 Profits: $-2,318	Int'l Ticker: Int'l Exchange:
2007 Sales: $82,578	2007 Profits: $-6,929	Employees: 277
2006 Sales: $64,060	2006 Profits: $4,969	Fiscal Year Ends: 6/30
2005 Sales: $44,902	2005 Profits: $-32,780	Parent Company:

SALARIES/BENEFITS:

Pension Plan:	ESOP Stock Plan:	Profit Sharing:	Top Exec. Salary: $367,202	Bonus: $244,365
Savings Plan: Y	Stock Purch. Plan: Y		Second Exec. Salary: $200,180	Bonus: $130,825

OTHER THOUGHTS:

Apparent Women Officers or Directors:
Hot Spot for Advancement for Women/Minorities:

LOCATIONS: ("Y" = Yes)

West:	Southwest:	Midwest:	Southeast:	Northeast:	International:
Y				Y	

LEXICON PHARMACEUTICALS INC

www.lexicon-genetics.com

Industry Group Code: 325412 Ranks within this company's industry group: Sales: 121 Profits: 155

Drugs:		Other:		Clinical:		Computers:		Services:	
Discovery:	Y	AgriBio:		Trials/Services:		Hardware:		Specialty Services:	
Licensing:		Genetic Data:	Y	Labs:		Software:		Consulting:	
Manufacturing:		Tissue Replacement:		Equipment/Supplies:		Arrays:		Blood Collection:	
Generics:				Research/Development Svcs.:	Y	Database Management:	Y	Drug Delivery:	
				Diagnostics:				Drug Distribution:	

TYPES OF BUSINESS:

Drug Discovery & Development
Genetic Databases
Research & Development-Genetics

BRANDS/DIVISIONS/AFFILIATES:

Lexicon Genetics Incorporated
LX2931
LX1031
LX1032
LX7101
LX4211
OmniBank
Genentech Inc

CONTACTS: Note: Officers with more than one job title may be intentionally listed here more than once.

Arthur T. Sands, CEO
Arthur T. Sands, Pres.
Brian P. Zambrowicz, Chief Scientific Officer/Exec. VP
Jeffrey L. Wade, General Counsel/Exec. VP
Steven A. Tragash, Sr. VP-Corp. Affairs
James Tessmer, VP-Finance & Acct.
Philip M. Brown, Sr. VP-Clinical Dev.
Alan J. Main, Exec. VP-Pharmaceutical Research
Samuel L. Barker, Chmn.

Phone: 281-863-3000	Fax: 281-863-8088
Toll-Free:	
Address: 8800 Technology Forest Pl., The Woodlands, TX 77381-1160 US	

GROWTH PLANS/SPECIAL FEATURES:

Lexicon Pharmaceuticals, Inc., formerly Lexicon Genetics Inc., is a biopharmaceutical company that focuses on the discovery and development of new medical treatments for human diseases using its proprietary gene knockout technology, which disrupts the function of genes in mice in order to determine the proper pharmaceutical treatment for the physiological and behavioral functions of each gene. Lexicon is conducting Phase II trials of four advanced drug candidates: LX1031, LX4211, LX2931 and LX1032. LX1031 is an orally-delivered drug for the treatment of gastrointestinal disorders such as irritable bowel syndrome. LX4211 is being developed to treat Type II diabetes. LX1031 and LX4211 are Lexicon's two most advanced drug candidates, and the company has announced positive results for both. LX2931 could potentially treat rheumatoid arthritis and other autoimmune diseases. LX1032 is being developed for the treatment of symptoms associated with carcinoid syndrome. Lexicon also has one other drug candidate, LX7101, which is in preclinical development. LX7101 is a topically-delivered drug for the treatment of glaucoma. Lexicon has identified and validated in living animals more than 100 targets with promising profiles for drug discovery. The firm collaborates with such companies as Bristol-Myers Squibb for development in the neuroscience field; Genentech for the discovery of novel therapeutic proteins and antibody targets; Schering-Plough/Organon to develop and commercialize biotherapeutic drugs; and Takeda to discover drugs for high blood pressure treatment. Lexicon's OmniBank contains over 270,000 frozen gene knockout embryonic mouse stem cell clones.

Employees are offered medical, dental, and vision insurance; life, AD&D, and short- and long-term disability insurance; flexible spending accounts; discounts for dependent care with partnered providers; a 529 college investing plan; a 401(k) plan; an employee stock option plan; a health and fitness program; an educational assistance plan with tuition reimbursement; relocation assistance for selected positions; an employee assistance program; and discounts at a variety of local businesses.

FINANCIALS: Sales and profits are in thousands of dollars—add 000 to get the full amount. 2009 Note: Financial information for 2009 was not available for all companies at press time.

2009 Sales: $10,700	2009 Profits: $-82,780	**U.S. Stock Ticker: LXRX**
2008 Sales: $32,321	2008 Profits: $-76,860	**Int'l Ticker:** Int'l Exchange:
2007 Sales: $50,118	2007 Profits: $-58,794	Employees: 345
2006 Sales: $72,798	2006 Profits: $-54,311	Fiscal Year Ends: 12/31
2005 Sales: $75,680	2005 Profits: $-36,315	Parent Company:

SALARIES/BENEFITS:

Pension Plan:	ESOP Stock Plan:	Profit Sharing:	Top Exec. Salary: $560,000	Bonus: $210,000
Savings Plan: Y	Stock Purch. Plan:		Second Exec. Salary: $365,000	Bonus: $120,000

OTHER THOUGHTS:

Apparent Women Officers or Directors: 1
Hot Spot for Advancement for Women/Minorities:

LOCATIONS: ("Y" = Yes)

West:	Southwest:	Midwest:	Southeast:	Northeast:	International:
	Y			Y	

LIFE SCIENCES RESEARCH INC

www.lsrinc.net

Industry Group Code: 541712 Ranks within this company's industry group: Sales: Profits:

Drugs:	Other:	Clinical:		Computers:	Services:	
Discovery:	AgriBio:	Trials/Services:	Y	Hardware:	Specialty Services:	
Licensing:	Genetic Data:	Labs:	Y	Software:	Consulting:	
Manufacturing:	Tissue Replacement:	Equipment/Supplies:		Arrays:	Blood Collection:	
Generics:		Research/Development Svcs.:	Y	Database Management:	Drug Delivery:	
		Diagnostics:			Drug Distribution:	

TYPES OF BUSINESS:
Contract Research
Drug Development Services
Non-Clinical Safety Testing

BRANDS/DIVISIONS/AFFILIATES:
Huntingdon Life Sciences Group plc
Lion Holdings Inc

CONTACTS: Note: Officers with more than one job title may be intentionally listed here more than once.
Andrew H. Baker, CEO
Brian Cass, Pres.
Richard A. Michaelson, CFO
Mark L. Bibi, General Counsel/Corp. Sec.
Julian T. Griffiths, Dir.-Oper.
Andrew H. Baker, Chmn.

Phone: 732-649-9961	Fax: 732-649-0021
Toll-Free:	
Address: Mettlers Rd., P.O. Box 2360, East Millstone, NJ 08875 US	

GROWTH PLANS/SPECIAL FEATURES:

Life Sciences Research, Inc. (LSR) is a global contract research organization providing pre-clinical and non-clinical testing services for biological safety evaluation research to pharmaceutical, biotechnology, agrochemical and industrial chemical companies. LSR serves the regulatory and commercial requirements to perform safety evaluations on new pharmaceutical compounds and chemical compounds contained within the products that humans use, eat and to which are otherwise exposed. The company also tests the effect of such compounds on the environment and assesses the safety and efficacy of veterinary products. Pre-clinical testing helps evaluate both how the drug affects the body and how the body affects the drug in order to assess safe and appropriate dose regimens. Non-clinical testing can focus on identifying and avoiding the longer-term cancer implications of exposure to the compound, the potential for possible reproductive implications and the stability of pharmaceuticals under a variety of storage conditions. LSR has also actively pursued opportunities to extend its range of capabilities supporting late stage drug discovery, focused around in vitro and in vivo models for lead candidate drug characterization and optimization. LSR was formed specifically to acquire the former Huntingdon Life Sciences Group plc in order to relocate the company to the U.S. The firm operates one research facilities in the U.S. and two in the U.K., and its services are designed to address the regulatory requirements of governments around the world. The U.K. accounts for approximately 74.6% of revenues, while the U.S. accounts for the remaining 25.4%. In November 2009, the firm was acquired by Lion Holdings, Inc., a holding company owned by LSR's CEO and Chairman.

FINANCIALS: Sales and profits are in thousands of dollars—add 000 to get the full amount. 2009 Note: Financial information for 2009 was not available for all companies at press time.

2009 Sales: $	2009 Profits: $	U.S. Stock Ticker: Subsidiary
2008 Sales: $242,422	2008 Profits: $10,418	Int'l Ticker: Int'l Exchange:
2007 Sales: $236,800	2007 Profits: $-13,974	Employees: 1,648
2006 Sales: $192,217	2006 Profits: $-14,872	Fiscal Year Ends: 12/31
2005 Sales: $172,013	2005 Profits: $1,491	Parent Company: LION HOLDINGS INC

SALARIES/BENEFITS:

Pension Plan:	ESOP Stock Plan:	Profit Sharing:	Top Exec. Salary: $636,111	Bonus: $
Savings Plan: Y	Stock Purch. Plan:		Second Exec. Salary: $636,111	Bonus: $

OTHER THOUGHTS:
Apparent Women Officers or Directors:
Hot Spot for Advancement for Women/Minorities:

LOCATIONS: ("Y" = Yes)

West:	Southwest:	Midwest:	Southeast:	Northeast:	International:
				Y	Y

LIFE TECHNOLOGIES CORP

www.lifetechnologies.com

Industry Group Code: 325413 Ranks within this company's industry group: Sales: 1 Profits: 1

Drugs:	Other:		Clinical:		Computers:		Services:	
Discovery:	AgriBio:		Trials/Services:		Hardware:		Specialty Services:	
Licensing:	Genetic Data:	Y	Labs:		Software:		Consulting:	
Manufacturing:	Tissue Replacement:		Equipment/Supplies:	Y	Arrays:	Y	Blood Collection:	
Generics:			Research/Development Svcs.:	Y	Database Management:		Drug Delivery:	
			Diagnostics:				Drug Distribution:	

TYPES OF BUSINESS:
Equipment-Gene Cloning Kits
Microarrays
Reagents
RNA & DNA Libraries
Mass Spectrometry

BRANDS/DIVISIONS/AFFILIATES:
Cell Culture Systems (CCS)
BioDiscovery
CellzDirect, Inc.
Applied Biosystems, Inc.
Invitrogen Corp.
BioTrove, Inc.
Invitrogen MAGic Sample Processor
Applied Biosystems ViiA 7 Real-Time PCR System

CONTACTS: Note: Officers with more than one job title may be intentionally listed here more than once.
Gregory T. Lucier, CEO
Mark P. Stevenson, COO
Mark P. Stevenson, Pres.
David F. Hoffmeister, CFO/Sr. VP
Amanda Clardy, Chief Mktg. Officer
Peter Leddy, Sr. VP-Global Human Resources
Brian Pollok, Chief Scientific Officer/Head-Global R&D
Joe Beery, CIO/Sr. VP
John A. Cottingham, Chief Legal Officer
Mark O'Donnell, Sr. VP-Global Oper. & Svcs.
Paul Grossman, Sr. VP-Strategy & Corp. Dev.
Farnaz Khadem, Sr. Dir.-Corp. Comm.
Eileen Pattinson, Sr. Dir.-Investor Rel.
Kelli Richard, VP-Finance/Chief Acct. Officer
Nicolas M. Barthelemy, Pres., Cell Systems
John Miller, Pres., Genetic Systems
Peter Dansky, Pres., Molecular Biology Systems
Claude D. Benchimol, Sr. VP-R&D, Genetic Systems
Gregory T. Lucier, Chmn.

Phone: 760-603-7200	Fax: 760-603-6500
Toll-Free: 800-955-6288	
Address: 5791 Van Allen Way, Carlsbad, CA 92008 US	

GROWTH PLANS/SPECIAL FEATURES:

Life Technologies Corp., formerly Invitrogen Corporation, is a global biotechnology tools company with operations in over 100 companies. The company's product portfolio includes technologies for capillary electrophoresis based sequencing, next generation sequencing, mass spectrometry, sample preparation, cell culture, RNA interference analysis, functional genomics research, proteomics and cell biology applications, as well as clinical diagnostic applications and water testing analysis. The company divides its operations into three segments: molecular biology systems (MBS); cell systems (CS); and genetic systems (GS). The MBS segment includes the molecular biology-based technologies including basic and real-time PCR; RNAi; DNA synthesis; sample prep; transfection; cloning; and protein expression profiling and protein analysis. The CS segment includes all product lines used in the study of cell function, including cell culture media and sera; stem cells and related tools; cellular imaging products; antibodies; drug discovery services; and cell therapy-related products. The GS division includes sequencing systems and reagents, including capillary electrophoresis and the SOLiD system; and reagent kits developed specifically for applied markets such as forensics, food and water safety and pharmaceutical quality monitoring. Additional products include mass spectrometry systems, Ambion RNA reagents and specialized applied markets products and services. The company has a presence in 160 countries, and its intellectual property portfolio contains 3,900 patents and exclusive licenses. In April 2010, the company debuted the Invitrogen MAGic Sample Processor, the life science industry's first proteomic and epigenetic sample preparation system. Also in April 2010, Life Technologies introduced the Applied Biosystems ViiA 7 Real-Time PCR System, a product that integrates a variety of genotyping and quantitative PCR applications.

The company offers its employees medical, dental and vision insurance; a 401(k) plan; life insurance; short- and long-term disability benefits; an employee stock purchase plan; educational reimbursement; and an employee assistance plan.

FINANCIALS: Sales and profits are in thousands of dollars—add 000 to get the full amount. 2009 Note: Financial information for 2009 was not available for all companies at press time.

2009 Sales: $3,280,344	2009 Profits: $144,594	U.S. Stock Ticker: LIFE
2008 Sales: $1,620,323	2008 Profits: $5,714	Int'l Ticker: Int'l Exchange:
2007 Sales: $1,281,747	2007 Profits: $119,149	Employees: 9,700
2006 Sales: $1,151,175	2006 Profits: $-191,049	Fiscal Year Ends: 12/31
2005 Sales: $1,198,452	2005 Profits: $132,046	Parent Company:

SALARIES/BENEFITS:

Pension Plan:	ESOP Stock Plan:	Profit Sharing:	Top Exec. Salary: $1,116,346	Bonus: $3,349,039
Savings Plan: Y	Stock Purch. Plan: Y		Second Exec. Salary: $650,000	Bonus: $1,505,000

OTHER THOUGHTS:
Apparent Women Officers or Directors: 5
Hot Spot for Advancement for Women/Minorities: Y

LOCATIONS: ("Y" = Yes)

West:	Southwest:	Midwest:	Southeast:	Northeast:	International:
Y	Y	Y	Y	Y	Y

LIFECELL CORPORATION

www.lifecell.com

Industry Group Code: 325414 **Ranks within this company's industry group:** Sales: Profits:

Drugs:		Other:		Clinical:	Computers:	Services:
Discovery:	Y	AgriBio:		Trials/Services:	Hardware:	Specialty Services:
Licensing:		Genetic Data:		Labs:	Software:	Consulting:
Manufacturing:	Y	Tissue Replacement:	Y	Equipment/Supplies:	Arrays:	Blood Collection:
Generics:				Research/Development Svcs.:	Database Management:	Drug Delivery:
				Diagnostics:		Drug Distribution:

TYPES OF BUSINESS:

Tissue Replacement Products
Skin Replacement Technology
Regenerative Medicine

BRANDS/DIVISIONS/AFFILIATES:

Kinetic Concepts Inc
AlloDerm
Strattice
Cymetra

CONTACTS: *Note: Officers with more than one job title may be intentionally listed here more than once.*

Lisa Colleran, Pres.
Catherine M. Burzik, CEO-Kinetic Concepts

Phone: 908-947-1100	Fax:
Toll-Free:	
Address: 1 Millennium Way, Branchburg, NJ 08876 US	

GROWTH PLANS/SPECIAL FEATURES:

LifeCell Corporation, a wholly-owned subsidiary of Kinetic Concepts, Inc. (KCI), specializes in regenerative medicine, developing and manufacturing products geared toward the repair, replacement and preservation of human tissues. The firm's products are used in a variety of reconstructive, orthopedic and urogynecologic surgical procedures. Of the company's products, AlloDerm is the flagship. AlloDerm is a tissue matrix derived from human skin. Originally used as a skin graft for deep second and third degree burns, it is now commonly used as a soft tissue replacement in a variety of procedures, including plastic, reconstructive, general surgical, burn and periodontal. Additional applications include abdominal wall reconstruction, post-mastectomy breast reconstruction, ENT/head and neck plastic reconstruction and grafting. The firm also offers its AlloDerm tissue in an injectable form. Marketed as Cymetra Micronized AlloDerm, this product offers a minimally invasive alternative for soft tissue defect correction. Additionally, LifeCell offers Strattice Reconstructive Tissue Matrix, a sterile reconstructive tissue matrix that supports tissue regeneration. Strattice is used for soft tissue reinforcement in reconstructive and general surgical procedures, including hernia repair and breast reconstruction or plastic surgery. LifeCell also maintains an interest in the researching of new biologics for multiple therapeutic areas, including reconstructive surgery, orthopedics and cardiovascular applications.

Employee benefits include: medical, dental and vision insurance; life/AD&D insurance; short and long-term disability; paid holidays; tuition reimbursement; 401(K) and stock purchase plans; and alternative summer working hours.

FINANCIALS: Sales and profits are in thousands of dollars—add 000 to get the full amount. 2009 Note: Financial information for 2009 was not available for all companies at press time.

2009 Sales: $	2009 Profits: $	**U.S. Stock Ticker: Subsidiary**
2008 Sales: $	2008 Profits: $	**Int'l Ticker:** Int'l Exchange:
2007 Sales: $191,130	2007 Profits: $26,883	Employees:
2006 Sales: $141,680	2006 Profits: $20,469	Fiscal Year Ends: 12/31
2005 Sales: $94,398	2005 Profits: $12,044	Parent Company: KINETIC CONCEPTS INC

SALARIES/BENEFITS:

Pension Plan:	ESOP Stock Plan:	Profit Sharing:	Top Exec. Salary: $	Bonus: $
Savings Plan: Y	Stock Purch. Plan:		Second Exec. Salary: $	Bonus: $

OTHER THOUGHTS:

Apparent Women Officers or Directors: 2
Hot Spot for Advancement for Women/Minorities:

LOCATIONS: ("Y" = Yes)

West:	Southwest:	Midwest:	Southeast:	Northeast:	International:
				Y	

LIFECORE BIOMEDICAL INC

www.lifecore.com

Industry Group Code: 325414 Ranks within this company's industry group: Sales: Profits:

Drugs:	Other:	Clinical:	Computers:	Services:
Discovery:	AgriBio:	Trials/Services:	Hardware:	Specialty Services:
Licensing:	Genetic Data:	Labs:	Software:	Consulting:
Manufacturing: Y	Tissue Replacement: Y	Equipment/Supplies:	Arrays: Y	Blood Collection:
Generics:		Research/Development Svcs.:	Database Management:	Drug Delivery:
		Diagnostics:		Drug Distribution:

TYPES OF BUSINESS:

Biomaterials Manufacturing

BRANDS/DIVISIONS/AFFILIATES:

Landec Corporation
Lurocoat Ophthalmic Viscoelastic
Corgel BioHydrogel
Ortholure Orthopedic Viscosupplement
Hyalose LLC
Nano HA
Oligo HA
Select HA

CONTACTS: Note: Officers with more than one job title may be intentionally listed here more than once.

Dennis J. Allingham, CEO
Dennis J. Allingham, Pres.
David M. Noel, CFO
James G. Hall, VP-Technical Oper.
Kipling Thacker, VP-New Bus. Dev.
David M. Noel, VP-Finance
Larry D. Hiebert, VP/Gen. Mgr.-Hyaluronan Div.

Phone: 952-368-4300	Fax: 952-368-3411
Toll-Free:	
Address: 3515 Lyman Blvd., Chaska, MN 55318 US	

GROWTH PLANS/SPECIAL FEATURES:

Lifecore Biomedical, Inc. develops and manufactures biomaterials and medical devices with applications in various surgical markets. The firm has two divisions, hyaluronan and hyaluronan finished products. The company's hyaluronan division is principally involved in the development and manufacture of products utilizing hyaluronan, a naturally occurring polysaccharide that is widely distributed in the extra-cellular matrix of connective tissues in both animals and humans. This division sells primarily to three medical segments: ophthalmic, orthopedic and veterinary. Lifecore also supplies hyaluronan to customers pursuing other medical applications, such as wound care, aesthetic surgery, medical device coatings, tissue engineering, drug delivery and pharmaceuticals. In the hyaluronan finished products division, the company manufactures Hyaluronan into several finished products for various medical applications. These applications are Lurocoat Ophthalmic Viscoelastic, for use in ophthalmic surgical procedures such as, cataract extraction and intraocular lens implantation; and Ortholure Orthopedic Viscosupplement used for pain relief of degenerative and traumatic changes in synovial joints. In May 2009, the firm commercially launched its Corgel BioHydrogel research kits; Corgel hydrogel serves as a tissue bulking agent and drug delivery matrix, and its biocompatibility allows for cells or bioactive agents to be included directly in the gel. In July 2009, the company signed an agreement with Hyalose LLC, to distribute Hyaloses's Nano HA, Oligo HA Select HA, Biotinylated Select HA and Hyalose Ladders (different sizes of hyaluronic acid). In May 2010, private equity firm Warburg Pincus sold the company to the Landec Corporation for $40 million.

Employees of the firm are offered medical and dental coverage; life insurance; paid time off; short and long-term disability; an employee assistant program; exercise room; flexible spending accounts and a 401(k)program.

FINANCIALS: Sales and profits are in thousands of dollars—add 000 to get the full amount. 2009 Note: Financial information for 2009 was not available for all companies at press time.

2009 Sales: $	2009 Profits: $	U.S. Stock Ticker: Private
2008 Sales: $	2008 Profits: $	Int'l Ticker: Int'l Exchange:
2007 Sales: $69,629	2007 Profits: $7,719	Employees:
2006 Sales: $63,097	2006 Profits: $7,040	Fiscal Year Ends: 6/30
2005 Sales: $55,695	2005 Profits: $17,511	Parent Company: LANDEC CORPORATION

SALARIES/BENEFITS:

Pension Plan:	ESOP Stock Plan:	Profit Sharing:	Top Exec. Salary: $	Bonus: $
Savings Plan: Y	Stock Purch. Plan: Y		Second Exec. Salary: $	Bonus: $

OTHER THOUGHTS:

Apparent Women Officers or Directors:
Hot Spot for Advancement for Women/Minorities:

LOCATIONS: ("Y" = Yes)

West:	Southwest:	Midwest: Y	Southeast:	Northeast:	International:

LIGAND PHARMACEUTICALS INC

www.ligand.com

Industry Group Code: 325412 Ranks within this company's industry group: Sales: 91   Profits: 77

Drugs:		Other:		Clinical:	Computers:	Services:
Discovery:	Y	AgriBio:		Trials/Services:	Hardware:	Specialty Services:
Licensing:	Y	Genetic Data:		Labs:	Software:	Consulting:
Manufacturing:		Tissue Replacement:		Equipment/Supplies:	Arrays:	Blood Collection:
Generics:				Research/Development Svcs.:	Database Management:	Drug Delivery:
				Diagnostics:		Drug Distribution:

TYPES OF BUSINESS:

Drugs-Diversified
Small-Molecule Drugs

BRANDS/DIVISIONS/AFFILIATES:

Eltrombopag/PROMACTA
AVINZA
Bazedoxifene/VIVIANT
Lasofoxifene/FABLYN
Dinaciclib
PS291822
PS540446
Neurogen Corporation

CONTACTS: *Note: Officers with more than one job title may be intentionally listed here more than once.*

John L. Higgins, CEO
John L. Higgins, Pres.
John P. Sharp, CFO
Audrey Warfield-Graham, VP-Human Resources
Charles Berkman, General Counsel/VP/Sec.
Syed Kazmi, VP-Bus. Dev. & Strategic Planning
John P. Sharp, VP-Finance
Lin Zhi, Sr. Dir.-Chemistry & Pharmaceutical Dev.
Keith Marschke, Senior Dir.-Molecular Sciences
John W. Kozarich, Chmn.

Phone: 858-550-7500	Fax: 858-550-1826
Toll-Free:	
Address: 11085 N. Torrey Pines Road, Ste. 300, La Jolla, CA 92037 US	

GROWTH PLANS/SPECIAL FEATURES:

Ligand Pharmaceuticals, Inc. develops drugs to address a variety of medical needs in areas including thrombocytopenia, anemia, cancer, hormone-related diseases, osteoporosis and inflammatory diseases. The firm uses intracellular receptor (IR) technology, which targets intracellular receptors in order to change cell function by selectively turning specific genes on or off. Ligand has five research programs: the selective androgen receptor modulators (SARMs) program, which is indicated for muscle wasting and frailty and is currently in Phase I clinical trials; the thyroid receptor beta antagonists program, which is used to treat hyperlipidemia and is in preclinical and Phase I clinical trials; the small molecule erythropoietin (EPO) receptor antagonists program, which is used to treat chemotherapy-induced anemia and is in the preclinical phase; the glucagon receptor antagonists program, which is used to treat diabetes and is in the preclinical phase; and the histamine 3 (H3) receptor antagonists program, which is used to treat cognitive disorders. Lead collaborative drug candidates include Eltrombopag (PROMACTA) for chronic immune thrombocytopenic purpura; AVINZA for chronic pain; Bazedoxifene (VIVIANT) for menopausal symptoms; Lasofoxifene (FABLYN) for osteoporosis; Dinaciclib for various types of cancer; PS291822 for COPD and asthma; and PS540446 for moderate to severe psoriasis. The company's collaborative partners include Bristol-Myers Squibb; GlaxoSmithKline; Pfizer; Merck & Co.; Roche; Cephalon; and Celgene. Royalties from King Pharmaceuticals' sale of AVINZA and GlaxoSmithKline's sales of PROMACTA generated 21% and 74% of Ligand's 2009 revenue, respectively. In December 2009, the firm acquired Neurogen Corporation for $11 million. In January 2010, the company acquired Metabasis Therapeutics, Inc. for $1.6 million. As a result of the transaction, Ligand gained a fully-funded partnership with Roche, additional pipeline assets and drug discovery technologies and resources. In May 2010, Ligand acquired intellectual property and royalties for MEDI-528, an IL-9 antibody program for moderate to severe asthma, from the Genaera Liquidating Trust for $2.75 million.

FINANCIALS: Sales and profits are in thousands of dollars—add 000 to get the full amount. 2009 Note: Financial information for 2009 was not available for all companies at press time.

2009 Sales: $38,940	2009 Profits: $-1,948	**U.S. Stock Ticker: LGND**
2008 Sales: $27,315	2008 Profits: $-98,114	**Int'l Ticker:** Int'l Exchange:
2007 Sales: $12,894	2007 Profits: $281,688	Employees: 72
2006 Sales: $140,960	2006 Profits: $-31,743	Fiscal Year Ends: 12/31
2005 Sales: $123,010	2005 Profits: $-36,399	Parent Company:

SALARIES/BENEFITS:

Pension Plan:	ESOP Stock Plan:	Profit Sharing:	Top Exec. Salary: $420,000	Bonus: $252,000
Savings Plan:	Stock Purch. Plan: Y		Second Exec. Salary: $320,256	Bonus: $128,102

OTHER THOUGHTS:

Apparent Women Officers or Directors: 1
Hot Spot for Advancement for Women/Minorities:

LOCATIONS: ("Y" = Yes)

West:	Southwest:	Midwest:	Southeast:	Northeast:	International:
Y					

LONZA GROUP

www.lonza.com

Industry Group Code: 325 Ranks within this company's industry group: Sales: 5 Profits: 6

Drugs:		Other:		Clinical:		Computers:		Services:	
Discovery:		AgriBio:	Y	Trials/Services:		Hardware:		Specialty Services:	Y
Licensing:		Genetic Data:		Labs:		Software:		Consulting:	
Manufacturing:	Y	Tissue Replacement:		Equipment/Supplies:	Y	Arrays:		Blood Collection:	
Generics:				Research/Development Svcs.:	Y	Database Management:		Drug Delivery:	
				Diagnostics:				Drug Distribution:	

TYPES OF BUSINESS:

Chemicals Manufacturing
Fine Chemicals & Pharmaceutical Intermediates
Biocides
Materials Research
Active Drug Ingredients
Monoclonal Antibody Drugs

BRANDS/DIVISIONS/AFFILIATES:

Alusuisse-Lonza Group
Life Science Ingredients
Custom Manufacturing
Bioscience
Algonomics NV
MODA Technology Partners
Applied Protein Services Platform

CONTACTS: Note: Officers with more than one job title may be intentionally listed here more than once.

Stefan Borgas, CEO
Toralf Haag, CFO
Uwe H. Bohlke, Chief Human Resources Officer
Stephan Kutzer, COO-Custom Mfg. Div.
Dominick Werner, Head-Media Rel.
Alexandre Pasini, Mgr.-Investor Rel.
Lukas Utiger, COO-Bioscience Div.
Rolf Soiron, Chmn.

Phone: 41-61-316-8111	Fax: 41-61-316-9111
Toll-Free:	
Address: Muenchensteinerstrasse 38, Basel, 4002 Switzerland	

GROWTH PLANS/SPECIAL FEATURES:

The Lonza Group brings together a global portfolio of companies engaged in the production and supply of active pharmaceutical ingredients both biotechnologically and chemically for a range of clients in the healthcare, pharmaceutical and life science industries. The group structure was established in 1999 through the de-merger of its core business units from Swiss aluminum and industrial conglomerate Alusuisse-Lonza Group. Lonza Group has strong capabilities in peptides, amino acids, niche bioproducts and small and large molecules. It is also a leader in cell-based research, cell therapy manufacturing and endotoxin detection. The company operates in three segments: Life Science Ingredients; Custom Manufacturing; and Bioscience. The Life Science Ingredients division is focused on ingredients used in nutrition, microbial control and select industrial markets. This division creates products and solutions that include active biocides for hospital disinfectants, complex chemical intermediates for the agricultural industry and nutritional ingredients for improving health. In the Custom Manufacturing segment, the firm partners with biopharmaceutical and pharmaceutical companies for their manufacturing needs. This segment manufactures intermediates and active ingredients for healthcare companies, ultimately used in critical drugs to treat cardiovascular disease, cancer, neurological and infectious diseases. Its product capabilities include small and large molecules resulting from chemical synthesis, biotransformation, peptide synthesis, mammalian cell culture and microbial fermentation. Lastly, the Bioscience segment supports pharmaceutical and biotechnology companies in the research, development and commercialization of human therapeutics. Lonza Group business units will oversee production through to market commercialization. In November 2009, the company acquired Belgium-based Algonomics NV, a contract research organization. In May 2010, Lonza Group acquired MODA Technology Partners, a software company that provides paperless quality control solutions. In June 2010, the firm launched the Applied Protein Services platform, a set of stability, immunogenicity and protein engineering services.

FINANCIALS: Sales and profits are in thousands of dollars—add 000 to get the full amount. 2009 Note: Financial information for 2009 was not available for all companies at press time.

2009 Sales: $2,536,130	2009 Profits: $152,730	**U.S. Stock Ticker:**
2008 Sales: $2,712,290	2008 Profits: $386,940	**Int'l Ticker: LONN** Int'l Exchange: Zurich-SWX
2007 Sales: $2,650,410	2007 Profits: $277,970	Employees: 8,386
2006 Sales: $2,193,600	2006 Profits: $213,100	Fiscal Year Ends: 12/31
2005 Sales: $1,818,200	2005 Profits: $180,500	Parent Company:

SALARIES/BENEFITS:

Pension Plan: Y	ESOP Stock Plan:	Profit Sharing:	Top Exec. Salary: $	Bonus: $
Savings Plan:	Stock Purch. Plan:		Second Exec. Salary: $	Bonus: $

OTHER THOUGHTS:

Apparent Women Officers or Directors: 1
Hot Spot for Advancement for Women/Minorities:

LOCATIONS: ("Y" = Yes)

West:	Southwest:	Midwest:	Southeast:	Northeast:	International:
		Y		Y	Y

LORUS THERAPEUTICS INC

www.lorusthera.com

Industry Group Code: 325412 Ranks within this company's industry group: Sales: 150   Profits: 96

Drugs:		Other:	Clinical:	Computers:	Services:
Discovery:	Y	AgriBio:	Trials/Services:	Hardware:	Specialty Services:
Licensing:		Genetic Data:	Labs:	Software:	Consulting:
Manufacturing:	Y	Tissue Replacement:	Equipment/Supplies:	Arrays:	Blood Collection:
Generics:			Research/Development Svcs.:	Database Management:	Drug Delivery:
			Diagnostics:		Drug Distribution:

TYPES OF BUSINESS:

Drugs-Cancer
Antisense Compounds
Low-Molecular-Weight Compounds

BRANDS/DIVISIONS/AFFILIATES:

LOR-2040
LOR-1284
LOR-253
LOR-220
Virulizin
Interleukin-17E (IL-17E)
LOR-264
LOR-500

CONTACTS: Note: Officers with more than one job title may be intentionally listed here more than once.

Aiping H. Young, CEO
Aiping H. Young, Pres.
Elizabeth Williams, Acting CFO
Yoon Lee, VP-Research
Elizabeth Williams, Dir.-Admin
Saeid Babaei, VP-Bus. Dev.
Elizabeth Williams, Dir.-Finance
Peter Murray, Dir.-Clinical Dev.
Denis R. Burger, Chmn.

Phone: 416-798-1200	Fax: 416-798-2200
Toll-Free:	
Address: 2 Meridian Rd., Toronto, ON M9W 4Z7 Canada	

GROWTH PLANS/SPECIAL FEATURES:

Lorus Therapeutics, Inc. is a life sciences company focused on the discovery, research and development of effective anticancer therapies. Lorus has product candidates in three classes of anticancer therapies: RNA-targeted therapies (antisense and siRNA); small molecule therapies; and immunotherapy. The company's RNA-targeted therapeutics include LOR-2040 and LOR-1284. LOR-2040 decreases expression of the R2 subunit of ribonucleotide reductase (RNR) and is used to treat acute myeloid leukemia. It has completed a Phase II clinical trial and is also in a Phase I trial as a single agent in patients with high grade myelodysplastic syndromes and acute leukemias. LOR-1284 is based on siRNA-mediated inhibition of R2 expression and is in preclinical development. Lorus' small molecule compounds include LOR-253 and LOR-220. LOR-253 is used to treat solid tumors, colon carcinoma and non-small cell lung cancer. The company plans to commence a Phase I clinical trial for LOR-253 in solid tumors. LOR-220 is in preclinical development and is active against multi-drug resistant Gram-positive bacteria including methicillin-resistant Staphylococcus aureus and vancomycin-resistant enterococci (VRE). The company's immunotherapy product candidates are Virulizin and Interleukin-17E (IL-17E). Virulizin is used to treat pancreatic cancer. It recently completed a Phase III clinical trial and is approved for the treatment of malignant melanoma in Mexico. IL-17E is an inflammatory cytokine with anticancer activity against colon cancer, melanoma and pancreatic cancer. It is in preclinical development. The firm's other preclinical drug candidates include LOR-264, an oral second-generation derivative of LOR-253 with anticancer activity; and LOR-500, which targets multikinases, including tyrosine kinase family members and a member of the calmodulin/calcium-dependent protein kinase family. The company holds an 80% interest in NuChem Pharmaceuticals, Inc.

FINANCIALS: Sales and profits are in thousands of dollars—add 000 to get the full amount. 2009 Note: Financial information for 2009 was not available for all companies at press time.

2009 Sales: $ 170	2009 Profits: $-8,610	U.S. Stock Ticker: LRUSF
2008 Sales: $ 40	2008 Profits: $-11,680	Int'l Ticker: LOR Int'l Exchange: Toronto-TSX
2007 Sales: $ 100	2007 Profits: $-8,910	Employees:
2006 Sales: $	2006 Profits: $-17,900	Fiscal Year Ends: 5/31
2005 Sales: $	2005 Profits: $-17,600	Parent Company:

SALARIES/BENEFITS:

Pension Plan:	ESOP Stock Plan:	Profit Sharing:	Top Exec. Salary: $335,236	Bonus: $112,320
Savings Plan:	Stock Purch. Plan:		Second Exec. Salary: $157,269	Bonus: $23,670

OTHER THOUGHTS:

Apparent Women Officers or Directors: 2
Hot Spot for Advancement for Women/Minorities: Y

LOCATIONS: ("Y" = Yes)

West:	Southwest:	Midwest:	Southeast:	Northeast:	International:
					Y

LUMINEX CORPORATION

www.luminexcorp.com

Industry Group Code: 325413 Ranks within this company's industry group: Sales: 12 Profits: 7

Drugs:	Other:	Clinical:		Computers:	Services:
Discovery:	AgriBio:	Trials/Services:		Hardware:	Specialty Services:
Licensing:	Genetic Data:	Labs:		Software:	Consulting:
Manufacturing:	Tissue Replacement:	Equipment/Supplies:	Y	Arrays:	Blood Collection:
Generics:		Research/Development Svcs.:		Database Management:	Drug Delivery:
		Diagnostics:	Y		Drug Distribution:

TYPES OF BUSINESS:

Medical Diagnostics
Bioassays
Software
xMAP Testing

BRANDS/DIVISIONS/AFFILIATES:

xMAP
Luminex 100 IS System
Luminex Molecular Diagnostics
Luminex 200 System
xPONENT
MagPlex Magnetic Microspheres
xTAG
Luminex Bioscience Group

CONTACTS: Note: Officers with more than one job title may be intentionally listed here more than once.

Patrick J. Balthrop, CEO
Patrick J. Balthrop, Pres.
Harriss T. Currie, CFO
Darin S. Leigh, VP-Mktg. & Sales
Timothy R. Dehne, VP-Systems R&D
Andrew D. Ewing, VP-Luminex Tech. Oper.
Steve Back, VP-Mfg.
David S. Reiter, General Counsel/VP/Corp. Sec.
Michael F. Pintek, Sr. VP-Oper.
Russell W. Bradley, VP-Bus. Dev. & Strategic Planning
Harriss T. Currie, VP-Finance/Treas.
Nancy Krunic, VP-Luminex Molecular Diagnostics
Gregory J. Gosch, VP-Luminex Bioscience Group
Jeremy Bridge-Cook, Sr. VP-Assay Group
G. Walter Loewenbaum, Chmn.

Phone: 512-219-8020	Fax: 512-219-5195
Toll-Free: 888-219-8020	
Address: 12212 Technology Blvd., Austin, TX 78727 US	

GROWTH PLANS/SPECIAL FEATURES:

Luminex Corporation develops, manufactures and markets biological testing technologies with applications for the life science and diagnostic industries. The company operates in two segments: the technology segment and the assay segment. The technology segment is responsible for the company's 68 strategic partnerships. Luminex licenses its technology to its partners who then develop reagent-based products utilizing that technology. Luminex also manufactures and sells xMAP instrumentation and microshperes to its partners. The partners then sell these products to end users, such as testing laboratories, and Luminex receives royalties from those sales. The assay segment consists of Luminex Bioscience Group (LLBG) and Luminex Molecular Diagnostics (LMD). This segment is primarily involved in the development and sale of assays utilizing the firm's proprietary xMAP technology. The xMAP system makes use of microspheres (microscopic polystyrene beads), lasers, digital signal processing and proprietary software in order to run various diagnostic tests. It can perform up to 500 bioassays on a single drop of fluid. xMAP technology is used within various segments of the life sciences industry including drug discovery and development; clinical diagostics; genetic analysis; bio-defense; protein analysis; and biomedical research. Additional products include xPONENT software and MagPlex magnetic microspheres, which enhance ease-of-use and automation capabilities in the xMAP technologies. The firm's xTAG technology, developed by LMD, is an assay cleared by the FDA to simultaneously detect and identify 12 viruses and viral subtypes that together are responsible for more than 85% of respiratory infections. Products designed for use in the Luminex Systems include the Luminex 100 IS System and the Luminex 200 System, compact analyzers that integrate fluidics, optics and digital signal processing to perform up to 100 bioassays simultaneously. In June 2009, Luminex opened an office in Shanghai, which will serve as its Asia-Pacific headquarters.

The firm offers its employees a 401(k) plan; and health, dental and life insurance.

FINANCIALS: Sales and profits are in thousands of dollars—add 000 to get the full amount. 2009 Note: Financial information for 2009 was not available for all companies at press time.

2009 Sales: $120,643	2009 Profits: $17,729	U.S. Stock Ticker: LMNX
2008 Sales: $104,447	2008 Profits: $3,057	Int'l Ticker: Int'l Exchange:
2007 Sales: $75,010	2007 Profits: $-2,711	Employees:
2006 Sales: $52,989	2006 Profits: $1,507	Fiscal Year Ends: 12/31
2005 Sales: $42,313	2005 Profits: $-2,666	Parent Company:

SALARIES/BENEFITS:

Pension Plan:	ESOP Stock Plan:	Profit Sharing:	Top Exec. Salary: $445,500	Bonus: $470,542
Savings Plan: Y	Stock Purch. Plan:		Second Exec. Salary: $325,600	Bonus: $257,253

OTHER THOUGHTS:

Apparent Women Officers or Directors: 1
Hot Spot for Advancement for Women/Minorities:

LOCATIONS: ("Y" = Yes)

West:	Southwest:	Midwest:	Southeast:	Northeast:	International:
	Y				Y

MANHATTAN PHARMACEUTICALS INC   www.manhattanpharma.com

Industry Group Code: 325412 Ranks within this company's industry group: Sales: Profits: 82

Drugs:		Other:	Clinical:	Computers:	Services:	
Discovery:	Y	AgriBio:	Trials/Services:	Hardware:	Specialty Services:	
Licensing:		Genetic Data:	Labs:	Software:	Consulting:	
Manufacturing:		Tissue Replacement:	Equipment/Supplies:	Arrays:	Blood Collection:	
Generics:			Research/Development Svcs.:	Database Management:	Drug Delivery:	Y
			Diagnostics:		Drug Distribution:	

TYPES OF BUSINESS:

Pharmaceuticals Discovery & Development
Drug Delivery Systems

BRANDS/DIVISIONS/AFFILIATES:

Hedrin
AST-726
AST-915
Ariston Pharmaceuticals, Inc.

CONTACTS: *Note: Officers with more than one job title may be intentionally listed here more than once.*

Michael G. McGuinness, COO
Michael G. McGuinness, CFO
Michelle Carroll, VP-Corp. Dev.
Mary C. Spellman, Head-Dermatology & Drug Dev.
Douglas Abel, Chmn.

Phone: 212-582-3950	Fax: 212-582-3957
Toll-Free:	
Address: 48 Wall St., Ste. 1100, New York, NY 10005 US	

GROWTH PLANS/SPECIAL FEATURES:

Manhattan Pharmaceuticals, Inc. is a specialty healthcare product company focused on developing and commercializing treatments primarily in the areas of dermatologic and immune disorders. With a pipeline consisting of four product candidates, the company is developing potential therapeutics for large, underserved patient populations. The company's portfolio consists of Hedrin; AST-726; AST-915; and a topical GEL for the treatment of mild psoriasis. Hedrin is a novel, non-insecticide one-hour treatment for pediculosis (head lice) and is currently being developed in the U.S. as a prescription medical device. It is one of the top selling head lice products in Europe and is marketed in over 27 countries. The company has a joint venture agreement with Nordic Biotech Advisors ApS to develop and commercialize Hedrin for the North American Market. AST-726 is a nasally-delivered form of hydroxocobalamin for the treatment of Vitamin B deficiency. It has completed Phase II clinical trials, and the company is planning a Phase III Vitamin B replacement study in the U.S. AST-915 is an orally-delivered treatment for essential tremor, a neurological disorder characterized by involuntary shaking of the hands, arms, head, voice and upper body. It is currently in a Phase I clinical study. Lastly, the company is developing a topical GEL for psoriasis which has completed Phase IIa studies. The firm owns global rights to the GEL and is exploring the possibility of developing it as an over-the-counter (OTC) product for mild psoriasis. In March 2010, Manhattan Pharmaceuticals acquired Delaware-based Ariston Pharmaceuticals, Inc. Through the acquisition, the company acquired global rights to AST-726 and AST-915.

FINANCIALS: Sales and profits are in thousands of dollars—add 000 to get the full amount. 2009 Note: Financial information for 2009 was not available for all companies at press time.

2009 Sales: $	2009 Profits: $-2,793	U.S. Stock Ticker: MHAN
2008 Sales: $	2008 Profits: $-4,269	Int'l Ticker: Int'l Exchange:
2007 Sales: $	2007 Profits: $-12,032	Employees: 4
2006 Sales: $	2006 Profits: $-9,695	Fiscal Year Ends: 12/31
2005 Sales: $	2005 Profits: $-19,141	Parent Company:

SALARIES/BENEFITS:

Pension Plan:	ESOP Stock Plan:	Profit Sharing:	Top Exec. Salary: $277,500	Bonus: $
Savings Plan:	Stock Purch. Plan:		Second Exec. Salary: $164,053	Bonus: $

OTHER THOUGHTS:

Apparent Women Officers or Directors: 2
Hot Spot for Advancement for Women/Minorities:

LOCATIONS: ("Y" = Yes)

West:	Southwest:	Midwest:	Southeast:	Northeast:	International:
				Y	

MAXYGEN INC

www.maxygen.com

Industry Group Code: 325412 Ranks within this company's industry group: Sales: 94 Profits: 137

Drugs:		Other:		Clinical:	Computers:		Services:	
Discovery:		AgriBio:		Trials/Services:	Hardware:		Specialty Services:	
Licensing:		Genetic Data:	Y	Labs:	Software:		Consulting:	
Manufacturing:	Y	Tissue Replacement:		Equipment/Supplies:	Arrays:		Blood Collection:	
Generics:				Research/Development Svcs.:	Database Management:		Drug Delivery:	
				Diagnostics:			Drug Distribution:	

TYPES OF BUSINESS:

Drug Discovery & Development
Research Services
Research & Development-Molecular Evolution

BRANDS/DIVISIONS/AFFILIATES:

MolecularBreeding
DNAShuffling
MaxyScan
Maxy-G34
MAXY-4
Codexis
Astellas Pharma, Inc.
Perseid Therapeutics LLC

CONTACTS: Note: Officers with more than one job title may be intentionally listed here more than once.

James Sulat, CEO
James Sulat, CFO
Grant Yonehiro, Sr. VP/Pres./CEO-Perseid Therapeutics LLC
Isaac Stein, Chmn.

Phone: 650-298-5300	Fax: 650-364-2715
Toll-Free:	
Address: 301 Galveston Dr., Redwood City, CA 94063 US	

GROWTH PLANS/SPECIAL FEATURES:

Maxygen, Inc. is a biotechnology company that, with its wholly-owned subsidiaries, Maxygen ApS and Maxygen Holdings, Ltd., works on the development of improved versions of protein pharmaceuticals for the treatment of diseases and serious medical conditions. Technologies developed by Maxygen include MolecularBreeding, a process that mimics the natural events of evolution using a recombination process called DNAShuffling that generates a diverse library of DNA sequences. MolecularBreeding allows the company to rapidly move from product concept to IND (investigational new drug)-ready drug candidate. The company's MaxyScan screening system selects individual proteins with desired characteristics from gene variants within the library for additional experimentation. Maxygen currently has two product candidates in different clinical study phases: Maxy-G34, a neutropenia treatment also undergoing clinical testing as a treatment for acute radiation syndrome (ARS); and MAXY-4, a rheumatoid arthritis and transplant rejection reduction treatment. Maxygen additionally has an HIV vaccine research program and a minority investment in Codexis, a biotechnology company focused on developing biocatalytic process technologies for pharmaceutical, energy and industrial chemical applications. In May 2009, the company announced an exclusive licensing agreement with Cangene Corporation for MAXY-G34's use in treating ARS. In September 2009, Maxygen established a joint venture, Perseid Therapeutics LLC, with Astellas Pharma, Inc., intended to further the discovery, research and development of multiple protein pharmaceutical programs. Astellas has the option to acquire all of Maxygen's ownership interests in the joint venture within three years. Substantially all of Maxygen's research and development operations are now under the control of Perseid.

Employees are offered stock options; financial planning services; medical and dental plans; tuition reimbursement; health club reimbursement; a commuter voucher program; concierge services; credit union membership; backup childcare services; notary services; computer training courses; flexible time off; and a 401(k) plan.

FINANCIALS: Sales and profits are in thousands of dollars—add 000 to get the full amount. 2009 Note: Financial information for 2009 was not available for all companies at press time.

2009 Sales: $36,376	2009 Profits: $-32,157	**U.S. Stock Ticker: MAXY**
2008 Sales: $100,709	2008 Profits: $30,325	**Int'l Ticker:** Int'l Exchange:
2007 Sales: $23,157	2007 Profits: $-49,315	Employees: 64
2006 Sales: $25,021	2006 Profits: $-16,482	Fiscal Year Ends: 12/31
2005 Sales: $14,501	2005 Profits: $-18,436	Parent Company:

SALARIES/BENEFITS:

Pension Plan:	ESOP Stock Plan:	Profit Sharing:	Top Exec. Salary: $520,520	Bonus: $200,000
Savings Plan: Y	Stock Purch. Plan: Y		Second Exec. Salary: $439,828	Bonus: $98,961

OTHER THOUGHTS:

Apparent Women Officers or Directors:
Hot Spot for Advancement for Women/Minorities:

LOCATIONS: ("Y" = Yes)

West:	Southwest:	Midwest:	Southeast:	Northeast:	International:
Y					

Note: Financial information, benefits and other data can change quickly and may vary from those stated here.

MDRNA INC

www.mdrnainc.com

Industry Group Code: 325412A Ranks within this company's industry group: Sales: 12   Profits: 10

Drugs:		Other:		Clinical:		Computers:		Services:	
Discovery:	Y	AgriBio:		Trials/Services:		Hardware:		Specialty Services:	
Licensing:	Y	Genetic Data:		Labs:		Software:		Consulting:	
Manufacturing:	Y	Tissue Replacement:		Equipment/Supplies:		Arrays:		Blood Collection:	
Generics:				Research/Development Svcs.:		Database Management:		Drug Delivery:	Y
				Diagnostics:				Drug Distribution:	

TYPES OF BUSINESS:

Drug Delivery Systems
RNA Interference Technology
Tight Junction Biology

BRANDS/DIVISIONS/AFFILIATES:

Cequent Pharmaceuticals
UNA-based Diagnostics

CONTACTS: Note: Officers with more than one job title may be intentionally listed here more than once.

J. Michael French, CEO
J. Michael French, Pres.
Peter S. Garcia, CFO
Barry Polinsky, Chief Scientific Officer
June D. Ameen, VP-Corp. Dev.
Bruce R. Thaw, Chmn.

Phone: 425-908-3600	Fax: 425-908-3650
Toll-Free:	
Address: 3830 Monte Villa Pkwy., Bothell, WA 98021 US	

GROWTH PLANS/SPECIAL FEATURES:

MDRNA, Inc. is a biotechnology company focused on the discovery, development and commercialization of therapies using RNA interference (RNAi) to down-regulate specific protein expressions that lead to disease without altering the DNA itself.   This technology may have applications for potential treatments in oncology, inflammation, metabolic disorders and viral infections.   A challenge involved with RNAi technology is that specific targeting is extremely important, and off-target deliveries can result in undesirable side effects. MDRNA's research is designed to eliminate the chance for off-target effects by using a specific type of small interfering RNAs (siRNAs). MDRNA focuses its pipeline on the discovery and development of siRNA therapeutics to treat forms of cancer such as hepatocellular carcinoma (liver cancer) and bladder cancer, as well as other types of diseases. The firm collaborates with a number of research partners, including the University of Helsinki, University of Michigan and University of British Columbia. The company also licenses its technology to F. Hoffmann-La Roche Inc., F. Hoffmann-La Roche Ltd. and Novartis. MDRNA owns or controls 15 issued or allowed patents and has 37 pending patent applications, 126 pending foreign patent applications and 7 PCT applications. In March 2010, the firm acquired intellectual property for bridged nucleic acids from Valeant Pharmaceuticals North America. In April 2010, MDRNA acquired Cequent Pharmaceuticals, a firm focused on the development of novel products to deliver RNAi-based therapeutics, for approximately $46 million. In June 2010, the company obtained exclusive rights to develop, manufacture, use and sell UNA (Unlocked Nucleobase Analog)-based diagnostics from RiboTask ApS.

MDRNA offers its employees medical, dental and vision insurance; flexible spending accounts; life insurance; short- and long-term disability coverage; a 401(k) plan; financial support for conferences, publications and seminars; and an employee stock purchase plan.

FINANCIALS: Sales and profits are in thousands of dollars—add 000 to get the full amount. 2009 Note: Financial information for 2009 was not available for all companies at press time.

2009 Sales: $14,732	2009 Profits: $-8,046	**U.S. Stock Ticker: MRNA**
2008 Sales: $2,609	2008 Profits: $-59,220	**Int'l Ticker:** Int'l Exchange:
2007 Sales: $18,137	2007 Profits: $-52,372	Employees: 46
2006 Sales: $28,490	2006 Profits: $-26,877	Fiscal Year Ends: 12/31
2005 Sales: $7,449	2005 Profits: $-32,163	Parent Company:

SALARIES/BENEFITS:

Pension Plan:	ESOP Stock Plan:	Profit Sharing:	Top Exec. Salary: $375,000	Bonus: $
Savings Plan: Y	Stock Purch. Plan: Y		Second Exec. Salary: $340,000	Bonus: $

OTHER THOUGHTS:

Apparent Women Officers or Directors: 1
Hot Spot for Advancement for Women/Minorities:

LOCATIONS: ("Y" = Yes)

West:	Southwest:	Midwest:	Southeast:	Northeast:	International:
Y					

MDS INC

www.mdsnordion.com

Industry Group Code: 6215 Ranks within this company's industry group: Sales: 3 Profits: 5

Drugs:		Other:		Clinical:		Computers:		Services:	
Discovery:	Y	AgriBio:		Trials/Services:		Hardware:		Specialty Services:	Y
Licensing:		Genetic Data:	Y	Labs:	Y	Software:		Consulting:	
Manufacturing:		Tissue Replacement:		Equipment/Supplies:	Y	Arrays:		Blood Collection:	
Generics:				Research/Development Svcs.:	Y	Database Management:		Drug Delivery:	
				Diagnostics:	Y			Drug Distribution:	

TYPES OF BUSINESS:

Drug Discovery & Development Services
Medical Isotopes
Imaging Agents
Sterilization Products
Irradiation Systems
Health Care Product Distribution

BRANDS/DIVISIONS/AFFILIATES:

MDS Pharma Services
MDS Analytical Technologies
MDS Nordion
TheraSphere
Nordion, Inc.

CONTACTS: Note: Officers with more than one job title may be intentionally listed here more than once.

Steven M. West, CEO
G. Peter Dans, CFO
Kevin Brooks, VP-Mktg.
Mary E. Federau, Exec. VP-Global Human Resources
Thomas E. Gernon, CIO/Exec. VP-IT
Peter Covitz, Sr. VP-Innovation, MDS Nordion
Kenneth L. Horton, General Counsel
Scott McIntosh, VP-Oper.
Kenneth L. Horton, Exec. VP-Corp. Dev.
Janet Ko, Sr. VP-Comm.
Kim Lee, Sr. Dir.-Investor Rel.
Douglas S. Prince, Exec. VP-Finance
Peter Brent, Sr. VP-Legal/Corp. Sec.
Christopher Ashwood, Sr. VP-Human Resources & IT, MDS Nordion
Jill Chitra, Sr. VP-Strategic Technologies & Regulatory Affairs
Chris Wagner, Sr. VP-Sales & Corp. Dev., MDS Nordion
William Anderson, Chmn.

Phone: 613-592-2790	Fax: 613-592-6937
Toll-Free:	
Address: 447 March Rd., Ottawa, ON K2K 1X8 Canada	

GROWTH PLANS/SPECIAL FEATURES:

MDS, Inc., founded in 1969 as Medical Data Sciences Limited, is a global biotechnology firm providing products and services for drug development and disease treatment and diagnosis. Its customers include a broad range of manufacturers of medical products such as, pharmaceutical manufacturers, biotechnology companies, manufacturers of medical supplies and devices, plus academic and government institutions. Following a dramatic restructuring process in which MDS sold a substantial portion of its assets and facilities, the firm established its focus on its MDS Nordion program. Nordion's focus continues to be the development and manufacture of technologies for use in medical imaging, radiotherapeutics and radiopharmaceuticals, as well as sterilization technologies for medical products and food safety. The firm supplies medical isotopes used in the diagnosis and treatment of diseases, while also offering specific treatments for various cancers, including liver cancer and non-Hodgkin's lymphoma. Its TheraSphere treatment for liver cancer offers an innovative and more precise way to combat instances of inoperability with external radiotherapy. Other fields supplied by Nordion's products include, cardiology and neurology. Its sterilization products, such as the isotope cobalt-60, use gamma radiation to ensure the destruction of harmful micro-organisms that can reside in food and on medical equipment. From late 2009 through early 2010, MDS announced and finalized plans to divest its programs in Pharma Services and Analytical Technologies. The Analytic Technologies segment, a supplier of drug discovery and life sciences research tools, was sold to Danaher Corporation, while its Pharma Services portion, which consisted of 38 centers of operation in 29 countries, was split into several independent portions between different purchasers.

FINANCIALS: Sales and profits are in thousands of dollars—add 000 to get the full amount. 2009 Note: Financial information for 2009 was not available for all companies at press time.

2009 Sales: $231,000	2009 Profits: $-135,000	U.S. Stock Ticker: MDZ
2008 Sales: $296,000	2008 Profits: $-553,000	Int'l Ticker: MDS Int'l Exchange: Toronto-TSX
2007 Sales: $290,000	2007 Profits: $781,000	Employees: 3,600
2006 Sales: $1,060,000	2006 Profits: $120,000	Fiscal Year Ends: 10/31
2005 Sales: $1,296,908	2005 Profits: $26,999	Parent Company:

SALARIES/BENEFITS:

Pension Plan:	ESOP Stock Plan:	Profit Sharing:	Top Exec. Salary: $703,563	Bonus: $
Savings Plan:	Stock Purch. Plan:		Second Exec. Salary: $363,322	Bonus: $

OTHER THOUGHTS:

Apparent Women Officers or Directors: 3
Hot Spot for Advancement for Women/Minorities: Y

LOCATIONS: ("Y" = Yes)

West:	Southwest:	Midwest:	Southeast:	Northeast:	International:
					Y

MEDICINES CO (THE)

www.themedicinescompany.com

Industry Group Code: 325412 Ranks within this company's industry group: Sales: 48 Profits: 152

Drugs:		Other:		Clinical:	Computers:	Services:
Discovery:		AgriBio:		Trials/Services:	Hardware:	Specialty Services:
Licensing:	Y	Genetic Data:		Labs:	Software:	Consulting:
Manufacturing:	Y	Tissue Replacement:		Equipment/Supplies:	Arrays:	Blood Collection:
Generics:				Research/Development Svcs.:	Database Management:	Drug Delivery:
				Diagnostics:		Drug Distribution:

TYPES OF BUSINESS:

Pharmaceuticals Acquisition & Development
Acute Care Hospital Products
Anticoagulants
Blood Pressure Control

BRANDS/DIVISIONS/AFFILIATES:

Angiomax
Cleviprex (Clevidine)
Cangrelor
Oritavancin
CU-2010
ApoA-I Milano

CONTACTS: *Note: Officers with more than one job title may be intentionally listed here more than once.*

Clive A. Meanwell, CEO
John P. Kelley, COO
John P. Kelley, Pres.
Glenn Sblendorio, CFO/Exec. VP
Leslie C. Rohrbacker, Chief Human Strategy Officer/VP
Paul M. Antinori, General Counsel/Sr. VP
Bill O'Connor, Chief Acct. Officer
Clive A. Meanwell, Chmn.

Phone: 973-290-6000	Fax: 973-656-9898
Toll-Free: 800-388-1183	
Address: 8 Sylvan Way, Parsippany, NJ 07054 US	

GROWTH PLANS/SPECIAL FEATURES:

The Medicines Co. (TMC) is a global pharmaceutical company specializing in acute care hospital products. The firm acquires, develops and commercializes pharmaceutical products in late stages of development. The company has two marketed products: Angiomax and Cleviprex. Angiomax is an intravenous direct thrombin inhibitor approved for use in patients undergoing percutaneous coronary intervention and, in Europe, for adult patients with acute coronary syndrome. Cleviprex is an intravenous drug intended for the control of blood pressure in anesthesiology and surgery, critical care and emergency conditions. The firm currently has two products in the Phase III stage of development: Cangrelor, an intravenous antiplatelet agent that prevents platelet activation and aggregation, which is believed to have potential advantages in the treatment of vascular disease; and Oritavancin, an intravenous antibiotic being developed for the treatment of gram-positive bacterial infections. The firm currently has one product in Phase I stage of production, CU-2010, a small molecule serine protease inhibitor being developed for the prevention of blood loss during surgery. In December 2009, the firm acquired the exclusive licensing rights to ApoA-I Milano, a potential atherosclerotic plaque development reversal drug, from Pfizer, Inc.

TMC offers its employees medical, dental, prescription and vision insurance; an employee assistance plan; tuition reimbursement; an employee stock purchase plan and associate stock options; and credit union membership.

FINANCIALS: Sales and profits are in thousands of dollars—add 000 to get the full amount. 2009 Note: Financial information for 2009 was not available for all companies at press time.

2009 Sales: $404,241	2009 Profits: $-76,229	U.S. Stock Ticker: MDCO
2008 Sales: $348,157	2008 Profits: $-8,504	Int'l Ticker: Int'l Exchange:
2007 Sales: $257,534	2007 Profits: $-18,272	Employees:
2006 Sales: $213,952	2006 Profits: $63,726	Fiscal Year Ends: 12/31
2005 Sales: $150,207	2005 Profits: $-7,753	Parent Company:

SALARIES/BENEFITS:

Pension Plan:	ESOP Stock Plan: Y	Profit Sharing:	Top Exec. Salary: $588,640	Bonus: $441,480
Savings Plan: Y	Stock Purch. Plan:		Second Exec. Salary: $434,534	Bonus: $201,710

OTHER THOUGHTS:

Apparent Women Officers or Directors: 2
Hot Spot for Advancement for Women/Minorities:

LOCATIONS: ("Y" = Yes)

West:	Southwest:	Midwest:	Southeast:	Northeast:	International:
				Y	Y

MEDICIS PHARMACEUTICAL CORP

www.medicis.com

Industry Group Code: 325412 Ranks within this company's industry group: Sales: 45 Profits: 47

Drugs:	Other:	Clinical:	Computers:	Services:
Discovery:	AgriBio:	Trials/Services:	Hardware:	Specialty Services:
Licensing:	Genetic Data:	Labs:	Software:	Consulting:
Manufacturing: Y	Tissue Replacement:	Equipment/Supplies:	Arrays:	Blood Collection:
Generics:		Research/Development Svcs.:	Database Management:	Drug Delivery:
		Diagnostics:		Drug Distribution:

TYPES OF BUSINESS:

Dermatological, Aesthetic & Podiatric Conditions Drugs
Acne Treatment
Topical Creams
Wrinkle Treatment

BRANDS/DIVISIONS/AFFILIATES:

Perlane
Restylane
Solodyn
Triaz
Vanos
Ziana
LipoSonix
Dysport

CONTACTS: *Note: Officers with more than one job title may be intentionally listed here more than once.*

Jonah Shacknai, CEO
Mark A. Prygocki, Sr., COO/Exec. VP
Richard D. Peterson, CFO/Exec. VP
Vincent Ippolito, Exec. VP-Sales & Mktg.
Mitchell S. Wortzman, Chief Scientific Officer/Exec. VP
Joseph P. Cooper, Exec. VP-Prod. Dev.
Jason Hanson, General Counsel/Exec. VP/Corp. Sec.
Joseph P. Cooper, Exec. VP-Corp. Dev.
Richard D. Peterson, Treas.
Jonah Shacknai, Chmn.

Phone: 602-808-8800	Fax: 602-808-0822
Toll-Free:	
Address: 7720 North Dobson Rd., Scottsdale, AZ 85256 US	

GROWTH PLANS/SPECIAL FEATURES:

Medicis Pharmaceutical Corp. is a specialty pharmaceutical company that develops and markets products for treatment of dermatological and aesthetic conditions. The company offers a range of products addressing various conditions or aesthetic improvement, including facial wrinkles; acne; fungal infections; rosacea; hyperpigmentation; photoaging; psoriasis; seborrheic dermatitis; and cosmesis (improvement in the texture and appearance of skin). The firm offers 17 branded products. Its primary brands include Dysport, an injection used to temporarily remove the look of moderate to severe frown lines between the eyebrows; the Liposonix system, a nonsurgical, non-invasive fat reduction technology that is used to permanently remove abdominal fat; Perlane, an injectable gel for implantation into the deep dermis to superficial subcutis for the correction of facial wrinkles and folds; Restylane, an injectable gel that is indicated for mid-to-deep implantation for treatment of moderate to severe facial wrinkles and folds such as nasolabial folds; Solodyn, an extended-release oral antibiotic for moderate to severe acne; Triaz acne cleansers, gels, pads and foaming cloths; Vanos, a corticosteroid to relieve inflammation and itching caused by certain skin conditions; and Ziana, a topical treatment for mild, moderate and severe acne vulgaris. Medicis customers include wholesale pharmaceutical distributors such as Cardinal Health, Inc. (which generated 37.1% of 2009 revenues); McKesson Corp. (40.8% of 2009 revenues); and other major drug chains. McKesson is the company's sole distributor of Restylane, Perlane and Dysport products in the U.S. In January 2010, the firm's Restylane-L and Perlane-L dermal filler products were approved by the U.S. Food and Drug Administration.

The firm offers employees medical, dental and vision coverage; a prescription plan; educational assistance; a 401(k) plan; and flexible spending accounts.

FINANCIALS: Sales and profits are in thousands of dollars—add 000 to get the full amount. 2009 Note: Financial information for 2009 was not available for all companies at press time.

2009 Sales: $571,915	2009 Profits: $75,951	**U.S. Stock Ticker: MRX**
2008 Sales: $517,750	2008 Profits: $10,276	**Int'l Ticker:** Int'l Exchange:
2007 Sales: $457,394	2007 Profits: $70,436	Employees: 578
2006 Sales: $393,165	2006 Profits: $-48,152	Fiscal Year Ends: 12/31
2005 Sales: $376,899	2005 Profits: $64,990	Parent Company:

SALARIES/BENEFITS:

Pension Plan:	ESOP Stock Plan:	Profit Sharing:	Top Exec. Salary: $1,100,000	Bonus: $1,039,500
Savings Plan: Y	Stock Purch. Plan:		Second Exec. Salary: $485,000	Bonus: $381,938

OTHER THOUGHTS:

Apparent Women Officers or Directors: 1
Hot Spot for Advancement for Women/Minorities:

LOCATIONS: ("Y" = Yes)

West:	Southwest:	Midwest:	Southeast:	Northeast:	International:
Y	Y	Y	Y	Y	

MEDIDATA SOLUTIONS INC

www.mdsol.com

Industry Group Code: 511210D Ranks within this company's industry group: Sales: 2 Profits: 3

Drugs:	Other:	Clinical:	Computers:		Services:	
Discovery:	AgriBio:	Trials/Services:	Hardware:		Specialty Services:	Y
Licensing:	Genetic Data:	Labs:	Software:	Y	Consulting:	
Manufacturing:	Tissue Replacement:	Equipment/Supplies:	Arrays:		Blood Collection:	
Generics:		Research/Development Svcs.:	Database Management:	Y	Drug Delivery:	
		Diagnostics:			Drug Distribution:	

TYPES OF BUSINESS:

Clinical Trial Software, Online

BRANDS/DIVISIONS/AFFILIATES:

Medidata Rave
Medidata Grants Manager
Medidata CRO Contractor
Medidata Designer
MedidatA Balance

CONTACTS: *Note: Officers with more than one job title may be intentionally listed here more than once.*

Tarek A. Sherif, CEO
Glen M. de Vries, Pres.
Bruce D. Dalziel, CFO
Lineene Krasnow, Exec. VP-Prod. & Mktg.
Arden Schneider, Sr. VP-Human Resources
Glenn Watt, VP-Global Info. Security & Privacy
Richard J. Piazza, VP-Prod. Mgmt.
Michael Otner, General Counsel
Lori Shields, VP-Oper., Trial Planning
Keith Howells, Sr. VP-Dev.
Cory Douglas, Controller/VP
Steven Hirschfield, Exec. VP-Sales and Alliances
Earl Hulihan, Sr. VP-Regulatory Compliance
Vik Shah, Sr. VP-Svcs.
Shih-Yin Ho, VP-Corp. Strategy
Tarek A. Sherif, Chmn.
Steve Heath, Head-EMEA/VP-Channel Sales

Phone: 212-918-1800	Fax: 212-918-1818
Toll-Free: 877-511-4200	
Address: 79 5th Ave., 8th Fl., New York, NY 10003 US	

GROWTH PLANS/SPECIAL FEATURES:

Medidata Solutions, Inc. offers hosted clinical trial software used by medical research and development firms worldwide. Its customers include pharmaceutical, biotechnology and medical device companies, as well as academic institutions, contract research organizations (CROs) and other organizations engaged in clinical trials. Included among these customers are companies such as Johnson & Johnson; AstraZeneca; Roche; Amgen; and Takeda Pharmaceutical. Its principal product is Medidata Rave, a comprehensive platform that integrates electronic data capture (EDC) with a clinical data management system (CDMS) in a single solution that replaces traditional paper-based methods of capturing and managing clinical data. Medidata Rave allows users to post real-time updates to clinical data and input and retrieve data in multiple languages simultaneously. Additionally, because it is an online program, customers can easily coordinate data from around the world in a single platform. Most of the revenue generated by this product derives from multi-study agreements made for a pre-determined number of studies. The company's other products similarly strive to help customers streamline the design, planning and management of key aspects of the clinical development process, including protocol development, CRO negotiation, investigator contracting, the capture and management of clinical trial data and the analysis and reporting of that data on a worldwide basis. These other products include Medidata Grants Manager, Medidata CRO Contractor and Medidata Designer (for designing protocols). In addition to its products, Medidata also offers professional services, including global consulting, implementation, technical support and training for customers and investigators. Application services (those relating to software) generated 73% of 2009 revenues, with professional services generating the remainder. In June 2010, the company introduced Medidata Balance, a randomization and trial supply (RTSM) offering for use in clinical research trials.

FINANCIALS: Sales and profits are in thousands of dollars—add 000 to get the full amount. 2009 Note: Financial information for 2009 was not available for all companies at press time.

2009 Sales: $140,400	2009 Profits: $5,182	U.S. Stock Ticker: MDSO
2008 Sales: $105,724	2008 Profits: $-18,272	Int'l Ticker: Int'l Exchange:
2007 Sales: $62,983	2007 Profits: $-23,662	Employees: 574
2006 Sales: $	2006 Profits: $	Fiscal Year Ends: 12/31
2005 Sales: $	2005 Profits: $	Parent Company:

SALARIES/BENEFITS:

Pension Plan:	ESOP Stock Plan:	Profit Sharing:	Top Exec. Salary: $360,000	Bonus: $524,958
Savings Plan: Y	Stock Purch. Plan: Y		Second Exec. Salary: $340,000	Bonus: $374,970

OTHER THOUGHTS:

Apparent Women Officers or Directors: 3
Hot Spot for Advancement for Women/Minorities: Y

LOCATIONS: ("Y" = Yes)

West:	Southwest:	Midwest:	Southeast:	Northeast:	International:
Y	Y			Y	Y

MEDIMMUNE INC

www.medimmune.com

Industry Group Code: 325412 Ranks within this company's industry group: Sales: Profits:

Drugs:	Other:	Clinical:	Computers:	Services:
Discovery:	AgriBio:	Trials/Services:	Hardware:	Specialty Services:
Licensing:	Genetic Data:	Labs:	Software:	Consulting:
Manufacturing: Y	Tissue Replacement:	Equipment/Supplies:	Arrays:	Blood Collection:
Generics:		Research/Development Svcs.:	Database Management:	Drug Delivery:
		Diagnostics:		Drug Distribution:

TYPES OF BUSINESS:

Pharmaceuticals Development & Manufacturing
Drugs-Cancer, Infectious & Immune Diseases

BRANDS/DIVISIONS/AFFILIATES:

MedImmune Ventures, Inc.
Ethyol
Synagis
FluMist

CONTACTS: *Note: Officers with more than one job title may be intentionally listed here more than once.*

Peter Greenleaf, Pres.
Tim Pearson, CFO/Sr. VP
Max Donley, VP-Human Resources
Bahija Jallal, Sr. VP-R&D
William C. Bertrand, Jr., General Counsel/Exec. VP-Legal Affairs
Bernardus N. M. Machielse, Exec. VP-Oper.
William C. Bertrand, Jr., Corporate Compliance Officer
Alexander A. Zukiwski, Sr. VP-Clinical Research/Chief Medical Officer

Phone: 301-398-0000	Fax: 301-398-9000
Toll-Free: 877-633-4411	
Address: 1 MedImmune Way, Gaithersburg, MD 20878 US	

GROWTH PLANS/SPECIAL FEATURES:

MedImmune, Inc., a subsidiary of AstraZeneca plc, is a biotechnology company engaged in the development and production of pharmaceuticals for infectious diseases, cancer and inflammatory disease. The company markets three products: Synagis, Ethyol and FluMist. Synagis is a treatment for respiratory syncytial virus (RSV). The drug was the first monoclonal antibody approved for an infectious disease and has become a primary pediatric product for the prevention of RSV, a major cause of viral pneumonia and bronchiolitis in infants and children. Ethyol is marketed as a treatment for the reduction of toxicities from cancer chemotherapy and radiotherapy. FluMist is a nasally administered flu vaccine for influenzas A and B. In its product candidate pipeline, MedImmune focuses its research and development efforts in the therapeutic areas of infectious disease, inflammatory diseases and cancer. The company has 11 candidate drugs in the infectious disease category, five in the neuroscience category and 10 in the oncology category. In recent years, the company opened a biologics research and development facility in Cambridge, U.K.

The firm offers employees benefits including health, dental, company-provided, supplemental life and short and long-term insurance; employee assistance program; adoption reimbursement program; vacation; paid holidays; and flexible spending accounts.

FINANCIALS: Sales and profits are in thousands of dollars—add 000 to get the full amount. 2009 Note: Financial information for 2009 was not available for all companies at press time.

2009 Sales: $	2009 Profits: $	U.S. Stock Ticker: Subsidiary
2008 Sales: $	2008 Profits: $	Int'l Ticker: Int'l Exchange:
2007 Sales: $	2007 Profits: $	Employees:
2006 Sales: $1,276,800	2006 Profits: $48,700	Fiscal Year Ends: 12/31
2005 Sales: $1,243,900	2005 Profits: $-16,600	Parent Company: ASTRAZENECA PLC

SALARIES/BENEFITS:

Pension Plan:	ESOP Stock Plan:	Profit Sharing:	Top Exec. Salary: $	Bonus: $
Savings Plan: Y	Stock Purch. Plan:		Second Exec. Salary: $	Bonus: $

OTHER THOUGHTS:

Apparent Women Officers or Directors: 1
Hot Spot for Advancement for Women/Minorities:

LOCATIONS: ("Y" = Yes)

West:	Southwest:	Midwest:	Southeast:	Northeast:	International:
Y				Y	Y

MEDTOX SCIENTIFIC INC

www.medtox.com

Industry Group Code: 6215 Ranks within this company's industry group: Sales: 4   Profits: 3

Drugs:	Other:	Clinical:		Computers:		Services:	
Discovery:	AgriBio:	Trials/Services:	Y	Hardware:		Specialty Services:	Y
Licensing:	Genetic Data:	Labs:	Y	Software:		Consulting:	
Manufacturing:	Tissue Replacement:	Equipment/Supplies:	Y	Arrays:		Blood Collection:	
Generics:		Research/Development Svcs.:		Database Management:		Drug Delivery:	
		Diagnostics:	Y			Drug Distribution:	

TYPES OF BUSINESS:

Diagnostic Device Manufacturing
Forensic & Clinical Lab Services
Diagnostic Drug Screening Devices

BRANDS/DIVISIONS/AFFILIATES:

MEDTOX Diagnostics, Inc.
MEDTOX Laboratories, Inc.
PROFILE-II
VERDICT-II
ClearCourse
SURE-SCREEN
Drug Abuse Recognition System (DARS)

CONTACTS: *Note: Officers with more than one job title may be intentionally listed here more than once.*

Richard J. Braun, CEO
Richard J. Braun, Pres.
Kevin J. Wiersma, CFO
James A. Schoonover, Chief Mktg. Officer/VP-Sales & Mktg.
Charlotte I. Sebastian, VP-Human Resources
Kevin J. Wiersma, VP/COO-Forensic Laboratory Oper.
Kevin J. Wiersma, Chief Admin. Officer
Susan E. Puskas, VP-Quality & Regulatory/COO-Clinical Lab Oper.
Steven J. Schmidt, VP-Finance
Kevin J. Wiersma, CFO-Medtox Scientific/COO-Forensic Lab Oper.
B. Mitchell Owens, VP/COO-Diagnostics Div.
Richard J. Braun, Chmn.

Phone: 651-636-7466	Fax: 651-636-5351
Toll-Free: 800-832-3244	
Address: 402 W. County Rd. D, St. Paul, MN 55112 US	

GROWTH PLANS/SPECIAL FEATURES:

MEDTOX Scientific, Inc. manufactures and distributes diagnostic devices and provides forensic and clinical laboratory services. The firm operates through two divisions, laboratory services and product sales. The laboratory services division is operated by the company's subsidiary MEDTOX Laboratories, Inc., which provides laboratory drug testing services for corporations, medical facilities and the federal government. MEDTOX Laboratories derives the majority of its revenues, 55% in 2009, from workplace drugs-of-abuse testing. It also offers clinical and other laboratory services accounting for 35% of revenues, including clinical toxicology; medical diagnostics; general clinical testing for the pharmaceutical industry; heavy metal, trace element and solvent analysis; it also offers logistics, data and program management services through its WEBTOX and eChain online services. The firm's other subsidiary, MEDTOX Diagnostics, Inc., which operates the product sales division, is based in Burlington, North Carolina and manufactures on-site drug screening products for the corporate, health care, criminal justice, temporary service and drug rehabilitation markets. MEDTOX Diagnostics' point-of-collection testing (POCT) products, which account for 90% of its sales, include the PROFILE line and MEDTOXScan, which are sold to hospitals, and the VERDICT-II and SURE-SCREEN lines, which are sold within the criminal justice and drug rehabilitation markets. These products test for a variety of drugs including amphetamines, marijuana, cocaine, opiates, phencyclidine (PCP), benzoylecgonine, morphine and methamphetamine. The company also offers its ClearCourse comprehensive drug testing program, which consists of Drug Abuse Recognition System (DARS) training; SURE-SCREEN drug testing devices; WEBTOX online data management; and MEDTOX's own laboratory assistance for confirmation testing. This division also offers contract manufacturing services. In July 2009, the company received FDA clearance to market its PROFILE-V MEDTOXScan system with additional capability to detect Oxycodone, Propoxyphene and Tricyclic Antidepressants.

MEDTOX offers its employees medical, dental and life insurance; tuition reimbursement; and flexible spending accounts.

FINANCIALS: Sales and profits are in thousands of dollars—add 000 to get the full amount. 2009 Note: Financial information for 2009 was not available for all companies at press time.

2009 Sales: $84,108	2009 Profits: $1,299	**U.S. Stock Ticker: MTOX**
2008 Sales: $85,813	2008 Profits: $5,572	Int'l Ticker: Int'l Exchange:
2007 Sales: $80,285	2007 Profits: $6,690	Employees:
2006 Sales: $69,804	2006 Profits: $4,548	Fiscal Year Ends: 12/31
2005 Sales: $63,047	2005 Profits: $3,318	Parent Company:

SALARIES/BENEFITS:

Pension Plan:	ESOP Stock Plan:	Profit Sharing:	Top Exec. Salary: $350,500	Bonus: $
Savings Plan: Y	Stock Purch. Plan:		Second Exec. Salary: $222,376	Bonus: $

OTHER THOUGHTS:

Apparent Women Officers or Directors: 2
Hot Spot for Advancement for Women/Minorities:

LOCATIONS: ("Y" = Yes)

West:	Southwest:	Midwest:	Southeast:	Northeast:	International:
		Y		Y	

Note: Financial information, benefits and other data can change quickly and may vary from those stated here.

MERCK & CO INC

www.merck.com

Industry Group Code: 325412 Ranks within this company's industry group: Sales: 9 Profits: 1

Drugs:		Other:	Clinical:	Computers:	Services:
Discovery:	Y	AgriBio:	Trials/Services:	Hardware:	Specialty Services:
Licensing:		Genetic Data:	Labs:	Software:	Consulting:
Manufacturing:	Y	Tissue Replacement:	Equipment/Supplies:	Arrays:	Blood Collection:
Generics:			Research/Development Svcs.:	Database Management:	Drug Delivery:
			Diagnostics:		Drug Distribution:

TYPES OF BUSINESS:

Drugs-Diversified
Anti-Infective & Anti-Cancer Drugs
Dermatologicals
Cardiovascular Drugs
Animal Health Products
Over-the-Counter Drugs
Foot & Sun Care Products

BRANDS/DIVISIONS/AFFILIATES:

Schering-Plough Corporation
AstraZeneca LP
Avecia Biologics
Rosetta Inpharmatics LLC
Sinra Therapeutics

CONTACTS: *Note: Officers with more than one job title may be intentionally listed here more than once.*

Richard T. Clark, CEO
Kenneth C. Frazier, Pres.
Peter N. Kellogg, CFO/Exec. VP
Mirian M. Graddick-Weir, Exec. VP-Human Resources
Peter S. Kim, Exec. VP/Pres., Merck Research Laboratories
J. Chris Scalet, CIO/Exec. VP-Global Svcs.
William E. Deese, Exec. VP/Pres., Merck Mfg. Div.
Bruce N. Kuhlik, General Counsel/Exec. VP
Mervyn Turner, Chief Strategy Officer/Sr. VP-Emerging Markets R&D
Richard S. Bowles III, Chief Compliance Officer/Exec. VP
Michael Rosenblatt, Chief Medical Officer/Exec. VP
Adam H. Schechter, Pres., Global Human Health
Bridgette P. Heller, Exec. VP/Pres., Consumer Care
Richard T. Clark, Chmn.

Phone: 908-423-1000	Fax: 908-735-1224
Toll-Free:	
Address: 1 Merck Dr., Whitehouse Station, NJ 08889-0100 US	

GROWTH PLANS/SPECIAL FEATURES:

Merck & Co., Inc., formerly Schering-Plough Corporation, is a global healthcare company that develops and manufactures medicines, vaccines, biologic therapies and consumer and animal products. The firm operates through four segments: pharmaceutical, animal health, consumer care and alliances. Pharmaceutical, the company's primary segment, markets human health pharmaceutical and vaccine products either directly or through joint ventures. Merck & Co. markets and develops human health pharmaceutical products for the treatment of bone, respiratory, dermatology, immunology, cardiovascular, diabetes, obesity, infectious disease, neuroscience, ophthalmology and oncology conditions. These products are sold primarily to drug wholesalers and retailers, hospitals, government agencies and managed health care providers such as health maintenance organizations (HMOs), pharmacy benefit managers and other institutions. Vaccine products are primarily sold to physicians, wholesalers, physician distributors and government entities. This segment also offers certain women's health products, including contraceptives and fertility treatments. The animal health segment offers vaccines, anti-infective and antiparasitic products for disease prevention, treatment and control in farm and companion animals. The consumer care segment markets allergy, foot care and sun care products. The alliances segment consists of revenue derived from the company's relationship with AstraZeneca LP, a global biopharmaceutical firm. In November 2009, the company, operating as Schering-Plough, merged with Merck & Co., Inc. Schering-Plough emerged as the parent company and subsequently renamed itself Merck & Co., Inc. In December 2009, Merck & Co. agreed to acquire Avecia Biologics, a contract manufacturing organization. In March 2010, the firm and Sanofi-aventis agreed to create an animal health joint venture company. In July 2010, the company and Sinopharm announced plans to jointly develop Human Papillomavirus (HPV) vaccine products in China.

Merck offers its employees medical, dental, vision and prescription drug coverage; flexible spending accounts; education assistance; an employee assistance program; fitness and wellness centers; a pension plan; and a 401(k) plan.

FINANCIALS: Sales and profits are in thousands of dollars—add 000 to get the full amount. 2009 Note: Financial information for 2009 was not available for all companies at press time.

2009 Sales: $27,428,300	2009 Profits: $13,024,200	U.S. Stock Ticker: MRK
2008 Sales: $23,850,300	2008 Profits: $7,932,300	Int'l Ticker: Int'l Exchange:
2007 Sales: $24,197,700	2007 Profits: $3,396,800	Employees: 100,000
2006 Sales: $10,594,000	2006 Profits: $1,143,000	Fiscal Year Ends: 12/31
2005 Sales: $9,508,000	2005 Profits: $269,000	Parent Company:

SALARIES/BENEFITS:

Pension Plan: Y	ESOP Stock Plan:	Profit Sharing:	Top Exec. Salary: $1,800,000	Bonus: $2,854,737
Savings Plan: Y	Stock Purch. Plan:		Second Exec. Salary: $1,467,542	Bonus: $

OTHER THOUGHTS:

Apparent Women Officers or Directors: 5
Hot Spot for Advancement for Women/Minorities: Y

LOCATIONS: ("Y" = Yes)

West:	Southwest:	Midwest:	Southeast:	Northeast:	International:
Y	Y	Y	Y	Y	Y

Note: Financial information, benefits and other data can change quickly and may vary from those stated here.

MERCK KGAA

www.merck.de

Industry Group Code: 325412 **Ranks within this company's industry group:** Sales: 16   Profits: 26

Drugs:		Other:		Clinical:	Computers:	Services:
Discovery:	Y	AgriBio:	Y	Trials/Services:	Hardware:	Specialty Services:
Licensing:	Y	Genetic Data:	Y	Labs:	Software:	Consulting:
Manufacturing:	Y	Tissue Replacement:	Y	Equipment/Supplies:	Arrays:	Blood Collection:
Generics:				Research/Development Svcs.:	Database Management:	Drug Delivery:
				Diagnostics:		Drug Distribution:

TYPES OF BUSINESS:

Pharmaceuticals
Over-the-Counter Drugs & Vitamins
Generic Drugs
Chemicals
LCD Components
Reagents & Diagnostics
Nanotechnology Research

BRANDS/DIVISIONS/AFFILIATES:

Merck Serono S.A.
Millipore Corporation
EMD Chemicals

CONTACTS: Note: Officers with more than one job title may be intentionally listed here more than once.

Karl-Ludwig Kley, Chmn.-Exec. Board
Michael Becker, CFO
Karl-Ludwig Kley, Dir.-Human Resources
Bernd Reckmann, Dir.-Corp. Info. Svcs.
Bernd Reckmann, Dir.-Prod.
Bernd Reckmann, Dir.-Eng.
Karl-Ludwig Kley, Dir.-Legal
Karl-Ludwig Kley, Dir.-Strategic Planning
Karl-Ludwig Kley, Dir.-Corp. Comm.
Markus Launer, Head-Investor Rel.
Michael Becker, Dir.-Acct., Finance, Controlling & Tax
Bernd Reckmann, Mgr.-Darmstadt & Gernsheim Sites
Elmar Schnee, Dir.-Pharmaceuticals Bus. Sector
Bernd Reckmann, Dir.-Chemicals Bus. Sector
Rolf Krebs, Chmn.-Supervisory Board
Michael Becker, Dir.-Purchasing

Phone: 49-6151-72-0	Fax: 49-6151-72-2000
Toll-Free:	
Address: Frankfurter St. 250, Darmstadt, 64293 Germany	

GROWTH PLANS/SPECIAL FEATURES:

Merck KGaA is a global pharmaceuticals and chemicals company with 192 companies in 61 countries. Outside Germany, Merck maintains research sites in France, Spain, the U.K., the U.S. and Japan. It has two main businesses: Pharmaceuticals and Chemicals. The Pharmaceuticals business manufactures prescription drugs and over-the-counter products through two divisions, Merck Serano and Consumer Health Care. Merck Serono's products target areas including oncology, neuro-degenerative diseases, fertility, endocrinology, cardio metabolic care and new specialist therapies. Products include colorectal cancer drug Erbitux; multiple sclerosis treatment Rebif; and Type 2 diabetes treatment Glucophage. The Consumer Health Care division focuses on four areas: mobility; everyday health protection; women's and children's health; and cough and cold. Specific products include nasal decongestant Nasivin; Bion probiotic and vitamin supplements; and Seven Seas/Flexagil products for joint health. The Chemical business has two segments: liquid crystals; and performance and life science chemicals. Merck's liquid crystals are used in a variety of devices including LCD televisions, computer monitors, digital cameras and mobile phones. The segment is also working on new display and lighting technologies including organic light emitting diodes (OLEDs), more energy efficient LEDs and solar powered technologies. The performance and life science chemicals segment operates in three business units: laboratory business, life science solutions and pigments. The laboratory business produces chemicals according to various specifications along with certificates of analysis and safety data sheets. The life science solutions unit manufactures products utilizing chemical and biotechnical processes for the pharmaceutical industry. The pigments unit produces pigments for applications in coatings, packaging and product design. American subsidiaries include EMD Serono Chemicals, Inc., based in New Jersey. In February 2010, Merck agreed to acquire Millipore Corporation, a Massachusetts-based life science company, for $7.2 billion.

FINANCIALS: Sales and profits are in thousands of dollars—add 000 to get the full amount. 2009 Note: Financial information for 2009 was not available for all companies at press time.

2009 Sales: $10,268,000	2009 Profits: $485,500	U.S. Stock Ticker:
2008 Sales: $10,000,300	2008 Profits: $501,600	Int'l Ticker: MRK Int'l Exchange: Frankfurt-Euronext
2007 Sales: $9,337,530	2007 Profits: $4,657,720	Employees: 33,062
2006 Sales: $8,352,950	2006 Profits: $1,335,880	Fiscal Year Ends: 12/31
2005 Sales: $7,697,680	2005 Profits: $898,150	Parent Company:

SALARIES/BENEFITS:

Pension Plan:	ESOP Stock Plan:	Profit Sharing:	Top Exec. Salary: $	Bonus: $
Savings Plan:	Stock Purch. Plan:		Second Exec. Salary: $	Bonus: $

OTHER THOUGHTS:

Apparent Women Officers or Directors: 2
Hot Spot for Advancement for Women/Minorities:

LOCATIONS: ("Y" = Yes)

West:	Southwest:	Midwest:	Southeast:	Northeast:	International:
Y		Y	Y	Y	Y

Note: Financial information, benefits and other data can change quickly and may vary from those stated here.

MERCK SERONO SA

www.merckserono.net

Industry Group Code: 325412 Ranks within this company's industry group: Sales: Profits:

Drugs:		Other:		Clinical:	Computers:		Services:	
Discovery:	Y	AgriBio:		Trials/Services:	Hardware:		Specialty Services:	
Licensing:		Genetic Data:	Y	Labs:	Software:		Consulting:	
Manufacturing:	Y	Tissue Replacement:		Equipment/Supplies:	Arrays:		Blood Collection:	
Generics:				Research/Development Svcs.:	Database Management:		Drug Delivery:	Y
				Diagnostics:			Drug Distribution:	

TYPES OF BUSINESS:

Pharmaceuticals Development
Fertility Drugs
Neurology Drugs
Growth & Metabolism Drugs
Dermatology Drugs
Oncology Research

BRANDS/DIVISIONS/AFFILIATES:

GONAL-f
Ovidrel/Ovitrelle
Luveris
Crinone
Cetrotide
Rebif
Saizen
EMD Serono Inc

CONTACTS: Note: Officers with more than one job title may be intentionally listed here more than once.

Elmar Schnee, Pres.
Dietmar Eidens, VP-Human Resources
Bernhard Kirschbaum, Exec. VP-Global R&D
Hanns-Eberhard Erle, Head-Tech. Oper.
Vincent Aurentz, Head-Portfolio Mgmt. & Bus. Dev.
Dorothea Wenzel, Head-Controlling
Fereydoun Firouz, Head-Fertility & Metabolic Endocrinology
Roberto Gradnik, Head-Neurodegenerative & Autoimmune Diseases
Wolfgang Wein, Head-Oncology
Franck Latrille, Head-CardioMetabolic Care & Gen. Medicine

Phone: 41-22-414-3000	Fax: 41-22-414-2179
Toll-Free:	
Address: 9, chemin des Mines, Case postale 54, Geneva 20, CH-1211 Switzerland	

GROWTH PLANS/SPECIAL FEATURES:

Merck Serono SA, formerly Serono SA, is a global biotechnology company that focuses on six therapeutic areas: oncology, neurodegenerative diseases, fertility, endocrinology, cardio-metabolic care and new specialist therapies. The company was formed when Merck KGaA acquired Serono in a $13 billion transaction. The company researches both traditional pharmaceutical drugs, consisting of small molecules treating the symptoms of a disease, and biopharmaceuticals, comprising large molecules such as proteins that target the underlying mechanism of a disease. For colorectal cancer, Merck Serono has developed Erbitux, a monoclonal antibody also sold for the treatment of head and neck cancer, and UFT, an oral 5-FU (5-fluorouracil) therapy. Rebif is Merck Serono's disease modifying drug, used to treat relapsing forms of multiple sclerosis. Infertility treatments developed by the company include GONAL-f, Pergoveris, Luveris, Ovidrel/Ovitrelle, Crinone and Cetrotide. Merck Serono offers a range of specialized endocrinology products. For childhood and adult growth hormone deficiency, as well as for children born small for gestational age (SGA), or with chronic renal failure or Turner's syndrome, the company has developed Saizen. Serostim, meanwhile, is used to treat HIV-associated wasting, and Zorbtive has been developed for the treatment of short bowel syndrome. Cardio-metabolic care treatments developed by Merck Serono include Glucophage/GlucophageXR and Glucovance for diabetes; Concor/ConcorCor for cardiovascular disease; and Euthyrox for hypothyroidism, euthyroid goiter and suppressive therapy of differentiated thyroid cancer. As part of its effort to develop products in new specialist areas with unmet medical needs, the company partnered with Genentech to develop and commercialize the psoriasis drug Raptiva. Merck Serono also has a number of products in development for the treatment of such diseases as lung, gastric and breast cancer; Parkinson's; arthritis and lupus; and hormone deficiencies. In Canada and the U.S., the Merck Serono division operates as EMD Serono, Inc.

FINANCIALS: Sales and profits are in thousands of dollars—add 000 to get the full amount. 2009 Note: Financial information for 2009 was not available for all companies at press time.

2009 Sales: $	2009 Profits: $	U.S. Stock Ticker: Subsidiary
2008 Sales: $	2008 Profits: $	Int'l Ticker: Int'l Exchange:
2007 Sales: $	2007 Profits: $	Employees:
2006 Sales: $2,804,900	2006 Profits: $735,400	Fiscal Year Ends: 12/31
2005 Sales: $2,586,400	2005 Profits: $-105,300	Parent Company: MERCK KGAA

SALARIES/BENEFITS:

Pension Plan:	ESOP Stock Plan:	Profit Sharing:	Top Exec. Salary: $	Bonus: $
Savings Plan:	Stock Purch. Plan:		Second Exec. Salary: $	Bonus: $

OTHER THOUGHTS:

Apparent Women Officers or Directors: 1
Hot Spot for Advancement for Women/Minorities:

LOCATIONS: ("Y" = Yes)

West:	Southwest:	Midwest:	Southeast:	Northeast:	International:
				Y	Y

MERIDIAN BIOSCIENCE INC

www.meridianbioscience.com

Industry Group Code: 325413 Ranks within this company's industry group: Sales: 9   Profits: 6

Drugs:	Other:	Clinical:	Computers:	Services:
Discovery:	AgriBio:	Trials/Services:	Hardware:	Specialty Services:
Licensing:	Genetic Data:	Labs:	Software:	Consulting:
Manufacturing: Y	Tissue Replacement:	Equipment/Supplies: Y	Arrays:	Blood Collection:
Generics:		Research/Development Svcs.:	Database Management:	Drug Delivery:
		Diagnostics: Y		Drug Distribution:

TYPES OF BUSINESS:

Diagnostic Test Kits
Contract Manufacturing
Bulk Antigens, Antibodies & Reagents

BRANDS/DIVISIONS/AFFILIATES:

Biodesign
OEM Concepts
Viral Antigens
ImmunoCard STAT Campy

CONTACTS: *Note: Officers with more than one job title may be intentionally listed here more than once.*

William J. Motto, CEO
John A. Kraeutler, COO
John A. Kraeutler, Pres.
Melissa A. Lueke, CFO/VP
Gregory S. Ballish, VP-Mktg. & Sales
Kenneth J. Kozak, VP-R&D
Lawrence J. Baldini, Exec. VP-Info. Systems
Melissa A. Leuke, Sec.
Lawrence J. Baldini, Exec. VP-Oper.
Susan D. Rolih, VP-Regulatory Affairs & Quality Systems
Richard L. Eberly, Exec. VP/Pres., Meridian Life Science
William J. Motto, Chmn.
Antonio A. Interno, Sr. VP/Pres./Managing Dir.-Europe

Phone: 513-271-3700	Fax: 513-271-3762
Toll-Free: 800-543-1980	
Address: 3471 River Hills Dr., Cincinnati, OH 45244 US	

GROWTH PLANS/SPECIAL FEATURES:

Meridian Biosciences, Inc. is a life science company that develops, manufactures, sells and distributes diagnostic test kits, primarily for respiratory, gastrointestinal, viral and parasitic infectious diseases. It also manufactures and distributes bulk antigens, antibodies and reagents; and contract manufactures proteins and other biologicals. The company operates in three segments: U.S. diagnostics, European diagnostics and life science. The U.S. diagnostics segment focuses on the development, manufacture, sale and distribution of diagnostic test kits, which utilize immunodiagnostic technologies that test samples of body fluids or tissue for the presence of antigens and antibodies of specific infectious diseases. Products also include transport media that store and preserve specimen samples from patient collection to laboratory testing.   The European diagnostics segment focuses on the sale and distribution of diagnostic test kits.   Its sales and distribution network consists of direct sales forces in Belgium, France, Holland and Italy, and independent distributors in other European, African and Middle Eastern countries.   The life sciences segment focuses on the development, manufacture, sale and distribution of bulk antigens, antibodies and reagents, as well as contract development and manufacturing services. The segment is represented by four product-line brands: Biodesign, which represents monoclonal and polyclonal antibodies and assay reagents; OEM (original equipment manufacturer) Concepts, which represents contract ascites and antibody production services; and Viral Antigens, which represents viral proteins.   In June 2009, the company received FDA clearance to market the ImmunoCard STAT! Campy, a new rapid test for detecting the Campylobacter bacteria.   This offering comes in addition to the firm's February 2009 PREMIER CAMPY culture test.

Meridian offers its employees medical, dental and life insurance; short- and long-term disability; stock options; a 401(k) plan; profit sharing; and tuition reimbursement.

FINANCIALS: Sales and profits are in thousands of dollars—add 000 to get the full amount. 2009 Note: Financial information for 2009 was not available for all companies at press time.

2009 Sales: $148,274	2009 Profits: $32,759	U.S. Stock Ticker: VIVO
2008 Sales: $139,639	2008 Profits: $30,202	Int'l Ticker: Int'l Exchange:
2007 Sales: $122,963	2007 Profits: $26,721	Employees:
2006 Sales: $108,413	2006 Profits: $18,333	Fiscal Year Ends: 9/30
2005 Sales: $92,965	2005 Profits: $12,565	Parent Company:

SALARIES/BENEFITS:

Pension Plan:	ESOP Stock Plan:	Profit Sharing: Y	Top Exec. Salary: $511,538	Bonus: $
Savings Plan: Y	Stock Purch. Plan: Y		Second Exec. Salary: $524,999	Bonus: $

OTHER THOUGHTS:

Apparent Women Officers or Directors: 2
Hot Spot for Advancement for Women/Minorities: Y

LOCATIONS: ("Y" = Yes)

West:	Southwest:	Midwest:	Southeast:	Northeast:	International:
		Y	Y	Y	Y

MILLENNIUM PHARMACEUTICALS INC

www.mlnm.com

Industry Group Code: 325412 Ranks within this company's industry group: Sales: Profits:

Drugs:		Other:		Clinical:	Computers:	Services:
Discovery:	Y	AgriBio:		Trials/Services:	Hardware:	Specialty Services:
Licensing:		Genetic Data:	Y	Labs:	Software:	Consulting:
Manufacturing:	Y	Tissue Replacement:		Equipment/Supplies:	Arrays:	Blood Collection:
Generics:				Research/Development Svcs.:	Database Management:	Drug Delivery:
				Diagnostics:		Drug Distribution:

TYPES OF BUSINESS:

Pharmaceuticals Discovery & Development
Gene-Based Drug Discovery Platform
Small-Molecule Drugs

BRANDS/DIVISIONS/AFFILIATES:

Takeda Pharmaceutical Company Ltd
VELCADE
MLN9708
1000CranesofHope.com
TAK-700

CONTACTS: Note: Officers with more than one job title may be intentionally listed here more than once.

Deborah Dunsire, CEO
Deborah Dunsire, Pres.
Todd Shegog, CFO
Stephen M. Gansler, Sr. VP-Human Resources
Joseph B. Bolen, Chief Scientific Officer
Laurie B. Keating, General Counsel/Sr. VP
Anna Protopapas, Sr. VP-Corp. Dev.
Lisa Adler, VP-Corp. Comm.
Todd Shegog, Sr. VP-Finance
Christophe Bianchi, Exec. VP
Peter F. Smith, Sr. VP-Non-Clinical Dev. Sciences
Nancy Simonian, Chief Medical Officer

Phone: 617-679-7000	Fax: 617-374-7788
Toll-Free: 800-390-5663	
Address: 40 Landsdowne St., Cambridge, MA 02139 US	

GROWTH PLANS/SPECIAL FEATURES:

Millennium Pharmaceuticals, a subsidiary of Takeda Pharmaceuticals Company, Ltd., researches and manufactures therapeutic products for the treatment of patients with cancer and inflammatory diseases. The company's development platform focuses on combing knowledge of genomics and protein homeostasis, a set of particular molecular pathways that affect the establishment and progression of diseases, in order to produce drug candidates. The firm currently has one marketed product called VELCADE. VELCADE is an oncology drug that is injected into patients for the treatment of multiple myeloma, a cancer of the blood, and mantle cell lymphoma. VELCADE is also being tested in Phase II and III trials for the treatment of follicular B-cell non-Hodgkin's lymphoma, hematologic malignancies and solid tumors including lung, breast, prostate and ovarian cancers. In July 2009, potential cancer treatment CBP501, which was jointly developed by Millennium, Takeda Pharmaceuticals and CanBas Co., Ltd., entered Phase II clinical trials for the treatment of non-small cell lung cancer. In September 2009, TAK-700, developed by the firm and Takeda Pharmaceuticals, entered Phase II clinical trials for the treatment of prostate cancer. In November 2009, the company's second-generation proteasome inhibitor candidate, MLN9708, entered Phase I clinical trials in its oral form. In December 2009, Millennium established 1000CranesofHope.com, a web site aimed at offering emotional support to cancer patients.

Millennium offers employees medical, dental and vision coverage; a 401(k); flexible spending accounts; life and disability insurance; and education and transportation assistance.

FINANCIALS: Sales and profits are in thousands of dollars—add 000 to get the full amount. 2009 Note: Financial information for 2009 was not available for all companies at press time.

2009 Sales: $	2009 Profits: $	U.S. Stock Ticker: Subsidiary	
2008 Sales: $	2008 Profits: $	Int'l Ticker: Int'l Exchange:	
2007 Sales: $527,525	2007 Profits: $14,909	Employees:	
2006 Sales: $486,830	2006 Profits: $-43,953	Fiscal Year Ends: 12/31	
2005 Sales: $558,308	2005 Profits: $-198,249	Parent Company: TAKEDA PHARMACEUTICAL COMPANY LTD	

SALARIES/BENEFITS:

Pension Plan:	ESOP Stock Plan:	Profit Sharing:	Top Exec. Salary: $	Bonus: $
Savings Plan: Y	Stock Purch. Plan:		Second Exec. Salary: $	Bonus: $

OTHER THOUGHTS:

Apparent Women Officers or Directors: 5
Hot Spot for Advancement for Women/Minorities: Y

LOCATIONS: ("Y" = Yes)

West:	Southwest:	Midwest:	Southeast:	Northeast:	International:
				Y	

MILLIPORE CORP

www.millipore.com

Industry Group Code: 3345 Ranks within this company's industry group: Sales: 2 Profits: 1

Drugs:	Other:	Clinical:		Computers:	Services:	
Discovery:	AgriBio:	Trials/Services:		Hardware:	Specialty Services:	Y
Licensing:	Genetic Data:	Labs:	Y	Software:	Consulting:	
Manufacturing:	Tissue Replacement:	Equipment/Supplies:	Y	Arrays:	Blood Collection:	
Generics:		Research/Development Svcs.:		Database Management:	Drug Delivery:	
		Diagnostics:			Drug Distribution:	

TYPES OF BUSINESS:

Biotechnology Instruments
Fluid Analysis, Identification & Purification Equipment
Chromatography Technologies

BRANDS/DIVISIONS/AFFILIATES:

Guava Technologies
Merck KGaA

CONTACTS: Note: Officers with more than one job title may be intentionally listed here more than once.

Martin D. Madaus, CEO
Martin D. Madaus, Pres.
Charles F. Wagner, Jr., CFO/VP
Bruce Bonnevier, VP-Global Human Resources
Dennis W. Harris, Chief Scientific Officer/VP
Peter C. Kershaw, VP-Worldwide Mfg. Oper.
Jeffrey Rudin, General Counsel/VP/Corp. Sec.
Peter C. Kershaw, VP-Global Oper.
Wei Zhang, VP-Strategic & Corp. Dev.
Karen Marinella Hall, Dir.-Corp. Comm.
Joshua S. Young, Dir.-Investor Rel.
Jon DiVincenzo, VP/Pres., Bioscience Div.
Gregory J. Sam, VP-Quality
Jean-Paul Mangeolle, VP/Pres., Bioprocess Div.
Martin D. Madaus, Chmn.
Geoffrey F. Ide, VP-Millipore Int'l

Phone: 978-715-4321	Fax: 800-645-5439
Toll-Free: 800-645-5476	
Address: 290 Concord Rd., Billerica, MA 01821 US	

GROWTH PLANS/SPECIAL FEATURES:

Millipore Corp. is a multinational bioscience company that provides technologies, tools and services for research, development and production. The company's products and services are based on technologies such as filtration, chromatography, cell culture supplements, antibodies and cell lines. The firm's products are offered through its two segments, the Bioscience division and the Bioprocess division. Millipore's Bioscience division, which accounted for 44% of 2009 revenue, is organized around four specific market segments: biotools for the separation, isolation and purification of biological samples; research reagents such as antibodies, dyes and biochemical reagents; drug discovery reagent for the analysis of drug candidates; and laboratory water purification systems that remove contaminants for critical laboratory analysis. The Bioprocess division, which accounted for 56% of 2009 revenue, provides bio-products and technologies for the manufacturing of biologic drugs in mammalian cell cultures; filtration, purification and chromatography technologies to clarify, concentrate, purify and remove viruses; process monitoring tools for the sampling and testing of drugs and intermediate products and advanced manufacturing systems for use in sterile biomanufacturing environments. The firm operates 11 manufacturing sites in Massachusetts, New Hampshire, Missouri, Illinois, California, France and the U.K., and 47 offices worldwide. In February 2009, the firm acquired Guava Technologies and opened a biomanufacturing sciences and training facility in Singapore. In June 2009, the firm opened a new bioprocess manufacturing facility in Massachusetts. In July 2010, Millipore launched the new ProRes-S media, an expansion to its downstream purification suite. In July 2010, the company was acquired by German firm Merck KGaA.

Millipore offers employees flexible spending accounts, tuition reimbursement, employee assistance programs and adoption assistance.

FINANCIALS: Sales and profits are in thousands of dollars—add 000 to get the full amount. 2009 Note: Financial information for 2009 was not available for all companies at press time.

2009 Sales: $1,654,410	2009 Profits: $179,190	U.S. Stock Ticker: Subsidiary
2008 Sales: $1,602,138	2008 Profits: $145,801	Int'l Ticker: Int'l Exchange:
2007 Sales: $1,531,555	2007 Profits: $136,472	Employees: 6,100
2006 Sales: $1,255,371	2006 Profits: $96,984	Fiscal Year Ends: 12/31
2005 Sales: $991,031	2005 Profits: $80,168	Parent Company: MERCK KGAA

SALARIES/BENEFITS:

Pension Plan:	ESOP Stock Plan:	Profit Sharing:	Top Exec. Salary: $850,962	Bonus: $655,666
Savings Plan: Y	Stock Purch. Plan:		Second Exec. Salary: $375,587	Bonus: $217,096

OTHER THOUGHTS:

Apparent Women Officers or Directors: 3
Hot Spot for Advancement for Women/Minorities: Y

LOCATIONS: ("Y" = Yes)

West:	Southwest:	Midwest:	Southeast:	Northeast:	International:
Y	Y	Y	Y	Y	Y

Note: Financial information, benefits and other data can change quickly and may vary from those stated here.

MONSANTO CO

www.monsanto.com

Industry Group Code: 11511 Ranks within this company's industry group: Sales: 1 Profits: 1

Drugs:	Other:		Clinical:	Computers:	Services:
Discovery:	AgriBio:	Y	Trials/Services:	Hardware:	Specialty Services:
Licensing:	Genetic Data:		Labs:	Software:	Consulting:
Manufacturing:	Tissue Replacement:		Equipment/Supplies:	Arrays:	Blood Collection:
Generics:			Research/Development Svcs.:	Database Management:	Drug Delivery:
			Diagnostics:		Drug Distribution:

TYPES OF BUSINESS:

Agricultural Biotechnology Products & Chemicals Manufacturing
Herbicides
Seeds
Genetic Products
Lawn & Garden Products

BRANDS/DIVISIONS/AFFILIATES:

Asgrow
Roundup Ready
Agroeste Sementes
Delta and Pine Land Company
Seminis
De Ruiter Seeds Group BV
Roundup
WestBred, LLC

CONTACTS: Note: Officers with more than one job title may be intentionally listed here more than once.

Hugh Grant, CEO
Hugh Grant, Pres.
Carl M. Casale, CFO/Exec. VP
Steven C. Mizell, Exec. VP-Human Resources
Robert T. Fraley, CTO/Exec. VP
Mark J. Leidy, Exec. VP-Mfg.
Janet M. Holloway, Chief of Staff/Sr. VP-Comm. Rel.
David F. Snively, General Counsel/Sr. VP/Sec.
Gerald A. Steiner, Exec. VP-Corp. Affairs & Sustainability
Scarlett Lee Foster, VP-Investor Rel.
Nicole M. Ringenberg, VP/Controller
Tom D. Hartley, VP/Treas.
Consuelo E. Madere, VP-Vegetable Bus.
Kerry J. Preete, VP-Crop Protection
Brett D. Begemann, Exec. VP-Seeds & Traits
Hugh Grant, Chmn.

Phone: 314-694-1000	Fax: 314-694-8394
Toll-Free:	
Address: 800 N. Lindbergh Blvd., St. Louis, MO 63167 US	

GROWTH PLANS/SPECIAL FEATURES:

Monsanto Co. is a global provider of agricultural products for farmers. The company operates in two principal business segments: Seeds and Genomics; and Agricultural Productivity. The Seeds and Genomics segment, representing 56% of sales, is responsible for producing seed brands and patenting genetic traits that enable seeds to resist insects, disease, drought and weeds. Major brands for row crop seeds produced by Monsanto include DEKALB and Channel Bio corn seeds; Asgrow soybean seeds; and Deltapine cotton seeds. Vegetable seeds such as tomato, pepper, eggplant, melon, cucumber, pumpkin, squash, beans, broccoli, onions and lettuce are sold under the Seminis and De Ruiter brands. The company's genetic trait products include Roundup Ready and Roundup Ready 2 Yield for soybeans; SmartStax, YieldGard, and YieldGardVT for corn; Bollgard and Bollgard II for cotton; and Genuity for multiple products. With the recent acquisition of WestBred, LLC, the segment also focuses on cereal grain seeds and biotech wheat products. The Agricultural Productivity segment, accounting for 44% of sales, produces herbicide products. Selective herbicides consist of the harness brand for corn and cotton. Other products include glyphosate-based herbicides, for weed control in nonselective agricultural, industrial, ornamental and turf applications; and lawn and garden herbicides for weed control in residential applications. Both glyphosate-based and lawn and garden herbicides are available under the Roundup brand. Monsanto market its seeds and commercial herbicides through a variety of channels and directly to farmers. Residential herbicides are marketed through the Scotts Miracle-Gro Company. Subsidiaries include Delta and Pine Land Company, a developer of cotton and soybean seeds; and Agroeste Sementes, a Brazilian corn seed company. In July 2009, the company acquired the assets of WestBred, LLC, a specialist in the genetic material of wheat seed.

Employees are offered medical, dental and vision insurance; life insurance; disability coverage; flexible spending accounts; adoption assistance; and relocation assistance.

FINANCIALS: Sales and profits are in thousands of dollars—add 000 to get the full amount. 2009 Note: Financial information for 2009 was not available for all companies at press time.

2009 Sales: $11,724,000	2009 Profits: $2,109,000	**U.S. Stock Ticker: MON**
2008 Sales: $11,365,000	2008 Profits: $2,024,000	**Int'l Ticker:** Int'l Exchange:
2007 Sales: $8,563,000	2007 Profits: $993,000	Employees: 22,900
2006 Sales: $7,294,000	2006 Profits: $689,000	Fiscal Year Ends: 8/31
2005 Sales: $6,275,000	2005 Profits: $255,000	Parent Company:

SALARIES/BENEFITS:

Pension Plan: Y	ESOP Stock Plan:	Profit Sharing:	Top Exec. Salary: $1,391,356	Bonus: $1,070,382
Savings Plan: Y	Stock Purch. Plan: Y		Second Exec. Salary: $593,173	Bonus: $300,000

OTHER THOUGHTS:

Apparent Women Officers or Directors: 5
Hot Spot for Advancement for Women/Minorities: Y

LOCATIONS: ("Y" = Yes)

West:	Southwest:	Midwest:	Southeast:	Northeast:	International:
Y	Y	Y	Y	Y	Y

MYLAN INC

www.mylan.com

Industry Group Code: 325412 Ranks within this company's industry group: Sales: 21 Profits: 33

Drugs:		Other:	Clinical:	Computers:	Services:
Discovery:	Y	AgriBio:	Trials/Services:	Hardware:	Specialty Services:
Licensing:		Genetic Data:	Labs:	Software:	Consulting:
Manufacturing:	Y	Tissue Replacement:	Equipment/Supplies:	Arrays:	Blood Collection:
Generics:			Research/Development Svcs.:	Database Management:	Drug Delivery:
			Diagnostics:		Drug Distribution:

TYPES OF BUSINESS:
Drugs-Generic
Generic Pharmaceuticals
Active Pharmaceutical Ingredients

BRANDS/DIVISIONS/AFFILIATES:
UDL Laboratories, Inc.
Mylan Pharmaceuticals, Inc.
Mylan Technologies, Inc.
Matrix Laboratories Limited
Dey
Mylan S.A.S.
Mylan-Nifedipine Extended Release Tablets
Mylan Pharmaceuticals S.L

CONTACTS: Note: Officers with more than one job title may be intentionally listed here more than once.
Robert J. Coury, CEO
Rajiv Malik, COO/Exec. VP
Heather Bresch, Pres.
John D. Sheehan, CFO/Exec. VP
Carolyn Meyers, Pres., Dey Pharma, L.P.
Harry A. Korman, Pres., North America
S. Srinivasan, CEO/Managing Dir.-Matrix
Robert J. Coury, Chmn.
Didier Barret, Pres., EMEA

Phone: 724-514-1800	Fax: 724-514-1870
Toll-Free:	
Address: 1500 Corporate Dr., Canonsburg, PA 15317 US	

GROWTH PLANS/SPECIAL FEATURES:

Mylan, Inc., formerly Mylan Laboratories, Inc., develops, licenses, manufactures, distributes and markets generic/branded generic pharmaceutical products, specialty pharmaceuticals and active pharmaceutical ingredients (APIs). The company has two operating segments: generics and specialty. The generics segment operates in the U.S., Canada, Europe, the Middle East, Africa, Australia, New Zealand, India and Japan. In the U.S., this segment markets over 200 products through three subsidiaries: Mylan Pharmaceuticals Inc. (MPI), which sells solid oral dosage products; UDL Laboratories, Inc., which re-packages and markets products obtained from MPI and third parties; and Mylan Technologies, Inc., which develops and markets transdermal patches. The firm also manufactures and markets generic pharmaceuticals in Canada through Mylan Pharmaceuticals ULC, which offers 120 products. Sales in EMEA are generated through operations in 25 countries. In France, Mylan markets 160 products through its wholly-owned subsidiaries Mylan S.A.S. and Qualimed S.A.S.; in Italy, 110 products through Mylan S.p.A; in Spain, 80 products through Mylan Pharmaceuticals S.L.; and in Germany, 160 products through Mylan dura. The company also markets 170 products in the U.K. In Asia Pacific, the firm markets products in Australia through Alphapharm; in New Zealand through Mylan New Zealand; in India through Matrix, which additionally has eight API and intermediate manufacturing facilities; and in Japan through Mylan Seiyaku. The firm also has a manufacturing facility located in Japan. The specialty segment is conducted through Dey, which focuses on the respiratory, psychiatry and severe allergy markets. Its principal products are EpiPen, which is used for severe allergic reactions; and Perforomist Inhalation Solution, a formoterol fumarate inhalation solution used to treat bronchoconstriction in chronic obstructive pulmonary disease patients. In May 2010, the company launched Mylan-Nifedipine Extended Release tablets, one of the only generic alternatives to Adalat XL in Canada.

FINANCIALS: Sales and profits are in thousands of dollars—add 000 to get the full amount. 2009 Note: Financial information for 2009 was not available for all companies at press time.

2009 Sales: $5,092,785	2009 Profits: $247,748	**U.S. Stock Ticker: MYL**
2008 Sales: $5,137,585	2008 Profits: $-181,215	**Int'l Ticker:** Int'l Exchange:
2007 Sales: $1,611,819	2007 Profits: $217,284	Employees: 15,500
2006 Sales: $1,257,164	2006 Profits: $184,542	Fiscal Year Ends: 12/31
2005 Sales: $1,253,374	2005 Profits: $203,592	Parent Company:

SALARIES/BENEFITS:

Pension Plan: Y	ESOP Stock Plan:	Profit Sharing:	Top Exec. Salary: $1,566,184	Bonus: $4,250,000
Savings Plan: Y	Stock Purch. Plan:		Second Exec. Salary: $633,173	Bonus: $1,450,000

OTHER THOUGHTS:
Apparent Women Officers or Directors: 3
Hot Spot for Advancement for Women/Minorities: Y

LOCATIONS: ("Y" = Yes)

West:	Southwest:	Midwest:	Southeast:	Northeast:	International:
	Y	Y	Y	Y	Y

Note: Financial information, benefits and other data can change quickly and may vary from those stated here.

MYRIAD GENETICS INC

www.myriad.com

Industry Group Code: 325412 Ranks within this company's industry group: Sales: 54   Profits: 44

Drugs:		Other:		Clinical:		Computers:		Services:	
Discovery:	Y	AgriBio:		Trials/Services:		Hardware:		Specialty Services:	
Licensing:		Genetic Data:	Y	Labs:		Software:		Consulting:	
Manufacturing:		Tissue Replacement:		Equipment/Supplies:		Arrays:		Blood Collection:	
Generics:				Research/Development Svcs.:		Database Management:		Drug Delivery:	
				Diagnostics:	Y			Drug Distribution:	

TYPES OF BUSINESS:

Pharmaceuticals Discovery & Development
Cancer Treatments
HIV Treatments
Cancer Diagnostics

BRANDS/DIVISIONS/AFFILIATES:

Myriad Pharmaceuticals, Inc.
TheraGuide 5-FU
Prezeon
BRACAnalysis
COLARIS
MELARIS
COLARIS AP
Azixa

CONTACTS: *Note: Officers with more than one job title may be intentionally listed here more than once.*

Peter D. Meldrum, CEO
Peter D. Meldrum, Pres.
James S. Evans, CFO
Jerry Lanchbury, Chief Scientific Officer
Robert G. Harrison, CIO
Richard Marsh, General Counsel/Exec. VP/Corp. Sec.
William A. Hockett, III, Exec. VP-Corp. Comm.
Mark C. Capone, Pres., Myriad Genetic Laboratories, Inc.
John T. Henderson, Chmn.

Phone: 801-584-3600	Fax: 801-584-3640
Toll-Free:	
Address: 320 Wakara Way, Salt Lake City, UT 84108 US	

GROWTH PLANS/SPECIAL FEATURES:

Myriad Genetics, Inc. is a biopharmaceutical company that develops and markets novel therapeutic and molecular diagnostic products. The company develops a number of proprietary technologies that are aimed at understanding the genetic basis of human disease in order to effectively treat them. Myriad's molecular diagnostic business is focused on both predictive and personalized medicine. Predictive medicine analyzes genes to assess an individual's potential risk of developing while personalized medicine works to assess a patient's risk in disease progression, recurrence and drug response. The company currently owns seven commercial predictive medicine products: BRACAnalysis for breast and ovarian cancer; COLARIS for colon and uterine cancer; COLARIS AP for polyp-forming syndromes of colon cancer; MELARIS for melanoma; TheraGuide 5-FU for determining the risk of toxic reaction to 5-FU chemotherapy; Prezeon for assessing PTEN function in cancer patients; and OnDose for optimizing 5-FU chemotherapy dosage. Prezeon and OnDose are relatively new additions to the company's product line and were launched in late 2008 and April 2009, respectively. The company works with its wholly-owned subsidiary Myriad Pharmaceuticals, Inc. to develop novel small molecule drugs. The lead product pipeline includes Azixa for the treatment of melanoma, glioblastoma and anaplastic glioma; MPC-3100 for inhibiting Heat shock protein 90 (Hsp90) in cancer patients; MPC-9528 for the treatment of cancer; as well as other preclinical treatments for cancer and HIV.

Myriad offers its employees medical, dental, life and AD&D insurance; an employee stock purchase plan; short- and long-term disability coverage; a 401(k) plan; and tax-free reimbursements.

FINANCIALS: Sales and profits are in thousands of dollars—add 000 to get the full amount. 2009 Note: Financial information for 2009 was not available for all companies at press time.

2009 Sales: $326,527	2009 Profits: $84,615	**U.S. Stock Ticker: MYGN**	
2008 Sales: $222,855	2008 Profits: $47,845	**Int'l Ticker:** Int'l Exchange:	
2007 Sales: $145,285	2007 Profits: $-34,962	Employees: 869	
2006 Sales: $114,279	2006 Profits: $-38,189	Fiscal Year Ends: 6/30	
2005 Sales: $82,406	2005 Profits: $-39,978	Parent Company:	

SALARIES/BENEFITS:

Pension Plan:	ESOP Stock Plan:	Profit Sharing:	Top Exec. Salary: $800,552	Bonus: $1,000,812
Savings Plan: Y	Stock Purch. Plan: Y		Second Exec. Salary: $551,465	Bonus: $415,812

OTHER THOUGHTS:

Apparent Women Officers or Directors: 2
Hot Spot for Advancement for Women/Minorities:

LOCATIONS: ("Y" = Yes)

West:	Southwest:	Midwest:	Southeast:	Northeast:	International:
Y					

NABI BIOPHARMACEUTICALS

www.nabi.com

Industry Group Code: 325412 Ranks within this company's industry group: Sales: 122   Profits: 114

Drugs:		Other:	Clinical:	Computers:	Services:
Discovery:	Y	AgriBio:	Trials/Services:	Hardware:	Specialty Services:
Licensing:		Genetic Data:	Labs:	Software:	Consulting:
Manufacturing:		Tissue Replacement:	Equipment/Supplies:	Arrays:	Blood Collection:
Generics:			Research/Development Svcs.:	Database Management:	Drug Delivery:
			Diagnostics:		Drug Distribution:

TYPES OF BUSINESS:

Drugs-Infectious Disease & Autoimmune Disorders
Vaccines
Addiction Treatments

BRANDS/DIVISIONS/AFFILIATES:

EnteroVAX
NicVAX
RENs

CONTACTS: *Note: Officers with more than one job title may be intentionally listed here more than once.*

Raafat E. F. Fahim, CEO
Raafat E. F. Fahim, Pres.
Raafat E. Fahim, Acting CFO
Mary Booth, Dir.-Human Resources
Ali E. Fattom, VP-R&D
Matthew Kalnik, VP-Bus. Oper.
Matthew Kalnik, VP-Strategic Planning
Greg Fries, Mgr.-Corp. Comm.
Greg Fries, Mgr.-Investor Rel.
Paul Kessler, Sr. VP-Clinical, Medical & Regulatory Affairs
Geoffrey F. Cox, Chmn.

Phone: 301-770-3099	Fax: 301-770-3097
Toll-Free: 800-685-5579	
Address: 12276 Wilkins Ave., Rockville, MD 20852 US	

GROWTH PLANS/SPECIAL FEATURES:

Nabi Biopharmaceuticals is a development level biopharmaceutical company that focuses on vaccines and therapies for the treatment and prevention of infectious diseases and nicotine addiction. The firm's two leading product in development are NicVax, a vaccine to treat nicotine addiction, and EnteroVAX, a vaccine for Enterococcous hospital-acquired bloodstream infections. The NicVax vaccine is designed to work by stimulating the immune system to produce antibodies that bind to nicotine, effectively keeping nicotine from entering the brain and triggering pleasure sensations. This vaccine is currently in Phase III clinical trials. The firm's other product in development is the RENs (Ring Expanded Nucleosides and Nucleotides) platform, under exclusive worldwide license from the University of Maryland Baltimore County, which is designed to develop analogs used to treat viral infections and cancer. Possibilities for this research include treatments for hepatitis B and C, respiratory syncytial virus, Epstein-Barr virus, West Nile virus and rhinovirus, as well as numerous cancers. In November 2009, the company sold PentaStaph, its pentavalent vaccine designed to prevent staphylococcus aureus infections, to GlaxoSmithKlineBiologicals (GSK) for $20 million up front with a potential of $26 million in additional payments. In March 2010, Nabi closed an option and license agreement for NicVax with GSK, under which Nabi will receive $40 million in upfront payments and up to $460 million in potential option fees and regulatory, development and sales milestones. GSK will be able to in-license NicVax globally and develop novel nicotine vaccines using Nabi's intellectual property.

Nabi offers its employees medical, dental, vision, prescription drug and life insurance; short- and long-term disability; an employee assistance program; a flexible spending account; a stock purchase plan; tuition assistance; access to a credit union; and employee referral incentives.

FINANCIALS: Sales and profits are in thousands of dollars—add 000 to get the full amount. 2009 Note: Financial information for 2009 was not available for all companies at press time.

2009 Sales: $10,489	2009 Profits: $-18,727	U.S. Stock Ticker: NABI
2008 Sales: $	2008 Profits: $-18,738	Int'l Ticker: Int'l Exchange:
2007 Sales: $	2007 Profits: $38,386	Employees: 42
2006 Sales: $117,852	2006 Profits: $-58,703	Fiscal Year Ends: 12/31
2005 Sales: $94,149	2005 Profits: $-128,449	Parent Company:

SALARIES/BENEFITS:

Pension Plan:	ESOP Stock Plan:	Profit Sharing:	Top Exec. Salary: $476,306	Bonus: $623,100
Savings Plan: Y	Stock Purch. Plan: Y		Second Exec. Salary: $320,769	Bonus: $181,500

OTHER THOUGHTS:

Apparent Women Officers or Directors: 1
Hot Spot for Advancement for Women/Minorities:

LOCATIONS: ("Y" = Yes)

West:	Southwest:	Midwest:	Southeast:	Northeast:	International:
				Y	

NANOBIO CORPORATION

www.nanobio.com

Industry Group Code: 325412A Ranks within this company's industry group: Sales: Profits:

Drugs:	Other:	Clinical:	Computers:	Services:
Discovery: Y	AgriBio:	Trials/Services:	Hardware:	Specialty Services:
Licensing:	Genetic Data:	Labs:	Software:	Consulting:
Manufacturing:	Tissue Replacement:	Equipment/Supplies:	Arrays:	Blood Collection:
Generics:		Research/Development Svcs.:	Database Management:	Drug Delivery:
		Diagnostics:		Drug Distribution:

TYPES OF BUSINESS:

Drug Delivery Systems
Nano-Emulsion Technology
Cold Sore Treatments
Drugs-Antimicrobial

BRANDS/DIVISIONS/AFFILIATES:

NanoStat
NanoHPX
NanoTxt

CONTACTS: Note: Officers with more than one job title may be intentionally listed here more than once.

James R. Baker, CEO
David Peralta, COO
David Peralta, CFO/VP
James R. Baker, Chief Scientific Officer
John Coffey, VP-Bus. Dev.
Mary R. Flack, VP-Clinical Research
Joyce Sutcliffe, VP-Research
Stephen Gracon, VP-Regulatory Affairs
James R. Baker, Exec. Chmn.

Phone: 734-302-4000	Fax: 734-302-9150
Toll-Free:	
Address: 2311 Green Rd., Ste. A, Ann Arbor, MI 48105 US	

GROWTH PLANS/SPECIAL FEATURES:

NanoBio Corporation is a privately-held biopharmaceutical company that develops and markets products for the prevention and treatment of infectious diseases. The company's main technology platform is NanoStat, which implements oil-in-water emulsions that are roughly 150-400 nanometers in size. Nanoemulsion particles can rapidly penetrate the skin through pores and hair shafts to the site of an infection, where it physically kills lipid-containing organisms by penetrating the outer membrane of pathogenic organisms. NanoStat emulsions have the added benefit of being selectively lethal to targeted microbes without affecting any surrounding skin and mucous membranes. The firm's NanoStat technology has shown activity against viruses, spores, bacteria and fungi. NanoBio's current commercial product pipeline consists of five anti-infective pharmaceuticals and two mucosal vaccines. The firm's anti-infective pharmaceuticals include NB-00X, a treatment that can be used in conjunction with systemic anti-virals for the treatment of genital herpes; and NB-401, a treatment that can help cystic fibrosis patients control bacterial airway infections. NanoStat mucosal vaccines are designed to treat influenza, H5N1, hepatitis B, pneumonia, tuberculosis, small pox anthrax and other various viral and bacterial diseases. NanoBio also holds exclusive intellectual property rights for virucidal, fungicidal, sporicidal and bactericidal applications in a wide spectrum of products that include personal care products, medical products and anti-bioterrorism applications. In October 2009, NanoBio and the University of Michigan received $9.3 million from the National Institute of Health to fund mucosal vaccine research. In December 2009, NanoBio and GlaxoSmithKline plc announced an exclusive licensing agreement in the U.S. and Canada for NB-001, a treatment for cold sores. In April 2010, the company launched a program to develop an intranasal vaccine for the treatment of hepatitis B.

NanoBio offers employees medical and dental coverage; medical and child care flexible spending accounts; a 401(k) plan; life insurance; and annual performance bonuses.

FINANCIALS: Sales and profits are in thousands of dollars—add 000 to get the full amount. 2009 Note: Financial information for 2009 was not available for all companies at press time.

2009 Sales: $	2009 Profits: $	**U.S. Stock Ticker: Private**
2008 Sales: $	2008 Profits: $	**Int'l Ticker:** Int'l Exchange:
2007 Sales: $	2007 Profits: $	Employees:
2006 Sales: $	2006 Profits: $	Fiscal Year Ends: 12/31
2005 Sales: $	2005 Profits: $	Parent Company:

SALARIES/BENEFITS:

Pension Plan:	ESOP Stock Plan:	Profit Sharing: Y	Top Exec. Salary: $	Bonus: $
Savings Plan: Y	Stock Purch. Plan:		Second Exec. Salary: $	Bonus: $

OTHER THOUGHTS:

Apparent Women Officers or Directors: 2
Hot Spot for Advancement for Women/Minorities:

LOCATIONS: ("Y" = Yes)

West:	Southwest:	Midwest:	Southeast:	Northeast:	International:
		Y			

NEKTAR THERAPEUTICS

www.nektar.com

Industry Group Code: 325412A **Ranks within this company's industry group:** Sales: 6   Profits: 20

Drugs:		Other:		Clinical:		Computers:		Services:	
Discovery:		AgriBio:		Trials/Services:		Hardware:		Specialty Services:	
Licensing:	Y	Genetic Data:		Labs:		Software:		Consulting:	
Manufacturing:		Tissue Replacement:		Equipment/Supplies:	Y	Arrays:		Blood Collection:	
Generics:				Research/Development Svcs.:	Y	Database Management:		Drug Delivery:	Y
				Diagnostics:				Drug Distribution:	

TYPES OF BUSINESS:

Drug Delivery Systems
PEG-Based Delivery Systems
Molecular & Particle Engineering
Equipment-Inhalers

BRANDS/DIVISIONS/AFFILIATES:

Nektar PEGylation Technology
PEG-INTRON
Macugen
PEGASYS
Somavert
Neulasta
NKTR-118
NKTR-119

CONTACTS: *Note: Officers with more than one job title may be intentionally listed here more than once.*

Howard W. Robin, CEO
Bharatt Chowrira, COO/Sr. VP
Howard W. Robin, Pres.
John Nicholson, CFO/Sr. VP
Dorian Rinella, VP-Human Resources & Facilities
Stephen K. Doberstein, Chief Scientific Officer/Sr. VP
Gil M. Labrucherie, General Counsel/Sec./Sr. VP
Rinko Ghosh, Chief Business Officer/Sr. VP
Jillian B Thomsen, VP-Finance/Chief Acct. Officer
Lorianne Masuoka, Chief Medical Officer/Sr. VP
Timothy A. Riley, Sr. VP-Global Research
Robert B. Chess, Chmn.

Phone: 650-631-3100	Fax: 650-631-3150
Toll-Free:	
Address: 201 Industrial Rd., San Carlos, CA 94070 US	

GROWTH PLANS/SPECIAL FEATURES:

Nektar Therapeutics is a biopharmaceutical company focused on developing technology for unmet medical needs. Nektar Therapeutics creates potential breakthrough products in two ways: by developing products in collaboration with pharmaceutical and biotechnology companies that seek to improve and differentiate their products; and by applying its technologies to already approved drugs. The company has two leading technology platforms: PEGylation technology and advanced polymer conjugate technology. PEGylation technology is a chemical process designed to enhance the performance of most drug classes with the potential to improve solubility and stability; increase drug half-life; reduce immune responses to an active drug; and improve the efficacy and/or safety of a molecule in certain instances. Leading products approved by the U.S. Food and Drug Administration (FDA) include Macugen treatment for macular degeneration; PEGASYS for chronic Hepatitis C; Somavert, a human growth hormone receptor antagonist; PEG-INTRON for treating Hepatitis C; and Neulasta for the treatment of neutropenia, a condition where the body produces too few white blood cells. Most of these have been developed in collaboration with other pharmaceutical companies, including Amgen; Novartis Pharma; Bayer; Pfizer; and Merck. The firm's drug portfolio also contains NKTR-118 and NKTR-119 to treat pain without the side effect of constipation associated with opioid therapy; BAY41-6551 to treat Gram-negative pneumonias; NKTR-102 for cancer indications such as breast, colorectal and ovarian; and NKTR-105, a chemotherapy agent, among others. In September 2009, the company entered into a license agreement with AstraZeneca AB for the global development and commercialization of Oral NKTR-118 and NKTR-119.

Nektar offers its employees medical, dental and vision plans; accident and life insurance; retirement and investment plans; and life enrichment programs, including tuition reimbursement and fitness membership discounts.

FINANCIALS: Sales and profits are in thousands of dollars—add 000 to get the full amount. 2009 Note: Financial information for 2009 was not available for all companies at press time.

2009 Sales: $71,931	2009 Profits: $-102,519	**U.S. Stock Ticker: NKTR**
2008 Sales: $90,185	2008 Profits: $-34,336	**Int'l Ticker:** Int'l Exchange:
2007 Sales: $273,027	2007 Profits: $-32,761	Employees: 335
2006 Sales: $217,718	2006 Profits: $-154,761	Fiscal Year Ends: 12/31
2005 Sales: $126,279	2005 Profits: $-185,111	Parent Company:

SALARIES/BENEFITS:

Pension Plan:	ESOP Stock Plan:	Profit Sharing:	Top Exec. Salary: $730,417	Bonus: $1,000,000
Savings Plan: Y	Stock Purch. Plan:		Second Exec. Salary: $492,417	Bonus: $415,000

OTHER THOUGHTS:

Apparent Women Officers or Directors: 4
Hot Spot for Advancement for Women/Minorities: Y

LOCATIONS: ("Y" = Yes)

West:	Southwest:	Midwest:	Southeast:	Northeast:	International:
Y			Y		Y

NEOGEN CORPORATION

www.neogen.com

Industry Group Code: 325413 Ranks within this company's industry group: Sales: 13 Profits: 8

Drugs:		Other:		Clinical:		Computers:		Services:	
Discovery:		AgriBio:	Y	Trials/Services:		Hardware:		Specialty Services:	
Licensing:		Genetic Data:		Labs:		Software:		Consulting:	
Manufacturing:	Y	Tissue Replacement:		Equipment/Supplies:	Y	Arrays:		Blood Collection:	
Generics:				Research/Development Svcs.:		Database Management:		Drug Delivery:	
				Diagnostics:	Y			Drug Distribution:	

TYPES OF BUSINESS:

Sanitary & Livestock Diagnostic Products
Food Safety Test Kits
Animal Health Test Kits
Pharmacology Test Kits
Agricultural Test Kits
Veterinary Instruments
Veterinary Pharmaceuticals

BRANDS/DIVISIONS/AFFILIATES:

GeneSeek, Inc.
Veratox for T-2/HT-2 Test
Neogen do Brasil
AmVet
PanaKare
RenaKare
NeogenVet
Vita-15

CONTACTS: *Note: Officers with more than one job title may be intentionally listed here more than once.*

James L. Herbert, CEO
Lon M. Bohannon, COO
Lon M. Bohannon, Pres.
Richard R. Current, CFO/VP
Mark A. Mozola, VP-R&D
Kenneth V. Kodilla, VP-Mfg.
Richard R. Current, Sec.
Edward L. Bradley, VP-Food Safety Oper.
Anthony E. Maltese, VP-Corp. Dev.
Terri A. Morrical, VP-Animal Safety Oper.
Joseph M. Madden, VP-Scientific Affairs
Paul S. Satoh, VP-Basic & Exploratory Research
James L. Herbert, Chmn.

Phone: 517-372-9200	Fax: 517-372-2006
Toll-Free: 800-234-5333	
Address: 620 Lesher Pl., Lansing, MI 48912 US	

GROWTH PLANS/SPECIAL FEATURES:

Neogen Corporation and its subsidiaries develop, manufacture and market a diverse line of products for food and animal safety. Its two operating segments are food safety and animal safety. The company's food safety segment primarily sells diagnostic test kits and complementary products marketed to food and feed producers and processors to detect food-borne pathogens, spoilage organisms, natural toxins, food allergens, genetic modifications, ruminant by-products, drug residues, pesticide residues and general sanitation concerns. Many of its food safety test kits use immunoassay technology to rapidly detect target substances. This segment also includes bioluminescence-based diagnostic technology. Revenue from the food safety segment generated 51.4% of the firm's total 2009 revenue. Neogen's animal safety segment, which generated approximately 48.6% of its 2009 revenue, primarily develops, manufactures and markets pharmaceuticals, rodenticides, disinfectants, vaccines, veterinary instruments, topicals and diagnostic products to the worldwide animal safety market. This segment's brands include AmVet, PanaKare, RenaKare, NeogenVet, Vita-15 and Liver 7. Its line of approximately 100 drug detection immunoassay test kits are sold worldwide for the detection of approximately 300 abused and therapeutic drugs in racing animals, such as horses, greyhounds and camels, as well as for testing farm animals. The company has a sales network of over 120 distributors in 100 countries. The firm manufactures its products in Lansing, Michigan; Lexington, Kentucky; Randolph, Wisconsin; and Ayr, Scotland. In September 2009, Neogen launched its Veratox for T-2/HT-2 test to detect T-2 and HT-2 mycotoxins. In October 2009, the firm formed Neogen do Brasil to distribute its food safety products throughout Brazil. In December 2009, the company acquired Gen-Probe Inc.'s BioKits food safety business. The acquisition includes more than 50 test kits for food allergens, plant genetics and meat and fish speciation. In April 2010, Neogen acquired GeneSeek, Inc., a commercial agricultural genetics laboratory, for $13.8 million.

FINANCIALS: Sales and profits are in thousands of dollars—add 000 to get the full amount. 2009 Note: Financial information for 2009 was not available for all companies at press time.

2009 Sales: $118,721	2009 Profits: $13,874	U.S. Stock Ticker: NEOG
2008 Sales: $102,418	2008 Profits: $12,098	Int'l Ticker: Int'l Exchange:
2007 Sales: $86,138	2007 Profits: $9,125	Employees: 515
2006 Sales: $72,433	2006 Profits: $7,941	Fiscal Year Ends: 5/31
2005 Sales: $62,756	2005 Profits: $5,916	Parent Company:

SALARIES/BENEFITS:

Pension Plan:	ESOP Stock Plan:	Profit Sharing:	Top Exec. Salary: $310,000	Bonus: $150,000
Savings Plan: Y	Stock Purch. Plan:		Second Exec. Salary: $215,000	Bonus: $80,000

OTHER THOUGHTS:

Apparent Women Officers or Directors: 1
Hot Spot for Advancement for Women/Minorities:

LOCATIONS: ("Y" = Yes)

West:	Southwest:	Midwest:	Southeast:	Northeast:	International:
		Y			Y

NEOPHARM INC

www.neopharm.com

Industry Group Code: 325412A **Ranks within this company's industry group:** Sales: Profits: 9

Drugs:		Other:	Clinical:	Computers:	Services:	
Discovery:	Y	AgriBio:	Trials/Services:	Hardware:	Specialty Services:	
Licensing:		Genetic Data:	Labs:	Software:	Consulting:	
Manufacturing:		Tissue Replacement:	Equipment/Supplies:	Arrays:	Blood Collection:	
Generics:			Research/Development Svcs.:	Database Management:	Drug Delivery:	Y
			Diagnostics:		Drug Distribution:	

TYPES OF BUSINESS:

Drug Delivery Systems
Liposomal Drug Delivery System
Tumor-Targeting Toxins
Drugs-Cancer
Drugs-Autoimmune Diseases
Drugs-Lung Disease

BRANDS/DIVISIONS/AFFILIATES:

NeoLipid
IL13-PE38QQR
LEP-ETU
LE-DT
LE-rafAON

CONTACTS: *Note: Officers with more than one job title may be intentionally listed here more than once.*

Aquilur Rahman, CEO
Aquilur Rahman, Pres.
Shahid Ali, Exec. VP-R&D
John N. Kapoor, Chmn.

Phone: 847-887-0800	Fax: 847-406-1764
Toll-Free:	
Address: 101 Waukegan Rd., Ste. 970, Lake Bluff, IL 60044 US	

GROWTH PLANS/SPECIAL FEATURES:

NeoPharm, Inc. is a biopharmaceutical company engaged in the research, development and commercialization of drugs for the treatment of cancer and other diseases. The company has built its drug portfolio based on two novel proprietary technology platforms: the NeoLipid drug delivery system and a tumor-targeting toxin platform. One of NeoPharm's leading drug candidates is IL13-PE38QQR (cintredekin besudotox), a tumor-targeting toxin being developed for the treatment of brain cancer and idiopathic pulmonary fibrosis (IPF). Currently, patients diagnosed with glioblastoma multiformes have few effective treatment options due to the physical layout of the brain, specifically the blood-brain barrier. IL13-PE38QQR is designed to overcome these obstacles by selectively target cancer cells in the brain without destroying normal tissue. Idiopathic Pulmonary Fibrosis, an indication related to immune-related damage and scarring in the lungs, also presents conditions that may be effectively treated with the drug's selective targeting. The company's other product candidates are based on its NeoLipid drug delivery platform, which combines drugs or other compounds with NeoPharm's proprietary lipids and allows for the creation of a stable liposome. Potential advantages of this technology include higher anticancer potency, the reduction of toxicity and fewer side effects of anticancer drugs. These products include LEP-ETU, in Phase II development for the treatment of metastatic breast cancer; LE-DT, a metastatic solid cancer treatment; and LE-rafAON, for pancreatic cancer. In May 2010, NeoPharm announced that the FDA granted IL13-PE38QQR orphan drug status for the treatment of IPF.

FINANCIALS: Sales and profits are in thousands of dollars—add 000 to get the full amount. 2009 Note: Financial information for 2009 was not available for all companies at press time.

2009 Sales: $	2009 Profits: $-7,461	**U.S. Stock Ticker:** NEOL	
2008 Sales: $	2008 Profits: $-8,218	**Int'l Ticker:** Int'l Exchange:	
2007 Sales: $	2007 Profits: $-11,001	Employees: 20	
2006 Sales: $ 11	2006 Profits: $-33,208	Fiscal Year Ends: 8/31	
2005 Sales: $ 543	2005 Profits: $-38,724	Parent Company:	

SALARIES/BENEFITS:

Pension Plan:	ESOP Stock Plan:	Profit Sharing:	Top Exec. Salary: $	Bonus: $
Savings Plan:	Stock Purch. Plan:		Second Exec. Salary: $	Bonus: $

OTHER THOUGHTS:

Apparent Women Officers or Directors:
Hot Spot for Advancement for Women/Minorities:

LOCATIONS: ("Y" = Yes)

West:	Southwest:	Midwest:	Southeast:	Northeast:	International:
		Y			

NEOPROBE CORPORATION

www.neoprobe.com

Industry Group Code: 325413 Ranks within this company's industry group: Sales: 18 Profits: 20

Drugs:	Other:	Clinical:		Computers:		Services:	
Discovery:	AgriBio:	Trials/Services:		Hardware:		Specialty Services:	
Licensing:	Genetic Data:	Labs:		Software:		Consulting:	
Manufacturing:	Tissue Replacement:	Equipment/Supplies:	Y	Arrays:		Blood Collection:	
Generics:		Research/Development Svcs.:		Database Management:		Drug Delivery:	
		Diagnostics:	Y			Drug Distribution:	

TYPES OF BUSINESS:

Supplies-Intraoperative Cancer Diagnosis
Gamma-Guided Surgical Instruments
Blood Flow Measurement Devices
Targeting Agents

BRANDS/DIVISIONS/AFFILIATES:

Neoprobe GDS Gamma Detection Systems
Activated Cellular Therapy
RIGScan CR
Lymphoseek
CIRA Biosciences, Inc.

CONTACTS: Note: Officers with more than one job title may be intentionally listed here more than once.

David C. Bupp, CEO
David C. Bupp, Pres.
Brent L. Larson, CFO
Douglas Rash, VP-Mktg.
Frederick O. Cope, VP-Pharmaceutical Research & Clinical Dev.
Anthony K. Blair, VP-Mfg. Oper.
Brent L. Larson, Sec.
Brent Larson, Contact-Investor Rel.
Brent L. Larson, VP-Finance/Treas.
Rodger A. Brown, VP-Regulatory Affairs & Quality Assurance
Carl J. Aschinger Jr., Chmn.

Phone: 614-793-7500	Fax: 614-793-7520
Toll-Free: 800-793-0079	
Address: 425 Metro Pl. N., Ste. 300, Dublin, OH 43017 US	

GROWTH PLANS/SPECIAL FEATURES:

Neoprobe Corporation is focused on developing and commercializing oncology products. The company focuses on improving cancer surgery outcomes using its gamma detection device in conjunction with radiopharmaceutical detection agents. Its lead product is the Neoprobe GDS gamma detection device, which uses its wireless gamma detection probes in Intraoperative Lymphatic Mapping (ILM) or sentinel lymph node biopsy (SNLB), enabling physicians to assess the potential spread of cancer to lymph node tissue or vital organs. Cancer utilizes the lymphatic system to rapidly spread through a patient's system, but the Neoprobe GDS allows surgeons and oncologists to biopsy sentinel nodes that are possible pathway locations, which makes it easier to evaluate if the cancer has spread. Other possible medical applications include thyroid function evaluation and parathyroid surgery. Neoprobe also has a majority-owned subsidiary, Cira Biosciences, Inc., focused on the development and commercialization of an activated cellular therapy technology that has shown promising early stage patient-specific treatment potential in oncology, viral (HIV/AIDS and hepatitis) and autoimmune diseases. Neoprobe's investigational activities are currently focused on three particular developmental product initiatives: Lymphoseek, RIGScan CR and Activated Cellular Therapy (ACT). Lymphoseek is a proprietary drug compound, under exclusive worldwide license from the University of California that works in complement with Neoprobe GDS as a sentinel lymph node targeting agent. RIGScan CR is an intraoperative targeting agent for use in the attempted removal of colorectal cancerous tumors. ACT is a technology platform intended to boost the patient's own immune system by removing lymph nodes identified during surgery, then activating and expanding T-cells found in those nodes. The company's gamma detection systems are marketed and distributed globally through Ethicon Endo-Surgery, Inc. In August 2009, the company divested its subsidiary Cardiosonix, Ltd. and its operations in the blood flow measurement device segment.

FINANCIALS: Sales and profits are in thousands of dollars—add 000 to get the full amount. 2009 Note: Financial information for 2009 was not available for all companies at press time.

2009 Sales: $9,518	2009 Profits: $-39,606	U.S. Stock Ticker: NEOP
2008 Sales: $7,590	2008 Profits: $-5,166	Int'l Ticker: Int'l Exchange:
2007 Sales: $7,125	2007 Profits: $-5,088	Employees: 34
2006 Sales: $6,051	2006 Profits: $-4,741	Fiscal Year Ends: 12/31
2005 Sales: $5,919	2005 Profits: $-4,929	Parent Company:

SALARIES/BENEFITS:

Pension Plan:	ESOP Stock Plan:	Profit Sharing:	Top Exec. Salary: $325,000	Bonus: $45,000
Savings Plan: Y	Stock Purch. Plan:		Second Exec. Salary: $184,000	Bonus: $15,313

OTHER THOUGHTS:

Apparent Women Officers or Directors:
Hot Spot for Advancement for Women/Minorities:

LOCATIONS: ("Y" = Yes)

West:	Southwest:	Midwest:	Southeast:	Northeast:	International:
		Y			

Note: Financial information, benefits and other data can change quickly and may vary from those stated here.

NEXMED INC

www.nexmed.com

Industry Group Code: 325412A **Ranks within this company's industry group:** Sales: 15   Profits: 17

Drugs:		Other:		Clinical:		Computers:		Services:	
Discovery:		AgriBio:		Trials/Services:		Hardware:		Specialty Services:	
Licensing:	Y	Genetic Data:		Labs:		Software:		Consulting:	
Manufacturing:	Y	Tissue Replacement:		Equipment/Supplies:	Y	Arrays:		Blood Collection:	
Generics:				Research/Development Svcs.:		Database Management:		Drug Delivery:	Y
				Diagnostics:				Drug Distribution:	

TYPES OF BUSINESS:

Drug Delivery Systems
Transdermal Drug Delivery Systems
Sexual Dysfunction Products
Medical Devices

BRANDS/DIVISIONS/AFFILIATES:

NexACT
Vitaros
Femprox
Bio-Quant Inc
NM100060

CONTACTS: *Note: Officers with more than one job title may be intentionally listed here more than once.*

Bassam B. Damaj, CEO
Hem Pandya, COO/VP
Bassam B. Damaj, Pres.
Mark Westgate, CFO/VP
Mark S. Wilson, VP-Tech. Dev., NexMed USA, Inc.
Edward Cox, Corp. Sec.
Edward Cox, VP-Corp. Dev.
Edward Cox, VP-Investor Rel.
Mark Westgate, Treas.
Vivian Liu, Exec. VP
Linda Smibert, VP-Bus. Dev., NexMed USA, Inc.
Richard Martin, VP-Chemistry Dev., NexMed USA, Inc.
Terry Ladd, VP-Bus. Dev., Bio-Quant, Inc.
Vivian Liu, Chmn.

Phone: 858-222-8041	Fax: 858-866-0482
Toll-Free:	
Address: 6330 Nancy Ridge Dr., Ste.103, San Diego, CA 92121 US	

GROWTH PLANS/SPECIAL FEATURES:

NexMed, Inc. is a pharmaceutical and medical technology company with a focus on developing and commercializing therapeutic products based on proprietary delivery systems. The firm operates in two divisions: pharmaceutical products and Bio-Quant contract research organization (CRO). The pharmaceutical products segment currently focuses on new and patented topical pharmaceutical products based on a penetration enhancement drug delivery technology called NexACT. NexACT facilitates an active drug's absorption through the skin. It develops topical treatments in forms including cream, gel, patch and tape, and has 10 U.S. patents in connection with its NexACT technology. NexMed's principal product candidates include Vitaros for the treatment of male erectile dysfunction (ED); Femprox cream for female sexual arousal disorder; and NM100060 (co-developed with Novartis International Pharmaceutical Ltd.), a topical treatment for nail fungus that has completed Phase III clinical trials. The Bio-Quant CRO division, which was formed after the December 2009 acquisition of Bio-Quant, Inc., offers in vitro/in vivo pharmacology, pharmacokinetics and toxicology studies regarding early stage drug development and discovery that support pre-IND enabling packages to approximately 300 international clients. In early 2009, NexMed sold the exclusive U.S. rights to Vitaros to Warner Chilcott Company, Inc., a subsidiary of former Vitaros licensing partner Warner Chilcott, Ltd.

The company offers its employees medical, dental and vision coverage; a 401(k); life, AD&D, and disability insurance; and an employee stock plan.

FINANCIALS: Sales and profits are in thousands of dollars—add 000 to get the full amount. 2009 Note: Financial information for 2009 was not available for all companies at press time.

2009 Sales: $2,974	2009 Profits: $-32,043	**U.S. Stock Ticker:** NEXM
2008 Sales: $5,957	2008 Profits: $-5,171	**Int'l Ticker:** Int'l Exchange:
2007 Sales: $1,270	2007 Profits: $-8,787	Employees: 34
2006 Sales: $1,867	2006 Profits: $-8,043	Fiscal Year Ends: 12/31
2005 Sales: $2,399	2005 Profits: $-15,442	Parent Company:

SALARIES/BENEFITS:

Pension Plan:	ESOP Stock Plan:	Profit Sharing:	Top Exec. Salary: $285,114	Bonus: $
Savings Plan: Y	Stock Purch. Plan:		Second Exec. Salary: $223,248	Bonus: $

OTHER THOUGHTS:

Apparent Women Officers or Directors: 3
Hot Spot for Advancement for Women/Minorities: Y

LOCATIONS: ("Y" = Yes)

West:	Southwest:	Midwest:	Southeast:	Northeast:	International:
Y					Y

NOVARTIS AG

www.novartis.com

Industry Group Code: 325412 Ranks within this company's industry group: Sales: 4 Profits: 6

Drugs:		Other:		Clinical:		Computers:		Services:	
Discovery:	Y	AgriBio:		Trials/Services:		Hardware:		Specialty Services:	
Licensing:	Y	Genetic Data:		Labs:		Software:		Consulting:	
Manufacturing:	Y	Tissue Replacement:		Equipment/Supplies:		Arrays:		Blood Collection:	
Generics:				Research/Development Svcs.:		Database Management:		Drug Delivery:	
				Diagnostics:				Drug Distribution:	

TYPES OF BUSINESS:

Drugs-Diversified
Therapeutic Drug Discovery
Therapeutic Drug Manufacturing
Generic Drugs
Over-the-Counter Drugs
Ophthalmic Products
Nutritional Products
Veterinary Products

BRANDS/DIVISIONS/AFFILIATES:

CIBA Vision
Chiron Corp
Sandoz
Novartis Institute for Biomedical Research Inc
Novartis Oncology
Alcon Inc
Corthera Inc

CONTACTS: Note: Officers with more than one job title may be intentionally listed here more than once.

Joseph Jimenez, CEO
Jon Symonds, CFO
Jurgen Brokatzky-Geiger, Head-Human Resources
Thomas Werlen, General Counsel
Sheldon Jones, Head-Corp. Comm.
George Gunn, Head-Consumer Health & Animal Health Div.
David Epstein, Head-Pharmaceuticals
Mark C. Fishman, Pres., Novartis Institute for Biomedical Research
Jeffrey George, Head-Sandoz Div.
Daniel Vasella, Chmn.

Phone: 11-4161-324-2745	Fax: 11-41-61-324-8001
Toll-Free:	
Address: Lichtstrasse 35, Basel, 4056 Switzerland	

GROWTH PLANS/SPECIAL FEATURES:

Novartis AG researches and develops pharmaceuticals as well as a large number of consumer and animal health products. It operates in four segments: pharmaceuticals; vaccines and diagnostics; consumer health; and Sandoz. The pharmaceuticals division, which accounts for 65% of sales, develops, manufactures, distributes and sells prescription medications in a variety of areas, which include cardiovascular and metabolism, oncology and hematology, neuroscience and ophthalmics, respiratory, immunology and infectious diseases. The segment is organized into global business franchises responsible for the marketing of various products as well as a business unit called Novartis Oncology, responsible for the global development and marketing of oncology products. The vaccines and diagnostics division (5% of sales) is focused on the development of preventive vaccine treatments and diagnostic tools. It has two activities: Novartis Vaccines, whose key products include meningococcal and travel vaccines; and Chiron, a blood testing and molecular diagnostics activity dedicated to preventing the spread of infectious diseases through the development of blood-screening tools that protect the world's blood supply. The Sandoz division (17% of sales) is a global generic pharmaceuticals company that develops, produces and markets drugs along with pharmaceutical and biotechnological active substances. The segment has activities in retail generics, anti-infectives and biopharmaceuticals. Sandoz offers roughly1,000 compounds in more than 130 countries. The most important product groups include antibiotics, gastrointestinal medicines, cardiovascular treatments and hormone therapies. The consumer health division (17% of sales) consists of three business units: over-the-counter medicines; animal health, which provides veterinary products for farm and companion animals; and CIBA Vision, which markets contact lenses and lens care products. In September 2009, the company's Sandoz division acquired EBEWE Pharma. In December 2009, Novartis agreed to acquire biopharmaceutical firm Corthera, Inc. In January 2010, the company announced plans to acquire Nestle S.A.'s remaining stake in U.S. eye-care firm, Alcon Inc., for $28.1 billion, bringing Novartis' ownership to 77%.

FINANCIALS: Sales and profits are in thousands of dollars—add 000 to get the full amount. 2009 Note: Financial information for 2009 was not available for all companies at press time.

2009 Sales: $45,075,000	2009 Profits: $8,400,000	U.S. Stock Ticker: NVS
2008 Sales: $42,535,000	2008 Profits: $8,195,000	Int'l Ticker: NOVN Int'l Exchange: Zurich-SWX
2007 Sales: $38,947,000	2007 Profits: $11,946,000	Employees: 99,834
2006 Sales: $34,393,000	2006 Profits: $7,202,000	Fiscal Year Ends: 12/31
2005 Sales: $31,005,000	2005 Profits: $6,141,000	Parent Company:

SALARIES/BENEFITS:

Pension Plan: Y	ESOP Stock Plan:	Profit Sharing:	Top Exec. Salary: $	Bonus: $
Savings Plan:	Stock Purch. Plan:		Second Exec. Salary: $	Bonus: $

OTHER THOUGHTS:

Apparent Women Officers or Directors: 4
Hot Spot for Advancement for Women/Minorities: Y

LOCATIONS: ("Y" = Yes)

West:	Southwest:	Midwest:	Southeast:	Northeast:	International:
Y		Y	Y	Y	Y

NOVAVAX INC

www.novavax.com

Industry Group Code: 325412A **Ranks within this company's industry group:** Sales: 18   Profits: 18

Drugs:		Other:	Clinical:		Computers:		Services:	
Discovery:	Y	AgriBio:	Trials/Services:		Hardware:		Specialty Services:	
Licensing:		Genetic Data:	Labs:		Software:		Consulting:	
Manufacturing:	Y	Tissue Replacement:	Equipment/Supplies:		Arrays:		Blood Collection:	
Generics:			Research/Development Svcs.:	Y	Database Management:		Drug Delivery:	Y
			Diagnostics:				Drug Distribution:	

TYPES OF BUSINESS:

Drug Delivery Systems
Drugs-Bacterial & Viral Infection
Hormone Replacement Therapies
Contract Research Services

BRANDS/DIVISIONS/AFFILIATES:

VLP Technology

CONTACTS: *Note: Officers with more than one job title may be intentionally listed here more than once.*

Rahul Singhvi, CEO
Rahul Singhvi, Pres.
Frederick W. Driscoll, CFO
Gale Smith, VP-Vaccine Dev.
John A. Herrmann, III, Exec. Dir.-Legal Affairs/Corp. Sec.
Raymond J. Hage, Jr., Sr. VP-Commercial Oper.
John J. Trizzino, Sr. VP-Bus. Dev.
Frederick W. Driscoll, Treas./VP
Thomas S. Johnston, VP-Strategy
Stanley C. Erck, Chmn.
John Trizzino, Sr. VP-Int'l & Gov't Alliances

Phone: 240-268-2000	Fax:
Toll-Free:	
Address: 9920 Belward Campus Dr., Rockville, MD 20850 US	

GROWTH PLANS/SPECIAL FEATURES:

Novavax, Inc. is a clinical-stage biopharmaceutical company focused on creating recombinant vaccines that improve upon current preventive options for a range of infectious diseases. These vaccines leverage the company's virus-like particle (VLP) platform technology. VLPs are genetically engineered three-dimensional nanostructures that imitate the structures of targeted viruses but are composed of recombinant proteins believed to be incapable of replicating and causing infection and disease. The company is focused in particular on pandemic influenza viruses and seasonal flu vaccine development programs. Novavax has several products in various stages of development, including seasonal influenza vaccine; varicella zoster (chicken pox and shingles) vaccine; and other undisclosed disease target vaccines. Rather than using chicken eggs or mammalian cells, both of which are commonly employed in the vaccine production process, the firm's proprietary production technology utilizes insect cells for virus gestation. Though newer and less established than egg-based production methods, Novavax's technology has potential benefits that may include higher batch yields; improved time frames for commissioning of new production sites; lower infrastructure costs; shorter overall time-frames for vaccine production; and a relatively portable production process that may make rapid response to pandemic outbreaks more feasible than with traditional methods. These systems are also being explored by the company for the potential treatment of infectious diseases including HIV and Severe Acute Respiratory Syndrome (SARS). During 2009, the company began producing a vaccine candidate targeted at H1N1 influenza (swine flu), with initial clinical trials started in October 2009.

The firm offers employees medical, dental and vision coverage; a 401(k) plan with company match; flexible spending accounts; life and disability insurance; stock options; paid holidays and vacation time; and an employee referral program, among other benefits.

FINANCIALS: Sales and profits are in thousands of dollars—add 000 to get the full amount. 2009 Note: Financial information for 2009 was not available for all companies at press time.

2009 Sales: $ 325	2009 Profits: $-38,374	**U.S. Stock Ticker: NVAX**
2008 Sales: $1,064	2008 Profits: $-36,049	**Int'l Ticker:** Int'l Exchange:
2007 Sales: $1,513	2007 Profits: $-34,765	Employees: 93
2006 Sales: $1,738	2006 Profits: $-23,068	Fiscal Year Ends: 12/31
2005 Sales: $5,343	2005 Profits: $-11,174	Parent Company:

SALARIES/BENEFITS:

Pension Plan:	ESOP Stock Plan:	Profit Sharing:	Top Exec. Salary: $434,141	Bonus: $100,000
Savings Plan: Y	Stock Purch. Plan:		Second Exec. Salary: $261,643	Bonus: $

OTHER THOUGHTS:

Apparent Women Officers or Directors: 1
Hot Spot for Advancement for Women/Minorities:

LOCATIONS: ("Y" = Yes)

West:	Southwest:	Midwest:	Southeast:	Northeast:	International:
				Y	

NOVEN PHARMACEUTICALS

www.noven.com

Industry Group Code: 325412A Ranks within this company's industry group: Sales: Profits:

Drugs:		Other:	Clinical:	Computers:	Services:	
Discovery:	Y	AgriBio:	Trials/Services:	Hardware:	Specialty Services:	
Licensing:		Genetic Data:	Labs:	Software:	Consulting:	
Manufacturing:	Y	Tissue Replacement:	Equipment/Supplies:	Arrays:	Blood Collection:	
Generics:			Research/Development Svcs.:	Database Management:	Drug Delivery:	Y
			Diagnostics:		Drug Distribution:	

TYPES OF BUSINESS:

Drug Delivery Systems
Hormone Replacement Products
Pain Management Products
Central Nervous System Products
Transdermal Drug Delivery Systems

BRANDS/DIVISIONS/AFFILIATES:

Hisamitsu Pharmaceutical Co., Inc.
Menorest
Daytrana
DOT Matrix
Estalis
Noven Therapeutics
Estradot
Stavzor

CONTACTS: Note: Officers with more than one job title may be intentionally listed here more than once.

Jeffrey F. Eisenberg, CEO
Jeffrey F. Eisenberg, Pres.
Michael D. Price, CFO/VP
Andrew C. Panagy, VP-Mktg. & Sales
Natalie P. Estrella, VP-Human Resources
Steven F. Dinh, Chief Scientific Officer/VP
Juan A. Mantelle, CTO/VP
Joel S. Lippman, VP-Clinical Dev./Chief Medical Officer
Jeff T. Mihm, Chief Admin. Officer
Jeff T. Mihm, General Counsel/VP
Richard Gilbert, VP-Oper.
Patrick Gallagher, VP-Bus. Dev.
Joseph C. Jones, VP-Corp. Affairs
Peter G. Amanatides, VP-Quality Assurance & Quality Control

Phone: 305-253-5099	Fax: 305-251-1887
Toll-Free:	
Address: 11960 S.W. 144th St., Miami, FL 33186 US	

GROWTH PLANS/SPECIAL FEATURES:

Noven Pharmaceuticals, Inc. develops and manufactures advanced transdermal drug delivery systems and prescription transdermal products. Its principal commercialized products are transdermal drug delivery systems designed with its DOT Matrix technology for use in hormone replacement therapy. The firm's first product was an estrogen patch for the treatment of menopausal symptoms, marketed under the name Vivelle in the U.S. and Canada, and under the name Menorest in Europe and other markets. The company also launched the smallest transdermal estrogen patch ever approved by the U.S. FDA, Vivelle-Dot, and markets the product in several foreign countries under the name Estradot. The firm also markets its Lidocaine/DentiPatch for dental pain associated with dental procedures. Other Noven products include Pexeva for depression, panic disorder, OCD and GAD; Lithobid for bipolar disorder; Stavzor for bipolar disorder, migraine and epilepsy; Daytrana for ADHD; and its fentanyl transdermal system for chronic pain. The company also has four products in development: Lithium QD for bipolar disorder; Stavzor ER (extended release) for bipolar disorder, migraines and epilepsy; Mesafem for hot flashes; and a treatment for ADHD. Noven has partial ownership of a joint venture formed with Novartis Pharmaceuticals, Vivelle Ventures, doing business under the name Novogyne Pharmaceuticals. Novogyne markets Vivelle, Vivelle-Dot and CombiPatch (a combination estrogen/progestin transdermal patch for the treatment of menopausal symptoms) in the U.S. Beyond Novartis, Noven collaborates with a number of other pharmaceutical companies, such as Aventis, Shire, and Procter & Gamble. In August 2009, the firm was acquired by Hisamitsu Pharmaceuticals Co., Inc., and it now operates as a stand-alone, wholly owned subsidiary of Hisamitsu.

FINANCIALS: Sales and profits are in thousands of dollars—add 000 to get the full amount. 2009 Note: Financial information for 2009 was not available for all companies at press time.

2009 Sales: $	2009 Profits: $	U.S. Stock Ticker: Subsidiary
2008 Sales: $108,175	2008 Profits: $21,412	Int'l Ticker: Int'l Exchange:
2007 Sales: $83,161	2007 Profits: $-45,376	Employees: 610
2006 Sales: $60,689	2006 Profits: $15,988	Fiscal Year Ends: 12/31
2005 Sales: $52,532	2005 Profits: $9,972	Parent Company: HISAMITSU PHARMACEUTICAL CO INC

SALARIES/BENEFITS:

Pension Plan:	ESOP Stock Plan:	Profit Sharing:	Top Exec. Salary: $483,524	Bonus: $182,500
Savings Plan: Y	Stock Purch. Plan:		Second Exec. Salary: $422,500	Bonus: $365,625

OTHER THOUGHTS:

Apparent Women Officers or Directors: 1
Hot Spot for Advancement for Women/Minorities:

LOCATIONS: ("Y" = Yes)

West:	Southwest:	Midwest:	Southeast:	Northeast:	International:
			Y	Y	

NOVO-NORDISK AS

www.novonordisk.com

Industry Group Code: 325412 Ranks within this company's industry group: Sales: 17 Profits: 17

Drugs:		Other:	Clinical:	Computers:	Services:	
Discovery:	Y	AgriBio:	Trials/Services:	Hardware:	Specialty Services:	
Licensing:		Genetic Data:	Labs:	Software:	Consulting:	
Manufacturing:	Y	Tissue Replacement:	Equipment/Supplies:	Arrays:	Blood Collection:	
Generics:			Research/Development Svcs.:	Database Management:	Drug Delivery:	Y
			Diagnostics:		Drug Distribution:	

TYPES OF BUSINESS:

Drugs-Diabetes
Hormone Replacement Therapy
Growth Hormone Drugs
Hemophilia Drugs
Insulin Delivery Systems
Educational & Training Services

BRANDS/DIVISIONS/AFFILIATES:

Levemir
NovoRapid
NovoNorm
NovoFine
Norditropin
NovoSeven
Activella
Victoza

CONTACTS: Note: Officers with more than one job title may be intentionally listed here more than once.

Lars R. Sorensen, CEO
Kare Schultz, COO/Exec. VP
Lars R. Sorensen, Pres.
Jesper Brandgaard, CFO/Exec. VP
Patrick Loustau, Dir.-Global Mktg.
Mads Krogsgaard Thomsen, Chief Science Officer/Exec. VP
Lars Fruergaard Jorgensen, Dir.-IT
Lise Kingo, Chief of Staff/Exec. VP
Ole Ramsby, Dir.-Legal Affairs
Peter Kristensen, Dir.-Global Dev.
Mike Rulis, Contact-Media
Mads Veggerby Lausten, Contact-Investor Rel.
Lars Green, Dir.-Corp. Finance
Goran A. Ando, Vice Chmn.
Jerzy Gruhn, Dir.-North American Bus.
Flemming Dahl, Dir.-Biopharmaceuticals
Peter Kurtzhals, Dir.-Diabetes Research Unit
Sten Scheibye, Chmn.
Jesper Hoiland, Dir.-Int'l Oper.

Phone: 45-4444-8888	Fax: 45-4449-0555
Toll-Free:	
Address: Novo Alle, Bagsvaerd, 2880 Denmark	

GROWTH PLANS/SPECIAL FEATURES:

Novo Nordisk A/S is a healthcare company that focuses on developing treatments for diabetes, hemostasis management and hormone therapy. It has two segments: diabetes care, which generated roughly 73% of 2009 sales; and biopharmaceuticals, 27%. The diabetes care segment manages the firm's insulin franchise, including modern insulins, human insulins, insulin-related sales and oral anti-diabetic drugs. Specific products include Levemir and NovoRapid modern insulin; NovoNorm, an oral anti-diabetic drug; FlexPen and Innolet insulin injectors; and GlucaGen and NovoFine (needles) diabetic devices. The biopharmaceuticals segment covers hemostasis management, growth hormone therapy, hormone replacement therapy, inflammation therapy and other therapy areas. Specific products include the following. Norditropin is a premixed liquid growth hormone designed to provide a very flexible and accurate dosing, while NordiFlex, NordiFlex PenMate and NordiLet are human growth hormone injection systems. NovoSeven is a hemostasis management product, a hemophilia treatment consisting of a recombinant coagulation factor that enables coagulation to proceed in the absence of natural blood factors. Lastly, the firm's post-menopausal hormone replacement therapy products include Activella and Vagifem. Novo Nordisk has employees in 76 countries and its products are marketed in 179 countries. Besides its products, the company offers educational services and training materials for both patients and health care professionals. Novo Nordisk owns and operates dedicated research centers in Denmark, China and the U.S., as well as production and processing facilities in Denmark, the U.S., France, Japan, China and Brazil. U.S. locations include Princeton, New Jersey; Clayton, North Carolina; Hayward, California; and Seattle, Washington. In February 2010, the company launched its Victoza injection pen in the U.S.; Victoza is a daily non-insulin medication approved for the treatment of type-2 diabetes.

Novo Nordisk offers its U.S. employees health, life, dental and supplemental insurance, as well as tuition reimbursement, among other benefits.

FINANCIALS: Sales and profits are in thousands of dollars—add 000 to get the full amount. 2009 Note: Financial information for 2009 was not available for all companies at press time.

2009 Sales: $9,335,490	2009 Profits: $1,968,060	**U.S. Stock Ticker: NVO**
2008 Sales: $8,239,030	2008 Profits: $1,744,460	**Int'l Ticker: NOVO B** Int'l Exchange: Copenhagen-CSE
2007 Sales: $8,190,000	2007 Profits: $1,670,000	Employees: 28,809
2006 Sales: $6,913,700	2006 Profits: $1,126,020	Fiscal Year Ends: 12/31
2005 Sales: $5,446,472	2005 Profits: $946,073	Parent Company:

SALARIES/BENEFITS:

Pension Plan:	ESOP Stock Plan:	Profit Sharing:	Top Exec. Salary: $	Bonus: $
Savings Plan: Y	Stock Purch. Plan:		Second Exec. Salary: $	Bonus: $

OTHER THOUGHTS:

Apparent Women Officers or Directors: 3
Hot Spot for Advancement for Women/Minorities: Y

LOCATIONS: ("Y" = Yes)

West:	Southwest:	Midwest:	Southeast:	Northeast:	International:
Y				Y	Y

Note: Financial information, benefits and other data can change quickly and may vary from those stated here.

NOVOZYMES

www.novozymes.com

Industry Group Code: 325414 Ranks within this company's industry group: Sales: 3 Profits: 3

Drugs:	Other:		Clinical:	Computers:	Services:
Discovery:	AgriBio:	Y	Trials/Services:	Hardware:	Specialty Services:
Licensing:	Genetic Data:	Y	Labs:	Software:	Consulting:
Manufacturing:	Tissue Replacement:		Equipment/Supplies:	Arrays:	Blood Collection:
Generics:			Research/Development Svcs.:	Database Management:	Drug Delivery:
			Diagnostics:		Drug Distribution:

TYPES OF BUSINESS:

Industrial Enzyme & Microorganism Production
Biopharmaceuticals
Enzymes
Microbiology

BRANDS/DIVISIONS/AFFILIATES:

Mannaway
Stainzyme Plus
Spirizyme
Sucrozyme
Acrylaway
Saczyme
Cellic CTec2
Met52

CONTACTS: Note: Officers with more than one job title may be intentionally listed here more than once.

Steen Riisgaard, CEO
Steen Riisgaard, Pres.
Benny Loft, CFO/Exec. VP
Henrik Meyer, VP-Mktg.
Per Falholt, Chief Scientific Officer/Exec. VP
Rasmus von Gottberg, VP-Bus. Dev. & Acquisitions
Thomas Nagy, Exec. VP-Stakeholder Rel.
Mads Bodenhoff, VP-Finance
Thomas Videbaek, Exec. VP-Bio Bus.
Peder Holk Nielsen, Exec. VP/Head-Enzyme Bus.
Michael Fredskov Christiansen, Regional Pres., China
Erik Gormsen, VP-R&D
Henrik Gurtler, Chmn.
Pedro Luiz Fernandes, Regional Pres., Brazil
Anders Spohr, VP-Supply Chain Oper.

Phone: 45-44-46-00-00	Fax: 45-44-46-99-99
Toll-Free:	
Address: Krogshoejvej 36, Bagsvaerd, 2880 Denmark	

GROWTH PLANS/SPECIAL FEATURES:

Novozymes is a biotechnology company that specializes in microbiology and enzymes. The firm currently sells over 700 products in 130 countries worldwide. It splits its business into two main areas: Enzyme Business and BioBusiness. The Enzyme Business, which accounts for over 90% of sales, is split into detergent, technical, food and feed enzymes. Enzymes are used in detergents, such as Mannaway and Stainzyme Plus enzymes break down water-insoluble stains into matersoluble molecules that can be rinsed away. Technical enzymes transform starch in sugar for starch and fuel industries and are applied in the textile, leather, forestry and alcohol industries. Technical enzymes include the brand name enzymes Spirizyme. The company's food enzymes increase the quality or production efficiency in the production of food products such as bread, wine, juice, beer, noodles, alcohol and pasta. Food enzyme products such as Acrylaway, Saczyme and Viscoferm are designed to improve the quality or production speed of foods such as bread, wine, juice, beer and pasta. Feed enzymes, such as Ronozyme NP, are designed to increase the nutritional value of feed and improve phosphorus absorption in animals. This leads to faster growth of animals, while improving the environment by decreasing the phosphorus released through manure. The firm's BioBusiness includes microorganisms and biopharmaceutical ingredients. Microorganisms have three main applications: wastewater treatment, cleaning products and natural growth enhancements for plants and turf grass. Biopharmaceutical products, such as proteins, are provided by the company as a replacement of human/animal substances, lowering the risk of disease transfer in pharmaceutical products. In February 2010, Novozymes began sales of Cellic CTec2 enzymes, which produce biofuel from agricultural waste. In March 2010, the company introduced a new enzyme, Met52, a bioinsecticide derived from a naturally occurring soil fungus and designed to protect over 100 plant species from Black Vine and Strawberry Root Weevil larvae.

FINANCIALS: Sales and profits are in thousands of dollars—add 000 to get the full amount. 2009 Note: Financial information for 2009 was not available for all companies at press time.

2009 Sales: $1,544,030	2009 Profits: $218,230	**U.S. Stock Ticker:**
2008 Sales: $1,453,610	2008 Profits: $189,510	**Int'l Ticker: NZYM** Int'l Exchange: Copenhagen-CSE
2007 Sales: $1,327,270	2007 Profits: $185,940	Employees: 4,968
2006 Sales: $1,251,490	2006 Profits: $167,610	Fiscal Year Ends: 12/31
2005 Sales: $1,079,840	2005 Profits: $148,025	Parent Company:

SALARIES/BENEFITS:

Pension Plan: Y	ESOP Stock Plan:	Profit Sharing:	Top Exec. Salary: $	Bonus: $
Savings Plan: Y	Stock Purch. Plan:		Second Exec. Salary: $	Bonus: $

OTHER THOUGHTS:

Apparent Women Officers or Directors:
Hot Spot for Advancement for Women/Minorities:

LOCATIONS: ("Y" = Yes)

West:	Southwest:	Midwest:	Southeast:	Northeast:	International:
Y				Y	Y

NPS PHARMACEUTICALS INC

www.npsp.com

Industry Group Code: 325412 **Ranks within this company's industry group:** Sales: 78   Profits: 112

Drugs:		Other:		Clinical:	Computers:	Services:
Discovery:	Y	AgriBio:		Trials/Services:	Hardware:	Specialty Services:
Licensing:	Y	Genetic Data:		Labs:	Software:	Consulting:
Manufacturing:	Y	Tissue Replacement:		Equipment/Supplies:	Arrays:	Blood Collection:
Generics:				Research/Development Svcs.:	Database Management:	Drug Delivery:
				Diagnostics:		Drug Distribution:

TYPES OF BUSINESS:

Small Molecule Drugs & Recombinant Proteins
Drugs-Gastrointestinal Diseases
Drugs-Hyperparathyroidism

BRANDS/DIVISIONS/AFFILIATES:

Preotact
Preos
Sensipar
Mimpara
GATTEX
NPSP558
NPSP156
REGPARA

CONTACTS: *Note: Officers with more than one job title may be intentionally listed here more than once.*

Francois Nader, CEO
Francois Nader, Pres.
Luke M. Beshar, CFO/Sr. VP
Nancy K. Bryan, Sr. VP-Commercial Oper.
Roger J. Garceau, Chief Medical Officer/Sr. VP
Joseph J. Rogus, VP-Tech. Oper.
Edward H. Stratemeier, General Counsel/Sr. VP-Legal Affairs
Sandra C. Cottrell, Sr. VP-Regulatory Affairs & Drug Safety
Peter G. Tombros, Chmn.
Joseph J. Rogus, VP-Supply Chain Mgmt.

Phone: 908-450-5300	Fax: 908-450-5351
Toll-Free:	
Address: 550 Hills Drive, 3rd Fl., Bedminster, NJ 07921 US	

GROWTH PLANS/SPECIAL FEATURES:

NPS Pharmaceuticals, Inc. is a clinical-stage biopharmaceutical company focused on the development and commercialization of drugs for the treatment of rare gastrointestinal and endocrine disorders and serious unmet medical needs. The company's lead pipeline products include GATTEX (teduglutide), which is in Phase III trials for the treatment of the short bowel syndrome (SBS), Phase II for Crohn's Disease and preclinical trials for pediatric SBS (caused by necrotizing enterocolitis) and gastrointestinal mucositis; NPSP558, which is in Phase III trials as a treatment for hypoparathyroidism; and a glycine reuptake inhibitor. Other products include Preos, for the treatment of post-menopausal osteoporosis, which is approved for marketing in Europe under the name Preotact and is awaiting FDA approval for commercialization in the U.S.; and NPSP156, in preclinical trials for the treatment of epilepsy, neuropathic pain and other central nervous system disorders. The firm's licensees, Amgen and Kyowa Kirin, developed the FDA-approved Cinacalcet HCl for the treatment of hyperparathyroidism, marketed under the trademark Sensipar in the U.S., Mimpara in Europe and REGPARA in Asia. Furthermore, Nycomed licenses Preotact (the firm's parathyroid hormone 1-84 injection) for distribution outside of the U.S., excluding Japan and Israel. NPS Pharmaceuticals has collaborative research and license agreements with several companies, including with Amgen concerning Sensipar; with GlaxoSmithKline concerning Ronacaleret for osteoporosis; with Kirin concerning REGPARA; with Ortho-McNeil concerning Tapentadol; with Hoffman-La Roche, Inc., concerning products related to the modulation of NMDA receptor activity; and with Nycomed concerning Preotact and GATTEX. The company has been issued roughly 191 patents in the U.S. and additional patents in other countries. In March 2010, NPS sold the royalty rights from REGPARA sales to a fund managed by DRI Capital, Inc.

FINANCIALS: Sales and profits are in thousands of dollars—add 000 to get the full amount. 2009 Note: Financial information for 2009 was not available for all companies at press time.

2009 Sales: $84,147	2009 Profits: $-17,862	**U.S. Stock Ticker:** NPSP
2008 Sales: $102,279	2008 Profits: $-31,726	**Int'l Ticker:** Int'l Exchange:
2007 Sales: $86,248	2007 Profits: $- 657	Employees: 53
2006 Sales: $48,502	2006 Profits: $-112,668	Fiscal Year Ends: 12/31
2005 Sales: $12,825	2005 Profits: $-169,723	Parent Company:

SALARIES/BENEFITS:

Pension Plan:	ESOP Stock Plan:	Profit Sharing:	Top Exec. Salary: $476,827	Bonus: $281,437
Savings Plan:	Stock Purch. Plan:		Second Exec. Salary: $345,586	Bonus: $114,398

OTHER THOUGHTS:

Apparent Women Officers or Directors: 3
Hot Spot for Advancement for Women/Minorities: Y

LOCATIONS: ("Y" = Yes)

West:	Southwest:	Midwest:	Southeast:	Northeast:	International:
				Y	

NYCOMED

www.nycomed.com

Industry Group Code: 325412 Ranks within this company's industry group: Sales: 25   Profits: 28

Drugs:		Other:	Clinical:	Computers:	Services:
Discovery:	Y	AgriBio:	Trials/Services:	Hardware:	Specialty Services:
Licensing:	Y	Genetic Data:	Labs:	Software:	Consulting:
Manufacturing:	Y	Tissue Replacement:	Equipment/Supplies: ·	Arrays:	Blood Collection:
Generics:			Research/Development Svcs.:	Database Management:	Drug Delivery:
			Diagnostics:		Drug Distribution:

TYPES OF BUSINESS:
Pharmaceuticals

BRANDS/DIVISIONS/AFFILIATES:
Alvesco
Curosurf
Ibumetin
Preotact
Matrifen
Neosaldina
Altana Pharma
Bradley Pharmaceuticals

CONTACTS: *Note: Officers with more than one job title may be intentionally listed here more than once.*
Hakan Bjorklund, CEO
Runar Bjorklund, CFO
Charles Depasse, Exec. VP-Human Resources
Anders Ullman, Exec. VP-R&D
Michael Kuner, General Counsel/Exec. VP
Barthold Piening, Exec. VP-Oper.
Kerstin Valinder, Exec. VP-Bus. Dev.
Beatrix Benz, Dir.-Media Rel.
Christian B. Seidelin, VP-Investor Rel.
Guido Oelkers, Exec. VP-Commercial Oper.
Toni Weitzberg, Chmn.

Phone: 41-44-55-51-000	Fax: 41-44-55-51-001
Toll-Free:	
Address: 130 Thurgauerstrasse, Zurich, 8152 Switzerland	

GROWTH PLANS/SPECIAL FEATURES:

Nycomed, also known as Nycomed International Management GmbH, is a European-based pharmaceutical company engaged in the research, licensing, manufacturing and marketing a wide array of pharmaceutical products. The company's focus is in the therapeutic areas of cardiology, gastroenterology, osteoporosis, respiratory, pain and tissue management. Some of the firm's products include: Zymelin, Xymelin, Omnaris, Amol, Febrisan, Alvesco and Curosurf for respiratory treatment; TachoSil for tissue management; Calcium D3 and Preotact for treatment of osteoporosis; Actovegin, Ibumetin, Matrifen, Neosaldina and Xefo Rapid for pain management; and Eparema, Hepatalgina, Pantoprazole and Riopan for gastroenterology problems. The firm's research and development program is led out of Konstanz, Germany with additional 15 facilities located in 12 countries. Nycomed has four reaserch and development sites located in Europe and India. The company is privately-owned and has operates in over 50 markets, Europe, Russia, Latin America, Asia Pacific, Africa, Middle East. Its subsidiary, Altana Pharma, researches therapeutic drugs for the treatment of gastrointestinal and respiratory diseases. It markets Alvesco, a respiratory drug, in 26 countries. The firm's subsidiary, Bradley Pharmaceuticals, a company focused on niche therapeutic markets in the U.S. The company's two biggest shareholders are Nordic Capital, and DLJ Merchant Banking, both private equity firms. In March 2009, the company agreed to purchase 20 branded generic products from Sanofi-Aventis and Zentiva. This will strengthen Nycomed's position in the Czech Republic and Slovakia. In July 2009, the European Commission granted marketing authorization to Nycomed for its new medication, Instanyl (intranasal fentanyl spray), which provides pain relief in adult cancer patients already receiving opioid therapy for chronic pain. In July 2010, the company announced plans to enter Turkish pharmaceutical market.

FINANCIALS: Sales and profits are in thousands of dollars—add 000 to get the full amount. 2009 Note: Financial information for 2009 was not available for all companies at press time.

2009 Sales: $4,278,460	2009 Profits: $308,430	U.S. Stock Ticker: Private
2008 Sales: $4,540,310	2008 Profits: $-110,860	Int'l Ticker: Int'l Exchange:
2007 Sales: $4,908,150 ·	2007 Profits: $334,900	Employees:
2006 Sales: $	2006 Profits: $	Fiscal Year Ends: 12/31
2005 Sales: $	2005 Profits: $	Parent Company:

SALARIES/BENEFITS:

Pension Plan:	ESOP Stock Plan:	Profit Sharing:	Top Exec. Salary: $	Bonus: $
Savings Plan:	Stock Purch. Plan:		Second Exec. Salary: $	Bonus: $

OTHER THOUGHTS:
Apparent Women Officers or Directors: 2
Hot Spot for Advancement for Women/Minorities: Y

LOCATIONS: ("Y" = Yes)

West:	Southwest:	Midwest:	Southeast:	Northeast:	International:
				Y	Y

ONCOGENEX PHARMACEUTICALS

www.oncogenex.com

Industry Group Code: 325412A Ranks within this company's industry group: Sales: 9   Profits: 7

Drugs:		Other:	Clinical:	Computers:	Services:	
Discovery:	Y	AgriBio:	Trials/Services:	Hardware:	Specialty Services:	
Licensing:		Genetic Data:	Labs:	Software:	Consulting:	
Manufacturing:		Tissue Replacement:	Equipment/Supplies:	Arrays:	Blood Collection:	
Generics:			Research/Development Svcs.:	Database Management:	Drug Delivery:	Y
			Diagnostics:		Drug Distribution:	

TYPES OF BUSINESS:

Drug Delivery Systems
Drugs-Cancer

BRANDS/DIVISIONS/AFFILIATES:

OGX-011
OGX-427
OGX-225
SN2310
CSP-9222
OncoGenex Technologies Inc.

CONTACTS: *Note: Officers with more than one job title may be intentionally listed here more than once.*

Scott Cormac, CEO
Scott Cormack, Pres.
Cameron Lawrence, Principal Financial Officer
Martin Gleave, Chief Scientific Advisor
Cindy Jacobs, Chief Medical Officer/Exec. VP
Monica Krieger, VP-Regulatory Affairs
Jack Goldstein, Chmn.

Phone: 425-686-1500	Fax: 425-686-1600
Toll-Free:	
Address: 1522 217th Place S.E., Ste. 100, Bothell, WA 98021 US	

GROWTH PLANS/SPECIAL FEATURES:

OncoGenex Pharmaceuticals is a biopharmaceutical company focused on the development and commercialization of new cancer therapies that address unmet needs in the treatment of cancer. The firm has five product candidates in its pipeline: OGX-011; OGX-427; OGX-225; SN2310; and CSP-9222. OGX-011 focuses on mechanisms of treatment resistance in cancer patients and is designed to address treatment resistance by blocking the production of specific proteins which are believed to promote survival of tumor cells. The firm recently finished five Phase II clinical trials to evaluate the ability of OGX-011 to enhance the effects of therapy in prostate, non-small cell lung and breast cancer. OncoGenex plans to initiate three Phase III clinical trials of OGX-011 with first- and second-line prostate cancer and first-line non-small cell lung cancer. OGX-427, an inhibitor of heat shock protein 27, is being evaluated in a Phase I clinical trial to evaluate safety for OGX-427 administered alone, as well as in combination with docetaxel chemotherapy, in patients with various types of cancer. OGX-225 is a second generation antisense drug which, in preclinical experiments, inhibits production of both Insulin Growth Factor Binding Protein 2 (IGFBP-2) and Insulin Growth Factor Binding Protein-5 (IGFBP-5). SN2310 is a novel camptothecin for the treatment of cancer. It recently completed a Phase I clinical trial, and OncoGenex is exploring options to license it. Lastly, CSP-9222, which is in pre-clinical development, is the lead compound from a family of compounds that have been in-licensed from Bayer Healthcare LLC. It demonstrates activation of programmed cell death in pre-clinical models. The firm has one wholly-owned subsidiary, OncoGenex Technologies Inc., which collaborates with Teva Pharmaceutical Industries Ltd. for the development and commercialization of OGX-011 and related compounds targeting clusterin, excluding OGX-427 and OGX-225.

FINANCIALS: Sales and profits are in thousands of dollars—add 000 to get the full amount. 2009 Note: Financial information for 2009 was not available for all companies at press time.

2009 Sales: $25,539	2009 Profits: $-5,476	**U.S. Stock Ticker:** OGXI
2008 Sales: $	2008 Profits: $-4,204	**Int'l Ticker:** Int'l Exchange:
2007 Sales: $	2007 Profits: $-8,536	Employees: 26
2006 Sales: $22,392	2006 Profits: $-11,594	Fiscal Year Ends: 12/31
2005 Sales: $8,254	2005 Profits: $-21,097	Parent Company:

SALARIES/BENEFITS:

Pension Plan:	ESOP Stock Plan:	Profit Sharing:	Top Exec. Salary: $360,000	Bonus: $108,000
Savings Plan: Y	Stock Purch. Plan:		Second Exec. Salary: $302,632	Bonus: $121,053

OTHER THOUGHTS:

Apparent Women Officers or Directors: 3
Hot Spot for Advancement for Women/Minorities: Y

LOCATIONS: ("Y" = Yes)

West:	Southwest:	Midwest:	Southeast:	Northeast:	International:
Y					Y

ONCOTHYREON INC

www.oncothyreon.com

Industry Group Code: 325412 Ranks within this company's industry group: Sales: 137   Profits: 109

Drugs:		Other:		Clinical:	Computers:	Services:
Discovery:	Y	AgriBio:		Trials/Services:	Hardware:	Specialty Services:
Licensing:	Y	Genetic Data:		Labs:	Software:	Consulting:
Manufacturing:		Tissue Replacement:		Equipment/Supplies:	Arrays:	Blood Collection:
Generics:				Research/Development Svcs.:	Database Management:	Drug Delivery:
				Diagnostics:		Drug Distribution:

TYPES OF BUSINESS:

Drugs-Cancer
Cancer Vaccines
Drugs-Immunological
Target Small Molecule Inhibitors

BRANDS/DIVISIONS/AFFILIATES:

Stimuvax
ONT0-10
PX-478
PX-866

CONTACTS: *Note: Officers with more than one job title may be intentionally listed here more than once.*

Robert L. Kirkman, CEO
Gary Christianson, COO
Robert L. Kirkman, Pres.
Scott Peterson, VP-R&D
Shashi Karan, Corp. Controller
Diana Hausman, VP-Clinical Dev.
Christopher S. Henney, Chmn.

Phone: 206-801-2100	Fax: 206-801-2101
Toll-Free:	
Address: 2601 4th Ave., Ste. 500, Seattle, WA 98121 US	

GROWTH PLANS/SPECIAL FEATURES:

Oncothyreon, Inc. is a clinical-stage biopharmaceutical company focused primarily on the development of cancer therapeutics. The company develops and commercializes novel synthetic vaccines, which stimulate a patient's immune system to recognize and fight malignant cells; and targeted small molecules for treatments of cancer, which inhibit the activity of specific cancer-related proteins. Oncothyreon's lead product candidate currently under phase 3 clinical development, Stimuvax, is a therapeutic vaccine designed to stimulate the immune system to recognize cancer cells and control the growth and spread of cancer. Stimuvax incorporates a 25 amino acid sequence of the cancer antigen MUC1, and the therapeutic is designed to induce an immune response to destroy cancer cells that express MUC1, a protein antigen expressed in many common cancers, such as lung, breast and colorectal cancer. The company is collaborates with Merck KGaA, a German pharmaceutical company, to pursue joint global product research, clinical development and commercialization of Stimuvax. Oncothyreon's ONT-10 is a completely synthetic MUC1-based liposomal vaccine, currently in pre-clinical development, for use in multiple forms of cancer, including breast, thyroid, colon, stomach, pancreas and prostate, as well as certain types of lung cancer. The company's targeted small molecule therapeutics, PX-866 and PX-478, are both in Phase 1 trials. PX866 is an inhibitor of the PI-3-kinase pathway, an important survival signaling pathway that is activated in many types of cancer, particularly ovarian, colon, urinary tract, cervical, and head and neck cancers. PX-478 is an inhibitor of HIF-1a, which encourages cell survival and growth. Animal tests have thus far show significant tumor regressions and growth inhibition. In March 2010, Oncothyreon announced that Merck temporarily suspended clinical development of Stimuvax because of adverse patient reactions.

The company offers employees medical, dental and vision benefits; and life and long-term disability insurance.

FINANCIALS: Sales and profits are in thousands of dollars—add 000 to get the full amount. 2009 Note: Financial information for 2009 was not available for all companies at press time.

2009 Sales: $2,100	2009 Profits: $-17,200	**U.S. Stock Ticker: ONTY**
2008 Sales: $40,300	2008 Profits: $7,400	**Int'l Ticker: BRA** Int'l Exchange: Toronto-TSX
2007 Sales: $3,798	2007 Profits: $-20,340	Employees: 16
2006 Sales: $4,199	2006 Profits: $-16,591	Fiscal Year Ends: 12/31
2005 Sales: $3,800	2005 Profits: $-16,400	Parent Company:

SALARIES/BENEFITS:

Pension Plan:	ESOP Stock Plan:	Profit Sharing:	Top Exec. Salary: $375,000	Bonus: $131,250
Savings Plan: Y	Stock Purch. Plan:		Second Exec. Salary: $250,000	Bonus: $70,000

OTHER THOUGHTS:

Apparent Women Officers or Directors: 1
Hot Spot for Advancement for Women/Minorities:

LOCATIONS: ("Y" = Yes)

West:	Southwest:	Midwest:	Southeast:	Northeast:	International:
Y	Y				

ONYX PHARMACEUTICALS INC
www.onyx-pharm.com

Industry Group Code: 325412 Ranks within this company's industry group: Sales: 62   Profits: 107

Drugs:		Other:	Clinical:	Computers:	Services:
Discovery:	Y	AgriBio:	Trials/Services:	Hardware:	Specialty Services:
Licensing:		Genetic Data:	Labs:	Software:	Consulting:
Manufacturing:	Y	Tissue Replacement:	Equipment/Supplies:	Arrays:	Blood Collection:
Generics:			Research/Development Svcs.:	Database Management:	Drug Delivery:
			Diagnostics:		Drug Distribution:

TYPES OF BUSINESS:
Pharmaceuticals Discovery & Development
Small-Molecule Drugs
Cancer Treatments

BRANDS/DIVISIONS/AFFILIATES:
Nexavar
ONX 0801
ONX 0803
ONX 0805
ONX 0912
ONX 0914
Proteolix Inc
Carfilzomib

CONTACTS: *Note: Officers with more than one job title may be intentionally listed here more than once.*
N. Anthony Coles, CEO
Laura A. Brege, COO/Exec. VP
N. Anthony Coles, Pres.
Matthew K. Fust, CFO/Exec. VP
Judy Batlin, VP-Human Resources
Ted W. Love, Exec. VP/Head-R&D
Suzanne M. Shema, General Counsel/Sr. VP
Juergen Lasowski, Sr. VP-Corp. Dev.
Julianna Wood, VP-Corp. Comm.
Julianna Wood, VP-Investor Rel.
Michael G. Kauffman, Chief Medical Officer
Judy Batlin, VP-Organizational Learning Dev.

Phone: 510-597-6500	Fax: 510-597-6600
Toll-Free:	
Address: 2100 Powell St., Emeryville, CA 94608 US	

GROWTH PLANS/SPECIAL FEATURES:
Onyx Pharmaceuticals, Inc. is engaged in the discovery, development and commercialization of innovative products that target oncological molecular mechanisms. The company's flagship drug, Nexavar, is the result of a partnership with Bayer Pharmaceuticals and is aimed at blocking inappropriate growth signals in tumor cells by inhibiting the responsible active enzymes that induce cancer cell growth. Nexavar is the first and only oral targeted therapy to improve overall survival in patients with liver and kidney cancer by inhibiting both tumor cell proliferation and angiogenesis. The drug is also approved in Europe for the treatment of advanced renal cell carcinoma in patients that have failed to respond or are unsuited for other forms of therapy. Nexavar is also being tested in Phase II and Phase III clinical trials in combination with standard chemotherapeutic and other anticancer agents for additional treatments of liver, lung, kidney, breast, colorectal and ovarian cancers. Onyx is also developing a cell cycle kinase inhibitor with Warner-Lambert Company, which regulates a cell's ability to replicate itself and case cancer. ONX 0801 is a preclinical anti-cancer development drug aimed at receptor-mediated targeting of tumor cells and the inhibition of thymidylate synthase. ONX 0803 and ONX 0805, in collaborative development with S*BIO, are drugs that are designed to inhibit JAK2, a major factor in solid tumors and rheumatoid arthritis. In November 2009, the firm acquired Proteolix, Inc., a privately held biopharmaceutical company, for an undisclosed amount. With the Proteolix acquisition, the company gained Carfilzomib, a selective proteasome inhibitor, and it is being developed for the treatment of patients with multiple myeloma and solid tumors; and two early-stage compounds ONX 0912, an oral proteasome inhibitor; and ONX 0914, a selective inhibitor of the immunoproteasome. The company is still developing both compounds, ONX 912 and ONX 914.

FINANCIALS: Sales and profits are in thousands of dollars—add 000 to get the full amount. 2009 Note: Financial information for 2009 was not available for all companies at press time.
2009 Sales: $250,390	2009 Profits: $-16,161	U.S. Stock Ticker: ONXX
2008 Sales: $194,343	2008 Profits: $1,948	Int'l Ticker: Int'l Exchange:
2007 Sales: $90,429	2007 Profits: $-34,167	Employees: 271
2006 Sales: $29,524	2006 Profits: $-92,681	Fiscal Year Ends: 12/31
2005 Sales: $1,000	2005 Profits: $-95,174	Parent Company:

SALARIES/BENEFITS:
Pension Plan:	ESOP Stock Plan:	Profit Sharing:	Top Exec. Salary: $471,134	Bonus: $825,000
Savings Plan: Y	Stock Purch. Plan: Y		Second Exec. Salary: $425,000	Bonus: $187,425

OTHER THOUGHTS:
Apparent Women Officers or Directors: 4
Hot Spot for Advancement for Women/Minorities: Y

LOCATIONS: ("Y" = Yes)
West:	Southwest:	Midwest:	Southeast:	Northeast:	International:
Y					

ORASURE TECHNOLOGIES INC

www.orasure.com

Industry Group Code: 325413 Ranks within this company's industry group: Sales: 14 Profits: 14

Drugs:	Other:	Clinical:		Computers:		Services:	
Discovery:	AgriBio:	Trials/Services:		Hardware:		Specialty Services:	
Licensing:	Genetic Data:	Labs:		Software:		Consulting:	
Manufacturing:	Tissue Replacement:	Equipment/Supplies:	Y	Arrays:		Blood Collection:	
Generics:		Research/Development Svcs.:	Y	Database Management:		Drug Delivery:	
		Diagnostics:	Y			Drug Distribution:	

TYPES OF BUSINESS:

Medical Devices Manufacturing
Oral Fluid Collection Devices
Cryosurgical Products

BRANDS/DIVISIONS/AFFILIATES:

OraQuick
OraSure
Histofreezer
Intercept
Q.E.D. Saliva Alcohol Test
MICRO-PLATE
Western Blot HIV-1 Confirmatory Test
AUTO-LYTE

CONTACTS: Note: Officers with more than one job title may be intentionally listed here more than once.

Douglas A. Michels, CEO
Ronald H. Spair, COO
Douglas A. Michels, Pres.
Ronald H. Spair, CFO
Henry B. Cohen, Sr. VP-Human Resources
Stephen Lee, Chief Science Officer/Exec. VP
Jack E. Jerrett, General Counsel/Sr. VP/Sec.
Nancy J. McLane, Sr. VP-Oper.
Ron Ticho, VP-Corp. Comm.
Mark K. Luna, Sr. VP-Finance/Controller
P. Michael Formica, Exec. VP/Gen. Mgr.-Cryosurgical Systems Div.
Debra Y. Fraser-Howze, VP-Gov't & External Affairs
Robert A. Gregg, Sr. VP-Regulatory Affairs & Quality Assurance
Douglas Watson, Chmn.

Phone: 610-882-1820	Fax: 610-882-1830
Toll-Free: 800-869-3538	
Address: 220 E. First St., Bethlehem, PA 18015 US	

GROWTH PLANS/SPECIAL FEATURES:

OraSure Technologies, Inc. develops, manufactures, markets and sells oral fluid diagnostic products and specimen collection devices using proprietary oral fluid technologies. The company is also interested in additional diagnostic products including immunoassays and in-vitro diagnostic tests that are used on other specimen types and medical devices used for the removal of benign skin lesions. The company's diagnostic products include tests that are processed in a laboratory and on a rapid basis at the point of care. The firm's primary products include the OraQuick rapid HIV test, which tests oral fluid, whole blood, plasma and serum samples; the OraSure and Intercept oral fluid collection devices; the Histofreezer cryosurgical removal system; the MICRO-PLATE and AUTO-LYTE immunoassay tests; the Western blot HIV-1 confirmatory test; and the Q.E.D. Saliva Alcohol Test. OraSure is also developing an oral test for Hepatitis C; OTC HIV antibody tests; and fully automated drug testing using OraQuick, currently in preclinical and clinical trials. OraSure Technologies' products are sold in the U.S. and internationally to various clinical laboratories, hospitals, clinics, community-based organizations and other public health organizations, distributors, government agencies, physicians' offices and commercial and industrial entities, with international sales generating 19% of revenue in 2009. In June 2010, the company received FDA approval for its OraQuick Hepatitis C Rapid Antibody Test.

The company offers its employees health, vision and dental coverage; life insurance; short- and long-term disability; flexible spending accounts; a 401(k) plan; a stock award plan; service recognition; employee referrals; free will preparation; discount plans; and an employee assistance plan.

FINANCIALS: Sales and profits are in thousands of dollars—add 000 to get the full amount. 2009 Note: Financial information for 2009 was not available for all companies at press time.

2009 Sales: $77,026	2009 Profits: $-7,813	U.S. Stock Ticker: OSUR
2008 Sales: $71,104	2008 Profits: $-31,275	Int'l Ticker: Int'l Exchange:
2007 Sales: $82,686	2007 Profits: $2,473	Employees: 280
2006 Sales: $68,155	2006 Profits: $5,268	Fiscal Year Ends: 12/31
2005 Sales: $69,366	2005 Profits: $27,448	Parent Company:

SALARIES/BENEFITS:

Pension Plan:	ESOP Stock Plan:	Profit Sharing:	Top Exec. Salary: $475,500	Bonus: $303,500
Savings Plan: Y	Stock Purch. Plan:		Second Exec. Salary: $380,500	Bonus: $202,500

OTHER THOUGHTS:

Apparent Women Officers or Directors: 2
Hot Spot for Advancement for Women/Minorities:

LOCATIONS: ("Y" = Yes)

West:	Southwest:	Midwest:	Southeast:	Northeast:	International:
				Y	

ORCHID CELLMARK INC

www.orchid.com

Industry Group Code: 6215 Ranks within this company's industry group: Sales: 5   Profits: 4

Drugs:	Other:	Clinical:	Computers:	Services:
Discovery:	AgriBio:	Trials/Services:	Hardware:	Specialty Services: Y
Licensing:	Genetic Data: Y	Labs:	Software:	Consulting:
Manufacturing:	Tissue Replacement:	Equipment/Supplies:	Arrays:	Blood Collection:
Generics:		Research/Development Svcs.:	Database Management:	Drug Delivery:
		Diagnostics: Y		Drug Distribution:

TYPES OF BUSINESS:

Research-Bioinformatics
Genomics Services
Diagnostic Products & Services
Genetic Databases
DNA Testing
Forensic Testing

BRANDS/DIVISIONS/AFFILIATES:

Orchid Cellmark
STRs
SNPs
Strand Analytical Laboratories LLC

CONTACTS: Note: Officers with more than one job title may be intentionally listed here more than once.

Thomas A. Bologna, CEO
Thomas A. Bologna, Pres.
James F. Smith, CFO/VP
Jeffrey S. Boschwitz, VP-North America Sales & Mktg.
William J. Thomas, General Counsel/VP
James Beery, Chmn.

Phone: 609-750-2200	Fax: 609-750-6405
Toll-Free:	
Address: 4390 US Route 1, Princeton, NJ 08540 US	

GROWTH PLANS/SPECIAL FEATURES:

Orchid Cellmark, Inc. (Orchid) provides identity genomics services for the forensic and public health markets, as well as paternity DNA testing, animal DNA testing and non-DNA forensic laboratory services. Forensic DNA testing is primarily used to establish and maintain DNA profile databases of individuals arrested or convicted of crimes, to analyze and compare evidence from crime scenes or, in paternity cases, to determine if a man has fathered a particular child. In agricultural applications, DNA testing services are available for selective trait breeding. Technologies utilized by Orchid include short tandem repeats (STRs) for forensic, security and paternal testing and single nucleotide polymorphisms (SNPs) for DNA agricultural applications and in some forensic DNA testing applications. Agricultural projects currently pursued by the company include scrapie genotyping, a U.K. government project designed to reduce the animal disease, scrapie, on sheep farms. Through four laboratories in the U.S. and one in the U.K., the firm provides DNA testing services to various government agencies, private individuals and commercial companies. Orchid's services have been utilized by major police departments in the U.S. and London's Metropolitan Police Force (Scotland Yard). Orchid Cellmark owns or has exclusive licenses to over 100 patents (36 U.S., 72 international) and has applied for an additional four. Recently, the firm announced plans to consolidate its Michigan paternity testing operations into its Ohio facility, and its Tennessee forensic testing operations into its Texas facility. In April 2010, the company acquired the paternity and immigration DNA testing business of Strand Analytical Laboratories LLC.

The company offers employees medical, dental and vision insurance; disability, life and AD&D insurance; a 401(k) plan; an employee assistance program; a credit union; and flexible spending accounts.

FINANCIALS: Sales and profits are in thousands of dollars—add 000 to get the full amount. 2009 Note: Financial information for 2009 was not available for all companies at press time.

2009 Sales: $59,062	2009 Profits: $-1,542	**U.S. Stock Ticker: ORCH**
2008 Sales: $57,595	2008 Profits: $-4,481	**Int'l Ticker:** Int'l Exchange:
2007 Sales: $60,303	2007 Profits: $-2,967	Employees: 466
2006 Sales: $56,854	2006 Profits: $-11,271	Fiscal Year Ends: 12/31
2005 Sales: $61,609	2005 Profits: $-9,439	Parent Company:

SALARIES/BENEFITS:

Pension Plan:	ESOP Stock Plan:	Profit Sharing:	Top Exec. Salary: $565,000	Bonus: $
Savings Plan: Y	Stock Purch. Plan:		Second Exec. Salary: $247,100	Bonus: $

OTHER THOUGHTS:

Apparent Women Officers or Directors: 1
Hot Spot for Advancement for Women/Minorities:

LOCATIONS: ("Y" = Yes)

West:	Southwest:	Midwest:	Southeast:	Northeast:	International:
	Y	Y	Y	Y	Y

ORE PHARMACEUTICALS HOLDINGS INC

www.orepharma.com

Industry Group Code: 541712 Ranks within this company's industry group: Sales: 18 Profits: 10

Drugs:	Other:		Clinical:		Computers:		Services:	
Discovery:	AgriBio:		Trials/Services:		Hardware:		Specialty Services:	
Licensing:	Genetic Data:	Y	Labs:		Software:		Consulting:	
Manufacturing:	Tissue Replacement:		Equipment/Supplies:		Arrays:		Blood Collection:	
Generics:			Research/Development Svcs.:	Y	Database Management:	Y	Drug Delivery:	
			Diagnostics:				Drug Distribution:	

TYPES OF BUSINESS:

Drug Development
Drug Repositioning

BRANDS/DIVISIONS/AFFILIATES:

Ore Pharmaceuticals, Inc.
ORE1001
ORE10002
Tiapamil
Romazarit
p-Value Capital Management LLC

CONTACTS: *Note: Officers with more than one job title may be intentionally listed here more than once.*

Mark J. Gabrielson, CEO
Mark J. Gabrielson, Pres.
Benjamin Palleiko, CFO
Stephen Donahue, Sr. VP-Clinical Dev.
Benjamin Palleiko, Sec./Sr. VP
Geoffrey Wilson, Dir.-Strategy
Benjamin Palleiko, Treas.
J. Stark Thompson, Chmn.

Phone: 617-649-2001	Fax:
Toll-Free:	
Address: 1 Main St., Ste. 300, Cambridge, MA 02142 US	

GROWTH PLANS/SPECIAL FEATURES:

Ore Pharmaceutical Holdings, Inc. is a pharmaceutical asset management company operating principally through subsidiary Ore Pharmaceuticals, Inc. The company is currently focusing on the development and monetization of its pharmaceutical asset portfolio, which includes four compounds in-licensed from pharmaceutical companies. Ore applies its experience in integrative pharmacology to identify potential new uses for drug candidates that have failed clinical development for reasons other than safety. All of its drug candidates have undergone extensive preclinical safety testing and early-stage human clinical trials, and has been observed to be well-tolerated in clinical trials. ORE1001, its lead drug candidate, is being developed for the potential treatment of Inflammatory Bowel Disease (IBD). In early 2009, this candidate entered a Phase Ib/IIa clinical trial for ulcerative colitis patients. ORE1001 is also being evaluated for the treatment of radiation enteritis, a side effect of radiation therapy. Other drug candidates include ORE10002, being evaluated for treatment of several inflammation-based conditions; Tiapamil, being evaluated for the treatment of certain central nervous system diseases; and Romazarit, being evaluated for metabolic therapeutic properties. In October 2009, the firm created Ore Pharmaceutical Holdings as a holding company and reorganized its business around the development of its drug candidates. In April 2010, the company entered a Management Services Agreement with p-Value Capital Management LLC, a private pharmaceutical asset management firm.

FINANCIALS: Sales and profits are in thousands of dollars—add 000 to get the full amount. 2009 Note: Financial information for 2009 was not available for all companies at press time.

2009 Sales: $ 175	2009 Profits: $-8,383	U.S. Stock Ticker: ORXE	
2008 Sales: $1,950	2008 Profits: $-22,461	Int'l Ticker: Int'l Exchange:	
2007 Sales: $1,596	2007 Profits: $-34,688	Employees: 7	
2006 Sales: $24,346	2006 Profits: $-54,710	Fiscal Year Ends: 12/31	
2005 Sales: $57,190	2005 Profits: $-48,304	Parent Company:	

SALARIES/BENEFITS:

Pension Plan:	ESOP Stock Plan:	Profit Sharing:	Top Exec. Salary: $295,000	Bonus: $44,625
Savings Plan:	Stock Purch. Plan:		Second Exec. Salary: $166,667	Bonus: $

OTHER THOUGHTS:

Apparent Women Officers or Directors:
Hot Spot for Advancement for Women/Minorities:

LOCATIONS: ("Y" = Yes)

West:	Southwest:	Midwest:	Southeast:	Northeast:	International:
				Y	

ORGANOGENESIS INC

www.organogenesis.com

Industry Group Code: 325414 Ranks within this company's industry group: Sales: Profits:

Drugs:	Other:	Clinical:	Computers:	Services:
Discovery:	AgriBio:	Trials/Services:	Hardware:	Specialty Services:
Licensing:	Genetic Data:	Labs:	Software:	Consulting:
Manufacturing: Y	Tissue Replacement: Y	Equipment/Supplies: Y	Arrays:	Blood Collection:
Generics:		Research/Development Svcs.: Y	Database Management:	Drug Delivery:
		Diagnostics:		Drug Distribution:

TYPES OF BUSINESS:

Tissue Replacement Products
Wound Dressing Products

BRANDS/DIVISIONS/AFFILIATES:

Apligraf
BioSTAR
FortaPerm
FortaGen
CuffPatch
CelTx
NanoMatrix, Inc.

CONTACTS: Note: Officers with more than one job title may be intentionally listed here more than once.

Geoff MacKay, CEO
Gary S. Gillheeney, Sr., COO/Exec. VP
Geoff MacKay, Pres.
Gary S. Gillheeney, Sr., CFO/Exec. VP
Santino Costanzo, VP-Sales, Bio-active Wound Healing
Houda Samaha, Dir.-Human Resources
Vincent Ronfard, VP-Research
Phillip Nolan, VP-Mfg. Oper.
Richard Shaw, VP-Finance
Dario Eklund, VP-Bio-Surgery & Oral Regeneration
Patrick Bilbo, VP-Regulatory
Damien Bates, Chief Medical Officer

Phone: 781-575-0775	Fax: 781-575-0440
Toll-Free:	
Address: 150 Dan Rd., Canton, MA 02021 US	

GROWTH PLANS/SPECIAL FEATURES:

Organogenesis, Inc. is a regenerative medicine firm that designs, develops and manufactures products containing living cells or natural connective tissue. It specializes in bio-active wound healing, oral regeneration and bio-surgery. The firm's bio-active wound healing products include: Apligraf, which has been used by over 200,000 patients, is designed for the treatment of venous leg ulcers and diabetic foot ulcers; and VCT01, a bio-engineered skin substitute currently in late stage development. Organogenesis is testing Apligraf in clinical studies for the treatment of burns and epidermolysis bullosa, a rare genetic disease characterized by extremely fragile skin. CelTx, the company's oral soft tissue regeneration product, has completed Phase III testing. Bio-surgery products include: CuffPatch for rotator cuff surgery, which repairs the tendons that connect the upper arm bones to the shoulder blade; FortaGen, used for tissue repair in cases of vaginal prolapse, cystocele, rectocele and abdominal hernias; FortaPerm, used as a sling to raise a dropped bladder neck back into place, fixing stress urinary incontinence; and BioSTAR, an implant technology for the treatment of cardiac sources of migraine headaches, strokes and other potential brain attacks. Organogenesis is also researching the potential connection between a common heart defect called patent foramen ovale and the aforementioned brain attacks. The company's subsidiary NanoMatrix, Inc., is a firm that uses electrospinning to create designer scaffolds for the purposes of regenerative medicine. The firm is based in Canton, Massachusetts and has an office located in Switzerland. In January 2009, the firm received a $7.4 million grant from The Massachusetts Life Sciences Center; Organogenesis plans to use this funding to build a new medical research, development and manufacturing plant. The company plans to add 280 employees with this expansion.

FINANCIALS: Sales and profits are in thousands of dollars—add 000 to get the full amount. 2009 Note: Financial information for 2009 was not available for all companies at press time.

2009 Sales: $	2009 Profits: $	U.S. Stock Ticker: Private
2008 Sales: $	2008 Profits: $	Int'l Ticker: Int'l Exchange:
2007 Sales: $	2007 Profits: $	Employees:
2006 Sales: $	2006 Profits: $	Fiscal Year Ends: 12/31
2005 Sales: $	2005 Profits: $	Parent Company:

SALARIES/BENEFITS:

Pension Plan:	ESOP Stock Plan:	Profit Sharing:	Top Exec. Salary: $	Bonus: $
Savings Plan: Y	Stock Purch. Plan:		Second Exec. Salary: $	Bonus: $

OTHER THOUGHTS:

Apparent Women Officers or Directors: 1
Hot Spot for Advancement for Women/Minorities:

LOCATIONS: ("Y" = Yes)

West:	Southwest:	Midwest:	Southeast:	Northeast:	International:
				Y	Y

OSI PHARMACEUTICALS INC

www.osip.com

Industry Group Code: 325412 Ranks within this company's industry group: Sales: Profits:

Drugs:		Other:	Clinical:	Computers:	Services:
Discovery:	Y	AgriBio:	Trials/Services:	Hardware:	Specialty Services:
Licensing:		Genetic Data:	Labs:	Software:	Consulting:
Manufacturing:	Y	Tissue Replacement:	Equipment/Supplies:	Arrays:	Blood Collection:
Generics:			Research/Development Svcs.:	Database Management:	Drug Delivery:
			Diagnostics:		Drug Distribution:

TYPES OF BUSINESS:

Drugs-Cancer
Drugs-Small-Molecule
Drugs-Diabetes

BRANDS/DIVISIONS/AFFILIATES:

Tarceva
OSI Oncology
Astella US Holding Inc
PSN-821
OSI-906
OSI-027
OSI-930
PSN-602

CONTACTS: Note: Officers with more than one job title may be intentionally listed here more than once.

Colin Goddard, CEO
Pierre Legault, CFO/Exec. VP
Linda E. Amper, Sr. VP-Human Resources
Anker Lundemose, Exec. VP-Diabetes & Obesity R&D
Robert L. Simon, Exec. VP-Pharmaceutical Dev. & Mfg.
Barbara A. Wood, General Counsel/Sr. VP/Sec.
Kathy Galante, Sr. Dir.-Public Rel.
Kathy Galante, Sr. Dir.-Investor Rel.
Pierre Legault, Treas.
Gabriel Leung, Exec. VP/Pres., Pharmaceuticals Bus.
Robert A. Ingram, Chmn.
Anker Lundemose, Pres., OSIP U.K.

Phone: 631-962-2000	Fax: 631-752-3880
Toll-Free:	
Address: 41 Pinelawn Rd., Melville, NY 11747 US	

GROWTH PLANS/SPECIAL FEATURES:

OSI Pharmaceuticals, Inc. is a biotechnology company that discovers, develops and commercializes molecular targeted therapies addressing the areas of oncology, obesity and diabetes. The firm's largest area of focus is oncology where its leading product is Tarceva, a small-molecule inhibitor of the epidermal growth factor receptor, which plays a role in the abnormal growth of many cancer cells. The drug is an oral, once-a-day pharmaceutical with approved indication for advanced non-small cell lung cancer (NSCLC) and for locally advanced and metastatic pancreatic cancer. It has been approved for treatment for advanced NSCLC after failed chemotherapy treatments in 109 countries, and has been approved in 80 countries for the treatment of pancreatic cancer. Tarceva is also in trials for the treatment of other tumor types, including hepatocellular carcinoma and ovarian and colorectal cancers. The firm collaborates with Genentech, Inc. and Roche for the continued development and commercialization of Tarceva. The company's other oncology products are all in clinical or late-stage pre-clinical development. They include OSI-906, an oral small molecule IGF-1 receptor inhibitor, which shows potential in treating NSCLC, breast, pancreas, prostate and colorectal cancers; PSN821, a potential anti-diabetic and appetite suppressing drug; OSI-027 a small molecule TORC1/TORC2 inhibitor; and OSI-930, which is designed to target both cancer cell proliferation and blood vessel growth in selected tumors. The firm's OSI Prosidion business unit focuses on the remaining drug candidates, including diabetes and obesity products. PSN821, an agonist with potential anti-diabetic and appetite suppressing effects, is in clinical trials. In late 2009, OSI Pharmaceuticals discontinued clinical trials on PSN602, a novel dual serotonin and noradrenaline reuptake inhibitor for the long-term treatment of obesity. In June 2010, the firm was acquired by Japan-based Astellas Pharma, Inc. It is now a wholly-owned subsidiary of Astellas U.S. Holding, Inc.

FINANCIALS: Sales and profits are in thousands of dollars—add 000 to get the full amount. 2009 Note: Financial information for 2009 was not available for all companies at press time.

2009 Sales: $	2009 Profits: $	U.S. Stock Ticker: Subsidiary
2008 Sales: $379,388	2008 Profits: $471,485	Int'l Ticker: Int'l Exchange:
2007 Sales: $341,030	2007 Profits: $66,319	Employees: 514
2006 Sales: $241,037	2006 Profits: $-582,184	Fiscal Year Ends: 12/31
2005 Sales: $174,194	2005 Profits: $-157,123	Parent Company: ASTELLAS PHARMA INC

SALARIES/BENEFITS:

Pension Plan:	ESOP Stock Plan:	Profit Sharing:	Top Exec. Salary: $652,214	Bonus: $705,000
Savings Plan: Y	Stock Purch. Plan: Y		Second Exec. Salary: $438,300	Bonus: $235,000

OTHER THOUGHTS:

Apparent Women Officers or Directors: 4
Hot Spot for Advancement for Women/Minorities: Y

LOCATIONS: ("Y" = Yes)

West:	Southwest:	Midwest:	Southeast:	Northeast:	International:
Y				Y	Y

Note: Financial information, benefits and other data can change quickly and may vary from those stated here.

OXIGENE INC

www.oxigene.com

Industry Group Code: 325412 Ranks within this company's industry group: Sales: Profits: 127

Drugs:		Other:	Clinical:	Computers:	Services:
Discovery:		AgriBio:	Trials/Services:	Hardware:	Specialty Services:
Licensing:	Y	Genetic Data:	Labs:	Software:	Consulting:
Manufacturing:	Y	Tissue Replacement:	Equipment/Supplies:	Arrays:	Blood Collection:
Generics:			Research/Development Svcs.:	Database Management:	Drug Delivery:
			Diagnostics:		Drug Distribution:

TYPES OF BUSINESS:

Pharmaceuticals Acquisition & Development
Drugs-Cancer
Anti-Inflammatory Agents
Ocular Disease Treatments
Ophthalmology Therapies

BRANDS/DIVISIONS/AFFILIATES:

Zybrestat
OXi4503

CONTACTS: Note: Officers with more than one job title may be intentionally listed here more than once.

Peter J. Langecker, CEO
James B. Murphy, CFO/VP
Suman Sharma, Dir.-CMC Oper.
Jodie Crowley, Dir.-Human Resources
Dai Chaplin, Chief Scientific Officer/VP-R&D
Jacqueline Moore, Sr. Dir.-Clinical Oper.
Michelle Edwards, Contact-Investor Rel.
Zelanna Goldberg, Dir.-Medical
Jai Balissoon, VP/Dir.-Clinical Research & Clinical Oper.
William Shiebler, Chmn.

Phone: 650-635-7000	Fax: 650-635-7001
Toll-Free:	
Address: 710 Gateway Blvd., Ste 210, San Francisco, CA 94080 US	

GROWTH PLANS/SPECIAL FEATURES:

OXiGENE, Inc. is a clinical stage biopharmaceutical company that researches and develops products to treat cancer and certain ocular diseases. It in-licenses complementary compounds from academic institutions in order to lead them through early-stage clinical trials and negotiate contracts with pharmaceutical companies to develop, market and manufacture resulting commercial drugs and clinical products. OXiGENE's primary drug development programs are based on a series of natural products called Combretastatins, which were originally isolated from the African bush willow tree (Combretum caffrum) by researchers at Arizona State University (ASU). ASU has granted the company a worldwide license for the use of the Combretastatins. OXiGENE has developed its primary technologies based on Combretastatins, called vascular disrupting agents (VDAs). The company's VDA compound, ZYBRESTAT (fosbretabulin, formerly known as combretastatin A4 phosphate or CA4P), attacks certain solid tumors and other diseases by selectively destroying their characteristically abnormal blood vessels. ZYBRESTAT is currently in Phase III clinical testing for anaplastic thyroid cancer; near the end of Phase II testing for platinum resistant ovarian cancer; and in the middle of Phase II testing for non-small cell lung cancer. Additionally, the firm is developing OXi4503, which is a structural analog to fosbretabulin, but differs in that it also possesses direct cytotoxic activity. This drug is currently in Phase I testing for solid and hepatic tumors. The company is also developing ZYBRESTAT for topical use in ophthalmological diseases and conditions that are characterized by abnormal blood vessel growth within the eye that results in a loss vision. This delivery method is preferable to current treatments, which require direct injection into the eye.

OXiGENE offers its employees medical and dental coverage and flexible spending accounts.

FINANCIALS: Sales and profits are in thousands of dollars—add 000 to get the full amount. 2009 Note: Financial information for 2009 was not available for all companies at press time.

2009 Sales: $	2009 Profits: $-24,728	**U.S. Stock Ticker: OXGN**
2008 Sales: $ 12	2008 Profits: $-21,401	**Int'l Ticker:** Int'l Exchange:
2007 Sales: $ 12	2007 Profits: $-20,389	Employees: 42
2006 Sales: $	2006 Profits: $-15,457	Fiscal Year Ends: 12/31
2005 Sales: $ 1	2005 Profits: $-11,909	Parent Company:

SALARIES/BENEFITS:

Pension Plan:	ESOP Stock Plan:	Profit Sharing:	Top Exec. Salary: $282,220	Bonus: $
Savings Plan: Y	Stock Purch. Plan: Y		Second Exec. Salary: $245,000	Bonus: $

OTHER THOUGHTS:

Apparent Women Officers or Directors: 1
Hot Spot for Advancement for Women/Minorities:

LOCATIONS: ("Y" = Yes)

West:	Southwest:	Midwest:	Southeast:	Northeast:	International:
Y				Y	Y

PACIFIC BIOMARKERS INC

www.pacbio.com

Industry Group Code: 541712 Ranks within this company's industry group: Sales: 15 Profits: 6

Drugs:	Other:	Clinical:		Computers:		Services:	
Discovery:	AgriBio:	Trials/Services:	Y	Hardware:		Specialty Services:	
Licensing:	Genetic Data:	Labs:	Y	Software:		Consulting:	Y
Manufacturing:	Tissue Replacement:	Equipment/Supplies:		Arrays:		Blood Collection:	
Generics:		Research/Development Svcs.:	Y	Database Management:		Drug Delivery:	
		Diagnostics:	Y			Drug Distribution:	

TYPES OF BUSINESS:

Clinical Trials
Laboratory Services
Contract Research & Development
Diagnostic Tests
DNA Amplification Systems

BRANDS/DIVISIONS/AFFILIATES:

PBI Technology Inc
Pacific Biometrics, Inc.
Cholesterol Reference Method Laboratory Network

CONTACTS: *Note: Officers with more than one job title may be intentionally listed here more than once.*

Ronald R. Helm, CEO
Elizabeth Teng Leary, Chief Scientific Officer
Amar A. Sethi, VP-Science & Tech
Michael Murphy, Sr. VP-Oper.
Patrick Jackle, Dir.-Bus. Dev.
Kari Charbonnel, Investor Rel.
John Jensen, Controller/VP-Finance
Ken Waters, Dir.-Strategic Planning
Kristin Walsh, Mgr.-Oper.
Timothy Carlson, Dir.-Laboratory
Tonya Aggoune, Dir.-Project Svcs.
Ronald R. Helm, Chmn.

Phone: 206-298-0068	Fax: 206-298-9838
Toll-Free:	
Address: 220 W. Harrison St., Seattle, WA 98119 US	

GROWTH PLANS/SPECIAL FEATURES:

Pacific Biomarkers, Inc. (PBI) provides specialty laboratory and clinical research services to pharmaceutical, biotechnology and laboratory manufacturers. Laboratory services are offered primarily in support of clinical trials and diagnostic product development. In clinical trial support, tailored databases are customized for each clinical research study protocol while a data analyst facilitates adapted data management plans for all clients. Diagnostic product development services include the development and improvement of reagents and point-of-care devices for diagnostic companies. PBI's areas of specialty include cardiovascular disease, diabetes and bone and joint diseases. Services offered include the measurement of cardiovascular disease markers through lipoprotein components, cholesterol, triglycerides, phospholipids and apolipoproteins; testing for diabetes markers such as glucose, HbA1c, microalbumin and non-esterified fatty acids; measurements of hormone and biochemical markers such as pyridonolines, procollagens and osteocalcin; and bone-specific alkaline phosphatase and cartilage oligomeric matrix protein for osteoporosis, bone and cartilage metabolism. PBI's wholly-owned subsidiary, PBI Technology, Inc., develops and commercializes molecular diagnostic technologies, non-invasive diagnostic devices and early-stage drug candidates. PBI Technology also owns DNA-based proprietary technologies, processes and equipment, including a proprietary isothermal DNA amplification method (LIDA) and a genetic method for distinguishing live cells from dead cells (Cell Viability). The Pacific Biometrics Research Foundation, an affiliated non-profit organization, is certified by the U.S. Centers for Disease Control and Prevention (CDC) as part of the Cholesterol Reference Method Laboratory Network (CRMLN), enabling it to perform official testing related to cholesterol measurement. In 2009, the company changed its name from Pacific Biometrics, Inc. to Pacific Biomarkers, Inc.

Employees of the firm are offered medical, dental and vision plans, flexible spending accounts and two weeks paid vacation, two paid floating personal days, eight days paid holidays, competitive salary, life insurance, 100% paid parking, stock purchase plan and a 401k plan.

FINANCIALS: Sales and profits are in thousands of dollars—add 000 to get the full amount. 2009 Note: Financial information for 2009 was not available for all companies at press time.

2009 Sales: $10,881	2009 Profits: $1,236	U.S. Stock Ticker: PBME
2008 Sales: $8,265	2008 Profits: $- 571	Int'l Ticker: Int'l Exchange:
2007 Sales: $8,480	2007 Profits: $-1,213	Employees: 64
2006 Sales: $10,750	2006 Profits: $ 179	Fiscal Year Ends: 6/30
2005 Sales: $3,230	2005 Profits: $-2,993	Parent Company:

SALARIES/BENEFITS:

Pension Plan:	ESOP Stock Plan: Y	Profit Sharing:	Top Exec. Salary: $247,500	Bonus: $47,750
Savings Plan: Y	Stock Purch. Plan:		Second Exec. Salary: $158,129	Bonus: $30,245

OTHER THOUGHTS:

Apparent Women Officers or Directors: 3
Hot Spot for Advancement for Women/Minorities: Y

LOCATIONS: ("Y" = Yes)

West:	Southwest:	Midwest:	Southeast:	Northeast:	International:
Y					

PAIN THERAPEUTICS INC

Industry Group Code: 325412 Ranks within this company's industry group: Sales: 108   Profits: 84

Drugs:		Other:	Clinical:	Computers:	Services:	
Discovery:	Y	AgriBio:	Trials/Services:	Hardware:	Specialty Services:	
Licensing:	Y	Genetic Data:	Labs:	Software:	Consulting:	
Manufacturing:		Tissue Replacement:	Equipment/Supplies:	Arrays:	Blood Collection:	
Generics:			Research/Development Svcs.:	Database Management:	Drug Delivery:	Y
			Diagnostics:		Drug Distribution:	

TYPES OF BUSINESS:

Drugs, Opioids
Abuse-Resistant Drug Delivery
Drugs-Metastatic Melanoma
Drugs-Hemophilia

BRANDS/DIVISIONS/AFFILIATES:

Oxytrex
Remoxy
PTI-202
PTI-721
PTI-188

CONTACTS: Note: Officers with more than one job title may be intentionally listed here more than once.

Remi Barbier, CEO
Nadav Friedmann, COO
Remi Barbier, Pres.
Peter S. Roddy, CFO/VP
Grant L. Schoenhard, Chief Scientific Officer
George Ben Thornton, Sr. VP-Tech.
Roger Fu, VP-Pharmaceutical Dev.
Judy Ishida, Mgr.-Admin.
Michael J. O'Donnell, Sec.
Michael Zamloot, Sr. VP-Tech. Oper.
Annelies de Kater, VP-Nonclinical Dev.
Nadav Friedmann, Chief Medical Officer
Michael Marsman, VP-Regulatory Affairs
Remi Barbier, Chmn.

Phone: 650-624-8200	Fax: 650-624-8222
Toll-Free:	
Address: 2211 Bridgepointe Pkwy., Ste. 500, San Mateo, CA 94404 US	

GROWTH PLANS/SPECIAL FEATURES:

Pain Therapeutics, Inc. is a biopharmaceutical company that develops novel drugs, mainly for oncology (cancer) and severe pain, particularly opioids. The company's lead drug candidate is Remoxy, which is currently in Phase III trials. Remoxy is an abuse-deterrent long-acting oral version of oxycodone designed to foil abusers who attempt to use the drug recreationally and to prevent accidental overdose by releasing just a small amount of oxycodone over time. Pain Therapeutics has a collaboration agreement with King Pharmaceuticals, Inc. to develop and commercialize Remoxy and other abuse-resistant opioid painkillers. The firm has two additional abuse-resistant opioid painkillers, PTI-202 and PTI-721, which have completed Phase I clinical studies. PTI-188, a novel radio-labeled monoclonal antibody for metastatic melanoma (a rare but deadly form of skin cancer), has completed two Phase I clinical studies, the most recent of which was completed in early 2010. Pain Therapeutics obtained the rights to PTI-188 from the Albert Einstein College of Medicine. It is also working on a Factor IX replacement for hemophilia, which is in pre-clinical study. Pain Therapeutics was previously working on Oxytrex, another non-addictive oral opioid, but work on this project was discontinued to focus on other developmental biotech products. In recent years, the FDA declined to approve Remoxy, citing the ease of possible abuse of Remoxy's active ingredient.

FINANCIALS: Sales and profits are in thousands of dollars—add 000 to get the full amount. 2009 Note: Financial information for 2009 was not available for all companies at press time.

2009 Sales: $20,563	2009 Profits: $-3,467	U.S. Stock Ticker: PTIE
2008 Sales: $63,725	2008 Profits: $15,347	Int'l Ticker: Int'l Exchange:
2007 Sales: $65,984	2007 Profits: $20,305	Employees: 27
2006 Sales: $53,918	2006 Profits: $6,188	Fiscal Year Ends: 12/31
2005 Sales: $5,080	2005 Profits: $-30,670	Parent Company:

SALARIES/BENEFITS:

Pension Plan:	ESOP Stock Plan:	Profit Sharing:	Top Exec. Salary: $588,333	Bonus: $250,000
Savings Plan: Y	Stock Purch. Plan: Y		Second Exec. Salary: $450,000	Bonus: $225,000

OTHER THOUGHTS:

Apparent Women Officers or Directors: 2
Hot Spot for Advancement for Women/Minorities:

LOCATIONS: ("Y" = Yes)

West:	Southwest:	Midwest:	Southeast:	Northeast:	International:
Y					

PALATIN TECHNOLOGIES INC

www.palatin.com

Industry Group Code: 325412 Ranks within this company's industry group: Sales: 120 Profits: 88

Drugs:		Other:		Clinical:		Computers:		Services:	
Discovery:	Y	AgriBio:		Trials/Services:		Hardware:		Specialty Services:	
Licensing:		Genetic Data:	Y	Labs:		Software:		Consulting:	
Manufacturing:		Tissue Replacement:		Equipment/Supplies:		Arrays:		Blood Collection:	
Generics:				Research/Development Svcs.:		Database Management:		Drug Delivery:	
				Diagnostics:	Y			Drug Distribution:	

TYPES OF BUSINESS:

Drugs-Diversified
Sexual Dysfunction Drugs
Inflammation Drugs
Peptide Technology
Diagnostic Imaging Products
Obesity Treatments

BRANDS/DIVISIONS/AFFILIATES:

Bremelanotide
MIDAS
PL-6983
PL-3994

CONTACTS: Note: Officers with more than one job title may be intentionally listed here more than once.

Carl Spana, CEO
Carl Spana, Pres.
Stephen T. Wills, CFO
Trevor Hallam, Exec. VP-R&D
Stephen T. Wills, Sec.
Stephen T. Wills, Exec. VP-Oper.
John Prendergast, Chmn.

Phone: 609-495-2200	Fax: 609-495-2201
Toll-Free:	
Address: 4-C Cedar Brook Dr., Cranbury, NJ 08512 US	

GROWTH PLANS/SPECIAL FEATURES:

Palatin Technologies, Inc. is a development-stage biopharmaceutical company committed to the development of peptide, peptide mimetic and small molecule agonist compounds. Its primary focus is discovering and developing natriuretic peptide receptor systems and melanocortin (MC)-based therapeutics. Palatin's patented MIDAS (Metal Ion-induced Distinctive Array of Structures) platform allows the company to synthesize pharmaceuticals that mimic the activity of peptides. MIDAS can generate both receptor antagonists and agonists to either block or promote metabolic responses. Palatin has active development programs targeting melanocortin and natriuretic receptors, including development of proposed products for treatment of heart failure, sexual dysfunction, obesity, diabetes and metabolic syndrome. The company has four active drug development programs: bremelanotide, a peptide melanocortin receptor antagonist used for treatment of sexual dysfunction, including female sexual dysfunction (FSD) and erectile dysfunction (ED) in patients non-responsive to current therapies; PL-6983, a peptide melanocortin receptor antagonist for treatment of sexual dysfunction; PL-3994, a peptide mimetic natriuretic peptide receptor A (NPRA) antagonist for the treatment of heart failure; and melanocortin receptor-based compounds for the treatment of obesity, diabetes and related metabolic syndromes pursuant to an ongoing research collaboration and global license with AstraZeneca AB. In May 2010, Palatin received a patent allowance on seven patents from the U.S. Patent and Trademark Office.

Employees are offered medical, vision and dental insurance; life and disability insurance; a 401(k) savings plan; incentive stock option plans; educational assistance; and tuition reimbursement.

FINANCIALS: Sales and profits are in thousands of dollars—add 000 to get the full amount. 2009 Note: Financial information for 2009 was not available for all companies at press time.

2009 Sales: $11,352	2009 Profits: $-4,802	U.S. Stock Ticker: PTN	
2008 Sales: $11,483	2008 Profits: $-14,384	Int'l Ticker:	Int'l Exchange:
2007 Sales: $14,406	2007 Profits: $-27,752	Employees: 43	
2006 Sales: $19,749	2006 Profits: $-28,959	Fiscal Year Ends: 6/30	
2005 Sales: $17,957	2005 Profits: $-14,358	Parent Company:	

SALARIES/BENEFITS:

Pension Plan:	ESOP Stock Plan:	Profit Sharing:	Top Exec. Salary: $390,000	Bonus: $25,000
Savings Plan: Y	Stock Purch. Plan: Y		Second Exec. Salary: $321,000	Bonus: $25,000

OTHER THOUGHTS:

Apparent Women Officers or Directors:
Hot Spot for Advancement for Women/Minorities:

LOCATIONS: ("Y" = Yes)

West:	Southwest:	Midwest:	Southeast:	Northeast:	International:
				Y	

Note: Financial information, benefits and other data can change quickly and may vary from those stated here.

PAR PHARMACEUTICAL COMPANIES INC   www.parpharm.com

Industry Group Code: 325412 Ranks within this company's industry group: Sales: 41   Profits: 46

Drugs:		Other:	Clinical:	Computers:	Services:
Discovery:	Y	AgriBio:	Trials/Services:	Hardware:	Specialty Services:
Licensing:		Genetic Data:	Labs:	Software:	Consulting:
Manufacturing:	Y	Tissue Replacement:	Equipment/Supplies:	Arrays:	Blood Collection:
Generics:			Research/Development Svcs.:	Database Management:	Drug Delivery:
			Diagnostics:		Drug Distribution:

TYPES OF BUSINESS:

Drugs-Generic & Branded
Pharmaceutical Intermediates

BRANDS/DIVISIONS/AFFILIATES:

Pharmaceutical Resources, Inc.
Par Pharmaceutical, Inc.
Megace ES
Nascobal
Strativa Pharmaceuticals
Androgel

CONTACTS: *Note: Officers with more than one job title may be intentionally listed here more than once.*

Patrick G. LePore, CEO
Gerard A. Martino, COO/Exec. VP
Patrick G. LePore, Pres.
Michael A. Tropiano, CFO/Exec. VP
Stephen Montalto, Sr. VP-Human Resources
Thomas J. Haughey, Chief Admin. Officer
Thomas Haughey, General Counsel/Corp. Sec./Exec. VP
John A. Neczesny, VP-Corp. Dev.
Allison Wey, VP-Corp. Affairs
Allison Wey, VP-Investor Rel.
Paul V. Campanelli, Exec. VP/Pres., Par Pharmaceutical
John A. MacPhee, Exec. VP/Pres., Strativa Pharmaceuticals
Patrick G. LePore, Chmn.

Phone: 201-802-4000	Fax: 201-802-4600
Toll-Free:	
Address: 300 Tice Blvd., Woodcliff Lake, NJ 07677 US	

GROWTH PLANS/SPECIAL FEATURES:

Par Pharmaceutical Companies, Inc. (formerly Pharmaceutical Resources, Inc.) develops, manufactures and markets branded and generic pharmaceuticals through its principal wholly-owned subsidiary, Par Pharmaceutical, Inc. The firm operates in two segments: Par Pharmaceutical, the company's generic products division; and Strativa Pharmaceuticals, the firm's branded products division. Par Pharmaceutical's product line comprises prescription drugs consisting of approximately 50 product names (molecules), each with an associated Abbreviated New Drug Application (ANDA) approved by the U.S. Food and Drug Administration and approximately 175 SKUs (packaging sizes). Par Pharmaceutical's products are manufactured principally in solid oral dosage form (tablet, caplet and two-piece hard shell capsule). In addition, it also markets several oral suspension products and products in the semi-solid form of a cream. Some products are the result of license agreements with the branded drug's manufacturer, including generics of Glucophage and Glucovance through Bristol-Meyers Squibb Company; Flonase and Zantac through GlaxoSmithKline plc; and Toprol XL through AstraZeneca. Par markets its generic products primarily to wholesalers, drug store chains, supermarket chains, mass merchandisers, distributors, managed health care organizations, mail order accounts and government, principally through its internal staff. Strativa Pharmaceuticals markets three products: Megace ES, which is approved for the treatment of anorexia, cachexia or any unexplained, significant weight loss in patients with a diagnosis of AIDS; Nascobal Nasal Spray; a prescription vitamin B12 treatment indicated for maintenance of remission in certain pernicious anemia patients, as well as a supplement for a variety of B12 deficiencies; and Solvay Pharmaceuticals' Androgel, a testosterone 1% gel indicated for replacement therapy in males for conditions associated with a deficiency or absence of endogenous testosterone. In May 2010, Par Pharmaceutical entered into an exclusive licensing agreement with Glenmark Generics Inc., USA and Glenmark Generics Limited to market and distribute a generic version of Merck & Co. Inc.'s Zetia, a cholesterol modifying agent.

FINANCIALS: Sales and profits are in thousands of dollars—add 000 to get the full amount. 2009 Note: Financial information for 2009 was not available for all companies at press time.

2009 Sales: $1,193,159	2009 Profits: $76,928	**U.S. Stock Ticker: PRX**
2008 Sales: $578,115	2008 Profits: $-55,171	Int'l Ticker: Int'l Exchange:
2007 Sales: $769,666	2007 Profits: $49,898	Employees: 616
2006 Sales: $725,168	2006 Profits: $5,847	Fiscal Year Ends: 12/31
2005 Sales: $432,256	2005 Profits: $-15,309	Parent Company:

SALARIES/BENEFITS:

Pension Plan:	ESOP Stock Plan:	Profit Sharing:	Top Exec. Salary: $830,769	Bonus: $1,400,000
Savings Plan: Y	Stock Purch. Plan: Y		Second Exec. Salary: $372,372	Bonus: $350,000

OTHER THOUGHTS:

Apparent Women Officers or Directors: 1
Hot Spot for Advancement for Women/Minorities:

LOCATIONS: ("Y" = Yes)

West:	Southwest:	Midwest:	Southeast:	Northeast:	International:
				Y	

PAREXEL INTERNATIONAL CORP

www.parexel.com

Industry Group Code: 541712 Ranks within this company's industry group: Sales: 4   Profits: 5

Drugs:	Other:	Clinical:		Computers:		Services:	
Discovery:	AgriBio:	Trials/Services:	Y	Hardware:	Y	Specialty Services:	Y
Licensing:	Genetic Data:	Labs:		Software:	Y	Consulting:	Y
Manufacturing:	Tissue Replacement:	Equipment/Supplies:		Arrays:		Blood Collection:	
Generics:		Research/Development Svcs.:	Y	Database Management:	Y	Drug Delivery:	
		Diagnostics:				Drug Distribution:	

TYPES OF BUSINESS:

Clinical Trial & Data Management
Biostatistical Analysis & Reporting
Medical Communications Services
Clinical Pharmacology Services
Consulting Services

BRANDS/DIVISIONS/AFFILIATES:

Clinical Research Services
PAREXEL Consulting & Medical Communications Svcs.
Perceptive Informatics Inc
Expert Office (The)

CONTACTS: *Note: Officers with more than one job title may be intentionally listed here more than once.*

Josef H. von Rickenbach, CEO
Mark A. Goldberg, COO
James F. Winschel, Jr., CFO/Sr. VP
Ulf Schneider, Chief Admin. Officer/Sr. VP
Douglas A. Batt, General Counsel/Sr. VP/Corp. Sec.
Jill L. Baker, VP-Investor Rel.
Kurt A. Brykman, Pres., PAREXEL Consulting & Medical Comm. Svcs.
Steve Kent, Pres., Perceptive Informatics
Josef H. von Rickenbach, Chmn.

Phone: 781-487-9900	Fax: 781-768-5512
Toll-Free:	
Address: 195 West St., Waltham, MA 02451 US	

GROWTH PLANS/SPECIAL FEATURES:

PAREXEL International is a biopharmaceutical services company. The firm provides clinical research, medical communications services, consulting and informatics and advanced technology products and services to the international biotechnology, pharmaceutical and medical device industries. Operating in approximately 70 locations throughout 52 countries, PAREXEL has three business segments: Clinical Research Services (CRS); PAREXEL Consulting and Medical Communications Services (PCMS); and Perceptive Informatics, Inc. PAREXEL's core business, CRS, accounts for roughly 76.5% of its revenues. It provides clinical trials management and biostatistics; data management; clinical pharmacology; and related medical advisory, patient recruitment and investigator site services. PCMS (11.6% of revenues) provides technical expertise and advice for drug development, regulatory affairs and biopharmaceutical process consulting; offers product launch support, including market development, product development and targeted communications services; identifies alternatives and solutions regarding product development, registration and commercialization; and provides health policy consulting and strategic reimbursement services. Lastly, Perceptive Informatics (11.9% of revenues) provides information technology designed to improve clients' product development processes, including medical imaging services, integrated voice response products, CTMS, web-based portals, systems integration, and patient diary applications. In November 2009, PAREXEL introduced The Expert Office, a program that provides its clients with a forum of specialists in the areas of therapeutic, medical, regulatory and clinical operations. In February 2010, the firm opened a new early phase research facility in South Africa.

FINANCIALS: Sales and profits are in thousands of dollars—add 000 to get the full amount. 2009 Note: Financial information for 2009 was not available for all companies at press time.

2009 Sales: $1,050,755	2009 Profits: $39,307	U.S. Stock Ticker: PRXL
2008 Sales: $964,283	2008 Profits: $64,640	Int'l Ticker: Int'l Exchange:
2007 Sales: $741,955	2007 Profits: $37,289	Employees: 9,275
2006 Sales: $614,947	2006 Profits: $23,544	Fiscal Year Ends: 6/30
2005 Sales: $544,726	2005 Profits: $-35,177	Parent Company:

SALARIES/BENEFITS:

Pension Plan:	ESOP Stock Plan:	Profit Sharing:	Top Exec. Salary: $650,000	Bonus: $
Savings Plan:	Stock Purch. Plan:		Second Exec. Salary: $469,165	Bonus: $

OTHER THOUGHTS:

Apparent Women Officers or Directors: 2
Hot Spot for Advancement for Women/Minorities:

LOCATIONS: ("Y" = Yes)

West:	Southwest:	Midwest:	Southeast:	Northeast:	International:
Y				Y	Y

PDL BIOPHARMA

www.pdl.com

Industry Group Code: 325412 Ranks within this company's industry group: Sales: 56   Profits: 387

Drugs:		Other:	Clinical:	Computers:	Services:
Discovery:	Y	AgriBio:	Trials/Services:	Hardware:	Specialty Services:
Licensing:	Y	Genetic Data:	Labs:	Software:	Consulting:
Manufacturing:		Tissue Replacement:	Equipment/Supplies:	Arrays:	Blood Collection:
Generics:			Research/Development Svcs.:	Database Management:	Drug Delivery:
			Diagnostics:		Drug Distribution:

TYPES OF BUSINESS:

Pharmaceuticals Development
Drugs-Kidney Transplant Rejection
Humanized Monoclonal Antibodies
Oncology Drugs
Asthma Drugs
Autoimmune Disease Drugs

BRANDS/DIVISIONS/AFFILIATES:

Protein Design Labs, Inc.
Avastin
Herceptin
Xolair
Raptiva
Lucentis
Tysabri
Mylotarg

CONTACTS: Note: Officers with more than one job title may be intentionally listed here more than once.

John McLaughlin, CEO
John McLaughlin, Pres.
Christine Larson, CFO/VP
Christopher L. Stone, General Counsel/VP/Sec.
Karen J. Wilson, VP-Finance/Principal Acct. Officer
Frederick Frank, Lead Dir.

Phone: 775-832-8500	Fax: 775-832-8501
Toll-Free:	
Address: 932 Southwood Blvd., Incline Village, NV 89451 US	

GROWTH PLANS/SPECIAL FEATURES:

PDL BioPharma is a biopharmaceutical company focused on discovery and development of novel antibodies in oncology and immunologic diseases. PDL also receives royalties and other revenues through licensing agreements with numerous biotechnology and pharmaceutical companies based on its proprietary antibody-based platform. These licensing agreements have contributed to the development by its licensees of eight marketed products and cover several antibodies in clinical development. These products include Avastin, Herceptin, Xolair, Raptiva and Lucentis through licenses with Genentech, Inc.; Tysabri through a license with Elan Corporation, Plc; Mylotarg through a license with Wyeth Pharmaceuticals, Inc.; and Actemra/RoActemra through a license with Chugai Pharmaceutical Co., Ltd. PDL currently has several investigational compounds in clinical development for severe or life-threatening diseases, some in collaboration. PDL's Zenapax was developed for the prevention of kidney transplant rejection. The product is currently licensed to Hoffman-La Roche, Inc. for marketing in the U.S., Europe and other countries. The firm has also entered into licensing agreements for non-marketed products that are currently in Phase 3 clinical trials. These agreements include those with Eli Lilly and Company and Wyeth Pharmaceuticals, which have licensed antibodies for the treatment of Alzheimer's disease; and an agreement with Eli Lilly for teplizumab, which is being studied for the treatment of newly-diagnosed type 1 diabetes mellitus. In 2009, Genentech accounted for 71% of the firm's revenues; MedImmune, 13%; and Elan, 8%.

FINANCIALS: Sales and profits are in thousands of dollars—add 000 to get the full amount. 2009 Note: Financial information for 2009 was not available for all companies at press time.

2009 Sales: $318,184	2009 Profits: $189,660	U.S. Stock Ticker: PDLI
2008 Sales: $294,270	2008 Profits: $68,387	Int'l Ticker: Int'l Exchange:
2007 Sales: $224,913	2007 Profits: $-21,061	Employees: 9
2006 Sales: $187,343	2006 Profits: $-130,020	Fiscal Year Ends: 12/31
2005 Sales: $280,569	2005 Profits: $-166,577	Parent Company:

SALARIES/BENEFITS:

Pension Plan:	ESOP Stock Plan:	Profit Sharing:	Top Exec. Salary: $500,000	Bonus: $262,500
Savings Plan: Y	Stock Purch. Plan: Y		Second Exec. Salary: $350,000	Bonus: $148,750

OTHER THOUGHTS:

Apparent Women Officers or Directors: 3
Hot Spot for Advancement for Women/Minorities: Y

LOCATIONS: ("Y" = Yes)

West:	Southwest:	Midwest:	Southeast:	Northeast:	International:
Y					

PENWEST PHARMACEUTICALS CO

www.penw.com

Industry Group Code: 325412A **Ranks within this company's industry group:** Sales: 11   Profits: 5

Drugs:		Other:		Clinical:	Computers:	Services:	
Discovery:	Y	AgriBio:		Trials/Services:	Hardware:	Specialty Services:	
Licensing:	Y	Genetic Data:		Labs:	Software:	Consulting:	
Manufacturing:	Y	Tissue Replacement:		Equipment/Supplies:	Arrays:	Blood Collection:	
Generics:				Research/Development Svcs.:	Database Management:	Drug Delivery:	Y
				Diagnostics:		Drug Distribution:	Y

TYPES OF BUSINESS:
Drug Delivery Systems
Nervous System Disorders Drugs

BRANDS/DIVISIONS/AFFILIATES:
TIMERx
Geminex
SyncroDose
GastroDose
Opana ER
Procardia XL

CONTACTS: *Note: Officers with more than one job title may be intentionally listed here more than once.*
Jennifer L. Good, CEO
Jennifer L. Good, Pres.
Paul Hayes, Sr. VP-Strategic Mktg.
Amale Hawi, Sr. VP-Pharmaceutical Dev.
Anand R. Baichwal, Sr. VP-Licensing & Bus. Dev.
Frank P. Muscolo, Chief Acct. Officer/Controller
Thomas R. Sciascia, Chief Medical Officer/Sr. VP
Paula D. Buckley, VP-QA/QC
Paul E. Freiman, Chmn.

Phone: 877-736-9378	**Fax:** 845-878-3484
Toll-Free:	
Address: 39 Old Ridgebury Rd., Ste. 11, Danbury, CT 06810 US	

GROWTH PLANS/SPECIAL FEATURES:

Penwest Pharmaceuticals Co. develops pharmaceutical products based on proprietary drug delivery technologies with a focus on products that address disorders of the nervous system. The firm's products are based on oral drug delivery technologies. Penwest currently has four proprietary drug delivery technologies: TIMERx, a controlled-release technology; Geminex, a technology enabling drug release at two different rates; SyncroDose, a technology enabling controlled release at the appropriate site in the body; and the GastroDose system, a technology enabling drug delivery to the upper gastrointestinal tract. Opana ER is an extended release formulation of oxymorphone hydrochloride that the company developed with Endo Pharmaceuticals, Inc. using the TIMERx drug delivery technology. The drug is an oral extended release opioid analgesic approved for twice-a-day dosing in patients with moderate to severe pain requiring extended, continuous opioid treatment. Opana is currently being marketed by Endo in the U.S. and Valeant Pharmaceuticals in Canada, Australia and New Zealand. Other products that utilize Timerx technology include Procardia XL, Slofedipine and Cronodipin for Hypertension and Cystrin CR for Urinary Incontience. Products in clinical development include nalbuphine ER, a controlled release formulation of nalbuphine hydrochloride, for the treatment of pain (Penwest is currently seeking a collaborator to continue development of this product); and A0001, a second-generation coenzyme Q10 analog for respiratory chain diseases. The company owns 30 U.S. and 122 foreign patents. In February 2010, Penwest began a second Phrase IIa clinical trial of A0001 targeted for treatment of Friedreich's Ataxia and MELAS syndrome. In April 2010, the firm signed an agreement with Alvogen, Inc. to develop and commercialize several generic drugs using Penwest's TIMERx technology.

Employees are offered medical, dental and vision insurance; disability benefits; life and accident insurance; flexible spending accounts; education reimbursement; a 401(k) savings plan; and an employee stock purchase program.

FINANCIALS: Sales and profits are in thousands of dollars—add 000 to get the full amount. 2009 Note: Financial information for 2009 was not available for all companies at press time.

2009 Sales: $23,812	2009 Profits: $-1,500	**U.S. Stock Ticker:** PPCO
2008 Sales: $8,534	2008 Profits: $-26,734	**Int'l Ticker:** Int'l Exchange:
2007 Sales: $3,308	2007 Profits: $-34,465	**Employees:** 39
2006 Sales: $3,499	2006 Profits: $-31,312	**Fiscal Year Ends:** 12/31
2005 Sales: $6,213	2005 Profits: $-22,898	**Parent Company:**

SALARIES/BENEFITS:

Pension Plan:	ESOP Stock Plan:	Profit Sharing:	Top Exec. Salary: $398,130	Bonus: $139,000
Savings Plan: Y	Stock Purch. Plan: Y		Second Exec. Salary: $318,991	Bonus: $77,000

OTHER THOUGHTS:
Apparent Women Officers or Directors: 4
Hot Spot for Advancement for Women/Minorities: Y

LOCATIONS: ("Y" = Yes)

West:	Southwest:	Midwest:	Southeast:	Northeast:	International:
				Y	

PEREGRINE PHARMACEUTICALS INC

www.peregrineinc.com

Industry Group Code: 325412 Ranks within this company's industry group: Sales: 109   Profits: 108

Drugs:		Other:	Clinical:	Computers:		Services:
Discovery:	Y	AgriBio:	Trials/Services:	Hardware:		Specialty Services:
Licensing:		Genetic Data:	Labs:	Software:		Consulting:
Manufacturing:	Y	Tissue Replacement:	Equipment/Supplies:	Arrays:		Blood Collection:
Generics:			Research/Development Svcs.:	Y	Database Management:	Drug Delivery:
			Diagnostics:			Drug Distribution:

TYPES OF BUSINESS:
Cancer & Hepatitis C Drugs
Contract Manufacturing Services

BRANDS/DIVISIONS/AFFILIATES:
Avid Bioservices, Inc.
Cotara
Peregrine Beijing Pharmaceuticals Technology Dev.
Bavituximab

CONTACTS: *Note: Officers with more than one job title may be intentionally listed here more than once.*
Steven W. King, CEO
Truc Le, COO
Steven W. King, Pres.
Paul J. Lytle, CFO
Marvin R. Garovoy, Head-Clinical Science
Richard Richieri, Sr. VP-BioProcess Dev. & Mfg.
Christopher E. Eso, VP-Bus. Oper.
Mary J. Boyd, Head-Bus. Dev.
Amy Figueroa, Contact-Corp. Comm.
Amy Figueroa, Contact-Investor Rel.
Joseph Shan, Exec. Dir.-Clinical & Regulatory Affairs
Shelley P.M. Fussey, VP-Intellectual Property
John Quick, Head-Quality Systems

Phone: 714-508-6000	Fax: 714-838-5817
Toll-Free:	
Address: 14272 Franklin Ave., Tustin, CA 92780 US	

GROWTH PLANS/SPECIAL FEATURES:
Peregrine Pharmaceuticals, Inc. is a biopharmaceutical company that manufactures and develops monoclonal antibodies for the treatment of cancer and serious viral infections. The company's products fall under two technology platforms: Anti-phosphatidylserine (anti-PS) and tumor necrosis therapy (TNT). The anti-PS immunotherapeutics are monoclonal antibodies that target cell components normally found only on the inner surface of the cell membrane but which become external in cells that line tumor blood vessels. The TNT technology uses monoclonal antibodies that target and bind to dead and dying cells at the core of solid tumors. The firm's lead anti-PS product, bavituximab, is in separate clinical trials for the treatment of solid cancers and hepatitis C viral infection in patients co-infected with HIV. A compound that combines bavituximab with docetal is also in trials for the treatment of advanced breast cancer; another compound, combining bavituximab with carboplatin and paclitaxel, are in trials to treat breast cancer and non-small cell lung cancer (NSCLC). Under the TNT platform, the lead product candidate is Cotara, for treatment of brain cancer. Additionally, the firm is pursuing two preclinical stage technologies: Anti-Angiogenesis Agents, which inhibit Vascular Endothelial Growth Factor (VEGF) to halt formation of blood vessels in tumors; and Vascular Targeting Agents, which bind to tumor blood vessels to inhibit blood flow. Peregrine operates a wholly-owned contact manufacturing subsidiary, Avid Bioservices, Inc. Avid provides several functions for Peregrine such as antibody and protein manufacture, cell culture development, process development and biologics testing. Avid also provides contract manufacturing services for outside companies on a fee-for-service basis. The firm has a wholly-owned subsidiary in China called Peregrine Beijing Pharmaceuticals Technology Development, Ltd. In July 2009, Peregrine licensed development and commercialization rights for its anti-VEGF antibodies to Affitech A/S. In May 2010, the firm licensed development rights for its TNT technologies to Stason Pharmaceuticals.

FINANCIALS: Sales and profits are in thousands of dollars—add 000 to get the full amount. 2009 Note: Financial information for 2009 was not available for all companies at press time.

2009 Sales: $18,151	2009 Profits: $-16,524	U.S. Stock Ticker: PPHM
2008 Sales: $6,093	2008 Profits: $-23,176	Int'l Ticker: Int'l Exchange:
2007 Sales: $3,708	2007 Profits: $-20,796	Employees: 138
2006 Sales: $3,193	2006 Profits: $-17,061	Fiscal Year Ends: 4/30
2005 Sales: $4,959	2005 Profits: $-15,452	Parent Company:

SALARIES/BENEFITS:
Pension Plan:	ESOP Stock Plan:	Profit Sharing:	Top Exec. Salary: $407,557	Bonus: $
Savings Plan:	Stock Purch. Plan:		Second Exec. Salary: $318,198	Bonus: $

OTHER THOUGHTS:
Apparent Women Officers or Directors: 3
Hot Spot for Advancement for Women/Minorities: Y

LOCATIONS: ("Y" = Yes)
West:	Southwest:	Midwest:	Southeast:	Northeast:	International:
Y					

PERRIGO CO

www.perrigo.com

Industry Group Code: 325412 Ranks within this company's industry group: Sales: 33 Profits: 39

Drugs:	Other:	Clinical:	Computers:	Services:	
Discovery:	AgriBio:	Trials/Services:	Hardware:	Specialty Services:	Y
Licensing:	Genetic Data:	Labs:	Software:	Consulting:	
Manufacturing: Y	Tissue Replacement:	Equipment/Supplies:	Arrays:	Blood Collection:	
Generics:		Research/Development Svcs.:	Database Management:	Drug Delivery:	
		Diagnostics:		Drug Distribution:	Y

TYPES OF BUSINESS:

Generic Prescription Drugs
Over-the-Counter Pharmaceuticals
Nutritional Products
Active Pharmaceutical Ingredients
Consumer Products

BRANDS/DIVISIONS/AFFILIATES:

Perrigo U.K.
Wrafton Laboratories, Ltd.
Perrigo Co. of South Carolina
Perrigo New York, Inc.
Perrigo Israel Pharmaceuticals, Ltd.
Orion Laboratories Pty, Ltd.
Quimica Y Farmacia S.A. de C.V.
PBM Holdings, Inc.

CONTACTS: *Note: Officers with more than one job title may be intentionally listed here more than once.*

Joseph C. Papa, CEO
Joseph C. Papa, Pres.
Judy L. Brown, CFO/Exec. VP
James Tomshack, Sr. VP-Consumer Healthcare Sales
Michael Stewart, Sr. VP-Global Human Resources
Jatin Shah, Chief Scientific Officer/Sr. VP
Thomas M. Farrington, CIO
Todd W. Kingma, General Counsel/Exec. VP/Sec.
John T. Hendrickson, Exec. VP-Global Oper.
Jeffrey R. Needham, Sr. VP-Global Bus. Dev.
Arthur J. Shannon, VP-Corp. Comm.
Arthur J. Shannon, VP-Investor Rel.
Sharon Kochan, Exec. VP-U.S. Generics
Louis Yu, Sr. VP-Global Quality & Compliance
David T. Gibbons, Chmn.
Refael Lebel, Exec. VP/Gen. Mgr.-Perrigo Israel
John T. Hendrickson, Exec. VP-Supply Chain

Phone: 269-673-8451	Fax: 269-673-7534
Toll-Free:	
Address: 515 Eastern Ave., Allegan, MI 49010 US	

GROWTH PLANS/SPECIAL FEATURES:

Perrigo Co. is a global healthcare supplier and one of the world's largest manufacturers of over-the-counter pharmaceutical and nutritional products for the store brand market. The company also develops and manufactures generic prescription drugs, active pharmaceutical ingredients (API) and consumer products. The firm operates through three segments: consumer healthcare, prescription pharmaceuticals and API. The consumer healthcare segment makes a broad line of products including analgesics, cough/cold/allergy/sinus, gastrointestinal, smoking cessation, first aid, vitamin and nutritional supplements. The pharmaceuticals segment's primary activity is the development, manufacture and sale of generic prescription drug products, generally for the U.S. market. The company currently markets roughly 250 generic prescription products to approximately 110 customers. The API segment, through subsidiary Chemagis, develops, manufactures and markets API for the drug industry and branded pharmaceutical companies. Perrigo's operations also include the Israel pharmaceutical and diagnostic products segment, which includes the marketing and manufacturing of branded prescription drugs under long-term exclusive licenses and the importation of pharmaceutical, diagnostics and other medical products into Israel. Perrigo operates through several wholly-owned subsidiaries. In the U.S., these subsidiaries consist primarily of L. Perrigo Co.; Perrigo Co. of South Carolina; and Perrigo New York, Inc. International subsidiaries include, Perrigo Israel Pharmaceuticals, Ltd.; Quimica Y Farmacia S.A. de C.V.; Wrafton Laboratories, Ltd.; and Perrigo U.K., Ltd. In November 2009, the firm divested interests in its Israel Consumer Products business, selling it and related production assets to Emilia Group, subsidiary of O. Feller Holdings, Ltd. In March 2010, the company acquired infant formula maker, PBM Holdings, Inc. and Orion Laboratories Pty, Ltd., an Australian store brand OTC supplier. In addition, the company broadened its generic offerings with several new launches and exclusive acquisitions. Newly offered products include, Ciclopirox shampoo, Fexofenadine HCl, Pseudophedrine, Ketotifen Fumarate eye solution, Miconazole Nitrate and several others.

FINANCIALS: Sales and profits are in thousands of dollars—add 000 to get the full amount. 2009 Note: Financial information for 2009 was not available for all companies at press time.

2009 Sales: $2,006,862	2009 Profits: $144,049	U.S. Stock Ticker: PRGO
2008 Sales: $1,729,921	2008 Profits: $135,773	Int'l Ticker: Int'l Exchange:
2007 Sales: $1,368,351	2007 Profits: $73,797	Employees: 7,250
2006 Sales: $1,366,821	2006 Profits: $71,400	Fiscal Year Ends: 6/30
2005 Sales: $1,024,098	2005 Profits: $-325,983	Parent Company:

SALARIES/BENEFITS:

Pension Plan:	ESOP Stock Plan:	Profit Sharing:	Top Exec. Salary: $875,000	Bonus: $904,875
Savings Plan:	Stock Purch. Plan:		Second Exec. Salary: $449,023	Bonus: $237,084

OTHER THOUGHTS:

Apparent Women Officers or Directors: 3
Hot Spot for Advancement for Women/Minorities: Y

LOCATIONS: ("Y" = Yes)

West:	Southwest:	Midwest:	Southeast:	Northeast:	International:
				Y	Y

Note: Financial information, benefits and other data can change quickly and may vary from those stated here.

PFIZER INC

www.pfizer.com

Industry Group Code: 325412 Ranks within this company's industry group: Sales: 2   Profits: 4

Drugs:		Other:	Clinical:	Computers:	Services:
Discovery:	Y	AgriBio:	Trials/Services:	Hardware:	Specialty Services:
Licensing:		Genetic Data:	Labs:	Software:	Consulting:
Manufacturing:	Y	Tissue Replacement:	Equipment/Supplies:	Arrays:	Blood Collection:
Generics:			Research/Development Svcs.:	Database Management:	Drug Delivery:
			Diagnostics:		Drug Distribution:

TYPES OF BUSINESS:

Pharmaceutical Drugs
Prescription Pharmaceuticals
Veterinary Pharmaceuticals

BRANDS/DIVISIONS/AFFILIATES:

Norvasc
Viagra
Zoloft
Revolution/Stronghold
Sutent
Lipitor
Chantix
Rimadyl

CONTACTS: Note: Officers with more than one job title may be intentionally listed here more than once.

Jeffrey B. Kindler, CEO
Frank D'Amelio, CFO/Sr. VP
Mary S. McLeod, Sr. VP-Worldwide Human Resources
Martin Mackay, Pres., Pfizer Global R&D
Natale Ricciardi, Pres./Team Leader-Pfizer Global Mfg.
Amy Schulman, General Counsel/Corp. Sec./Sr. VP
William Ringo, Sr. VP-Strategy & Bus. Dev.
Sally Susman, Sr. VP-Policy, External Affairs & Comm.
Freda C. Lewis-Hall, Chief Medical Officer/Sr. VP
Mikael Dolsten, Pres., BioTherapeutics R&D/Sr. VP
Ian Read, Sr. VP/Pres., Worldwide Pharmaceutical Oper.
Jeffrey B. Kindler, Chmn.

Phone: 212-733-2323	Fax: 212-573-7851
Toll-Free:	
Address: 235 E. 42nd St., New York, NY 10017 US	

GROWTH PLANS/SPECIAL FEATURES:

Pfizer, Inc. is a research-based, global pharmaceutical company. It discovers, develops, manufactures and markets prescription medicines for humans and animals. The company operates in two segments: Biopharmaceutical and Diversified. The Biopharmaceutical segment includes four global units: Primary Care; Specialty Care; Established Products; and Emerging Markets and Oncology. This segment includes products that prevent and treat cardiovascular and metabolic diseases, central nervous system disorders, arthritis and pain, infectious and respiratory diseases, urogenital conditions, cancer, eye diseases and endocrine disorders. Some of the Biopharmaceutical products include Lipitor, Chantix/Champix, Lyrica, Aricept, Celebrex and Viagra. The Diversified segment includes four global units: Animal Health, Consumer Healthcare, Nutrition and Capsugel. The Animal Health unit deals with products that prevent and treat diseases in livestock and companion animals. The Consumer Healthcare unit deals with products that include over-the-counter health care products such as pain management therapies, cough/cold/allergy remedies, dietary supplements, hemorrhoidal care and personal care items. The Nutrition unit deals with products such as infant and toddler formula. The Capsugel unit deals with gelatin capsules. Some of the Diversified products include Cerenia, Excede, Zulvac, Centrum, Advil, ChapStick and Preparation H. Research and development is conducted internally and through contracts with third parties, through collaborations with universities and biotechnology companies and in cooperation with other pharmaceutical firms. In October 2009, the company completed its acquisition of Wyeth. In April 2010, Pfizer announced a collaboration and research licensing agreement with Stemgent, Inc.

Employees are offered medical, dental, and vision insurance; prescription drug benefits; a health care account; a retirement plan; a 401(k) plan; a group legal plan; a dependent care account; short-term and long-term disability; life insurance; adoption reimbursements; educational assistance programs; and a work/life program.

FINANCIALS: Sales and profits are in thousands of dollars—add 000 to get the full amount. 2009 Note: Financial information for 2009 was not available for all companies at press time.

2009 Sales: $50,009,000	2009 Profits: $8,644,000	**U.S. Stock Ticker:** PFE
2008 Sales: $48,296,000	2008 Profits: $8,104,000	**Int'l Ticker:** Int'l Exchange:
2007 Sales: $48,418,000	2007 Profits: $8,144,000	Employees: 116,500
2006 Sales: $48,371,000	2006 Profits: $19,337,000	Fiscal Year Ends: 12/31
2005 Sales: $47,405,000	2005 Profits: $8,085,000	Parent Company:

SALARIES/BENEFITS:

Pension Plan: Y	ESOP Stock Plan:	Profit Sharing:	Top Exec. Salary: $1,600,000	Bonus: $3,500,000
Savings Plan: Y	Stock Purch. Plan:		Second Exec. Salary: $1,139,500	Bonus: $2,157,000

OTHER THOUGHTS:

Apparent Women Officers or Directors: 6
Hot Spot for Advancement for Women/Minorities: Y

LOCATIONS: ("Y" = Yes)

West:	Southwest:	Midwest:	Southeast:	Northeast:	International:
Y	Y	Y	Y	Y	Y

PHARMACEUTICAL PRODUCT DEVELOPMENT INC www.ppdi.com

Industry Group Code: 541712 Ranks within this company's industry group: Sales: 2 Profits: 2

Drugs:	Other:	Clinical:	Computers:	Services:	
Discovery:	AgriBio:	Trials/Services:	Hardware:	Specialty Services:	Y
Licensing:	Genetic Data:	Labs:	Software: Y	Consulting:	Y
Manufacturing:	Tissue Replacement:	Equipment/Supplies:	Arrays:	Blood Collection:	
Generics:		Research/Development Svcs.: Y	Database Management:	Drug Delivery:	
		Diagnostics:		Drug Distribution:	

TYPES OF BUSINESS:

Contract Research
Drug Discovery & Development Services
Clinical Data Consulting Services
Medical Marketing & Information Support Services
Drug Discovery Services
Medical Device Development

BRANDS/DIVISIONS/AFFILIATES:

AbC.R.O., Inc.
Magen BioSciences, Inc.
Excel PharmaStudies, Inc.
BioDuro LLC

CONTACTS: Note: Officers with more than one job title may be intentionally listed here more than once.

David L. Grange, CEO
William J. Sharbaugh, COO
Daniel G. Darazsdi, CFO
Christine A. Dingivan, Chief Medical Officer/Exec. VP
B. Judd Hartman, General Counsel/Corp. Sec.
William W. Richardson, Sr. VP-Global Bus. Dev.
Louise Caudle, Dir.-Corp. Comm.
Luke Heagle, Dir.-Investor Rel.
Daniel G. Darazsdi, Treas./Assistant Sec.
Lee E. Babiss, Exec. VP-Global Laboratory Svcs.
Michael O. Wilkinson, Exec. VP-Global Clinical Dev.
Fred N. Eshelman, Chmn.

Phone: 910-251-0081	Fax: 910-762-5820
Toll-Free:	
Address: 929 N. Front St., Wilmington, NC 28401-3331 US	

GROWTH PLANS/SPECIAL FEATURES:

Pharmaceutical Product Development, Inc. (PPD) provides drug discovery and development services to pharmaceutical, biotechnology, medical device, academic and government organizations. PPD's services are divided into two segments: Discovery Sciences and Development. Through the combined services of these segments, PPD helps pharmaceutical companies through all stages of clinical testing. The stages of testing can be specifically divided into discovery and preclinical, phase I, phase II-IIIb and post-approval. In the discovery and preclinical stages of drug testing, PPD helps clients accelerate drug discovery and provides information concerning the pharmaceutical composition of a new drug, its safety, its formulaic design and how it will be administered. During phase I testing, PPD conducts healthy volunteer clinics, provides data and project management services and offers regulatory affairs assistance. In phase II and III tests, PPD oversees the later stages of product development and government approval, providing project management, clinical quality assurance, patient recruitment, medical device development and clinical monitoring. In the post-approval stage, PPD provides technology, approval and marketing services intended to maximize the new drug's lifecycle. PPD has experience conducting research and drug development in the areas of central nervous system diseases, cardiovascular diseases, critical care studies, dermatological disorders, endocrine/metabolic studies, infectious disease treatment, hematology/oncology studies and pulmonary/allergy treatment, among others. The firm conducts its operations through offices in 40 countries, and recently opened labs and offices in the Philippines, India, Singapore, Japan, Ireland and North Carolina. During 2009, the firm acquired AbC.R.O., Inc., a contract research organization; divested subsidiary Piedmont Research Center LLC, a preclinical research provider for anticancer therapies; acquired Magen BioSciences, Inc., a biotechnology company; announced plans to spin off its compound partnering business; agreed to invest in Celtic Therapeutics Holdings L.P.; acquired Excel PharmaStudies, Inc., a China-based contract research organizations; and acquired BioDuro LLC, a drug discovery outsourcing company.

FINANCIALS: Sales and profits are in thousands of dollars—add 000 to get the full amount. 2009 Note: Financial information for 2009 was not available for all companies at press time.

2009 Sales: $1,416,770	2009 Profits: $159,295	U.S. Stock Ticker: PPDI
2008 Sales: $1,551,384	2008 Profits: $187,519	Int'l Ticker: Int'l Exchange:
2007 Sales: $1,414,465	2007 Profits: $163,401	Employees: 10,860
2006 Sales: $1,247,682	2006 Profits: $156,652	Fiscal Year Ends: 12/31
2005 Sales: $1,037,090	2005 Profits: $119,897	Parent Company:

SALARIES/BENEFITS:

Pension Plan:	ESOP Stock Plan:	Profit Sharing:	Top Exec. Salary: $740,000	Bonus: $300,000
Savings Plan:	Stock Purch. Plan:		Second Exec. Salary: $380,519	Bonus: $150,000

OTHER THOUGHTS:

Apparent Women Officers or Directors: 3
Hot Spot for Advancement for Women/Minorities: Y

LOCATIONS: ("Y" = Yes)

West:	Southwest:	Midwest:	Southeast:	Northeast:	International:
Y	Y	Y		Y	Y

PHARMACYCLICS INC

www.pharmacyclics.com

Industry Group Code: 325412 **Ranks within this company's industry group:** Sales: Profits: 122

Drugs:		Other:	Clinical:	Computers:	Services:
Discovery:	Y	AgriBio:	Trials/Services:	Hardware:	Specialty Services:
Licensing:	Y	Genetic Data:	Labs:	Software:	Consulting:
Manufacturing:		Tissue Replacement:	Equipment/Supplies:	Arrays:	Blood Collection:
Generics:			Research/Development Svcs.:	Database Management:	Drug Delivery:
			Diagnostics:		Drug Distribution:

TYPES OF BUSINESS:

Pharmaceuticals Development
Drugs-Cancer, Cardiovascular Diseases & Autoimmune Diseases
Texaphyrins

BRANDS/DIVISIONS/AFFILIATES:

Xcytrin
PCI-24781
PCI-24783
PCI-32765
PCI-34051

CONTACTS: *Note: Officers with more than one job title may be intentionally listed here more than once.*

Robert W. Duggan, CEO
Maky Zanganeh, VP-Mktg.
David Loury, Chief Scientific Officer
Ramses Erdtmann, VP-Admin.
Maky Zanganeh, VP-Bus. Dev.
Ramses Erdtmann, VP-Finance
Ahmed Hamdy, Chief Medical Officer
Gregory Hemmi, VP-Chemical Oper.
Joseph J. Buggy, VP-Research
Robert W. Duggan, Chmn.

Phone: 408-774-0330	Fax: 408-774-0340
Toll-Free:	
Address: 995 E. Arques Ave., Sunnyvale, CA 94085-4521 US	

GROWTH PLANS/SPECIAL FEATURES:

Pharmacyclics, Inc. is a clinical-stage biopharmaceutical company focused on developing and commercializing innovative treatments for cancer and other diseases. The company is working on a range of clinical and laboratory programs and is pursuing the development of novel drugs aimed at specific pathways. Pharmacyclics has four product candidates in clinical development and two product candidates in pre-clinical development. The clinical development products include PCI-24781, PCI-24783, PCI-32765 and MGd. PCI-24781 is a histone deacetylase inhibitor that is used for advanced solid tumors, recurrent lymphomas and sarcoma. It is about to enter a Phase II clinical trial. PCI-24783 is an inhibitor of Factor VIIa and is used for cancer therapy. It is soon to be in a Phase I/II clinical trial. PCI-32765 is an inhibitor of Bruton's tyrosine kinase (Btk) and is used for B-Cell lymphomas, autoimmune diseases and Mast cell diseases. It is currently in a Phase I clinical trial targeting oncology applications. MGd (Motexafin Gadolinium) is a radiation and chemotherapy sensitizing agent that is used for primary brain tumors and childhood brain tumors. It has completed enrollment for a Phase II clinical trial. The pre-clinical development products include a series of Btk inhibitors in advanced preclinical lead optimization and testing targeting autoimmune and allergic indications; and HDAC8 inhibitors (i.e., PCI-34051 and others) that are currently being optimized for autoimmune and cancer indications. Pharmacyclics owns or licenses rights to 66 issued U.S. patents and 28 other pending U.S. applications. It also owns or licenses 70 issued foreign patents, six Patent Cooperation Treaty (PCT) patent applications and more than 81 pending non-U.S. patent applications filed with the European Patent Office and nationally in Canada, Japan, China, Australia and other countries.

FINANCIALS: Sales and profits are in thousands of dollars—add 000 to get the full amount. 2009 Note: Financial information for 2009 was not available for all companies at press time.

2009 Sales: $	2009 Profits: $-23,447	U.S. Stock Ticker: PCYC
2008 Sales: $	2008 Profits: $-24,298	Int'l Ticker: Int'l Exchange:
2007 Sales: $ 126	2007 Profits: $-26,217	Employees: 46
2006 Sales: $ 181	2006 Profits: $-42,158	Fiscal Year Ends: 6/30
2005 Sales: $	2005 Profits: $-31,048	Parent Company:

SALARIES/BENEFITS:

Pension Plan:	ESOP Stock Plan:	Profit Sharing:	Top Exec. Salary: $265,215	Bonus: $
Savings Plan: Y	Stock Purch. Plan: Y		Second Exec. Salary: $192,115	Bonus: $

OTHER THOUGHTS:

Apparent Women Officers or Directors: 2
Hot Spot for Advancement for Women/Minorities:

LOCATIONS: ("Y" = Yes)

West:	Southwest:	Midwest:	Southeast:	Northeast:	International:
Y					

PHARMANET DEVELOPMENT GROUP INC   www.pharmanet.com

Industry Group Code: 541712 Ranks within this company's industry group: Sales: Profits:

Drugs:	Other:	Clinical:		Computers:		Services:	
Discovery:	AgriBio:	Trials/Services:	Y	Hardware:		Specialty Services:	
Licensing:	Genetic Data:	Labs:	Y	Software:	Y	Consulting:	
Manufacturing:	Tissue Replacement:	Equipment/Supplies:		Arrays:		Blood Collection:	
Generics:		Research/Development Svcs.:	Y	Database Management:		Drug Delivery:	
		Diagnostics:				Drug Distribution:	

TYPES OF BUSINESS:

Contract Research
Clinical Trial Services
Bioanalytical Laboratory Services
Clinical Trial Software
Consulting Services
Data Management Tools

BRANDS/DIVISIONS/AFFILIATES:

JLL Holdings
Taylor Technology
Anapharm, Inc.
PharmaSoft
Azopharma Product Development Group LLC

CONTACTS: Note: Officers with more than one job title may be intentionally listed here more than once.

Jeffry P. McMullen, CEO
Jeffry P. McMullen, Pres.
George McMillan, CFO/Exec. VP
Anne-Marie Hess, VP-Mktg. & Corp. Comm.
Robin C. Sheldrick, Sr. VP-Human Resources
Steven Kasay, CIO
Christopher S. Brennan, General Counsel
Ian B. Holmes, Sr. VP-Corp. Dev.
Diane Mitchell, Mgr.-Corp. Comm.
Pablo Fernandez, Sr. VP-Medical Affairs
Robert Reekie, COO-Late Stage Dev.
Dalvir Gill, Pres., Late Stage Dev.
Thomas J. Newman, Pres., PharmaNet Dev. Group
Gregory Skalicky, Sr. VP-Worldwide Bus. Dev.

Phone: 609-951-6800	Fax: 609-514-0390
Toll-Free:	
Address: 504 Carnegie Ctr., Princeton, NJ 08540 US	

GROWTH PLANS/SPECIAL FEATURES:

PharmaNet Development Group, Inc. (PharmaNet), controlled by private equity firm JLL Partners, is a global drug development services company, which provides a broad range of both early and late stage clinical drug development services to branded pharmaceutical, biotechnology, generic drug and medical device companies around the world. PharmaNet also offers bioequivalency and pharmacodynamic studies, bioanalytical analyses, consulting services and clinical trial data management technology. The company separates its products and services into six categories. The Phase I/bioequivalence category offers early stage clinical development services, including conducting Phase I studies through subsidiary Anapharm, Inc.'s three clinical facilities in Canada. This segment also offers bioanalytical laboratory services. The bioanalytical laboratory services division, operating through Anapharm and subsidiary Taylor Technology, provides validated assays and sample analysis throughout the drug development process. The Phase II-IV segment provides late stage clinical development and project management services, through 2 GLP-compliant laboratories, in various therapeutic areas, including dermatology, general medicine, cardiovascular health, infectious disease, neuroscience, oncology and ophthalmology. The Phase IV development segment provides similar late stage services, as well as product selection, risk management, brand expansion and development solutions. The firm's clinical-research technology group markets the PharmaSoft suite of applications for the capture, verification and management of clinical trial data. Finally, PharmaNet offers a number of early- and late-phase consulting services related to international regulatory affairs; scientific and medical affairs; chemistry, manufacturing and controls; and biopharmaceutical investor interest. PharmaNet operates through 41 offices and facilities throughout more than 20 countries. In February 2009, the firm opened an office in Brazil. In March 2009, the company was acquired by JLL Partners. In October 2009, PharmaNet entered a strategic alliance with Azopharma Product Development Group LLC to jointly promote their preclinical services.

FINANCIALS: Sales and profits are in thousands of dollars—add 000 to get the full amount. 2009 Note: Financial information for 2009 was not available for all companies at press time.

2009 Sales: $	2009 Profits: $	U.S. Stock Ticker: Private
2008 Sales: $451,453	2008 Profits: $-251,093	Int'l Ticker: Int'l Exchange:
2007 Sales: $470,257	2007 Profits: $12,916	Employees: 2,400
2006 Sales: $406,956	2006 Profits: $-36,025	Fiscal Year Ends: 12/31
2005 Sales: $361,506	2005 Profits: $4,779	Parent Company:

SALARIES/BENEFITS:

Pension Plan:	ESOP Stock Plan:	Profit Sharing:	Top Exec. Salary: $741,932	Bonus: $
Savings Plan:	Stock Purch. Plan:		Second Exec. Salary: $534,867	Bonus: $

OTHER THOUGHTS:

Apparent Women Officers or Directors: 5
Hot Spot for Advancement for Women/Minorities: Y

LOCATIONS: ("Y" = Yes)

West:	Southwest:	Midwest:	Southeast:	Northeast:	International:
Y		Y		Y	Y

PHARMOS CORP

www.pharmoscorp.com

Industry Group Code: 325412 Ranks within this company's industry group: Sales: Profits: 83

Drugs:		Other:	Clinical:	Computers:	Services:	
Discovery:	Y	AgriBio:	Trials/Services:	Hardware:	Specialty Services:	
Licensing:		Genetic Data:	Labs:	Software:	Consulting:	
Manufacturing:		Tissue Replacement:	Equipment/Supplies:	Arrays:	Blood Collection:	
Generics:			Research/Development Svcs.:	Database Management:	Drug Delivery:	Y
			Diagnostics:		Drug Distribution:	

TYPES OF BUSINESS:

Neurological Diseases Drugs
Drug Delivery Systems
Synthetic Cannabinoids
Cannabinoids

BRANDS/DIVISIONS/AFFILIATES:

Dextofisopam
Cannabinor
PRS-639,058
Tianeptine

CONTACTS: *Note: Officers with more than one job title may be intentionally listed here more than once.*

S. Colin Neill, Pres.
S. Colin Neill, CFO
S. Colin Neill, Sec.
S. Colin Neill, Treas.
Robert F. Johnston, Exec. Chmn.

Phone: 732-452-9556	Fax: 732-452-9557
Toll-Free:	
Address: 99 Wood Ave. S., Ste. 311, Iselin, NJ 08830 US	

GROWTH PLANS/SPECIAL FEATURES:

Pharmos Corp. is a biopharmaceutical company that discovers and develops novel therapeutics focusing on specific diseases of the nervous system including disorders of the brain-gut axis (GI/IBS), pain/inflammation and autoimmune disorders.   The company's lead product, Dextofisopam, completed a Phase IIb clinical trial to evaluate safety and efficacy of the compound in irritable bowel syndrome. Dextofisopam is currently the only drug under active development. Pharmos' past discovery efforts were focused primarily on CB2-selective compounds, which are small molecule cannabinoid receptor agonists that bind preferentially to CB2 receptors found primarily in peripheral immune cells and peripheral nervous system. Cannabinator completed two Phase IIa tests for pain indications with an intravenous formulation, but failed to meet its endpoints. Another CB2-selective compound, PRS-639,058, was able to overcome some of the difficulties commonly occurring in cannabinoid drugs, including low solubility and metabolism, poor bioavailability and side effects. Pharmos also worked to develop Tianeptine, a possible follow-up to Dextofisopam in IBS, but this program is also currently inactive.

FINANCIALS: Sales and profits are in thousands of dollars—add 000 to get the full amount. 2009 Note: Financial information for 2009 was not available for all companies at press time.

2009 Sales: $	2009 Profits: $-3,100	U.S. Stock Ticker: PARS.PK
2008 Sales: $	2008 Profits: $-10,089	Int'l Ticker: Int'l Exchange:
2007 Sales: $	2007 Profits: $-15,626	Employees: 4
2006 Sales: $	2006 Profits: $-35,137	Fiscal Year Ends: 12/31
2005 Sales: $	2005 Profits: $-2,930	Parent Company:

SALARIES/BENEFITS:

Pension Plan:	ESOP Stock Plan:	Profit Sharing:	Top Exec. Salary: $300,000	Bonus: $
Savings Plan: Y	Stock Purch. Plan: Y		Second Exec. Salary: $	Bonus: $

OTHER THOUGHTS:

Apparent Women Officers or Directors:
Hot Spot for Advancement for Women/Minorities:

LOCATIONS: ("Y" = Yes)

West:	Southwest:	Midwest:	Southeast:	Northeast:	International:
				Y	

PHARSIGHT CORP

www.pharsight.com

Industry Group Code: 511210D Ranks within this company's industry group: Sales: Profits:

Drugs:	Other:	Clinical:	Computers:		Services:	
Discovery:	AgriBio:	Trials/Services:	Hardware:		Specialty Services:	
Licensing:	Genetic Data:	Labs:	Software:	Y	Consulting:	Y
Manufacturing: ·	Tissue Replacement:	Equipment/Supplies:	Arrays:		Blood Collection:	
Generics:		Research/Development Svcs.:	Database Management:	Y	Drug Delivery:	
		Diagnostics:			Drug Distribution:	

TYPES OF BUSINESS:

Health Care Data Management Software
Strategic Consulting Services
Reporting & Analysis Services

BRANDS/DIVISIONS/AFFILIATES:

Phoenix WinNonlin
Phoenix NLME
WinNonlin AutoPilot
IVIVC Toolkit
Trial Simulator
Tripos International
Pharsight Knowledgebase Server (PKS)
PKS Reporter

CONTACTS: Note: Officers with more than one job title may be intentionally listed here more than once.

Jim Hopkins, CEO
Patrick Flanagan, COO
John D. Yingling, CFO
Mark Hovde, Sr. VP-Mktg.
J.F. Marier, VP/Lead Scientist-Worldwide Consulting
Daniel L. Weiner, CTO/Sr. VP
James Hayden, Sr. VP-Global Sales
Nancy Risch, VP-Global Sales
Rene Bruno, Managing Dir.-Consulting Svcs., Europe

Phone: 314-951-3000	Fax: 314-647-9241
Toll-Free:	
Address: 1699 S. Hanley Rd., St. Louis, MO 63144 US	

GROWTH PLANS/SPECIAL FEATURES:

Pharsight Corp., a Certara company, is a provider of scientific consulting services and software products to help biotechnology and pharmaceutical companies improve their drug development processes, regulatory compliance and strategic decision-making. It provides solutions for pharmacokinetics, clinical medicine, biostatistics, clinical pharmacology, data management, marketing, regulatory affairs and formulation development. Pharsight offers two types of services: strategic consulting services for multiple therapeutic areas; and reporting and analysis services to assist with data analysis and report writing for pre-clinical and clinical studies to support new drug submission and approval. The strategic consulting services offer clinical program design; clinical utility modeling; QTc modeling; therapeutic domain expertise; modeling and simulation for competitive product positioning; treatment optimization; public source meta-databases for modeling; computer assisted trial design (CATD) and analysis; and decision support for in-licensing. The reporting and analysis services offer preclinical PK analysis; clinical PK study analysis; biostatistics and data management; PK-based formulation strategy; population PK modeling; and protocol writing and regulatory and scientific writing. Pharsight's products are offered in four categories: analysis, data management, reporting and validation. Its analysis products include Phoenix WinNonlin; Phoenix NLME; WinNonlin AutoPilot; IVIVC Toolkit; and Trial Simulator. Its data management products include Phoenix Connect and Pharsight Knowledgebase Server (PKS). Its reporting products include PKS Reporter and Drug Model Explorer. Its validation products include Phoenix Validation Suite and PKS Validation Suite. The firm has offices in Cary, North Carolina; Mountain View, California; Munich, Germany; Montreal, Canada; and staff throughout the U.S. and Europe.

FINANCIALS: Sales and profits are in thousands of dollars—add 000 to get the full amount. 2009 Note: Financial information for 2009 was not available for all companies at press time.

2009 Sales: $	2009 Profits: $	U.S. Stock Ticker: Subsidiary
2008 Sales: $	2008 Profits: $	Int'l Ticker: Int'l Exchange:
2007 Sales: $	2007 Profits: $	Employees:
2006 Sales: $	2006 Profits: $	Fiscal Year Ends:
2005 Sales: $	2005 Profits: $	Parent Company: CERTARA CORPORATION

SALARIES/BENEFITS:

Pension Plan:	ESOP Stock Plan:	Profit Sharing:	Top Exec. Salary: $	Bonus: $
Savings Plan:	Stock Purch. Plan:		Second Exec. Salary: $	Bonus: $

OTHER THOUGHTS:

Apparent Women Officers or Directors: 1
Hot Spot for Advancement for Women/Minorities:

LOCATIONS: ("Y" = Yes)

West:	Southwest:	Midwest:	Southeast:	Northeast:	International:
Y		Y		Y	Y

PIONEER HI-BRED INTERNATIONAL INC

www.pioneer.com

Industry Group Code: 11511 Ranks within this company's industry group: Sales: Profits:

Drugs:	Other:		Clinical:	Computers:	Services:	
Discovery:	AgriBio:	Y	Trials/Services:	Hardware:	Specialty Services:	Y
Licensing:	Genetic Data:		Labs:	Software:	Consulting:	
Manufacturing:	Tissue Replacement:		Equipment/Supplies:	Arrays:	Blood Collection:	
Generics:			Research/Development Svcs.:	Database Management:	Drug Delivery:	
			Diagnostics:		Drug Distribution:	

TYPES OF BUSINESS:

Seed Production & Distribution
Agronomics
Seed Genetics

BRANDS/DIVISIONS/AFFILIATES:

Optimum AcreMax 1
Plenish
AgVenture
Hoegemeyer Hybrids
NuTech Seed
PROaccess

CONTACTS: *Note: Officers with more than one job title may be intentionally listed here more than once.*

Paul Shickler, Pres.
Jeff Austin, CFO/VP
Susan Bunz, VP-Human Resources
John Soper, VP-R&D, Crop Genetics
Lane Arthur, CIO/VP
Judith E. McKay, General Counsel/VP
Peter Hemken, VP-Strategic Planning
Doyle Karr, Dir.-Comm.
Arun Baral, Regional Dir.-Asia Pacific
Daniel Glat, Regional Dir.-Latin America & Africa
Gyula Kovacs, Regional Dir.-Europe
Alejandro Munoz, VP/Regional Dir.-North America
William S. Niebur, VP-China

Phone: 515-270-3200	Fax: 515-334-4415
Toll-Free:	
Address: P.O. Box 1000, Johnston, IA 50131 US	

GROWTH PLANS/SPECIAL FEATURES:

Pioneer Hi-Bred International, Inc., a DuPont subsidiary, develops and supplies advanced plant genetics to farmers in nearly 70 countries. Research and development at Pioneer focuses on discovery and delivery of elite seed genetics with the goal of increasing crop yield, sustainability and pest and disease resistance. The company's products include seeds of many different varieties such as, corn, soybeans, alfalfa, corn silage, canola, sorghum, sunflower and wheat. Pioneer's seeds are bred to ensure higher yields as well as disease and pest protection. Its Plenish high oleic soybean seed varieties offer greater nutritional value when compared to other soybeans used in the production of soybean oil for consumer use. Also offered are pesticides including Optimum AcreMax 1, (the first product of its kind supported by the EPA) designed to maximize field-by-field productivity in fields with corn rootworm (CRW) pressure by placing in-plant protection. Pioneer also offers hybrid seeds to grain farmers supplying crops to ethanol producers. The firm's HTF (high total fermentable) hybrids increase the maximum ethanol yield potential (EYP) of their crop, assisting growers in meeting ethanol processors demands. Pioneer currently offers over 170 HTF hybrids. In May 2010, the firm expanded its research and development facilities by adding a second research center in Puerto Rico, making it one of over 100 international research facilities operated by Pioneer. Additionally, it opened two new facilities in the U.S. In June 2010, through an acquisition carried out by DuPont, Pioneer absorbed three new seed companies as part of its PROaccess distribution strategy. The companies acquired, AgVenture, Hoegemeyer Hybrids and NuTech Seed, will now sell seeds under the Pioneer brand.

Pioneer offers its employees: Health and wellness benefits including; Medical, dental, vision and mental health coverage, disability and long-term care; financial and retirement security packages; flexible spending accounts and tuition reimbursement.

FINANCIALS: Sales and profits are in thousands of dollars—add 000 to get the full amount. 2009 Note: Financial information for 2009 was not available for all companies at press time.

2009 Sales: $	2009 Profits: $	**U.S. Stock Ticker: Subsidiary**
2008 Sales: $	2008 Profits: $	**Int'l Ticker:** Int'l Exchange:
2007 Sales: $	2007 Profits: $	Employees:
2006 Sales: $	2006 Profits: $	Fiscal Year Ends:
2005 Sales: $	2005 Profits: $	Parent Company: E I DU PONT DE NEMOURS & CO (DUPONT)

SALARIES/BENEFITS:

Pension Plan: Y	ESOP Stock Plan:	Profit Sharing:	Top Exec. Salary: $	Bonus: $
Savings Plan: Y	Stock Purch. Plan:		Second Exec. Salary: $	Bonus: $

OTHER THOUGHTS:

Apparent Women Officers or Directors: 2
Hot Spot for Advancement for Women/Minorities:

LOCATIONS: ("Y" = Yes)

West:	Southwest:	Midwest:	Southeast:	Northeast:	International:
Y	Y	Y	Y	Y	Y

POLYDEX PHARMACEUTICALS

www.polydex.com

Industry Group Code: 325414 Ranks within this company's industry group: Sales: 6   Profits: 5

Drugs:		Other:	Clinical:	Computers:	Services:	
Discovery:		AgriBio:	Trials/Services:	Hardware:	Specialty Services:	
Licensing:		Genetic Data:	Labs:	Software:	Consulting:	
Manufacturing:	Y	Tissue Replacement:	Equipment/Supplies:	Arrays:	Blood Collection:	
Generics:			Research/Development Svcs.: Y	Database Management:	Drug Delivery:	Y
			Diagnostics:		Drug Distribution:	Y

TYPES OF BUSINESS:

Drugs-Raw Ingredients
Contraceptive Development
HIV & STD Preventatives
Anemia Prevention-Veterinary

BRANDS/DIVISIONS/AFFILIATES:

Dextran Products Limited
Chemdex Inc.
Iron Dextran
Dextran Sulphate
Ushercell
Usherdex 4

CONTACTS: *Note: Officers with more than one job title may be intentionally listed here more than once.*

George G. Usher, CEO
Sharon L. Wardlaw, COO
George G. Usher, Pres.
John A. Luce, CFO
Sharon L. Wardlaw, Sec.
Sharon L. Wardlaw, Treas.
George G. Usher, Chmn.

Phone: 416-755-2231	Fax:
Toll-Free:	
Address: 421 Comstock Rd., Toronto, ON M1L 2H5 Canada	

GROWTH PLANS/SPECIAL FEATURES:

Polydex Pharmaceuticals, Ltd. is engaged in the research, development, manufacture and marketing of biotechnology-based products for the human pharmaceutical market and also manufactures bulk pharmaceutical intermediates for the global veterinary pharmaceutical industry. Polydex conduct business through two wholly-owned subsidiaries: Canada-based Dextran Products Limited, which manufactures and sells dextran and derivative products; and Kansas-based Chemdex, Inc., which has U. S. Food and Drug Administration (FDA) approval to manufacture and sell Iron Dextran for veterinary use. Dextran is the generic name applied to polymer compounds formed by bacterial growth on sucrose. Iron dextran is a derivative of dextran produced by complexing iron with dextran and is injected into most pigs at birth as a treatment for anemia. Polydex sells iron dextran to independent wholesalers primarily in Europe, the U.S. and Canada, with increasing sales in China. Dextran sulphate, another of the company's products, is a specialty chemical derivative of dextran used in research applications by the pharmaceutical industry and other chemical research centers. It is sold primarily to independent distributors and wholesalers in the U.S. as analytical chemical applications. Polydex is developing its Ushercell product, a high molecular weight cellulose sulphate, as a topical vaginal product primarily for the prevention of AIDS transmission and other sexually transmitted diseases. The company is also researching and developing Usherdex 4, a special form of dextran, to treat cystic fibrosis. Usherdex 4 could be effective in preventing the colonization of bacteria in the mouth and in stimulating the macrophages in the lungs to remove present bacteria and lessen secondary infections.

FINANCIALS: Sales and profits are in thousands of dollars—add 000 to get the full amount. 2009 Note: Financial information for 2009 was not available for all companies at press time.

2009 Sales: $4,825	2009 Profits: $-1,326	**U.S. Stock Ticker:** POLXF
2008 Sales: $5,735	2008 Profits: $- 885	**Int'l Ticker:** Int'l Exchange:
2007 Sales: $6,499	2007 Profits: $- 261	Employees: 20
2006 Sales: $5,265	2006 Profits: $-1,489	Fiscal Year Ends: 1/31
2005 Sales: $6,372	2005 Profits: $1,139	Parent Company:

SALARIES/BENEFITS:

Pension Plan:	ESOP Stock Plan:	Profit Sharing:	Top Exec. Salary: $234,700	Bonus: $
Savings Plan:	Stock Purch. Plan:		Second Exec. Salary: $100,000	Bonus: $

OTHER THOUGHTS:

Apparent Women Officers or Directors: 1
Hot Spot for Advancement for Women/Minorities:

LOCATIONS: ("Y" = Yes)

West:	Southwest:	Midwest:	Southeast:	Northeast:	International:
					Y

PONIARD PHARMACEUTICALS INC

www.poniard.com

Industry Group Code: 325412 Ranks within this company's industry group: Sales: Profits: 144

Drugs:		Other:	Clinical:	Computers:	Services:
Discovery:	Y	AgriBio:	Trials/Services:	Hardware:	Specialty Services:
Licensing:	Y	Genetic Data:	Labs:	Software:	Consulting:
Manufacturing:		Tissue Replacement:	Equipment/Supplies:	Arrays:	Blood Collection:
Generics:			Research/Development Svcs.:	Database Management:	Drug Delivery:
			Diagnostics:		Drug Distribution:

TYPES OF BUSINESS:
Drugs-Cancer

BRANDS/DIVISIONS/AFFILIATES:
Picoplatin

CONTACTS: *Note: Officers with more than one job title may be intentionally listed here more than once.*
Ronald A. Martell, CEO
Michael S. Perry, Pres.
Greg Weaver, CFO
Anna Lewak Wight, VP-Legal/Corp. Sec.
Cheni Kwok, Sr. VP-Corp. Dev.
Greg Weaver, Sr. VP-Finance
Michael S. Perry, Chief Medical Officer
Gerald McMahon, Chmn.

Phone: 650-583-3774	Fax: 650-583-3789
Toll-Free:	
Address: 7000 Shoreline Ct., Ste. 270, South San Francisco, CA 94080 US	

GROWTH PLANS/SPECIAL FEATURES:

Poniard Pharmaceuticals, Inc. is a biopharmaceutical company focused on the development and commercialization of cancer therapeutics. Poniard's lead product candidate is picoplatin, a platinum-based cancer therapy that can treat multiple cancer indications, including small cell lung, colorectal, prostate and ovarian cancers. It is an intravenous chemotherapeutic designed to treat solid tumors that are resistant to existing platinum-based cancer therapies. It targets solid tumors by binding to DNA and interfering with its replication and transcription processes, thereby causing apoptosis (cell death). Potential benefits of picoplatin are that it is broadly applicable for the treatment of solid tumors; the compound may have less severe side effects than other platinum treatments, and it may be possible to use this product as treatment for platinum sensitive, resistant and refractory disease. Poniard is conducting two separate Phase II trials evaluating picoplatin as a first-line treatment of metastatic colorectal cancer and castration-resistant prostate cancer. The firm has also completed a Phase I cardiac safety trial of picoplatin and a Phase I study evaluating an oral formulation of picoplatin in solid tumors. Picoplatin has received orphan drug designation from the FDA and recently completed enrollment and initial statistical analysis of a Phase III SPEAR trial of picoplatin in the second-line treatment of patients with small cell lung cancer. The company has entered into agreements with W.C Heraeus GmbH for the manufacture of picoplatin active pharmaceutical ingredient for use in its clinical studies and for commercial purposes. It has also entered into an agreement with Baxter Oncology GmbH for the bulk production and distribution of finished picoplatin drug product for clinical and commercial use. In February 2010, the company reduced its workforce by approximately 57%.

FINANCIALS: Sales and profits are in thousands of dollars—add 000 to get the full amount. 2009 Note: Financial information for 2009 was not available for all companies at press time.

2009 Sales: $	2009 Profits: $-45,715	U.S. Stock Ticker: PARD
2008 Sales: $	2008 Profits: $-48,565	Int'l Ticker: Int'l Exchange:
2007 Sales: $	2007 Profits: $-32,782	Employees: 22
2006 Sales: $	2006 Profits: $-23,294	Fiscal Year Ends: 12/31
2005 Sales: $ 15	2005 Profits: $-20,997	Parent Company:

SALARIES/BENEFITS:

Pension Plan:	ESOP Stock Plan:	Profit Sharing:	Top Exec. Salary: $446,719	Bonus: $
Savings Plan: Y	Stock Purch. Plan:		Second Exec. Salary: $355,364	Bonus: $

OTHER THOUGHTS:
Apparent Women Officers or Directors: 2
Hot Spot for Advancement for Women/Minorities: Y

LOCATIONS: ("Y" = Yes)

West:	Southwest:	Midwest:	Southeast:	Northeast:	International:
Y				Y	

POZEN INC

www.pozen.com

Industry Group Code: 325412 Ranks within this company's industry group: Sales: 99 Profits: 70

Drugs:		Other:	Clinical:	Computers:	Services:
Discovery:	Y	AgriBio:	Trials/Services:	Hardware:	Specialty Services:
Licensing:		Genetic Data:	Labs:	Software:	Consulting:
Manufacturing:		Tissue Replacement:	Equipment/Supplies:	Arrays:	Blood Collection:
Generics:			Research/Development Svcs.:	Database Management:	Drug Delivery:
			Diagnostics:		Drug Distribution:

TYPES OF BUSINESS:

Drugs, Analgesic
Migraine Therapy
Gastric Ulcer Prevention
Cardiovascular Protection

BRANDS/DIVISIONS/AFFILIATES:

Treximet
VIMOVO
PA

CONTACTS: Note: Officers with more than one job title may be intentionally listed here more than once.

John R. Plachetka, CEO
John R. Plachetka, Pres.
William L. Hodges, CFO
Elizabeth A. Cermak, Chief Commercial Officer/Exec. VP
John G. Fort, Chief Medical Officer
Everadus Orlemans, Exec. VP-Prod. Dev.
William L. Hodges, Sr. VP-Admin.
Gilda M. Thomas, General Counsel/Sr. VP
William L. Hodges, Sr. VP-Finance
John R. Plachetka, Chmn.

Phone: 919-913-1030	Fax: 919-913-1039
Toll-Free:	
Address: 1414 Raleigh Rd., Chapel Hill, NC 27517 US	

GROWTH PLANS/SPECIAL FEATURES:

Pozen, Inc. is a pharmaceutical company focused primarily on products that can provide improved efficacy, safety or patient convenience in the treatment of migraine, acute and chronic pain and other pain-related indications. Pozen's primary product is Treximet, developed in collaboration with GlaxoSmithKline (GSK), for the treatment of acute migraines. Treximet is a single tablet containing sumatriptan succinate, a 5-HT 1B/1D agonist formulated with GSK's RT Technology, and naproxen sodium, a non-steroidal anti-inflammatory drug (NSAID). The company also has a handful of other drugs in development. Under Pozen's PN program, the company has completed formulation development and clinical studies for combining a proton pump inhibitor (PPI) and an NSAID into a single tablet, meant to manage pain associated with chronic osteoarthritis, rheumatoid arthritis and ankylosing spondylitis in patients who are prone to developing gastric ulcers caused by NSAIDs. The drug, known as VIMOVO, was developed in collaboration with AstraZeneca and has now been approved by the FDA. Similarly, the company's PA program combines aspirin with a PPI for cardiovascular protection and to reduce the risk of colorectal cancer and adenomas with fewer gastrointestinal side effects. Pozen is also examining other indications in which it will develop drugs, including exploratory initiatives in cardiovascular and oncology areas and proof of concept trials in gastroenterology and cardiovascular studies. In April 2010, Pozen's VIMOVO was approved by the FDA for treatment of arthritis patients at risk of developing NSAID-associated ulcers.

FINANCIALS: Sales and profits are in thousands of dollars—add 000 to get the full amount. 2009 Note: Financial information for 2009 was not available for all companies at press time.

2009 Sales: $32,187	2009 Profits: $1,959	U.S. Stock Ticker: POZN
2008 Sales: $66,133	2008 Profits: $-5,976	Int'l Ticker: Int'l Exchange:
2007 Sales: $53,444	2007 Profits: $4,666	Employees: 31
2006 Sales: $13,517	2006 Profits: $-19,310	Fiscal Year Ends: 12/31
2005 Sales: $28,647	2005 Profits: $1,959	Parent Company:

SALARIES/BENEFITS:

Pension Plan:	ESOP Stock Plan:	Profit Sharing:	Top Exec. Salary: $528,865	Bonus: $1,121,746
Savings Plan: Y	Stock Purch. Plan:		Second Exec. Salary: $281,845	Bonus: $111,600

OTHER THOUGHTS:

Apparent Women Officers or Directors: 1
Hot Spot for Advancement for Women/Minorities:

LOCATIONS: ("Y" = Yes)

West:	Southwest:	Midwest:	Southeast:	Northeast:	International:
				Y	

PRA INTERNATIONAL

www.prainternational.com

Industry Group Code: 541712 Ranks within this company's industry group: Sales: Profits:

Drugs:	Other:	Clinical:		Computers:		Services:	
Discovery:	AgriBio:	Trials/Services:	Y	Hardware:		Specialty Services:	
Licensing:	Genetic Data:	Labs:		Software:		Consulting:	
Manufacturing:	Tissue Replacement:	Equipment/Supplies:		Arrays:		Blood Collection:	
Generics:		Research/Development Svcs.:	Y	Database Management:	Y	Drug Delivery:	
		Diagnostics:				Drug Distribution:	

TYPES OF BUSINESS:

Clinical Research & Testing Services
Clinical Development Services
Clinical Trials
Data Management Services

BRANDS/DIVISIONS/AFFILIATES:

PRA E-TMF

CONTACTS: Note: Officers with more than one job title may be intentionally listed here more than once.

Colin Shannon, CEO
Colin Shannon, Pres.
Linda Baddour, CFO/Exec. VP
Roger Boutin, Dir.-Mktg. Sales & Support
Jeff Chambers, Sr. VP-Human Resources
Willem Jan Drijfhout, Sr. VP-Early Dev. Svcs.
Tami Klerr-Naivar, Sr. VP-Bus. Dev.
William M. Walsh III, Exec. VP-Corp. Svcs.
Steve Powell, Sr. VP-Clinical Informatics & Late Phase Svcs.
Kent Thoelke, Sr. VP/Head-Scientific & Medical Affairs
Bruce Teplitzky, Exec. VP-Strategic Bus. Dev.
Melvin Booth, Chmn.
Susan C. Stansfield, Exec. VP-Prod. Registration, Europe, Africa & APAC

Phone: 919-786-8200	Fax: 919-786-8201
Toll-Free:	
Address: 4130 Park Lake Ave., Ste. 400, Raleigh, NC 27612 US	

GROWTH PLANS/SPECIAL FEATURES:

PRA International is a contract research organization (CRO) that provides clinical drug development services to pharmaceutical and biotechnology companies around the world. CROs typically assist companies in developing drug compounds, biologics, drug delivery devices and the attainment of certain regulatory approvals necessary to market these technologies. PRA International specializes in oncology, CNS, respiratory/allergy, cardiovascular and infectious diseases. In addition, PRA provides a broad array of services in clinical development programs, including the creation of drug development and regulatory strategy plans, the utilization of bioanalytical laboratory testing and the development of integrated global clinical databases. Bioanalytical sample testing includes the use of LC-MS Machines, the HPLC System and the Immunoassays suite. PRA International also provides data management services such as electronic data capture, data monitoring and database development. The company supports its data services through the use of its proprietary PRA Electronic Trial Master File (E-TMF) enterprise-wide electronic document management system, which provides clients with faster document handling and archival services. Clinical trials in the U.S. are largely centered around PRA International's facilities in Kansas. The company has performed more than 2,700 studies in 75 countries on six continents. In October 2009, the company opened a new office in Milan, Italy. In June 2010, the firm opened a Phase I facility In Budapest, Hungary.

PRA International offers employees comprehensive medial benefits, life insurance, retirements programs, holidays and paid time off, tuition advance payment programs and a scholarship program for employee dependents.

FINANCIALS: Sales and profits are in thousands of dollars—add 000 to get the full amount. 2009 Note: Financial information for 2009 was not available for all companies at press time.

2009 Sales: $	2009 Profits: $	**U.S. Stock Ticker: Private**
2008 Sales: $	2008 Profits: $	**Int'l Ticker:** Int'l Exchange:
2007 Sales: $	2007 Profits: $	Employees:
2006 Sales: $338,166	2006 Profits: $26,845	Fiscal Year Ends: 12/31
2005 Sales: $326,244	2005 Profits: $32,223	Parent Company: GENSTAR CAPITAL LLC

SALARIES/BENEFITS:

Pension Plan:	ESOP Stock Plan:	Profit Sharing:	Top Exec. Salary: $	Bonus: $
Savings Plan:	Stock Purch. Plan:		Second Exec. Salary: $	Bonus: $

OTHER THOUGHTS:

Apparent Women Officers or Directors: 2
Hot Spot for Advancement for Women/Minorities: Y

LOCATIONS: ("Y" = Yes)

West:	Southwest:	Midwest:	Southeast:	Northeast:	International:
Y	Y	Y		Y	Y

PROGENICS PHARMACEUTICALS

www.progenics.com

Industry Group Code: 325412   Ranks within this company's industry group: Sales: 85   Profits: 135

Drugs:		Other:		Clinical:	Computers:	Services:
Discovery:	Y	AgriBio:	Y	Trials/Services:	Hardware:	Specialty Services:
Licensing:	Y	Genetic Data:		Labs:	Software:	Consulting:
Manufacturing:	Y	Tissue Replacement:		Equipment/Supplies:	Arrays:	Blood Collection:
Generics:				Research/Development Svcs.:	Database Management:	Drug Delivery:
				Diagnostics:		Drug Distribution:

TYPES OF BUSINESS:

Pharmaceuticals Development
Drugs-HIV
Drugs-Cancer
Small-Molecule Drugs
Viral Entry Inhibitors
Vaccines

BRANDS/DIVISIONS/AFFILIATES:

RELISTOR
PRO 140

CONTACTS: Note: Officers with more than one job title may be intentionally listed here more than once.

Paul J. Maddon, CEO
Mark R. Baker, Pres.
Robert A. McKinney, CFO
William C. Olson, Sr. VP-R&D
Thomas A. Boyd, Sr. VP-Prod. Dev.
Nitya G. Ray, Sr. VP-Mfg.
Mark R. Baker, General Counsel/Exec. VP
Robert A. McKinney, Sr. VP-Oper.
Walter M. Capone, VP-Commercial Dev. & Oper.
Dory A. Kurowski, Contact-Investor Rel.
Robert A. McKinney, Sr. VP-Finance/Treas.
Paul J. Maddon, Chief Science Officer
Robert J. Isreal, Sr. VP-Medical Affairs
Benedict Osorio, Sr. VP-Quality
Tage Ramakrishna, VP-Clinical Research
Peter J. Crowley, Chmn.

Phone: 914-789-2800	Fax: 914-789-2817
Toll-Free:	
Address: 777 Old Saw Mill River Rd., Tarrytown, NY 10591 US	

GROWTH PLANS/SPECIAL FEATURES:

Progenics Pharmaceuticals, Inc. develops and commercializes therapeutic products to treat patients with debilitating conditions and life-threatening diseases. The firm's principal programs are directed toward supportive care, oncology and virology. The firm's lead product in the area of symptom management is RELISTOR, which is designed to reverse the side effects of opioid pain medications while maintaining pain relief. RELISTOR is approved for use as a subcutaneous injection in 40 countries, including the U.S., Australia, Canada, Brazil and the countries of the E.U. The company is also developing oral delivery methods for RELISTOR. In the area of virology, the firm has developed viral-entry inhibitors, which are molecules designed to inhibit a virus' ability to enter certain types of immune system cells. Viral products in development include the PRO 140 viral entry inhibitor, a potential treatment for HIV infection, currently in Phase II development trials. The firm's virology division is also evaluating candidates for therapies that block entry of the Hepatitis C virus into cells. The oncology division is focused on developing immunotherapies for prostate cancer, including monoclonal antibodies directed against prostate specific membrane antigen (PSMA), a protein found on the surface of prostate cancer cells. Progenics is also developing therapeutic prostate cancer vaccines designed to enhance the body's own defense mechanisms against PSMA. In May 2009, the firm finalized its plans to cut 10% of its staff in order to streamline production. In October 2009, the company regained the worldwide rights to RELISTOR from former co-developer Wyeth Pharmaceuticals. In March 2010, the firm announced plans to commence a Phase IIb/III clinical trial with chronic pain patients for its oral methylnaltrexone candidate.

Progenics offers employees medical, vision, dental and life insurance; a 401(k) plan; stock options; an employee stock purchase plan; flexible spending accounts; tuition and seminar reimbursement; and an employee assistance plan.

FINANCIALS: Sales and profits are in thousands of dollars—add 000 to get the full amount. 2009 Note: Financial information for 2009 was not available for all companies at press time.

2009 Sales: $48,947	2009 Profits: $-30,612	U.S. Stock Ticker: PGNX
2008 Sales: $67,671	2008 Profits: $-44,672	Int'l Ticker: Int'l Exchange:
2007 Sales: $75,646	2007 Profits: $-43,688	Employees: 204
2006 Sales: $69,906	2006 Profits: $-21,618	Fiscal Year Ends: 12/31
2005 Sales: $9,486	2005 Profits: $-69,429	Parent Company:

SALARIES/BENEFITS:

Pension Plan:	ESOP Stock Plan: Y	Profit Sharing:	Top Exec. Salary: $618,000	Bonus: $250,000
Savings Plan: Y	Stock Purch. Plan: Y		Second Exec. Salary: $391,667	Bonus: $220,000

OTHER THOUGHTS:

Apparent Women Officers or Directors: 2
Hot Spot for Advancement for Women/Minorities:

LOCATIONS: ("Y" = Yes)

West:	Southwest:	Midwest:	Southeast:	Northeast:	International:
				Y	

PROMEGA CORP

www.promega.com

Industry Group Code: 325413 **Ranks within this company's industry group:** Sales: Profits:

Drugs:	Other:	Clinical:		Computers:		Services:	
Discovery:	AgriBio:	Trials/Services:		Hardware:		Specialty Services:	Y
Licensing:	Genetic Data:	Labs:	Y	Software:		Consulting:	
Manufacturing:	Tissue Replacement:	Equipment/Supplies:	Y	Arrays:		Blood Collection:	
Generics:		Research/Development Svcs.:	Y	Database Management:		Drug Delivery:	
		Diagnostics:	Y			Drug Distribution:	

TYPES OF BUSINESS:

Life Sciences Solutions & Technical Support
Specialty Biochemicals
DNA & RNA Analysis Systems
Protein Expression, Purification & Amalysis Systems
Cellular Analysis Systems
Genetic Identifier Systems
Automation & Robotics

BRANDS/DIVISIONS/AFFILIATES:

Terso Solutions
PowerPlex 16 HS
ProteasMAX
QuantiFluor-ST
Renilla-Glo
GloMax-Multi+
pmirGLO
GoTaq

CONTACTS: Note: Officers with more than one job title may be intentionally listed here more than once.

William A. Linton, CEO
William A. Linton, Pres.
Laura Francis, CFO
Randy Dimond, CTO/VP
Laura Francis, VP-Finance
William A. Linton, Chmn.

Phone: 608-274-4330	Fax: 608-277-2601
Toll-Free: 800-356-9526	
Address: 2800 Woods Hollow Rd., Madison, WI 53711 US	

GROWTH PLANS/SPECIAL FEATURES:

Promega Corp. provides solutions and technical support to the life sciences industry. The company's over 2,000 products aid scientists in life science research, especially genomics, proteomics and cellular analysis. The firm's products comprise kits and reagents, as well as integrated solutions for life sciences research and drug discovery. Reagents include DNA and RNA purification; genotype analysis; protein expression and analysis; and DNA sequencing. Integrated solutions provide customers with personal automation tools that combine instrumentation and reagents into one system. Promega's Terso Solutions subsidiary focuses on leveraging RFID technology to solve customers' needs for complete inventory management solutions. In addition, Promega's New Lab Setup program assists researchers begin new operations from the ground up. The company has nearly 900 U.S. and foreign patents and pending applications in areas including nucleic acid purification, human identification, bioluminescence, cell biology and coupled in vitro transcription and translation. The firm has branches in 14 countries and sells its products through over 50 global distributors. In 2009, Promega launched its StemElite ID system, which improves human cell line authentication; GloMax-Multi+ detection system; a 16-locus STR system to profile inhibited DNA samples; and the pmirGLO Dual-Luciferase miRNA Target Expression Vector for microRNA study. In November 2009, Promega's PowerPlex 16 HS system was approved for use by the FBI in generating DNA records for the National DNA Index System. The firm's product line grew again in 2010 with new additions including: GoTaq 2-Step RT-qPCR System, Renilla-Glo Luciferase Assay System, QuantiFluor-ST and QuantiFluor-P Single-Tube Fluorometers, in addition to others.

The company offers its employees health, life, disability, long-term care and dental insurance; flexible spending accounts; tuition assistance; and an employee assistance program. Additional perks include an on-campus Farmer's Market, convenience services and wellness programs, which include on-site fitness classes, exercise facilities, yoga classes and massage therapy.

FINANCIALS: Sales and profits are in thousands of dollars—add 000 to get the full amount. 2009 Note: Financial information for 2009 was not available for all companies at press time.

2009 Sales: $	2009 Profits: $	**U.S. Stock Ticker: Private**
2008 Sales: $	2008 Profits: $	**Int'l Ticker:** Int'l Exchange:
2007 Sales: $	2007 Profits: $	Employees:
2006 Sales: $200,000	2006 Profits: $	Fiscal Year Ends: 12/31
2005 Sales: $	2005 Profits: $	Parent Company:

SALARIES/BENEFITS:

Pension Plan:	ESOP Stock Plan:	Profit Sharing:	Top Exec. Salary: $	Bonus: $
Savings Plan: Y	Stock Purch. Plan:		Second Exec. Salary: $	Bonus: $

OTHER THOUGHTS:

Apparent Women Officers or Directors: 1
Hot Spot for Advancement for Women/Minorities:

LOCATIONS: ("Y" = Yes)

West:	Southwest:	Midwest:	Southeast:	Northeast:	International:
Y		Y			Y

PROTEIN POLYMER TECHNOLOGIES

www.ppti.com

Industry Group Code: 325411 Ranks within this company's industry group: Sales: Profits:

Drugs:		Other:		Clinical:		Computers:		Services:	
Discovery:		AgriBio:		Trials/Services:		Hardware:		Specialty Services:	
Licensing:		Genetic Data:		Labs:		Software:		Consulting:	
Manufacturing:	Y	Tissue Replacement:	Y	Equipment/Supplies:		Arrays:		Blood Collection:	
Generics:				Research/Development Svcs.:	Y	Database Management:		Drug Delivery:	Y
				Diagnostics:				Drug Distribution:	

TYPES OF BUSINESS:

Medical Products Development
Medical Products-Bodily Repair
Drug Delivery Devices
Protein Polymers

BRANDS/DIVISIONS/AFFILIATES:

Spine Wave, Inc.
NuCore
Polymer 47K
Investigational Device Exemption

CONTACTS: Note: Officers with more than one job title may be intentionally listed here more than once.

James B. McCarthy, CEO
James B. McCarthy, Interim Pres.
Joseph Cappello, VP-R&D
Joseph Cappello, CTO
Franco A. Ferrari, VP-Laboratory Oper. & Polymer Prod.
James B. McCarthy, Chmn.

Phone: 858-558-6064	Fax: 858-558-6477
Toll-Free:	
Address: 7660-H Fay Ave., Ste. 352, La Jolla, CA 92037 US	

GROWTH PLANS/SPECIAL FEATURES:

Protein Polymer Technologies, Inc. (PPT) is a development-stage company that uses its proprietary protein-based biomaterials technology to research, produce and clinically test medical products that aid in the natural process of bodily repair. Its product focus includes surgical adhesives and sealants, soft-tissue augmentation products, wound-healing matrices, drug delivery devices and surgical adhesion barriers. PPT uses biomaterials that can mimic the natural properties and functions of proteins and peptides, and interact with a cell's enzymatic reactions to spur protein formation. The company has developed surgical tissue sealants which combine the biocompatibility of fibrin glues, without the risks associated with use of blood-derived products, with high strength and fast setting times. The product adheres well to tissue to seal gas and fluid leaks and comes in resorbable and non-resorbable formulations. PPT has developed protein polymers for use on dermal wounds, particularly chronic wounds such as decubitous ulcers, which assist the tissue to heal in such a way that it regenerates as functional tissue instead of scar tissue. Polymer 47K is the company's urethral bulking agent for the treatment of female stress urinary incontinence and is given as an injection. The company's Investigational Device Exemption (IDE) allows the testing of the safety and effectiveness of the incontinence product in women over the age of 40 who have become incontinent due to the shifting of their bladder or the weakening of the muscle at its base. Another application for the firm's bodily repair technology is spine applications. PPT used the company's patented tissue adhesive technology to create Spine Wave's NuCore intervertebral disc repair material. PPT is a technical partner of Spine Wave, Inc., and the company manufactured the NuCore material for Spine Wave's clinical trials.

FINANCIALS: Sales and profits are in thousands of dollars—add 000 to get the full amount. 2009 Note: Financial information for 2009 was not available for all companies at press time.

2009 Sales: $	2009 Profits: $	U.S. Stock Ticker: PPTI.OB	
2008 Sales: $ 25	2008 Profits: $-2,938	Int'l Ticker: Int'l Exchange:	
2007 Sales: $ 287	2007 Profits: $-3,252	Employees: 2	
2006 Sales: $ 605	2006 Profits: $-7,878	Fiscal Year Ends: 12/31	
2005 Sales: $ 867	2005 Profits: $-5,822	Parent Company:	

SALARIES/BENEFITS:

Pension Plan:	ESOP Stock Plan:	Profit Sharing:	Top Exec. Salary: $	Bonus: $
Savings Plan:	Stock Purch. Plan:		Second Exec. Salary: $	Bonus: $

OTHER THOUGHTS:

Apparent Women Officers or Directors:
Hot Spot for Advancement for Women/Minorities:

LOCATIONS: ("Y" = Yes)

West:	Southwest:	Midwest:	Southeast:	Northeast:	International:
Y					

QIAGEN NV

www.qiagen.com

Industry Group Code: 325413 **Ranks within this company's industry group:** Sales: 3 Profits: 2

Drugs:	Other:		Clinical:		Computers:	Services:	
Discovery:	AgriBio:		Trials/Services:		Hardware:	Specialty Services:	Y
Licensing:	Genetic Data:	Y	Labs:		Software:	Consulting:	
Manufacturing:	Tissue Replacement:		Equipment/Supplies:	Y	Arrays:	Blood Collection:	
Generics:			Research/Development Svcs.:		Database Management:	Drug Delivery:	
			Diagnostics:	Y		Drug Distribution:	

TYPES OF BUSINESS:
Supplies-Plasmid Purification
Genomics Analysis Products
Diagnostic Products

BRANDS/DIVISIONS/AFFILIATES:
Qiagen, GmbH
Tianwei Times
Nextal Biotechnology, Inc.
artus GmbH
Shenzhen PG Biotech Co. Ltd.
Gentra Systems, Inc.
QIA Symphony SP
Corbett Life Science Pty Ltd

CONTACTS: Note: Officers with more than one job title may be intentionally listed here more than once.
Peer Schatz, CEO
Roland Sackers, CFO
Bernd Uder, Sr. VP-Global Sales
Gisela Orth, VP-Global Human Resources
Joachim Schorr, Sr. VP-Global R&D
Douglas Liu, VP-Global Oper.
Ulrich Schriek, VP-Corp. Bus. Dev.
Michael Collasius, VP-Automated Systems
Thomas Schweins, VP-Mktg. & Strategy
Detlev H. Riesner, Chmn.

Phone: 31-77-320-8400	Fax: 31-77-320-8409
Toll-Free:	
Address: Spoorstraat 50, Venlo, 5911 KJ The Netherlands	

GROWTH PLANS/SPECIAL FEATURES:
QIAGEN NV provides technologies and products for the separation and purification of nucleic acids, used in genomics, molecular diagnostics and genetic vaccination and gene therapy. Through its many subsidiaries, QIAGEN offers more than 500 products in over 40 countries worldwide. The firm's products and services find application in animal and veterinary research; biomedical research; biosecurity and biodefense; epigenetics; genetic identity and forensics; gene expression analysis; gene silencing; influenza research; molecular diagnostics; pharmacogenomics; plant research; and protein science fields. The company owns TianGen; Nextal Biotechnology, Inc.; Artus GmbH; and Shenzhen PG Biotech Co. Ltd. TianGen, based in Beijing, develops, manufactures and supplies nucleic acid sample preparation consumables in China. Nextal provides proprietary sample preparation tools that make protein crystallization more accessible. Artus GmbH is a leader in PCR molecular diagnostic systems. Shenzhen PG Biotech develops, manufactures and supplies PCR based molecular diagnostic kits in China. The company also features the QIAGENcares program, which aims to improve access to screening methods for infectious diseases in developing and emerging countries. In January 2010, the company launched the QIAsymphony AS, a module that provides fully automated setup of QIAGEN PCR kits and assays in a single platform. Also in January 2010, the firm acquired German-based ESE GmbH, a developer and manufacturer of UV and fluorescence optical measurement devices. In February 2010, QIAGEN announced an agreement with Celera Corporation, in which QIAGEN will exclusively distribute one of Celera's molecular multiplex assays.

FINANCIALS: Sales and profits are in thousands of dollars—add 000 to get the full amount. 2009 Note: Financial information for 2009 was not available for all companies at press time.

2009 Sales: $1,009,825	2009 Profits: $137,767	**U.S. Stock Ticker:** QGEN
2008 Sales: $892,975	2008 Profits: $89,033	**Int'l Ticker:** QIA Int'l Exchange: Frankfurt-Euronext
2007 Sales: $649,774	2007 Profits: $50,122	Employees: 3,495
2006 Sales: $465,778	2006 Profits: $70,539	Fiscal Year Ends: 12/31
2005 Sales: $298,395	2005 Profits: $62,225	Parent Company:

SALARIES/BENEFITS:
Pension Plan:	ESOP Stock Plan:	Profit Sharing:	Top Exec. Salary: $	Bonus: $
Savings Plan: Y	Stock Purch. Plan:		Second Exec. Salary: $	Bonus: $

OTHER THOUGHTS:
Apparent Women Officers or Directors: 1
Hot Spot for Advancement for Women/Minorities:

LOCATIONS: ("Y" = Yes)
West:	Southwest:	Midwest:	Southeast:	Northeast:	International:
Y				Y	Y

Note: Financial information, benefits and other data can change quickly and may vary from those stated here.

QLT INC

www.qltinc.com

Industry Group Code: 325412 Ranks within this company's industry group: Sales: 89 Profits: 42

Drugs:		Other:	Clinical:	Computers:	Services:	
Discovery:	Y	AgriBio:	Trials/Services:	Hardware:	Specialty Services:	
Licensing:	Y	Genetic Data:	Labs:	Software:	Consulting:	
Manufacturing:	Y	Tissue Replacement:	Equipment/Supplies:	Arrays:	Blood Collection:	
Generics:			Research/Development Svcs.:	Database Management:	Drug Delivery:	Y
			Diagnostics:		Drug Distribution:	

TYPES OF BUSINESS:

Drugs-Photodynamic
Eye Disease Treatments
Drug Delivery-Controlled & Extended Release
Punctual Plug Delivery Systems

BRANDS/DIVISIONS/AFFILIATES:

Visudyne
Novartis Pharma AG
Latanoprost
Olopatadine
QLT Ophthalmics, Inc.

CONTACTS: Note: Officers with more than one job title may be intentionally listed here more than once.

Robert L. Butchofsky, CEO
Robert L. Butchofsky, Pres.
Cameron Nelson, CFO
Linda Lupini, Sr. VP-Human Resources
Dipak Panigrahi, Sr. VP-R&D
Alexander R. Lussow, Sr. VP-Bus. Dev. & Commercial Oper.
Therese Hayes, VP-Corp. Comm.
Therese Hayes, VP-Investor Rel.
Cameron Nelson, VP-Finance
Linda Lupini, Sr. VP-Organizational Dev.
Dipak Panigrahi, Chief Medical Officer
C. Boyd Clarke, Chmn.

Phone: 604-707-7000	Fax: 604-707-7001
Toll-Free: 800-663-5486	
Address: 887 Great Northern Way, Vancouver, BC V5T 4T5 Canada	

GROWTH PLANS/SPECIAL FEATURES:

QLT, Inc. is a global biopharmaceutical company that focuses on the discovery, development and commercialization of pharmaceutical products in the fields of ophthalmology. The firm's key product in the field of ophthalmology, Visudyne, uses photodynamic therapy to stifle and kill abnormally growing tissue. By administering drugs to target proteins then activating the drugs through externally provided light, Visudyne can destroy abnormal blood vessel cells by releasing singlet oxygen. It is one of the few approved treatments for wet age-related macular degeneration (AMD), the leading cause of blindness in people over the age of 55. Visudyne, which was co-developed by Swiss firm Novartis Pharma AG, is also approved for treatment of subfoveal choroidal neovascularization (CNV) and is sold in 77 countries. QLT is also developing punctal plug delivery (PPD) drug systems, a minimally invasive system that delivers a variety of drugs to the eye through controlled sustained release to the tear film. The firm is researching the use of this platform for a broad range of ocular diseases that are currently being treated with eye drops. Its leading PPD candidate for market approval, now in phase II clinical development, is Latanoprost, for use in patients with either ocular hypertension or open-angle glaucoma. In addition, QLT currently has several other drugs in clinical development. QLT091001, now in phase I of development, is a synthetic retinoid targeted for the treatment of Leber Congenital Amourosis (LCA). Other drugs, still in preclinical development include, Olopatadine and QLT091568. In order to focus entirely on the field of ophthalmology, in recent years QLT has sold its holdings in its non-ocular based product lines. In December 2009, QLT established a market presence in the U.S. through a newly formed subsidiary, QLT Ophthalmics, Inc. The new subsidiary sells and markets Visudyne to U.S. customers.

FINANCIALS: Sales and profits are in thousands of dollars—add 000 to get the full amount. 2009 Note: Financial information for 2009 was not available for all companies at press time.

2009 Sales: $42,100	2009 Profits: $99,400	U.S. Stock Ticker: QLTI
2008 Sales: $48,300	2008 Profits: $134,891	Int'l Ticker: QLT Int'l Exchange: Toronto-TSX
2007 Sales: $67,700	2007 Profits: $-109,997	Employees: 142
2006 Sales: $174,136	2006 Profits: $-101,605	Fiscal Year Ends: 12/31
2005 Sales: $241,973	2005 Profits: $-325,412	Parent Company:

SALARIES/BENEFITS:

Pension Plan: Y	ESOP Stock Plan:	Profit Sharing:	Top Exec. Salary: $430,599	Bonus: $153,939
Savings Plan:	Stock Purch. Plan:		Second Exec. Salary: $299,027	Bonus: $105,800

OTHER THOUGHTS:

Apparent Women Officers or Directors: 3
Hot Spot for Advancement for Women/Minorities: Y

LOCATIONS: ("Y" = Yes)

West:	Southwest:	Midwest:	Southeast:	Northeast:	International:
Y					Y

Note: Financial information, benefits and other data can change quickly and may vary from those stated here.

QUEST DIAGNOSTICS INC

www.questdiagnostics.com

Industry Group Code: 6215 Ranks within this company's industry group: Sales: 1 Profits: 1

Drugs:	Other:	Clinical:		Computers:	Services:
Discovery:	AgriBio:	Trials/Services:		Hardware:	Specialty Services:
Licensing:	Genetic Data:	Labs:	Y	Software:	Consulting:
Manufacturing:	Tissue Replacement:	Equipment/Supplies:		Arrays:	Blood Collection:
Generics:		Research/Development Svcs.:		Database Management:	Drug Delivery:
		Diagnostics:	Y		Drug Distribution:

TYPES OF BUSINESS:

Services-Testing & Diagnostics
Clinical Laboratory Testing
Clinical Trials Testing
Esoteric Testing Laboratories

BRANDS/DIVISIONS/AFFILIATES:

Cardio CRP
HEPTIMAX
ImmunoCap
Nichols Institute
Risk Management Services
Wellness Testing

CONTACTS: Note: Officers with more than one job title may be intentionally listed here more than once.

Surya N. Mohapatra, CEO
Robert A. Hagemann, CFO/Sr. VP
David W. Norgard, VP-Human Resources
Jon R. Cohen, Chief Medical Officer/Sr. VP
David Evans, VP-IT
Michael E. Prevoznik, General Counsel/VP-Legal & Compliance
Wayne R. Simmons, VP-Oper.
Laura E. Park, VP-Comm.
Laura E. Park, VP-Investor Rel.
Joan E. Miller, Sr. VP-Pathology & Hospital Svcs.
Richard L. Bevan, VP-Insurer & Employer Svcs.
Steve Burton, VP-Health & Wellness Svcs.
Stephen C. Suffin, Chief Laboratory Officer/VP
Surya N. Mohapatra, Chmn.
Ken Finnegan, VP-Int'l

Phone: 201-393-5000	Fax: 201-462-4169
Toll-Free: 800-222-0446	
Address: 3 Giralda Farms, Madison, NJ 07940 US	

GROWTH PLANS/SPECIAL FEATURES:

Quest Diagnostics, Inc. is a U.S. clinical laboratory testing company, offering diagnostic testing and related services to the health care industry. The firm's operations consist of routine, esoteric and clinical trials testing. Quest operates through its national network of over 2,000 patient service centers, principal laboratories in 30 major metropolitan areas, approximately 150 rapid-response laboratories, outpatient anatomic pathology centers, hospital-based laboratories and esoteric testing laboratories on both coasts. Routine tests measure various important bodily health parameters. Tests in this category include blood cholesterol level tests; complete blood cell counts; HIV-related tests; urinalyses; pregnancy and prenatal tests; substance-abuse tests; and allergy tests such as the ImmunoCap test. The company also provides cancer diagnostics, including anatomic pathology services in the U.S. Gene-based and other esoteric tests require more sophisticated technology and highly skilled personnel. The firm's tests in this field include Cardio CRP and HEPTIMAX. Quest's two esoteric testing laboratories, comprising the Nichols Institute, are among the leading esoteric clinical testing laboratories in the world. Esoteric tests involve endocrinology, genetics, immunology, microbiology, oncology, serology, endocrinology, hematology and toxicology. Clinical trial testing primarily involves assessing the safety and efficacy of new drugs to meet FDA requirements. The company has clinical trials testing centers in the U.S. and the U.K., and also provides clinical trials testing in Australia, China, Singapore and South America through affiliated laboratories. Additionally, Quest provides risk management services to the life insurance industry in the U.S. and Canada as well as many other countries. The firm also provides clinical testing to employers for the detection of employee drug use. Other employer services include wellness testing and analytic services to employers who strive to take an active roll in improving the health of themselves and their employees.

Employees are offered medical, dental and life insurance; an employee assistance program; disability coverage; flexible spending accounts; and an employee assistance program.

FINANCIALS: Sales and profits are in thousands of dollars—add 000 to get the full amount. 2009 Note: Financial information for 2009 was not available for all companies at press time.

2009 Sales: $7,455,243	2009 Profits: $729,111	**U.S. Stock Ticker: DGX**
2008 Sales: $7,249,447	2008 Profits: $581,490	**Int'l Ticker:** Int'l Exchange:
2007 Sales: $6,704,907	2007 Profits: $339,939	Employees: 43,000
2006 Sales: $6,268,659	2006 Profits: $586,421	Fiscal Year Ends: 12/31
2005 Sales: $5,456,726	2005 Profits: $546,277	Parent Company:

SALARIES/BENEFITS:

Pension Plan:	ESOP Stock Plan:	Profit Sharing:	Top Exec. Salary: $1,201,095	Bonus: $1,567,670
Savings Plan: Y	Stock Purch. Plan: Y		Second Exec. Salary: $541,009	Bonus: $634,928

OTHER THOUGHTS:

Apparent Women Officers or Directors: 7
Hot Spot for Advancement for Women/Minorities: Y

LOCATIONS: ("Y" = Yes)

West:	Southwest:	Midwest:	Southeast:	Northeast:	International:
Y	Y	Y	Y	Y	Y

QUESTCOR PHARMACEUTICALS

www.questcor.com

Industry Group Code: 325412   Ranks within this company's industry group: Sales: 77   Profits: 49

Drugs:		Other:	Clinical:	Computers:	Services:	
Discovery:	Y	AgriBio:	Trials/Services:	Hardware:	Specialty Services:	
Licensing:		Genetic Data:	Labs:	Software:	Consulting:	
Manufacturing:		Tissue Replacement:	Equipment/Supplies:	Arrays:	Blood Collection:	
Generics:			Research/Development Svcs.:	Database Management:	Drug Delivery:	
			Diagnostics:		Drug Distribution:	Y

TYPES OF BUSINESS:

Drugs-Neurology
Multiple Sclerosis Treatment
Kidney Proteinuria (Nephrotic Syndrome) Treatment

BRANDS/DIVISIONS/AFFILIATES:

HP Acthar Gel
Doral
QSC-001

CONTACTS: Note: Officers with more than one job title may be intentionally listed here more than once.

Don M. Bailey, CEO
Don M. Bailey, Pres.
Gary Sawka, CFO/Sr. VP
Eldon Mayer, VP-Commercial Oper.
David Young, Chief Scientific Officer
Timothy O'Neill, VP-Contract Mfg.
Michael H. Mulroy, Sec.
Dave Medieros, Sr. VP-Pharmaceutical Oper.
Steve Cartt, Chief Business Officer/Exec. VP
Gary Sawka, Sr. VP-Finance
Sian Bigora, Sr. VP-Regulatory Affairs
Jason Zielonka, Chief Medical Officer/Sr. VP
Virgil D. Thompson, Chmn.

Phone: 510-400-0700	Fax: 510-400-0799
Toll-Free:	
Address: 3260 Whipple Rd., Union City, CA 94587 US	

GROWTH PLANS/SPECIAL FEATURES:

Questcor Pharmaceuticals, Inc. is a biopharmaceutical company that focuses on novel therapeutics for the treatment of diseases and disorders of the central nervous system (CNS). The company currently owns and markets two commercial products: H.P. Acthar Gel (Acthar) and Doral. Acthar is a natural source, highly purified preparation of the adrenal corticotropin hormone, which is specially formulated to provide prolonged release after intramuscular or subcutaneous injection. The product is indicated for use in acute exacerbations of multiple sclerosis (MS) and is prescribed for patients that have MS and experience painful, episodic flares. It is also FDA-approved for treatment in patients with nephrotic syndrome. The company is in the process of getting Acthar approved for infantile spasms (IS), a condition for which no other drug has been approved. Doral (quazepam) is a non-narcotic, selective benzodiazepine receptor agonist that is indicated for the treatment of insomnia. Sleep disturbance and insomnia are very common side effects of many neurological diseases and disorders such as multiple sclerosis, Epilepsy, Parkinson's disease and Alzheimer's disease. The company, in conjunction with Eurand, is developing QSC-001, an orally dissolving tablet for severe pain in patients with swallowing difficulties

Employees are offered medical, dental and vision benefits; short- and long-term disability plans; life insurance; flexible spending accounts; and an employee assistance program.

FINANCIALS: Sales and profits are in thousands of dollars—add 000 to get the full amount. 2009 Note: Financial information for 2009 was not available for all companies at press time.

2009 Sales: $88,320	2009 Profits: $26,629	U.S. Stock Ticker: QSC
2008 Sales: $95,248	2008 Profits: $40,532	Int'l Ticker: Int'l Exchange:
2007 Sales: $49,768	2007 Profits: $36,449	Employees: 46
2006 Sales: $12,788	2006 Profits: $-10,109	Fiscal Year Ends: 12/31
2005 Sales: $14,162	2005 Profits: $7,392	Parent Company:

SALARIES/BENEFITS:

Pension Plan:	ESOP Stock Plan:	Profit Sharing:	Top Exec. Salary: $546,000	Bonus: $301,665
Savings Plan: Y	Stock Purch. Plan: Y		Second Exec. Salary: $364,000	Bonus: $170,170

OTHER THOUGHTS:

Apparent Women Officers or Directors: 1
Hot Spot for Advancement for Women/Minorities:

LOCATIONS: ("Y" = Yes)

West:	Southwest:	Midwest:	Southeast:	Northeast:	International:
Y					

QUINTILES TRANSNATIONAL CORP

www.quintiles.com

Industry Group Code: 541712 Ranks within this company's industry group: Sales: Profits:

Drugs:		Other:	Clinical:		Computers:		Services:	
Discovery:		AgriBio:	Trials/Services:	Y	Hardware:		Specialty Services:	Y
Licensing:	Y	Genetic Data:	Labs:	Y	Software:		Consulting:	Y
Manufacturing:		Tissue Replacement:	Equipment/Supplies:		Arrays:		Blood Collection:	
Generics:			Research/Development Svcs.:	Y	Database Management:		Drug Delivery:	
			Diagnostics:				Drug Distribution:	

TYPES OF BUSINESS:

Contract Research
Pharmaceutical, Biotech & Medical Device Research
Consulting & Training Services
Sales & Marketing Services

BRANDS/DIVISIONS/AFFILIATES:

Novaquest
Innovex
Movetis NV
Resolor

CONTACTS: *Note: Officers with more than one job title may be intentionally listed here more than once.*

Dennis Gillings, CEO
John Ratliff, COO
Mike Troullis, CFO
Millie Tan, Sr. VP-Global Mktg.
Christopher Cabell, Chief Medical & Scientific Officer
William R. Deam, CIO/Exec. VP
John Goodacre, General Counsel/Exec. VP
Ron Wooten, Exec. VP-Corp. Dev.
Oren Cohen, Sr. VP-Clinical Research Strategies
Dennis Gillings, Chmn.

Phone: 919-998-2000	Fax: 919-998-9113
Toll-Free:	
Address: 4820 Emperor Blvd., Durham, NC 27703 US	

GROWTH PLANS/SPECIAL FEATURES:

Quintiles Transnational Corp. provides full-service contract research, sales and marketing services to the global pharmaceutical, biotechnology and medical device industries. The company is one of the world's top contract research organizations, and it provides a broad range of contract services to speed the process from development to peak sales of a new drug or medical device. The firm operates through offices in 50 countries, organized in three primary business segments, including the product development group, the Innovex commercialization group and the Novaquest strategic partnering solutions group. The product development group provides a full range of drug development services from strategic planning and preclinical services to regulatory submission and approval. The commercial services group, which operates under the Innovex brand, engages in sales solutions such as recruitment, training and deployment of Innovex managed sales teams; and medical communications, which provides and promotes physician education. The Novaquest strategic partnering solutions segment attempts to optimize portfolio development, company growth and profits for pharmaceutical and biotech companies through a variety of solutions such as structured finance, strategic resourcing and eBio. The company also provides a consulting service that focus on product commercialization, market intelligence, market access, regulatory and quality consulting. In January 2010, the company contracted with the European pharmaceutical firm Movetis NV to commercialize its new product, Resolor. In March 2010, the company opened a new facility in Tokyo, Japan, as well as a new office in Nairobi, Kenya. Also in March 2010, the company opened a newly expanded research facility in London.

The company offers its employees a comprehensive benefits package including on-the-job training, recreational activities and community support activities.

FINANCIALS: Sales and profits are in thousands of dollars—add 000 to get the full amount. 2009 Note: Financial information for 2009 was not available for all companies at press time.

2009 Sales: $	2009 Profits: $	U.S. Stock Ticker: Private
2008 Sales: $2,800,000	2008 Profits: $	Int'l Ticker: Int'l Exchange:
2007 Sales: $	2007 Profits: $	Employees: 23,000
2006 Sales: $	2006 Profits: $	Fiscal Year Ends: 12/31
2005 Sales: $2,398,583	2005 Profits: $ 648	Parent Company:

SALARIES/BENEFITS:

Pension Plan:	ESOP Stock Plan:	Profit Sharing:	Top Exec. Salary: $	Bonus: $
Savings Plan:	Stock Purch. Plan:		Second Exec. Salary: $	Bonus: $

OTHER THOUGHTS:

Apparent Women Officers or Directors:
Hot Spot for Advancement for Women/Minorities:

LOCATIONS: ("Y" = Yes)

West:	Southwest:	Midwest:	Southeast:	Northeast:	International:
Y	Y	Y	Y	Y	Y

RANBAXY LABORATORIES LIMITED

www.ranbaxy.com

Industry Group Code: 325412 Ranks within this company's industry group: Sales: 36 Profits: 41

Drugs:		Other:		Clinical:		Computers:		Services:	
Discovery:		AgriBio:		Trials/Services:		Hardware:		Specialty Services:	
Licensing:	Y	Genetic Data:		Labs:		Software:		Consulting:	
Manufacturing:	Y	Tissue Replacement:		Equipment/Supplies:		Arrays:		Blood Collection:	
Generics:				Research/Development Svcs.:		Database Management:		Drug Delivery:	Y
				Diagnostics:	Y			Drug Distribution:	

TYPES OF BUSINESS:

Generic Pharmaceuticals
Active Pharmaceutical Ingredients
Drugs-Anti-Retroviral & HIV/AIDS
Drug Delivery Systems
Veterinary Pharmaceuticals
Specialty Chemicals
Diagnostics

BRANDS/DIVISIONS/AFFILIATES:

Valacyclovir Hydrochloride
SEBIFIN Terbinafine
Simvastatin
Olvance
Ran-Simvastatin
Ran-Amlodipine
Covance
Ketorolac

CONTACTS: *Note: Officers with more than one job title may be intentionally listed here more than once.*

Atul Sobti, CEO
Omesh Sethi, Pres.
Omesh Sethi, CFO
Bhagwat Yagnik, Head-Global Human Resources
Sudarshan K. Arora, Pres., R&D
Arun Sawhney, Pres., Global Pharmaceuticals Bus.
Dale Adkisson, Head-Global Quality
Hiroyuki Okuzawa, Head-Global Synergy Project
Tsutomu Une, Chmn.

Phone: 91-124-413-5000	Fax: 91-124-413-5001
Toll-Free:	
Address: Plot 90, Sector 32, Gurgaon, 122001 India	

GROWTH PLANS/SPECIAL FEATURES:

Ranbaxy Laboratories Limited is a leading Indian manufacturer and marketer of generic pharmaceuticals, branded generics, over-the-counter medications and active pharmaceutical ingredients. It has manufacturing operations in seven countries, a ground presence in 46 countries and its products are available in over 125 nations worldwide. The firm offers its products in six segments: anti-infectives; cardiovasculars, musculoskeletal, central nervous system, gastrointestinals and dermatologicals. The anti-infectives products include Valacyclovir Hydrochloride tablets in the U.S. and SEBIFIN Terbinafine tablets in Australia. The cardiovasculars product line includes Simvastatin; Olvance (Olmesartan Medoxomil) and its fixed dose combination with Amlodipine (Ol-Vamlo) in India; Ran-Simvastatin (Simvastatin) and Ran-Amlodipine (Amlodipine) in Canada; and Covance (Losartan) in Malaysia. The musculoskeletal product line includes Ketorolac; Evista (Raloxifene) in Romania; and Volini in India. The central nervous system product line includes Gabapentin; Sertraline; Oxcarbazepine Suspension and Sumatriptan tablets; Topiramate tablets in the US; Risperidone tablets in Australia; and Ropinirole tablets in Canada. The gastrointestinals product line includes Famotidine tablets; Glycopyrrolate tablets; and Ondansetron tablets in Canada. Lastly, the dermatologicals segment includes Derma products in India; Neuronox, which is purified Botulinum Toxin Type-A; and Lulifin (Luliconazole), which was launched under license from Summit Pharmaceuticals, Japan. In 2009, the company launched 42 products, including seven First-to-Launch products for the domestic market. Rambaxy conducts major research collaborations with GlaxoSmithKline (GSK). The firm has over 50 subsidiaries in numerous countries including India, Brazil, Canada, Egypt, Ireland, Italy, Malaysia, Mexico, Nigeria, Poland, Portugal, China, Romania, South Africa, Spain, Thailand and the U.S., among others. In May 2010, the firm gained approval from Health Canada to market RAN-Atorvastatin tablets in the Canadian Healthcare System. In June 2010, the company launched Prasita, a generic version of Prasugrel, in India; and Lipogen, a generic version of Lipitor, in South Africa.

Ranbaxy offers employees group life insurance, medical insurance and pension plans.

FINANCIALS: Sales and profits are in thousands of dollars—add 000 to get the full amount. 2009 Note: Financial information for 2009 was not available for all companies at press time.

2009 Sales: $1,619,700	2009 Profits: $103,328	**U.S. Stock Ticker: RBXLF.PK**
2008 Sales: $1,531,700	2008 Profits: $-196,320	**Int'l Ticker: 500359** Int'l Exchange: Bombay-BSE
2007 Sales: $1,399,530	2007 Profits: $159,870	Employees: 12,995
2006 Sales: $1,444,400	2006 Profits: $123,710	Fiscal Year Ends: 12/31
2005 Sales: $1,244,190	2005 Profits: $60,200	Parent Company:

SALARIES/BENEFITS:

Pension Plan: Y	ESOP Stock Plan:	Profit Sharing:	Top Exec. Salary: $	Bonus: $
Savings Plan:	Stock Purch. Plan:		Second Exec. Salary: $	Bonus: $

OTHER THOUGHTS:

Apparent Women Officers or Directors:
Hot Spot for Advancement for Women/Minorities:

LOCATIONS: ("Y" = Yes)

West:	Southwest:	Midwest:	Southeast:	Northeast:	International:
				Y	Y

Note: Financial information, benefits and other data can change quickly and may vary from those stated here.

RAPTOR PHARMACEUTICAL CORP   www.raptorpharmaceuticals.com

Industry Group Code: 325412 Ranks within this company's industry group: Sales: Profits:

Drugs:		Other:	Clinical:	Computers:	Services:
Discovery:	Y	AgriBio:	Trials/Services:	Hardware:	Specialty Services:
Licensing:	Y	Genetic Data:	Labs:	Software:	Consulting:
Manufacturing:		Tissue Replacement:	Equipment/Supplies:	Arrays:	Blood Collection:
Generics:			Research/Development Svcs.:	Database Management:	Drug Delivery:
			Diagnostics:		Drug Distribution:

TYPES OF BUSINESS:

Pharmaceutical Acquisition & Development
Biopharmaceuticals
Drug Development
Breast Cancer Treatment Products
Cystinosis Treatment Products
Huntington's Disease Treatment Products
Migraine Treatment Products
Hepatitis C Treatment Products

BRANDS/DIVISIONS/AFFILIATES:

DR Cysteamine
HepTide
NGX426
NeuroTrans
Convivia
WntTide
Tezampanel

CONTACTS: Note: Officers with more than one job title may be intentionally listed here more than once.

Christopher M. Starr, CEO
Thomas E. Daley, Pres.
Kim R. Tscuchimoto, CFO/Sec.
Todd C. Zankel, Chief Scientific Officer
Paul R. Schneider, General Counsel/VP
Kim R. Tsuchimoto, Treas.
Patrice P. Rioux, Chief Medical Officer

Phone: 415-382-8111	Fax: 415-382-1386
Toll-Free: 877-727-8679	
Address: 9 Commercial Blvd., Ste. 200, Novato, CA 94949 US	

GROWTH PLANS/SPECIAL FEATURES:

Raptor Pharmaceuticals Corporation engages in the discovery, development, and commercialization of novel small molecules to treat diseases and disorders of the central nervous system. Its therapeutic focus on chronic diseases including dry mouth, migraine and neuropathic pain; and cognitive disorders, including cognitive impairment associated with schizophrenia and Alzheimer's disease. Raptor has three products in active clinical development, all of which utilize delayed release, enterically coated cysteamine bitartrate: DR Cysteamine for the treatment of Cystinosis; DR Cysteamine for the treatment of Non-Alcoholic Steatohepatitis (NASH); and DR Cysteamine for the treatment of Huntington's disease. The company also has three product candidates for which it is seeking business development partners but are not actively developing: Convivia, for the management of acetaldehyde toxicity due to alcohol consumption by individuals with aldehyde dehydrogenase, an inherited metabolic disorder; Tezampanel, which has undergone a Phase IIb trial for the abortive treatment of migraines; and its oral prodrug, NGX426, which recently completed its Phase I clinical trial for chronic pain treatment. Raptor's preclinical product candidates include two therapeutics based on the company's receptor-associated protein (RAP): HepTide, for the potential treatment of primary liver cancer and hepatitis C; and NeuroTrans for the potential delivery of therapeutics across the blood-brain barrier for treatment if various neurological diseases. The third, WntTide, is based on Raptor's mesoderm development protein, Mesd, and is being developed for the potential treatment of breast cancer. In June 2010, the company announced that it had reached an agreement with Uni Pharma Co., Ltd. to commercialize Convivia in Taiwan.

FINANCIALS: Sales and profits are in thousands of dollars—add 000 to get the full amount. 2009 Note: Financial information for 2009 was not available for all companies at press time.

2009 Sales: $	2009 Profits: $	U.S. Stock Ticker: RPTP
2008 Sales: $6,071	2008 Profits: $-22,785	Int'l Ticker: Int'l Exchange:
2007 Sales: $9,850	2007 Profits: $-23,369	Employees: 10
2006 Sales: $9,850	2006 Profits: $-25,377	Fiscal Year Ends: 12/31
2005 Sales: $7,967	2005 Profits: $-11,542	Parent Company:

SALARIES/BENEFITS:

Pension Plan:	ESOP Stock Plan:	Profit Sharing:	Top Exec. Salary: $266,987	Bonus: $
Savings Plan:	Stock Purch. Plan:		Second Exec. Salary: $215,115	Bonus: $

OTHER THOUGHTS:

Apparent Women Officers or Directors: 2
Hot Spot for Advancement for Women/Minorities: Y

LOCATIONS: ("Y" = Yes)

West:	Southwest:	Midwest:	Southeast:	Northeast:	International:
Y					

REGENERON PHARMACEUTICALS INC

www.regeneron.com

Industry Group Code: 325412 Ranks within this company's industry group: Sales: 50 Profits: 148

Drugs:		Other:		Clinical:	Computers:	Services:
Discovery:	Y	AgriBio:		Trials/Services:	Hardware:	Specialty Services:
Licensing:	Y	Genetic Data:	Y	Labs:	Software:	Consulting:
Manufacturing:	Y	Tissue Replacement:		Equipment/Supplies:	Arrays:	Blood Collection:
Generics:				Research/Development Svcs.:	Database Management:	Drug Delivery:
				Diagnostics:		Drug Distribution:

TYPES OF BUSINESS:

Drugs-Diversified
Protein-Based Drugs
Small-Molecule Drugs
Genetics & Transgenic Mouse Technologies

BRANDS/DIVISIONS/AFFILIATES:

VelocImmune
VelociGene
VelociMouse
VEGF Trap
VEGF Trap-Eye
VelociMab
ARCALYST
REGN88

CONTACTS: Note: Officers with more than one job title may be intentionally listed here more than once.

Leonard S. Schleifer, CEO
Leonard S. Schleifer, Pres.
Murray A. Goldberg, CFO
George D. Yancopoulos, Chief Scientific Officer/Exec. VP
Murray A. Goldberg, Sr. VP-Admin.
Stuart A. Kolinski, General Counsel/Sr. VP/Sec.
Daniel Van Plew, Sr. VP/Gen. Mgr.-Industrial Oper.
Murray A. Goldberg, Sr. VP-Finance/Treas.
George D. Yancopoulos, Pres., Regeneron Research Laboratories
Neil Stahl, Sr. VP-R&D Sciences
Peter Powchik, Sr. VP-Clinical Dev.
Robert J. Terifay, Sr. VP-Commercial
P. Roy Vagelos, Chmn.
Daniel Van Plew, Gen. Mgr.-Prod. Supply

Phone: 914-345-7400	Fax:
Toll-Free:	
Address: 777 Old Saw Mill River Rd., Tarrytown, NY 10591 US	

GROWTH PLANS/SPECIAL FEATURES:

Regeneron Pharmaceuticals, Inc. is a biopharmaceutical company that discovers, develops and commercializes pharmaceutical drugs for the treatment of serious medical conditions. The company's sole marketed product is Arcalyst, a therapy for the treatment of Cryopyrin-Associated Periodic Syndromes (CAPS), a rare, inherited inflammatory condition. The firm currently has three late-stage clinical development programs: aflibercept (VEGF Trap) in oncology; VEGF Trap eye formulation (VEGF Trap-Eye) in eye diseases using intraocular delivery; and rilonacept (IL-1 Trap) in gout. The VEGF Trap oncology development program is being developed jointly with the sanofi-aventis Group. Also in collaboration with sanofi-aventis is REGN88, an antibody to the Interleukin-6 receptor (IL-6R) for treatment of rheumatoid arthritis, developed using VelocImmune technologies. Regeneron's preclinical research programs are in the areas of oncology, angiogenesis, ophthalmology, metabolic and related diseases, muscle diseases and disorders, inflammation and immune diseases, bone and cartilage, pain, metabolic diseases, infectious diseases and cardiovascular diseases. The company expects that its next generation of product candidates will be based on its proprietary technologies for developing Traps and Human Monoclonal Antibodies. Regeneron's proprietary technologies include VelociGene, VelociMouse and VelocImmune, among others. The VelociGene technology allows precise DNA manipulation and gene staining, helping to identify where a particular gene is active in the body. VelociMouse technology allows for the direct and immediate generation of genetically altered mice from ES cells, avoiding the lengthy process involved in generating and breeding knock-out mice from chimeras. VelocImmune is a novel mouse technology platform for producing fully human monoclonal antibodies. Additionally, the VelociMab suite of technologies allows rapid screenings of therapeutic antibodies and eliminates the need for slower development techniques.

Regeneron employee benefits include medical, dental, vision, prescription, life and AD&D insurance; short- and long-term disability; flexible spending accounts, and tuition reimbursement.

FINANCIALS: Sales and profits are in thousands of dollars—add 000 to get the full amount. 2009 Note: Financial information for 2009 was not available for all companies at press time.

2009 Sales: $379,268	2009 Profits: $-67,830	U.S. Stock Ticker: REGN
2008 Sales: $238,457	2008 Profits: $-86,249	Int'l Ticker: Int'l Exchange:
2007 Sales: $125,024	2007 Profits: $-115,373	Employees: 1,029
2006 Sales: $63,447	2006 Profits: $-102,337	Fiscal Year Ends: 12/31
2005 Sales: $66,193	2005 Profits: $-95,446	Parent Company:

SALARIES/BENEFITS:

Pension Plan:	ESOP Stock Plan:	Profit Sharing:	Top Exec. Salary: $734,400	Bonus: $2,054,720
Savings Plan: Y	Stock Purch. Plan:		Second Exec. Salary: $609,900	Bonus: $1,709,900

OTHER THOUGHTS:

Apparent Women Officers or Directors:
Hot Spot for Advancement for Women/Minorities:

LOCATIONS: ("Y" = Yes)

West:	Southwest:	Midwest:	Southeast:	Northeast:	International:
				Y	

Note: Financial information, benefits and other data can change quickly and may vary from those stated here.

REPLIGEN CORPORATION

www.repligen.com

Industry Group Code: 325412 **Ranks within this company's industry group:** Sales: 101   Profits: 65

Drugs:		Other:	Clinical:		Computers:	Services:	
Discovery:	Y	AgriBio:	Trials/Services:		Hardware:	Specialty Services:	
Licensing:		Genetic Data:	Labs:		Software:	Consulting:	
Manufacturing:	Y	Tissue Replacement:	Equipment/Supplies:		Arrays:	Blood Collection:	
Generics:			Research/Development Svcs.:		Database Management:	Drug Delivery:	
			Diagnostics:	Y		Drug Distribution:	

TYPES OF BUSINESS:

Neuropsychiatric Drugs
Protein A Drugs

BRANDS/DIVISIONS/AFFILIATES:

SecreFlo
ELISA
RPA50
IPA300
IPA400HC
CaptivA
Uridine

CONTACTS: *Note: Officers with more than one job title may be intentionally listed here more than once.*

Walter C. Herlihy, CEO
Walter C. Herlihy, Pres.
William J. Kelly, CFO
Laura Whitehouse, VP-Market Dev.
James R. Rusche, Sr. VP-R&D
William J. Kelly, VP-Admin.
William J. Kelly, Sec.
Daniel P. Witt, VP-Oper.
Howard Benjamin, VP-Bus. Dev.
William J. Kelly, VP-Finance
Stephen Tingley, VP-Bioprocessing Bus. Dev.
Alexander Rich, Chmn.

Phone: 781-250-0111	Fax: 781-250-0115
Toll-Free: 800-622-2259	
Address: 41 Seyon St., Bldg. 1, Ste. 100, Waltham, MA 02453 US	

GROWTH PLANS/SPECIAL FEATURES:

Repligen Corp. develops therapies that harness biological pathways for the treatment of neurological and gastroenterological diseases and disorders, as well as rare and under-addressed, or orphan, diseases. The company develops drugs for diseases such as pancreatitis, bipolar disorder, Friedrich's ataxia and spinal muscular atrophy. Repligen manufactures a line of products based on protein A, which is used in the production of many therapeutic monoclonal antibodies and other biopharmaceutical manufacturing applications. Its protein A products include recombinant protein A ligands (rPA50, srP450), immobilized protein A (IPA300and IPA400HC), protein A ELISA kits, CaptivA affinity resins, and Opus Pre-Packed Run Ready chromatography columns. Protein A products have a wide variety of end-uses including applications in drugs for colon cancer, infection, Crohn's disease and arthritis. In addition to protein A products, the firm market SecreFlo, a synthetic form of the hormone secretin which is used as an aid in the diagnosis chronic pancreatitis and gastrinoma. Repligen also has products in the development stage for neuropsychiatric disorders and other applications: secretin, a hormone produced in the small intestine that regulates the function of the pancreas as part of the process of digestion, evaluated by the company for improvement of MRI imaging of the pancreas; Uridine, a biological compound essential for the synthesis of DNA and RNA, being tested by the firm under an oral formulation for the treatment of Bipolar Depression; and Histone Deacetylase inhibitors for Friedreich's Ataxia. In March 2010, the company announced the acquisition of the assets of BioFlash Partners, LLC, which include a technology platform for the production of pre-packed chromatography packs, for $1.8 million.

Repligen offers its employees benefits that include health and dental insurance; life and long-term disability coverage; and equity participation.

FINANCIALS: Sales and profits are in thousands of dollars—add 000 to get the full amount. 2009 Note: Financial information for 2009 was not available for all companies at press time.

2009 Sales: $29,362	2009 Profits: $5,746	U.S. Stock Ticker: RGEN
2008 Sales: $19,296	2008 Profits: $37,107	Int'l Ticker: Int'l Exchange:
2007 Sales: $14,074	2007 Profits: $- 889	Employees: 69
2006 Sales: $12,911	2006 Profits: $ 697	Fiscal Year Ends: 3/31
2005 Sales: $9,360	2005 Profits: $-2,984	Parent Company:

SALARIES/BENEFITS:

| Pension Plan: | ESOP Stock Plan: Y | Profit Sharing: | Top Exec. Salary: $365,000 | Bonus: $149,013 |
| Savings Plan: Y | Stock Purch. Plan: | | Second Exec. Salary: $246,000 | Bonus: $61,313 |

OTHER THOUGHTS:

Apparent Women Officers or Directors: 2
Hot Spot for Advancement for Women/Minorities:

LOCATIONS: ("Y" = Yes)

West:	Southwest:	Midwest:	Southeast:	Northeast:	International:
				Y	

REPROS THERAPEUTICS INC

www.reprosrx.com

Industry Group Code: 325412 **Ranks within this company's industry group: Sales: 145   Profits: 130**

Drugs:		Other:	Clinical:	Computers:	Services:
Discovery:	Y	AgriBio:	Trials/Services:	Hardware:	Specialty Services:
Licensing:	Y	Genetic Data:	Labs:	Software:	Consulting:
Manufacturing:		Tissue Replacement:	Equipment/Supplies:	Arrays:	Blood Collection:
Generics:			Research/Development Svcs.:	Database Management:	Drug Delivery:
			Diagnostics:		Drug Distribution:

TYPES OF BUSINESS:

Drugs-Fertility & Sexual Dysfunction
Reproductive System Disorder Treatment
Hormonal Disorder Treatment

BRANDS/DIVISIONS/AFFILIATES:

Proellex
Androxal
VASOMAX
Z-Max

CONTACTS: Note: Officers with more than one job title may be intentionally listed here more than once.

Joseph S. Podolski, CEO
Joseph S. Podolski, Pres.
Ronald Wiehle, VP-R&D
Katherine A. Anderson, Corp. Sec.
Katherine A. Anderson, Chief Acct. Officer
Nola E. Masterson, Chmn.

Phone: 281-719-3400	Fax: 281-719-3446
Toll-Free:	
Address: 2408 Timberloch Pl., Ste. B-7, The Woodlands, TX 77380 US	

GROWTH PLANS/SPECIAL FEATURES:

Repros Therapeutics, Inc., formerly Zonagen, Inc., develops oral small molecule drugs for the treatment of male and female reproductive disorders. Proellex, its leading product candidate, is a selective progesterone receptor blocker for the treatment of uterine fibroids and the treatment of endometriosis. The National Uterine Fibroids Foundation estimates that as many as 80% of all women in the U.S. have uterine fibroids, and one in four of these women require treatment. According to The Endometriosis Association, endometriosis affects 7.3 million women in the U.S. and Canada. During 2009, the FDA placed Proellex under a full clinical hold, in accordance with which all clinical testing was halted. Proellex's status was changed from full to partial clinical hold in June 2010. Under this new status, Repros will be allowed to run a single low-dose study. The company's other product candidate, Androxal, is an orally available small molecule compound. Androxal targets adult-onset idiopathic hypogonadotrophic hypogonadism with concomitant plasma glucose and lipid elevations; and improves or maintains fertility and/or sperm function in men being treated for low testosterone. The drug is also being evaluated for use in treating Type II diabetes. Androxal has completed Phase IIb testing in the U.S., and Repros has received FDA approval to conduct a Phase IIa trial. The company has a third, phentolamine-based product, VASOMAX, which is out-licensed in certain Latin American countries for the treatment of male erectile dysfunction (ED) under the Z-Max brand. VASOMAX is currently on partial clinical hold in the U.S. In January 2009, the firm reported positive results in its Proellex endometriosis trials, with statistically significant pain reductions compared to a placebo.

FINANCIALS: Sales and profits are in thousands of dollars—add 000 to get the full amount. 2009 Note: Financial information for 2009 was not available for all companies at press time.

2009 Sales: $ 551	2009 Profits: $-27,234	U.S. Stock Ticker: RPRX
2008 Sales: $ 433	2008 Profits: $-25,202	Int'l Ticker: Int'l Exchange:
2007 Sales: $1,508	2007 Profits: $-13,700	Employees: 5
2006 Sales: $ 596	2006 Profits: $-14,195	Fiscal Year Ends: 12/31
2005 Sales: $ 634	2005 Profits: $-7,391	Parent Company:

SALARIES/BENEFITS:

Pension Plan:	ESOP Stock Plan:	Profit Sharing:	Top Exec. Salary: $353,682	Bonus: $
Savings Plan:	Stock Purch. Plan:		Second Exec. Salary: $285,111	Bonus: $

OTHER THOUGHTS:

Apparent Women Officers or Directors: 2
Hot Spot for Advancement for Women/Minorities:

LOCATIONS: ("Y" = Yes)

West:	Southwest:	Midwest:	Southeast:	Northeast:	International:
	Y				

RIGEL PHARMACEUTICALS INC

www.rigel.com

Industry Group Code: 325412 Ranks within this company's industry group: Sales: 142   Profits: 160

Drugs:		Other:	Clinical:	Computers:	Services:
Discovery:	Y	AgriBio:	Trials/Services:	Hardware:	Specialty Services:
Licensing:	Y	Genetic Data:	Labs:	Software:	Consulting:
Manufacturing:		Tissue Replacement:	Equipment/Supplies:	Arrays:	Blood Collection:
Generics:			Research/Development Svcs.:	Database Management:	Drug Delivery:
			Diagnostics:		Drug Distribution:

TYPES OF BUSINESS:

Biopharmaceuticals Development
Small-Molecule Drugs
Drugs-Cancer & Inflammatory Diseases
Drugs-Viral Diseases
Drugs-Autoimmune Diseases

BRANDS/DIVISIONS/AFFILIATES:

R788
R763/AS703569
R343

CONTACTS: Note: Officers with more than one job title may be intentionally listed here more than once.

James M. Gower, CEO
Raul R. Rodriguez, COO
Raul R. Rodriguez, Pres.
Ryan Maynard, CFO/Exec. VP
Donald G. Payan, Exec. VP/Pres., Research & Discovery
Dolly Vance, General Counsel/Corp. Sec.
Dolly Vance, Exec. VP-Corp. Affairs
Elliott B. Grossbard, Chief Medical Officer/Exec. VP
James M. Gower, Chmn.

Phone: 650-624-1100	Fax: 650-624-1101
Toll-Free:	
Address: 1180 Veterans Blvd., South San Francisco, CA 94080 US	

GROWTH PLANS/SPECIAL FEATURES:

Rigel Pharmaceuticals, Inc. is a biotechnology company focused on developing small-molecule drugs in the fields of inflammatory/autoimmune diseases, metabolic diseases and certain cancers. The firm's lead candidate, R788, is a potential drug for the treatment of rheumatoid arthritis that functions by inhibiting IgG receptor signaling in macrophages and B-cells. It recently completed a Phase II clinical trial, and Rigel is currently designing a global Phase III clinical trial for R788. R788 is also being tested by the firm in various stages of clinical trials for indications of B-Cell lymphoma, T-Cell lymphoma, certain solid tumors and immune thrombocytopenia purpura. The company's cancer treatment candidate, R763/AS703569, is a specific inhibitor of aurora kinase, shown to block proliferation of trigger apoptosis in several tumor cell lines. It is currently in Phase I clinical trials for the treatment of patients with refractory solid tumors and hematological malignancies. R343, an oral Syk kinase inhibitor and the company's asthma and allergy treatment candidate, functions by inhibiting the IgE receptor signaling in respiratory tract mast cells. The company holds a partnership with Pfizer for the development of R343, which is currently in Phase I development. Rigel also has a partnership with Daiichi Sankyo for ubiquitin ligase oncology targets, which are used to minimize tumor growth.

Rigel offers its employees medical, dental and vision coverage; domestic partner benefits; a 401(k) plan; commuter benefits; free shuttle services; employee stock options; an employee stock purchase plan; flexible spending accounts; an employee assistance program; life and AD&D insurance; business travel accident insurance; an employee referral plan; pre-paid legal services and identity theft monitoring services; short- and long-term disability; a subsidized cafeteria; a tuition reimbursement program; a wellness benefit; and employee discounts.

FINANCIALS: Sales and profits are in thousands of dollars—add 000 to get the full amount. 2009 Note: Financial information for 2009 was not available for all companies at press time.

2009 Sales: $ 750	2009 Profits: $-111,547	U.S. Stock Ticker: RIGL
2008 Sales: $	2008 Profits: $-132,435	Int'l Ticker: Int'l Exchange:
2007 Sales: $12,600	2007 Profits: $-74,272	Employees: 142
2006 Sales: $33,473	2006 Profits: $-37,637	Fiscal Year Ends: 12/31
2005 Sales: $16,526	2005 Profits: $-45,256	Parent Company:

SALARIES/BENEFITS:

Pension Plan:	ESOP Stock Plan:	Profit Sharing:	Top Exec. Salary: $600,000	Bonus: $504,000
Savings Plan: Y	Stock Purch. Plan: Y		Second Exec. Salary: $483,000	Bonus: $338,100

OTHER THOUGHTS:

Apparent Women Officers or Directors: 1
Hot Spot for Advancement for Women/Minorities:

LOCATIONS: ("Y" = Yes)

West:	Southwest:	Midwest:	Southeast:	Northeast:	International:
Y					

ROCHE HOLDING LTD

www.roche.com

Industry Group Code: 325412 Ranks within this company's industry group: Sales: 3   Profits: 8

Drugs:		Other:	Clinical:		Computers:	Services:
Discovery:	Y	AgriBio:	Trials/Services:		Hardware:	Specialty Services:
Licensing:		Genetic Data:	Labs:		Software:	Consulting:
Manufacturing:	Y	Tissue Replacement:	Equipment/Supplies:		Arrays:	Blood Collection:
Generics:			Research/Development Svcs.:		Database Management:	Drug Delivery:
			Diagnostics:	Y		Drug Distribution:

TYPES OF BUSINESS:

Pharmaceuticals Manufacturing
Antibiotics
Diagnostics
Cancer Drugs
Virology Products
HIV/AIDS Treatments
Transplant Drugs

BRANDS/DIVISIONS/AFFILIATES:

Genentech Inc
454 Life Sciences
NimbleGen Systems Inc
Therapeutic Human Polyclonals Inc
Ventana Medical Systems Inc
Chugai Pharmaceuticals
Gilead Sciences
Memory Pharmaceuticals Corp

CONTACTS: Note: Officers with more than one job title may be intentionally listed here more than once.

Severin Schwan, CEO
Erich Hunziker, CFO
Sylvia Ayyoubi, Head-Human Resources
Jonathan Knowles, Head-Group Research
Erich Hunziker, CIO
Gottlieb Keller, General Counsel/Head-Corp. Svcs.
Pascal Soriot, Head-Commercial Oper., Pharmaceuticals
Per-Olof Attinger, Head-Global Corp. Comm.
William M. Burns, CEO-Roche Pharmaceuticals
Jurgen Schwiezer, CEO-Roche Diagnostics
Osamu Nagayama, Pres./CEO-Chugai
Franz B. Humer, Chmn.

Phone: 41-61-688-1111	Fax: 41-61-691-9391
Toll-Free:	
Address: Grenzacherstrasse 124, Basel, 4070 Switzerland	

GROWTH PLANS/SPECIAL FEATURES:

Roche Holding, Ltd. is one of the world's largest health care companies, occupying an industry-leading position in the global diagnostics market and ranking as one of the top producers of pharmaceuticals, with particular recognition in the areas of cancer drugs, autoimmune disease and metabolic disorder treatments, virology and transplantation medicine. The company's operations currently extend to over 150 countries, with additional alliances and research and development agreements with corporate and institutional partners, furthering Roche's collective reach. The firm operates via two divisions, Pharmaceuticals and Diagnostics. The Pharmaceuticals division includes Roche Pharma; Genentech; and Chugai. The Diagnostics division consists of Roche Applied Science; Roche Molecular Diagnostics; Roche Professional Diagnostics; Roche Tissue Diagnostics; and Roche Diabetes Care. The company's primary drug products include the cancer drugs Avastin, Bondronat, Herceptin and Tarceva; the antibiotic Rocephin; the HIV/AIDS treatments Viracept, Invirase and Fuzeon; and Tamiflu, which is used to prevent and treat influenza. Roche companies control proprietary diagnostic technologies across a range of areas, including advanced DNA tests, leading consumer diabetes monitoring devices and applied sciences methodologies for laboratory research. As part of the mobilization of Tamiflu, the company has maintained a long-term strategic partnership with Gilead Sciences to coordinate the licensing and manufacture of the drug, important in the case of a flu pandemic. In February 2010, the U.S. FDA approved the company's Adult Leukemia treatment drug, Rituxan/Mabthera, for use within the U.S. In March 2010, the company announced that its Colon Cancer treatment drug, Xeloda, had been approved by the European Commission for sale in Europe. Also in March 2010, the company acquired the MAUI system from Nanostart AG; the MAUI allows for bio-chip analysis of DNA. In April 2010, the company acquired Medingo, Ltd., a subsidiary of the Elron Group.

FINANCIALS: Sales and profits are in thousands of dollars—add 000 to get the full amount. 2009 Note: Financial information for 2009 was not available for all companies at press time.

2009 Sales: $48,287,100	2009 Profits: $7,348,180	U.S. Stock Ticker: RHHBY
2008 Sales: $41,676,500	2008 Profits: $7,803,000	Int'l Ticker: RO Int'l Exchange: Zurich-SWX
2007 Sales: $40,650,000	2007 Profits: $8,600,000	Employees: 81,507
2006 Sales: $34,851,500	2006 Profits: $7,116,030	Fiscal Year Ends: 12/31
2005 Sales: $27,385,668	2005 Profits: $5,189,777	Parent Company:

SALARIES/BENEFITS:

Pension Plan: Y	ESOP Stock Plan:	Profit Sharing:	Top Exec. Salary: $	Bonus: $
Savings Plan:	Stock Purch. Plan:		Second Exec. Salary: $	Bonus: $

OTHER THOUGHTS:

Apparent Women Officers or Directors: 4
Hot Spot for Advancement for Women/Minorities: Y

LOCATIONS: ("Y" = Yes)

West:	Southwest:	Midwest:	Southeast:	Northeast:	International:
Y	Y	Y	Y	Y	Y

Note: Financial information, benefits and other data can change quickly and may vary from those stated here.

RULES BASED MEDICINE INC
www.rulesbasedmedicine.com

Industry Group Code: 6215 Ranks within this company's industry group: Sales: Profits:

Drugs:	Other:	Clinical:		Computers:		Services:	
Discovery:	AgriBio:	Trials/Services:		Hardware:		Specialty Services:	Y
Licensing:	Genetic Data:	Labs:	Y	Software:		Consulting:	
Manufacturing:	Tissue Replacement:	Equipment/Supplies:		Arrays:		Blood Collection:	
Generics:		Research/Development Svcs.:		Database Management:		Drug Delivery:	
		Diagnostics:	Y			Drug Distribution:	

TYPES OF BUSINESS:
Diagnostic Biomarker Testing

BRANDS/DIVISIONS/AFFILIATES:
HumanMAP
PsyMAP
CardiovascularMAP
InflammationMAP
MetabolicMAP
KidneyMAP
RodentMAP
EDI GmbH

CONTACTS: *Note: Officers with more than one job title may be intentionally listed here more than once.*
Craig Benson, CEO
Craig Benson, Pres.
Patirck McClain, CFO
Peter Amatulli, VP-Sales & Mktg.
Joshua L. Kemp, Dir.-R&D
Laurie Stephen, Dir.-Assay Dev.
Samuel T. LaBrie, VP-Corp. Dev.
Michael Spain, Chief Medical Officer
James P. Mapes, Chief Scientific Officer
Margaret Sandefur, Dir.-Quality Assurance
Anthony Barnes, Sr. VP-Diagnostics

Phone: 512-835-8026	Fax: 512-835-4687
Toll-Free: 866-726-6277	
Address: 3300 Duval Rd., Austin, TX 78759 US	

GROWTH PLANS/SPECIAL FEATURES:

Rules-Based Medicine, Inc. (RBM) is a leading multiplexed biomarker testing laboratory in Austin, TX. Based on its Multi-Analyte Profiling (MAP) technology, RBM offers testing services to preclinical and clinical researchers for hundreds of biomarkers. The effects of both diseases and drugs are typically manifested in abnormal levels of certain biomarkers in a blood sample, and by providing tests for these biomarkers, RBM is able to assist researchers in identifying patients most likely to respond to a given therapy and the biochemical reason for that response. The smallness of the amount of a sample required for these tests allows research that was previously either unavailable or expensive. The initial focus of its testing was on psychiatric-based conditions such as schizophrenia, bipolar disorder and major depressive disorder, where diagnosis is primarily based on patient self-reports and clinically observed behavior. Biomarker-based diagnosis could offer a decrease in the ambiguity of current diagnostics. RBM has since expanded its MAP development into other fields of the health sector. The firm now offers its biomarker MAPs in two divisions; human and rodent. The human MAPs include; HumanMAP, PsyMAP, CardiovascularMAP, InflammationMAP, MetabolicMAP, CytokineMAP A and B and KidneyMAP. Rodent offerings include; RodentMAP for antigens, Rat MetabolicMAP, KidneyMAP and Mouse Cytokine MAPs A, B and C. In addition it offers custom MAP development. A secondary facility in Lake Placid, NY performs custom multiplex assay development and manufacturing. RBM also runs a wholly owned subsidiary, EDI GmbH, from a facility in Germany. Through this facility RBM provides Human Organo-Typic cell culture systems, which are ideal for ex vivo studies of drug safety and efficacy. In December 2009, RBM had an initial public offering (IPO). In June 2010, RBM announced it had received a grant in order to further develop its new OncologyMAP program.

FINANCIALS: Sales and profits are in thousands of dollars—add 000 to get the full amount. 2009 Note: Financial information for 2009 was not available for all companies at press time.

2009 Sales: $	2009 Profits: $	U.S. Stock Ticker: RULE
2008 Sales: $	2008 Profits: $	Int'l Ticker: Int'l Exchange:
2007 Sales: $	2007 Profits: $	Employees:
2006 Sales: $	2006 Profits: $	Fiscal Year Ends:
2005 Sales: $	2005 Profits: $	Parent Company:

SALARIES/BENEFITS:

Pension Plan:	ESOP Stock Plan:	Profit Sharing:	Top Exec. Salary: $257,500	Bonus: $125,000
Savings Plan:	Stock Purch. Plan:		Second Exec. Salary: $242,565	Bonus: $117,000

OTHER THOUGHTS:
Apparent Women Officers or Directors: 6
Hot Spot for Advancement for Women/Minorities: Y

LOCATIONS: ("Y" = Yes)

West:	Southwest:	Midwest:	Southeast:	Northeast:	International:
	Y			Y	Y

Note: Financial information, benefits and other data can change quickly and may vary from those stated here.

S&W SEED CO

www.swseedco.com

Industry Group Code: 11511 Ranks within this company's industry group: Sales: Profits:

Drugs:	Other:		Clinical:		Computers:	Services:
Discovery:	AgriBio:	Y	Trials/Services:	Y	Hardware:	Specialty Services:
Licensing:	Genetic Data:		Labs:		Software:	Consulting:
Manufacturing:	Tissue Replacement:		Equipment/Supplies:		Arrays:	Blood Collection:
Generics:			Research/Development Svcs.:		Database Management:	Drug Delivery:
			Diagnostics:			Drug Distribution:

TYPES OF BUSINESS:
Alfalfa Seed Research & Development

BRANDS/DIVISIONS/AFFILIATES:
Stevia

CONTACTS: Note: Officers with more than one job title may be intentionally listed here more than once.
Mark S. Grewal, CEO
Mark S. Grewal, Pres.
Matthew K. Szot, CFO
Daniel Z. Karsten, VP-Oper.
Matthew K. Szot, VP-Finance
Grover T. Wickersham, Chmn.

Phone: 559-884-2535	**Fax:** 559-884-2750
Toll-Free:	
Address: 25552 S. Butte Ave., Five Points, CA 93624 US	

GROWTH PLANS/SPECIAL FEATURES:

S&W Seed Co. is a leader in breeding and developing proprietary alfalfa seed varieties that grow in warm climates and which can thrive on poor, saline soils. Although sale of its proprietary alfalfa seed varieties has been a mainstay of its business for decades, the firm has in the past derived material revenue from processing wheat and other small grains, which it continue to pursue on a limited scale. The company contracts with a small network of growers for its seeds; processes the seed purchased from its growers in its 40-acre cleaning and processing plant; and then sells the seed through a network of dealers and distributors, as well directly to farmers. S&W's alfalfa seed varieties are sold in dormancies 4, 6, 7, 8, 9 and 10. For nearly 30 years, S&W operated as a general partnership of five partners; however, the firm became incorporated in October 2009 and is now publicly traded. Three customers, including largest client Genetics International, Inc., account for roughly 64% of the firm's sales. In May 2010, the company implemented a stevia growing trial program through a partnership with stevia producer PureCircle; the stevia plant is a source of an all natural, non-caloric sweetener.

FINANCIALS: Sales and profits are in thousands of dollars—add 000 to get the full amount. 2009 Note: Financial information for 2009 was not available for all companies at press time.

2009 Sales: $	2009 Profits: $	**U.S. Stock Ticker: SANWU**
2008 Sales: $	2008 Profits: $	**Int'l Ticker:** Int'l Exchange:
2007 Sales: $	2007 Profits: $	Employees:
2006 Sales: $	2006 Profits: $	Fiscal Year Ends: 12/31
2005 Sales: $	2005 Profits: $	Parent Company:

SALARIES/BENEFITS:

Pension Plan:	ESOP Stock Plan:	Profit Sharing:	Top Exec. Salary: $	Bonus: $
Savings Plan:	Stock Purch. Plan:		Second Exec. Salary: $	Bonus: $

OTHER THOUGHTS:
Apparent Women Officers or Directors:
Hot Spot for Advancement for Women/Minorities:

LOCATIONS: ("Y" = Yes)

West:	Southwest:	Midwest:	Southeast:	Northeast:	International:
Y					

SALIX PHARMACEUTICALS

www.salix.com

Industry Group Code: 325412 Ranks within this company's industry group: Sales: 64   Profits: 143

Drugs:	Other:	Clinical:	Computers:	Services:
Discovery:	AgriBio:	Trials/Services:	Hardware:	Specialty Services:
Licensing:	Genetic Data:	Labs:	Software:	Consulting:
Manufacturing: Y	Tissue Replacement:	Equipment/Supplies:	Arrays:	Blood Collection:
Generics:		Research/Development Svcs.:	Database Management:	Drug Delivery:
		Diagnostics:		Drug Distribution:

TYPES OF BUSINESS:

Pharmaceuticals Development & Manufacturing
Drugs-Gastroenterology

BRANDS/DIVISIONS/AFFILIATES:

Colazal
Azasan
Proctocort
Anusol-HC
OsmoPrep
Xifaxan
Visicol
Lupin Ltd

CONTACTS: *Note: Officers with more than one job title may be intentionally listed here more than once.*

Carolyn J. Logan, CEO
Carolyn J. Logan, Pres.
Adam C. Derbyshire, CFO
William P. Forbes, Exec. VP-R&D
Adam C. Derbyshire, Sr. VP-Admin.
William P. Forbes, Chief Dev. Officer
G. Michael Freeman, VP-Corp. Comm.
G. Michael Freeman, VP-Investor Rel.
Adam C. Derbyshire, Exec. VP-Finance
John F. Chappell, Chmn.

Phone: 919-862-1000	Fax: 919-862-1095
Toll-Free: 888-802-9956	
Address: 1700 Perimeter Park Dr., Morrisville, NC 27560 US	

GROWTH PLANS/SPECIAL FEATURES:

Salix Pharmaceuticals is a specialty pharmaceutical company dedicated to acquiring, developing and commercializing prescription drugs used in the treatment of a variety of gastrointestinal diseases. The company seeks to identify late-stage or approved proprietary therapeutics for in-licensing, which subsequently advances the new drugs through regulatory procedures and final product development stages. The firm's products are divided into three segments: GI, Colonoscopy preps and other Salix products. In the GI segment, the company sells Metozolv ODT, a orally indigested metoclopramide tablet; Apriso, a once-a-day tablet for ulcerative colitis; Colazal (balsalazide disodium) treats ulcerative colitis; Xifaxan, an antibiotic for the gut; and Pepcid Oral Suspension, short-term treatment of gastroesophageal reflux disease. In the Colonoscopy prep segment, Salix sells OsmoPrep tablets; MoviPrep oral solution a gastrointestinal-specific oral antibiotic; and Visicol, a product indicated for cleansing of the bowel as a preparation for colonoscopy. Other Salix Products include, Azasan (azathioprine tablets), is intended to suppress immune response in organ transplant recipients; Anusol-HC rectal suppositories; Proctocort, which is available in a cream form that is indicated for the relief of the inflammation and in a suppository form, which is indicated for use in inflamed hemorrhoids and postirradiation proclitis; and Diuril Oral Suspension, treats high blood pressure. The primary product candidates Salix is developing are balsalazide disodium tablets, which the company intends to sell for the treatment of ulcerative colitis; a patented, granulated formula of mesalamine, which it intends to sell for the treatment of ulcerative colitis; Metoclopramide, for short-term therapy following gastroesophageal reflux; budesonide, a foam preparation for the treatment of mild to moderate forms of distal ulcerative colitis; and Vapreotide Acetate Powder, for the treatment of acute esophageal variceal bleeding. In October 2009, the firm entered into a partnership with Lupin Ltd., to develop and commercialize rifaximin using Lupin's proprietary technology.

FINANCIALS: Sales and profits are in thousands of dollars—add 000 to get the full amount. 2009 Note: Financial information for 2009 was not available for all companies at press time.

2009 Sales: $232,890	2009 Profits: $-43,619	**U.S. Stock Ticker:** SLXP
2008 Sales: $178,766	2008 Profits: $-47,037	**Int'l Ticker:** Int'l Exchange:
2007 Sales: $235,792	2007 Profits: $8,225	Employees: 395
2006 Sales: $208,533	2006 Profits: $31,510	Fiscal Year Ends: 12/31
2005 Sales: $154,903	2005 Profits: $-60,585	Parent Company:

SALARIES/BENEFITS:

Pension Plan:	ESOP Stock Plan:	Profit Sharing: Y	Top Exec. Salary: $741,500	Bonus: $650,683
Savings Plan: Y	Stock Purch. Plan:		Second Exec. Salary: $410,000	Bonus: $235,750

OTHER THOUGHTS:

Apparent Women Officers or Directors: 1
Hot Spot for Advancement for Women/Minorities:

LOCATIONS: ("Y" = Yes)

West:	Southwest:	Midwest:	Southeast:	Northeast:	International:
				Y	

SANGAMO BIOSCIENCES INC
www.sangamo.com

Industry Group Code: 541712 Ranks within this company's industry group: Sales: 13   Profits: 12

Drugs:		Other:		Clinical:		Computers:		Services:	
Discovery:	Y	AgriBio:		Trials/Services:		Hardware:		Specialty Services:	
Licensing:		Genetic Data:	Y	Labs:		Software:		Consulting:	
Manufacturing:		Tissue Replacement:	Y	Equipment/Supplies:		Arrays:		Blood Collection:	
Generics:				Research/Development Svcs.:	Y	Database Management:		Drug Delivery:	
				Diagnostics:				Drug Distribution:	

TYPES OF BUSINESS:
Drug Research & Development
Gene Expression Regulation Therapies
Transcription Factor Technology

BRANDS/DIVISIONS/AFFILIATES:
ZFP Therapeutic
ZFP Transcription Factors (ZFP TFs)
ZFP Nucleases (ZFNs)
Zinc Finger DNA-Binding Proteins
SB-509
SB-728-T

CONTACTS:
Note: Officers with more than one job title may be intentionally listed here more than once.
Edward O. Lanphier, II, CEO
Edward O. Lanphier, II, Pres.
H. Ward Wolff, CFO/Exec. VP
Phillip D. Gregory, Chief Scientific Officer/VP
Edward J. Rebar, VP-Tech.
Gregory S. Zante, VP-Admin.
David G. Ichikawa, Sr. VP-Bus. Dev.
Gregory S. Zante, VP-Finance
Ely Benaim, VP-Clinical Affairs
Dale G. Ando, Chief Medical Officer/VP-Therapeutic Dev.
Shirley Clift, VP-Regulatory Affairs
Winson Tang, VP-Clinical Research

Phone: 510-970-6000	Fax: 510-236-8951
Toll-Free:	
Address: 501 Canal Blvd., Ste. A100, Richmond, CA 94804 US	

GROWTH PLANS/SPECIAL FEATURES:
Sangamo BioSciences, Inc. is a biotechnology company that develops, researches and commercializes zinc finger DNA-binding proteins (ZFPs), a naturally occurring class of proteins. ZFPs can be engineered to make ZFP transcription factors (ZFP TFs), which can be used to turn genes on or off, and ZFP nucleases (ZFNs), which enable the modification of DNA sequences in a variety of ways. Sangamo's lead ZFP Therapeutic, SB-509, is a plasmid formulation of a ZFP TF activator of the vascular endothelial growth factor-A gene, and is currently in a Phase IIb clinical trial for the treatment of diabetic neuropathy (DN) and a Phase II clinical trial for sever DN. SB-509 is also in a Phase II trial for amyotrophic lateral sclerosis (ALS), or Lou Gehrig's Disease. Sangamo expects to have clinical data from its Phase II trials in severe DN and ALS in 2010. The firm has additional research-stage programs in X-linked severe combined immunodeficiency (X-linked SCID), hemophilia and hemoglobinopathies. The company filed an application for a Phase I trial to evaluate a ZFN-based therapeutic for the treatment of glioblastoma multiforme, a type of brain cancer, and hopes to begin the trial in 2010. As ZFPs act at the DNA level, they have broad potential applications in several areas including human therapeutics, plant agriculture, research reagents and cell-line engineering. Sangamo has attempted to capitalize on the broad potential commercial applications of its ZFPs by facilitating the sale and licensing of its ZFP TFs and ZFNs to companies working in fields outside of human therapeutics, including Dow AgroSciences; Johnson & Johnson; Pfizer; Sigma-Aldrich; Medarex; Novo Nordisk; Novartis A/G; and Genentech. In October 2009, Sangamo and Sigma-Aldrich Corporation announced an expansion of its ZFP technology license.

Sangamo offers its employees health, dental, vision and life insurance; a 401(k) plan; an employee stock purchase plan; and a cafeteria plan.

FINANCIALS:
Sales and profits are in thousands of dollars—add 000 to get the full amount. 2009 Note: Financial information for 2009 was not available for all companies at press time.

2009 Sales: $22,187	2009 Profits: $-18,587	U.S. Stock Ticker: SGMO
2008 Sales: $16,186	2008 Profits: $-24,302	Int'l Ticker: Int'l Exchange:
2007 Sales: $9,098	2007 Profits: $-21,480	Employees: 74
2006 Sales: $7,885	2006 Profits: $-17,864	Fiscal Year Ends: 12/31
2005 Sales: $2,484	2005 Profits: $-13,293	Parent Company:

SALARIES/BENEFITS:
Pension Plan:	ESOP Stock Plan:	Profit Sharing:	Top Exec. Salary: $510,000	Bonus: $216,750
Savings Plan: Y	Stock Purch. Plan: Y		Second Exec. Salary: $385,000	Bonus: $98,175

OTHER THOUGHTS:
Apparent Women Officers or Directors: 1
Hot Spot for Advancement for Women/Minorities:

LOCATIONS: ("Y" = Yes)
West:	Southwest:	Midwest:	Southeast:	Northeast:	International:
Y					

SANOFI-AVENTIS SA

www.en.sanofi-aventis.com

Industry Group Code: 325412 Ranks within this company's industry group: Sales: 6 Profits: 9

Drugs:		Other:		Clinical:		Computers:		Services:	
Discovery:	Y	AgriBio:		Trials/Services:		Hardware:		Specialty Services:	
Licensing:		Genetic Data:		Labs:		Software:		Consulting:	
Manufacturing:	Y	Tissue Replacement:		Equipment/Supplies:		Arrays:		Blood Collection:	
Generics:				Research/Development Svcs.:		Database Management:		Drug Delivery:	
				Diagnostics:				Drug Distribution:	

TYPES OF BUSINESS:

Pharmaceuticals Development & Manufacturing
Over-the-Counter Drugs
Cardiovascular Drugs
CNS Drugs
Oncology Drugs
Diabetes Drugs
Generics
Vaccines

BRANDS/DIVISIONS/AFFILIATES:

Aprovel
Plavix
Allegra
Depakine
Stilnox
Sanofi Pasteur
Eloxatin
Lantus

CONTACTS: Note: Officers with more than one job title may be intentionally listed here more than once.

Chris Viehbacher, CEO
Jerome Contamine, CFO/Exec. VP
Roberto Pucci, Sr. VP-Human Resources
Marc Cluzel, Sr. VP-R&D
Karen Linehan, General Counsel/Sr. VP-Legal Affairs
Hanspeter Spek, Exec. VP-Global Oper.
Laurence Debroux, Chief Strategic Officer
Michel Labie, Sr. VP-Comm./VP-Institutional & Professional Rel.
Philippe Peyre, Sr. VP-Corp. Affairs
Olivier Charmeil, Sr. VP-Pharmaceutical Oper., Asia-Pacific & Japan
Philippe Luscan, Sr. VP-Industrial Affairs
Gregory Irace, Sr. VP-Pharmaceutical Oper., U.S.
Jean-Francois Dehecq, Chmn.
Antoine Ortoli, Sr. VP-Pharmaceuticals Oper., Int'l

Phone: 33-1-53-77-4000	Fax:
Toll-Free:	
Address: 174 Ave. de France, Paris, 75365 France	

GROWTH PLANS/SPECIAL FEATURES:

Sanofi-Aventis SA is an international pharmaceutical group engaged in the research, development, manufacturing and marketing of healthcare products. The company has two main business activities: pharmaceuticals and human vaccines through subsidiary Sanofi Pasteur. The firm's pharmaceutical business includes specialties in three main areas: diabetes; oncology; and thrombosis and cardiovascular. Diabetes products include Lantus, a long-acting analog of human insulin; apidra, a fast-acting analog of human insulin; and Amaryl, an oral once-daily sulfonylurea. Oncology treatments include Taxotere, a taxane derivative and therapy for several cancer types, and Eloxatine, a platinum agent, which is a leading treatment of colorectal cancer. Thrombosis and cardiovascular medicines include two leading drugs in their categories: Plavix, an anti-platelet agent indicated for a number of atherothrombotic conditions, and Lovenox, a low molecular weight heparin indicated for prophylaxis, deep vein thrombosis, unstable angina and myocardial infarction. Cardiovascular medicines include Multaq, an anti-arrhythmic agent and two major hypertension treatments: Aprovel/Co-Aprovel and Tritace. Other therapeutic areas include central nervous system medicines, such as Stilnow and Ambien CR, a sleep disorder medication, and internal medicine, including respiratory/allergy products Allegra and Nasacort; urology medicines such as Xatral, a treatment for benign prosatic hypertrophy; and osteoporosis treatment Actonel. The vaccines segment provides pediatric combination vaccines for such diseases as pertussis, diphtheria, tetanus, Haemophilus Influenzae type B infection and polio with products such as Daptacel and Tripedia; influenza vaccines for seasonal flu vaccinations in both hemispheres with Intanza and Fluzone; adult and adolescent booster vaccines protecting against petrussis, tetanus, diphtheria and polio with Adacel, Adacel Polio and Decavac; meningitis vaccines, Menactra and Menomune; and travel and endemic vaccines guarding against an array of diseases including hepatitis A, typhoid, rabies, yellow fever and cholera with Imovax, Verorab and Typhim Vi.

FINANCIALS: Sales and profits are in thousands of dollars—add 000 to get the full amount. 2009 Note: Financial information for 2009 was not available for all companies at press time.

2009 Sales: $41,776,800	2009 Profits: $7,153,240	U.S. Stock Ticker: SNY
2008 Sales: $36,751,700	2008 Profits: $5,721,790	Int'l Ticker: SAN Int'l Exchange: Paris-Euronext
2007 Sales: $37,397,000	2007 Profits: $7,574,840	Employees: 104,867
2006 Sales: $38,722,100	2006 Profits: $6,003,540	Fiscal Year Ends: 12/31
2005 Sales: $37,272,700	2005 Profits: $3,538,800	Parent Company:

SALARIES/BENEFITS:

Pension Plan:	ESOP Stock Plan:	Profit Sharing:	Top Exec. Salary: $	Bonus: $
Savings Plan:	Stock Purch. Plan:		Second Exec. Salary: $	Bonus: $

OTHER THOUGHTS:

Apparent Women Officers or Directors: 5
Hot Spot for Advancement for Women/Minorities: Y

LOCATIONS: ("Y" = Yes)

West:	Southwest:	Midwest:	Southeast:	Northeast:	International:
Y	Y	Y	Y	Y	Y

Note: Financial information, benefits and other data can change quickly and may vary from those stated here.

SAVIENT PHARMACEUTICALS INC

www.savientpharma.com

Industry Group Code: 325412 Ranks within this company's industry group: Sales: 134 Profits: 157

Drugs:			Other:	Clinical:	Computers:	Services:
Discovery:		Y	AgriBio:	Trials/Services:	Hardware:	Specialty Services:
Licensing:		Y	Genetic Data:	Labs:	Software:	Consulting:
Manufacturing:			Tissue Replacement:	Equipment/Supplies:	Arrays:	Blood Collection:
Generics:				Research/Development Svcs.:	Database Management:	Drug Delivery:
				Diagnostics:		Drug Distribution:

TYPES OF BUSINESS:

Pharmaceuticals Discovery & Development
Weight Gain Products
Hormone Therapy

BRANDS/DIVISIONS/AFFILIATES:

Oxandrolone
KRYSTEXXA

CONTACTS: Note: Officers with more than one job title may be intentionally listed here more than once.

Paul Hamelin, Pres.
David Gionco, CFO/Sr. VP
Philip K. Yachmetz, General Counsel/Sr. VP/Sec.
David Gionco, Treas.
Stephen Jaeger, Chmn.

Phone: 732-418-9300	Fax: 732-418-0570
Toll-Free:	
Address: 1 Tower Ctr., 14th Fl., E. Brunswick, NJ 08816 US	

GROWTH PLANS/SPECIAL FEATURES:

Savient Pharmaceuticals, Inc. is engaged in the development, manufacture and marketing of both genetically engineered and niche-focused specialty pharmaceutical products. Currently, the primary product marketed worldwide by Savient is Oxandrin, an oral anabolic agent primarily used to promote weight gain following involuntary weight loss. Oxandrin is currently used for patients suffering from weight loss due to a number of medical conditions, including HIV/AIDS, surgery recovery, cancer and chronic diseases. Savient also distributes oxandrolone as a generic form of Oxandrin. The firm is currently developing the drug KRYSTEXXA, also referred to as pegloticase, an infused genetically engineered enzyme conjugate being studied for the elimination of excess uric acid in individuals suffering from severe gout. The rights to this technology were purchased from Duke University and Mountain View Pharmaceuticals, Inc. Current treatments and research endeavors aim to block uric acid production, but do little to reduce levels present in the body. The drug is designed to convert the acid to allantoin, which the body can remove by itself. The firm has completed several Phase III clinical trials and is currently awaiting the FDA's approval to market KRYSTEXXA.

Savient offers its employees medical, vision, dental, life and AD&D insurance; a flexible spending account; tuition reimbursement; short- and long-term disability; a 401(k) plan; and an employee referral program.

FINANCIALS: Sales and profits are in thousands of dollars—add 000 to get the full amount. 2009 Note: Financial information for 2009 was not available for all companies at press time.

2009 Sales: $2,960	2009 Profits: $-90,853	U.S. Stock Ticker: SVNT	
2008 Sales: $3,181	2008 Profits: $-84,169	Int'l Ticker: Int'l Exchange:	
2007 Sales: $14,024	2007 Profits: $-48,668	Employees: 43	
2006 Sales: $47,514	2006 Profits: $60,325	Fiscal Year Ends: 12/31	
2005 Sales: $49,495	2005 Profits: $5,968	Parent Company:	

SALARIES/BENEFITS:

Pension Plan:	ESOP Stock Plan:	Profit Sharing:	Top Exec. Salary: $517,274	Bonus: $
Savings Plan: Y	Stock Purch. Plan: Y		Second Exec. Salary: $406,000	Bonus: $

OTHER THOUGHTS:

Apparent Women Officers or Directors:
Hot Spot for Advancement for Women/Minorities:

LOCATIONS: ("Y" = Yes)

West:	Southwest:	Midwest:	Southeast:	Northeast:	International:
				Y	

SCICLONE PHARMACEUTICALS

www.sciclone.com

Industry Group Code: 325412 Ranks within this company's industry group: Sales: 81 Profits: 57

Drugs:		Other:	Clinical:	Computers:	Services:
Discovery:	Y	AgriBio:	Trials/Services:	Hardware:	Specialty Services:
Licensing:	Y	Genetic Data:	Labs:	Software:	Consulting:
Manufacturing:	Y	Tissue Replacement:	Equipment/Supplies:	Arrays:	Blood Collection:
Generics:			Research/Development Svcs.:	Database Management:	Drug Delivery:
			Diagnostics:		Drug Distribution:

TYPES OF BUSINESS:

Pharmaceuticals Acquisition & Development
Immune System Enhancers
Infectious Disease Therapies
Cancer Therapies

BRANDS/DIVISIONS/AFFILIATES:

SciClone Pharmaceuticals International, Ltd.
ZADAXIN
SCV-07
RapidFilm
DC Bead
SciClone China

CONTACTS: *Note: Officers with more than one job title may be intentionally listed here more than once.*

Friedhelm Blobel, CEO
Friedhelm Blobel, Pres.
Gary Titus, CFO
Cynthia W. Tuthill, Chief Scientific Officer
Jeffery Lange, VP-Bus. Dev.
Gary Titus, Sr. VP-Finance
Israel Rios, Chief Medical Officer/Sr. VP-Medical Affairs
Cynthia W. Tuthill, Sr. VP-Scientific Affairs
Jon S. Saxe, Chmn.
Hans P. Schmid, Pres., SciClone Pharmaceuticals Int'l, Ltd.

Phone: 650-358-3456	Fax: 650-358-3469
Toll-Free:	
Address: 950 Tower Ln., Ste. 900, Foster City, CA 94404 US	

GROWTH PLANS/SPECIAL FEATURES:

SciClone Pharmaceuticals, Inc. develops and commercializes pharmaceutical and biological therapeutic compounds for oncology and infectious diseases. ZADAXIN (thymosin alpha 1), the firm's lead primary product, is used for treatment of the hepatitis B and hepatitis C viruses, certain cancers and as a vaccine adjuvant. ZADAXIN is approved for sale in over 30 countries, primarily in Asia, Eastern Europe, the Middle East and Latin America, with 96% of the company's sales of ZADAXIN coming from China through SciClone China. The company also markets DC Bead, a novel treatment for advanced liver cancer that is currently approved in 40 countries, including Europe and the U.S. The firm obtained the Chinese marketing rights for the treatment from Biocompatibles International PLC. RapidFilm, the firm's other marketed product, is an oral thin film formulation of ondansetron, commonly used to treat and prevent nausea and vomiting caused by chemotherapy, radiotherapy and surgery. RapidFilm is currently being evaluated for regulatory approval in Europe, and the company has commercialization rights for RapidFilm in China and Vietnam. SciClone's proprietary drug development candidate is SCV-07, which is used to treat infectious diseases and in oncology applications. SCV-07 is a synthetic dipeptide that has demonstrated immunomodulatory activity by increasing T-cell differentiation and function, biological processes that are necessary for the body to fight infection. The firm is currently conducting a Phase II clinical trial to assess the safety and efficacy of SCV-07 for the severity of severe oral mucositis in subjects receiving chemoradiation therapy for head and neck cancers. The company acquired exclusive worldwide rights outside of Russia to SCV-07 from Verta, Ltd. In March 2010, the company received European approval for RapidFilm in 16 European countries.

SciClone offers employees medical, dental, vision and life insurance; short- and long-term disability coverage; flexible spending accounts; a 401(k) plan; and an employee stock purchase program.

FINANCIALS: Sales and profits are in thousands of dollars—add 000 to get the full amount. 2009 Note: Financial information for 2009 was not available for all companies at press time.

2009 Sales: $72,411	2009 Profits: $11,945	**U.S. Stock Ticker:** SCLN
2008 Sales: $54,113	2008 Profits: $-8,348	**Int'l Ticker:** Int'l Exchange:
2007 Sales: $37,058	2007 Profits: $-9,948	Employees: 223
2006 Sales: $32,662	2006 Profits: $ 727	Fiscal Year Ends: 12/31
2005 Sales: $28,334	2005 Profits: $-7,713	Parent Company:

SALARIES/BENEFITS:

Pension Plan:	ESOP Stock Plan:	Profit Sharing:	Top Exec. Salary: $442,000	Bonus: $176,800
Savings Plan: Y	Stock Purch. Plan: Y		Second Exec. Salary: $340,000	Bonus: $120,000

OTHER THOUGHTS:

Apparent Women Officers or Directors: 1
Hot Spot for Advancement for Women/Minorities:

LOCATIONS: ("Y" = Yes)

West:	Southwest:	Midwest:	Southeast:	Northeast:	International:
Y					Y

SCIOS INC

www.sciosinc.com

Industry Group Code: 325412 Ranks within this company's industry group: Sales: Profits:

Drugs:		Other:	Clinical:	Computers:		Services:	
Discovery:	Y	AgriBio:	Trials/Services:	Hardware:		Specialty Services:	
Licensing:		Genetic Data:	Labs:	Software:		Consulting:	
Manufacturing:	Y	Tissue Replacement:	Equipment/Supplies:	Arrays:		Blood Collection:	
Generics:			Research/Development Svcs.:	Database Management:	Y	Drug Delivery:	
			Diagnostics:			Drug Distribution:	

TYPES OF BUSINESS:
Cardiovascular Drugs

BRANDS/DIVISIONS/AFFILIATES:
Natrecor
ADHERE
Johnson & Johnson

CONTACTS: Note: Officers with more than one job title may be intentionally listed here more than once.
William C. Weldon, CEO-Johnson & Johnson
William C. Weldon, Chmn.-Johnson & Johnson

Phone: 732-524-0400	Fax:
Toll-Free: 877-462-873267	
Address: 1 Johnson & Johnson Plz., New Brunswick, NJ 08933 US	

GROWTH PLANS/SPECIAL FEATURES:

Scios, Inc., a subsidiary of Johnson & Johnson, is a biopharmaceutical company developing treatments for cardiovascular diseases. The firm's technology platform fuses classical medicinal chemistry with the most recent advances in disease-based gene array, bioinformatics and computational chemistry. The company's lead product, Natrecor, is an intravenous cardiovascular drug approved to treat acutely decompensated congestive heart failure (ADHF) for patients who have dyspnea (shortness of breath) at rest or with minimal activity, such as talking, eating or bathing. Natrecor is a recombinant form of the human B-type natriuretic peptide (hBNP), which is normally produced by the heart. The firm's current research and development program focuses on the discovery of new therapeutics for other cardiovascular diseases. Scios also has a program call ADHERE, Acute Decompensated Heart Failure National Registry, which is an observational registry that lists data on heart failure patient treatment and associated outcomes.

The firm offers employees benefits including life, disability, medical, dental, long-term care and accident insurance; paid time off; savings and pension plans; and health and wellness services.

FINANCIALS: Sales and profits are in thousands of dollars—add 000 to get the full amount. 2009 Note: Financial information for 2009 was not available for all companies at press time.

2009 Sales: $	2009 Profits: $	U.S. Stock Ticker: Subsidiary	
2008 Sales: $	2008 Profits: $	Int'l Ticker: Int'l Exchange:	
2007 Sales: $98,600	2007 Profits: $	Employees:	
2006 Sales: $	2006 Profits: $	Fiscal Year Ends: 12/31	
2005 Sales: $	2005 Profits: $	Parent Company: JOHNSON & JOHNSON	

SALARIES/BENEFITS:

Pension Plan: Y	ESOP Stock Plan:	Profit Sharing:	Top Exec. Salary: $	Bonus: $
Savings Plan:	Stock Purch. Plan:		Second Exec. Salary: $	Bonus: $

OTHER THOUGHTS:
Apparent Women Officers or Directors:
Hot Spot for Advancement for Women/Minorities:

LOCATIONS: ("Y" = Yes)

West:	Southwest:	Midwest:	Southeast:	Northeast:	International:
				Y	

SEATTLE GENETICS

www.seagen.com

Industry Group Code: 325412 **Ranks within this company's industry group:** Sales: 84   Profits: 154

Drugs:		Other:		Clinical:		Computers:		Services:	
Discovery:	Y	AgriBio:		Trials/Services:		Hardware:		Specialty Services:	
Licensing:	Y	Genetic Data:		Labs:		Software:		Consulting:	
Manufacturing:		Tissue Replacement:		Equipment/Supplies:		Arrays:		Blood Collection:	
Generics:				Research/Development Svcs.:	Y	Database Management:		Drug Delivery:	
				Diagnostics:				Drug Distribution:	

TYPES OF BUSINESS:

Biopharmaceuticals Development
Cancer Treatments
Monoclonal Antibodies
Autoimmune Diseases

BRANDS/DIVISIONS/AFFILIATES:

SGN-40 (Dacetuzumab)
SGN-75
SGN-33 (Lintuzumab)
SGN-35 (Brentuximab Vedotin)
SGN-70
Antibody-Drug Conjugates (ADCs)
Sugar Engineered Antibody (SEA)

CONTACTS: *Note: Officers with more than one job title may be intentionally listed here more than once.*

Clay B. Siegall, CEO
Clay B. Siegall, Pres.
Todd Simpson, CFO
Charles R. Romp, VP-Sales
Christopher Pawlowicz, Sr. VP-Human Resources
Jonathan Drachman, Sr. VP-Research & Translational Medicine
Vaughn B. Himes, Dir.-Mfg.
Kirk D. Schumacher, General Counsel/VP-Legal Affairs & Compliance
Vaughn B. Himes, Exec. VP-Tech. Oper.
Eric L. Dobmeier, Dir.-Corp. Comm.
Thomas C. Reynolds, Chief Medical Officer
Morris Z. Rosenberg, Exec. VP-Process Sciences
Eric L. Dobmeier, Chief Bus. Officer
Peter D. Senter, VP-Chemistry
Clay B. Siegall, Chmn.
Vaughn B. Himes, Dir.-Supply Chain

Phone: 425-527-4000	Fax: 425-527-4001
Toll-Free:	
Address: 21823 30th Dr. SE, Bothell, WA 98021 US	

GROWTH PLANS/SPECIAL FEATURES:

Seattle Genetics develops monoclonal antibody (mAb)-based therapies for the treatment of cancer and autoimmune diseases. Its research and development activities focus on mAb-based (monoclonal antibodies) therapies for human cancers including lung, renal cell, Hodgkin's disease, non-Hodgkin's lymphoma, multiple myeloma and melanoma. Products in the developmental stage include SGN-33 (lintuzumab), in testing for acute myeloid leukemia (AML) and myelodysplastic syndromes (MDS); SGN-40 (dacetuzumab), in testing for multiple myeloma, diffuse large B-cell lymphoma and non-Hodgkin's lymphoma; SGN-35, in testing for relapsed Hodgkin's lymphoma and systemic anaplastic large cell lymphoma; SGN-70, an IND for autoimmune diseases; SGN-75, a trial drug for non-Hodgkin lymphoma and renal cell carcinoma; SGN-19A, a preclinical candidate for the treatment of hematological malignancies; ASG-5ME, a preclinical drug for the treatment of prostate and pancreatic cancer, developed jointly with Agensys. These product candidates represent applications of Seattle Genetics' primary platform technologies: genetically engineered monoclonal antibodies and antibody-drug conjugates (ADC). Each of these platforms is designed to support therapies that are able to identify and kill cancer cells while limiting damage to normal tissue. Seattle Genetics currently has license agreements for its proprietary ADC technology with Genentech; Bayer Pharmaceuticals, Inc.; Progenics; Daiichi Sankyo; GlaxoSmithKline; Millennium Pharmaceuticals; and MedImmune, Inc. The company has a co-development agreement with Agensys to research and develop ADC products. In September 2009, the company introduced its sugar engineered antibody (SEA) technology, which can increase the potency of monoclonal antibodies through enhanced effector function.

Seattle Genetics offers its employees medical, dental and vision coverage; a flexible spending plan; short and long-term disability; and life and AD&D coverage.

FINANCIALS: Sales and profits are in thousands of dollars—add 000 to get the full amount. 2009 Note: Financial information for 2009 was not available for all companies at press time.

2009 Sales: $51,965	2009 Profits: $-81,683	**U.S. Stock Ticker: SGEN**
2008 Sales: $35,236	2008 Profits: $-85,501	**Int'l Ticker:** Int'l Exchange:
2007 Sales: $22,420	2007 Profits: $-48,932	Employees:
2006 Sales: $10,005	2006 Profits: $-36,015	Fiscal Year Ends: 12/31
2005 Sales: $9,757	2005 Profits: $-29,433	Parent Company:

SALARIES/BENEFITS:

Pension Plan:	ESOP Stock Plan:	Profit Sharing:	Top Exec. Salary: $600,071	Bonus: $346,351
Savings Plan: Y	Stock Purch. Plan: Y		Second Exec. Salary: $369,217	Bonus: $176,014

OTHER THOUGHTS:

Apparent Women Officers or Directors:
Hot Spot for Advancement for Women/Minorities:

LOCATIONS: ("Y" = Yes)

West:	Southwest:	Midwest:	Southeast:	Northeast:	International:
Y					

Note: Financial information, benefits and other data can change quickly and may vary from those stated here.

SENETEK PLC

www.senetekplc.com

Industry Group Code: 325412 Ranks within this company's industry group: Sales: 139 Profits: 89

Drugs:		Other:		Clinical:	Computers:	Services:
Discovery:	Y	AgriBio:	Y	Trials/Services:	Hardware:	Specialty Services:
Licensing:	Y	Genetic Data:		Labs:	Software:	Consulting:
Manufacturing:	Y	Tissue Replacement:		Equipment/Supplies:	Arrays:	Blood Collection:
Generics:				Research/Development Svcs.:	Database Management:	Drug Delivery:
				Diagnostics:		Drug Distribution:

TYPES OF BUSINESS:
Drugs-Sexual Dysfunction
Anti-Aging Products
Skin Care Drugs
Cancer Treatment

BRANDS/DIVISIONS/AFFILIATES:
Invicorp
Kinetin
Zeatin
Reliaject

CONTACTS: *Note: Officers with more than one job title may be intentionally listed here more than once.*
John P. Ryan, CEO
Phillip J. Rose, COO
Howard Crosby, Pres.
Jan-Elo Jorgensen, Dir.-Research
John P. Ryan, Chmn.
Jan-Elo Jorgensen, Managing Dir.-Senetek Denmark ApS

Phone: 707-226-3900	Fax: 707226-3999
Toll-Free:	
Address: 831A Latour Ct., Napa, CA 94558 US	

GROWTH PLANS/SPECIAL FEATURES:
Senetek PLC develops and markets products that target the science of healthy aging. Its business is divided in two segments, skincare and pharmaceuticals, which market anti-aging products, products for the treatment of sexual dysfunction and cancer therapy. In the anti-aging market, the company offers Kinetin, an antioxidant skin cream that has been shown to reduce fine lines and blotchiness and increase the amount of moisture retained by the skin. Senetek developed an entire line of Kinetin products, as well as Zeatin, an analog of Kinetin that is indicated for dermatological applications including the treatment of psoriasis. Kinetin products are marketed worldwide by Valeant Pharmaceuticals International. For erectile dysfunction, the company markets Invicorp, a self-administered injection that delivers a combination therapy of two specific drugs, vasoactive intestinal polypeptide and phentolamine mesylate. Invicorp is meant to be a second-line treatment, after oral therapies have failed. Plethora Holdings is the North American marketer of Invicorp. Aside from drugs, Senetek developed Reliaject, an auto-injector that aids in the accuracy of self-administered shots, which it sold to Ranbaxy Pharmaceuticals. The firm owns a royalty-based license for development of tumor treatments based on the use of RNAi therapy. Treatment of cancer using interference RNA works to inhibit the production of tenascin-c, which has shown a correlation with brain tumor formation. In addition to its Napa headquarters, Senetek runs a research facility in Denmark. In March 2010, the company sold its skincare segment to Skinvera LLC as part of a restructuring process. The firm retained its Kinetin and Zeatin product lines.

FINANCIALS: Sales and profits are in thousands of dollars—add 000 to get the full amount. 2009 Note: Financial information for 2009 was not available for all companies at press time.

2009 Sales: $1,907	2009 Profits: $-5,105	U.S. Stock Ticker: SNTK
2008 Sales: $1,845	2008 Profits: $-3,759	Int'l Ticker: Int'l Exchange:
2007 Sales: $26,471	2007 Profits: $18,632	Employees: 9
2006 Sales: $8,431	2006 Profits: $1,883	Fiscal Year Ends: 12/31
2005 Sales: $5,871	2005 Profits: $-1,739	Parent Company:

SALARIES/BENEFITS:
Pension Plan:	ESOP Stock Plan:	Profit Sharing:	Top Exec. Salary: $185,000	Bonus: $
Savings Plan:	Stock Purch. Plan:		Second Exec. Salary: $165,000	Bonus: $

OTHER THOUGHTS:
Apparent Women Officers or Directors:
Hot Spot for Advancement for Women/Minorities:

LOCATIONS: ("Y" = Yes)
West:	Southwest:	Midwest:	Southeast:	Northeast:	International:
Y					Y

SEPRACOR INC

www.sepracor.com

Industry Group Code: 325412 **Ranks within this company's industry group:** Sales: Profits:

Drugs:		Other:		Clinical:	Computers:	Services:
Discovery:	Y	AgriBio:	Y	Trials/Services:	Hardware:	Specialty Services:
Licensing:	Y	Genetic Data:		Labs:	Software:	Consulting:
Manufacturing:	Y	Tissue Replacement:		Equipment/Supplies:	Arrays:	Blood Collection:
Generics:				Research/Development Svcs.:	Database Management:	Drug Delivery:
				Diagnostics:		Drug Distribution:

TYPES OF BUSINESS:

Pharmaceuticals Discovery & Development
Respiratory Treatments
Central Nervous System Disorder Treatments
Insomnia

BRANDS/DIVISIONS/AFFILIATES:

LUNESTA
XOPENEX
STEDESA
OMNARIS
ALVESCO
BROVANA
Dainippon Sumitomo Pharma Co. Ltd.
Dainippon Sumitomo Pharma America

CONTACTS: Note: Officers with more than one job title may be intentionally listed here more than once.

Saburo Hamanaka, CEO
Mark Iwicki, COO
Mark Iwicki, Pres.
Nobuhiko Tamura, Chief Science Officer/Exec. VP
Susan Adler, VP-Corp. Planning
Susan Adler, VP-Corp. Comm.
Saburo Hamanaka, Chmn.

Phone: 508-481-6700	Fax:
Toll-Free:	
Address: 84 Waterford Dr., Marlborough, MA 01752 US	

GROWTH PLANS/SPECIAL FEATURES:

Sepracor, Inc., a subsidiary of Japan-based Dainippon Sumitomo Pharma Co., Ltd. (DSP), is a research-based pharmaceutical company whose goal is to discover, develop and market products directed toward the treatment of respiratory and central nervous system (CNS) disorders. Sepracor's lead product is an adult insomnia treatment, Lunesta (eszopiclone). It also markets Xopenex Inhalation Solution for use in nebulizer machines, which turn liquids into fine sprays; it is indicated for patients with asthma or chronic obstructive pulmonary disease (COPD). The company also markets Brovana for COPD; Omnaris for allergic rhinitis; and Alvesco for asthma. Wholly owned subsidiary Sepracor Pharmaceuticals, Inc. markets additional products in Canada for cardiovascular, CNS, pain and infectious diseases. Due to holding certain patents, the company also has out-licensing agreements for various allergy medications, including agreements with Schering-Plough for Clarinex (desloratadine); sanofi-aventis for Allegra (fexofenadine); and UCB Farchim SA for its Xusal/Xyzal products (levocetirizine). Additionally, the firm is developing new treatments for depression, pain, asthma, COPD and allergic rhinitis. In January 2009, Sepracor announced it would cut 20% of its workforce, or about 530 jobs. In March 2009, the firm formally submitted a new drug application for Stedesa, a treatment for epilepsy, to the FDA. In May 2009, the company withdrew plans to market Lunesta in Europe after the European Medical Agency declined to grant the drug a new active substance status. In October 2009, the firm was acquired by DSP. In April 2010, Sepracor merged with another DSP subsidiary, Dainippon Sumitomo Pharma America, and was the surviving company.

Sepracor offers its employees medical, dental and vision coverage; short- and long-term disability; life and AD&D insurance; flexible spending accounts; an employee assistance program; a stock purchase plan; a 401(k) plan tuition and adoption assistance; and childcare services.

FINANCIALS: Sales and profits are in thousands of dollars—add 000 to get the full amount. 2009 Note: Financial information for 2009 was not available for all companies at press time.

2009 Sales: $	2009 Profits: $	**U.S. Stock Ticker: Subsidiary**
2008 Sales: $1,292,289	2008 Profits: $515,110	Int'l Ticker: Int'l Exchange:
2007 Sales: $1,225,230	2007 Profits: $58,333	Employees:
2006 Sales: $1,183,133	2006 Profits: $171,161	Fiscal Year Ends: 12/31
2005 Sales: $820,928	2005 Profits: $3,927	Parent Company: DAINIPPON SUMITOMO PHARMA CO LTD

SALARIES/BENEFITS:

Pension Plan:	ESOP Stock Plan:	Profit Sharing:	Top Exec. Salary: $1,050,000	Bonus: $682,500
Savings Plan: Y	Stock Purch. Plan: Y		Second Exec. Salary: $648,442	Bonus: $

OTHER THOUGHTS:

Apparent Women Officers or Directors: 1
Hot Spot for Advancement for Women/Minorities:

LOCATIONS: ("Y" = Yes)

West:	Southwest:	Midwest:	Southeast:	Northeast:	International:
				Y	Y

Note: Financial information, benefits and other data can change quickly and may vary from those stated here.

SEQUENOM INC

www.sequenom.com

Industry Group Code: 325413 Ranks within this company's industry group: Sales: 16 Profits: 21

Drugs:	Other:		Clinical:	Computers:		Services:	
Discovery:	AgriBio:		Trials/Services:	Hardware:		Specialty Services:	
Licensing:	Genetic Data:	Y	Labs:	Software:	Y	Consulting:	
Manufacturing:	Tissue Replacement:		Equipment/Supplies:	Arrays:	Y	Blood Collection:	
Generics:			Research/Development Svcs.:	Database Management:	Y	Drug Delivery:	
			Diagnostics:			Drug Distribution:	

TYPES OF BUSINESS:

Chips-DNA Arrays
Genotype Analysis Software
Cell Research Database

BRANDS/DIVISIONS/AFFILIATES:

MassARRAY
iPLEX Gold
SEQureDx
Center for Molecular Medicine
AttoSense
SensiGen LLC

CONTACTS: Note: Officers with more than one job title may be intentionally listed here more than once.

Harry F. Hixson, Jr., Interim CEO
Paul Maier, Interim CFO
Michael Monko, Sr. VP-Sales & Mktg.
Alisa Judge, VP-Human Resources
Charles R. Cantor, Chief Scientific Officer
Clarke Neumann, General Counsel/VP
Larry Myres, VP-Oper.
Gary S. Riordan, VP-Regulatory Affairs & Quality
Ronald M. Lindsay, Interim Sr. VP-R&D
Shawn Marcell, VP-Molecular Diagnostics
Allan T. Lombard, Chief Medical Officer
Harry F. Hixson, Jr., Chmn.

Phone: 858-202-9000	Fax: 858-202-9001
Toll-Free:	
Address: 3595 John Hopkins Ct., San Diego, CA 92121-1331 US	

GROWTH PLANS/SPECIAL FEATURES:

Sequenom, Inc. is a genetics and molecular diagnostic company providing genetic analysis products, services and diagnostic applications for the development of noninvasive diagnostics in prenatal, oncology and infectious diseases and other disorders. The company's main source of revenue is derived from MassARRAY, a hardware and software application with consumable chips and reagents that analyzes high performance nucleic acids and quantitatively measures genetic target material and variations. The MassARRAY system offers cost-effective methods for numerous types of DNA analysis applications, which can range from SNP genotyping and allelotyping; quantitative gene expression analysis; quantitative methylation marker analysis; epigenomics; and pathogen typing. Sequenom's iPLEX Gold assay provides for multiplexed DNA sample analysis that enables the user to perform multiple sample genotyping analyses using a similar amount of reagents and chip surface area as used for a single DNA sample analysis. Customers of MassARRAY include clinical research laboratories, biotechnology companies, academic institutions and government agencies worldwide. Sequenom is developing various molecular diagnostic tests in prenatal genetic disorders, oncology and infectious diseases under the brand name SEQureDx. In the agricultural market, MassARRAY is utilized to provide farm-of-origin verification, country-of-origin verification, age verification and national ID programs for traceability analysis of livestock. In 2009, the firm introduced three new products: MassARRAY Compact 96 System, which offers genetic analysis solutions; and SensiGene Cystic Fibrosis Carrier Screening test, which screens for 103 mutants and five variants through genomic DNA and DNA control testing to identify carriers of the genetic condition cystic fibrosis. In February 2010, the company introduced two new products: SensiGene Fetal RHD Genotyping test, a noninvasive Rhesus D (RHD) screening test for pregnant women; and SensiGene Fetal (XY) test, a fetal sex determination test.

FINANCIALS: Sales and profits are in thousands of dollars—add 000 to get the full amount. 2009 Note: Financial information for 2009 was not available for all companies at press time.

2009 Sales: $37,863	2009 Profits: $-71,012	**U.S. Stock Ticker:** SQNM
2008 Sales: $47,149	2008 Profits: $-44,154	**Int'l Ticker:** Int'l Exchange:
2007 Sales: $41,002	2007 Profits: $-21,983	Employees: 234
2006 Sales: $28,496	2006 Profits: $-17,577	Fiscal Year Ends: 12/31
2005 Sales: $19,421	2005 Profits: $-26,537	Parent Company:

SALARIES/BENEFITS:

Pension Plan:	ESOP Stock Plan: Y	Profit Sharing:	Top Exec. Salary: $441,000	Bonus: $181,912
Savings Plan: Y	Stock Purch. Plan:		Second Exec. Salary: $319,020	Bonus: $82,247

OTHER THOUGHTS:

Apparent Women Officers or Directors: 2
Hot Spot for Advancement for Women/Minorities: Y

LOCATIONS: ("Y" = Yes)

West:	Southwest:	Midwest:	Southeast:	Northeast:	International:
Y				Y	Y

Note: Financial information, benefits and other data can change quickly and may vary from those stated here.

SERACARE LIFE SCIENCES INC

www.seracare.com

Industry Group Code: 325414 Ranks within this company's industry group: Sales: 4   Profits: 7

Drugs:		Other:		Clinical:		Computers:		Services:	
Discovery:		AgriBio:		Trials/Services:		Hardware:		Specialty Services:	Y
Licensing:		Genetic Data:		Labs:		Software:	Y	Consulting:	
Manufacturing:	Y	Tissue Replacement:		Equipment/Supplies:	Y	Arrays:	Y	Blood Collection:	Y
Generics:				Research/Development Svcs.:	Y	Database Management:		Drug Delivery:	
				Diagnostics:	Y			Drug Distribution:	

TYPES OF BUSINESS:

Diagnostics Products
Blood & Plasma Collection & Processing
Research Support Services
Plasma-based Products
Biological Product Database

BRANDS/DIVISIONS/AFFILIATES:

ACCURUN
ACCUTYPE H1N1
ACCURUN 106 Series 1000 Positive Control
AccuCell Human PBMC-Basic
AccuCell Human PBMC-Characterized

CONTACTS: *Note: Officers with more than one job title may be intentionally listed here more than once.*

Susan L. N. Vogt, CEO
Susan L. N. Vogt, Pres.
Gregory A. Gould, CFO
Bill Smutny, VP-Mktg. & Sales
Ron Dilling, VP-Mfg.
Gregory A. Gould, Sec.
Katheryn E. Shea, VP-BioServices Oper.
Gregory A. Gould, Treas.
David M. Olsen, VP-Corp. Quality
Eugene I. Davis, Chmn.

Phone: 508-244-6400	Fax: 508-634-3394
Toll-Free: 800-676-1881	
Address: 37 Birch St., Milford, MA 01757 US	

GROWTH PLANS/SPECIAL FEATURES:

SeraCare Life Sciences, Inc. develops, manufactures and sells a broad range of biological based materials and services, which facilitate the discovery, development and production of diagnostic and therapeutic products. The company's portfolio includes plasma-based reagents; diagnostic controls; and molecular biomarkers, biobanking and contract research services. SeraCare's business is divided into two segments: the Diagnostic and Biopharmaceutical Products division and the BioServices division. The Diagnostic and Biopharmaceutical Products division includes two product lines: controls and panels; and reagents and bioprocessing products. Controls and panels include the manufacture of products used for the evaluation and quality control of infectious disease testing in hospital and clinical testing labs and blood banks, and by in vitro diagnostic ("IVD") manufacturers. The firm offers over 100 control and panel products for infectious diseases including HIV, hepatitis A, HBV, HCV, West Nile Virus, Chagas and HPV. Most of the control products are sold under the ACCURUN brand name, and its panel products are called seroconversion and performance panels. The reagents and bioprocessing products line includes the manufacture and supply of biological materials used in the research, development and manufacturing of human and animal diagnostics, therapeutics and vaccines. Its products include diagnostic intermediates, therapeutic grade albumin, purified viable human cells and cell culture additives and media. The BioServices segment includes biobanking, sample processing and testing services for research and clinical trials; and contract research services in molecular biology, virology, immunology and biochemistry. In January 2010, SeraCare added five products to its portfolio: the ACCURUN 1 Series 2700 Multi-Market Positive Control; the ACCURUN 106 Series 1000 Positive Control; the AccuCell Human PBMC-Basic and AccuCell Human PBMC-Characterized products; and the ACCUTYPE H1N1.

SeraCare offers its employees health and dental coverage; short- and long-term disability; life and travel insurance; a 401(k) plan; a flexible spending account; and an employee assistance program.

FINANCIALS: Sales and profits are in thousands of dollars—add 000 to get the full amount. 2009 Note: Financial information for 2009 was not available for all companies at press time.

2009 Sales: $44,434	2009 Profits: $-15,379	**U.S. Stock Ticker: SRLS**
2008 Sales: $48,967	2008 Profits: $-11,963	**Int'l Ticker:** Int'l Exchange:
2007 Sales: $47,304	2007 Profits: $-13,165	Employees: 193
2006 Sales: $49,176	2006 Profits: $-24,278	Fiscal Year Ends: 9/30
2005 Sales: $50,300	2005 Profits: $-21,097	Parent Company:

SALARIES/BENEFITS:

Pension Plan:	ESOP Stock Plan:	Profit Sharing:	Top Exec. Salary: $378,000	Bonus: $49,825
Savings Plan: Y	Stock Purch. Plan:		Second Exec. Salary: $270,000	Bonus: $70,284

OTHER THOUGHTS:

Apparent Women Officers or Directors: 3
Hot Spot for Advancement for Women/Minorities: Y

LOCATIONS: ("Y" = Yes)

West:	Southwest:	Midwest:	Southeast:	Northeast:	International:
				Y	Y

SHIONOGI INC

www.shionogi-inc.com

Industry Group Code: 325412 Ranks within this company's industry group: Sales: Profits:

Drugs:		Other:	Clinical:	Computers:	Services:
Discovery:	Y	AgriBio:	Trials/Services:	Hardware:	Specialty Services:
Licensing:		Genetic Data:	Labs:	Software:	Consulting:
Manufacturing:		Tissue Replacement:	Equipment/Supplies:	Arrays:	Blood Collection:
Generics:			Research/Development Svcs.:	Database Management:	Drug Delivery:
			Diagnostics:		Drug Distribution:

TYPES OF BUSINESS:

Pharmaceuticals Discovery & Development
Research & Development

BRANDS/DIVISIONS/AFFILIATES:

Shionogi & Co Ltd
Shionogi Pharma Inc
Sular
Fortamet
Allegra
Triglide
Altoprev
Nitrolingual

CONTACTS: *Note: Officers with more than one job title may be intentionally listed here more than once.*

Sapan Shah, CEO
Joseph J. Ciaffoni, COO/Exec. VP
Sapan Shah, Pres.
Gregg Siefert, Exec. VP-Human Resources
Leslie Zacks, Chief Legal & Compliance Officer/Exec. VP
Donald Manning, Chief Medical Officer/Exec. VP
Susan Witham, Exec. VP-Regulatory Affairs & Quality Assurance

Phone: 973-966-6900	Fax:
Toll-Free: 800-849-9707	
Address: 100 Campus Dr., Florham Park, NJ 07932 US	

GROWTH PLANS/SPECIAL FEATURES:

Shionogi, Inc., is the U.S. branch of Japanese company Shionogi & Co. Ltd. It oversees both the commercial, and the research and development activities of Shionogi that take place in the U.S. The firm is responsible for the clinical development, regulatory affairs and corporate strategy, while it's wholly-owned subsidiary Shionogi Pharma, Inc., (formerly Sciele Pharma Inc.), is responsible for sales, marketing and distribution. The firm focuses on cardiology, diabetes, women's health, and pediatric treatment. Some of the top products, include: Sular for hypertension; Fortamet, a medication that lowers blood glucose in type 2 diabetes patients; Altoprev, for cholesterol reduction and coronary heart disease; Triglide, for hyperchlesterolemia and hypertriglyceridemia; Nitrolingual, a pumpspray that provides acute relief during attacks of angina pectoris due to coronary artery disease; Prenate Elite and Prenate DHA, prenatal vitamins; Allegra, a pediatric allergy and chronic idiopathic urticaria treatment; Methylin chewable tablets, for attention deficit/hyperactivity disorder; and Orapred, whose applications include severe allergy relief for patients with asthma. The company's other products include treatments for swimmer's ear infection, tension headaches, peptic ulcers, dementia, urinary tract infections and seasonal allergies. In addition to its marketed drugs, the company has products under development for a range of indications including chronic drooling, premature ejaculation, head lice, obesity, HIV Infection, diabetes and hypertension. The firm enlists third-party manufacturers for all its products. In January 2010, Shionogi & Co. Ltd changed the name of Sciele Pharma Inc., to Shionogi Pharma Inc. In July 2010, Shionogi Inc. was established as the U.S. group headquarters. Shionogi & Co. Ltd plans to merge Shionogi Pharma Inc., into Shionogi Inc.

Employees of the firm are offered medical, dental, vision and Rx coverage; short and long-term disability; accidental death and dismemberment; wellness programs; voluntary life insurance; and a 401(k) plan.

FINANCIALS: Sales and profits are in thousands of dollars—add 000 to get the full amount. 2009 Note: Financial information for 2009 was not available for all companies at press time.

2009 Sales: $	2009 Profits: $	**U.S. Stock Ticker:** Subsidiary
2008 Sales: $	2008 Profits: $	**Int'l Ticker:** Int'l Exchange:
2007 Sales: $	2007 Profits: $	Employees:
2006 Sales: $	2006 Profits: $	Fiscal Year Ends: 3/31
2005 Sales: $	2005 Profits: $	Parent Company: SHIONOGI & CO LTD

SALARIES/BENEFITS:

Pension Plan:	ESOP Stock Plan:	Profit Sharing:	Top Exec. Salary: $	Bonus: $
Savings Plan: Y	Stock Purch. Plan:		Second Exec. Salary: $	Bonus: $

OTHER THOUGHTS:

Apparent Women Officers or Directors: 1
Hot Spot for Advancement for Women/Minorities:

LOCATIONS: ("Y" = Yes)

West:	Southwest:	Midwest:	Southeast:	Northeast:	International:
			Y	Y	Y

SHIRE CANADA INC

www.shirecanada.com

Industry Group Code: 325412 Ranks within this company's industry group: Sales: Profits:

Drugs:		Other:	Clinical:	Computers:	Services:
Discovery:	Y	AgriBio:	Trials/Services:	Hardware:	Specialty Services:
Licensing:		Genetic Data:	Labs:	Software:	Consulting:
Manufacturing:	Y	Tissue Replacement:	Equipment/Supplies:	Arrays:	Blood Collection:
Generics:			Research/Development Svcs.:	Database Management:	Drug Delivery:
			Diagnostics:		Drug Distribution:

TYPES OF BUSINESS:

Pharmaceuticals Discovery & Development
Influenza Vaccine
Drugs-HIV

BRANDS/DIVISIONS/AFFILIATES:

Shire Human Genetic Therapies (Canada)
Estrace
Shire PLC
Vyvanse
Aderall XR
Agrylin
Alertec
Replagal

CONTACTS: *Note: Officers with more than one job title may be intentionally listed here more than once.*

Claude Perron, Gen. Mgr./VP

Phone: 514-787-2300	Fax: 514-787-2427
Toll-Free:	
Address: 2250 Alfred-Nobel Blvd., Ste. 500, Saint-Laurent, QC 24 Canada	

GROWTH PLANS/SPECIAL FEATURES:

Shire Canada, Inc., formerly Shire BioChem, Inc., is a Canadian company that focuses on the research, development and commercialization of innovative products for the prevention and treatment of human diseases. The company is also in charge of marketing products in Canada. The firm is a subsidiary of Shire PLC, a global specialty pharmaceutical company. Shire Canada specializes in marketing products related to attention deficit hyperactivity disorder (ADHD), gastrointestinal/renal diseases and human genetic therapies. Through a partnership with GlaxoSmithKline, Inc., the firm also offers antiretroviral treatment options for Canadians diagnosed with Hepatitis B and/or HIV. Shire Canada currently markets Adderall XR for ADHD; Agrylin, a treatment of essential thrombocythaemia; Amatine, a cardiovascular aid; Elaprase, an enzyme replacement therapy for Hunter syndrome patients; Fosrenol, which reduces serum phosphate in patients that suffer from end-stage renal disease; Alertec, a symptomatic treatment of excessive sleepiness in adult patients with obstructive sleep apnea/hypopnea syndrome (OSAHS), narcolepsy, and shift work sleep disorder (SWSD); Estrace, a menopausal relief drug; Mezavant, a treatment for patients suffering from active mild to moderate ulcerative colitis; Vyvanse (which became available in Canada in February 2010), an ADHD treatment specifically for children 6-12 years of age; and Replagal, a possible treatment for Fabry disease. The firm's sister company, Shire Human Genetic Therapies (Canada), specializes in the development of treatments for rare and orphan genetic diseases such as Fabry disease and Hunter syndrome (MPS II).

FINANCIALS: Sales and profits are in thousands of dollars—add 000 to get the full amount. 2009 Note: Financial information for 2009 was not available for all companies at press time.

2009 Sales: $	2009 Profits: $	U.S. Stock Ticker: Subsidiary
2008 Sales: $	2008 Profits: $	Int'l Ticker: Int'l Exchange:
2007 Sales: $	2007 Profits: $	Employees:
2006 Sales: $	2006 Profits: $	Fiscal Year Ends: 12/31
2005 Sales: $	2005 Profits: $	Parent Company: SHIRE PLC

SALARIES/BENEFITS:

Pension Plan:	ESOP Stock Plan:	Profit Sharing:	Top Exec. Salary: $	Bonus: $
Savings Plan:	Stock Purch. Plan:		Second Exec. Salary: $	Bonus: $

OTHER THOUGHTS:

Apparent Women Officers or Directors:
Hot Spot for Advancement for Women/Minorities:

LOCATIONS: ("Y" = Yes)

West:	Southwest:	Midwest:	Southeast:	Northeast:	International:
					Y

SHIRE PLC

www.shire.com

Industry Group Code: 325412   Ranks within this company's industry group: Sales: 29   Profits: 25

Drugs:		Other:	Clinical:	Computers:	Services:	
Discovery:	Y	AgriBio:	Trials/Services:	Hardware:	Specialty Services:	
Licensing:	Y	Genetic Data:	Labs:	Software:	Consulting:	
Manufacturing:	Y	Tissue Replacement:	Equipment/Supplies:	Arrays:	Blood Collection:	
Generics:			Research/Development Svcs.:	Database Management:	Drug Delivery:	Y
			Diagnostics:		Drug Distribution:	

TYPES OF BUSINESS:

Drugs-Diversified
Drug Delivery Technology
Small-Molecule Drugs

BRANDS/DIVISIONS/AFFILIATES:

Vyvanse
Intuniv
Equasym XL
Daytrana
Adderall XR
Pentasa
Lialda/Mezavant
Replagal

CONTACTS: Note: Officers with more than one job title may be intentionally listed here more than once.

Angus Russell, CEO
Graham Hetherington, CFO
Anita Graham, Chief Admin. Officer
Tatjana May, General Counsel/Exec. VP-Global Legal Affairs/Sec.
Barbara Deptula, Chief Corp. Dev. Officer/Exec. VP
Sylvie Gregoire, Pres., Human Genetic Therapies
Mike Cola, Pres., Specialty Pharmaceuticals
Anita Graham, Exec. VP-Corp. Bus. Svcs.
Matthew Emmens, Chmn.

Phone: 353-1-429-7700	Fax: 353-1-429-7701
Toll-Free:	
Address: 5 Riverwalk, Citywest Business Campus, Dublin 24, UK	

GROWTH PLANS/SPECIAL FEATURES:

Shire plc, formerly Shire Ltd., is an international specialty pharmaceutical company. The firm is focused on three therapeutic areas: attention deficit and hyperactivity disorder (ADHD), gastrointestinal (GI) diseases and human genetic therapies (HGT). Shire's products for the treatment of ADHD include Vyvanse, a pro-drug stimulant; Intuniv, an alpha-2A receptor agonist; Equasym XL; Daytrana, a methylphenidate transdermal delivery system; and Adderall XR, an extended release treatment that uses MICROTROL drug delivery technology. The firm's treatments for GI diseases include Pentasa controlled release capsules for the treatment of patients with mild to moderately active ulcerative colitis; and Lialda/Mezavant for the induction of remission in patients with ulcerative colitis. Shire's HGT products include Replagal, a treatment for Fabry disease; Elaprase, a treatment for Hunter syndrome; and Firazyr, a peptide-based therapeutic developed for the symptomatic treatment of acute attacks of HAE, a debilitating genetic disease. The firm also offers Fosrenol, a phosphate binder for use in chronic kidney disease patients; the Calcichew range of calcium and calcium/vitamin D3 supplements for the adjunctive treatment of osteoporosis; Carbatrol, an anti-convulsant for individuals with epilepsy; Reminyl for the symptomatic treatment of mild to moderately severe dementia; and Xagrid, which is used for the reduction of elevated platelet counts in at-risk essential thrombocythemia patients. The firm markets products in the U.S., Canada, the U.K., Germany, France, Italy, Ireland and Spain. In June 2010, Shire purchased the Lexington Technology Park campus in Lexington, Massachusetts for $165 million. In July 2010, the company received FDA approval of Daytrana for the treatment of ADHD in adolescents aged 13-17.

Employees are offered health, dental and vision insurance; flexible spending accounts; life and AD&D insurance; disability benefits; an employee assistance program; a healthy lifestyle reimbursement; educational assistance; an employee stock purchase plan; employee referral incentives; a 401(k) plan; and adoption assistance.

FINANCIALS: Sales and profits are in thousands of dollars—add 000 to get the full amount. 2009 Note: Financial information for 2009 was not available for all companies at press time.

2009 Sales: $3,007,700	2009 Profits: $491,600	U.S. Stock Ticker: SHPGY
2008 Sales: $3,022,200	2008 Profits: $156,000	Int'l Ticker: SHP.L Int'l Exchange: London-LSE
2007 Sales: $2,436,300	2007 Profits: $-1,451,800	Employees: 3,875
2006 Sales: $1,796,500	2006 Profits: $278,200	Fiscal Year Ends: 12/31
2005 Sales: $1,599,300	2005 Profits: $-578,400	Parent Company:

SALARIES/BENEFITS:

Pension Plan:	ESOP Stock Plan:	Profit Sharing:	Top Exec. Salary: $	Bonus: $
Savings Plan: Y	Stock Purch. Plan: Y		Second Exec. Salary: $	Bonus: $

OTHER THOUGHTS:

Apparent Women Officers or Directors: 6
Hot Spot for Advancement for Women/Minorities: Y

LOCATIONS: ("Y" = Yes)

West:	Southwest:	Midwest:	Southeast:	Northeast:	International:
		Y		Y	Y

SIEMENS HEALTHCARE DIAGNOSTICS

diagnostics.siemens.com

Industry Group Code: 325413 Ranks within this company's industry group: Sales: Profits:

Drugs:	Other:	Clinical:		Computers:	Services:
Discovery:	AgriBio:	Trials/Services:		Hardware:	Specialty Services:
Licensing:	Genetic Data:	Labs:		Software:	Consulting:
Manufacturing:	Tissue Replacement:	Equipment/Supplies:	Y	Arrays:	Blood Collection:
Generics:		Research/Development Svcs.:		Database Management:	Drug Delivery:
		Diagnostics:	Y		Drug Distribution:

TYPES OF BUSINESS:

Supplies-Immunodiagnostic Kits
Nonisotopic Diagnostic Tests
Immunoassay Analyzers
Allergy Testing

BRANDS/DIVISIONS/AFFILIATES:

IMMULITE
ADVIA
Bayer Diagnostics
Dimension EXL
CLINITEK AUWi
RAPIDPoint
Siemens AG
Diagnostic Products Corp.

CONTACTS: Note: Officers with more than one job title may be intentionally listed here more than once.

Donal Quinn, CEO
Donal Quinn, Pres.
Denice Kronau, CFO/Exec. VP

Phone: 847-267-5300	Fax:
Toll-Free:	
Address: 1717 Deerfield Rd., Deerfield, IL 60015-0778 US	

GROWTH PLANS/SPECIAL FEATURES:

Siemens Healthcare Diagnostics (formerly Siemens Medical Solutions Diagnostics), formed from the acquisition of Diagnostic Products Corp. (DPC) and Bayer Diagnostics, is a wholly-owned subsidiary of Siemens Healthcare. The firm is focused on molecular imaging diagnostic and in vitro (laboratory) diagnostic products. Its portfolio includes products and services designed to optimize efficiency, improve workflow and help ensure patient safety. The company offers a broad spectrum of testing systems such as immunoassay, integrated chemistry, routine chemistry, automation, hematology, microbiology, hemostasis, molecular, urinalysis, diabetes and blood gas. It also offers automation, informatics and consulting services. The firm's proprietary products include the IMMULITE immunoassay systems; ADVIA hematology systems; RapidLab systems for blood gas testing; the Clinitek family of products for urinalysis and urine chemistry testing; the DCA Vantage Analyzer, an immunoassay testing system for diabetes management. In April 2008, the company added the RAPIDPoint 340 and 350 blood gas analyzers to its product inventory. These analyzers are designed for low - to mid-volume testing locations. In May of the same year, the firm launched the CLINITEK AUWi System, which is the combination of the CLINITEK Atlas Automated Urine Chemistry Analyzer and the Sysmex UF-1000i Urine Cell Analyzer. This combined system is designed to minimize follow-up, as well as workflow with the automated transferring of samples to different workstations. Also in May 2009, the company introduced the Dimension EXL, an immunoassay testing product that uses LOCI chemiluminescence technology.

FINANCIALS: Sales and profits are in thousands of dollars—add 000 to get the full amount. 2009 Note: Financial information for 2009 was not available for all companies at press time.

2009 Sales: $	2009 Profits: $	U.S. Stock Ticker: Subsidiary
2008 Sales: $	2008 Profits: $	Int'l Ticker: Int'l Exchange:
2007 Sales: $	2007 Profits: $	Employees:
2006 Sales: $	2006 Profits: $	Fiscal Year Ends: 12/31
2005 Sales: $	2005 Profits: $	Parent Company: SIEMENS AG

SALARIES/BENEFITS:

Pension Plan:	ESOP Stock Plan:	Profit Sharing:	Top Exec. Salary: $	Bonus: $
Savings Plan:	Stock Purch. Plan:		Second Exec. Salary: $	Bonus: $

OTHER THOUGHTS:

Apparent Women Officers or Directors:
Hot Spot for Advancement for Women/Minorities:

LOCATIONS: ("Y" = Yes)

West:	Southwest:	Midwest:	Southeast:	Northeast:	International:
Y				Y	Y

SIGA TECHNOLOGIES INC   www.siga.com

Industry Group Code: 325412 Ranks within this company's industry group: Sales: 115   Profits: 111

Drugs:		Other:	Clinical:	Computers:	Services:	
Discovery:	Y	AgriBio:	Trials/Services:	Hardware:	Specialty Services:	
Licensing:		Genetic Data:	Labs:	Software:	Consulting:	
Manufacturing:		Tissue Replacement:	Equipment/Supplies:	Arrays:	Blood Collection:	
Generics:			Research/Development Svcs.:	Database Management:	Drug Delivery:	Y
			Diagnostics:		Drug Distribution:	

TYPES OF BUSINESS:

Drugs-Infectious Diseases
Vaccines
Antibiotics
Biothreat Rapid-Response Therapeutics
Mucosal Drug Delivery

BRANDS/DIVISIONS/AFFILIATES:

ST-246
ST-669
ST-610
ST-148
ST-294
ST-193
Orthopox Antiviral
New World Arenavirus Antiviral

CONTACTS: Note: Officers with more than one job title may be intentionally listed here more than once.

Eric A. Rose, CEO
Ayelet Dugary, CFO
Dennis E. Hruby, Chief Scientific Officer
Ayelet Dugary, Sec.
Eric A. Rose, Chmn.

Phone: 212-672-9100	Fax: 212-697-3130
Toll-Free:	
Address: 35 E. 62nd St., New York, NY 10065 US	

GROWTH PLANS/SPECIAL FEATURES:

SIGA Technologies is a development-stage biotechnology company focused on the discovery, development and commercialization of novel products for the prevention and treatment of serious infectious diseases, including products for use in defense against biological warfare agents such as smallpox and Arenaviruses. Its lead product, ST-246, is an orally administered anti-viral drug targeting the smallpox virus that has been granted fast-track orphan drug status by the FDA. SIGA's biological warfare defense product portfolio includes ST-246, which also targets vaccinia, cowpox, mousepox, monkeypox and camelpox; ST-669, a broad-spectrum antiviral used to target viruses in the Flaviviridae, Togaviridae, Retroviridae and Picornaviridae families; ST-610 and ST-148 for the treatment of Dengue fever, dengue hemorrhagic fever and dengue shock syndrome; and ST-294 and ST-193, which are used against Arenaviruses. SIGA's antiviral programs are designed to prevent or limit the replication of the viral pathogen. Its anti-infectives programs are aimed at the increasingly serious problem of drug resistance. Its antivirals product portfolio includes the Orthopox antiviral; the New World Arenavirus antiviral; the Old World Arenavirus antiviral; the Filovirus (Ebola and Marburg) antivirals; the Dengue Fever virus antiviral; and Bunyavirus antivirals. The firm's other technology platforms include its Strep Bacterial Commensal Vector program as a subunit vaccine delivery system. The company has collaborations with organizations including the National Institutes of Health, the National Institute of Allergy and Infectious Diseases and TransTech Pharma, Inc. In February 2010, the company was awarded a $2.8 million contract by the Department of Defense for the development of ST-669.

FINANCIALS: Sales and profits are in thousands of dollars—add 000 to get the full amount. 2009 Note: Financial information for 2009 was not available for all companies at press time.

2009 Sales: $13,812	2009 Profits: $-17,618	U.S. Stock Ticker: SIGA
2008 Sales: $8,066	2008 Profits: $-8,599	Int'l Ticker: Int'l Exchange:
2007 Sales: $6,699	2007 Profits: $-5,639	Employees: 55
2006 Sales: $7,258	2006 Profits: $-9,899	Fiscal Year Ends: 12/31
2005 Sales: $8,477	2005 Profits: $-2,288	Parent Company:

SALARIES/BENEFITS:

Pension Plan:	ESOP Stock Plan:	Profit Sharing:	Top Exec. Salary: $400,000	Bonus: $
Savings Plan:	Stock Purch. Plan:		Second Exec. Salary: $275,000	Bonus: $275,000

OTHER THOUGHTS:

Apparent Women Officers or Directors: 1
Hot Spot for Advancement for Women/Minorities:

LOCATIONS: ("Y" = Yes)

West:	Southwest:	Midwest:	Southeast:	Northeast:	International:
Y				Y	

SIGMA-ALDRICH CORP

www.sigmaaldrich.com

Industry Group Code: 325 Ranks within this company's industry group: Sales: 6   Profits: 5

Drugs:	Other:	Clinical:		Computers:	Services:
Discovery:	AgriBio:	Trials/Services:		Hardware:	Specialty Services:
Licensing:	Genetic Data:	Labs:		Software:	Consulting:
Manufacturing:	Tissue Replacement:	Equipment/Supplies:	Y	Arrays:	Blood Collection:
Generics:		Research/Development Svcs.:		Database Management:	Drug Delivery:
		Diagnostics:	Y		Drug Distribution:

TYPES OF BUSINESS:

Chemicals Manufacturing
Biotechnology Equipment
Pharmaceutical Ingredients
Fine Chemicals
Chromatography Products

BRANDS/DIVISIONS/AFFILIATES:

Research Essentials
Research Specialties
Research Biotech
SAFC
ChemInformatics, Inc.

CONTACTS: *Note: Officers with more than one job title may be intentionally listed here more than once.*

Jai Nagarkatti, CEO
Jai Nagarkatti, Pres.
Rakesh Sachdev, CFO/VP
Gerrit van den Dool, VP-Sales
Doug Rau, VP-Human Resources
Rakesh Sachdev, Chief Admin. Officer
Rakesh Sachdev, Sec.
Karen Miller, VP-Strategy & Corp. Dev.
Kirk Richter, Treas.
Gilles Cottier, Pres., SAFC
Frank Wicks, Pres., Research Specialties & Essentials
David Smoller, Pres., Research Biotech
Steve Walton, VP-Quality & Safety
Jai Nagarkatti, Chmn.
Dave Julien, Pres., Supply Chain

Phone: 314-771-5765	**Fax:** 314-771-5757
Toll-Free: 800-521-8956	
Address: 3050 Spruce St., St. Louis, MO 63103 US	

GROWTH PLANS/SPECIAL FEATURES:

Sigma-Aldrich Corp. is a life science and technology company that develops, manufactures, purchases and distributes a broad range of biochemicals and organic chemicals. The company offers roughly 130,000 chemicals (including 48,000 chemicals manufactured in-house) and 40,000 equipment products used for scientific and genomic research; biotechnology; pharmaceutical development; disease diagnosis; and pharmaceutical and high technology manufacturing. Sigma-Aldrich is structured into four units: research essentials, research specialties, research biotech and SAFC. The research essentials unit sells biological buffers; cell culture reagents; biochemicals; chemicals; solvents; and other reagents and kits. The research specialties unit provides organic chemicals, biochemicals, analytical reagents, chromatography consumables, reference materials and high-purity products. The research biotech unit supplies immunochemical, molecular biology, cell signaling and neuroscience biochemicals and kits used in biotechnology, genomic, proteomic and other life science research applications. The SAFC fine chemicals unit offers large-scale organic chemicals and biochemicals used in development and production by pharmaceutical, biotechnology, industrial and diagnostic companies. The company operates in 38 countries, selling its products in nearly 160 countries and servicing over 1 million customers. Customers include commercial laboratories; pharmaceutical and industrial companies; universities; diagnostics, chemical and biotechnology companies and hospitals; non-profit organizations; and governmental institutions. In August 2009, the firm acquired ChemNavigator, Inc., a company that designs and develops chemical informatics software.

FINANCIALS: Sales and profits are in thousands of dollars—add 000 to get the full amount. 2009 Note: Financial information for 2009 was not available for all companies at press time.

2009 Sales: $2,147,600	2009 Profits: $346,700	**U.S. Stock Ticker:** SIAL
2008 Sales: $2,200,700	2008 Profits: $341,500	**Int'l Ticker:** Int'l Exchange:
2007 Sales: $2,038,100	2007 Profits: $311,100	Employees: 7,740
2006 Sales: $1,797,500	2006 Profits: $276,800	Fiscal Year Ends: 12/31
2005 Sales: $1,666,500	2005 Profits: $258,300	Parent Company:

SALARIES/BENEFITS:

Pension Plan: Y	ESOP Stock Plan:	Profit Sharing:	Top Exec. Salary: $750,000	Bonus: $384,000
Savings Plan: Y	Stock Purch. Plan:		Second Exec. Salary: $475,000	Bonus: $447,600

OTHER THOUGHTS:

Apparent Women Officers or Directors: 2
Hot Spot for Advancement for Women/Minorities: Y

LOCATIONS: ("Y" = Yes)

West:	Southwest:	Midwest:	Southeast:	Northeast:	International:
Y	Y	Y	Y	Y	Y

Note: Financial information, benefits and other data can change quickly and may vary from those stated here.

SIMCERE PHARMACEUTICAL GROUP

www.simcere.com

Industry Group Code: 325412 Ranks within this company's industry group: Sales: 61   Profits: 69

Drugs:		Other:	Clinical:	Computers:	Services:
Discovery:	Y	AgriBio:	Trials/Services:	Hardware:	Specialty Services:
Licensing:		Genetic Data:	Labs:	Software:	Consulting:
Manufacturing:	Y	Tissue Replacement:	Equipment/Supplies:	Arrays:	Blood Collection:
Generics:			Research/Development Svcs.:	Database Management:	Drug Delivery:
			Diagnostics:		Drug Distribution:

TYPES OF BUSINESS:

Branded Generic Pharmaceuticals
Drug Research

BRANDS/DIVISIONS/AFFILIATES:

Endu
Bicun
Zailin
Yingtaiqing
Anqi
Biqi
Master Luck Corporation Limited
Zanamivir

CONTACTS: Note: Officers with more than one job title may be intentionally listed here more than once.

Jinsheng Ren, CEO
Yehong Zhang, Pres.
Frank Zhigang Zhao, CFO
Huaping Fu, VP-Commercial Sales
Quanfu Feng, VP-Human Resources
Peng Wang, Chief Scientific Officer
Haibo Qian, Corp. Sec.
Jindong Zhou, Exec. VP
Xiaohua Yang, VP-Hospital Sales
Xiaojin Yin, Sr. VP-R&D
Qinseng Li, VP-Training & Dev.
Jinsheng Ren, Chmn.

Phone: 86-25-8556-6666	Fax: 85-25-8547-1729
Toll-Free:	
Address: No. 699-18 Xuan Wu Ave., Nanjing, 210042 China	

GROWTH PLANS/SPECIAL FEATURES:

Simcere Pharmaceutical Group is a Chinese manufacturer and supplier of branded generic pharmaceuticals. The company manufactures and sells 46 pharmaceutical products and is the exclusive distributor of two additional pharmaceutical products marketed under the firm's brand name. The firm's products are used for treatment of a wide range of diseases including cancer; cerebrovascular and cardiovascular diseases; infections; arthritis; diarrhea; allergies; respiratory conditions; and urinary conditions. Simcere's products include Bicun, an anti-stroke medication and the first synthetic-free radical scavenger sold in China; Zailin, a generic amoxicillin granule antibiotic; Endu, an anticancer medication and recombinant human endostatin injection; Yingtaiqing, a generic diclofenac sodium sustained-release capsule for inflammation and pain relief; Anqi, a generic amoxicillin with clavulanate potassium antibiotic; and Biqi, an over-the-counter generic smectite powder for diarrhea. In addition, the Chinese FDA has approved the sale of more than 222 other company products. The firm also has 12 product candidates in various stages of development. Simcere Pharmaceutical has seven manufacturing facilities, two nationwide sales and marketing subsidiaries and one research and development center. In recent years, the firm acquired Master Luck Corporation Limited, which is a majority-owner of Nanjing Chit Pharmaceutical, a quick-growing manufacturer and supplier of anti-cancer drugs in China. In October 2009, Simcere agreed to acquire a 74.49% stake in ChinaVax. In November 2009, the firm agreed to acquire the manufacturing license in China of Rosuvastatin from Tianjin Tianda Pharmaceutical Co., Ltd. In February 2010, Simcere received approval from the State Food and Drug Administration (SFDA) to manufacture and sell Zanamivir (an inhalant used to prevent and treat Influenza A and Influenza B) in China. In June 2010, the company agreed to acquire an 80% stake in Xiangao Investment Company Ltd.

The company offers employee benefits that include health insurance and a housing plan.

FINANCIALS: Sales and profits are in thousands of dollars—add 000 to get the full amount. 2009 Note: Financial information for 2009 was not available for all companies at press time.

2009 Sales: $272,062	2009 Profits: $3,871	U.S. Stock Ticker: SCR
2008 Sales: $225,206	2008 Profits: $51,323	Int'l Ticker: Int'l Exchange:
2007 Sales: $187,638	2007 Profits: $41,300	Employees: 3,974
2006 Sales: $	2006 Profits: $	Fiscal Year Ends: 12/31
2005 Sales: $91,300	2005 Profits: $12,700	Parent Company:

SALARIES/BENEFITS:

Pension Plan:	ESOP Stock Plan:	Profit Sharing:	Top Exec. Salary: $	Bonus: $
Savings Plan:	Stock Purch. Plan:		Second Exec. Salary: $	Bonus: $

OTHER THOUGHTS:

Apparent Women Officers or Directors:
Hot Spot for Advancement for Women/Minorities:

LOCATIONS: ("Y" = Yes)

West:	Southwest:	Midwest:	Southeast:	Northeast:	International: Y

SKYEPHARMA PLC

www.skyepharma.com

Industry Group Code: 325412A Ranks within this company's industry group: Sales: 5 Profits: 6

Drugs:		Other:		Clinical:	Computers:	Services:	
Discovery:	Y	AgriBio:		Trials/Services:	Hardware:	Specialty Services:	
Licensing:	Y	Genetic Data:		Labs:	Software:	Consulting:	
Manufacturing:	Y	Tissue Replacement:		Equipment/Supplies:	Arrays:	Blood Collection:	
Generics:				Research/Development Svcs.:	Database Management:	Drug Delivery:	Y
				Diagnostics:		Drug Distribution:	

TYPES OF BUSINESS:

Drug Delivery Systems
Generic Drugs

BRANDS/DIVISIONS/AFFILIATES:

Sular
Xatral
Lodotra
ZYFLO CR
Paxil CR
Corumo
Pulmicort HFA-MDI
Solaraze

CONTACTS: Note: Officers with more than one job title may be intentionally listed here more than once.

Ken Cunningham, CEO
Peter Grant, CFO
John Murphy, General Counsel/Sec.
Tim McBride, Exec. VP-Commercial
Kirsten Kaiser, Exec. VP-Medical & Regulatory Dev.
Anne Brindley, Exec. VP-Pharmaceutical Dev. & Project Mgmt.
Frank Condella, Chmn.

Phone: 44-20-7491-1777	Fax: 44-20-7491-3338
Toll-Free:	
Address: 105 Piccadilly, London, W1J 7NJ UK	

GROWTH PLANS/SPECIAL FEATURES:

SkyePharma PLC is a worldwide provider of drug delivery technologies. Its drug delivery products and new drugs are manufactured under the SkyePharma name, which is contracted to several global pharmaceutical companies, including Abbott Laboratories; AstraZeneca; Novartis; and GlaxoSmithKline. SkyePharma specializes in three technologies: oral technologies, inhalation technologies and solubilisation technologies. Each of these areas is supported by dedicated research facilities located primarily in Switzerland, with additional operations in Lyon, France. The oral technologies segment uses the company's Geomatrix technology, enabling the formulation of oral products that provide control over the timing, location and amount of the release of drug compounds in the human body. The inhalation technologies segment develops advanced formulations of inhaled products in hydrofluoroalkane (HFA) metered dose inhalers (MDI) and dry powder inhalers (DPI). The solubilisation technology platform consists of three technologies that address formulation and delivery problems caused by poor compound solubility: insoluble drug delivery technology; dissocubes; and solid lipid nanoparticles (SLN). The firm's oral treatment products include Sular for high blood pressure; Coruno for throat diseases; Madopar DR for Parkinson's disease; and ZYFLO CR for asthma. Pulmicort HFA-MDI, SkyePharma's FDA-approved dry powder inhaler, is an asthma treatment for adults and children. Solaraze, the firm's topical drug, is used to treat Actinic Keratosis. SkyePharma is currently developing oral drugs Lodotra for rheumatoid arthritis and SKP-1041 for sleep maintenance. SkyePharma's Flutiform HFA-MDI, an inhalation product for the treatment of asthma, is currently under review in both the U.S. and Europe.

FINANCIALS: Sales and profits are in thousands of dollars—add 000 to get the full amount. 2009 Note: Financial information for 2009 was not available for all companies at press time.

2009 Sales: $80,470	2009 Profits: $-2,020	U.S. Stock Ticker:
2008 Sales: $94,710	2008 Profits: $-43,700	Int'l Ticker: SKP Int'l Exchange: London-LSE
2007 Sales: $63,340	2007 Profits: $-36,540	Employees: 121
2006 Sales: $85,100	2006 Profits: $-156,600	Fiscal Year Ends: 12/31
2005 Sales: $100,600	2005 Profits: $-100,800	Parent Company:

SALARIES/BENEFITS:

Pension Plan:	ESOP Stock Plan:	Profit Sharing:	Top Exec. Salary: $	Bonus: $
Savings Plan:	Stock Purch. Plan:		Second Exec. Salary: $	Bonus: $

OTHER THOUGHTS:

Apparent Women Officers or Directors: 2
Hot Spot for Advancement for Women/Minorities:

LOCATIONS: ("Y" = Yes)

West:	Southwest:	Midwest:	Southeast:	Northeast:	International:
					Y

SOLIGENIX INC
www.soligenix.com

Industry Group Code: 325412A **Ranks within this company's industry group:** Sales: 16 Profits: 8

Drugs:		Other:	Clinical:	Computers:	Services:
Discovery:	Y	AgriBio:	Trials/Services:	Hardware:	Specialty Services:
Licensing:		Genetic Data:	Labs:	Software:	Consulting:
Manufacturing:		Tissue Replacement:	Equipment/Supplies:	Arrays:	Blood Collection:
Generics:			Research/Development Svcs.:	Database Management:	Drug Delivery:
			Diagnostics:		Drug Distribution:

TYPES OF BUSINESS:
Drug Delivery Technologies
Oral Formulations
Biodefense Vaccines

BRANDS/DIVISIONS/AFFILIATES:
orBec
RiVax
BT-VACC
Leuprolide

CONTACTS: *Note: Officers with more than one job title may be intentionally listed here more than once.*
Christopher J. Schaber, CEO
Christopher J. Schaber, Pres.
Evan Myrianthopoulos, CFO/Sr. VP
Robert N. Brey, Chief Scientific Officer/Sr. VP
Christopher P. Schnittker, VP-Admin.
James Clavijo, Corp. Sec.
Christopher P. Schnittker, Controller
Brian L. Hamilton, Chief Medical Officer/Sr. VP
Christopher J. Schaber, Chmn.

Phone: 786-425-3848	Fax: 786-425-3853
Toll-Free:	
Address: 29 Emmons Dr., Ste. C-10, Princeton, NJ 08540 US	

GROWTH PLANS/SPECIAL FEATURES:
Soligenix, Inc., formerly DOR BioPharma, Inc., develops products to treat the life-threatening side effects of cancer treatment, serious gastrointestinal diseases, as well as biodefense vaccines and therapeutics. The firm operates in two business units, biotherapeutics and biodefense. In the biotherapeutics segment, the company develops orBec, its leading product, which is used to treat gastrointestinal Graft-versus-Host-Disease (GVHD). The firm has completed Phase III trials of orBec, which has been granted orphan fast track drug status by the FDA. Another drug in the biotherapeutic segment, in pre-clinical development, is Lipid Polymer Micelle (LPM) Leuprolide, a treatment for prostate cancer, endometriosis, and precocious puberty. The biodefense division, working in collaboration with the University of Texas Southwest Medical Center and Thomas Jefferson University, has vaccines under development for the treatment of two toxins, ricin and botulinum. RiVax, a vaccine for ricin, has completed Phase I trials and has received support from the NIH and the National Institute of Allergy and Infectious Diseases. In February 2010, the company announced the issuance of a Hong patent for its LPM oral drug delivery technology. In May 2010, Soligenix was granted a United States patent for the use of oral beclomethasone dipropionate (BDP), the active ingredient in orBec, to treat irritable bowel syndrome. In June 2010, the company received a European patent for the use of BDP in conjunction with prednisone to treat GVHD and Leukemia.

FINANCIALS: Sales and profits are in thousands of dollars—add 000 to get the full amount. 2009 Note: Financial information for 2009 was not available for all companies at press time.

2009 Sales: $2,800	2009 Profits: $-6,000	**U.S. Stock Ticker: SNGX.OB**	
2008 Sales: $2,310	2008 Profits: $-3,422	**Int'l Ticker:** Int'l Exchange:	
2007 Sales: $1,258	2007 Profits: $-6,165	Employees: 12	
2006 Sales: $2,313	2006 Profits: $-8,163	Fiscal Year Ends: 12/31	
2005 Sales: $3,076	2005 Profits: $-4,720	Parent Company:	

SALARIES/BENEFITS:

Pension Plan:	ESOP Stock Plan:	Profit Sharing:	Top Exec. Salary: $300,000	Bonus: $100,000
Savings Plan:	Stock Purch. Plan:		Second Exec. Salary: $200,000	Bonus: $50,000

OTHER THOUGHTS:
Apparent Women Officers or Directors:
Hot Spot for Advancement for Women/Minorities:

LOCATIONS: ("Y" = Yes)

West:	Southwest:	Midwest:	Southeast:	Northeast:	International:
				Y	Y

SPECIALTY LABORATORIES INC

www.specialtylabs.com

Industry Group Code: 6215 Ranks within this company's industry group: Sales: Profits:

Drugs:	Other:	Clinical:		Computers:	Services:
Discovery:	AgriBio:	Trials/Services:		Hardware:	Specialty Services:
Licensing:	Genetic Data:	Labs:	Y	Software:	Consulting:
Manufacturing:	Tissue Replacement:	Equipment/Supplies:		Arrays:	Blood Collection:
Generics:		Research/Development Svcs.:		Database Management:	Drug Delivery:
		Diagnostics:	Y		Drug Distribution:

TYPES OF BUSINESS:

Clinical Laboratory Tests
Assays

BRANDS/DIVISIONS/AFFILIATES:

DataPassport MD
Quest Diagnostics Inc

CONTACTS: Note: Officers with more than one job title may be intentionally listed here more than once.

Christopher Lockhart, Laboratory Dir.
Surya N. Mohapatra, Chmn./CEO-Quest Diagnostics
Jon R. Cohen, Chief Medical Officer/Sr. VP-Quest Diagnostics
Robert A. Hagemann, CFO/Sr. VP-Quest Diagnostics
David W. Norgard, VP-Human Resources Quest Diagnostics

Phone: 661-799-6543	Fax: 661-799-6634
Toll-Free: 800-421-7110	
Address: 27027 Tourney Rd., Valencia, CA 91355 US	

GROWTH PLANS/SPECIAL FEATURES:

Specialty Laboratories, Inc. (SL), a subsidiary of Quest Diagnostics, Inc., is a research-based clinical laboratory, predominantly focused on developing and performing esoteric clinical laboratory tests, referred to as assays. The firm offers a comprehensive menu of thousands of assays, many of which it developed through internal research and development efforts, that are used to diagnose, evaluate and monitor patients in the areas of endocrinology; genetics; infectious diseases; neurology; pediatrics; urology; allergy and immunology; cardiology and coagulation; hepatology; microbiology; oncology; rheumatology; women's health; dermatopathology; gastroenterology; nephrology; pathology; and toxicology. Some of the company's assays include evaluations for H1N1, SARS, vitamin D deficiency, potassium and chlorine levels and protein chemistry, and the ovarian cancer marker HE4. The company also collaborates with the Molecular Profiling Institute, Inc. (MPI), a subsidiary of Translational Genomics Research Institute (TGen), to offer molecular-based assays. In addition, SL owns proprietary information technology that accelerates and automates test ordering and results reporting with customers. Products include DataPassportMD, a web-based laboratory order entry and resulting system. The company's primary customers are hospitals, independent clinical laboratories and physicians.

Employees of the firm are offered health and dental plans; a 401(k) plan; a stock purchase plan; short- and long-term disability plans; flexible spending accounts; an employee assistance program; and an education reimbursement program.

FINANCIALS: Sales and profits are in thousands of dollars—add 000 to get the full amount. 2009 Note: Financial information for 2009 was not available for all companies at press time.

2009 Sales: $	2009 Profits: $	U.S. Stock Ticker: Subsidiary
2008 Sales: $	2008 Profits: $	Int'l Ticker: Int'l Exchange:
2007 Sales: $	2007 Profits: $	Employees:
2006 Sales: $	2006 Profits: $	Fiscal Year Ends: 12/31
2005 Sales: $	2005 Profits: $	Parent Company: QUEST DIAGNOSTICS INC

SALARIES/BENEFITS:

Pension Plan:	ESOP Stock Plan: Y	Profit Sharing:	Top Exec. Salary: $	Bonus: $
Savings Plan: Y	Stock Purch. Plan:		Second Exec. Salary: $	Bonus: $

OTHER THOUGHTS:

Apparent Women Officers or Directors:
Hot Spot for Advancement for Women/Minorities:

LOCATIONS: ("Y" = Yes)

West:	Southwest:	Midwest:	Southeast:	Northeast:	International:
Y					

SPECTRAL DIAGNOSTICS INC

www.spectraldx.com

Industry Group Code: 325413 Ranks within this company's industry group: Sales: 20 Profits: 13

Drugs:	Other:	Clinical:		Computers:		Services:
Discovery:	AgriBio:	Trials/Services:		Hardware:		Specialty Services:
Licensing:	Genetic Data:	Labs:		Software:		Consulting:
Manufacturing:	Tissue Replacement:	Equipment/Supplies:	Y	Arrays:		Blood Collection:
Generics:		Research/Development Svcs.:		Database Management:		Drug Delivery:
		Diagnostics:	Y			Drug Distribution:

TYPES OF BUSINESS:

Medical Diagnostics Products
West Nile Virus Diagnostics
Sepsis Diagnostics
Antibodies

BRANDS/DIVISIONS/AFFILIATES:

Toraymyxin
Endotoxin Activity Assays

CONTACTS:

Note: Officers with more than one job title may be intentionally listed here more than once.

Paul M. Walker, CEO
Paul M. Walker, Pres.
Tony Businskas, CFO/Exec. VP
Robert Verhagen, VP-Bus. Dev.
Debra M. Foster, Dir.-Sepsis Program

Phone: 416-626-3233	Fax: 416-626-7383
Toll-Free: 888-426-4264	
Address: 135-2 The West Mall, Toronto, ON M9C 1C2 Canada	

GROWTH PLANS/SPECIAL FEATURES:

Spectral Diagnostics, Inc. is a medical diagnostics company that focuses on developing, manufacturing and selling rapid assays for the identification of life-threatening illnesses. Its activities are primarily involved with diagnostics products for sepsis. Spectral makes Endotoxin Activity Assays (EAA), which are rapid, whole blood tests used to identify patients who are at risk for sepsis. With the development of the EAA test as a rapid indicator of endotoxin, physicians are assisted in stratifying patients into those who are at low risk of severe sepsis and those who require directed anti-sepsis or anti-infection therapy. The company's other principal product, Toraymyxin, is a therapeutic hemoperfusion device that removes endotoxin from the bloodstream. Spectral also develops, produces and markets recombinant cardiac proteins, antibodies and calibrators. These products are sold for use in research and development, as well as in products manufactured by other diagnostic companies. Spectral has successfully engineered and produced over 30 antigens in bacteria and has created several patented novel molecules.

FINANCIALS:

Sales and profits are in thousands of dollars—add 000 to get the full amount. 2009 Note: Financial information for 2009 was not available for all companies at press time.

2009 Sales: $3,283	2009 Profits: $-2,764	**U.S. Stock Ticker:** DIAGF
2008 Sales: $3,011	2008 Profits: $-1,502	**Int'l Ticker:** SDI Int'l Exchange: Toronto-TSX
2007 Sales: $2,800	2007 Profits: $-1,500	Employees:
2006 Sales: $2,900	2006 Profits: $3,800	Fiscal Year Ends: 12/31
2005 Sales: $9,700	2005 Profits: $-7,100	Parent Company:

SALARIES/BENEFITS:

Pension Plan:	ESOP Stock Plan:	Profit Sharing:	Top Exec. Salary: $268,252	Bonus: $
Savings Plan:	Stock Purch. Plan:		Second Exec. Salary: $201,189	Bonus: $

OTHER THOUGHTS:

Apparent Women Officers or Directors: 1
Hot Spot for Advancement for Women/Minorities:

LOCATIONS: ("Y" = Yes)

West:	Southwest:	Midwest:	Southeast:	Northeast:	International:
					Y

SPECTRUM PHARMACEUTICALS INC

www.spectrumpharm.com

Industry Group Code: 325412 Ranks within this company's industry group: Sales: 92   Profits: 116

Drugs:		Other:		Clinical:	Computers:	Services:
Discovery:	Y	AgriBio:	Y	Trials/Services:	Hardware:	Specialty Services:
Licensing:	Y	Genetic Data:	Y	Labs:	Software:	Consulting:
Manufacturing:	Y	Tissue Replacement:		Equipment/Supplies:	Arrays:	Blood Collection:
Generics:				Research/Development Svcs.:	Database Management:	Drug Delivery:
				Diagnostics:		Drug Distribution:

TYPES OF BUSINESS:
Oncology & Urology Drugs
Cancer Treatments

GROWTH PLANS/SPECIAL FEATURES:

Spectrum Pharmaceuticals, Inc. is a biopharmaceutical company that develops, acquires and advances a diversified portfolio of drug candidates, with a focus on oncology, urology and other critical health challenges. The company has two drugs currently on the U.S. market and several others in development. Fusilev (levoleucovorin), its first approved drug, is a novel folate analog formulation that is marketed for patients with osteosarcoma after high-dose methotrexate therapy, a common cancer treatment that can cause toxicity or inadvertent overdose of folic acid antagonists. Fusilev has also been submitted for a new drug application (NDA) for the treatment of colorectal cancer. Spectrum's other marketed drug, Zevalin, is a treatment for patients with relapsed or refractory low-grade or follicular B-cell non-Hodgkin's lymphoma (NHL). The drug has also been submitted as an NDA for first-line therapy in NHL. The company currently has 10 drugs in various developmental stages. The firm's late-stage development products are EOquin, an anti-cancer agent for the treatment of non-invasive bladder cancer and Belinostat, a histone deacetylase (HDAC) inhibitor for peripheral t-cell lymphoma. The company's Phase I and II drug candidates include: Ozarelix, an antagonist that is being investigated for indications in non muscle-invasive bladder cancer, hormone dependent prostate cancer and endometriosis; Ortataxel, a treatment for solid tumors; Satraplatin, a treatment for non-small cell lung cancer; SPI-1620, an adjunct to chemotherapy; Elsamitrucin, an anti-tumor antibiotic; Lucanthone, a chemo sensitizer in malignant brain tumors; RenaZorb, a second-generation lanthanum-based treatment for hyperphosphatemia; and SPI-205, a drug with possible benefits in treating chemotherapy-induced pheripheral neuropathy. In May 2009, Spectrum obtained 100% control of RIT Oncology, LLC, a joint company formed with Cell Therapeutics, Inc. to market Zevalin. In February 2010, the firm entered a co-development and commercialization agreement with TopoTarget A/S for Belinostat.

BRANDS/DIVISIONS/AFFILIATES:
Fusilev
Elsamitrucin
Lucanthone
EOquin
Ortataxel
Ozarelix
Satraplatin
RenaZorb

CONTACTS: *Note: Officers with more than one job title may be intentionally listed here more than once.*
Rajesh C. Shrotriya, CEO
Rajesh C. Shrotriya, Pres.
George Uy, VP-Mktg. & Sales
Andrew Sandler, Chief Medical Officer
William Pedranti, General Counsel/VP
Michael Adam, Sr. VP-Pharmaceutical Oper.
Russell L. Skibsted, Chief Business Officer/Sr. VP
Shyam Kumaria, VP-Finance
James E. Shields, Chief Commercial Officer/ Sr. VP
Rajesh C. Shrotriya, Chmn.

Phone: 949-788-6700	Fax: 949-788-6706
Toll-Free:	
Address: 157 Technology Dr., Irvine, CA 92618 US	

FINANCIALS: Sales and profits are in thousands of dollars—add 000 to get the full amount. 2009 Note: Financial information for 2009 was not available for all companies at press time.

2009 Sales: $38,025	2009 Profits: $-19,046	**U.S. Stock Ticker: SPPI**
2008 Sales: $28,725	2008 Profits: $-14,196	**Int'l Ticker:** Int'l Exchange:
2007 Sales: $7,672	2007 Profits: $-21,981	Employees: 158
2006 Sales: $5,673	2006 Profits: $-23,284	Fiscal Year Ends: 12/31
2005 Sales: $ 577	2005 Profits: $-18,642	Parent Company:

SALARIES/BENEFITS:

Pension Plan:	ESOP Stock Plan:	Profit Sharing:	Top Exec. Salary: $600,000	Bonus: $1,000,000
Savings Plan: Y	Stock Purch. Plan:		Second Exec. Salary: $275,000	Bonus: $60,000

OTHER THOUGHTS:
Apparent Women Officers or Directors:
Hot Spot for Advancement for Women/Minorities:

LOCATIONS: ("Y" = Yes)

West:	Southwest:	Midwest:	Southeast:	Northeast:	International:
Y					Y

STEMCELLS INC

www.stemcellsinc.com

Industry Group Code: 325414 Ranks within this company's industry group: Sales: 8 Profits: 9

Drugs:		Other:		Clinical:	Computers:	Services:
Discovery:	Y	AgriBio:		Trials/Services:	Hardware:	Specialty Services:
Licensing:	Y	Genetic Data:	Y	Labs:	Software:	Consulting:
Manufacturing:		Tissue Replacement:	Y	Equipment/Supplies:	Arrays:	Blood Collection:
Generics:				Research/Development Svcs.:	Database Management:	Drug Delivery:
				Diagnostics:		Drug Distribution:

TYPES OF BUSINESS:

Cell-Based Therapeutics

BRANDS/DIVISIONS/AFFILIATES:

HuCNS-SC
hLEC
SC Proven
iSTEM
GSI-RTM
GS2-MTM
RHB-A
Stem Cell Sciences (UK) Ltd.

CONTACTS: Note: Officers with more than one job title may be intentionally listed here more than once.

Martin McGlynn, CEO
Ann Tsukamoto, COO
Martin McGlynn, Pres.
Rodney Young, CFO
Ann Tsukamoto, Exec. VP-R&D
Rodney Young, VP-Admin.
Ken Stratton, General Counsel
Stewart Craig, Sr. VP-Oper.
Stewart Craig, Sr. VP-Dev.
Rodney Young, VP-Finance
Stephen Huhn, VP/Head-CNS Program
Nobuko Uchida, VP-Stem Cell Biology
Maria Millan, VP/Head-Liver Program
John J. Schwartz, Chmn.

Phone: 650-475-3100	Fax: 650-475-3101
Toll-Free:	
Address: 3155 Porter Dr., Palo Alto, CA 94304 US	

GROWTH PLANS/SPECIAL FEATURES:

StemCells, Inc. is focused on the discovery and development of stem cell therapeutics that will create the foundation of therapies to support or replace cells that have been destroyed or damaged through disease or genetic defect. The company seeks to identify multiple types of human stem and progenitor cells with therapeutic and commercial importance, to develop techniques and processes to purify those cells, expand and bank them as transplantable cells and then advance them into clinical development with the ultimate aim of commercializing them as cell-based therapeutic products. StemCells has completed a Phase I clinical trial to evaluate the safety and efficacy of its lead product, the human neural stem cell (HuCNS-SC), as a treatment for neuronal ceroid lipofuscinosis, also known as Batten Disease, which is a brain disorder in children. It is now conducting a Phase I trial of HuCNS-SC in Pelizaeus-Merzbacher disease, a fatal myelination disorder of the brain that afflicts male children. In addition, the company is in preclinical development with its human liver engrafting cells (hLEC), to evaluate them as a potential cellular therapy for liver diseases. The company also markets a range of proprietary cell culture technologies for academic and industrial laboratories conducting stem cell research. They are marketed under the SC Proven name, including iSTEM, GSI-RTM, GS2-MTM, RHB-A, RHB-Basal, NDiff N2B27, NDiff 2 and 27, HEScGROTM, and ESGRO complete. The firm owns 50 issued U.S. patents and over 200 foreign patents. In February 2010, the company announced that its HuCNS-SC stem cells had been used in its Phase I clinical trials to perform the first human neural stem cell transplant on a patient with the most severe form of Pelizeus-Merzbacher Disease.

FINANCIALS: Sales and profits are in thousands of dollars—add 000 to get the full amount. 2009 Note: Financial information for 2009 was not available for all companies at press time.

2009 Sales: $ 608	2009 Profits: $-27,026	**U.S. Stock Ticker: STEM**
2008 Sales: $ 232	2008 Profits: $-29,087	**Int'l Ticker:** Int'l Exchange:
2007 Sales: $ 57	2007 Profits: $-25,023	Employees: 75
2006 Sales: $ 93	2006 Profits: $-18,948	Fiscal Year Ends: 12/31
2005 Sales: $ 206	2005 Profits: $-11,738	Parent Company:

SALARIES/BENEFITS:

Pension Plan:	ESOP Stock Plan:	Profit Sharing:	Top Exec. Salary: $540,885	Bonus: $202,125
Savings Plan: Y	Stock Purch. Plan:		Second Exec. Salary: $311,538	Bonus: $52,500

OTHER THOUGHTS:

Apparent Women Officers or Directors: 3
Hot Spot for Advancement for Women/Minorities: Y

LOCATIONS: ("Y" = Yes)

West:	Southwest:	Midwest:	Southeast:	Northeast:	International:
Y				Y	Y

Note: Financial information, benefits and other data can change quickly and may vary from those stated here.

STIEFEL LABORATORIES INC

www.stiefel.com

Industry Group Code: 325412 Ranks within this company's industry group: Sales: Profits:

Drugs:	Other:	Clinical:	Computers:	Services:	
Discovery:	AgriBio:	Trials/Services:	Hardware:	Specialty Services:	Y
Licensing:	Genetic Data:	Labs:	Software:	Consulting:	
Manufacturing: Y	Tissue Replacement:	Equipment/Supplies:	Arrays:	Blood Collection:	
Generics:		Research/Development Svcs.:	Database Management:	Drug Delivery:	
		Diagnostics:		Drug Distribution:	

TYPES OF BUSINESS:

Dermatological & Skin Care Products

BRANDS/DIVISIONS/AFFILIATES:

Evoclin
Duac
Olux
MimyX
LUXIQ Foam
Verdeso
GlaxoSmithKline plc

CONTACTS: *Note: Officers with more than one job title may be intentionally listed here more than once.*

Charles W. Stiefel, CEO
Bill Humphries, Pres.
Alfonso Ugarte, VP-Global Mktg.
Gavin Corcoran, Chief Scientific Officer
Jeff Klimaski, VP/Global Corp. Compliance Officer
Wayne Wilson, VP-U.S. Sales
Charles W. Stiefel, Chmn.
Richard MacKay, Pres., Stiefel Canada, Inc.

Phone: 305-443-3800	Fax: 305-443-3467
Toll-Free:	
Address: 20 T. W. Alexander Dr., Research Triangle Park, NC 33134-7412 US	

GROWTH PLANS/SPECIAL FEATURES:

Stiefel Laboratories, Inc., a subsidiary of GlaxoSmithKline plc, is a specialized pharmaceutical company that focuses on the advancement of dermatology and skin care.   The company's sells hundreds of products that treat a wide range of dermatological ailments including acne, psoriasis, fungal infections, eczema, dry skin, oily skin, rosacea, seborrhea, pruritus and sun damaged and aging skin. Leading products include Evoclin and Duac, acne skin care products; MimyX, a cream that relieves dry and waxy skin for a variety of dermatoses; LUXIQ Foam, which reduces the signs of scalp dermatoses, such as scaling, redness, and plaques; Olux, also used for scalp treatment; and Verdeso, a treatment for eczema.   The firm also offers pharmaceutical contract manufacturing of gels, creams, ointments, topical solutions and liquid orals, both prescription and over-the-counter. The company has subsidiaries in more than 30 countries and its products are available in more than 100 countries. In July 2009, the firm was acquired by GlaxoSmithKline plc for $2.9 billion.

FINANCIALS: Sales and profits are in thousands of dollars—add 000 to get the full amount. 2009 Note: Financial information for 2009 was not available for all companies at press time.

2009 Sales: $	2009 Profits: $	U.S. Stock Ticker: Private
2008 Sales: $	2008 Profits: $	Int'l Ticker: Int'l Exchange:
2007 Sales: $	2007 Profits: $	Employees:
2006 Sales: $	2006 Profits: $	Fiscal Year Ends:
2005 Sales: $	2005 Profits: $	Parent Company:

SALARIES/BENEFITS:

Pension Plan:	ESOP Stock Plan:	Profit Sharing:	Top Exec. Salary: $	Bonus: $
Savings Plan:	Stock Purch. Plan:		Second Exec. Salary: $	Bonus: $

OTHER THOUGHTS:

Apparent Women Officers or Directors: 3
Hot Spot for Advancement for Women/Minorities: Y

LOCATIONS: ("Y" = Yes)

West:	Southwest:	Midwest:	Southeast:	Northeast:	International:
Y			Y	Y	Y

SUPERGEN INC

www.supergen.com

Industry Group Code: 325412 **Ranks within this company's industry group:** Sales: 90   Profits: 67

Drugs:		Other:	Clinical:	Computers:	Services:
Discovery:	Y	AgriBio:	Trials/Services:	Hardware:	Specialty Services:
Licensing:	Y	Genetic Data:	Labs:	Software:	Consulting:
Manufacturing:		Tissue Replacement:	Equipment/Supplies:	Arrays:	Blood Collection:
Generics:			Research/Development Svcs.:	Database Management:	Drug Delivery:
			Diagnostics:		Drug Distribution:

TYPES OF BUSINESS:

Pharmaceuticals Acquisition & Development
Oncology Drugs

BRANDS/DIVISIONS/AFFILIATES:

SGI-1776
Dacogen
Amuvatinib (MP470)
SGI-110
JAK2
Axl
ETK/BMX
CLIMB

CONTACTS: Note: Officers with more than one job title may be intentionally listed here more than once.

James S. Manuso, CEO
James S. Manuso, Pres.
Michael Molkentin, CFO/Corp. Sec.
David Bearss, Scientific Advisor
Sanjeev Redkar, VP-Mfg. & Pre-Clinical Dev.
Shu Lee, VP-Legal Affairs & Intellectual Property
Michael McCullar, VP-Discovery Oper. & Strategy
Timothy L. Enns, Sr. VP-Bus. Dev.
Timothy L. Enns, Sr. VP-Corp. Comm.
Gavin Choy, VP-Clinical Oper.
David S. Smith, VP-Regulatory & Quality Affairs
Mohammad Azab, Chief Medical Officer
James S. Manuso, Chmn.

Phone: 925-560-0100	Fax: 925-560-0101
Toll-Free:	
Address: 4140 Dublin Blvd., Ste. 200, Dublin, CA 94568 US	

GROWTH PLANS/SPECIAL FEATURES:

SuperGen Inc. develops and commercializes pharmaceutical products intended to treat patients with cancer. The firm's development platform is based on its proprietary CLIMB process that merges rapid screening of compound libraries with computational chemistry and systems biology techniques to identify small-molecule drug candidates. SuperGen currently possesses a portfolio of five oncological drugs candidates that treat a variety of solid tumors and hematological malignancies. The company's leading product, Amuvatinib (MP-470), is a tyrosine kinase inhibitor/rad 51 suppressor, which has completed Phase I clinical trials. Amuvatinib is being tested on both healthy persons and cancer patients. SGI-1776, an oral pim kinase inhibitor, blocks the pro-survival activity of potentially malignant cells (pim kinases), allowing them to self-abort. The drug has completed Phase I trials, and a Phase I/II clinical trial is planned for launch in mid 2010. SuperGen's preclinical stage product, SGI-110, is a compound which targets and blocks the mechanism by which DNA methylation occurs, thus allowing tumor suppressor genes to function. Products in the company's pipeline that have not yet reached preclinical stages include three kinase inhibitors: the JAK2, Axl and ETK/BMX (epithelial and endothelial tyrosine kinase/ bone marrow X kinase). In addition to drug candidates, SuperGen continues to generate royalties from the 2004 sale of its drug Dacogen to MGI Pharma, Inc. MGI was acquired by Eisai Corporation, which now has worldwide rights to develop, manufacture and distribute Dacogen and pays the firm at specific regulatory and commercialization milestones. Dacogen is a therapeutic product that decreases the degree of methylation, the replacement of hydrogen with methyl, at certain DNA sites to assist patients with myelodysplastic syndrome. In October 2009, SuperGen agreed to partner with GlaxoSmithKline to jointly discover and develop cancer therapeutics based on epigenetic processes.

Employee benefits include life, AD&D, disability, medical, dental and vision insurance; and paid vacation.

FINANCIALS: Sales and profits are in thousands of dollars—add 000 to get the full amount. 2009 Note: Financial information for 2009 was not available for all companies at press time.

2009 Sales: $41,253	2009 Profits: $4,737	**U.S. Stock Ticker:** SUPG
2008 Sales: $38,422	2008 Profits: $-9,111	**Int'l Ticker:** Int'l Exchange:
2007 Sales: $22,954	2007 Profits: $13,081	Employees: 80
2006 Sales: $38,083	2006 Profits: $-16,487	Fiscal Year Ends: 12/31
2005 Sales: $30,169	2005 Profits: $-14,482	Parent Company:

SALARIES/BENEFITS:

Pension Plan:	ESOP Stock Plan:	Profit Sharing:	Top Exec. Salary: $574,237	Bonus: $500,000
Savings Plan: Y	Stock Purch. Plan: Y		Second Exec. Salary: $326,375	Bonus: $119,000

OTHER THOUGHTS:

Apparent Women Officers or Directors:
Hot Spot for Advancement for Women/Minorities:

LOCATIONS: ("Y" = Yes)

West:	Southwest:	Midwest:	Southeast:	Northeast:	International:
Y					

SYNBIOTICS CORP

www.synbiotics.com

Industry Group Code: 325412B Ranks within this company's industry group: Sales: Profits:

Drugs:		Other:		Clinical:		Computers:		Services:	
Discovery:		AgriBio:		Trials/Services:		Hardware:		Specialty Services:	Y
Licensing:		Genetic Data:		Labs:		Software:	Y	Consulting:	
Manufacturing:	Y	Tissue Replacement:		Equipment/Supplies:	Y	Arrays:		Blood Collection:	
Generics:				Research/Development Svcs.:		Database Management:		Drug Delivery:	
				Diagnostics:	Y			Drug Distribution:	Y

TYPES OF BUSINESS:

Veterinary Products Manufacturing
Animal Health Diagnostics
Animal Pregnancy Tests
Breeding Services
Flock Management Software

BRANDS/DIVISIONS/AFFILIATES:

ReproCHEK
Fungassay
Ovassay
ViraCHEK
WITNESS
OVUCHECK
TiterCHEK
Synbiotics Europe SAS

CONTACTS: Note: Officers with more than one job title may be intentionally listed here more than once.

Paul R. Hays, CEO
Paul R. Hays, Pres.
Daniel O'Rourke, CFO
Kevin Hayes, VP-Sales & Mktg.
Chinta Lamichhane, Chief Scientific Officer/VP-R&D
Daniel O'Rourke, VP-Finance
Clifford J. Frank, VP-Strategic Projects
Kraig Stemme, Dir.-Reproduction Sales
Thomas A. Donelan, Chmn.
John Trabucco, Dir.-National Sales, US & Canada

Phone: 816-464-3500	Fax: 816-464-3521
Toll-Free: 800-228-4305	
Address: 12200 NW Ambassador Dr., Ste. 101, Kansas City, MO 64163 US	

GROWTH PLANS/SPECIAL FEATURES:

Synbiotics Corp. develops, manufactures and markets animal diagnostic products to veterinarians and breeders. The firm produces products for poultry, cows, swine, dogs, horses, cats and non-human primates, and is able to treat animal coagulation, dermatophytes, gastrointestinal parasites, tuberculosis, heartworm, parvovirus and arthritis. Synbiotics' brand names include D-Tec to diagnose Brucella canis infections in dogs; ReproCHEK, a canine pregnancy test; TiterCHEK, a canine distemper and parvovirus antibody test; Fungassay for dermatophyte infections; Ovassay for identifying gastrointestinal parasites; CRF, a test for canine rheumatoid factor; ViraCHEK, a coronavirus identifier; OVUCHECK, a pregnancy test for mares; and WITNESS, the company's heartworm identification kit. Synbiotics also markets domestic animal pregnancy tests and the Profile flock management software. The firm distributes its products through a number of companies in the U.S, Canada, Latin America, Asia, New Zealand and Australia. The company offers canine reproduction services that include a referral network and a freezing center that allows breeders to conduct long-distance breeding. The firm's specialized use of frozen semen allows dog breeders to preserve valuable genetic lines and to conduct long-distance breeding without shipping the dogs themselves. The firm also runs a wholly-owned affiliate, Synbiotics Europe SAS, which handles the sales and services in Europe, the Middle East and Africa. Synbiotics manufactures most of its products at its facilities located in San Diego, California and Lyon, France.

FINANCIALS: Sales and profits are in thousands of dollars—add 000 to get the full amount. 2009 Note: Financial information for 2009 was not available for all companies at press time.

2009 Sales: $	2009 Profits: $	U.S. Stock Ticker: SYNB.PK
2008 Sales: $	2008 Profits: $	Int'l Ticker: Int'l Exchange:
2007 Sales: $	2007 Profits: $	Employees:
2006 Sales: $	2006 Profits: $	Fiscal Year Ends: 12/31
2005 Sales: $	2005 Profits: $	Parent Company:

SALARIES/BENEFITS:

Pension Plan:	ESOP Stock Plan:	Profit Sharing:	Top Exec. Salary: $	Bonus: $
Savings Plan:	Stock Purch. Plan:		Second Exec. Salary: $	Bonus: $

OTHER THOUGHTS:

Apparent Women Officers or Directors:
Hot Spot for Advancement for Women/Minorities:

LOCATIONS: ("Y" = Yes)

West:	Southwest:	Midwest:	Southeast:	Northeast:	International:
Y				Y	Y

SYNGENTA AG

www.syngenta.com

Industry Group Code: 11511 Ranks within this company's industry group: Sales: 2 Profits: 2

Drugs:	Other:		Clinical:	Computers:	Services:
Discovery:	AgriBio:	Y	Trials/Services:	Hardware:	Specialty Services:
Licensing:	Genetic Data:		Labs:	Software:	Consulting:
Manufacturing:	Tissue Replacement:		Equipment/Supplies:	Arrays:	Blood Collection:
Generics:			Research/Development Svcs.:	Database Management:	Drug Delivery:
			Diagnostics:		Drug Distribution:

TYPES OF BUSINESS:

Agricultural Biotechnology Products & Chemicals Manufacturing
Crop Protection Products
Seeds

BRANDS/DIVISIONS/AFFILIATES:

Dual Gold
Avistar
Acanto
Score
Touchdown
Cruiser
Maxim
Dividend

CONTACTS: *Note: Officers with more than one job title may be intentionally listed here more than once.*

Michael Mack, CEO
John Ramsay, CFO
Alejandro Aruffo, Head-R&D
Christoph Mader, Head-Legal & Taxes/Corp. Sec.
Mark Peacock, Head-Global Oper.
Robert Berendes, Head-Bus. Dev.
John Atkin, COO-Crop Protection
Davor Pisk, COO-Syngenta Seeds
Martin Taylor, Chmn.

Phone: 41-61-323-9094	Fax: 41-61-323-2324
Toll-Free:	
Address: Schwarzwaldallee 215, Basel, 4058 Switzerland	

GROWTH PLANS/SPECIAL FEATURES:

Syngenta AG is an international agrochemical companies and a leading worldwide supplier of conventional and bioengineered crop protection and seeds. The firm's products designed for crop protection include seed treatments to control weeds, insects and diseases, herbicides, fungicides and insecticides. Additionally, the firm produces seeds for field crops, vegetables and flowers. Its leading marketed products include the following: Dual Gold, Axial and Fusilade selective herbicides; Touchdown, Reglone and Granoxone non-selective herbicides; Bravo, Score and Amistar fungicides; Proclaim, Match and Actara insecticides; and Dividend, Apron, Maxim and Cruiser seed care treatments. Syngenta has a seed portfolio of over 200 product lines and more than 6,800 proprietary varieties. The seeds that the company markets are for field crops, such as corn, soybeans, sugar beets, sunflowers and oilseed rape (canola); vegetables, including tomatoes, lettuce, melons, squash, cabbages, peppers, beans and radishes; and garden plants such as begonias, violas, petunias and many other seasonal flowers and herbs, some of which can also be purchased as plugs or full-grown plants. The firm spends roughly $368 million annually on research and development at its laboratories in the U.S., Sweden, Chile, China, France, India, Singapore and the Netherlands. Syngenta's research and development division is engaged in collaborations with several companies and universities, including Anhui Rice Research Institute of China, Dow AgroSciences and Chromatin, Inc. The company's gene technology has become so refined that single genes can be isolated from a type of plant material and transferred to the DNA of another. This process allows manipulation of such traits as nutrient composition, appearance, and even the specifics of the taste of a certain crop. In 2009, Syngenta acquired antitoxin crop protection technology developer Circle One Global, Inc.; the global hybrid seed operations of Monsanto; and lettuce seed firms Synergene Seed & Technology, Inc. and Pybas Vegetable Seed Co.

FINANCIALS: Sales and profits are in thousands of dollars—add 000 to get the full amount. 2009 Note: Financial information for 2009 was not available for all companies at press time.

2009 Sales: $10,992,000	2009 Profits: $1,374,000	U.S. Stock Ticker: SYT
2008 Sales: $11,624,000	2008 Profits: $1,385,000	Int'l Ticker: SYNN Int'l Exchange: Zurich-SWX
2007 Sales: $9,240,000	2007 Profits: $1,109,000	Employees: 25,900
2006 Sales: $8,046,000	2006 Profits: $634,000	Fiscal Year Ends: 12/31
2005 Sales: $8,104,000	2005 Profits: $622,000	Parent Company:

SALARIES/BENEFITS:

Pension Plan: Y	ESOP Stock Plan:	Profit Sharing:	Top Exec. Salary: $	Bonus: $
Savings Plan: Y	Stock Purch. Plan: Y		Second Exec. Salary: $	Bonus: $

OTHER THOUGHTS:

Apparent Women Officers or Directors:
Hot Spot for Advancement for Women/Minorities:

LOCATIONS: ("Y" = Yes)

West:	Southwest:	Midwest:	Southeast:	Northeast:	International:
Y		Y	Y	Y	Y

Note: Financial information, benefits and other data can change quickly and may vary from those stated here.

SYNOVIS LIFE TECHNOLOGIES INC

www.synovislife.com

Industry Group Code: 33911 **Ranks within this company's industry group:** Sales: 10   Profits: 9

Drugs:	Other:	Clinical:		Computers:	Services:
Discovery:	AgriBio:	Trials/Services:		Hardware:	Specialty Services:
Licensing:	Genetic Data:	Labs:		Software:	Consulting:
Manufacturing:	Tissue Replacement:	Equipment/Supplies:	Y	Arrays:	Blood Collection:
Generics:		Research/Development Svcs.:	Y	Database Management:	Drug Delivery:
		Diagnostics:			Drug Distribution:

TYPES OF BUSINESS:

Surgical & Interventional Treatment Products
Implantable Biomaterials

BRANDS/DIVISIONS/AFFILIATES:

Synovis Surgical Innovations
Synovis Micro Companies Alliance
Biover Microvascular Clamp
Microvascular Anastomotis COUPLER System
Peri-Strips
Veritas
Pegasus Biologics Inc
Synovis Orthopedic and Woundcare

CONTACTS: *Note: Officers with more than one job title may be intentionally listed here more than once.*

Richard W. Kramp, CEO
Richard W. Kramp, Pres.
Brett A. Reynolds, CFO
Daniel L. Mooradian, VP-R&D
Brett A. Reynolds, Corp. Sec.
Tim Floeder, VP-Corp. Dev.
Brett A. Reynolds, VP-Finance
Michael K. Campbell, Pres., Micro Companies Alliance, Inc.
Mary L. Frick, VP-Regulatory Affairs & Quality Assurance
Mary L. Frick, VP-Clinical Affairs
Timothy M. Scanlan, Chmn.

Phone: 651-796-7300	**Fax:** 651-642-9018
Toll-Free: 800-255-4018	
Address: 2575 University Ave., St. Paul, MN 55114 US	

GROWTH PLANS/SPECIAL FEATURES:

Synovis Life Technologies, Inc. is a diversified medical device firm engaged in developing, manufacturing and bringing to market products for the surgical treatment of disease. Its products are marketed for use in bariatric, cardiac, neurologic, thoracic, urogynecologic, vascular and general surgeries. The company operates through three subsidiaries: Synovis Surgical Innovations; Synovis Micro Companies Alliance; and Synovis Orthopedic and Woundcare. Synovis Surgical Innovations is focused on developing implantable biomaterial products designed to improve a patient's outcome in critical surgeries. The products are made from bovine pericardium, which easily assimilates into host tissue. Products offered by the firm include Peri-Strips, which are stapling buttresses used to reinforce surgical staple lines and reduce potential leakage; Dura-Guard, which is used by neurosurgeons in the treatment of tumors, brain trauma, cerebrovascular disorders and congenital disorders; and Veritas, for use in pelvic floor reconstruction, stress urinary incontinence treatment, vaginal and rectal prolapse repair, hernia repair as well as soft tissue repair. Synovis Micro Companies Alliance provides products to the niche microsurgery market. Products offered include the GEM Microvascular Anastomotic COUPLER System, which is used in small vessel anastomases; GEM Neurotube, which is used in primary or secondary nerve repair; GEM MicroClip, a hemostatic clip; GEM03 Focus Headlight, a rechargeable medical lamp; and the Biover Microvascular Clamp, a single use clamp for arteries and veins. Synovis Orthopedic and Woundcare is a medical device manufacturer that develops advanced biologic solutions for the repair, reinforcement, augmentation and reconstruction of soft tissues; and advanced wound management. Its products include Unite Biomatrix for chronic wounds and the OrthADAPT Bioimplant collagen scaffold. In July 2009 Synovis Life acquired medical device manufacturer Pegasus Biologics, Inc. for roughly $12.1 million. In January 2010, Synovis Micro acquired the exclusive worldwide marketing rights to the GEM SuperFine MicroClip.

FINANCIALS: Sales and profits are in thousands of dollars—add 000 to get the full amount. 2009 Note: Financial information for 2009 was not available for all companies at press time.

2009 Sales: $58,211	2009 Profits: $2,706	**U.S. Stock Ticker:** SYNO
2008 Sales: $49,800	2008 Profits: $11,485	**Int'l Ticker:** Int'l Exchange:
2007 Sales: $37,691	2007 Profits: $3,810	**Employees:** 290
2006 Sales: $27,743	2006 Profits: $-1,481	**Fiscal Year Ends:** 10/31
2005 Sales: $60,256	2005 Profits: $ 883	**Parent Company:**

SALARIES/BENEFITS:

Pension Plan:	ESOP Stock Plan:	Profit Sharing:	Top Exec. Salary: $362,000	Bonus: $109,396
Savings Plan: Y	Stock Purch. Plan:		Second Exec. Salary: $225,000	Bonus: $49,050

OTHER THOUGHTS:

Apparent Women Officers or Directors: 2
Hot Spot for Advancement for Women/Minorities: Y

LOCATIONS: ("Y" = Yes)

West:	Southwest:	Midwest:	Southeast:	Northeast:	International:
Y		Y	Y		

SYNTHETIC GENOMICS

www.syntheticgenomics.com

Industry Group Code: 325199 **Ranks within this company's industry group: Sales:** Profits:

Drugs:	Other:		Clinical:	Computers:	Services:
Discovery:	AgriBio:	Y	Trials/Services:	Hardware:	Specialty Services:
Licensing:	Genetic Data:	Y	Labs:	Software:	Consulting:
Manufacturing:	Tissue Replacement:		Equipment/Supplies:	Arrays:	Blood Collection:
Generics:			Research/Development Svcs.:	Database Management:	Drug Delivery:
			Diagnostics:		Drug Distribution:

TYPES OF BUSINESS:

Biofuels
Genomics-Based Technologies

BRANDS/DIVISIONS/AFFILIATES:

ExxonMobil Research and Engineering Co
Draper Fisher Juvetson
Plenus SA de CV
Biotechonomy LLC
BP plc
ACGT Sdn Bhd
Meteor Group
Asiatic Centre for Genome Technology

CONTACTS: *Note: Officers with more than one job title may be intentionally listed here more than once.*

J. Craig Venter, CEO
Aristides A. N. Patrinos, Pres.
Chuck McBride, CFO
Tina Jones, VP-Human Resources
Hamilton O. Smith, Co-Chief Scientific Officer
Fernanda Gandara, VP-Bus. Dev.
J. Craig Venter, Co-Chief Scientific Officer
Paul Roessler, VP-Renewable Fuels & Chemicals
Eric J. Mathur, VP-Genomic Research
Toby Richardson, VP-Bioinformatics
J. Craig Venter, Chmn.

Phone: 858-754-2900	Fax: 858-754-2988
Toll-Free:	
Address: 11149 N. Torrey Pines Rd., La Jolla, CA 92037 US	

GROWTH PLANS/SPECIAL FEATURES:

Synthetic Genomics is involved in the commercialization of genomic-driven technologies. Founded in 2005, the firm seeks to develop marketable genomic-driven solutions that address energy and environment challenges. The company is focused on developing new biological solutions to increase production and/or recovery rates of subsurface hydrocarbons; harnessing photosynthetic organisms to produce energy directly from sunlight and carbon dioxide; designing advanced biofuels with superior properties compared to ethanol and biodiesel; and developing high-yielding, economic and more disease-resistant feedstocks. Synthetic Genomics' extensive scientific team allows the firm the capability to pursue research and development in several areas, including genome engineering, microbiology, synthetic biology, plant genomics, biochemistry, environmental genomics, bioinformatics and climate change. The firm is a sponsor of fundamental research at the J. Craig Venter Institute, a non-profit organization of over 400 scientists and staff that conduct a variety of genomic research. Synthetic Genomics has received funding from several investors, including Draper Fisher Juvetson, Plenus, S.A. de C.V., Biotechonomy LLC, BP plc, ACGT Sdn Bhd. and Meteor Group. The firm was founded by Dr. J. Craig Venter and Nobel Laureate Dr. Hamilton O. Smith, both of whom remain Synthetic Genomics executives and research leaders. In May 2009, the company and joint venture partner Asiatic Centre for Genome Technology completed the first draft, 10X assembly of a jatropha genome. Jatrophas are very high yielding tropical oilseed plants that can be grown on generally non-food producing lands and have a very short generation time. Most importantly for Synthetic Genomics, the plant can produce oil for 30 to 40 years and its seed oil and biomass are ideal for biofuel production. In July 2009, ExxonMobil Research and Engineering Company agreed to back Synthetic Genomic's research and development of next generation biofuels from photosynthetic algae. ExxonMobil will initially invest up to $300 million in the effort. The firm's technology enables it to cultivate and harvest algae which utilize energy from sunlight to convert carbon dioxide into oils that can be processed into fuels and chemicals.

FINANCIALS: Sales and profits are in thousands of dollars—add 000 to get the full amount. 2009 Note: Financial information for 2009 was not available for all companies at press time.

2009 Sales: $	2009 Profits: $	**U.S. Stock Ticker: Private**
2008 Sales: $	2008 Profits: $	**Int'l Ticker:** Int'l Exchange:
2007 Sales: $	2007 Profits: $	Employees:
2006 Sales: $	2006 Profits: $	Fiscal Year Ends:
2005 Sales: $	2005 Profits: $	Parent Company:

SALARIES/BENEFITS:

Pension Plan:	ESOP Stock Plan:	Profit Sharing:	Top Exec. Salary: $	Bonus: $
Savings Plan:	Stock Purch. Plan:		Second Exec. Salary: $	Bonus: $

OTHER THOUGHTS:

Apparent Women Officers or Directors: 2
Hot Spot for Advancement for Women/Minorities: Y

LOCATIONS: ("Y" = Yes)

West:	Southwest:	Midwest:	Southeast:	Northeast:	International:
Y				Y	

TAKEDA PHARMACEUTICAL COMPANY LTD

www.takeda.com

Industry Group Code: 325412 Ranks within this company's industry group: Sales: 12 Profits: 14

Drugs:		Other:	Clinical:	Computers:	Services:
Discovery:	Y	AgriBio:	Trials/Services:	Hardware:	Specialty Services:
Licensing:	Y	Genetic Data:	Labs:	Software:	Consulting:
Manufacturing:	Y	Tissue Replacement:	Equipment/Supplies:	Arrays:	Blood Collection:
Generics:			Research/Development Svcs.:	Database Management:	Drug Delivery:
			Diagnostics:		Drug Distribution:

TYPES OF BUSINESS:

Pharmaceuticals Discovery & Development
Over-the-Counter Drugs
Vitamins

BRANDS/DIVISIONS/AFFILIATES:

Takeda America Holdings, Inc.
IDM Pharma Inc.
Millennium Pharmaceuticals, Inc.
Lupron Depot
Enantone
Blopress
Alinamin
Benza

CONTACTS: Note: Officers with more than one job title may be intentionally listed here more than once.

Yasuchika Hasegawa, CEO
Yasuchika Hasegawa, Pres.
Yasuhiko Yamanaka, Gen. Mgr.-Pharmaceutical Mktg. Div.
Shigenori Ohkawa, Chief Scientific Officer/Exec. VP
Takashi Inkyo, Gen. Mgr.-Pharmaceutical Prod. Div.
Toyoji Yoshida, Managing Dir./Chief Admin. Officer
Masumitsu Inoue, Gen. Mgr.-Corp. Strategy & Planning Dept.
Hiroshi Ohtsuki, Gen. Mgr.-Corp. Comm.
Hiroshi Takahara, Gen. Mgr.-Finance & Acct.
Kanji Negi, Gen. Mgr.-Admin. Mgmt., Pharmaceutical Affairs
Makoto Yamaoka, Pres., Takeda Pharmaceuticals Int'l, Inc.
Alan MacKenzie, Exec. VP-Int'l Oper.

Phone: 81-6-6204-2111	Fax: 81-6-6204-2880
Toll-Free:	
Address: 1-1, Doshomachi 4-chome, Chuo-ku,, Osaka, 540-8645 Japan	

GROWTH PLANS/SPECIAL FEATURES:

Takeda Pharmaceutical Company Ltd., based in Japan, is an international research-based global pharmaceuticals company. One of the largest pharmaceutical companies in Japan, it operates research and development facilities in four countries and production facilities in five countries. Takeda discovers, develops, manufactures and markets pharmaceutical products in two categories: ethical and consumer health care drugs. Ethical drugs, as the firm denominates them, are marketed in about 90 countries worldwide. This segment includes the anti-prostatic cancer agent leuprolide acetate, marketed as Lupron Depot, Enantone, Prostap and Leuplin; the anti-peptic ulcer agent lansoprazole, marketed as Prevacid, Ogast, Takepron and other brands; the anti-hypertensive agent candesartan cilexetil, marketed as Blopress, Kenzen and Amias; and the anti-diabetic agent pioglitazone hydrochloride, marketed as Actos. Ethical drugs account for over 80% of Takeda's total sales. The company's consumer health care division focuses on the over-the-counter drug market. Takeda's main consumer brands include Alinamin, a vitamin B1 derivative; and Benza, a cold remedy. Within research and development, the firm focuses on the lifestyle-related diseases, oncology, urologic diseases, central nervous system diseases and gastroenterological diseases. Outside Japan, Takeda maintains subsidiaries and affiliates in the U.S., Canada, the U.K., France, Italy, Germany, Austria, Switzerland, Spain, Portugal, Ireland, Netherlands, Sweden, Belgium, Turkey, China, Taiwan, Philippines, Thailand, Indonesia, Singapore, Mexico and Brazil. The company pursues alliances with other pharmaceutical manufacturers, biotechnology companies, universities and other research institutions. During 2009, Takeda established and expanded its presence in Canada, Spain, Ireland, Portugal, Mexico, Turkey, Sweden, Norway, Denmark, Belgium and Luxembourg; and integrated its two manufacturing subsidiaries in Ireland. In May 2009, subsidiary Takeda America Holdings, Inc. agreed to acquire IDM Pharma, a developer of cancer products. In April 2010, Takeda received regulatory approval to market five new products, designed to treat diabetes, insomnia, cancer and hypertension, in Japan.

FINANCIALS: Sales and profits are in thousands of dollars—add 000 to get the full amount. 2009 Note: Financial information for 2009 was not available for all companies at press time.

2009 Sales: $16,813,800	2009 Profits: $2,561,790	**U.S. Stock Ticker: TKPHF**
2008 Sales: $13,748,020	2008 Profits: $3,554,540	**Int'l Ticker: 4502** Int'l Exchange: Tokyo-TSE
2007 Sales: $11,060,737	2007 Profits: $2,845,805	Employees: 15,717
2006 Sales: $10,360,744	2006 Profits: $2,677,342	Fiscal Year Ends: 3/31
2005 Sales: $10,441,300	2005 Profits: $2,579,600	Parent Company:

SALARIES/BENEFITS:

Pension Plan:	ESOP Stock Plan:	Profit Sharing:	Top Exec. Salary: $	Bonus: $
Savings Plan:	Stock Purch. Plan:		Second Exec. Salary: $	Bonus: $

OTHER THOUGHTS:

Apparent Women Officers or Directors:
Hot Spot for Advancement for Women/Minorities:

LOCATIONS: ("Y" = Yes)

West:	Southwest:	Midwest:	Southeast:	Northeast:	International:
Y		Y		Y	Y

Note: Financial information, benefits and other data can change quickly and may vary from those stated here.

TAMIR BIOTECHNOLOGY INC

www.alfacell.com

Industry Group Code: 325412 Ranks within this company's industry group: Sales: Profits: 86

Drugs:		Other:	Clinical:	Computers:	Services:
Discovery:	Y	AgriBio:	Trials/Services:	Hardware:	Specialty Services:
Licensing:		Genetic Data:	Labs:	Software:	Consulting:
Manufacturing:	Y	Tissue Replacement:	Equipment/Supplies:	Arrays:	Blood Collection:
Generics:			Research/Development Svcs.:	Database Management:	Drug Delivery:
			Diagnostics:		Drug Distribution:

TYPES OF BUSINESS:

Cancer & Pathological Conditions Drugs
RNase-based Drugs

BRANDS/DIVISIONS/AFFILIATES:

Onconase
Amphinases

CONTACTS: *Note: Officers with more than one job title may be intentionally listed here more than once.*

Charles Muniz, CEO
Charles Muniz, Pres.
Charles Muniz, CFO
Diane Scudiery, Dir.-Clinical & Regulatory Oper.
David Sidransky, Chmn.

Phone: 732-652-4525	Fax: 732-652-4575
Toll-Free:	
Address: 300 Atrium Dr., Somerset, NJ 08873 US	

GROWTH PLANS/SPECIAL FEATURES:

Tamir Biotechnology, Inc., formerly known as Alfacell Corporation, is a biopharmaceutical company engaged in the research, development and commercialization of drugs for life-threatening diseases, such as malignant mesothelioma and other cancers. The company's drug discovery and development program consists of novel therapeutics that are being developed from amphibian ribonucleases (RNases). RNases are biologically active enzymes that split RNA molecules. The firm uses RNases for the development of therapeutics for cancer and other life-threatening diseases, including HIV and autoimmune diseases, that require anti-proliferative and apoptotic, or programmed cell death, properties. The company's proprietary product is Onconase, which targets solid tumors that have become resistant to other chemotherapeutic drugs. Onconase affects primarily exponentially growing malignant cells, with activity controlled through specific molecular mechanisms. The drug is being evaluated as a treatment for inoperable malignant mesothelioma, a rare cancer primarily affecting the pleura (lining of the lungs) usually caused by exposure to asbestos. The drug received orphan drug designation for malignant mesothelioma in Australia and the U.S., as well as from the European Agency for the Evaluation of Medicinal Products (EMEA). The company is also evaluating Onconase in Phase I/II clinical development for applications in treating lung cancer and other solid tumors. The company also has another amphibian RNases product, known as Amphinases, in pre-clinical research and development. The firm was awarded a U.S. patent for a methodology for synthesizing gene sequences of ranpirnase, the active component of Onconase. The firm owns 20 U.S. patents, four patents in Europe, three in Japan and one in Singapore.

FINANCIALS: Sales and profits are in thousands of dollars—add 000 to get the full amount. 2009 Note: Financial information for 2009 was not available for all companies at press time.

2009 Sales: $	2009 Profits: $-4,539	U.S. Stock Ticker: ACEL
2008 Sales: $	2008 Profits: $-12,321	Int'l Ticker: Int'l Exchange:
2007 Sales: $	2007 Profits: $-8,755	Employees: 6
2006 Sales: $	2006 Profits: $-7,810	Fiscal Year Ends: 7/31
2005 Sales: $ 152	2005 Profits: $-6,462	Parent Company:

SALARIES/BENEFITS:

Pension Plan:	ESOP Stock Plan:	Profit Sharing:	Top Exec. Salary: $207,692	Bonus: $
Savings Plan: Y	Stock Purch. Plan:		Second Exec. Salary: $109,615	Bonus: $

OTHER THOUGHTS:

Apparent Women Officers or Directors: 1
Hot Spot for Advancement for Women/Minorities:

LOCATIONS: ("Y" = Yes)

West:	Southwest:	Midwest:	Southeast:	Northeast:	International:
				Y	

TARGETED GENETICS CORP

www.targen.com

Industry Group Code: 325412   Ranks within this company's industry group: Sales: 119   Profits: 61

Drugs:		Other:		Clinical:	Computers:	Services:	
Discovery:	Y	AgriBio:		Trials/Services:	Hardware:	Specialty Services:	
Licensing:		Genetic Data:	Y	Labs:	Software:	Consulting:	
Manufacturing:		Tissue Replacement:		Equipment/Supplies:	Arrays:	Blood Collection:	
Generics:				Research/Development Svcs.:	Database Management:	Drug Delivery:	Y
				Diagnostics:		Drug Distribution:	

TYPES OF BUSINESS:
Gene Therapeutics

BRANDS/DIVISIONS/AFFILIATES:
Adeno-Associated Viral (AAV) vectors
Celladon Corporation
Sirna Therapeutics

CONTACTS: Note: Officers with more than one job title may be intentionally listed here more than once.
B. G. Susan Robinson, CEO
B. G. Susan Robinson, Pres.
David J. Poston, CFO
Richard W. Peluso, VP-Mfg. & Process Sciences
David J. Poston, Corp. Sec.
David J. Poston, VP-Finance/Treas.
Jeremy Curnock Cook, Chmn.

Phone: 206-623-7612	Fax: 206-223-0288
Toll-Free:	
Address: 1100 Olive Way, Ste. 100, Seattle, WA 98101 US	

GROWTH PLANS/SPECIAL FEATURES:

Targeted Genetics Corp. is a clinical-stage biotechnology company that develops gene therapeutics. The company's gene therapeutics consists of a delivery vehicle, called a vector, and genetic material. The role of the vector is to carry the genetic material into a target cell. Once delivered into the cell, the gene can express or direct production of the specific proteins encoded by the gene. The firm develops and manufactures adeno-associated viral (AAV) vectors, or AAV-based gene therapeutics. The company has products in development, including treatments for inflammatory arthritis, HIV/AIDS, congestive heart failure and Huntington's disease. TGAAC94, the inflammatory arthritis candidate, works to administer an inhibitor of TNF-alpha (Tumor Necrosis Factor) in order to modulate inflammation. TGAAC09, a potential HIV/AIDS vaccine based on HIV subtype C, is designed to use the body's immune system to counteract the disease. The congestive heart failure program, being developed with Celladon Corporation, uses the AAV delivery method with SERCA2a and phospholamban gene variants in order to modify the heart's ability to contract. Targeted Genetics' Huntington's disease program was acquired from Sirna Therapeutics and aims to use the AAV vector with siRNA (short interfering RNA) to destroy specific RNA targets. The company recently realigned product development priorities and narrowed activities to three programs. The programs include a clinical stage AAV-based product candidate for treatment of Leber's congenital amaurosis (LCA); a preclinical AAV-based Huntington's disease (HD) product candidate; and a preclinical small molecule product candidate to treat amyotrophic lateral sclerosis (ALS). In addition, it maintained the development and manufacturing collaboration with Celladon; and its suspended self-funded efforts to advance our inflammatory arthritis product candidate, TGAAC94, as the company would require a partner that has the resources to take the product into Phase II development. Recently, the company cut about 10% of its workforce and scaled back executive pay.

FINANCIALS: Sales and profits are in thousands of dollars—add 000 to get the full amount. 2009 Note: Financial information for 2009 was not available for all companies at press time.

2009 Sales: $12,171	2009 Profits: $7,948	U.S. Stock Ticker: TGEN.PK
2008 Sales: $8,718	2008 Profits: $-20,720	Int'l Ticker: Int'l Exchange:
2007 Sales: $10,332	2007 Profits: $-16,127	Employees: 58
2006 Sales: $9,864	2006 Profits: $-33,990	Fiscal Year Ends: 12/31
2005 Sales: $6,874	2005 Profits: $-19,198	Parent Company:

SALARIES/BENEFITS:

Pension Plan:	ESOP Stock Plan:	Profit Sharing:	Top Exec. Salary: $374,000	Bonus: $
Savings Plan:	Stock Purch. Plan:		Second Exec. Salary: $257,500	Bonus: $

OTHER THOUGHTS:
Apparent Women Officers or Directors: 1
Hot Spot for Advancement for Women/Minorities:

LOCATIONS: ("Y" = Yes)

West:	Southwest:	Midwest:	Southeast:	Northeast:	International:
Y					

TARO PHARMACEUTICAL INDUSTRIES

www.taro.com

Industry Group Code: 325412 Ranks within this company's industry group: Sales: Profits:

Drugs:		Other:	Clinical:	Computers:	Services:
Discovery:	Y	AgriBio:	Trials/Services:	Hardware:	Specialty Services:
Licensing:	Y	Genetic Data:	Labs:	Software:	Consulting:
Manufacturing:	Y	Tissue Replacement:	Equipment/Supplies:	Arrays:	Blood Collection:
Generics:			Research/Development Svcs.:	Database Management:	Drug Delivery:
			Diagnostics:		Drug Distribution:

TYPES OF BUSINESS:

Drugs-Generic & Proprietary
Over-the-Counter Analgesics
Vitamins
Anti-Cancer Drugs
Dermatological Drugs

BRANDS/DIVISIONS/AFFILIATES:

Taro Pharmaceuticals U.S.A., Inc.
Kusch Manual
RxDesktop
Lustra
Ovide
Clotrimazole/Betamethasomne
Tarodex
Taro International

CONTACTS: Note: Officers with more than one job title may be intentionally listed here more than once.

Tal Levitt, Corp. Sec.
Tal Levitt, Sr. VP-Corp. Affairs
Tal Levitt, Treas.
Barrie Levitt, Chmn.

Phone: 972-9-971-1821	Fax: 972-9-955-7443
Toll-Free:	
Address: 14 Hakitor St., Haifa Bay, 26110 Israel	

GROWTH PLANS/SPECIAL FEATURES:

Taro Pharmaceutical Industries is a multinational pharmaceutical company that discovers, develops, manufactures and markets a wide range of both proprietary and generic health care products. These range from over-the-counter and prescription analgesics to vitamins. Its products address illnesses varying from the common cold to cancer, and are mainly used in dermatology, cardiology, neurology and pediatrics. The firm has five subsidiaries for distribution and manufacturing, which include Taro U.S.A, Taro Canada, Taro Israel, Taro UK and Taro International. Taro produces more than 200 pharmaceutical products, including topical preparations such as creams, ointments, gels and solutions; oral medications such as tablets, capsules, powders and liquids; and sterile products such as injectables, ophthalmic drops and powders. Some of the top products manufactured by Taro include: Lustra, a treatment for dyschromia (discolored skin); Ovide, a lotion for the treatment of head lice; Warfarin, sodium tablets for certain heart conditions; Clotrimazole and BetamethasoneDipropionate Creams (generic equivalents of Lotrisone), for the treatment of the topical effects of fungus; Uramox tablets, a diuretics; Avipur tablets and Calcimore tablets, for nutritional supplements; Tarocidin, for eye drops; Tarodex, adult and pediatric syrup for treating coughs; Jungborn Granules, a laxative; and Terconazole Vaginal Cream, for treating yeast infections. Taro also funds the Kusch Manual, a dermatology diagnosis manual for physicians; and RxDesktop, a program for health care professionals that contains conversation tools, dosage calculators and other information. The company is in the process of establishing pharmaceutical research and manufacturing operations for its sixth subsidiary, in Roscrea, Ireland.

FINANCIALS: Sales and profits are in thousands of dollars—add 000 to get the full amount. 2009 Note: Financial information for 2009 was not available for all companies at press time.

2009 Sales: $ 2009 Profits: $
2008 Sales: $ 2008 Profits: $
2007 Sales: $ 2007 Profits: $
2006 Sales: $ 2006 Profits: $
2005 Sales: $ 2005 Profits: $

U.S. Stock Ticker: TAROF.PK
Int'l Ticker: Int'l Exchange:
Employees:
Fiscal Year Ends: 12/31
Parent Company:

SALARIES/BENEFITS:

Pension Plan:	ESOP Stock Plan:	Profit Sharing:	Top Exec. Salary: $	Bonus: $
Savings Plan:	Stock Purch. Plan:		Second Exec. Salary: $	Bonus: $

OTHER THOUGHTS:

Apparent Women Officers or Directors: 1
Hot Spot for Advancement for Women/Minorities:

LOCATIONS: ("Y" = Yes)

West:	Southwest:	Midwest:	Southeast:	Northeast:	International:
				Y	Y

Note: Financial information, benefits and other data can change quickly and may vary from those stated here.

TECHNE CORP

www.techne-corp.com

Industry Group Code: 325413 Ranks within this company's industry group: Sales: 6   Profits: 4

Drugs:	Other:	Clinical:		Computers:		Services:	
Discovery:	AgriBio:	Trials/Services:		Hardware:		Specialty Services:	
Licensing:	Genetic Data:	Labs:		Software:		Consulting:	
Manufacturing:	Tissue Replacement:	Equipment/Supplies:	Y	Arrays:		Blood Collection:	
Generics:		Research/Development Svcs.:		Database Management:		Drug Delivery:	
		Diagnostics:	Y			Drug Distribution:	

TYPES OF BUSINESS:

Biotechnology Products
Reagents, Antibodies & Assay Kits
Hematology Products

BRANDS/DIVISIONS/AFFILIATES:

Research and Diagnostic Systems, Inc.
R&D Systems Europe, Ltd.
R&D Systems GmbH
R&D Systems China Co. Ltd.
Quantikine
BiosPacific

CONTACTS: Note: Officers with more than one job title may be intentionally listed here more than once.

Thomas E. Oland, CEO
Thomas E. Oland, Pres.
Gregory J. Melsen, CFO/VP
Richard A. Krzyzek, VP-Biotech Div.
Marcel Veronneau, VP-Hematology Oper.
Gregory J. Melsen, VP-Finance
Thomas E. Oland, Chmn.
Wendy Shao, Gen. Mgr.-R&D Systems China Co. Ltd.

Phone: 612-379-8854	Fax: 612-379-6580
Toll-Free: 800-343-7475	
Address: 614 McKinley Pl. NE, Minneapolis, MN 55413-2610 US	

GROWTH PLANS/SPECIAL FEATURES:

Techne Corp. is a holding company involved in biotechnology and hematology products. The firm operates through two subsidiaries: Research and Diagnostic Systems, Inc. (R&D Systems) and R&D Systems Europe, Ltd. (R&D Europe). R&D Systems manufactures biological products in two major segments: hematology controls, which are used in clinical and hospital laboratories to monitor the accuracy of blood analysis instruments; and biotechnology products, which include purified proteins and antibodies that are sold exclusively to the research market and assay kits (under the trade name Quantikine) that are sold to the research and clinical diagnostic markets. R&D Europe distributes biotechnology products throughout Europe and operates sales offices in France and Germany through its German subsidiary, R&D Systems GmbH. The company also operates under its BiosPacific subsidiary, which supplies biologics to manufacturers of in vitro diagnostic systems and immunodiagnostic kits. In recent years, R&D Systems has also expanded its product portfolio to include enzymes and intracellular cell signaling reagents such as kinases, proteases and phosphatases; these reagents detect diseases such as cancer, Alzheimer's, arthritic, autoimmunity, diabetes, hypertension, obesity, AIDS and SARS. Techne also produces controls and calibrators for a variety of medical brands such as Abbott Diagnostics, Beckman Coulter, Siemens Healthcare Diagnostics and Sysmex. In the hematology sector, the company's Whole Blood Flow Cytometry Controls are used to identify and quantify white blood cells by their surface antigens while linearity and reportable range controls assess the linearity of hematology analyzers for white blood cells, red blood cells, platelets and reticulocytes. R&D Systems China Co. Ltd. is a wholly-owned subsidiary established to improve the level of service offered to R&D Systems' distributors and customers in Shanghai, China. The company operates as a warehouse/distribution hub for R&D Systems' cell biology research products and offers technical services and marketing support for the Chinese market.

FINANCIALS: Sales and profits are in thousands of dollars—add 000 to get the full amount. 2009 Note: Financial information for 2009 was not available for all companies at press time.

2009 Sales: $263,956	2009 Profits: $105,242	**U.S. Stock Ticker:** TECH
2008 Sales: $257,420	2008 Profits: $103,558	**Int'l Ticker:** Int'l Exchange:
2007 Sales: $223,482	2007 Profits: $85,111	Employees: 746
2006 Sales: $202,617	2006 Profits: $73,351	Fiscal Year Ends: 6/30
2005 Sales: $178,700	2005 Profits: $66,100	Parent Company:

SALARIES/BENEFITS:

Pension Plan:	ESOP Stock Plan:	Profit Sharing: Y	Top Exec. Salary: $275,000	Bonus: $12,705
Savings Plan: Y	Stock Purch. Plan:		Second Exec. Salary: $254,100	Bonus: $

OTHER THOUGHTS:

Apparent Women Officers or Directors: 2
Hot Spot for Advancement for Women/Minorities: Y

LOCATIONS: ("Y" = Yes)

West:	Southwest:	Midwest:	Southeast:	Northeast:	International:
Y		Y			Y

TELIK INC

www.telik.com

Industry Group Code: 325412 Ranks within this company's industry group: Sales: Profits: 124

Drugs:		Other:	Clinical:	Computers:	Services:
Discovery:	Y	AgriBio:	Trials/Services:	Hardware:	Specialty Services:
Licensing:	Y	Genetic Data:	Labs:	Software:	Consulting:
Manufacturing:		Tissue Replacement:	Equipment/Supplies:	Arrays:	Blood Collection:
Generics:			Research/Development Svcs.:	Database Management:	Drug Delivery:
			Diagnostics:		Drug Distribution:

TYPES OF BUSINESS:
Drugs-Cancer
Small-Molecule Drugs

BRANDS/DIVISIONS/AFFILIATES:
Target-Related Affinity Profiling (TRAP)
TELCYTA
TELINTRA
TLK58747
TLK60596

CONTACTS: *Note: Officers with more than one job title may be intentionally listed here more than once.*
Michael M. Wick, CEO
Cynthia M. Butitta, COO
Michael M. Wick, Pres.
Cynthia M. Butitta, CFO
Gail L. Brown, Chief Medical Officer/Sr. VP
William P. Kaplan, General Counsel/Corp. Sec./VP
Marc L. Steuer, Sr. VP-Bus. Dev.
Michael M. Wick, Chmn.

Phone: 650-845-7700	Fax: 650-845-7800
Toll-Free:	
Address: 3165 Porter Dr., Palo Alto, CA 94304 US	

GROWTH PLANS/SPECIAL FEATURES:
Telik, Inc. discovers, develops and commercializes small-molecule pharmaceuticals, mainly for the treatment of specific cancers. The company's proprietary Target-Related Affinity Profiling (TRAP) chemoinformatics technology is used by Telik and its partners to rapidly identify promising chemicals for development. TRAP can select a small sample from its large compound library and effectively identify small, biologically active molecules. The company currently has two product candidates in clinical trials. Telcyta is a novel tumor-activated compound that is currently in Phase III clinical trials for use against ovarian and non-small cell lung cancers and in Phase II of development for colorectal, breast, and certain forms of lung and ovarian cancers. The drug product candidate is designed to be activated in cancer cells through binding to the GST P1-1 protein, which is elevated in many human cancers, even more so in patients who have already been treated with other standard chemotherapy drugs. Once bound to the protein inside a cancer cell, a chemical reaction occurs that releases Telcyta and causes programmed cancer death. Telintra, a small-molecule bone marrow stimulant, is in Phase II trials for myelodysplastic syndrome (MDS), a form of pre-leukemia. The product is used in the treatment of blood disorders with low blood cell levels, such as neutropenia or anemia. The company has a variety of potential products in the preclinical pipeline, including TLK58747, which has shown significant anti-tumor activity in human breast, pancreatic and colon tumors while in preclinical study; and TLK60596, a small-molecule dual inhibitor that has shown significantly reduced tumor growth in standard animal models of human colon cancer.

FINANCIALS: Sales and profits are in thousands of dollars—add 000 to get the full amount. 2009 Note: Financial information for 2009 was not available for all companies at press time.

2009 Sales: $	2009 Profits: $-23,693	U.S. Stock Ticker: TELK
2008 Sales: $	2008 Profits: $-31,763	Int'l Ticker: Int'l Exchange:
2007 Sales: $	2007 Profits: $-55,215	Employees: 43
2006 Sales: $	2006 Profits: $-79,624	Fiscal Year Ends: 12/31
2005 Sales: $ 19	2005 Profits: $-75,542	Parent Company:

SALARIES/BENEFITS:

Pension Plan:	ESOP Stock Plan:	Profit Sharing:	Top Exec. Salary: $514,000	Bonus: $
Savings Plan:	Stock Purch. Plan: Y		Second Exec. Salary: $379,000	Bonus: $

OTHER THOUGHTS:
Apparent Women Officers or Directors: 2
Hot Spot for Advancement for Women/Minorities: Y

LOCATIONS: ("Y" = Yes)

West:	Southwest:	Midwest:	Southeast:	Northeast:	International:
Y					

TENGION INC

www.tengion.com

Industry Group Code: 33911 Ranks within this company's industry group: Sales: Profits:

Drugs:		Other:		Clinical:	Computers:	Services:
Discovery:	Y	AgriBio:		Trials/Services:	Hardware:	Specialty Services:
Licensing:		Genetic Data:		Labs:	Software:	Consulting:
Manufacturing:	Y	Tissue Replacement:	Y	Equipment/Supplies:	Arrays:	Blood Collection:
Generics:				Research/Development Svcs.:	Database Management:	Drug Delivery:
				Diagnostics:		Drug Distribution:

TYPES OF BUSINESS:

Medical Implant & Regenerative Therapy

BRANDS/DIVISIONS/AFFILIATES:

Organ Regeneration Platform
Neo-Urinary Conduit (The)
Tengion Neo-Bladder Augment (The)
Tengion Neo-GI Augment (The)
Tengion Neo-Kidney Augment (The)
Tengion Neo-Vessel Replacement
Tengion Neo-Bladder Replacement

CONTACTS: Note: Officers with more than one job title may be intentionally listed here more than once.

Steven Nichtberger, CEO
Steven Nichtberger, Pres.
Tim Bertram, Sr. VP-Science
Tim Bertram, Sr. VP-Tech.
Joseph W. La Barge, Corp. Counsel/Exec. Dir.
Jason Krentz, Exec. Dir.-Tech. Oper.
Linda Hearne, VP-Finance
Sunita B. Sheth, VP-Clinical & Regulatory Affairs
Sunita B. Sheth, Chief Medical Officer
Francois Dubois, VP-Quality
Mark Stejbach, VP/Chief Commercial Officer
David Scheer, Chmn.

Phone: 610-292-8364	Fax:
Toll-Free:	
Address: 2900 Potshop Ln., Ste. 100, East Norriton, PA 19403 US	

GROWTH PLANS/SPECIAL FEATURES:

Tengion, Inc. is a regenerative medicine firm. The company is focused on the discovery, development, manufacturing and marketing of replacement human neo-organs and neo-tissues derived from autologous cells (a patient's own cells). The firm's Organ Regeneration Platform allows for the creation of proprietary product candidates intended to harness the intrinsic regenerative pathways of the body to produce a range of native-like organs and tissues. Tengion's lead product candidate, the Neo-Urinary Conduit, is a potential treatment for bladder cancer. The candidate is currently in a Phase 1 clinical trial for patients with bladder cancer who require a total cystectomy. The company's other candidate, the Tengion Neo-Bladder Augment, is a potential alternative to intestinal surgery and entercystoplasty (using a portion of intestine as a graft); it is intended to increase bladder capacity and decrease bladder pressure while integrating into the patient's body. The candidate has completed Phase II clinical trials for spina bifida, spinal cord injury and urge incontinence. The firm's other candidates, which are in preclinical development, include the Tengion Neo-GI Augment, a possible alternative to an esophagectomy (surgical removal of the esophagus); Tengion Neo-Kidney Augment, for the treatment of chronic kidney disease; Tengion Neo-Vessel Replacement, a potential alternative to coronary artery bypass, vascular access graft and peripheral arterial bypass surgery; and Tengion Neo-Bladder Replacement for patients with bladder cancer. In April 2010, Tengion became a publicly traded company.

The firm offers employees benefits including medical, dental, vision, disability, life and AD&D insurance; a 401(k); paid time off; and flexible spending accounts.

FINANCIALS: Sales and profits are in thousands of dollars—add 000 to get the full amount. 2009 Note: Financial information for 2009 was not available for all companies at press time.

2009 Sales: $	2009 Profits: $	U.S. Stock Ticker: TNGN
2008 Sales: $	2008 Profits: $	Int'l Ticker: Int'l Exchange:
2007 Sales: $	2007 Profits: $	Employees:
2006 Sales: $	2006 Profits: $	Fiscal Year Ends: 12/31
2005 Sales: $	2005 Profits: $	Parent Company:

SALARIES/BENEFITS:

Pension Plan:	ESOP Stock Plan:	Profit Sharing:	Top Exec. Salary: $	Bonus: $
Savings Plan: Y	Stock Purch. Plan:		Second Exec. Salary: $	Bonus: $

OTHER THOUGHTS:

Apparent Women Officers or Directors: 4
Hot Spot for Advancement for Women/Minorities: Y

LOCATIONS: ("Y" = Yes)

West:	Southwest:	Midwest:	Southeast:	Northeast:	International:
				Y	

TEVA PARENTERAL MEDICINES INC

www.tevausa.com

Industry Group Code: 325412 **Ranks within this company's industry group: Sales: Profits:**

Drugs:		Other:	Clinical:		Computers:		Services:	
Discovery:		AgriBio:	Trials/Services:		Hardware:		Specialty Services:	
Licensing:	Y	Genetic Data:	Labs:	Y	Software:		Consulting:	
Manufacturing:	Y	Tissue Replacement:	Equipment/Supplies:		Arrays:		Blood Collection:	
Generics:			Research/Development Svcs.:	Y	Database Management:		Drug Delivery:	
			Diagnostics:				Drug Distribution:	

TYPES OF BUSINESS:
Drugs-Diversified
Bulk Pharmaceutical Ingredients
Contract Manufacturing & Research
Injectable Pharmaceuticals

BRANDS/DIVISIONS/AFFILIATES:
Teva Pharmaceuticals Industries, Ltd.
Teva Pharmaceuticals USA
Teva Biopharmaceuticals USA
Teva Speciality Pharmaceuticals, LLC
TAPI
Gate Pharmaceuticals
Teva Animal Health
Barr Laboratories

CONTACTS: *Note: Officers with more than one job title may be intentionally listed here more than once.*
Frank Becker, CEO
Shlomo Yanai, CEO/Pres., Teva Pharmaceutical Industries Ltd.
Eyal Desheh, CFO-Teva Pharmaceutical Industries, Ltd.
Eli Hurvitz, Chmn., Teva Pharmaceutical Industries, Ltd.

Phone: 949-455-4700	Fax: 949-855-8210
Toll-Free:	
Address: 19 Hughes, Irvine, CA 92618 US	

GROWTH PLANS/SPECIAL FEATURES:

Teva Parenteral Medicines, Inc. is a wholly-owned subsidiary of Teva Pharmaceutical Industries and a global leader in the development and marketing of generic injectable drugs. The company uses its internal research and development capabilities, together with its operational manufacturing and regulatory experience, to develop a wide range of pharmaceutical products. The firm markets oncolytic and specialty injectable products, particularly in the oncology and anesthesiology care areas. It concentrates on products and technologies that face significant barriers to entering the worldwide market. A few of the firm's products include Acyclovir powder for injection; Cholografin Meglumine injection; Leucovorin Clacium injection USP; Octreotide Acetate injection; and Sulfamethoxazole and Trimethoprim injection. Teva Parenteral Medicines has concentrated its expansion toward securing marketing approval for new drugs and marketing them first in developing nations, then in Europe and finally in the U.S. The firm is one of 23 business units controlled by Teva Pharmaceutical Industries. The business units include Teva Pharmaceuticals USA; Teva Biopharmaceuticals USA; Teva Parenteral Medicines, Inc.; TAPI; Teva Speciality Pharmaceuticals, LLC; Gate Pharmaceuticals; Teva Animal Health; Barr Laboratories; Duramed; Global Generic R&D; and Global Respiratory R&D.

The firm offers employees medical and dental coverage; life insurance; long and short term disability; a child and elder care referral program; flexible spending accounts; tuition reimbursement; an employee assistance program; and professional development programs.

FINANCIALS: Sales and profits are in thousands of dollars—add 000 to get the full amount. 2009 Note: Financial information for 2009 was not available for all companies at press time.

2009 Sales: $	2009 Profits: $	**U.S. Stock Ticker: Subsidiary**
2008 Sales: $	2008 Profits: $	**Int'l Ticker:** Int'l Exchange:
2007 Sales: $	2007 Profits: $	Employees:
2006 Sales: $	2006 Profits: $	Fiscal Year Ends:
2005 Sales: $	2005 Profits: $	Parent Company: TEVA PHARMACEUTICAL INDUSTRIES

SALARIES/BENEFITS:

Pension Plan:	ESOP Stock Plan:	Profit Sharing: Y	Top Exec. Salary: $	Bonus: $
Savings Plan: Y	Stock Purch. Plan: Y		Second Exec. Salary: $	Bonus: $

OTHER THOUGHTS:
Apparent Women Officers or Directors:
Hot Spot for Advancement for Women/Minorities:

LOCATIONS: ("Y" = Yes)

West:	Southwest:	Midwest:	Southeast:	Northeast:	International:
Y					Y

TEVA PHARMACEUTICAL INDUSTRIES

www.tevapharm.com

Industry Group Code: 325412 Ranks within this company's industry group: Sales: 14   Profits: 16

Drugs:		Other:	Clinical:	Computers:		Services:
Discovery:		AgriBio:	Trials/Services:	Hardware:		Specialty Services:
Licensing:		Genetic Data:	Labs:	Software:		Consulting:
Manufacturing:	Y	Tissue Replacement:	Equipment/Supplies:	Y	Arrays:	Blood Collection:
Generics:			Research/Development Svcs.:	Database Management:		Drug Delivery:
			Diagnostics:			Drug Distribution:

TYPES OF BUSINESS:

Drugs-Generic
Active Pharmaceutical Ingredients

BRANDS/DIVISIONS/AFFILIATES:

Teva Pharmaceuticals USA
Pharmachemie BV
Copaxone
Ivax Corporation
Bentley Pharmaceuticals, Inc.
Azilect
CoGenesys, Inc.
Barr Pharmaceuticals

CONTACTS: Note: Officers with more than one job title may be intentionally listed here more than once.

Shlomo Yanai, CEO
Shlomo Yanai, Pres.
Eyal Desheh, CFO
Isaac Abravanel, Corp. VP-Human Resources
Ben-Zion Weiner, Chief R&D Officer
Richard S. Egosi, Chief Legal Officer/Corp. VP
Itzhak Krinsky, Corp. VP-Bus. Dev.
William S. Marth, CEO/Pres., Teva North America
Gerard Van Odlijk, Group VP-Europe
Moshe Manor, Group VP-Global Branded Prod.
Aharon Yaari, Group VP-Teva Generics System
Phillip Frost, Chmn.
Chaim Hurvitz, Pres., Int'l

Phone: 972-3-926-7267	Fax: 972-3-923-4050
Toll-Free:	
Address: 5 Basel St., Petach Tikva, 49131 Israel	

GROWTH PLANS/SPECIAL FEATURES:

Teva Pharmaceutical Industries, Ltd., based in Israel, is a pharmaceutical company that produces, distributes and sells pharmaceutical products internationally. Approximately 85% of the firm's sales are generated from North America and Europe. The firm specializes in the development of generic, also known as human pharmaceuticals (HP), and active pharmaceutical ingredients (API). The HP segment produces generic drugs in all major therapeutics in a variety of dosage forms, including capsules, tablets, creams, ointments and liquids. The API segment distributes ingredients to manufacturers worldwide, in addition to supporting its own pharmaceutical products. Teva also manufactures innovative drugs in niche markets. Currently, the firm focuses on products for neurological disorders and auto-immune diseases such as Alzheimer's disease, Amyotrophic Lateral Sclerosis, Lupus and Parkinson's disease. Subsidiary Teva Pharmaceuticals USA is a dominant figure in the manufacture and distribution of generic drugs in the U.S. The firm's two chief products are Copaxone, a branded treatment for multiple sclerosis (MS), currently marketed and sold in 42 countries including the U.S., and Azilect, a treatment for Parkinson's. Teva has direct operations in more than 60 countries, including 38 finished dosage pharmaceutical manufacturing sites in 17 countries, 15 generic research and development centers within the firm's operating sites and 21 API manufacturing sites worldwide. In August 2009, the company announced the approval and launch of Oxaliplatin Injection, a drug for the adjuvant treatment of stage III colon cancer. In November 2009, Teva launched its Fexofenadine Hydrochloride 60mg and Pseudoephedrine Hydrochloride 120mg extended release allergy tablets. In March 2010, the firm agreed to acquire German generics company Ratiopharm Group for $5 billion. In April 2010, Teva launched its generic version of Flomax, a product used for the treatment of benign prostatic hyperplasia symptoms.

FINANCIALS: Sales and profits are in thousands of dollars—add 000 to get the full amount. 2009 Note: Financial information for 2009 was not available for all companies at press time.

2009 Sales: $13,899,000	2009 Profits: $2,000,000	**U.S. Stock Ticker: TEVA**
2008 Sales: $11,085,000	2008 Profits: $635,000	**Int'l Ticker: TEVA** Int'l Exchange: Tel Aviv-TASE
2007 Sales: $9,408,000	2007 Profits: $1,952,000	Employees: 35,089
2006 Sales: $8,408,000	2006 Profits: $546,000	Fiscal Year Ends: 12/31
2005 Sales: $5,250,000	2005 Profits: $1,072,000	Parent Company:

SALARIES/BENEFITS:

Pension Plan:	ESOP Stock Plan:	Profit Sharing:	Top Exec. Salary: $	Bonus: $
Savings Plan:	Stock Purch. Plan:		Second Exec. Salary: $	Bonus: $

OTHER THOUGHTS:

Apparent Women Officers or Directors: 3
Hot Spot for Advancement for Women/Minorities: Y

LOCATIONS: ("Y" = Yes)

West:	Southwest:	Midwest:	Southeast:	Northeast:	International:
Y	Y	Y	Y	Y	Y

THERAGENICS CORP

www.theragenics.com

Industry Group Code: 33911 Ranks within this company's industry group: Sales: 9 Profits: 8

Drugs:	Other:	Clinical:		Computers:	Services:
Discovery:	AgriBio:	Trials/Services:		Hardware:	Specialty Services:
Licensing:	Genetic Data:	Labs:		Software:	Consulting:
Manufacturing:	Tissue Replacement:	Equipment/Supplies:	Y	Arrays:	Blood Collection:
Generics:		Research/Development Svcs.:	Y	Database Management:	Drug Delivery:
		Diagnostics:			Drug Distribution:

TYPES OF BUSINESS:

Medical Devices
Surgical Products

BRANDS/DIVISIONS/AFFILIATES:

TheraSeed
Galt Medical Corp.
CP Medical Corp.
NeedleTech Products, Inc.
I-Seed

CONTACTS: Note: Officers with more than one job title may be intentionally listed here more than once.

M. Christine Jacobs, CEO
M. Christine Jacobs, Pres.
Frank J. Tarallo, CFO
C. Russell Small, Exec. VP-Sales & Mktg., Surgical Prod.
Bruce W. Smith, Corp. Sec.
Bruce W. Smith, Exec. VP-Strategy & Bus. Dev.
Frank J. Tarallo, Treas.
Joseph Plante, Pres., NeedleTech
Janet E. Zeman, Pres., CP Medical
M. Christine Jacobs, Chmn.

Phone: 770-271-0233	Fax:
Toll-Free:	
Address: 5203 Bristol Industrial Way, Buford, GA 30518 US	

GROWTH PLANS/SPECIAL FEATURES:

Theragenics Corp. is a medical device company serving the cancer treatment and surgical markets. The company operates in two segments: the brachytherapy seed business and the surgical products business. The brachytherapy seed business segment produces, markets and sells TheraSeed, the firm's premier palladium-103 prostate cancer treatment device; I-seed, its iodine-125 based prostrate cancer treatment device; and other related products and services. Theragenics is one of the world's largest producers of palladium-103, the radioactive isotope that supplies the therapeutic radiation for its TheraSeed device. TheraSeed is an implant the size of a grain of rice that is used primarily in treating localized prostate cancer with a one-time, minimally invasive procedure. The implant emits radiation within the immediate prostate area, killing the tumor while sparing surrounding organs from significant radiation exposure. Physicians, hospitals and other healthcare providers, primarily located in the U.S., utilize the TheraSeed device. The majority of TheraSeed sales are channeled through one third-party distributor. The surgical products business segment consists of wound closure, vascular access and specialty needle products. Wound closure products include sutures, needles and other surgical products with applications in, among other areas, urology, veterinary medicine, cardiology, orthopedics, plastic surgery and dental surgery. Vascular access products include introducers, guidewires and related products. Specialty needles include coaxial, biopsy, spinal and disposable veress needles; access trocars; and other needle-based products. The surgical products business sells its devices and components primarily to original equipment manufacturers (OEMs) and a network of distributors. The firm's wholly-owned subsidiaries CP Medical Corp., Galt Medical Corp. and NeedleTech Products, Inc. comprise the surgical products business and accounted for 69% of revenue in 2009.

Employees are offered medical, dental and life insurance; short- and long-term disability coverage; a 401(k) plan; and an employee stock purchase plan.

FINANCIALS: Sales and profits are in thousands of dollars—add 000 to get the full amount. 2009 Note: Financial information for 2009 was not available for all companies at press time.

2009 Sales: $78,326	2009 Profits: $3,075	U.S. Stock Ticker: TGX
2008 Sales: $67,358	2008 Profits: $-58,540	Int'l Ticker: Int'l Exchange:
2007 Sales: $62,210	2007 Profits: $5,635	Employees:
2006 Sales: $54,096	2006 Profits: $6,865	Fiscal Year Ends: 12/31
2005 Sales: $44,270	2005 Profits: $-29,006	Parent Company:

SALARIES/BENEFITS:

Pension Plan:	ESOP Stock Plan:	Profit Sharing:	Top Exec. Salary: $545,000	Bonus: $619,680
Savings Plan: Y	Stock Purch. Plan: Y		Second Exec. Salary: $302,500	Bonus: $261,585

OTHER THOUGHTS:

Apparent Women Officers or Directors: 3
Hot Spot for Advancement for Women/Minorities: Y

LOCATIONS: ("Y" = Yes)

West:	Southwest:	Midwest:	Southeast:	Northeast:	International:
Y	Y		Y	Y	

THERMO FISHER SCIENTIFIC INC
www.thermofisher.com

Industry Group Code: 423450 Ranks within this company's industry group: Sales: 1   Profits: 1

Drugs:		Other:	Clinical:		Computers:		Services:	
Discovery:		AgriBio:	Trials/Services:		Hardware:	Y	Specialty Services:	Y
Licensing:		Genetic Data:	Labs:		Software:	Y	Consulting:	Y
Manufacturing:	Y	Tissue Replacement:	Equipment/Supplies:	Y	Arrays:		Blood Collection:	
Generics:			Research/Development Svcs.:	Y	Database Management:		Drug Delivery:	
			Diagnostics:	Y			Drug Distribution:	

TYPES OF BUSINESS:
Laboratory Equipment & Supplies Distribution
Contract Manufacturing
Equipment Calibration & Repair
Clinical Trial Services
Laboratory Workstations
Clinical Consumables
Diagnostic Reagents
Custom Chemical Synthesis

BRANDS/DIVISIONS/AFFILIATES:
B.R.A.H.M.S. AG
Ahura Scientific
Finnzymes
Proxeon A/S
Fermentas International, Inc.
ABGene
Dharmacon
Owl Separation Systems

CONTACTS: Note: Officers with more than one job title may be intentionally listed here more than once.
Marc N. Casper, CEO
Marc N. Casper, Pres.
Peter M. Wilver, CFO/Sr. VP
Stephen G. Sheehan, Sr. VP-Human Resources
Seth H. Hoogasian, General Counsel/Sec./Sr. VP
Kenneth J. Apicerno, VP-Investor Rel.
Peter E. Hornstra, Chief Acct. Officer/VP
Alan J. Malus, Sr. VP/Pres., Laboratory Products
Gregory J. Herrema, Sr. VP/Pres., Analytical Instruments
Edward A. Pesicka, Sr. VP/Pres., Customer Channels
Ken Berger, VP/Pres., Specialty Diagnostics
Jim Manzi, Chmn.

Phone: 781-622-1000	Fax: 781-622-1207
Toll-Free: 800-678-5599	
Address: 81 Wyman St., Waltham, MA 02454 US	

GROWTH PLANS/SPECIAL FEATURES:
Thermo Fisher Scientific Inc. is a distributor of products and services principally to the scientific-research and clinical laboratory markets. The firm serves over 350,000 customers including biotechnology and pharmaceutical companies; colleges and universities; medical-research institutions; hospitals; reference, quality control, process-control and research and development labs in various industries; and government agencies. It operates in two segments: Analytical Technologies and Laboratory Products and Services. The Analytical Technologies segment has three primary growth platforms: analytical instruments, specialty diagnostics and biosciences; and through them, provides a broad range of instruments, software and services, bioscience reagents, and diagnostic assays. Thermo Fisher's products include analytical instruments; automation and robotics; life science research supplies; chemicals; disposable laboratory supplies; diagnostic supplies; laboratory equipment; furniture; software; and custom products. The company's services include asset management, instrument standards compliance services, education, technical support and professional assistance. The company's brands include ABGene, BioImage, Cellomics, Consolidated Technologies, Fisher Diagnostics, Microgenics, Owl Separation Systems, Inc, and Nalgene Outdoor. The company has over 600 subsidiaries, including Alchematrix, Echochem N.B., Flux Instruments, LambTrack Ltd., and Thermo Electron Industries. In October 2009, Thermo Fisher completed the acquisition of German In-Vitro Diagnostics firm B.R.A.H.M.S. AG for roughly $470 million. In February 2010, the company acquired Ahura Scientific, a producer of spectroscopy instruments. In March 2010, Thermo Fisher completed the acquisition of Finnzymes, a provider of tools for polymerase chain reaction. In April 2010, the company acquired Danish proteomics firm Proxeon A/S. In May 2010, Thermo Fisher agreed to acquire Fermentas International, Inc. (a distributor and manufacturer of enzymes, reagents, and kits for molecular and cellular biology research) for $260 million.

ThermoFisher offers its employees comprehensive benefits, paid time off, tuition reimbursement, retirement plans and employee recognition awards.

FINANCIALS: Sales and profits are in thousands of dollars—add 000 to get the full amount. 2009 Note: Financial information for 2009 was not available for all companies at press time.
2009 Sales: $10,109,700	2009 Profits: $850,300	U.S. Stock Ticker: TMO
2008 Sales: $10,498,000	2008 Profits: $980,900	Int'l Ticker: Int'l Exchange:
2007 Sales: $9,746,400	2007 Profits: $748,400	Employees: 35,400
2006 Sales: $3,791,617	2006 Profits: $168,935	Fiscal Year Ends: 12/31
2005 Sales: $2,633,027	2005 Profits: $223,218	Parent Company: THERMO ELECTRON CORP

SALARIES/BENEFITS:
Pension Plan: Y	ESOP Stock Plan:	Profit Sharing:	Top Exec. Salary: $955,644	Bonus: $
Savings Plan: Y	Stock Purch. Plan:		Second Exec. Salary: $790,220	Bonus: $824,533

OTHER THOUGHTS:
Apparent Women Officers or Directors: 2
Hot Spot for Advancement for Women/Minorities: Y

LOCATIONS: ("Y" = Yes)
West:	Southwest:	Midwest:	Southeast:	Northeast:	International:
Y	Y	Y	Y	Y	Y

Note: Financial information, benefits and other data can change quickly and may vary from those stated here.

THERMO SCIENTIFIC

www.perbio.com

Industry Group Code: 325413 Ranks within this company's industry group: Sales: Profits:

Drugs:	Other:	Clinical:		Computers:		Services:
Discovery:	AgriBio:	Trials/Services:		Hardware:		Specialty Services:
Licensing:	Genetic Data:	Labs:		Software:	Y	Consulting:
Manufacturing:	Tissue Replacement:	Equipment/Supplies:	Y	Arrays:		Blood Collection:
Generics:		Research/Development Svcs.:	Y	Database Management:		Drug Delivery:
		Diagnostics:	Y			Drug Distribution:

TYPES OF BUSINESS:

Research Diagnostics Products
Software
Laboratory Equipment
Sample Handling

BRANDS/DIVISIONS/AFFILIATES:

Pierce
HyClone
Endogen
Cellomics
Thermo Fisher Scientific, Inc.

CONTACTS: Note: Officers with more than one job title may be intentionally listed here more than once.

Sonja Neerinckx, Mgr.-Human Resources
Marc N. Casper, CEO/Pres., ThermoFisher Scientific. Inc.
Peter M. Wilver, CFO/Sr. VP-ThermoFisher Scientific. Inc.
Jim P. Manzi, Chmn.-ThermoFisher Scientific. Inc.

Phone: 32-53-834-404	Fax: 32-53-837-638
Toll-Free:	
Address: Industrielaan 27, Erembodegem, 9320 Belgium	

GROWTH PLANS/SPECIAL FEATURES:

Thermo Scientific (formerly Perbio Science AB), a subsidiary of Thermo Fisher Scientific, Inc., focuses on protein production and research. The firm specializes in products, systems and services for the study and production of proteins, cell cultures and RNA, primarily for academic and research institutes, pharmaceutical companies and diagnostics companies. The company's portfolio of products includes innovative technologies for mass spectrometry, molecular spectroscopy, microbiology, elemental analysis, informatics, fine- and high-purity chemistry production, sample preparation, cell culture, protein analysis, RNA-interference techniques, immunodiagnostic testing, , as well as environmental monitoring and process control. The company operates in two segments: bioresearch and cell cultures. The bioresearch segment develops products used for research in protein chemistry and molecular biology. Products include kits, reagents and services for identifying, quantifying, purifying and modifying proteins, amidites and nucleotides for nucleic acid-based drugs and diagnostics. The cell culture segment offers proper nutrition for animal cells, process liquids for protein purification, sterile liquid handling systems and disposable sterile liquid processing systems. Products include animal sera such as fetal bovine serum, cosmic calf serum and equine serum, as well as cell culture media in powder and liquid forms. Leading brands include Pierce, a provider of research products for use in protein chemistry, sample handling, immunology and chromatography; Endogen, a provider of kits for the detection of infections, multiplex assays testing up to 16 types of antibodies and antibodies to cytokines and kinases; HyClone, a provider of cell-culture products and bioprocessing systems; and Cellomics, a provider of cellular measuring tools and screening instruments. In addition, the firm provides high-end analytical instruments; software; laboratory equipment; and consumables and reagents for research. Thermo Scientific has operating companies in the U.K., France, Germany, Switzerland, the Netherlands and Belgium.

FINANCIALS: Sales and profits are in thousands of dollars—add 000 to get the full amount. 2009 Note: Financial information for 2009 was not available for all companies at press time.

2009 Sales: $	2009 Profits: $	U.S. Stock Ticker: Subsidiary
2008 Sales: $	2008 Profits: $	Int'l Ticker: Int'l Exchange:
2007 Sales: $	2007 Profits: $	Employees:
2006 Sales: $	2006 Profits: $	Fiscal Year Ends: 12/31
2005 Sales: $	2005 Profits: $	Parent Company: THERMO FISHER SCIENTIFIC INC

SALARIES/BENEFITS:

Pension Plan:	ESOP Stock Plan:	Profit Sharing:	Top Exec. Salary: $	Bonus: $
Savings Plan:	Stock Purch. Plan:		Second Exec. Salary: $	Bonus: $

OTHER THOUGHTS:

Apparent Women Officers or Directors:
Hot Spot for Advancement for Women/Minorities:

LOCATIONS: ("Y" = Yes)

West:	Southwest:	Midwest:	Southeast:	Northeast:	International:
Y		Y	Y	Y	Y

TITAN PHARMACEUTICALS

www.titanpharm.com

Industry Group Code: 325412 Ranks within this company's industry group: Sales: Profits:

Drugs:		Other:	Clinical:	Computers:	Services:	
Discovery:	Y	AgriBio:	Trials/Services:	Hardware:	Specialty Services:	
Licensing:		Genetic Data:	Labs:	Software:	Consulting:	
Manufacturing:		Tissue Replacement:	Equipment/Supplies:	Arrays:	Blood Collection:	
Generics:			Research/Development Svcs.:	Database Management:	Drug Delivery:	Y
			Diagnostics:		Drug Distribution:	

TYPES OF BUSINESS:

Central Nervous System, Cardiovascular & Bone Diseases Therapeutics
Drug Delivery Systems

BRANDS/DIVISIONS/AFFILIATES:

Fanapt
Probuphine
ProNeura
Vanda Pharmaceutical Inc

CONTACTS: Note: Officers with more than one job title may be intentionally listed here more than once.

Sunil Bhonsle, Pres.
Kate Beebe, Sr. VP-Clinical Dev. & Medical Affairs
Sunil Bhonsle, Sec.
Marc Rubin, Exec. Chmn.

Phone: 650-244-4990	Fax: 650-244-4956
Toll-Free:	
Address: 400 Oyster Point Blvd., Ste. 505, South San Francisco, CA 94080 US	

GROWTH PLANS/SPECIAL FEATURES:

Titan Pharmaceuticals, Inc. is a biopharmaceutical company that develops proprietary therapeutics for the treatment of central nervous system disorders, cardiovascular disease, bone disease and other disorders. The firm's main product, Fanapt (also known as Iloperidone), which was developed in collaboration with Vanda Pharmaceutical, Inc., is an oral pill used to treat adult schizophrenia. The company is focused on clinical development of Probuphine, for the treatment of opioid dependence. Titan Pharmaceuticals utilizes the ProNeura technology in the Probuphine treatment. ProNeura is a drug delivery system that lasts approximately six months and is placed subcutaneously in the patient. ProNeura also has potential applications where conventional treatment is limited by poor patient compliance and variability in blood drug levels. The firm utilizes contract-manufacturing organizations to manufacture its products for pre-clinical studies and clinical trials. In May 2009, Fanapt received marketing approval from the U.S. FDA. In October 2009, Novartis Pharma AG acquired the U.S. and Canadian rights to co-market, commercialize and develop Fanapt. In June 2010, Titan announced that it has a patent for Probuphine.

FINANCIALS: Sales and profits are in thousands of dollars—add 000 to get the full amount. 2009 Note: Financial information for 2009 was not available for all companies at press time.

2009 Sales: $	2009 Profits: $	U.S. Stock Ticker: TTNP.PK	
2008 Sales: $	2008 Profits: $	Int'l Ticker: Int'l Exchange:	
2007 Sales: $ 24	2007 Profits: $-17,647	Employees:	
2006 Sales: $ 32	2006 Profits: $-15,737	Fiscal Year Ends: 12/31	
2005 Sales: $ 89	2005 Profits: $-22,462	Parent Company:	

SALARIES/BENEFITS:

Pension Plan:	ESOP Stock Plan:	Profit Sharing:	Top Exec. Salary: $402,487	Bonus: $
Savings Plan:	Stock Purch. Plan:		Second Exec. Salary: $384,326	Bonus: $

OTHER THOUGHTS:

Apparent Women Officers or Directors: 1
Hot Spot for Advancement for Women/Minorities:

LOCATIONS: ("Y" = Yes)

West:	Southwest:	Midwest:	Southeast:	Northeast:	International:
Y					

TOLMAR INC   www.tolmar.com

Industry Group Code: 325412 Ranks within this company's industry group: Sales: Profits:

Drugs:	Other:	Clinical:	Computers:	Services:
Discovery:	AgriBio:	Trials/Services:	Hardware:	Specialty Services:
Licensing:	Genetic Data:	Labs: Y	Software:	Consulting:
Manufacturing: Y	Tissue Replacement:	Equipment/Supplies:	Arrays:	Blood Collection:
Generics:		Research/Development Svcs.:	Database Management:	Drug Delivery:
		Diagnostics:		Drug Distribution:

TYPES OF BUSINESS:

Pharmaceuticals Manufacturing & Packaging
Generic Pharmaceutical Products-Dental, Oncology & Dermatology
Commercial Product Development
Contract Manufacturing

BRANDS/DIVISIONS/AFFILIATES:

ATRIDOX
ATRISORB
ATRISORB-D
Zila

CONTACTS: Note: Officers with more than one job title may be intentionally listed here more than once.

Mike Duncan, CEO

Phone: 970-212-4500	Fax:
Toll-Free: 877-986-5627	
Address: 701 Centre Ave., Fort Collins, CO 80526 US	

GROWTH PLANS/SPECIAL FEATURES:

TOLMAR Inc. is a pharmaceutical company that develops and manufactures proprietary and generic pharmaceutical products in the dental, oncology and dermatology therapeutic areas. TOLMAR has three operating segments: dental products; commercial product development; and contract manufacturing. The dental products segment offers three products: ATRIDOX, ATRISORB and ATRISORB-D. ATRIDOX (doxycycline hyclate) is used to treat periodontal disease. It is a locally applied antibiotic (LAA) that is placed below the gum line into periodontal pockets where bacteria exist. ATRISORB Free Flow Bioabsorbable (GTR) Barrier is a flowable gel that forms over a bone graft, creating a barrier at the Guided Tissue Regeneration (GTR) site that allows cell regeneration. ATRISORB-D Free Flow Bioabsorbable (GRT) Barrier contains doxycycline 4%, which provides a controlled release of doxycycline for 7 days and prevents bacterial colonization of the barrier. The commercial product development segment develops and manufactures generic topical dermatology products. Its products include creams, such as Lidocaine 2.5% and Mometasone Furoate Cream USP 0.1%; gels, such as Metronidazole Vaginal Gel USP, 0.75%; lotions, such as Mometasone Furoate Topical Solution USP, 0.17%; shampoos, such as Ketoconazole Shampoo, 2%; solutions, including Ciclopirox Topical Solution, 8%; and ointments, such as Mometasone Furoate Ointment USP, 0.1%. Lastly, the contract manufacturing segment's services include formulation and analytical development; primary package design and development; INA, ANDA and NDA CMC support; and Lyophilization cycle development. Zila, a division of TOLMAR, offers consumable home care products, acute care products, prevention tools and in-office diagnostics.

Employees are offered medical, dental and vision insurance; life and AD&D insurance; short- and long-term disability; a 401(k) plan; flexible spending accounts; education assistance; an employee assistance program; lactation rooms; health services; health and wellness discounts; travel assistance; and company sponsored events, including summer picnics, holiday parties, recreational leagues and philanthropic events.

FINANCIALS: Sales and profits are in thousands of dollars—add 000 to get the full amount. 2009 Note: Financial information for 2009 was not available for all companies at press time.

2009 Sales: $	2009 Profits: $	U.S. Stock Ticker: Private
2008 Sales: $	2008 Profits: $	Int'l Ticker: Int'l Exchange:
2007 Sales: $	2007 Profits: $	Employees:
2006 Sales: $	2006 Profits: $	Fiscal Year Ends:
2005 Sales: $	2005 Profits: $	Parent Company:

SALARIES/BENEFITS:

Pension Plan:	ESOP Stock Plan:	Profit Sharing:	Top Exec. Salary: $	Bonus: $
Savings Plan: Y	Stock Purch. Plan:		Second Exec. Salary: $	Bonus: $

OTHER THOUGHTS:

Apparent Women Officers or Directors:
Hot Spot for Advancement for Women/Minorities:

LOCATIONS: ("Y" = Yes)

West:	Southwest:	Midwest:	Southeast:	Northeast:	International:
Y		Y		Y	

TOYOBO CO LTD

www.toyobo.co.jp

Industry Group Code: 313 Ranks within this company's industry group: Sales: 1 Profits: 1

Drugs:		Other:	Clinical:	Computers:	Services:
Discovery:	Y	AgriBio:	Trials/Services:	Hardware:	Specialty Services:
Licensing:		Genetic Data:	Labs:	Software:	Consulting:
Manufacturing:	Y	Tissue Replacement:	Equipment/Supplies:	Arrays:	Blood Collection:
Generics:			Research/Development Svcs.:	Database Management:	Drug Delivery:
			Diagnostics:		Drug Distribution:

TYPES OF BUSINESS:

Textile & Fiber Manufacturing
Advanced Materials
Biomedical Products
Industrial Textiles
Plastics & Films
Engineering
Pharmaceuticals
Water Treatment Membranes

BRANDS/DIVISIONS/AFFILIATES:

Zylon
Dyneema
Tsunooga
Toyo Cloth Co., Ltd.
Miyuki Holdings Co., Ltd.
Toyobo Biologics, Inc.

CONTACTS: Note: Officers with more than one job title may be intentionally listed here more than once.

Ryuzo Sakamoto, COO
Ryuzo Sakakmoto, Pres.
Kenji Hayashi, Controlling Supervisor-Personnel & Labor Dept.
Kenji Hayashi, Controlling Supervisor-Gen. Admin. Dept.
Kenji Hayashi, Controlling Supervisor-Law Dept.
Masaaki Sekino, Controlling Supervisor-Bus. Dev. Planning Office
Fumishige Imamura, Dir.-Audit Dept. & Finance, Acct. & Control Dept.
Fumiaki Miyoshi, Head, Plastics Div./Gen. Mgr.-Tsuruga Center
Kazuo Kurita, Head-Bioscience & Medical Div.
Hiroyuki Kagawa, Head-Fibers & Textiles Div.
Kanji Aono, Head-Fine Chemicals Div.
Junji Tsumura, Chmn.
Masayuki Yoshikawa, Gen. Mgr.-Procurement Oper. Office

Phone: 81-6-6348-3111	Fax: 81-6-6348-3206
Toll-Free:	
Address: 2-8, Dojima Hama 2-chome, Kita-ku, Osaka, 530-8230 Japan	

GROWTH PLANS/SPECIAL FEATURES:

Toyobo Co. Ltd., founded as a textile business in 1882, has expanded into domains that utilize its core technologies of polymerization, modification, processing and biotechnology. Currently, the company operates in four segments: textiles; life science; films and functional polymers; and industrial materials. The textiles segment produces a range of fabric materials, functional textiles for sports clothing, underwear and uniforms, synthetic fibers and acrylic and acrylate fibers. The firm also operates trading companies in textiles. Toyobo's life science segment produces pharmaceuticals through contract manufacturing, including injections, pharmaceutical intermediaries and raw pharmaceuticals; bioproducts, including enzymes for diagnostics, diagnostic systems and reagents for research; medical membranes, equipment and devices, such as artificial kidney hollow fiber and anti-clotting materials; and water treatment membranes for seawater desalination. The films and functional polymers segment produces industrial films for LCD and optical use; synthetic paper; packaging films for food packaging, including PET, polyolefin and nylon; heat-shrink PET films; functional polymers, including engineering plastics, industrial adhesives and coatings, photosensitive printing plates, acrylate polymers and electronic materials; and also conducts rubber and coat processing. The industrial materials segment produces fibers for airbags and tire cords and a variety of air filter products, as well as high-performance fibers. These include Zylon, a fiber with high thermal resistance, and Dyneema, a super-strong fiber. In April 2009, the company developed a cut-resistant, high strength melt spinning polyethylene fiber called Tsunooga. In May 2009, Miyuki Holdings Co., Ltd., and Toyo Cloth Co. Ltd., became wholly owned subsidiaries of Toyobo. In February 2010, the firm agreed to form a joint venture in Saudi Arabia with ITOCHU Corporation and Arabian Company for Water & Power Development. In March 2010, the company announced plans to merge subsidiaries Toyobo Gene Analysis Co., Ltd., and Toyobo Biologics, Inc.

FINANCIALS: Sales and profits are in thousands of dollars—add 000 to get the full amount. 2009 Note: Financial information for 2009 was not available for all companies at press time.

2009 Sales: $4,083,660	2009 Profits: $-139,040	**U.S. Stock Ticker: TYOBY**
2008 Sales: $4,305,989	2008 Profits: $46,891	**Int'l Ticker: 3101** Int'l Exchange: Tokyo-TSE
2007 Sales: $3,614,282	2007 Profits: $114,121	Employees:
2006 Sales: $3,418,200	2006 Profits: $107,100	Fiscal Year Ends: 3/31
2005 Sales: $3,660,500	2005 Profits: $113,500	Parent Company:

SALARIES/BENEFITS:

Pension Plan:	ESOP Stock Plan:	Profit Sharing:	Top Exec. Salary: $	Bonus: $
Savings Plan:	Stock Purch. Plan:		Second Exec. Salary: $	Bonus: $

OTHER THOUGHTS:

Apparent Women Officers or Directors:
Hot Spot for Advancement for Women/Minorities:

LOCATIONS: ("Y" = Yes)

West:	Southwest:	Midwest:	Southeast:	Northeast:	International:
				Y	Y

Note: Financial information, benefits and other data can change quickly and may vary from those stated here.

TRIMERIS INC

www.trimeris.com

Industry Group Code: 325412 Ranks within this company's industry group: Sales: 113 Profits: 55

Drugs:		Other:	Clinical:	Computers:	Services:
Discovery:	Y	AgriBio:	Trials/Services:	Hardware:	Specialty Services:
Licensing:		Genetic Data:	Labs:	Software:	Consulting:
Manufacturing:		Tissue Replacement:	Equipment/Supplies:	Arrays:	Blood Collection:
Generics:			Research/Development Svcs.:	Database Management:	Drug Delivery:
			Diagnostics:		Drug Distribution:

TYPES OF BUSINESS:

Antivirals
HIV Drugs

BRANDS/DIVISIONS/AFFILIATES:

Fuzeon (T-20)
TRI-1144

CONTACTS: Note: Officers with more than one job title may be intentionally listed here more than once.

Martin A. Mattingly, CEO
Andrew L. Graham, CFO
Michael Alrutz, General Counsel
Andrew Graham, Corp. Sec.

Phone: 919-806-4682	Fax: 919-806-4770
Toll-Free:	
Address: 2530 Meridian Pkwy., Fl. 2, Durham, NC 27713 US	

GROWTH PLANS/SPECIAL FEATURES:

Trimeris, Inc. is a biopharmaceutical company primarily engaged in the development and commercialization of a class of antiviral drug treatments called fusion inhibitors. Fusion inhibitors impair viral fusion, a complex process by which viruses attach to, penetrate and infect host cells. The firm focuses on Fuzeon (T-20), an HIV fusion inhibitor that was developed in collaboration with F. Hoffman-La Roche, Ltd. (Roche). When used in combination with other anti-HIV drugs, Fuzeon has been shown to reduce the amount of HIV in the blood and increase the number of T-cells. The drug is approved for marketing by the FDA in combination with other anti-HIV drugs for the treatment of HIV-1 infection in treatment-experienced patients with evidence of HIV-1 replication despite ongoing anti-HIV therapy. It is also approved by the European Agency for the Evaluation of Medicinal Products in exceptional circumstances. Trimeris currently relies on Roche for the manufacture, sales, marketing and distribution of Fuzeon. Fuzeon is available through retail and specialty pharmacies across the U.S. and Canada. The company's other product, TRI-1144, is a next-generation fusion inhibitor peptide meant to suppress HIV by raising the genetic barrier to the development of resistance. TRI-1144 has completed a Phase I clinical trial. In October 2009, Trimeris entered into a merger agreement with Korea-based Arigene Co., Ltd. Under the terms of the agreement, Arigene agreed to acquire Trimeris for $81 million.

FINANCIALS: Sales and profits are in thousands of dollars—add 000 to get the full amount. 2009 Note: Financial information for 2009 was not available for all companies at press time.

2009 Sales: $15,180	2009 Profits: $12,296	U.S. Stock Ticker: TRMS
2008 Sales: $19,647	2008 Profits: $8,009	Int'l Ticker: Int'l Exchange:
2007 Sales: $47,870	2007 Profits: $27,425	Employees: 4
2006 Sales: $36,980	2006 Profits: $7,384	Fiscal Year Ends: 12/31
2005 Sales: $19,059	2005 Profits: $-8,106	Parent Company:

SALARIES/BENEFITS:

Pension Plan:	ESOP Stock Plan:	Profit Sharing:	Top Exec. Salary: $396,564	Bonus: $198,282
Savings Plan: Y	Stock Purch. Plan:		Second Exec. Salary: $180,264	Bonus: $54,079

OTHER THOUGHTS:

Apparent Women Officers or Directors:
Hot Spot for Advancement for Women/Minorities:

LOCATIONS: ("Y" = Yes)

West:	Southwest:	Midwest:	Southeast:	Northeast:	International:
				Y	

TRINITY BIOTECH PLC

www.trinitybiotech.com

Industry Group Code: 325413 Ranks within this company's industry group: Sales: 11   Profits: 9

Drugs:	Other:	Clinical:		Computers:	Services:
Discovery:	AgriBio:	Trials/Services:		Hardware:	Specialty Services:
Licensing:	Genetic Data:	Labs:		Software:	Consulting:
Manufacturing:	Tissue Replacement:	Equipment/Supplies:	Y	Arrays:	Blood Collection:
Generics:		Research/Development Svcs.:		Database Management:	Drug Delivery:
		Diagnostics:	Y		Drug Distribution:

TYPES OF BUSINESS:

Medical Diagnostics Products
Immunoassay Technology

BRANDS/DIVISIONS/AFFILIATES:

Trini
Destiny Max
ACE
Bile Acids
Lactate
Oxalate
UniGold HIV

CONTACTS: Note: Officers with more than one job title may be intentionally listed here more than once.

Ronan O'Caoimh, CEO
Rory Nealon, COO
Kevin Tansley, CFO
Kevin Tansley, Sec.
Ronan O'Caoimh, Chmn.

Phone: 353-1-276-9800	**Fax:** 353-1-276-9888
Toll-Free:	
Address: 1 Southern Cross, IDA Business Park, Bray, Ireland UK	

GROWTH PLANS/SPECIAL FEATURES:

Trinity Biotech plc develops, acquires, manufactures and markets medical diagnostic products for the clinical laboratory and point-of-care (POC) segments of the diagnostic market. Trinity has manufacturing sites in Ireland, Germany and the U.S. and sells over 500 products in 75 countries worldwide through its own sales force, as well as international distributors and strategic partners. The company's product offerings are used for testing and diagnosis in infectious diseases, sexually transmitted diseases, blood disorders and autoimmune disorders. The firm also provides raw materials to the life sciences industry. Trinity's clinical laboratory products can be divided into three lines, which include coagulation, infectious diseases and clinical chemistry. The coagulation product line includes test kits and instrumentation used in the detection of blood coagulation and clotting disorders. It offers its products under the Trini and Destiny Max brands. The infectious diseases product line includes a wide range of diagnostic kits for autoimmune diseases, including celiac, lupus and rheumatoid arthritis; hormonal imbalances; sexually transmitted diseases, including syphilis, chlamydia and herpes; intestinal infections; lung/bronchial infections; and cardiovascular diseases. Products are sold primarily in North America, Europe and Asia. Lastly, Trinity's clinical chemistry products have proven performance in the diagnosis of many disease states from liver and kidney disease to G6PDH deficiency, which is an indicator of hemolytic anemia. Its products include ACE, Bile Acids, Lactate and Oxalate. The firm also offers POC products which test for the presence of HIV antibodies. Trinity's principal product is UniGold HIV. In May 2010, the firm sold its coagulation business to the Stago Group for $90 million. The remaining portion of Trinity Biotech's coagulation operating segment is working to develop a new range of coagulation technologies for the point-of-care market in the near term.

FINANCIALS: Sales and profits are in thousands of dollars—add 000 to get the full amount. 2009 Note: Financial information for 2009 was not available for all companies at press time.

2009 Sales: $125,907	2009 Profits: $11,824	**U.S. Stock Ticker:** TRIB
2008 Sales: $140,139	2008 Profits: $-77,778	**Int'l Ticker:** TWU Int'l Exchange: Dublin-ISE
2007 Sales: $143,617	2007 Profits: $-35,372	Employees: 658
2006 Sales: $118,674	2006 Profits: $3,276	Fiscal Year Ends: 12/31
2005 Sales: $98,560	2005 Profits: $5,280	Parent Company:

SALARIES/BENEFITS:

Pension Plan:	ESOP Stock Plan:	Profit Sharing:	Top Exec. Salary: $	Bonus: $
Savings Plan:	Stock Purch. Plan:		Second Exec. Salary: $	Bonus: $

OTHER THOUGHTS:

Apparent Women Officers or Directors:
Hot Spot for Advancement for Women/Minorities:

LOCATIONS: ("Y" = Yes)

West:	Southwest:	Midwest:	Southeast:	Northeast:	International:
Y		Y		Y	Y

Note: Financial information, benefits and other data can change quickly and may vary from those stated here.

TRIPOS INTERNATIONAL

www.tripos.com

Industry Group Code: 511210D Ranks within this company's industry group: Sales: Profits:

Drugs:	Other:	Clinical:	Computers:		Services:	
Discovery:	AgriBio:	Trials/Services:	Hardware:		Specialty Services:	Y
Licensing:	Genetic Data:	Labs:	Software:	Y	Consulting:	Y
Manufacturing:	Tissue Replacement:	Equipment/Supplies:	Arrays:		Blood Collection:	
Generics:		Research/Development Svcs.:	Database Management:		Drug Delivery:	
		Diagnostics:			Drug Distribution:	

TYPES OF BUSINESS:

Biotech Software, Research Products & Services
Clinical Software
Consulting Services

BRANDS/DIVISIONS/AFFILIATES:

SYBYL
Benchware
Tripos Inc
Vector Capital
Certara Corporation
Pharsight Corp
Pantheon

CONTACTS: *Note: Officers with more than one job title may be intentionally listed here more than once.*

Jim Hopkins, CEO
Patrick Flanagan, COO
John D. Yingling, CFO
Diana O'Rourke, Dir.-Mktg.
Richard D. Cramer III, Chief Scientific Officer
Gregory B. Smith, CTO
William Cobert, VP-European Sales, Certara

Phone: 314-647-1099	**Fax:** 314-647-9241
Toll-Free: 800-323-2960	
Address: 1699 S. Hanley Rd., St. Louis, MO 63144-2319 US	

GROWTH PLANS/SPECIAL FEATURES:

Tripos International, formed from the Discovery Informatics business of Tripos, Inc., provides research products and services for biotechnology, pharmaceutical and other life science enterprises. Drug development companies use the firm's informatics technologies, including computational methods and software tools, to study products at a molecular level. The firm serves over 1,000 customers in 46 countries. The company's two primary products lines are the SYBYL suite of molecular modeling applications and Benchware software products. The SYBYL program offers complete computational chemistry and molecular modeling systems that are tailored to accelerate and ease the discovery process. It comes with its Standard Base, which offers tools and functions designed to optimize, visualize and compare the attributes of molecular models and structures. SYBYL also offers two optional base modules: MOLCAD, which creates graphical images that reveal the properties of molecules; and Advanced Computation, a fast systematic conformational searching algorithm. The SIBYL suite also includes additional programs that focus on ligands, receptors or protein structures to design new compounds and drugs. The Benchware line of products designed for laboratory scientists, digitally stores, searches and retrieves research information; supports chemical synthesis and testing decisions; and shares experimental results with project team members, including computational scientists. The firm's Pantheon software platform using cheminformatics allows scientists to organize, analyze, and visualize chemical and biological data. Additionally, Tripos International provides informatics services, including product support and training. Tripos is a subsidiary of Certara, which also owns Pharsight Corporation. Vector Capital owns Certara.

FINANCIALS: Sales and profits are in thousands of dollars—add 000 to get the full amount. 2009 Note: Financial information for 2009 was not available for all companies at press time.

2009 Sales: $	2009 Profits: $	**U.S. Stock Ticker: Subsidiary**
2008 Sales: $	2008 Profits: $	**Int'l Ticker:** Int'l Exchange:
2007 Sales: $	2007 Profits: $	Employees:
2006 Sales: $27,384	2006 Profits: $-38,593	Fiscal Year Ends: 12/31
2005 Sales: $27,981	2005 Profits: $-4,288	Parent Company: CERTARA CORPORATION

SALARIES/BENEFITS:

Pension Plan:	ESOP Stock Plan:	Profit Sharing:	Top Exec. Salary: $	Bonus: $
Savings Plan:	Stock Purch. Plan:		Second Exec. Salary: $	Bonus: $

OTHER THOUGHTS:

Apparent Women Officers or Directors: 1
Hot Spot for Advancement for Women/Minorities:

LOCATIONS: ("Y" = Yes)

West:	Southwest:	Midwest:	Southeast:	Northeast:	International:
		Y			Y

TRIUS THERAPUTICS INC

www.triusrx.com

Industry Group Code: 325412 Ranks within this company's industry group: Sales: 130   Profits: 120

Drugs:		Other:	Clinical:	Computers:	Services:
Discovery:	Y	AgriBio:	Trials/Services:	Hardware:	Specialty Services:
Licensing:		Genetic Data:	Labs:	Software:	Consulting:
Manufacturing:		Tissue Replacement:	Equipment/Supplies:	Arrays:	Blood Collection:
Generics:			Research/Development Svcs.:	Database Management:	Drug Delivery:
			Diagnostics:		Drug Distribution:

TYPES OF BUSINESS:

Drugs (Pharmaceuticals), Discovery & Manufacturing
Biodefense
Antibiotics

BRANDS/DIVISIONS/AFFILIATES:

FAST
SBDD
GyrB/ParE
Marine Natural Products

CONTACTS: Note: Officers with more than one job title may be intentionally listed here more than once.

Jeffrey Stein, CEO
Jeffrey Stein, Pres.
John P. Schmid, CFO
John Finn, Chief Science Officer
Neil Abdollahian, Sr. Dir.-Bus. Dev.
Karen E. Potts, VP-Regulatory Affairs
Karen Joy Shaw, Sr. VP-Biology
Philippe Prokocimer, Chief Medical Officer
Kenneth Bartizal, Chief Dev. Officer
David S. Kabakoff, Chmn.

Phone: 858-452-0370	Fax: 858-677-9975
Toll-Free:	
Address: 6310 Nancy Ridge Dr., Ste. 101, San Diego, CA 92121 US	

GROWTH PLANS/SPECIAL FEATURES:

Trius Therapeutics, Inc. is a biopharmaceutical company focused on the discovery, development and commercialization of innovative antibiotics for serious, life-threatening infections. Phase III clinical trials are currently proceeding for its torezolid phosphate (TR-701). TR-701 is an IV and orally administered antibiotic being tested for its usefulness in treatment of serious gram-positive bacterial infections, including those caused by methicillin-resistant Staphylococcus aureus. The drug is intended to treat acute bacterial skin and skin structure infections (ABSSSI). These infections involve deeper tissue than most antibiotics can adequately treat and often require surgical intervention. Further complicating traditional therapeutic responses to ABSSSI is the fact that they are frequently caused by underlying diseases such as, diabetes and systematic immunosuppression. In addition, Trius has two research platforms; focused antisense screening technology (FAST) and structure-based drug design (SBDD). The FAST program consists of engineered bacterial strains with antisense DNA fragments. The synthesis of these fragments can be controlled in order to inhibit production of targeted proteins. Trius has strains targeted at over 20 essential bacteria, selected for their likelihood of providing wide coverage antibacterial agents. The SBDD program is able to design drug compounds that bind to specific bacterial targets by identifying the structural patterns of the target enzymes that make up bacterial pathogens. The capabilities of these two research platforms have aided the company in developing two preclinical drug programs. The GyrB/ParE program is focused on developing novel antibiotics to treat serious gram-negative bacterial pathogen infections such as, respiratory tract, urinary tract and intra-abdominal infections. In April 2010, the company received a grant from an agency within the Department of Defense to support its Marine Natural Products program intended to develop antibiotics as bio-defense countermeasures. The company has filed a registration statement with the SEC for a proposed initial public offering of common stock.

FINANCIALS: Sales and profits are in thousands of dollars—add 000 to get the full amount. 2009 Note: Financial information for 2009 was not available for all companies at press time.

2009 Sales: $5,000	2009 Profits: $-22,700	**U.S. Stock Ticker: TSRX**
2008 Sales: $1,300	2008 Profits: $-20,800	**Int'l Ticker:** Int'l Exchange:
2007 Sales: $1,200	2007 Profits: $-8,700	Employees: 37
2006 Sales: $	2006 Profits: $	Fiscal Year Ends:
2005 Sales: $	2005 Profits: $	Parent Company:

SALARIES/BENEFITS:

Pension Plan:	ESOP Stock Plan:	Profit Sharing:	Top Exec. Salary: $315,000	Bonus: $
Savings Plan:	Stock Purch. Plan:		Second Exec. Salary: $305,000	Bonus: $

OTHER THOUGHTS:

Apparent Women Officers or Directors: 5
Hot Spot for Advancement for Women/Minorities: Y

LOCATIONS: ("Y" = Yes)

West:	Southwest:	Midwest:	Southeast:	Northeast:	International:
Y					

UCB SA

www.ucb-group.com

Industry Group Code: 325412 Ranks within this company's industry group: Sales: 26   Profits: 22

Drugs:		Other:	Clinical:		Computers:	Services:
Discovery:	Y	AgriBio:	Trials/Services:		Hardware:	Specialty Services:
Licensing:		Genetic Data:	Labs:		Software:	Consulting:
Manufacturing:	Y	Tissue Replacement:	Equipment/Supplies:	Y	Arrays:	Blood Collection:
Generics:			Research/Development Svcs.:		Database Management:	Drug Delivery:
			Diagnostics:			Drug Distribution:

TYPES OF BUSINESS:

Pharmaceuticals Development
Industrial Chemical Products
Allergy & Respiratory Treatments
Central Nervous System Disorder Treatments

BRANDS/DIVISIONS/AFFILIATES:

Keppra
Neupro
Cimzia
Lortab
Xyzal
Zyrtec
Cyclofluidic
Tussionex

CONTACTS: *Note: Officers with more than one job title may be intentionally listed here more than once.*

Roch Doliveux, CEO
Detlef Thielgen, CFO/Exec. VP
Fabrice Enderlin, Exec. VP-Human Resources
Iris Low-Friedrich, Exec. VP-Dev./Chief Medical Officer
Michele Antonelli, Exec. VP-Tech. Oper.
Bob Trainor, General Counsel/Exec. VP
Bill Robinson, Exec. VP-Global Oper.
Mark McDade, VP-Bus. Dev. & Corp. Strategy
Antje Witte, VP-Corp. Comm.
Antje Witte, VP-Investor Rel.
Michele Antonelli, Exec. VP-Quality Assurance
Karel Boone, Chmn.

Phone: 32-2-559-99-99	Fax: 32-2-599-99-00
Toll-Free:	
Address: Allee de la Recherche, 60, Brussels, 1070 Belgium	

GROWTH PLANS/SPECIAL FEATURES:

UCB S.A. is a Belgian biopharmaceutical firm with operations in more than 40 countries. The firm's focuses on severe diseases in the fields of the central nervous system (CNS), inflammation and oncology. Keppra, the firm's lead product, is used in both monotherapy and adjunctive therapy for the treatment of epilepsy. Another CNS product is Neupro, approved in Europe for the treatment of early-stage Parkinson's disease. UCB's anti-allergics include Xyzal, an anti-allergic designed to treat and prevent persistent rhinitis in children, characterized by severe and long-lasting allergic symptoms with a tendency to evolve towards allergic asthma; Cimzia, approved in the U.S. and Switzerland for the treatment of Crohn's disease; and Zyrtec, an antihistamine approved in over 100 countries for use in children from the age of six months. Other products include Nootropil, a cerebral function regulator used to treat adults and the elderly; Lortab, an analgesic approved in the U.S. for the relief of moderate to moderately severe pain; Tussionex, a 12-hour cough suppressant approved in the U.S.; Innovair, a combination therapy for asthma; Isoket, used to treat coronary heart disease; and BUP-4, a once-daily treatment for urinary incontinence.

FINANCIALS: Sales and profits are in thousands of dollars—add 000 to get the full amount. 2009 Note: Financial information for 2009 was not available for all companies at press time.

2009 Sales: $4,233,520	2009 Profits: $696,980	**U.S. Stock Ticker:**
2008 Sales: $4,860,380	2008 Profits: $-58,040	**Int'l Ticker: UCB** Int'l Exchange: Brussels-Euronext
2007 Sales: $4,894,120	2007 Profits: $295,590	Employees:
2006 Sales: $2,987,500	2006 Profits: $501,100	Fiscal Year Ends: 12/31
2005 Sales: $2,602,500	2005 Profits: $961,760	Parent Company:

SALARIES/BENEFITS:

Pension Plan: Y	ESOP Stock Plan:	Profit Sharing:	Top Exec. Salary: $	Bonus: $
Savings Plan:	Stock Purch. Plan:		Second Exec. Salary: $	Bonus: $

OTHER THOUGHTS:

Apparent Women Officers or Directors: 2
Hot Spot for Advancement for Women/Minorities:

LOCATIONS: ("Y" = Yes)

West:	Southwest:	Midwest:	Southeast:	Northeast:	International:
		Y	Y	Y	Y

UNIGENE LABORATORIES
www.unigene.com

Industry Group Code: 325412 Ranks within this company's industry group: Sales: 117   Profits: 104

Drugs:		Other:	Clinical:	Computers:	Services:	
Discovery:	Y	AgriBio:	Trials/Services:	Hardware:	Specialty Services:	
Licensing:	Y	Genetic Data:	Labs:	Software:	Consulting:	
Manufacturing:	Y	Tissue Replacement:	Equipment/Supplies:	Arrays:	Blood Collection:	
Generics:			Research/Development Svcs.:	Database Management:	Drug Delivery:	Y
			Diagnostics:		Drug Distribution:	

TYPES OF BUSINESS:
Peptides Research & Production
Drug Delivery Systems

BRANDS/DIVISIONS/AFFILIATES:
Fortical
Tarsa Therapeutics Inc
EnteriPep

CONTACTS: Note: Officers with more than one job title may be intentionally listed here more than once.
Ashleigh Palmer, CEO
Ashleigh Palmer, Pres.
Nozer M. Mehta, VP-Biological R&D
James P. Gilligan, VP-Prod. Dev.
Paul Shields, VP-Mfg. Oper.
William Steinhauer, VP-Finance
Ronald Levy, Exec. VP
Richard Levy, Chmn.

Phone: 973-265-1100	Fax: 973-335-0972
Toll-Free:	
Address: 81 Fulton St., Boonton, NJ 07005 US	

GROWTH PLANS/SPECIAL FEATURES:
Unigene Laboratories, Inc. is a biopharmaceutical company that focuses on the research, production and delivery of small proteins, or peptides, for medical use. The company uses its proprietary peptide manufacturing and delivery technology, branded EnteriPep for oral delivery. The firm's primary focus is on the development of salmon calcitonin and other peptide products, for the treatment of osteoporosis and other indications. Unigene has licensed worldwide rights to its manufacturing and delivery technologies for oral parathyroid hormone, which is used to regulate calcium and phosphate metabolism in the bones and kidneys, to GlaxoSmithKline. In the U.S., the company licensed its nasal salmon calcitonin product, trademarked Fortical, to Upsher-Smith Laboratories, Inc. Fortical, aimed at treating osteoporosis, was the company's first drug approved in the U.S. The firm licensed worldwide rights to its patented manufacturing technology for the production of calcitonin to Novartis Pharma AG. Unigene holds 13 U.S. patents. The firm is also developing site-directed bone-growth technology in conjunction with Yale University. In May 2009, the firm was issued its first patent for site-directed bone growth technology. In October 2009, Unigene sold its international rights (with the exception of China) to its oral calcitonin operations to Tarsa Therapeutics, Inc. In exchange, Unigene acquired roughly 26% ownership of Tarsa. In December 2009, the company agreed to layoff approximately 33% of its total workforce.

FINANCIALS: Sales and profits are in thousands of dollars—add 000 to get the full amount. 2009 Note: Financial information for 2009 was not available for all companies at press time.

2009 Sales: $12,792	2009 Profits: $-13,380	**U.S. Stock Ticker: UGNE**
2008 Sales: $19,229	2008 Profits: $-6,078	**Int'l Ticker:** Int'l Exchange:
2007 Sales: $20,423	2007 Profits: $-3,448	Employees: 72
2006 Sales: $6,059	2006 Profits: $-11,784	Fiscal Year Ends: 12/31
2005 Sales: $14,276	2005 Profits: $- 496	Parent Company:

SALARIES/BENEFITS:

Pension Plan:	ESOP Stock Plan:	Profit Sharing:	Top Exec. Salary: $345,000	Bonus: $
Savings Plan: Y	Stock Purch. Plan:		Second Exec. Salary: $295,000	Bonus: $

OTHER THOUGHTS:
Apparent Women Officers or Directors:
Hot Spot for Advancement for Women/Minorities:

LOCATIONS: ("Y" = Yes)

West:	Southwest:	Midwest:	Southeast:	Northeast:	International:
				Y	

UNITED THERAPEUTICS CORP

www.unither.com

Industry Group Code: 325412 **Ranks within this company's industry group:** Sales: 51 Profits: 52

Drugs:		Other:		Clinical:		Computers:		Services:	
Discovery:	Y	AgriBio:		Trials/Services:		Hardware:		Specialty Services:	
Licensing:		Genetic Data:		Labs:		Software:		Consulting:	
Manufacturing:	Y	Tissue Replacement:		Equipment/Supplies:	Y	Arrays:		Blood Collection:	
Generics:				Research/Development Svcs.:		Database Management:		Drug Delivery:	
				Diagnostics:				Drug Distribution:	

TYPES OF BUSINESS:

Cardiovascular, Cancer & Infectious Diseases Therapeutics
Dietary Supplements
Telecardiology Products

BRANDS/DIVISIONS/AFFILIATES:

Unither Pharma Inc
Medicomp Inc
Unither Biotech Inc
Unither Pharmaceuticals LLC
Unither Neuroscience Inc
Lung Rx LLC
Unither Telmed Ltd
LungRx Limited

CONTACTS: *Note: Officers with more than one job title may be intentionally listed here more than once.*

Martine Rothblatt, CEO
Roger Jeffs, COO
Roger Jeffs, Pres.
John Ferrari, CFO
Paul A. Mahon, General Counsel
Paul A. Mahon, Exec. VP-Strategic Planning
John Ferrari, Treas.
Dan Balda, Pres./COO-Medicomp, Inc.
Martine Rothblatt, Chmn.

Phone: 301-608-9292	**Fax:** 301-608-9291
Toll-Free:	
Address: 1110 Spring St., Silver Spring, MD 20910 US	

GROWTH PLANS/SPECIAL FEATURES:

United Therapeutics Corp. (UTC) is a biotechnology company focused on the development and commercialization of therapeutic products for patients with cancer, cardiovascular diseases and other infectious diseases. Remodulin, the company's lead product, is an FDA-approved, treprostinil-based compound designed for the treatment of pulmonary arterial hypertension (PAH) for patients with negative symptoms associated with exercise. UTC has commercial rights for Remodulin in most European Union countries, Israel, Australia, Mexico, Argentina and Peru. The firm manufactures the treprostinil-based compound in an inhaled form, TYVASO; while Adcirca is the tablet form. Both are used in the treatment of pulmonary arterial hypertension to improve exercise ability. In addition, the company has a joint venture with Toray Industries, Inc., to develop Beraprost-MR, a prostacyclin analog drug designed to treat cardiovascular diseases. Subsidiary Unither Pharma, Inc., markets the HeartBar line of products, which are arginine-enriched dietary supplements designed to help maintaining healthy circulatory function. Medicomp, Inc., another subsidiary of the company, manufactures and markets a variety of telecardiology services, including cardiac Holter monitoring, event monitoring and analysis and pacemaker monitoring. Lastly, Unither Biotech Inc. distributes and sells the company's products in Canada. The firm's other pharmaceutical subsidiaries include Unither Pharmaceuticals LLC.; Unither Neuroscience Inc.; Lung Rx, LLC; Unither Telmed, Ltd.; United Therapeutics Europe, Ltd.; Unither Therapeutik GmbH; Unither.com, Inc.; LungRx Limited; and Unither Virology, LLC. In February 2010, Lung Rx, a subsidiary of UTC, announced a development agreement with ImmuneWorks, Inc., to develop an oral solution for the treatment of Idiopathic Pulmonary Fibrosis (IPF) and Primary Graft Dysfunction (PGD) in patients receiving lung transplant.

FINANCIALS: **Sales and profits are in thousands of dollars—add 000 to get the full amount. 2009 Note: Financial information for 2009 was not available for all companies at press time.**

2009 Sales: $369,848	2009 Profits: $19,462	**U.S. Stock Ticker: UTHR**
2008 Sales: $281,497	2008 Profits: $-42,789	**Int'l Ticker:** Int'l Exchange:
2007 Sales: $210,943	2007 Profits: $19,859	Employees: 410
2006 Sales: $159,632	2006 Profits: $73,965	Fiscal Year Ends: 12/31
2005 Sales: $115,915	2005 Profits: $65,016	Parent Company:

SALARIES/BENEFITS:

Pension Plan:	ESOP Stock Plan:	Profit Sharing:	Top Exec. Salary: $798,203	Bonus: $440,125
Savings Plan: Y	Stock Purch. Plan:		Second Exec. Salary: $710,000	Bonus: $288,150

OTHER THOUGHTS:

Apparent Women Officers or Directors: 1
Hot Spot for Advancement for Women/Minorities:

LOCATIONS: ("Y" = Yes)

West:	Southwest:	Midwest:	Southeast:	Northeast:	International:
			Y	Y	Y

UNITED-GUARDIAN INC

www.u-g.com

Industry Group Code: 325412 Ranks within this company's industry group: Sales: 116 Profits: 48

Drugs:		Other:	Clinical:		Computers:		Services:	
Discovery:	Y	AgriBio:	Trials/Services:		Hardware:		Specialty Services:	
Licensing:		Genetic Data:	Labs:		Software:		Consulting:	
Manufacturing:	Y	Tissue Replacement:	Equipment/Supplies:		Arrays:		Blood Collection:	
Generics:			Research/Development Svcs.:	Y	Database Management:		Drug Delivery:	
			Diagnostics:				Drug Distribution:	

TYPES OF BUSINESS:

Cosmetic Ingredients & Pharmaceuticals
Personal & Healthcare Products
Organic Chemicals
Test Solutions
Indicators
Dyes
Reagents

BRANDS/DIVISIONS/AFFILIATES:

Guardian Laboratories
Orchid Complex
Razoride
Lubrajel
Renacidin
Clorpactin
Klensoft
Lubrasil

CONTACTS: *Note: Officers with more than one job title may be intentionally listed here more than once.*

Kenneth H. Globus, CEO
Kenneth H. Globus, Pres.
Robert S. Rubinger, CFO/Exec. VP
Joseph J. Vernice, VP/Mgr.-R&D
Joseph J. Vernice, Dir.-Tech. Svcs.
Robert S. Rubinger, Dir.-Prod. Dev./Sec.
Kenneth H. Globus, General Counsel
Charles W. Castanza, Sr. VP/Dir.-Plant Oper.
Cecile M. Brophy, Treas./Controller
Peter A. Hiltunen, VP/Prod. Mgr.
Kenneth H. Globus, Chmn.

Phone: 631-273-0900	Fax: 631-273-0858
Toll-Free: 800-645-5566	
Address: 230 Marcus Blvd., Hauppauge, NY 11788 US	

GROWTH PLANS/SPECIAL FEATURES:

United-Guardian, Inc., through Guardian Laboratories, conducts research, product development, manufacturing and marketing of cosmetic ingredients and other personal care products, pharmaceuticals, medical and health care products and specialty industrial products. Guardian Laboratories' two largest marketed product lines are the Lubrajel line of cosmetic ingredients and Renacidin Irrigation. The Lubrajel line of cosmetic ingredients are nondrying water-based moisturizing and lubricating gels that have applications in the cosmetic industry primarily as a moisturizer and as a base for other cosmetic ingredients, and in the medical field as a lubricant. The Lubrajel line accounted for roughly 78% of revenue in 2009. Renacidin Irrigation is a urological prescription drug, used primarily to prevent the formation of and to dissolve calcifications in catheters implanted in the urinary bladder. Other products include Clorpactin, a microbicidal product used primarily in urology and surgery as an antiseptic for treating a wide range of localized infections in the urinary bladder, the peritoneum, the abdominal cavity, the eye, ear, nose and throat, and the sinuses; Lubrasil, a form of Lubrajel into which silicone oil is incorporated; Lubrajel PF, a preservative-free form of Lubrajel; Unitwix, a cosmetic additive used as a thickener for oils and oil-based liquids; Orchid Complex, base for skin creams, lotions, cleansers, and other cosmetics; Razoride, a water-based shaving product; Klensoft, a surfactant that can be used in shampoos, shower gels, makeup removers and other cosmetic formulations; and Confetti Dermal Delivery Flakes, a product line that incorporates various functional oil-soluble ingredients into colorful flakes that can be added to, and suspended in, various water-based products. The company markets its products primarily through marketing partners, distributors, wholesalers, direct advertising, mailings and trade exhibitions.

FINANCIALS: Sales and profits are in thousands of dollars—add 000 to get the full amount. 2009 Note: Financial information for 2009 was not available for all companies at press time.

2009 Sales: $13,277	2009 Profits: $3,879	U.S. Stock Ticker: UG	
2008 Sales: $12,292	2008 Profits: $3,163	Int'l Ticker:	Int'l Exchange:
2007 Sales: $11,889	2007 Profits: $3,544	Employees: 37	
2006 Sales: $11,208	2006 Profits: $2,737	Fiscal Year Ends: 12/31	
2005 Sales: $12,135	2005 Profits: $2,617	Parent Company:	

SALARIES/BENEFITS:

Pension Plan:	ESOP Stock Plan:	Profit Sharing:	Top Exec. Salary: $246,779	Bonus: $68,200
Savings Plan: Y	Stock Purch. Plan:		Second Exec. Salary: $167,514	Bonus: $16,600

OTHER THOUGHTS:

Apparent Women Officers or Directors: 1
Hot Spot for Advancement for Women/Minorities:

LOCATIONS: ("Y" = Yes)

West:	Southwest:	Midwest:	Southeast:	Northeast:	International:
				Y	

URIGEN PHARMACEUTICALS INC

www.urigen.com

Industry Group Code: 325412 Ranks within this company's industry group: Sales: Profits: 79

Drugs:		Other:		Clinical:	Computers:		Services:	
Discovery:		AgriBio:		Trials/Services:	Hardware:		Specialty Services:	
Licensing:	Y	Genetic Data:		Labs:	Software:		Consulting:	
Manufacturing:	Y	Tissue Replacement:		Equipment/Supplies:	Arrays:		Blood Collection:	
Generics:				Research/Development Svcs.:	Database Management:		Drug Delivery:	
				Diagnostics:			Drug Distribution:	

TYPES OF BUSINESS:

Drugs-Urological

BRANDS/DIVISIONS/AFFILIATES:

URG101
URG301

CONTACTS: Note: Officers with more than one job title may be intentionally listed here more than once.

William J. Garner, CEO
William J. Garner, Pres.
Martin E. Shmagin, CFO
Lowell Parsons, Chief Medical Officer

Phone: 415-781-0350	Fax: 415-781-0385
Toll-Free:	
Address: 27 Maiden Ln., Ste. 595, San Francisco, CA 94108 US	

GROWTH PLANS/SPECIAL FEATURES:

Urigen Pharmaceuticals, Inc. specializes in the development and commercialization of products for the treatment and diagnosis of urological disorders. The firm focuses specifically on products for amelioration painful bladder syndrome/interstitial cystitis (PBS or PBS/IC), urethritis, nocturia and overactive bladder (OAB). Urigen has two clinical stage products: URG101, a bladder instillation for PBS/IC; and URG301, a female urethral suppository for urethritis and nocturia. URG101, the firm's leading product, is a combination of lidocaine, an FDA approved anesthetic, and heparin, an FDA approved anti-coagulant, which are delivered locally to the bladder for rapid relief of pain and urgency. It has undertaken two Phase II clinical trials for the treatment of painful bladder syndrome (PBS). URG301 has completed Phase I trials for both overactive bladder and urethritis. The product is also being evaluated as a treatment for Acute Urethral Discomfort (AUD). The treatment would be delivered as an intraurethral suppository, which would melt and distribute the drug to the urethral tissue, achieving a rapid anesthetic effect. Urigen is also developing an Investigational New Drug (IND) to initiate an exploratory study to evaluate the safety and efficacy of an intraurethral suppository to treat urethritis, nocturia and the symptoms of acute urinary urgency associated with overactive bladder. The company maintains strategic partnerships with Life Science Strategy Group and Kalium, Inc. In July 2010, Urigen announced its plans to conduct Phase IIb studies for URG101.

FINANCIALS: Sales and profits are in thousands of dollars—add 000 to get the full amount. 2009 Note: Financial information for 2009 was not available for all companies at press time.

2009 Sales: $	2009 Profits: $-2,257	**U.S. Stock Ticker:** URGP	
2008 Sales: $	2008 Profits: $-4,844	**Int'l Ticker:** Int'l Exchange:	
2007 Sales: $ 571	2007 Profits: $-3,733	Employees: 2	
2006 Sales: $ 727	2006 Profits: $-15,337	Fiscal Year Ends: 6/30	
2005 Sales: $2,177	2005 Profits: $-11,083	Parent Company:	

SALARIES/BENEFITS:

Pension Plan:	ESOP Stock Plan:	Profit Sharing:	Top Exec. Salary: $250,000	Bonus: $
Savings Plan:	Stock Purch. Plan:		Second Exec. Salary: $225,000	Bonus: $

OTHER THOUGHTS:

Apparent Women Officers or Directors: 1
Hot Spot for Advancement for Women/Minorities:

LOCATIONS: ("Y" = Yes)

West:	Southwest:	Midwest:	Southeast:	Northeast:	International:
Y					

VALEANT PHARMACEUTICALS INTERNATIONAL www.valeant.com

Industry Group Code: 325412 Ranks within this company's industry group: Sales: 43 Profits: 32

Drugs:		Other:		Clinical:	Computers:	Services:
Discovery:	Y	AgriBio:		Trials/Services:	Hardware:	Specialty Services:
Licensing:	Y	Genetic Data:		Labs:	Software:	Consulting:
Manufacturing:	Y	Tissue Replacement:		Equipment/Supplies:	Arrays:	Blood Collection:
Generics:				Research/Development Svcs.:	Database Management:	Drug Delivery:
				Diagnostics:		Drug Distribution:

TYPES OF BUSINESS:
Prescription & Non-Prescription Pharmaceuticals
Neurology Drugs
Dermatology Drugs
Infectious Diseases Drugs

BRANDS/DIVISIONS/AFFILIATES:
Efudex/Efudix
Dermatix
Oxsoralen-Ultra
Cloderm
Mestinon
Librax
Migranal
Tasmar

CONTACTS: *Note: Officers with more than one job title may be intentionally listed here more than once.*
J. Michael Pearson, CEO
Peter J. Blott, CFO/Exec. VP
Elisa A. Carlson, Chief Admin. Officer/Exec. VP
Steve T. Min, General Counsel/Exec. VP/Corp. Sec.
Rajiv De Silva, COO-Specialty Pharmaceuticals
J. Michael Pearson, Chmn.
Bhaskar Chaudhuri, Pres., Valeant Pharmaceuticals Int'l

Phone: 949-461-6000	Fax: 949-461-6609
Toll-Free: 800-548-5100	
Address: 1 Enterprise, Aliso Viejo, CA 92656 US	

GROWTH PLANS/SPECIAL FEATURES:
Valeant Pharmaceuticals International is a pharmaceutical company that develops, manufactures and markets a broad spectrum of prescription and non-prescription pharmaceuticals. The company focuses in the therapeutic areas of neurology, dermatology and infectious diseases. The firm's products also treat neuromuscular disorders, cancer, cardiovascular disease, diabetes and psychiatric disorders. Valeant's products are sold through three pharmaceutical segments: Specialty Pharmaceuticals, Branded Generics-Europe and Branded Generics-Latin America. The company's specialty pharmaceuticals segment includes product revenues primarily from the U.S., Canada and Australia. Its product portfolio comprises roughly 380 branded products with approximately 2,000 stock keeping units. Products in the neurology field include Mestinon, Librax, Migranal, Tasmar, Zelapar and Diastat AcuDial; in the dermatology field, Efudex/Efudix, Tetrix, Oxsoralen-Ultra and Cloderm. The Branded Generics-Europe segment includes revenues from products in Poland, Hungary, the Czech Republic and Slovakia. The Branded Generics-Latin America segment includes revenues from products in Mexico and Brazil. Valeant's research and development program focuses on preclinical and clinical development of identified molecules. The company is developing product candidates, including taribavirin and retigabine. In early 2009, the firm drastically cut its research budget from $100 million to $50 million. In October 2009, the company agreed to acquire several prescription and cosmetic dermatology products from a Poland-based specialty pharmaceutical company. In April 2010, the firm agreed to acquire a privately held pharmaceutical company based in Brazil for $56 million. In May 2010, the company acquired Anton Pharma, Inc, a New Jersey-based specialty pharmaceutical company. In June 2010, the company agreed to merge with Biovail Corporation, the combined company will retain the name Valeant Pharmaceuticals International, Inc.

FINANCIALS: Sales and profits are in thousands of dollars—add 000 to get the full amount. 2009 Note: Financial information for 2009 was not available for all companies at press time.

2009 Sales: $830,461	2009 Profits: $263,741	**U.S. Stock Ticker:** VRX
2008 Sales: $656,977	2008 Profits: $-23,714	**Int'l Ticker:** Int'l Exchange:
2007 Sales: $689,503	2007 Profits: $-6,186	Employees: 3,100
2006 Sales: $685,052	2006 Profits: $-57,568	Fiscal Year Ends: 12/31
2005 Sales: $823,886	2005 Profits: $-188,143	Parent Company:

SALARIES/BENEFITS:
Pension Plan:	ESOP Stock Plan:	Profit Sharing:	Top Exec. Salary: $1,000,000	Bonus: $3,000,000
Savings Plan: Y	Stock Purch. Plan:		Second Exec. Salary: $475,881	Bonus: $600,000

OTHER THOUGHTS:
Apparent Women Officers or Directors: 2
Hot Spot for Advancement for Women/Minorities: Y

LOCATIONS: ("Y" = Yes)
West:	Southwest:	Midwest:	Southeast:	Northeast:	International:
Y					Y

VAXGEN INC

www.vaxgen.com

Industry Group Code: 325412 Ranks within this company's industry group: Sales: Profits: 91

Drugs:		Other:	Clinical:		Computers:		Services:
Discovery:	Y	AgriBio:	Trials/Services:		Hardware:		Specialty Services:
Licensing:		Genetic Data:	Labs:		Software:		Consulting:
Manufacturing:		Tissue Replacement:	Equipment/Supplies:	Y	Arrays:		Blood Collection:
Generics:			Research/Development Svcs.:		Database Management:		Drug Delivery:
			Diagnostics:				Drug Distribution:

TYPES OF BUSINESS:

Biologic Products

BRANDS/DIVISIONS/AFFILIATES:

CONTACTS: *Note: Officers with more than one job title may be intentionally listed here more than once.*

James P. Panek, Pres.
James P. Panek, Chief Acct. Officer

Phone: 650-624-1000	**Fax:** 650-624-4785
Toll-Free:	
Address: 379 Oyster Point Blvd., Ste. 10, South San Francisco, CA 94080 US	

GROWTH PLANS/SPECIAL FEATURES:

VaxGen, Inc. is a biopharmaceutical company that historically developed, manufactured and commercialized biologic products for the prevention and treatment of human infectious diseases. VaxGen owns a facility in California with a 1,000-liter bioreactor that can be used to make cell culture or microbial biologic products and is capable of either microbial or mammalian fermentation. However, the company has ended all product development activities and sold or otherwise terminated its drug development programs. As a result of a proposed merger between VaxGen and Raven BioTechnologies, Inc., which was terminated, VaxGen recently underwent significant restructuring, which included reduction of workforce by 75%. VaxGen recently sold all of its assets and rights related to its anthrax vaccine candidate to Emergent BioSolutions, Inc. for $2 million with another possible $8 million in milestone payments. In May 2010, the company agreed to acquire diaDexus, a diagnostics company focused on the development and commercialization of patent-protected in vitro diagnostic products addressing cardiovascular disease.

FINANCIALS: Sales and profits are in thousands of dollars—add 000 to get the full amount. 2009 Note: Financial information for 2009 was not available for all companies at press time.

2009 Sales: $	2009 Profits: $-6,208	**U.S. Stock Ticker:** VXGN
2008 Sales: $ 293	2008 Profits: $-12,563	**Int'l Ticker:** Int'l Exchange:
2007 Sales: $5,011	2007 Profits: $-44,180	Employees: 3
2006 Sales: $14,836	2006 Profits: $37,592	Fiscal Year Ends: 12/31
2005 Sales: $29,939	2005 Profits: $-55,958	Parent Company:

SALARIES/BENEFITS:

Pension Plan:	ESOP Stock Plan:	Profit Sharing:	Top Exec. Salary: $211,250	Bonus: $
Savings Plan:	Stock Purch. Plan:		Second Exec. Salary: $160,417	Bonus: $

OTHER THOUGHTS:

Apparent Women Officers or Directors: 1
Hot Spot for Advancement for Women/Minorities:

LOCATIONS: ("Y" = Yes)

West:	Southwest:	Midwest:	Southeast:	Northeast:	International:
Y					

VENTRIA BIOSCIENCE

Industry Group Code: 325411 Ranks within this company's industry group: Sales: Profits:

Drugs:	Other:		Clinical:	Computers:	Services:
Discovery:	AgriBio:	Y	Trials/Services:	Hardware:	Specialty Services:
Licensing:	Genetic Data:		Labs:	Software:	Consulting:
Manufacturing: Y	Tissue Replacement:		Equipment/Supplies:	Arrays:	Blood Collection:
Generics:			Research/Development Svcs.:	Database Management:	Drug Delivery:
			Diagnostics:		Drug Distribution:

TYPES OF BUSINESS:

Medical Supplies, Manufacturing
Lactoferrin Manufacturing
Lysozyme Manufacturing

BRANDS/DIVISIONS/AFFILIATES:

ExpressPro
ExpressTec
ExpressMab
BioShare

CONTACTS: *Note: Officers with more than one job title may be intentionally listed here more than once.*

Scott E. Deeter, CEO
Scott E. Deeter, Pres.
Ning Huang, VP-R&D
Randy Semadeni, VP-Bus. Dev.
Randy Semadeni, VP-Finance
Greg Unruh, VP/Gen. Mgr.

Phone:	Fax: 970-472-0500
Toll-Free: 800 916-8311	
Address: P.O. Box 273330, Fort Collins, CO 80527 US	

GROWTH PLANS/SPECIAL FEATURES:

Ventria Bioscience focuses on human nutrition and human therapeutics products.   The company manufactures two products, lactoferrin and lysozyme. Lactoferrin is a globular multifunctional protein with antimicrobial activity. Lactoferrin protein has applications in the alleviation of fungal infections gastrointestinal health, dietary management of acute diarrhea and the treatment of topical infections and inflammations. Lysozyme is an enzyme that attacks the cell walls of many bacterias and has anti-bacterial, anti-viral and anti-fungal properties. Lysozyme enzyme has applications in the alleviation of fungal infections gastrointestinal health, dietary management of acute diarrhea and the treatment of topical infections and inflammations. These materials are generally found in human breast milk, as well as tears, nasogastric secretions, saliva and bronchial secretions. Ventria's ExpressTec platform uses self-pollinating crops specifically rice and barley, as the production host for these products. Ventria uses the ExpressTec production system as a basis for forming, specialized systems that produce specific molecules. These platforms include ExpressPro for production of proteins, ExpressTide for production of peptides and ExpressMab for production of monoclonal antibodies. The company has a program for researchers called BioShare, in which researchers submit requests for materials and granted access to a free supply of recombinant proteins and peptides.

Employees of the firm are offered health care, equity ownership, life insurance, long-term disability and a 401(K) plan.

FINANCIALS: Sales and profits are in thousands of dollars—add 000 to get the full amount. 2009 Note: Financial information for 2009 was not available for all companies at press time.

2009 Sales: $	2009 Profits: $	U.S. Stock Ticker: Private	
2008 Sales: $	2008 Profits: $	Int'l Ticker: Int'l Exchange:	
2007 Sales: $	2007 Profits: $	Employees:	
2006 Sales: $	2006 Profits: $	Fiscal Year Ends:	
2005 Sales: $	2005 Profits: $	Parent Company:	

SALARIES/BENEFITS:

Pension Plan:	ESOP Stock Plan:	Profit Sharing:	Top Exec. Salary: $	Bonus: $
Savings Plan: Y	Stock Purch. Plan:		Second Exec. Salary: $	Bonus: $

OTHER THOUGHTS:

Apparent Women Officers or Directors:
Hot Spot for Advancement for Women/Minorities:

LOCATIONS: ("Y" = Yes)

West:	Southwest:	Midwest:	Southeast:	Northeast:	International:
Y		Y			

VERENIUM CORPORATION

www.verenium.com

Industry Group Code: 541712 Ranks within this company's industry group: Sales: 8   Profits: 16

Drugs:	Other:		Clinical:	Computers:	Services:	
Discovery:	AgriBio:	Y	Trials/Services:	Hardware:	Specialty Services:	Y
Licensing:	Genetic Data:		Labs:	Software:	Consulting:	
Manufacturing:	Tissue Replacement:		Equipment/Supplies:	Arrays:	Blood Collection:	
Generics:			Research/Development Svcs.:	Database Management:	Drug Delivery:	
			Diagnostics:		Drug Distribution:	

TYPES OF BUSINESS:

Specialty Enzyme Production
Biofuels
Cellulosic Ethanol Technology

BRANDS/DIVISIONS/AFFILIATES:

Verenium Biofuels Corporation
Vercipia Biofuels
Fuelzyme-LF
Purifine
Luminase
Phyzyme XP
Cottonase

CONTACTS: *Note: Officers with more than one job title may be intentionally listed here more than once.*

Carlos A. Riva, CEO
Carlos A. Riva, Pres.
James Levine, CFO/Exec. VP
Nell Jones, Sr. VP-Human Resources
Nelson Barton, Sr. VP-R&D
J. Chris Terajewicz, Sr. VP-Eng. & Construction
Gerald M. Haines II, Chief Legal Officer/Sr. VP
Carey Buckles, VP-Oper.
William H. Baum, Exec. VP-Bus. Dev.
Kelly Lindenboom, VP-Corp. Comm.
Jeffrey G. Black, Chief Acct. Officer/Sr. VP
John B. Howe, VP-Public Affairs
Janet Roemer, Exec. VP-Specialty Enzymes Bus. Unit
James Cavanaugh, Chmn.

Phone: 617-674-5300	Fax:
Toll-Free:	
Address: 55 Cambridge Pkwy., Cambridge, MA 02142 US	

GROWTH PLANS/SPECIAL FEATURES:

Verenium Corporation is an alternative energy and bioenzymes company. The firm operates in two segments: Specialty Enzymes and Biofuels. It also operates a Research and Development business unit supporting both segments. The Specialty Enzymes segment manufactures chemicals for three target markets: alternative fuels, specialty industrial processes and animal health and nutrition. Its commercial enzymes for the biofuels market include Fuelzyme-LF, used to increase the efficiency of corn ethanol production, and Purifine, used to convert vegetable oils to biodiesel and oilseed into edible oils. For the industrial market, Luminase makes pulp more susceptible to bleaching and Cottonase is designed to provide better cleaning of cotton that traditional methods like chemical scouring. For the animal health & nutrition market, the Phzyme enzyme helps phosphorous and other nutrients in animal feed become more digestible. The Biofuels segment operates through the company's wholly-owned subsidiary, Verenium Biofuels Corporation, and focuses on cellulosic ethanol production, aiming to commercially produce ethanol fuel from almost any cellulose-based biomass, including agricultural waste, switch grass and wood pulp. It currently has one research and development ethanol plant in Jennings, Louisiana dedicated to improving Verenium's extraction process and extending it to new biomass materials; and one demonstration-scale plant, also in Jennings, that has the capacity to process 1.4 million gallons of cellulosic ethanol per year, at which the company works to test, refine and improve the processing technology. The firm partners with BP in a 50-50 joint-venture under the operating name Vercipia Biofuels in the development of a commercial-scale cellulosic ethanol production plant in Florida at a cost of about $300 million. In addition, BP invested $112.5 million in Verenium and received a 50% stake in Verenium's technology licensing business.

Employees are offered health insurance and stock options.

FINANCIALS: Sales and profits are in thousands of dollars—add 000 to get the full amount. 2009 Note: Financial information for 2009 was not available for all companies at press time.

2009 Sales: $65,911	2009 Profits: $-56,240	**U.S. Stock Ticker: VRNM**
2008 Sales: $69,659	2008 Profits: $-185,542	**Int'l Ticker:** Int'l Exchange:
2007 Sales: $46,273	2007 Profits: $-107,585	Employees: 270
2006 Sales: $49,198	2006 Profits: $-39,271	Fiscal Year Ends: 12/31
2005 Sales: $54,303	2005 Profits: $-89,718	Parent Company:

SALARIES/BENEFITS:

Pension Plan:	ESOP Stock Plan:	Profit Sharing:	Top Exec. Salary: $498,991	Bonus: $
Savings Plan: Y	Stock Purch. Plan: Y		Second Exec. Salary: $372,885	Bonus: $

OTHER THOUGHTS:

Apparent Women Officers or Directors: 3
Hot Spot for Advancement for Women/Minorities: Y

LOCATIONS: ("Y" = Yes)

West:	Southwest:	Midwest:	Southeast:	Northeast:	International:
Y			Y	Y	

VERMILLION INC

www.vermillion.com

Industry Group Code: 325413 Ranks within this company's industry group: Sales: Profits: 17

Drugs:	Other:		Clinical:		Computers:	Services:
Discovery:	AgriBio:		Trials/Services:		Hardware:	Specialty Services:
Licensing:	Genetic Data:	Y	Labs:	Y	Software:	Consulting:
Manufacturing:	Tissue Replacement:		Equipment/Supplies:	Y	Arrays:	Blood Collection:
Generics:			Research/Development Svcs.:	Y	Database Management:	Drug Delivery:
			Diagnostics:	Y		Drug Distribution:

TYPES OF BUSINESS:

Biomarkers
Translation Proteomics
Research Services
Specialty Diagnostics

BRANDS/DIVISIONS/AFFILIATES:

OVA1

CONTACTS: *Note: Officers with more than one job title may be intentionally listed here more than once.*

Gail S. Page, CEO
Sandra A. Gardiner, CFO/VP
William B. Creech, VP-Mktg. & Sales
Eric Fung, Chief Scientific Officer/Sr. VP
John H. Tran, VP-Finance/Chief Acct. Officer
Gail S. Page, Chmn.

Phone: 510-226-2800	Fax: 510-226-2801
Toll-Free:	
Address: 47350 Fremont Blvd., Fremont, CA 94538 US	

GROWTH PLANS/SPECIAL FEATURES:

Vermillion, Inc. discovers, develops and commercializes high-value diagnostic tests in the fields of oncology, hematology, cardiology and women's health.   Vermillion utilizes advanced protein separation methods to identify and resolve variants of specific biomarkers (known as translational proteomics) for developing a procedure to measure a property or concentration of an analyte (known as an assay) and commercializing novel diagnostic tests. It will also address clinical questions related to early disease detection, treatment response, monitoring of disease progression, prognosis and others through collaborations with leading academic and research institutions in addition to a strategic alliance agreement with Quest.   OVA1, the company's lead product, is an ovarian tumor triage test that helps to identify women who are at right high risk of having a malignant ovarian tumor, prior to them having surgery. It is a qualitative serum test that utilizes five biomarkers and proprietary FDA-cleared software to determine the likelihood of malignancy in women with a pelvic mass for whom surgery is planned. Academic and research institutions that Vermillion has collaborated with include The Johns Hopkins University School of Medicine; The University of Texas M.D. Anderson Cancer Center; University College London; The University of Texas Medical Branch; The Katholieke Universiteit Leuven; The Ohio State University Research Foundation; and Stanford University. The company has one diagnostic product commercially available for thrombotic thrombocytopenic purpura, a rare disorder of the blood coagulation system that is potentially life-threatening if left untreated.   In March 2009, Vermillion filed for Chapter 11 bankruptcy protection, from which it emerged in January 2010. In March 2010, the company announced Medicare coverage for its OVA1 test.

Vermillion offers its employees medical, dental, life, short-term and long-term disability insurance as well as flexible spending accounts.

FINANCIALS: Sales and profits are in thousands of dollars—add 000 to get the full amount. 2009 Note: Financial information for 2009 was not available for all companies at press time.

2009 Sales: $	2009 Profits: $-22,048	U.S. Stock Ticker: VRML
2008 Sales: $ 124	2008 Profits: $-18,330	Int'l Ticker: Int'l Exchange:
2007 Sales: $ 44	2007 Profits: $-21,282	Employees: 13
2006 Sales: $18,215	2006 Profits: $-22,066	Fiscal Year Ends: 12/31
2005 Sales: $27,246	2005 Profits: $-35,433	Parent Company:

SALARIES/BENEFITS:

Pension Plan:	ESOP Stock Plan:	Profit Sharing:	Top Exec. Salary: $	Bonus: $
Savings Plan: Y	Stock Purch. Plan: Y		Second Exec. Salary: $	Bonus: $

OTHER THOUGHTS:

Apparent Women Officers or Directors:
Hot Spot for Advancement for Women/Minorities:

LOCATIONS: ("Y" = Yes)

West:	Southwest:	Midwest:	Southeast:	Northeast:	International:
Y					

VERNALIS PLC

www.vernalis.com

Industry Group Code: 325412 Ranks within this company's industry group: Sales: 107   Profits: 115

Drugs:		Other:	Clinical:	Computers:	Services:	
Discovery:	Y	AgriBio:	Trials/Services:	Hardware:	Specialty Services:	Y
Licensing:	Y	Genetic Data:	Labs:	Software:	Consulting:	
Manufacturing:		Tissue Replacement:	Equipment/Supplies:	Arrays:	Blood Collection:	
Generics:			Research/Development Svcs.:	Database Management:	Drug Delivery:	
			Diagnostics:		Drug Distribution:	Y

TYPES OF BUSINESS:

Drugs-Neurology & Acute Pain
Drugs-Parkinson's Disease
Drugs-Migraine
Drugs-Obesity
Drugs-Oncology

BRANDS/DIVISIONS/AFFILIATES:

Frovatriptan
Menarini

CONTACTS: *Note: Officers with more than one job title may be intentionally listed here more than once.*

Ian Garland, CEO
John Slater, COO
David Mackney, CFO
Mike Wood, Dir.-Research
Alison Hood, General Counsel/Sec.
Nerida Scott, Dir.-Bus. Dev.
Steve Pawsey, Dir.-Dev.
Peter Fellner, Chmn.

Phone: 440-118-977-3133	Fax: 440-118-989-9300
Toll-Free:	
Address: Oakdene Ct., 613 Reading Rd., Winnersh, RG41 5UA UK	

GROWTH PLANS/SPECIAL FEATURES:

Vernalis plc discovers, researches and develops new compounds for central nervous system disorders. The company's three developmental franchises are for pain, including the treatment of migraines, neurological disorders, such as Parkinson's disease and cancer treatment. The firm actively seeks partnerships with pharmaceutical companies to complete the development and marketing of its compounds. Vernalis currently has one marketed product and 11 products in clinical development. Vernalis' marketed drug is frovatriptan, intended for the oral treatment of migraines. Frovatriptan has been approved in the U.S., Canada, Europe, seven Central American countries and South Korea. Included in the firm's products in Phase II clinical development are V2006, a drug hoped to aid in the restoration of motor function in patients with Parkinson's disease. Vernalis is investigating the potential use of certain receptor antagonists to treat Parkinson's disease and obesity at the level of the nervous system. The firm's Oncology program is working in collaboration with Novartis on Hsp90 inhibitors. Currently there are two products in its Oncology program entering clinical development; AUY922 is an intravenous Hsp90 inhibitor, which could possibly target a variety of solid and haematological cancers and a follow-up to AUY992, known as HSP990, which can be orally administered. Newer drugs in Vernalis' portfolio which have yet to make it into clinical development include; V158411, a Chk1 inhibitor used in conjunction with other cytotoxic drugs and V158866, a FAAH enzyme inhibitor that can possibly reduce the sensation of both pain and inflammation. In March 2010, the firm regained rights to 100% of the royalties receivable from European frovatriptan marketer Menarini. Resulting from the transaction, Vernalis now retains 25.25% of the royalties on Menarini's sales of frovatriptan.

FINANCIALS: Sales and profits are in thousands of dollars—add 000 to get the full amount. 2009 Note: Financial information for 2009 was not available for all companies at press time.

2009 Sales: $20,720	2009 Profits: $-18,810	**U.S. Stock Ticker:**	
2008 Sales: $89,900	2008 Profits: $- 100	**Int'l Ticker: VER** Int'l Exchange: London-LSE	
2007 Sales: $32,500	2007 Profits: $-74,500	Employees: 84	
2006 Sales: $20,400	2006 Profits: $-69,600	Fiscal Year Ends: 12/31	
2005 Sales: $28,000	2005 Profits: $-65,000	Parent Company:	

SALARIES/BENEFITS:

Pension Plan:	ESOP Stock Plan:	Profit Sharing:	Top Exec. Salary: $511,555	Bonus: $369,297
Savings Plan:	Stock Purch. Plan:		Second Exec. Salary: $343,570	Bonus: $

OTHER THOUGHTS:

Apparent Women Officers or Directors: 2
Hot Spot for Advancement for Women/Minorities: Y

LOCATIONS: ("Y" = Yes)

West:	Southwest:	Midwest:	Southeast:	Northeast:	International:
					Y

VERTEX PHARMACEUTICALS INC

www.vrtx.com

Industry Group Code: 325412 Ranks within this company's industry group: Sales: 73 Profits: 169

Drugs:		Other:	Clinical:	Computers:	Services:
Discovery:	Y	AgriBio:	Trials/Services:	Hardware:	Specialty Services:
Licensing:	Y	Genetic Data:	Labs:	Software:	Consulting:
Manufacturing:		Tissue Replacement:	Equipment/Supplies:	Arrays:	Blood Collection:
Generics:			Research/Development Svcs.:	Database Management:	Drug Delivery:
			Diagnostics:		Drug Distribution:

TYPES OF BUSINESS:

Small Molecule Drugs

BRANDS/DIVISIONS/AFFILIATES:

Telaprevir
Lexiva
Telzir
VX-770
VX-809
VX-509
VX-765

CONTACTS: Note: Officers with more than one job title may be intentionally listed here more than once.

Matthew W. Emmens, CEO
Matthew W. Emmens, Pres.
Ian F. Smith, CFO/Exec. VP
Lisa Kelly-Croswell, Sr. VP-Human Resources
Peter Mueller, Chief Scientific Officer
Kenneth S. Boger, General Counsel/Sr. VP
Kurt C. Graves, Exec. VP/Head-Strategic Dev.
Amit K. Sachdev, Sr. VP-Corp. Affairs & Public Policy
Peter Mueller, Exec. VP-Global R&D
Nancy J. Wysenski, Chief Commercial Officer/Exec. VP
Matthew W. Emmens, Chmn.

Phone: 617-444-6100	Fax: 617-444-6680
Toll-Free:	
Address: 130 Waverly St., Cambridge, MA 02139 US	

GROWTH PLANS/SPECIAL FEATURES:

Vertex Pharmaceuticals, Inc. discovers, develops and commercializes small molecule drugs for the treatment of serious diseases. Telaprevir, the firm's lead drug candidate, is an oral hepatitis C virus (HCV) protease inhibitor. It is being evaluated in a registration program focused on treatment-naïve and treatment-failure patients with genotype 1 HCV infection. The firm is also engaged in a number of other clinical development programs. VX-770, the lead drug candidate in the firm's cystic fibrosis (CF) program, is being evaluated in a registration program that focuses on patients with CF who have the G551D mutation in the gene responsible for CF. Vertex is also planning on beginning a number of Phase 2a clinical trials of its earlier-stage drug candidates. These clinical trials consist of a planned clinical trial that will evaluate telaprevir in combination with the HCV polymerase inhibitor VX-222; a planned clinical trial of VX-809 in combination with VX-770 in patients with the most common mutation in the gene responsible for CF; a clinical trial of VX-509 in patients with moderate to severe rheumatoid arthritis; and a clinical trial of VX-765 in patients with treatment-resistant epilepsy. Vertex's pipeline includes several drug candidates that are being developed by its collaborators. Collaborations include Cystic Fibrosis Foundation Therapeutics, Inc., with which it is developing VX-770; GlaxoSmithKline, for the development and commercialization of Lexiva/Telzir; and Janssen Pharmaceutica, N.V. and Mitsubishi Tanabe Pharma Corporation for the development of VX-950.

Employee benefits at Vertex include four weeks of vacation from the first day of employment; an employee stock ownership plan at a 15% discounted rate; a discount entertainment program, including discounts for movie theaters, movie rentals, Broadway theater shows, theme parks, ski tickets and special family events; and Living Well at Vertex, an employee health and wellness program.

FINANCIALS: Sales and profits are in thousands of dollars—add 000 to get the full amount. 2009 Note: Financial information for 2009 was not available for all companies at press time.

2009 Sales: $101,889	2009 Profits: $-642,178	U.S. Stock Ticker: VRTX
2008 Sales: $175,504	2008 Profits: $-459,851	Int'l Ticker: Int'l Exchange:
2007 Sales: $199,012	2007 Profits: $-391,279	Employees: 1,432
2006 Sales: $216,356	2006 Profits: $-206,891	Fiscal Year Ends: 12/31
2005 Sales: $160,890	2005 Profits: $-203,417	Parent Company:

SALARIES/BENEFITS:

Pension Plan:	ESOP Stock Plan:	Profit Sharing:	Top Exec. Salary: $1,002,693	Bonus: $2,846,251
Savings Plan: Y	Stock Purch. Plan: Y		Second Exec. Salary: $538,029	Bonus: $495,000

OTHER THOUGHTS:

Apparent Women Officers or Directors: 3
Hot Spot for Advancement for Women/Minorities: Y

LOCATIONS: ("Y" = Yes)

West:	Southwest:	Midwest:	Southeast:	Northeast:	International:
Y				Y	Y

VIACORD INC
www.viacord.com

Industry Group Code: 325414 Ranks within this company's industry group: Sales: Profits:

Drugs:	Other:	Clinical:	Computers:	Services:	
Discovery:	AgriBio:	Trials/Services:	Hardware:	Specialty Services:	Y
Licensing:	Genetic Data:	Labs: Y	Software:	Consulting:	
Manufacturing:	Tissue Replacement: Y	Equipment/Supplies:	Arrays:	Blood Collection:	Y
Generics:		Research/Development Svcs.:	Database Management:	Drug Delivery:	
		Diagnostics:		Drug Distribution:	

TYPES OF BUSINESS:
Human Cells-Based Medicine
Stem Cell Research

BRANDS/DIVISIONS/AFFILIATES:
ViaCord
ViaCord Research Institute
PerkinElmer Inc

CONTACTS: Note: Officers with more than one job title may be intentionally listed here more than once.
Jim Corbett, Pres.
Nadia Altomare, VP-Sales & Service
Karen Foster, VP-ViaCord Processing Laboratory
Morey Kraus, CTO
Christina Willwerth, VP-Prod. Dev.
Nadia Altomare, VP-Oper.
Mary T. Thistle, Sr. VP-Bus. Dev.
Christopher Stump, VP-Sales & Training
Karen Nichols, VP-Regulatory Affairs & Quality Systems

Phone: 617-914-3900	Fax: 866-565-2243
Toll-Free: 866-668-4895	
Address: 245 First St., 15th Fl., Cambridge, MA 02142 US	

GROWTH PLANS/SPECIAL FEATURES:

ViaCord, Inc., formerly ViaCell, Inc., is a leading provider of neonatal screening systems. PerkinElmer, a testing and diagnostic provider, owns the company. The firm's sole offering is the ViaCord cord blood-banking product, which includes the collection, testing, processing and preserving of umbilical cord blood. Currently, ViaCord stores approximately 145,000 cord blood units for its customers. Additional services are offered when a customer requests cord blood banking. These services include: comprehensive testing, which is conducted at the firm's laboratory and includes several tests that are necessary in the event the unit is ever needed for transplant purposes; processing, which is also done at the firm's laboratory and is designed to maximize the number of stem cells preserved; cryopreservation, which consists of the firm freezing the cord blood unit and storing it in liquid nitrogen; and provision of collection kits, which includes all required items for the collection of the newborn umbilical cord blood at the time of birth. In addition, ViaCord provides genetic screening on newborns to determine the risk of metabolic or other inherited disorder. Stem cells from umbilical cord blood are a potential treatment option for more than 80 diseases, including certain blood cancers and genetic diseases. The company also operates the ViaCord Research Institute and concentrates on five primary areas of research: cord blood technologies, emerging stem cell therapies, genetic screening, therapeutic development and related transplants. Recently, the firm discontinued its lead drug candidate, ViaCyte and sold the naming rights to Novocell, Inc.

FINANCIALS: Sales and profits are in thousands of dollars—add 000 to get the full amount. 2009 Note: Financial information for 2009 was not available for all companies at press time.

2009 Sales: $	2009 Profits: $	U.S. Stock Ticker: Private
2008 Sales: $	2008 Profits: $	Int'l Ticker: Int'l Exchange:
2007 Sales: $	2007 Profits: $	Employees:
2006 Sales: $54,426	2006 Profits: $-21,330	Fiscal Year Ends: 12/31
2005 Sales: $44,443	2005 Profits: $-14,677	Parent Company: PERKINELMER INC

SALARIES/BENEFITS:

Pension Plan:	ESOP Stock Plan:	Profit Sharing:	Top Exec. Salary: $	Bonus: $
Savings Plan:	Stock Purch. Plan:		Second Exec. Salary: $	Bonus: $

OTHER THOUGHTS:
Apparent Women Officers or Directors: 5
Hot Spot for Advancement for Women/Minorities: Y

LOCATIONS: ("Y" = Yes)

West:	Southwest:	Midwest:	Southeast:	Northeast:	International:
		Y		Y	

VICAL INC

www.vical.com

Industry Group Code: 325412 **Ranks within this company's industry group:** Sales: 118   Profits: 132

Drugs:		Other:		Clinical:	Computers:	Services:
Discovery:	Y	AgriBio:		Trials/Services:	Hardware:	Specialty Services:
Licensing:	Y	Genetic Data:		Labs:	Software:	Consulting:
Manufacturing:		Tissue Replacement:		Equipment/Supplies:	Arrays:	Blood Collection:
Generics:				Research/Development Svcs.:	Database Management:	Drug Delivery:
				Diagnostics:		Drug Distribution:

TYPES OF BUSINESS:
Drug Delivery Systems
Cancer, Infectious Disease & Metabolic Disease Drugs & Vaccines

BRANDS/DIVISIONS/AFFILIATES:
Allovectin-7
Vaxfectin
CyMVectin
TransVax

CONTACTS: *Note: Officers with more than one job title may be intentionally listed here more than once.*
Vijay B. Samant, CEO
Vijay B. Samant, Pres.
Jill M. Broadfoot, CFO/Sr. VP
Alain P. Rolland, Exec. VP-Prod. Dev.
Kevin R. Bracken, VP-Mfg.
Jill M. Broadfoot, Sec.
Alan Engbring, Exec. Dir.-Investor Rel.
Larry R. Smith, VP-Vaccine Research
Richard T. Kenney, P-Clinical Dev.
R. Gordon Douglas, Chmn.

Phone: 858-646-1100	Fax: 858-646-1150
Toll-Free:	
Address: 10390 Pacific Ctr. Ct., San Diego, CA 92121-4340 US	

GROWTH PLANS/SPECIAL FEATURES:
Vical, Inc. researches and develops biopharmaceutical products based on its patented DNA delivery technologies for the prevention and treatment of serious or life-threatening diseases. The company's research areas include vaccines for use in high-risk population for infectious disease targets; vaccines for general pediatric, adolescent and adult populations for infectious disease applications; and cancer vaccines or immunotherapies. The firm has five active independent development programs in the areas of infectious disease and cancer, which include a Phase III clinical trial using the Allovectin-7 immunotherapeutic in patients with metastatic melanoma; a Phase II clinical trial using the TransVax cytomegalovirus DNA vaccine in hematopoietic cell transplant patients; a Phase I clinical trial of a pandemic influenza DNA vaccine candidate using the proprietary Vaxfectin as an adjuvant; CyMVectin, a preclinical stage candidate to prevent cytomegalovirus infection by fetal transmission prior to or in the course of a pregnancy; and a preclinical program using the H1N1 pandemic influenza DNA vaccine formulated with Vaxfectin. Vical licenses its technologies to companies including Merck & Co., Inc.; the Sanofi-Aventis Groups; and AnGes. The firm also has two veterinary vaccine studies, licensed to Merial, Ltd. and Aqua Health, Ltd., for infectious diseases and cancer in household animals.

FINANCIALS: Sales and profits are in thousands of dollars—add 000 to get the full amount. 2009 Note: Financial information for 2009 was not available for all companies at press time.
2009 Sales: $12,686	2009 Profits: $-28,558	U.S. Stock Ticker: VICL
2008 Sales: $7,956	2008 Profits: $-36,896	Int'l Ticker: Int'l Exchange:
2007 Sales: $5,512	2007 Profits: $-35,894	Employees: 113
2006 Sales: $14,740	2006 Profits: $-23,148	Fiscal Year Ends: 12/31
2005 Sales: $12,003	2005 Profits: $-24,357	Parent Company:

SALARIES/BENEFITS:
Pension Plan:	ESOP Stock Plan:	Profit Sharing:	Top Exec. Salary: $470,000	Bonus: $250,000
Savings Plan: Y	Stock Purch. Plan:		Second Exec. Salary: $306,000	Bonus: $120,000

OTHER THOUGHTS:
Apparent Women Officers or Directors: 1
Hot Spot for Advancement for Women/Minorities:

LOCATIONS: ("Y" = Yes)
West:	Southwest:	Midwest:	Southeast:	Northeast:	International:
Y					

VIRBAC CORP

www.virbaccorp.com

Industry Group Code: 325412B Ranks within this company's industry group: Sales: Profits:

Drugs:		Other:	Clinical:	Computers:	Services:
Discovery:	Y	AgriBio:	Trials/Services:	Hardware:	Specialty Services:
Licensing:		Genetic Data:	Labs:	Software:	Consulting:
Manufacturing:	Y	Tissue Replacement:	Equipment/Supplies:	Arrays:	Blood Collection:
Generics:			Research/Development Svcs.:	Database Management:	Drug Delivery:
			Diagnostics:		Drug Distribution:

TYPES OF BUSINESS:
Drugs-Animal Health & Pet Care

BRANDS/DIVISIONS/AFFILIATES:
C.E.T.
IVERHART MAX
Preventic
Soloxine
Pancrezyme
Novifit NoviSAMe
ResiKetoChlor Leave-On Lotion
Anxitan Chewable Tablets

CONTACTS: Note: Officers with more than one job title may be intentionally listed here more than once.
Erik R. Martinez, CEO
Erik R. Martinez, Pres.
Christo White, CFO/VP
Dena Ware, Mgr.-Mktg.
George Stanly, Dir.-IT
Eric Maree, Chmn.

Phone: 817-831-5030	Fax: 817-831-8327
Toll-Free: 800-338-3659	
Address: 3200 Meacham Blvd., Fort Worth, TX 76137 US	

GROWTH PLANS/SPECIAL FEATURES:
Virbac Corporation, a subsidiary of Virbac SA, develops, manufactures and markets a variety of pet and companion animal health products. The firm specializes in the heartworm, tick/flea, dermatology, antibiotics, endocrinology and oral hygiene markets. The company's dermatology line markets treatments for allergic dermatitis, keratoseborrheic disorders, infectious dermatitis and otitis externa. The firm's leading brands include the C.E.T. line of at home pet oral hygiene products, which are designed specifically for the mouth of a dog or cat; IVERHART MAX, a medication that protects against heartworm disease and roundworm, hookworm and tapeworm infections; and Preventic, a line of products that prevent flea and tick infestation, including tick collars, shampoo, dips and environmental parasite control sprays and treatments for the house and yard. The firm also sells a line of endocrinology and urology products including Soloxine, sodium tablets, for management of canine hypothyroidism; Tumil-K tablets or gel, for potassium deficiencies; Pancrezyme tablets and powder, for the treatment for exocrine pancreatic insufficiency; Uroeze tablets or powder and Ammonil tablets, for proper urinary pH; Biomox for soft tissue infections; Rebound OES, a rehydrating solution; Novifit NoviSAMe, developed to enhance brain activity in aging pets; C.E.T VeggieDent dental chews; Anxitan Chewable Tablets for cat and dog anxiety; and ResiKetoChlor Leave-On Lotion, which treats dermatological infections. Additional products offered by Virbac include dewormers for dogs and natural fibers to ease constipation in dogs and cats. In addition to veterinary products, the company operates the Virbac University online web site, an interactive resource for veterinarians and their staff members. Virbac University provides up-to-date technical and treatment information in four major areas and provides the opportunity to earn CE Credits and Certification. Virbac also operates a Canadian branch in St Lazare, Quebec. In June 2009, Virbac and Orion Corporation agreed to develop a cardiovascular disease product for older dogs.

FINANCIALS: Sales and profits are in thousands of dollars—add 000 to get the full amount. 2009 Note: Financial information for 2009 was not available for all companies at press time.

2009 Sales: $	2009 Profits: $	U.S. Stock Ticker: Subsidiary
2008 Sales: $	2008 Profits: $	Int'l Ticker: Int'l Exchange:
2007 Sales: $80,800	2007 Profits: $	Employees:
2006 Sales: $	2006 Profits: $	Fiscal Year Ends: 10/31
2005 Sales: $80,778	2005 Profits: $3,873	Parent Company: VIRBAC SA

SALARIES/BENEFITS:

Pension Plan:	ESOP Stock Plan:	Profit Sharing:	Top Exec. Salary: $	Bonus: $
Savings Plan:	Stock Purch. Plan:		Second Exec. Salary: $	Bonus: $

OTHER THOUGHTS:
Apparent Women Officers or Directors: 1
Hot Spot for Advancement for Women/Minorities:

LOCATIONS: ("Y" = Yes)

West:	Southwest:	Midwest:	Southeast:	Northeast:	International:
	Y	Y			Y

Note: Financial information, benefits and other data can change quickly and may vary from those stated here.

VIROPHARMA INC

www.viropharma.com

Industry Group Code: 325412 Ranks within this company's industry group: Sales: 58 Profits: 101

Drugs:		Other:	Clinical:	Computers:	Services:
Discovery:		AgriBio:	Trials/Services:	Hardware:	Specialty Services:
Licensing:	Y	Genetic Data:	Labs:	Software:	Consulting:
Manufacturing:		Tissue Replacement:	Equipment/Supplies:	Arrays:	Blood Collection:
Generics:			Research/Development Svcs.:	Database Management:	Drug Delivery:
			Diagnostics:		Drug Distribution:

TYPES OF BUSINESS:

Oral Antibiotics
Hepatitis C Drugs
Infectious Diseases Drugs

BRANDS/DIVISIONS/AFFILIATES:

Cinryze
Vancocin
CINRYZESolutions
Ryze Above

CONTACTS: Note: Officers with more than one job title may be intentionally listed here more than once.

Vincent J. Milano, CEO
Daniel B. Soland, COO/VP
Vincent J. Milano, Pres.
Charles A Rowland, CFO/VP
Colin Broom, Chief Scientific Officer/VP
Thomas F. Doyle, VP-Strategic Initiatives
Will Roberts, VP-Corp. Comm.
Robert Doody, Assistant Dir.-Investor Rel.
Vincent J. Milano, Chmn.
Robert G. Pietrusko, VP-Global Regulatory Affairs & Quality

Phone: 610-458-7300	Fax: 610-458-7380
Toll-Free: 888-651-0201	
Address: 730 Stockton Dr., Exton, PA 19341 US	

GROWTH PLANS/SPECIAL FEATURES:

ViroPharma, Inc. is a biopharmaceutical company that develops and commercializes products that address serious infectious diseases, with a focus on products used by physician specialists or in hospital settings. The firm has two marketed products and two development programs. ViroPharma's marketed products include Cinryze and Vancocin. ViroPharma markets and sells Cinryze in the U.S. for routine prophylaxis against angioedema attacks in adolescent and adult patients with hereditary angioedema (HAE). Cinryze is a C1 esterase inhibitor therapy for routine prohylaxis against HAE, also known as C1 inhibitor (C1-INH), a life-threatening genetic disorder. The firm recently obtained expanded rights to commercialize Cinryze and future C1-INH derived products in certain European countries and other territories throughout the world, as well as rights to develop future C1-INH derived products for additional indications. ViroPharma intends to commercialize Cinryze in Europe in 2011. ViroPharma also markets and sells Vancocin HCl capsules in the U.S. Vancocin is an antibiotic that is used to treat pseudomembranous colitis and enterocolitis, including methicillin-resistant strains. ViroPharma's development programs include C1 esterase inhibitor to identify therapeutic uses and potential additional indications and other modes of administration for the treatment of HAE and other C1 mediated diseases; and a non-toxigenic strain of C. difficile (NTCD) for the treatment and prevention of CDI, currently in Phase I. The firm licensed the U.S. and Canadian rights for a further product development candidate, an intranasal formulation of pleconaril, to Merck & Co., Inc. for the treatment of picornavirus infections. In March 2010, the company launched Ryze Above, an exclusive patient resources program within CINRYZESolutions, the company's patient support program. In May 2010, ViroPharma announced that it has begun packaging Cinryze with West Pharmaceutical Services, Inc.'s needleless reconstitution system.

The company offers its employees medical and dental benefits; a 401(k) plan; and stock options.

FINANCIALS: Sales and profits are in thousands of dollars—add 000 to get the full amount. 2009 Note: Financial information for 2009 was not available for all companies at press time.

2009 Sales: $310,449	2009 Profits: $-11,077	U.S. Stock Ticker: VPHM
2008 Sales: $232,307	2008 Profits: $67,617	Int'l Ticker: Int'l Exchange:
2007 Sales: $203,770	2007 Profits: $95,353	Employees: 188
2006 Sales: $167,181	2006 Profits: $66,666	Fiscal Year Ends: 12/31
2005 Sales: $132,417	2005 Profits: $113,705	Parent Company:

SALARIES/BENEFITS:

Pension Plan:	ESOP Stock Plan:	Profit Sharing:	Top Exec. Salary: $500,000	Bonus: $244,972
Savings Plan: Y	Stock Purch. Plan: Y		Second Exec. Salary: $365,000	Bonus: $193,360

OTHER THOUGHTS:

Apparent Women Officers or Directors:
Hot Spot for Advancement for Women/Minorities:

LOCATIONS: ("Y" = Yes)

West:	Southwest:	Midwest:	Southeast:	Northeast:	International:
				Y	

VIVUS INC

www.vivus.com

Industry Group Code: 33911 Ranks within this company's industry group: Sales: 11 Profits: 14

Drugs:		Other:	Clinical:	Computers:	Services:	
Discovery:	Y	AgriBio:	Trials/Services:	Hardware:	Specialty Services:	
Licensing:		Genetic Data:	Labs:	Software:	Consulting:	
Manufacturing:		Tissue Replacement:	Equipment/Supplies:	Arrays:	Blood Collection:	
Generics:			Research/Development Svcs.:	Database Management:	Drug Delivery:	
			Diagnostics:		Drug Distribution:	Y

TYPES OF BUSINESS:

Pharmaceuticals Manufacturing
Female Sexual Dysfunction Treatments
Obesity Treatment
Erectile Dysfunction Therapies
Diabetes Treatment
Obstructive Sleep Apnea Therapy

BRANDS/DIVISIONS/AFFILIATES:

MUSE
Qnexa
Luramist
Avanafil

CONTACTS: *Note: Officers with more than one job title may be intentionally listed here more than once.*

Leland F. Wilson, CEO
Peter Y. Tam, COO
Peter Y. Tam, Pres.
Timothy E. Morris, CFO/Sr. VP-Finance
Michael P. Miller, Sr. VP/Chief Commercial Officer
Barbara Troupin, Sr. Dir.-Medical Affairs
Ted Broman, VP-Chemistry, Mfg. & Control
Guy Marsh, VP-Oper./Gen. Mgr.
Rob Janosky, Sr. Dir.-Comm. Dev.
Lee B. Perry, VP/Chief Acct. Officer
Wesley W. Day, VP-Clinical Dev.
Mark B. Logan, Chmn.

Phone: 650-934-5265	Fax: 650-934-5389
Toll-Free:	
Address: 1172 Castro St., Ste. 200, Mountain View, CA 94040 US	

GROWTH PLANS/SPECIAL FEATURES:

Vivus, Inc. is a biopharmaceutical company developing products to treat obesity, diabetes, obstructive sleep apnea, and sexual health. The firm currently develops and markets one FDA-approved drug: MUSE, which is a prescription treatment for erectile dysfunction (ED). The company's core technology, the transurethral system for erection, is based on the discovery that the urethra can absorb certain pharmacologic agents into the surrounding erectile tissues. MUSE is an applicator that delivers a tiny medicated pellet to the urethra in order to relax blood vessels and combat erectile dysfunction. Vivus has an international distribution agreement with Stockholm-based Meda AB to market and distribute MUSE internationally in all member states of the European Union. Vivus also has an investigational ED product, Avanafil, which is currently in Phase III clinical trials. The company is also developing Qnexa, an investigational drug with applications in the treatment of obesity, diabetes, obstructive sleep apnea, fatty liver disease and hyperlipidemia. Qnexa for Obesity is currently under review for marketing by the Endocrinologic and Metabolic Drugs Advisory Committee of the FDA. Another investigational Vivus product is Luramist, a therapy for the treatment of Hypoactive Sexual Desire Disorder (HSDD) which is the persistent or recurrent lack of interest in sexual activity resulting in personal distress for women who experience it. Vivus has been approved by the FDA to conduct Phase III trials for Luramist.

FINANCIALS: Sales and profits are in thousands of dollars—add 000 to get the full amount. 2009 Note: Financial information for 2009 was not available for all companies at press time.

2009 Sales: $50,041	2009 Profits: $-54,291	U.S. Stock Ticker: VVUS
2008 Sales: $102,233	2008 Profits: $-9,940	Int'l Ticker: Int'l Exchange:
2007 Sales: $54,698	2007 Profits: $-2,384	Employees: 126
2006 Sales: $	2006 Profits: $	Fiscal Year Ends: 12/31
2005 Sales: $	2005 Profits: $	Parent Company:

SALARIES/BENEFITS:

Pension Plan:	ESOP Stock Plan:	Profit Sharing:	Top Exec. Salary: $612,721	Bonus: $266,063
Savings Plan: Y	Stock Purch. Plan: Y		Second Exec. Salary: $384,204	Bonus: $159,051

OTHER THOUGHTS:

Apparent Women Officers or Directors: 2
Hot Spot for Advancement for Women/Minorities: Y

LOCATIONS: ("Y" = Yes)

West:	Southwest:	Midwest:	Southeast:	Northeast:	International:
Y				Y	

WARNER CHILCOTT PLC

www.wcrx.com

Industry Group Code: 325412 Ranks within this company's industry group: Sales: 39 Profits: 24

Drugs:	Other:	Clinical:	Computers:	Services:
Discovery:	AgriBio:	Trials/Services:	Hardware:	Specialty Services:
Licensing:	Genetic Data:	Labs:	Software:	Consulting:
Manufacturing: Y	Tissue Replacement:	Equipment/Supplies:	Arrays:	Blood Collection:
Generics:		Research/Development Svcs.:	Database Management:	Drug Delivery:
		Diagnostics:		Drug Distribution:

TYPES OF BUSINESS:

Pharmaceuticals Development & Manufacturing
Contraceptives
Hormone Therapies
Vitamins
Dermatology Treatments
Gastroenterological Treatments

BRANDS/DIVISIONS/AFFILIATES:

Loestrin 24
Actonel
femhrt
Ovcon
Estrace
Taclonex
Asacol
Doryx

CONTACTS: Note: Officers with more than one job title may be intentionally listed here more than once.

Roger Boissonneault, CEO
Roger Boissonneault, Pres.
Paul Herendeen, CFO/Exec. VP
Mahdi B. Fawzi, Pres., R&D
Leland H. Cross, Sr. VP-Tech. Oper.
Izumi Hara, General Counsel/Sr. VP/Corp. Sec.
Anthony D.Bruno, Exec. VP-Corp. Dev.
W. Carl Reichel, Pres., Pharmaceuticals
Alvin Howard, Sr. VP-Regulatory Affairs
Herman Ellman, Sr. VP-Clinical Dev.
John A. King, Chmn.
Marinus Johannes van Zoonen, Pres., Europe, Int'l & Global Mktg.

Phone: 973-442-3200	Fax: 973-442-3283
Toll-Free: 800-521-8813	
Address: 100 Enterprise Dr., Rockaway, NJ 07866 US	

GROWTH PLANS/SPECIAL FEATURES:

Warner Chilcott is a specialty pharmaceutical company that focuses on developing, manufacturing and marketing branded prescription pharmaceutical products in women's healthcare, gastroenterology, urology and dermatology. Its franchises comprise complementary portfolios of established, branded and development stage products. Its women's healthcare franchise is anchored by its strong presence in the hormonal contraceptive and hormone therapy categories and its dermatology franchise is built on its established positions in the markets for psoriasis and acne therapies. In hormonal oral contraceptives, its products include Loestrin 24 Fe, OVCON 35, OVCON 50, ESTROSTEP FE, and FEMCON FE. In the hormone therapy market, its products include femhrt, Femtrace, and Estrace cream and tablets. In dermatology, its products include psoriasis medications Dovonex and Taclonex and acne medication DORYX. Its gastrointestinal products include Asacol and Asacol HD, which are both used for the treatment of ulcerative colitis. In October 2009, the firm acquired the global branded pharmaceuticals business (PGP) of Procter & Gamble Co, for approximately $3 billion. As a result of the acquisition, Warner Chilcott is part of a global collaboration agreement with pharmaceutical company Sanofi, under which both co-develop and market Actonel, an osteoporosis medication.

FINANCIALS: Sales and profits are in thousands of dollars—add 000 to get the full amount. 2009 Note: Financial information for 2009 was not available for all companies at press time.

2009 Sales: $1,435,816	2009 Profits: $514,118	U.S. Stock Ticker: WCRX
2008 Sales: $938,125	2008 Profits: $-8,357	Int'l Ticker: Int'l Exchange:
2007 Sales: $899,561	2007 Profits: $28,875	Employees: 1,000
2006 Sales: $	2006 Profits: $	Fiscal Year Ends: 12/31
2005 Sales: $	2005 Profits: $	Parent Company:

SALARIES/BENEFITS:

Pension Plan: Y	ESOP Stock Plan:	Profit Sharing:	Top Exec. Salary: $960,000	Bonus: $1,700,000
Savings Plan: Y	Stock Purch. Plan:		Second Exec. Salary: $468,971	Bonus: $672,074

OTHER THOUGHTS:

Apparent Women Officers or Directors:
Hot Spot for Advancement for Women/Minorities:

LOCATIONS: ("Y" = Yes)

West:	Southwest:	Midwest:	Southeast:	Northeast:	International:
				Y	Y

WATSON PHARMACEUTICALS INC

www.watson.com

Industry Group Code: 325412 Ranks within this company's industry group: Sales: 30   Profits: 35

Drugs:		Other:	Clinical:	Computers:	Services:	
Discovery:	Y	AgriBio:	Trials/Services:	Hardware:	Specialty Services:	
Licensing:		Genetic Data:	Labs:	Software:	Consulting:	
Manufacturing:	Y	Tissue Replacement:	Equipment/Supplies:	Arrays:	Blood Collection:	
Generics:			Research/Development Svcs.:	Database Management:	Drug Delivery:	
			Diagnostics:		Drug Distribution:	Y

TYPES OF BUSINESS:

Generic Pharmaceuticals
Branded Drugs
Urology Drugs
Anti-Hypertensive Drugs
Nephrology Drugs
Anti-Inflammatory Drugs
Oral Contraceptive Drugs
Pain Management Drugs

BRANDS/DIVISIONS/AFFILIATES:

Watson Laboratories
Watson Pharma
Arrow Group
Anda Pharmaceuticals
Anda
Valmed
Trelstar Depot
Gelnique

CONTACTS: Note: Officers with more than one job title may be intentionally listed here more than once.

Paul M. Bisaro, CEO
Paul M. Bisaro, Pres.
R. Todd Joyce, CFO/Sr. VP
Clare Carmichael, Sr. VP-Human Resources
Charles D. Ebert, Sr. VP-R&D
Thomas R. Giordano, CIO/Sr. VP
David A. Buchen, General Counsel/Sr. VP/Sec.
Robert A. Stewart, Sr. VP-Global Oper.
Patricia L. Eisenhauer, VP-Corp. Comm.
Patricia L. Eisenhauer, VP-Investor Rel.
Edward F. Heimers, Jr., Exec. VP/Pres., Brand Division
Albert Paonessa III, Exec. VP/COO-Anda, Inc.
Gordon Munro, Sr. VP-Quality Assurance
Francois A. Menard, Sr. VP-Generics Research & Dev.
Andrew L. Turner, Chmn.

Phone: 951-493-5300	Fax:
Toll-Free:	
Address: 311 Bonnie Cir., Corona, CA 92880 US	

GROWTH PLANS/SPECIAL FEATURES:

Watson Pharmaceuticals, Inc. develops, manufactures, markets, sells and distributes over 30 branded and over 170 generic pharmaceutical products. The firm operates through three segments: global generics, global brands and distribution. The global generics segment includes pharmaceutical products that are bioequivalent to proprietary products. These generic products address the therapeutic areas of antibiotics, anti-inflammatories, depression, hypertension, oral contraceptives, pain management and smoking cessation. Generic products accounted for roughly 60% of net revenues in 2009. The global brands segment develops, manufactures, markets, sells and distributes products primarily through two core areas: specialty products and nephrology. The specialty products include urology and a number of non-promoted products. The nephrology product line concerns products for the treatment of iron deficiency anemia. Brands include Trelstar Depot and Trelstar LA, treatments for advanced prostate cancer; Gelnique, a treatment for overactive bladder symptoms; Rapaflo, a treatment for symptoms of benign prostatic hyperplasia; and Ferrlecit and INFeD, iron replacement therapies for patients with iron deficiency anemia. The company markets its brand products through 350 sales professionals. Brand products accounted for roughly 16% of total revenue in 2009. The distribution segment distributes generic products and select brand products to independent pharmacies, pharmacy chains, alternative care facilities and physicians' offices in the U.S. The company's distribution business subsidiaries include Anda, Anda Pharmaceuticals and Valmed. In December 2009, Watson acquired London-based Arrow Group for approximately $1.75 billion.

The company offers employees medical, dental and vision insurance; life and AD&D insurance; domestic partner coverage; flexible spending accounts; business travel accident insurance; short and long-term disability; pet insurance; and tuition reimbursement.

FINANCIALS: Sales and profits are in thousands of dollars—add 000 to get the full amount. 2009 Note: Financial information for 2009 was not available for all companies at press time.

2009 Sales: $2,793,000	2009 Profits: $222,000	U.S. Stock Ticker: WPI
2008 Sales: $2,535,501	2008 Profits: $238,379	Int'l Ticker: Int'l Exchange:
2007 Sales: $2,496,651	2007 Profits: $141,030	Employees: 5,830
2006 Sales: $1,979,244	2006 Profits: $-445,005	Fiscal Year Ends: 12/31
2005 Sales: $1,646,203	2005 Profits: $138,557	Parent Company:

SALARIES/BENEFITS:

Pension Plan:	ESOP Stock Plan:	Profit Sharing:	Top Exec. Salary: $1,038,462	Bonus: $1,250,000
Savings Plan: Y	Stock Purch. Plan:		Second Exec. Salary: $854,921	Bonus: $650,288

OTHER THOUGHTS:

Apparent Women Officers or Directors: 2
Hot Spot for Advancement for Women/Minorities: Y

LOCATIONS: ("Y" = Yes)

West:	Southwest:	Midwest:	Southeast:	Northeast:	International:
Y		Y	Y	Y	

WHATMAN PLC
www.whatman.com

Industry Group Code: 3345 Ranks within this company's industry group: Sales: Profits:

Drugs:	Other:	Clinical:		Computers:	Services:
Discovery:	AgriBio:	Trials/Services:		Hardware:	Specialty Services:
Licensing:	Genetic Data:	Labs:		Software:	Consulting:
Manufacturing:	Tissue Replacement:	Equipment/Supplies:	Y	Arrays:	Blood Collection:
Generics:		Research/Development Svcs.:		Database Management:	Drug Delivery:
		Diagnostics:	Y		Drug Distribution:

TYPES OF BUSINESS:
Equipment-Filtration Systems
Diagnostic Supplies

BRANDS/DIVISIONS/AFFILIATES:
UNIPLATE
UNIFILTER
GE Healthcare

CONTACTS: Note: Officers with more than one job title may be intentionally listed here more than once.
Kieran Murphy, CEO
Helen Evans, VP-Mktg.
Shari States, Group Mgr.-Mktg. Comm.
Joe Hogan, CEO/Pres., GE Healthcare

Phone: 440-162-267-6670	Fax: 440-162-269-1425

Toll-Free: 800-942-8626

Address: Springfield Mill, James Whatman Way, Maidstone, Kent ME14 2LE UK

GROWTH PLANS/SPECIAL FEATURES:
Whatman plc, operating under General Electric Co.'s GE Healthcare division, is a global leader in filtration technology and manufactures and distributes separation and filtration products used in laboratories, health care facilities and bioscience research. The firm offers over 100 different products for filtration including student kits, DNA purification and pH testers. Whatman has developed total sample preparation solutions through its robust line of filtration devices and membranes. The company's breakthrough protein array and FTA technology, which works to capture archive and purify DNA at room temperature, enables it to provide novel solutions for the analytical, healthcare and bioscience markets. Whatman products serve three markets: LabSciences, MedTech and BioScience. The LabSciences unit produces chromatography products used for purification, extraction products, filter papers, filtration devices, membrane filters and specialty products. The MedTech sector makes diagnostic components and medical devices such as purification and sterilization tools used by original equipment manufacturers. The BioScience unit produces the UNIPLATE and UNIFILTER lines of filter plates and products used for nucleic acid sample preparation.

FINANCIALS: Sales and profits are in thousands of dollars—add 000 to get the full amount. 2009 Note: Financial information for 2009 was not available for all companies at press time.
2009 Sales: $	2009 Profits: $	U.S. Stock Ticker: Subsidiary
2008 Sales: $	2008 Profits: $	Int'l Ticker: Int'l Exchange:
2007 Sales: $	2007 Profits: $	Employees:
2006 Sales: $	2006 Profits: $	Fiscal Year Ends: 12/31
2005 Sales: $	2005 Profits: $	Parent Company: GENERAL ELECTRIC CO (GE)

SALARIES/BENEFITS:
Pension Plan:	ESOP Stock Plan:	Profit Sharing:	Top Exec. Salary: $	Bonus: $
Savings Plan:	Stock Purch. Plan:		Second Exec. Salary: $	Bonus: $

OTHER THOUGHTS:
Apparent Women Officers or Directors:
Hot Spot for Advancement for Women/Minorities:

LOCATIONS: ("Y" = Yes)
West:	Southwest:	Midwest:	Southeast:	Northeast:	International:
				Y	Y

XECHEM INTERNATIONAL

www.xechem.com

Industry Group Code: 325412 Ranks within this company's industry group: Sales: Profits:

Drugs:		Other:	Clinical:	Computers:		Services:
Discovery:	Y	AgriBio:	Trials/Services:	Hardware:		Specialty Services:
Licensing:		Genetic Data:	Labs:	Software:		Consulting:
Manufacturing:	Y	Tissue Replacement:	Equipment/Supplies:	Arrays:		Blood Collection:
Generics:			Research/Development Svcs.:	Y	Database Management:	Drug Delivery:
			Diagnostics:			Drug Distribution:

TYPES OF BUSINESS:

Pharmaceuticals Development & Manufacturing
Drugs-Proprietary
Drugs-Generic
Research & Development Services
Fine Chemicals
Nutraceuticals

BRANDS/DIVISIONS/AFFILIATES:

Xechem Nigeria Pharmaceuticals, Ltd.
NICOSAN
5-HMF

CONTACTS: *Note: Officers with more than one job title may be intentionally listed here more than once.*

Robert Swift, Interim Pres.
Robert Swift, Chief Oversight Officer
Robert Swift, Interim Chmn.

Phone: 732-205-0500	Fax: 732-2474090
Toll-Free:	
Address: 379 Thornall, Edison, NJ 08818 US	

GROWTH PLANS/SPECIAL FEATURES:

Xechem International is a development stage biopharmaceutical company working on sickle cell disease. Recently, its focus has been the development of NICOSAN, as well as another sickle cell compound, 5-HMF, which it licensed from Virginia Commonwealth University. Xechem is restructuring its U.S. operations as well as the oversight of its Nigerian subsidiary, Xechem Nigeria Pharmaceuticals, Ltd., which produces and sells NICOSAN. As a part of this restructuring, the company recently closed its operations in New Brunswick, New Jersey; this included termination of personnel and the transfer of over $2 million in equipment to Nigeria. The company still maintains its Edison, New Jersey headquarters. Xechem, in partnership with Rutgers University, has developed a formulation of NICOSAN that can be made into tablets for sale in the U.S. as a nutraceutical. Recently, Xechem International and one of its subsidiaries, Xechem, Inc., filed for Chapter 11 protection; the firms are continuing restructuring efforts with the intent to emerge from bankruptcy.

FINANCIALS: Sales and profits are in thousands of dollars—add 000 to get the full amount. 2009 Note: Financial information for 2009 was not available for all companies at press time.

2009 Sales: $	2009 Profits: $	U.S. Stock Ticker: XKEM.Q
2008 Sales: $	2008 Profits: $	Int'l Ticker: Int'l Exchange:
2007 Sales: $	2007 Profits: $	Employees:
2006 Sales: $ 202	2006 Profits: $-11,130	Fiscal Year Ends: 12/31
2005 Sales: $ 6	2005 Profits: $-10,039	Parent Company:

SALARIES/BENEFITS:

Pension Plan:	ESOP Stock Plan:	Profit Sharing:	Top Exec. Salary: $	Bonus: $
Savings Plan:	Stock Purch. Plan: Y		Second Exec. Salary: $	Bonus: $

OTHER THOUGHTS:

Apparent Women Officers or Directors:
Hot Spot for Advancement for Women/Minorities:

LOCATIONS: ("Y" = Yes)

West:	Southwest:	Midwest:	Southeast:	Northeast:	International:
				Y	Y

XENOPORT INC

www.xenoport.com

Industry Group Code: 325412 **Ranks within this company's industry group:** Sales: 97   Profits: 147

Drugs:		Other:		Clinical:	Computers:	Services:	
Discovery:	Y	AgriBio:		Trials/Services:	Hardware:	Specialty Services:	
Licensing:	Y	Genetic Data:		Labs:	Software:	Consulting:	
Manufacturing:		Tissue Replacement:		Equipment/Supplies:	Arrays:	Blood Collection:	
Generics:				Research/Development Svcs.:	Database Management:	Drug Delivery:	Y
				Diagnostics:		Drug Distribution:	

TYPES OF BUSINESS:
Drug Development

BRANDS/DIVISIONS/AFFILIATES:
Transported Prodrugs
Horizant
XP21510
XP13512
Arbaclofen Placarbil
XP21279
XP20925

CONTACTS: *Note: Officers with more than one job title may be intentionally listed here more than once.*
Ronald W. Barrett, CEO
William J. Rieflin, Pres.
William G. Harris, CFO
Mark A. Gallop, Sr. VP-Research
Kenneth C. Cundy, Sr. VP-Preclinical Dev.
Gianna M. Bosko, Sec.
David R. Savello, Sr. VP-Dev.
William G. Harris, Sr. VP-Finance
David A. Stamler, Chief Medical Officer/Sr. VP
Vincent J. Angotti, Chief Commercialization Officer/Sr. VP

Phone: 408-616-7200	Fax:
Toll-Free:	
Address: 3410 Central Expwy., Santa Clara, CA 95051 US	

GROWTH PLANS/SPECIAL FEATURES:
XenoPort, Inc. develops and commercializes potential treatments of central nervous system (CNS) disorders. Its drug candidates, which it calls Transported Prodrugs, are designed to modify the chemical structure of currently marketed drugs, called the parent drug, in order to correct deficiencies in oral absorption, distribution and/or metabolism by utilizing the body's natural nutrient transporter mechanisms. The firm's most advanced product candidate, XP13512 (known as Horizant in the U.S.), has passed its Phase III clinical trial for moderate to severe restless leg syndrome (RLS) and passed Phase II trials for the treatment of post-herpetic neuralgia (PHN), a type of nerve damage associated with certain strains of the herpes virus. Partner company Astellas Pharma, Inc. is conducting two Phase II trials in Japan evaluating XP13512 for the treatment of painful diabetic neuropathy (PDN) and RLS; another partner, GlaxoSmithKline plc (GSK), is evaluating XP13512 for PHN, PDN and migraine prophylaxis. Astellas has the exclusive right to develop and commercialize XP13512 in Japan, Korea, the Philippines, Indonesia, Thailand and Taiwan; GSK has exclusive development and commercialization rights for all other territories. The company's second product candidate, Arbaclofen Placarbil (previously known as XP19986), is in separate Phase II trials for two indications: the potential treatment of gastroesophageal reflux disease (GERD), sometimes called acid reflux; and for the treatment of spasticity. The firm's third candidate, XP21279 (levodopa), is in Phase I trials for the treatment of Parkinson's disease. Its newest candidate, XP21510, is in the preclinical stage for the treatment of menorrhagia (heavy menstrual flow). In November 2009, the firm and Astellas filed a new drug application for XP13512 with Japan's Pharmaceuticals and Medical Device Agency. In March 2010, XenoPort agreed to layoff approximately 50% of its workforce as part of a restructuring plan to focus on its later-stage potential products.

FINANCIALS: Sales and profits are in thousands of dollars—add 000 to get the full amount. 2009 Note: Financial information for 2009 was not available for all companies at press time.

2009 Sales: $34,273	2009 Profits: $-66,334	**U.S. Stock Ticker:** XNPT
2008 Sales: $41,996	2008 Profits: $-62,540	**Int'l Ticker:** Int'l Exchange:
2007 Sales: $113,822	2007 Profits: $28,193	Employees: 219
2006 Sales: $10,606	2006 Profits: $-64,313	Fiscal Year Ends: 12/31
2005 Sales: $4,753	2005 Profits: $-42,909	Parent Company:

SALARIES/BENEFITS:

Pension Plan:	ESOP Stock Plan:	Profit Sharing:	Top Exec. Salary: $500,000	Bonus: $300,000
Savings Plan:	Stock Purch. Plan:		Second Exec. Salary: $385,000	Bonus: $188,843

OTHER THOUGHTS:
Apparent Women Officers or Directors: 2
Hot Spot for Advancement for Women/Minorities:

LOCATIONS: ("Y" = Yes)

West:	Southwest:	Midwest:	Southeast:	Northeast:	International:
Y					

XOMA LTD

www.xoma.com

Industry Group Code: 325412 Ranks within this company's industry group: Sales: 74 Profits: 72

Drugs:		Other:		Clinical:	Computers:	Services:
Discovery:	Y	AgriBio:		Trials/Services:	Hardware:	Specialty Services:
Licensing:	Y	Genetic Data:	Y	Labs:	Software:	Consulting:
Manufacturing:	Y	Tissue Replacement:		Equipment/Supplies:	Arrays:	Blood Collection:
Generics:				Research/Development Svcs.:	Database Management:	Drug Delivery:
				Diagnostics:		Drug Distribution:

TYPES OF BUSINESS:
Therapeutic Antibodies

BRANDS/DIVISIONS/AFFILIATES:
Raptiva
Lucentis
CIMZIA
XOMA 3AB
XOMA 052
HCD 122
Novartis AG
Takeda Pharmaceutical Company Ltd.

CONTACTS: Note: Officers with more than one job title may be intentionally listed here more than once.
Steven B. Engle, CEO
Fred Kurland, CFO
Charles C. Wells, VP-Human Resources
Charles C. Wells, VP-IT
Christopher J. Margolin, General Counsel/Corp. Sec./VP
Robert S. Tenerowicz, VP-Oper.
Jim Neal, VP-Bus. Dev.
Fred Kurland, VP-Finance
Patrick J. Scannon, Chief Medical Officer/Exec. VP
Daniel P. Cafaro, VP-Regulatory Affairs & Compliance
Susan Kramer, VP-Project & Alliance Mgmt.
Calvin C. McGoogan, VP-Quality & Facilities
Steven B. Engle, Chmn.

Phone: 510-204-7200	Fax: 510-644-2011
Toll-Free: 800-246-9662	
Address: 2910 7th St., Berkeley, CA 94710 US	

GROWTH PLANS/SPECIAL FEATURES:

Xoma, Ltd., is a biopharmaceutical company that discovers, develops and manufactures therapeutic antibodies designed to treat inflammatory, autoimmune, infectious and oncological diseases. Xoma's product development pipeline includes proprietary products and collaborative programs at various stages of preclinical and clinical development. Products in the development pipeline include XOMA 052, an anti-interleukin-1 beta (anti-IL-1 beta) antibody being evaluated as a treatment for type 1 and type 2 diabetes and cardiovascular disease; XOMA 3AB, a biodefense anti-botulism antibody candidate; HCD 122, a fully human antibody being developed with Novartis AG as a treatment for B-cell cancers such as lymphoma; and several antibodies in preclinical development. Xoma has collaborations with several companies, including Takeda Pharmaceutical Company Limited; Schering-Plough Research Institute; and Novartis. The company also has royalty interests in CIMZIA, approved for the treatment of Crohn's disease and rheumatoid arthritis. CIMZIA is in clinical trials for the treatment of psoriasis. In January 2009, Xoma announced a workforce reduction of about 42%. In April 2009, the firm began voluntary removal of Raptiva, developed to treat immune system disorders, from the U.S. market. In late 2009, Xoma sold its royalty interest in Lucentis, for the treatment of neovascular wet age-related macular degeneration, to Genentech. In September 2009, the firm announced a discovery collaboration agreement with Arana Therapeutics Limited; in November 2009, the company announced another antibody discovery collaboration with Kaketsuken. In February 2010, Xoma initiated a Phase IIb clinical trial for XOMA 052 for the treatment of type 2 diabetes. In March 2010, it began a Phase II trial with XOMA 052 for type 1 diabetes.

Xoma offers its employees medical, dental and vision plans, flexible spending accounts, short and long-term disability programs, life insurance, a stock purchase plan, an employee assistance program, educational assistance, a commuter checks program, concierge services, ticket discounts and credit union membership.

FINANCIALS: Sales and profits are in thousands of dollars—add 000 to get the full amount. 2009 Note: Financial information for 2009 was not available for all companies at press time.

2009 Sales: $98,430	2009 Profits: $ 550	**U.S. Stock Ticker: XOMA**
2008 Sales: $67,987	2008 Profits: $-45,245	Int'l Ticker: Int'l Exchange:
2007 Sales: $84,252	2007 Profits: $-12,326	Employees: 195
2006 Sales: $29,498	2006 Profits: $-51,841	Fiscal Year Ends: 12/31
2005 Sales: $18,669	2005 Profits: $2,779	Parent Company:

SALARIES/BENEFITS:

Pension Plan: Y	ESOP Stock Plan:	Profit Sharing:	Top Exec. Salary: $540,750	Bonus: $262,267
Savings Plan:	Stock Purch. Plan:		Second Exec. Salary: $389,340	Bonus: $123,811

OTHER THOUGHTS:
Apparent Women Officers or Directors: 1
Hot Spot for Advancement for Women/Minorities:

LOCATIONS: ("Y" = Yes)

West:	Southwest:	Midwest:	Southeast:	Northeast:	International:
Y					

Note: Financial information, benefits and other data can change quickly and may vary from those stated here.

ZIOPHARM ONCOLOGY INC

www.ziopharm.com

Industry Group Code: 325412 Ranks within this company's industry group: Sales: Profits: 94

Drugs:		Other:		Clinical:	Computers:	Services:
Discovery:		AgriBio:		Trials/Services:	Hardware:	Specialty Services:
Licensing:	Y	Genetic Data:		Labs:	Software:	Consulting:
Manufacturing:	Y	Tissue Replacement:		Equipment/Supplies:	Arrays:	Blood Collection:
Generics:				Research/Development Svcs.:	Database Management:	Drug Delivery:
				Diagnostics:		Drug Distribution:

TYPES OF BUSINESS:

Biopharmaceutical Development

BRANDS/DIVISIONS/AFFILIATES:

palifosfamide
Zymafos
ZIO-201
indibulin
Zybulin
ZIO-301
darinaparsin
Zinapar

CONTACTS: *Note: Officers with more than one job title may be intentionally listed here more than once.*

Jonathan Lewis, CEO
Richard Bagley, COO
Richard Bagley, Pres.
Richard Bagley, CFO/Treas.
Jonathan Lewis, Chief Medical Officer
Barbara Wallner, CTO
Steve Bloom, VP-Bus. Dev.
Tyler Cook, Sr. Dir.-Finance
Bob Morgan, Sr. VP-Regulatory Affairs, Quality & Clinical Dev.
Jan Stevens, VP-Clinical Dev.
Murray Brennan, Lead Dir.

Phone: 646-214-0700	Fax: 646-214-0711
Toll-Free:	
Address: 1180 Ave. of the Americas, 19th Fl., New York, NY 10036 US	

GROWTH PLANS/SPECIAL FEATURES:

ZIOPHARM Oncology, Inc. is a biopharmaceutical firm that seeks to develop and market a diverse portfolio of in-licensed cancer drugs that address unmet medical needs. It is focused on licensing and developing proprietary small molecule drug candidates in intravenous and/or oral forms and that are related to cancer therapeutics already on the market/in development. The firm has three product candidates: palifosfamide (Zymafos, ZIO-201); indibulin (Zybulin, ZIO-301); and darinaparsin (Zinapar, ZIO-101). Palifosfamide is an active metabolite of the pro-drug ifosfamide, an alkylating agent used to treat sarcoma, lymphoma and testicular cancer. ZIOPHARM's main priority is the ongoing randomized phase II trial with palifosfamide aimed at supporting a registration trial for palifosfamide in combination with doxorubicin in the front- and second-line treatment of soft tissue sarcoma. Darinaparsin is a novel anti-mitochondrial agent (organic arsenic) that has been shown to be less toxic than Trisenox, an inorganic arsenic currently on the market; it is a possible treatment of brain, ovarian, lung, colon, melanoma and kidney cancer. Indibulin is a novel, orally taken small molecular-weight tubulin (an anti-tumor target) polymerization inhibitor that the company acquired from Baxter Healthcare. The candidate is distinct from other tubulin inhibitors currently on the market in that it binds to a unique site on tubulin and is active in taxane-resistant and multi-drug-resistant tumors. Both palifosfamide and darinaparsin have completed Phase I and II clinical trials, while indibulin has completed two Phase I clinical trials in combination with Tarceva, a non-small cell lung cancer treatment, and Xeloda, an oral chemotherapy for both metastatic breast and colorectal cancer.

FINANCIALS: Sales and profits are in thousands of dollars—add 000 to get the full amount. 2009 Note: Financial information for 2009 was not available for all companies at press time.

2009 Sales: $	2009 Profits: $-7,649	U.S. Stock Ticker: ZIOP
2008 Sales: $	2008 Profits: $-25,231	Int'l Ticker: Int'l Exchange:
2007 Sales: $	2007 Profits: $-26,608	Employees: 17
2006 Sales: $	2006 Profits: $	Fiscal Year Ends: 12/31
2005 Sales: $	2005 Profits: $	Parent Company:

SALARIES/BENEFITS:

Pension Plan:	ESOP Stock Plan:	Profit Sharing:	Top Exec. Salary: $420,000	Bonus: $200,000
Savings Plan:	Stock Purch. Plan:		Second Exec. Salary: $315,000	Bonus: $100,000

OTHER THOUGHTS:

Apparent Women Officers or Directors: 2
Hot Spot for Advancement for Women/Minorities:

LOCATIONS: ("Y" = Yes)

West:	Southwest:	Midwest:	Southeast:	Northeast:	International:
				Y	

ZYMOGENETICS INC

www.zymogenetics.com

Industry Group Code: 325412 Ranks within this company's industry group: Sales: 68   Profits: 142

Drugs:		Other:		Clinical:	Computers:	Services:
Discovery:	Y	AgriBio:		Trials/Services:	Hardware:	Specialty Services:
Licensing:	Y	Genetic Data:	Y	Labs:	Software:	Consulting:
Manufacturing:	Y	Tissue Replacement:		Equipment/Supplies:	Arrays:	Blood Collection:
Generics:				Research/Development Svcs.:	Database Management:	Drug Delivery:
				Diagnostics:		Drug Distribution:

TYPES OF BUSINESS:

Therapeutic Proteins
Homeostasis, Inflammatory & Autoimmune Diseases Drugs
Cancer & Viral Infections Drugs

BRANDS/DIVISIONS/AFFILIATES:

Recothrom rThrombin
Interleukin-21
PEG-IFN
Atacicept
rFactor XIII
Bristol-Myers Squibb Company
Bayer Schering Pharma AG
Merck Serono

CONTACTS: Note: Officers with more than one job title may be intentionally listed here more than once.

Douglas E. Williams, CEO
Stephen W. Zaruby, Pres.
James A. Johnson, CFO/Exec. VP
Darren R. Hamby, Sr. VP-Human Resources
Dennis M. Miller, Sr. VP-Research & Preclinical Dev.
James A. Johnson, Corp. Sec.
Heather Franklin, Sr. VP-Bus. Dev.
James A. Johnson, Treas.
Eleanor L. Ramos, Chief Medical Officer/Sr. VP
Darren R. Hamby, Sr. VP-Corp. Svcs.
Bruce L. A. Carter, Chmn.

Phone: 206-442-6600	Fax: 206-442-6608
Toll-Free: 800-775-6686	
Address: 1201 Eastlake Ave. E., Seattle, WA 98102 US	

GROWTH PLANS/SPECIAL FEATURES:

ZymoGenetics, Inc. discovers, develops, manufactures and commercializes therapeutic proteins for the treatment of human diseases. The company's current therapeutic focus is in the areas of hemostasis; inflammatory and autoimmune diseases; cancer; and viral infections. The firm's first internally developed product candidate, Recothrom Thrombin (also referred to as rThrombin or recombinant thrombin), is FDA-approved as a topical hemostat to control moderate bleeding during surgical procedures and is now being marketed in the U.S. Outside of the U.S., the company has partnered with Bayer Schering Pharma AG to develop and commercialize Recothrom. Other in-house products include Interleukin-21 (IL-21), a cytokine with potential applications for the treatment of cancer; PEG-IFN (also known as IL-29 and PEG-Interferon lambda), a cytokine with potential applications for the treatment of viral infections; and monoclonal antibody IL-31 (mAb) for atopic dermatitis. ZymoGenetics collaborates with a number of external companies for developmental products, including Merck Serono for ataicept (formerly known as TACI-Ig), a soluble receptor with potential applications for the treatment of cancer and autoimmune diseases; with Novo Nordisk for rFactor XIII for congenital deficiency and cardiac surgery; with BioMimetic Therapeutics for Augment Bone Grafts (formerly GEM-OS1TM); and many others. The company also out-licenses a number of commercial products, including treatments for diabetes, hemophilia, wound healing, periodontal defects, hypoglycemia and myocardial infarctions. In April 2009, the firm announced it would be laying off about 32% of its workforce and shift focus from oncology development to immunology. In January 2009, the firm announced a partnership with Bristol-Myers Squibb Company for developing PEG-Interferon lambda.

Employees of the company are offered medical, dental and vision coverage; performance based bonuses; stock options; a 401(k) plan; retirement planning assistance; employee discounts; tuition reimbursements; life insurance; short- and long-term disability coverage; and parental leave.

FINANCIALS: Sales and profits are in thousands of dollars—add 000 to get the full amount. 2009 Note: Financial information for 2009 was not available for all companies at press time.

2009 Sales: $136,972	2009 Profits: $-42,981	U.S. Stock Ticker: ZGEN
2008 Sales: $73,989	2008 Profits: $-116,241	Int'l Ticker: Int'l Exchange:
2007 Sales: $38,477	2007 Profits: $-148,144	Employees: 323
2006 Sales: $25,380	2006 Profits: $-130,002	Fiscal Year Ends: 12/31
2005 Sales: $42,909	2005 Profits: $-78,027	Parent Company:

SALARIES/BENEFITS:

Pension Plan:	ESOP Stock Plan: Y	Profit Sharing:	Top Exec. Salary: $549,745	Bonus: $316,250
Savings Plan: Y	Stock Purch. Plan:		Second Exec. Salary: $398,485	Bonus: $222,000

OTHER THOUGHTS:

Apparent Women Officers or Directors: 2
Hot Spot for Advancement for Women/Minorities: Y

LOCATIONS: ("Y" = Yes)

West:	Southwest:	Midwest:	Southeast:	Northeast:	International:
Y					

Note: Financial information, benefits and other data can change quickly and may vary from those stated here.

ADDITIONAL INDEXES

Contents:

INDEX OF FIRMS NOTED AS HOT SPOTS FOR ADVANCEMENT FOR WOMEN & MINORITIES

NPS PHARMACEUTICALS INC
NYCOMED
ONCOGENEX PHARMACEUTICALS
ONYX PHARMACEUTICALS INC
OSI PHARMACEUTICALS INC
PACIFIC BIOMARKERS INC
PDL BIOPHARMA
PENWEST PHARMACEUTICALS CO
PEREGRINE PHARMACEUTICALS INC
PERRIGO CO
PFIZER INC
PHARMACEUTICAL PRODUCT DEVELOPMENT INC
PHARMANET DEVELOPMENT GROUP INC
PONIARD PHARMACEUTICALS INC
PRA INTERNATIONAL
QLT INC
QUEST DIAGNOSTICS INC
RAPTOR PHARMACEUTICAL CORP
ROCHE HOLDING LTD
RULES BASED MEDICINE INC
SANOFI-AVENTIS SA
SEQUENOM INC
SERACARE LIFE SCIENCES INC
SHIRE PLC
SIGMA-ALDRICH CORP
STEMCELLS INC
STIEFEL LABORATORIES INC
SYNOVIS LIFE TECHNOLOGIES INC
SYNTHETIC GENOMICS
TECHNE CORP
TELIK INC
TENGION INC
TEVA PHARMACEUTICAL INDUSTRIES
THERAGENICS CORP
THERMO FISHER SCIENTIFIC INC
TRIUS THERAPUTICS INC
VALEANT PHARMACEUTICALS INTERNATIONAL
VERENIUM CORPORATION
VERNALIS PLC
VERTEX PHARMACEUTICALS INC
VIACORD INC
VIVUS INC
WATSON PHARMACEUTICALS INC
ZYMOGENETICS INC

INDEX OF SUBSIDIARIES, BRAND NAMES AND AFFILIATIONS

Brand or subsidiary, followed by the name of the related corporation

1000CranesofHope.com; **MILLENNIUM PHARMACEUTICALS INC**
21Ventures; **BIONANOMATRIX**
3-Nitro; **ALPHARMA ANIMAL HEALTH**
454 Life Sciences; **ROCHE HOLDING LTD**
4MyHeart.com; **CELERA CORPORATION**
4SC-101; **4SC AG**
4SC-201; **4SC AG**
4SC-202; **4SC AG**
4SC-203; **4SC AG**
4SC-205; **4SC AG**
4SC-207; **4SC AG**
4SCan Technology; **4SC AG**
5-HMF; **XECHEM INTERNATIONAL**
Abbott Laboratories; **CELERA CORPORATION**
Abbott Medical Optics Inc; **ABBOTT LABORATORIES**
AbC.R.O., Inc.; **PHARMACEUTICAL PRODUCT DEVELOPMENT INC**
ABGene; **THERMO FISHER SCIENTIFIC INC**
Abilify; **BRISTOL-MYERS SQUIBB CO**
Abraxis Bioscience, Inc.; **CELGENE CORP**
Abreva; **AVANIR PHARMACEUTICALS**
AbsorbaTack; **COVIDIEN PLC**
Abthrax; **HUMAN GENOME SCIENCES INC**
AC Vaccine; **AVAX TECHNOLOGIES INC**
Acanto; **SYNGENTA AG**
Accellerase; **GENENCOR INTERNATIONAL INC**
ACCESS Oncology, Inc.; **KERYX BIOPHARMACEUTICALS INC**
Acclarent, Inc.; **JOHNSON & JOHNSON**
Accretropin; **CANGENE CORP**
AccuCell Human PBMC-Basic; **SERACARE LIFE SCIENCES INC**
AccuCell Human PBMC-Characterized; **SERACARE LIFE SCIENCES INC**
AccuDrain External Ventricular Drainage Systems; **INTEGRA LIFESCIENCES HOLDINGS CORP**
Accumin Diagnostics, Inc.; **KERYX BIOPHARMACEUTICALS INC**
ACCURUN; **SERACARE LIFE SCIENCES INC**
ACCURUN 106 Series 1000 Positive Control; **SERACARE LIFE SCIENCES INC**
ACCUTYPE H1N1; **SERACARE LIFE SCIENCES INC**
ACE; **TRINITY BIOTECH PLC**
Acetadote; **CUMBERLAND PHARMACEUTICALS**
Acetazolamide; **LANNETT COMPANY INC**
ACGT Sdn Bhd; **SYNTHETIC GENOMICS**
ACHN-490; **ISIS PHARMACEUTICALS INC**
Aciphex/Pariet; **EISAI CO LTD**
Acrylaway; **NOVOZYMES**

AcrySof; **ALCON INC**
ACTCellerate; **BIOTIME INC**
ACTEGA Coatings & Sealants; **ALTANA AG**
Actimmune; **INTERMUNE INC**
Actiq; **CEPHALON INC**
Activase; **GENENTECH INC**
Activated Cellular Therapy; **NEOPROBE CORPORATION**
Activated Checkpoint Therapy (ACT); **ARQULE INC**
Activella; **NOVO-NORDISK AS**
Actonel; **WARNER CHILCOTT PLC**
AcuForm; **DEPOMED INC**
Adagen; **ENZON PHARMACEUTICALS INC**
Adamis Labs; **ADAMIS PHARMACEUTICALS CORPORATION**
Adamis Viral Therapies; **ADAMIS PHARMACEUTICALS CORPORATION**
Adderall XR; **SHIRE PLC**
Adeno-Associated Viral (AAV) vectors; **TARGETED GENETICS CORP**
Aderall XR; **SHIRE CANADA INC**
ADHERE; **SCIOS INC**
ADL5945; **ADOLOR CORP**
ADL7445; **ADOLOR CORP**
Adoair; **GLAXOSMITHKLINE PLC**
AdVac; **CRUCELL NV**
AdvanSource Biomaterials Corp.; **ADVANSOURCE BIOMATERIALS CORPORATION**
ADVIA; **SIEMENS HEALTHCARE DIAGNOSTICS**
AEOL 10150; **AEOLUS PHARMACEUTICALS INC**
Aequus BioPharma, Inc.; **CELL THERAPEUTICS INC**
Aerosurf; **DISCOVERY LABORATORIES INC**
AERx; **ARADIGM CORPORATION**
AERx Essence; **ARADIGM CORPORATION**
AERx iDMS; **ARADIGM CORPORATION**
AEZS-108; **AETERNA ZENTARIS INC**
AEZS-120; **AETERNA ZENTARIS INC**
AEZS-130; **AETERNA ZENTARIS INC**
AG-707; **ANTIGENICS INC**
Agroeste Sementes; **MONSANTO CO**
Agroproducts Corey, S.A. de C.V.; **DU PONT AGRICULTURE & NUTRITION**
Agrylin; **SHIRE CANADA INC**
AgVenture; **PIONEER HI-BRED INTERNATIONAL INC**
Ahura Scientific; **THERMO FISHER SCIENTIFIC INC**
Akorn (New Jersey), Inc.; **AKORN INC**
Akorn-Strides, LLC; **AKORN INC**
AKuScreen; **CELSIS INTERNATIONAL PLC**
Albac; **ALPHARMA ANIMAL HEALTH**
Alcon Inc; **NOVARTIS AG**
Alcon Surgical; **ALCON INC**
ALD-201; **ALDAGEN INC**
ALD-301; **ALDAGEN INC**
ALD-401; **ALDAGEN INC**
Aldurazyme; **BIOMARIN PHARMACEUTICAL INC**

INDEX OF SUBSIDIARIES, BRAND NAMES AND AFFILIATIONS, CONT.

INDEX OF SUBSIDIARIES, BRAND NAMES AND AFFILIATIONS, CONT.

INDEX OF SUBSIDIARIES, BRAND NAMES AND AFFILIATIONS, CONT.

INDEX OF SUBSIDIARIES, BRAND NAMES AND AFFILIATIONS, CONT.

INDEX OF SUBSIDIARIES, BRAND NAMES AND AFFILIATIONS, CONT.

Cloderm; **VALEANT PHARMACEUTICALS INTERNATIONAL**
Clolar; **GENZYME ONCOLOGY**
Clorpactin; **UNITED-GUARDIAN INC**
Clotrimazole/Betamethasomne; **TARO PHARMACEUTICAL INDUSTRIES**
CML Healthcare, Inc.; **CML HEALTHCARE INCOME FUND**
Coag Dx; **IDEXX LABORATORIES INC**
Cobalamin; **ACCESS PHARMACEUTICALS INC**
Coban; **ELI LILLY & COMPANY**
Codexis; **MAXYGEN INC**
CoGenesys, Inc.; **TEVA PHARMACEUTICAL INDUSTRIES**
COLARIS; **MYRIAD GENETICS INC**
COLARIS AP; **MYRIAD GENETICS INC**
Colazal; **SALIX PHARMACEUTICALS**
Colby Pharmaceuticals Corporation; **ADAMIS PHARMACEUTICALS CORPORATION**
Colcys; **FLAMEL TECHNOLOGIES SA**
Colisure; **IDEXX LABORATORIES INC**
Complete Genomics, Inc.; **BIONANOMATRIX**
Convivia; **RAPTOR PHARMACEUTICAL CORP**
Copaxone; **TEVA PHARMACEUTICAL INDUSTRIES**
Corbett Life Science Pty Ltd; **QIAGEN NV**
Cordis Corp; **JOHNSON & JOHNSON**
Core Companion Animal Health; **HESKA CORP**
Coretec; **EISAI CO LTD**
Corgel BioHydrogel; **LIFECORE BIOMEDICAL INC**
Corgentin; **CARDIUM THERAPEUTICS INC**
Corian; **E I DU PONT DE NEMOURS & CO (DUPONT)**
Corthera Inc; **NOVARTIS AG**
Corumo; **SKYEPHARMA PLC**
CoSeal; **ANGIOTECH PHARMACEUTICALS INC**
Cotara; **PEREGRINE PHARMACEUTICALS INC**
Cottonase; **VERENIUM CORPORATION**
Covance; **RANBAXY LABORATORIES LIMITED**
CP Medical Corp.; **THERAGENICS CORP**
CR011-vcMMAE; **CURAGEN CORPORATION**
Crave-NX 7-Day Diet Aid Spray; **GENEREX BIOTECHNOLOGY**
Crestor; **ASTRAZENECA PLC**
Crinone; **COLUMBIA LABORATORIES INC**
Crinone; **MERCK SERONO SA**
Cruiser; **SYNGENTA AG**
CSL Behring LLC; **CSL LIMITED**
CSL Bioplasma; **CSL LIMITED**
CSL Biotherapies; **CSL LIMITED**
CSL Pharmaceutical; **CSL LIMITED**
CSP-9222; **ONCOGENEX PHARMACEUTICALS**
CU-2010; **MEDICINES CO (THE)**
CUBICIN; **CUBIST PHARMACEUTICALS INC**
CuffPatch; **ORGANOGENESIS INC**
Culex; **BIOANALYTICAL SYSTEMS INC**

Culex ACB; **BIOANALYTICAL SYSTEMS INC**
Culex-L ABC; **BIOANALYTICAL SYSTEMS INC**
Cumberland Emerging Technologies, Inc.; **CUMBERLAND PHARMACEUTICALS**
CuraGen Corporation; **CELLDEX THERAPEUTICS INC**
Curosurf; **NYCOMED**
Currenta GmbH & Co.; **BAYER AG**
Custom Manufacturing; **LONZA GROUP**
CustomExpress; **AFFYMETRIX INC**
CustomSeq; **AFFYMETRIX INC**
CX1739; **CORTEX PHARMACEUTICALS INC**
CX2007; **CORTEX PHARMACEUTICALS INC**
CX929; **CORTEX PHARMACEUTICALS INC**
Cyclofluidic; **UCB SA**
Cymetra; **LIFECELL CORPORATION**
CyMVectin; **VICAL INC**
CynerGene LLC; **BIORELIANCE CORP**
Cytosol Laboratories, Inc.; **BIOMET INC**
Dacogen; **SUPERGEN INC**
Daiichi Pharmaceutical Co., Ltd.; **DAIICHI SANKYO CO LTD**
Daiichi Sankyo Healthcare Co. Ltd.; **DAIICHI SANKYO CO LTD**
Daiichi Sankyo, Inc.; **DAIICHI SANKYO CO LTD**
Dainippon Sumitomo Pharma America; **SEPRACOR INC**
Dainippon Sumitomo Pharma Co. Ltd.; **SEPRACOR INC**
Danisco A/S; **GENENCOR INTERNATIONAL INC**
Darapladib; **HUMAN GENOME SCIENCES INC**
darinaparsin; **ZIOPHARM ONCOLOGY INC**
DataPassport MD; **SPECIALTY LABORATORIES INC**
Daytrana; **NOVEN PHARMACEUTICALS**
Daytrana; **SHIRE PLC**
DC Bead; **BIOCOMPATIBLES INTERNATIONAL PLC**
DC Bead; **SCICLONE PHARMACEUTICALS**
De Ruiter Seeds Group BV; **MONSANTO CO**
Deccox; **ALPHARMA ANIMAL HEALTH**
deCODE genetics ehf; **DECODE GENETICS LTD**
deCODE ProstateCancer; **DECODE GENETICS LTD**
deCODEme; **DECODE GENETICS LTD**
deCODET2; **DECODE GENETICS LTD**
Delta and Pine Land Company; **MONSANTO CO**
Delursan; **AXCAN PHARMA INC**
Dendrite International Inc; **CEGEDIM SA**
Denufosol Tetrasodium; **INSPIRE PHARMACEUTICALS INC**
Denville Scientific Inc.; **HARVARD BIOSCIENCE INC**
Depakine; **SANOFI-AVENTIS SA**
DePuy Inc; **JOHNSON & JOHNSON**
Dermatix; **VALEANT PHARMACEUTICALS INTERNATIONAL**
Destiny Max; **TRINITY BIOTECH PLC**
Devoe; **AKZO NOBEL NV**

INDEX OF SUBSIDIARIES, BRAND NAMES AND AFFILIATIONS, CONT.

INDEX OF SUBSIDIARIES, BRAND NAMES AND AFFILIATIONS, CONT.

INDEX OF SUBSIDIARIES, BRAND NAMES AND AFFILIATIONS, CONT.

Fungassay; **SYNBIOTICS CORP**
Fusilev; **SPECTRUM PHARMACEUTICALS INC**
Fuzeon (T-20); **TRIMERIS INC**
Gabapentin GR; **DEPOMED INC**
Gabitril; **CEPHALON INC**
Gadovist; **BAYER SCHERING PHARMA AG**
Galt Medical Corp.; **THERAGENICS CORP**
Gammagard; **BAXTER INTERNATIONAL INC**
Ganite; **GENTA INC**
GastroDose; **PENWEST PHARMACEUTICALS CO**
Gate Pharmaceuticals; **TEVA PARENTERAL MEDICINES INC**
GATTEX; **NPS PHARMACEUTICALS INC**
GDC-0449; **CURIS INC**
GE Healthcare; **HARVARD BIOSCIENCE INC**
GE Healthcare; **WHATMAN PLC**
GED-aPC; **CARDIOME PHARMA CORP**
Gelnique; **WATSON PHARMACEUTICALS INC**
Geminex; **PENWEST PHARMACEUTICALS CO**
GemStar; **HOSPIRA INC**
Gemzar; **ELI LILLY & COMPANY**
Genasense; **GENTA INC**
GeneChip; **AFFYMETRIX INC**
Genelabs Technologies; **GLAXOSMITHKLINE PLC**
Genentech Inc; **23ANDME INC**
Genentech Inc; **ROCHE HOLDING LTD**
Genentech Inc; **LEXICON PHARMACEUTICALS INC**
General Electric Co (GE); **GE HEALTHCARE**
Generx; **CARDIUM THERAPEUTICS INC**
GeneSeek, Inc.; **NEOGEN CORPORATION**
Genetech, Inc.; **IMMUNOGEN INC**
GeneXpert; **CEPHEID**
GeneXpert Infinity; **CEPHEID**
Genome Analyzer; **ILLUMINA INC**
Genopoietic, S.A.; **AVAX TECHNOLOGIES INC**
Gen-Probe; **CHIRON CORP**
Gensweet; **GENENCOR INTERNATIONAL INC**
Gentra Systems, Inc.; **QIAGEN NV**
Genvir; **FLAMEL TECHNOLOGIES SA**
Genzyme Corp; **GENZYME ONCOLOGY**
Genzyme Corp.; **GENZYME BIOSURGERY**
Genzyme Corporation; **BIOMARIN PHARMACEUTICAL INC**
Gilead Sciences; **ROCHE HOLDING LTD**
GlaxoSmithKline plc; **STIEFEL LABORATORIES INC**
Global Pharmaceuticals; **IMPAX LABORATORIES INC**
GloMax-Multi+; **PROMEGA CORP**
Gloucester Pharmaceuticals; **CELGENE CORP**
Glucose RapidSpray; **GENEREX BIOTECHNOLOGY**
Glumetza; **DEPOMED INC**
Gold Avenue Ltd.; **IMMTECH PHARMACEUTICALS INC**
GONAL-f; **MERCK SERONO SA**
Google Inc; **23ANDME INC**

GoTaq; **PROMEGA CORP**
G-Protein Coupled Receptors (GPCRs); **ACTELION LTD**
Green Cross Corp.; **BIOCRYST PHARMACEUTICALS INC**
Grifols SA; **CERUS CORPORATION**
GRN163L; **GERON CORPORATION**
GRNCM1; **GERON CORPORATION**
GRNOPC1; **GERON CORPORATION**
GRNVAC1; **GERON CORPORATION**
GS2-MTM; **STEMCELLS INC**
GSI-RTM; **STEMCELLS INC**
Guardian Laboratories; **UNITED-GUARDIAN INC**
Guava Technologies; **MILLIPORE CORP**
Gynazole-1; **KV PHARMACEUTICAL CO**
GyrB/ParE; **TRIUS THERAPUTICS INC**
Haemacure Corp; **ANGIOTECH PHARMACEUTICALS INC**
Hammerite; **AKZO NOBEL NV**
Harnal; **ASTELLAS PHARMA INC**
Harvard Apparatus; **HARVARD BIOSCIENCE INC**
HCD 122; **XOMA LTD**
Health Care Products Division; **HI-TECH PHARMACAL CO INC**
Hedrin; **MANHATTAN PHARMACEUTICALS INC**
Helinx; **CERUS CORPORATION**
Hemagen Diagnosticos Comerico; **HEMAGEN DIAGNOSTICS INC**
HEMATRUE Veterinary Hematology Analyzer; **HESKA CORP**
HepaGam B; **CANGENE CORP**
HepaSphere; **BIOSPHERE MEDICAL INC**
Hepavax-Gene; **CRUCELL NV**
Hepsera; **GILEAD SCIENCES INC**
HepTide; **RAPTOR PHARMACEUTICAL CORP**
HEPTIMAX; **QUEST DIAGNOSTICS INC**
Herceptin; **GENENTECH INC**
Herceptin; **PDL BIOPHARMA**
Herculex; **DOW AGROSCIENCES LLC**
hERGexpress; **AVIVA BIOSCIENCES CORP**
HESKA Feline UltraNasal FVRCP Vaccine; **HESKA CORP**
Hextend; **BIOTIME INC**
Hisamitsu Pharmaceutical Co., Inc.; **NOVEN PHARMACEUTICALS**
HiSeq 2000; **ILLUMINA INC**
Histofreezer; **ORASURE TECHNOLOGIES INC**
Histostat; **ALPHARMA ANIMAL HEALTH**
hLEC; **STEMCELLS INC**
Hoefer, Inc.; **HARVARD BIOSCIENCE INC**
Hoegemeyer Hybrids; **PIONEER HI-BRED INTERNATIONAL INC**
Horizant; **XENOPORT INC**
Horizontal Needle Box; **BRACHYTHERAPY SERVICES INC**

INDEX OF SUBSIDIARIES, BRAND NAMES AND AFFILIATIONS, CONT.

INDEX OF SUBSIDIARIES, BRAND NAMES AND AFFILIATIONS, CONT.

INDEX OF SUBSIDIARIES, BRAND NAMES AND AFFILIATIONS, CONT.

INDEX OF SUBSIDIARIES, BRAND NAMES AND AFFILIATIONS, CONT.

Menorest; **NOVEN PHARMACEUTICALS**
Menostar; **BAYER SCHERING PHARMA AG**
Mentor Corp; **JOHNSON & JOHNSON**
Merck & Co. Inc.; **CARDIOME PHARMA CORP**
Merck KGaA; **MILLIPORE CORP**
Merck Serono; **ZYMOGENETICS INC**
Merck Serono S.A.; **MERCK KGAA**
Merck Sharp & Dohme (Switzerland) GmbH;
CARDIOME PHARMA CORP
Meridian Medical Technologies, Inc.; **KING**
PHARMACEUTICALS INC
Merz Pharmaceuticals GmbH; **BIOCOMPATIBLES**
INTERNATIONAL PLC
Mestinon; **VALEANT PHARMACEUTICALS**
INTERNATIONAL
Met52; **NOVOZYMES**
MetabolicMAP; **RULES BASED MEDICINE INC**
Meteor Group; **SYNTHETIC GENOMICS**
Metformin XL; **FLAMEL TECHNOLOGIES SA**
Meyrin Photovoltaic Application Laboratory; **E I DU**
PONT DE NEMOURS & CO (DUPONT)
MGI Pharma Inc; **EISAI CO LTD**
Micardis; **ASTELLAS PHARMA INC**
Microbia Inc; **IRONWOOD PHARMACEUTICALS**
INC
Microbiological Associates; **BIORELIANCE CORP**
MICRODUR; **DURECT CORP**
MICRO-PLATE; **ORASURE TECHNOLOGIES INC**
Micropump; **FLAMEL TECHNOLOGIES SA**
Microvascular Anastomotis COUPLER System;
SYNOVIS LIFE TECHNOLOGIES INC
MIDAS; **IMS HEALTH INC**
MIDAS; **PALATIN TECHNOLOGIES INC**
Midlothian Laboratories; **HI-TECH PHARMACAL CO**
INC
Miglustat; **ACTELION LTD**
Migranal; **VALEANT PHARMACEUTICALS**
INTERNATIONAL
Milatuzumab; **IMMUNOMEDICS INC**
Millennium Pharmaceuticals, Inc.; **TAKEDA**
PHARMACEUTICAL COMPANY LTD
Millipore Corporation; **MERCK KGAA**
Miltex; **INTEGRA LIFESCIENCES HOLDINGS**
CORP
Mimotopes Pty Ltd; **COMMONWEALTH**
BIOTECHNOLOGIES INC
Mimpara; **NPS PHARMACEUTICALS INC**
MimyX; **STIEFEL LABORATORIES INC**
Mintop; **DR REDDY'S LABORATORIES LIMITED**
Mipomersen; **ISIS PHARMACEUTICALS INC**
MIR Preclinical Services; **CHARLES RIVER**
LABORATORIES INTERNATIONAL INC
Miyuki Holdings Co., Ltd.; **TOYOBO CO LTD**
MKC-1; **ENTREMED INC**

MLN9708; **MILLENNIUM PHARMACEUTICALS**
INC
MODA Technology Partners; **LONZA GROUP**
MolecularBreeding; **MAXYGEN INC**
Monarch Pharmaceuticals, Inc.; **KING**
PHARMACEUTICALS INC
MoRu-Viraten; **CRUCELL NV**
Movetis NV; **QUINTILES TRANSNATIONAL CORP**
Mozobil; **GENZYME CORP**
Mozobil; **GENZYME ONCOLOGY**
MuGard; **ACCESS PHARMACEUTICALS INC**
Multi-Betic; **HI-TECH PHARMACAL CO INC**
Multifect Protex; **GENENCOR INTERNATIONAL**
INC
MultiKine; **CEL-SCI CORPORATION**
Multi-Link Vision; **ABBOTT LABORATORIES**
MultiStem; **ANGIOTECH PHARMACEUTICALS INC**
Mundipharma; **BIOCRYST PHARMACEUTICALS**
INC
MUSE; **VIVUS INC**
Mvax; **AVAX TECHNOLOGIES INC**
My Study Portal; **ERESEARCH TECHNOLOGY INC**
Mycogen; **DOW AGROSCIENCES LLC**
Mylan Pharmaceuticals S.L; **MYLAN INC**
Mylan Pharmaceuticals, Inc.; **MYLAN INC**
Mylan S.A.S.; **MYLAN INC**
Mylan Technologies, Inc.; **MYLAN INC**
Mylan-Nifedipine Extended Release Tablets; **MYLAN**
INC
Mylotarg; **PDL BIOPHARMA**
MyoCell; **BIOHEART INC**
MyoCell SDF-1; **BIOHEART INC**
Myonal; **EISAI CO LTD**
Myriad Pharmaceuticals, Inc.; **MYRIAD GENETICS**
INC
Mytogen, Inc.; **ADVANCED CELL TECHNOLOGY**
INC
Naglazyme; **BIOMARIN PHARMACEUTICAL INC**
Namenda; **FOREST LABORATORIES INC**
Nano HA; **LIFECORE BIOMEDICAL INC**
NanoCrystal; **ELAN CORP PLC**
NanoHPX; **NANOBIO CORPORATION**
NanoMatrix, Inc.; **ORGANOGENESIS INC**
NanoStat; **NANOBIO CORPORATION**
NanoTxt; **NANOBIO CORPORATION**
Nascobal; **PAR PHARMACEUTICAL COMPANIES**
INC
National Human Genome Research Institute;
BIONANOMATRIX
National Institutes of Health; **BIONANOMATRIX**
Natrecor; **SCIOS INC**
Nav1.5; **AVIVA BIOSCIENCES CORP**
NeedleTech Products, Inc.; **THERAGENICS CORP**
Nektar PEGylation Technology; **NEKTAR**
THERAPEUTICS

INDEX OF SUBSIDIARIES, BRAND NAMES AND AFFILIATIONS, CONT.

INDEX OF SUBSIDIARIES, BRAND NAMES AND AFFILIATIONS, CONT.

INDEX OF SUBSIDIARIES, BRAND NAMES AND AFFILIATIONS, CONT.

INDEX OF SUBSIDIARIES, BRAND NAMES AND AFFILIATIONS, CONT.

INDEX OF SUBSIDIARIES, BRAND NAMES AND AFFILIATIONS, CONT.

INDEX OF SUBSIDIARIES, BRAND NAMES AND AFFILIATIONS, CONT.

INDEX OF SUBSIDIARIES, BRAND NAMES AND AFFILIATIONS, CONT.

INDEX OF SUBSIDIARIES, BRAND NAMES AND AFFILIATIONS, CONT.

INDEX OF SUBSIDIARIES, BRAND NAMES AND AFFILIATIONS, CONT.

Western Blot HIV-1 Confirmatory Test; **ORASURE TECHNOLOGIES INC**
WI Harper Group; **AVIVA BIOSCIENCES CORP**
WideStrike; **DOW AGROSCIENCES LLC**
WinNonlin AutoPilot; **PHARSIGHT CORP**
WinRho SDF; **CANGENE CORP**
Wintershall AG; **BASF SE**
Wipe Out Dairy Wipes; **IMMUCELL CORPORATION**
WITNESS; **SYNBIOTICS CORP**
WntTide; **RAPTOR PHARMACEUTICAL CORP**
Wrafton Laboratories, Ltd.; **PERRIGO CO**
WuXi PharmaTech; **CHARLES RIVER LABORATORIES INTERNATIONAL INC**
Xatral; **SKYEPHARMA PLC**
Xcytrin; **PHARMACYCLICS INC**
Xechem Nigeria Pharmaceuticals, Ltd.; **XECHEM INTERNATIONAL**
Xenazine; **BIOVAIL CORPORATION**
Xenerex; **AVANIR PHARMACEUTICALS**
Xifaxan; **SALIX PHARMACEUTICALS**
xMAP; **LUMINEX CORPORATION**
Xolair; **PDL BIOPHARMA**
XOMA 052; **XOMA LTD**
XOMA 3AB; **XOMA LTD**
XOPENEX; **SEPRACOR INC**
XP13512; **XENOPORT INC**
XP20925; **XENOPORT INC**
XP21279; **XENOPORT INC**
XP21510; **XENOPORT INC**
Xpert C; **CEPHEID**
Xpert vanA; **CEPHEID**
xPONENT; **LUMINEX CORPORATION**
xTAG; **LUMINEX CORPORATION**
Xyrem; **JAZZ PHARMACEUTICALS**
Xyzal; **UCB SA**
Yamanouchi Pharmaceutical Co., Ltd.; **ASTELLAS PHARMA INC**
Yasmin; **BAYER SCHERING PHARMA AG**
Yingtaiqing; **SIMCERE PHARMACEUTICAL GROUP**
Yttrium Y 90; **IMMUNOMEDICS INC**
Yttrium Y 90 epratuzumab tetraxetan; **IMMUNOMEDICS INC**
ZADAXIN; **SCICLONE PHARMACEUTICALS**
Zailin; **SIMCERE PHARMACEUTICAL GROUP**
ZALBIN; **HUMAN GENOME SCIENCES INC**
Zanamivir; **SIMCERE PHARMACEUTICAL GROUP**
Zantac; **GLAXOSMITHKLINE PLC**
Zavesca; **ACTELION LTD**
Zeatin; **SENETEK PLC**
Zebrafish; **EVOTEC AG**
Zentaris GmbH; **AETERNA ZENTARIS INC**
Zenvia; **AVANIR PHARMACEUTICALS**
Zerenex; **KERYX BIOPHARMACEUTICALS INC**
Zevalin; **CELL THERAPEUTICS INC**

ZFP Nucleases (ZFNs); **SANGAMO BIOSCIENCES INC**
ZFP Therapeutic; **SANGAMO BIOSCIENCES INC**
ZFP Transcription Factors (ZFP TFs); **SANGAMO BIOSCIENCES INC**
Ziana; **MEDICIS PHARMACEUTICAL CORP**
Zila; **TOLMAR INC**
Zinapar; **ZIOPHARM ONCOLOGY INC**
Zinc Finger DNA-Binding Proteins; **SANGAMO BIOSCIENCES INC**
ZIO-201; **ZIOPHARM ONCOLOGY INC**
ZIO-301; **ZIOPHARM ONCOLOGY INC**
Z-Max; **REPROS THERAPEUTICS INC**
Zoloft; **PFIZER INC**
Zostrix; **HI-TECH PHARMACAL CO INC**
Zovirax; **BIOVAIL CORPORATION**
Zybrestat; **OXIGENE INC**
Zybulin; **ZIOPHARM ONCOLOGY INC**
ZYFLO CR; **SKYEPHARMA PLC**
Zylon; **TOYOBO CO LTD**
Zymafos; **ZIOPHARM ONCOLOGY INC**
ZymeQuest; **CHIRON CORP**
Zyoptic; **BAUSCH & LOMB INC**
Zyprexa; **ELI LILLY & COMPANY**
Zyrtec; **UCB SA**

LaVergne, TN USA
09 March 2011
219368LV00003BB/3/P